AIRFRAME
STRESS ANALYSIS AND SIZING

About the Author

Mr. Michael C. Y. Niu is the president of AD Airframe Consulting Company and is a metallic and composite airframe consultant. He was a Senior Research and Development Engineer (management position) at Lockheed Aeronautical Systems Co. Mr. Niu has over 30 years experience in aerospace and airframe structural analysis and design. At Lockheed, he was the acting Department Manager and Program Manager in charge of various structural programs, including development of innovative structural design concepts for both metallic and composite materials that are applicable to current and/or future advanced tactical fighters and advanced transports. During his service at Lockheed, he was deeply involved in all aspects of preliminary design, including aircraft layout, integration, configuration selection, and airworthiness. He was lead engineer responsible for the L-1011 wide body derivative aircraft wing and empennage stress analysis. During 1966 and 1968, he served as stress engineer for the B727 and B747 at The Boeing Company.

Mr. Niu is the author of the texts, AIRFRAME STRUCTURAL DESIGN (1988) and COMPOSITE AIRFRAME STRUCTURES (1992). He has also written Lockheed's Composites Design Guide and Composites Drafting Handbook. He received the Lockheed Award of Achievement and Award of Product Excellence in 1973 and 1986, respectively. He is listed in Who's Who in Aviation, 1973.

He is a consulting professor at Beijing University of Aeronautics and Astronautics and a chair professor at Nanjing University of Aeronautics and Astronautics, The People's Republic of China.

He teaches the UCLA (University of California in Los Angeles) Engineering Short Courses "Airframe Structural Design and Repairs", "Composite Airframe Structures" and "Airframe Stress Analysis and Sizing".

Mr. Niu received a B.S. degree in Hydraulic Engineering from Chungyuan University, Taiwan, in 1962 and a M.S. degree in Civil Engineering from the University of Wyoming in 1966.

A MESSAGE FROM THE AUTHOR

In the universe, there is always a balance, typified by ying (negative) and yang (positive). This may help explain the Industrial Revolution which brought humankind civilization (positive) but also inflicted on us the problem of "pollution" (negative) which, if not effectively controlled, could be catastrophic. There are many such instances which require balance in this world and it all depends on how we handle them before it is too late. Diagnosing such problems in the early stages makes them much easier and less costly to handle.

Since the 1970's the calculation revolution instigated by the use of computer analysis has given the engineer quicker and more accurate answers. Computer analysis can even solve some problems of highly redundant structures that were practically impossible to calculate in the past. Computer analyses will replace all hand calculations in the near future. The merits of using computer hardware and software are not disputed; however, "Murphy's Law" is always operative everywhere and mistakes do happen in design. An engineer who only knows how to do inputs to computer analysis may not know whether or not the output is correct (the garbage-in and garbage-out process!). An engineer is not a robot or a machine; ingenuity comes from the engineer, not the machine. Unfortunately, industries ignore engineers' ambition to pursue the necessary experience due to cost considerations. But, engineers cannot learn valuable experience from computers; the computer is just a beautiful tool and is not everything!

A second issue is work experience; in particular, engineers now spend most of their time in front of computer screens and the importance of communication with colleagues from other disciplines, i.e. team work, is overlooked. Engineers are losing the opportunity to gain needed experience from the "old timers". The end result is that industries gradually lose the most valuable experience which passes away with the retiring engineers of the "old school". Industries should ask retiring engineers to write down their valuable experience in report form before they leave so that it will be passed on to those who follow. Big companies have their own handbooks, manuals, etc., for company use, but they are merely handbooks and do not provide enough experience. Industry management spends too much effort on figuring out how to save on costs, but, in the long run, they lose the bases of experience, and will end up as big losers. If this situation is not remedied soon, the aircraft manufacturing industries as well as other industries will walk into backlash.

Another issue concerns aging aircraft, an issue that has been debated since Aloha Airlines' old B737 aircraft disaster, but, as yet, nobody, including the aircraft manufacturers and users, wants to define the flyable life of the aircraft, i.e. limit flight hours or years of service, whichever is more appropriate. From the structural engineer's standpoint, we realize that any metallic structure has its own fatigue life, just like the human body, and no life can go on forever, even with vigilant maintenance work. When a vehicle becomes too old, even maintenance is of no use. If the aircraft has a problem in midair, it will fall and, unlike ground vehicles, cannot stop to wait for rescue. The ultimate goal of the aircraft manufacturers and the government certifying agency should be to determine a standard for a reasonable and affordable lifespan for aircraft to reduce peril to passengers. This should be done now, before more lives are lost.

AIRFRAME
STRESS ANALYSIS AND SIZING

Second Edition

Michael Chun-Yung Niu

HONG KONG CONMILIT PRESS LTD.

MICHAEL C. Y. NIU'S AIRFRAME BOOK SERIES:
  AIRFRAME STRUCTURAL DESIGN (1988) – Green Book
  COMPOSITE AIRFRAME STRUCTURES (1992) – Blue Book
  AIRFRAME STRESS ANALYSIS AND SIZING (1997) – Red Book

©1997 Conmilit Press Ltd. All right reserved.
No part of this book may be reproduced in any form or
by any electronic or mechanical means, including
information storage and retrieval devices or systems,
without prior written permission from the publisher.

First edition – Oct., 1997
Second edition – Jan., 1999

ISBN 962-7128-08-2

All inquiries should be directed to: HONG KONG CONMILIT PRESS LTD.
           Flat A, 10th Floor,   or   P. O. Box 23250
           Shing Loong Court,   Wanchai Post Office
           No.13, Dragon Terrace,   HONG KONG
           North Point,   Tel: (852) 2578 9529
           HONG KONG   Fax:(852) 2578 1183

U.S. order/inquiry: TECHNICAL BOOK COMPANY
           2056 Westwood Blvd.
           Los Angeles, CA 90025 U.S.A.
           Tel: (310) 475 5711
           Fax: (310) 470 9810

Please forward any suggestions or comments to:
    Prof. Michael C. Y. Niu
    P. O. Box 3552
    Granada Hills, CA 91394 U.S.A.
    Fax: (818) 701 0298
    E-mail: mniu@worldnet.att.net

DISCLAIMER

It is the intent of the author of this book "AIRFRAME STRESS ANALYSIS AND SIZING" to include all the pertinent structural design data required for airframe stress analysis and sizing into one convenient book. The author does not guarantee the contents provided in this book, and it is the responsibility of the reader to determine if the data and information in this book are in agreement with the latest design allowables and the reader's authoritative company source.

Printed in Hong Kong

PREFACE

This book has been prepared as a source of data and procedures for use in the sizing of both airframe and space vehicle structures. The material presented herein has been compiled largely from the published data of government agencies, such as NACA reports and technical publications. The reader will not, of course, be able to read this book with complete comprehension unless he is familiar with the basic concepts of strength of materials and structural analysis; it is assumed that the reader is familiar with these subjects and such material generally is not repeated herein. To maintain the compact size of this book, only data and information relevant to airframe structures are included. Since today's airframe structures are primarily constructed from metallic materials, this book focuses on metallic structural sizing. A few material allowables are included for reference and for in performing the sizing examples.

Step-by-step procedures are included whenever possible and, in many chapters, examples of numerical calculation are included to clarify either the method of analysis or the use of design data and/or design curves to give engineers a real-world feeling of how to achieve the most efficient structures. The intent of this book is to provide a fundamental understanding of the stress analysis required for airframe sizing. The emphasis is on practical application with input from both material strength and hands-on experience. A balance between theory and practical application in airframe structures depends extensively on test data which must be correlated with theory to provide an analytical procedure.

The structural problems of aircraft usually involve the buckling and crippling of thin-sheets (or shells) and stiffened panels. Thin sheet buckling design is one of the most important subjects of airframe analysis. The NACA reports from the 1940's have contributed tremendous amount of information and design data in this field and today's airframe engineers still use them and consider them to be the backbone of airframe stress analysis.

The careful selection of structural configurations and materials, which are combined to produce an optimized design while also considering the effects of static loads, fatigue, fail-safe requirements, damage tolerance and cost, is the most important issue in this book. A considerable amount of material on the sizing of metallic airframes is presented in tables, charts and curves that are based on past experience and/or test results. Another purpose of this book is to give airframe engineers a broad overview of data and information based on the experience and the lessons learned in the aircraft industry (including service of components) that can be used to design a weight-efficient structure which has structural integrity.

Structural sizing approaches and methods will be introduced for those who need to do rough estimations to support aircraft structural design during the preliminary aircraft design stage and also for those who are involved in airframe repair work.

Preface

There is a big gap between theoretical and practical design and this book attempts to bridge the two with real-life examples. The list of references at the end of each chapter complements the material presented in this book and provides the interested reader with additional sources of relevant information.

In order to reinforce the reader's knowledge of airframe sizing, it is strongly recommended that the reader use the author's book AIRFRAME STRUCTURAL DESIGN as an important complementary reference source because it contains a tremendous amount of design information and data that are generally not repeated in this book.

This book was put together in the style of an engineering report with clear sketches of good quality instead of expensive commercial illustrations like those in old-fashioned textbooks. Sincere appreciation and thanks to those who have contributed to correct many errors in the first edition. Special thanks to Mr. Lawrence W. Maxwell for his review in checking for errors throughout the entire book (first edtiion). It is inevitable that minor errors will occur in this book and the author welcomes any comments for future revisions.

Michael Chun-yung Niu
（牛春勻）

Los Angeles, California, U.S.A.
October, 1998 (second edition)

CONTENTS

PREFACE

ABBREVIATIONS, ACRONYMS AND NOMENCLATURE

Chapter 1.0	**GENERAL OVERVIEW**	1
	1.1 OVERVIEW	1
	1.2 DISCUSSION OF OPTIMUM DESIGN	6
	1.3 STRUCTURAL WEIGHT	9
	1.4 DESIGN FOR MANUFACTURING	11
Chapter 2.0	**SIZING PROCEDURES**	14
	2.1 OVERVIEW	14
	2.2 PRELIMINARY SIZING	15
	2.3 PRODUCTION STRESS ANALYSIS	16
	2.4 FORMAL STRESS REPORTS	16
	2.5 SIGN CONVENTIONS	19
	2.6 LOAD PATHS AND FREE-BODY DIAGRAMS	21
	2.7 DRAWING TOLERANCES	23
	2.8 MARGIN OF SAFETY (MS)	24
	2.9 STIFFNESS REQUIREMENTS	25
	2.10 DISCUSSION	26
Chapter 3.0	**EXTERNAL LOADS**	27
	3.1 OVERVIEW	27
	3.2 STRUCTURAL DESIGN CRITERIA	28
	3.3 WEIGHT AND BALANCE	32
	3.4 FLIGHT LOADS	36
	3.5 GROUND LOADS	38
	3.6 DYNAMIC LOADS	45
	3.7 CONTROLLABLE SURFACES	46
	3.8 FUSELAGE PRESSURE LOADS	46
	3.9 WING FUEL PRESSURE LOADS	49
	3.10 MISCELLANEOUS LOADS	52
	3.11 LOAD CONDITIONS SUMMARY FOR COMMERCIAL TRANSPORT	54
Chapter 4.0	**MATERIAL PROPERTIES**	58
	4.1 INTRODUCTION	58
	4.2 STRESS-STRAIN CURVES	61
	4.3 ALLOWABLES	67
	4.4 AIRWORTHINESS REQUIREMENTS	73
	4.5 TOUGHNESS AND CRACK GROWTH RATE	75
	4.6 MATERIAL USAGE	76
	4.7 MATERIAL SELECTION PROCEDURES	78

Chapter 5.0	**STRUCTURAL ANALYSIS**	80
5.1	REVIEW OF HOOKE'S LAW	80
5.2	PRINCIPAL STRESSES	83
5.3	EQUILIBRIUM AND COMPATIBILITY	85
5.4	DETERMINATE STRUCTURES	87
5.5	INDETERMINATE STRUCTURES	97
5.6	FINITE ELEMENT MODELING	124

Chapter 6.0	**BEAM STRESS**	138
6.1	BEAM THEORY	138
6.2	SECTION PROPERTIES	140
6.3	BENDING OF SYMMETRICAL AND UNSYMMETRIC SECTIONS	143
6.4	PLASTIC BENDING	146
6.5	SHEAR STRESS IN BEAMS	152
6.6	SHEAR CENTER	161
6.7	TAPERED BEAMS	164
6.8	TORSION	169
6.9	CRUSHING LOAD ON BOX BEAMS	178

Chapter 7.0	**PLATES AND SHELLS**	181
7.1	INTRODUCTION	181
7.2	PLATES	182
7.3	CYLINDRICAL SHELLS	197
7.4	HEMISPHERICAL SHELLS	199
7.5	HONEYCOMB PANELS	203

Chapter 8.0	**BOX BEAMS**	221
8.1	INTRODUCTION	221
8.2	SHEAR FLOWS DUE TO TORSION	226
8.3	SINGLE-CELL BOX (TWO-STRINGER SECTION)	236
8.4	SINGLE-CELL BOX (MULTI-STRINGER SECTION)	238
8.5	TWO-CELL BOXES	249
8.6	TAPERED CROSS-SECTIONS	256
8.7	SHEAR LAG	262

Chapter 9.0	**JOINTS AND FITTINGS**	273
9.1	INTRODUCTION	273
9.2	FASTENER INFORMATION	279
9.3	SPLICES	289
9.4	ECCENTRIC JOINTS	292
9.5	GUSSET JOINTS	298
9.6	WELDED JOINTS	305
9.7	BONDED JOINTS	313
9.8	LUG ANALYSIS (BOLT IN SHEAR)	321
9.9	TENSION FITTINGS (BOLT IN TENSION)	343
9.10	TENSION CLIPS	361
9.11	GAPS AND THE USE OF SHIMS	370
9.12	FATIGUE CONSIDERATIONS	374

Chapter 10.0	**COLUMN BUCKLING**	394
10.1	INTRODUCTION	394
10.2	EULER EQUATION (LONG COLUMN)	395
10.3	STEPPED COLUMNS	401
10.4	TAPERED COLUMN WITH VARIABLE CROSS-SECTION	406

10.5	LATERAL BUCKLING OF A BEAM BENDING	411
10.6	BEAM-COLUMNS	418
10.7	CRIPPLING STRESS	437
10.8	INTERACTION BETWEEN COLUMN AND CRIPPLING STRESS	448

Chapter 11.0 BUCKLING OF THIN SHEETS — 451
- 11.1 INTRODUCTION — 451
- 11.2 GENERAL BUCKLING FORMULAS — 453
- 11.3 FLAT PLATES — 458
- 11.4 CURVED PLATES — 464
- 11.5 COMBINED LOADINGS — 468

Chapter 12.0 SHEAR PANELS — 473
- 12.1 INTRODUCTION — 473
- 12.2 SHEAR RESISTANT WEBS — 474
- 12.3 PURE DIAGONAL TENSION WEBS — 483
- 12.4 DIAGONAL TENSION FLAT WEBS — 484
- 12.5 DIAGONAL TENSION CUREVED WEBS — 504
- 12.6 DIAGONAL TENSION EFFECT AT END BAY AND SPLICES — 523

Chapter 13.0 CUTOUTS — 530
- 13.1 INTRODUCTION — 530
- 13.2 UNSTIFFENED-WEB SHEAR BEAMS — 534
- 13.3 STIFFENED-WEB SHEAR BEAMS — 538
- 13.4 STIFFENED-WEB HAVING DOUBLER-REINFORCED HOLES — 546
- 13.5 WEB CUTOUT WITH BENT-DOUBLER — 548
- 13.6 FRAMING CUTOUTS IN DEEP SHEAR BEAM — 563
- 13.7 CUTOUTS IN SKIN-STRINGER PANEL UNDER AXIAL LOAD — 573
- 13.8 LARGE CUTOUTS IN CURVED SKIN-STRINGER PANEL (FUSELAGE) — 585

Chapter 14.0 COMPRESSION PANELS — 607
- 14.1 INTRODUCTION — 607
- 14.2 EFFECTIVE WIDTH — 610
- 14.3 INTER-RIVET BUCKLING — 613
- 14.4 SKIN-STRINGER PANELS — 616
- 14.5 STURDY INTEGRALLY-STIFFENED PANELS — 643
- 14.6 UNFLANGED INTEGRALLY-STIFFENED PANELS — 646

Chapter 15.0 DAMAGE TOLERANT PANELS (TENSION) — 655
- 15.1 INTRODUCTION — 655
- 15.2 THE STRESS CYCLE AND THE LOAD SPECTRA — 658
- 15.3 STRUCTURAL LIFE PREDICTION (SAFE-LIFE) — 660
- 15.4 STRUCTURAL CRACK GROWTH (INSPECTION INTERVAL) — 664
- 15.5 RESIDUAL STRENGTH (FAIL-SAFE DESIGN) — 668
- 15.6 RESIDUAL STRENGTH OF BEAM ASSEMBLY — 670

Chapter 16.0 STRUCTURAL SALVAGE AND REPAIRS — 677
- 16.1 INTRODUCTION — 677
- 16.2 STRUCTURAL SALVAGE — 679
- 16.3 SALVAGE EXAMPLES — 681
- 16.4 REPAIR CONSIDERATIONS — 686
- 16.5 REPAIR EXAMPLES — 696
- 16.6 REPAIRS FOR CORROSION DAMAGE — 713
- 16.7 REPAIR OF SKIN-STRINGER PANELS — 716
- 16.8 STRUCTURAL RETROFIT — 718

Chapter 17.0 STRUCTURAL TEST SETUP — 722
    17.1 INTRODUCTION — 722
    17.2 TESTING THE LOAD SPECTRUM — 722
    17.3 STRAIN GAUGES — 723
    17.4 FATIGUE TEST PANEL SPECIMEN — 726
    17.5 COMPRESSION TEST PANEL SPECIMEN — 737
    17.6 SHEAR TEST PANEL SPECIMEN — 738
    17.7 TEST SPECIMENS FOR PRIMARY JOINTS — 740

ASSIGNMENTS — 742

APPENDIX A   CONVERSION FACTORS — 755

APPENDIX B   FASTENER DATA — 756

APPENDIX C   COMMON PROPERTIES OF SECTIONS — 763

APPENDIX D   COMMON FORMULAS FOR BEAMS — 772

APPENDIX E   COMPUTER AIDED ENGINEERING (CAE) — 783

INDEX — 788

ABBREVIATIONS, ACRONYMS AND NOMENCLATURE

Aircraft Geometry

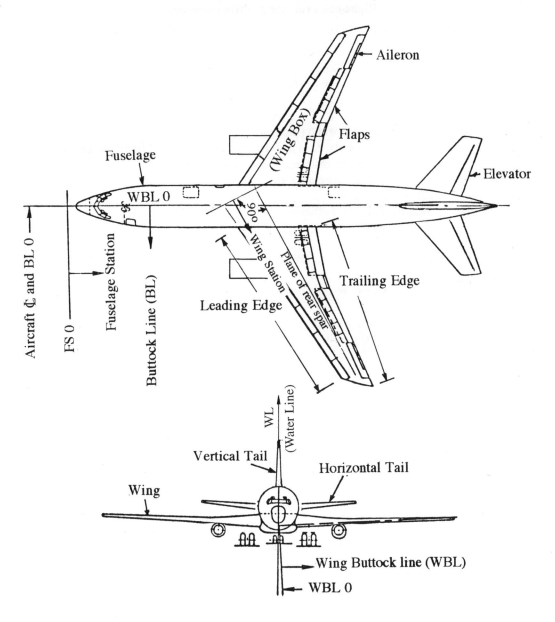

Airframe Stress Analysis and Sizing

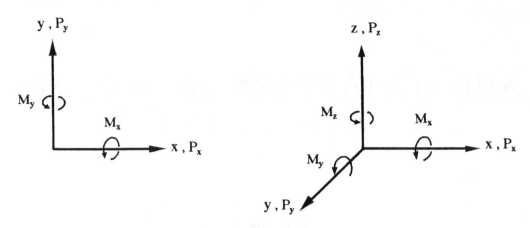

Right-hand rule for positive moments

(2-dimensional Coordinates)　　　　　　　　(3-dimensional Coordinates)

Coordinate Systems

a	–	Panel skin length (long side); crack length
A	–	Cross-sectional area
A/C	–	Aircraft
b	–	Width; panel skin width (short side)
BL	–	Buttock Line - distance in a horizontal plane measured from fuselage centerline
c	–	Column end-fixity; distance from N.A. to extreme fiber
C	–	Coefficient; fastener constant
c.g.	–	Center of gravity
C_N MAX	–	Maximum lifting coefficient
CSK	–	Countersunk
d	–	Diameter
D	–	Drag; diameter; plate stiffness
DM	–	drag force on main gear
DN	–	drag force on nose gear
Dia.	–	Diameter
e	–	Elongation; edge distance; shear center; eccentricity; effective
E	–	Modulus of elasticity (Young's modulus in elasticity range)
Eq.	–	Equation
E_t	–	Tangent modulus
f	–	Applied stress
F	–	Allowable stress; Farrar's efficiency factor
FAA	–	Federal Aviation Administration
FAR	–	Federal Aviation Requlations
fps	–	Feet per second
FS	–	Fuselage station – distance measured parallel to fuselage from a vertical plane forward of fuselage
ft.	–	Foot
Fwd	–	Forward
g	–	Load factor; gross
G	–	Modulus of rigidity (or shear modulus)
GAG	–	Ground-air-ground cyclic load spectrum
h	–	Height
HD	–	Head

H.T.	–	Heat treat
I	–	Moment of inertia; Angular inertia force necessary for equilibrium
in.	–	Inch
Inb'd	–	Inboard
in.-kips	–	Inch-(1,000) pounds
in.-lbs	–	Inch-pound
j	–	$\sqrt{\dfrac{EI}{P}}$
J	–	Torsional constant
JAR	–	(European) Joint Airworthiness Regulations
k	–	Buckling coefficient; section factor; kips; fastener spring constant; diagonal tension factor;
K	–	Modified buckling coefficient; shear lag parameter; stress intensity factor; stress concentration factor; fatigue quality index; efficiency factor
kip(s)	–	1,000 lbs.
ksi	–	(1,000 pounds) per square inch
L	–	Wing lift; length; longitudinal grain direction
LT	–	Long transverse grain direction
L'	–	Effective column length ($\dfrac{L}{\sqrt{c}}$)
lbs.	–	Pounds
lbs./in. or #/in.	–	Pounds per inch
m	–	Bending moment; meter; empirical material constant; mean value
M	–	Bending moment; Mach Number;
MAC	–	Mean Aerodynamic Chord
max.	–	Maximum
MFD	–	Manufactured
Min. or min.	–	Minimum
MS or M.S.	–	Margin of Safety
n	–	Material shape parameter; load factor; drag factor; number of cycles
N	–	Load intensity; number of cycles
NA or N.A.	–	Neutral axis
NACA	–	National Advisory Committee for Aeronautics
NASA	–	National Aeronautics and Space Administration
Outb'd	–	Outboard
P	–	Concentrated load; column load
psi	–	Pounds per square inch
pt.	–	Point
q	–	Shear flow
Q	–	First moment of area; concentrated load; shear load
R	–	Support reaction load; radius; load ratio; stress ratio
r	–	Radius
rad.	–	Radian
s	–	Fastener spacing
S	–	Stress
SF	–	Severtiy factor
S-N	–	Cyclic stress; stress vs. number of cycles
ST	–	Short transverse grain direction
T	–	Torsion load; total thickness
t	–	Thickness
Typ	–	Typical
V	–	Velocity; vertical load; shear load

W	–	Weight; width; concentrated load
w	–	Uniform loading; width
WBL	–	Wing Buttock Line – a plane normal to wing chord plane and parallel to fuselage and distance measured from intersection of wing chord plane and BL 0.0
WL	–	Water line – distance measured perpendicular from a horizontal plane located below bottom of fuselage
WS	–	Wing station – a plane perpendicular to wing chord plane and plane of rear spar, distance measured from intersection of leading edge and BL 0.0
x	–	x-axis
X	–	c.g. location in x-axis direction
y	–	y-axis
Y	–	c.g. location in y-axis direction
z	–	z-axis
Z	–	c.g. location in z-axis direction
"	–	Inch
α	–	Angle of oblique load; coefficient; angle of diagonal tension web; hole condition factor; correction factor
β	–	Coefficient; hole filling factor; reinforcement factor
γ	–	Shear strain; coefficient
θ	–	Twist angle; tangential angle; bearing distribution factor; petch angle
ϕ	–	Principal stress inclined angle; meridian angle; roll angle
ψ	–	Yaw angle
ρ	–	Radius of gyration ($\sqrt{\frac{I}{A}}$); density
μ	–	Poisson's ratio; torsional spring constant; coefficient
η	–	Plasticity reduction factor; fastener hole-out efficiency; plasticity correction factor
δ	–	Deflection
Δ	–	Deflection; increment
ε	–	Strain; rotational restraint
Σ	–	Summation
#	–	Pound
τ	–	Shear stress
κ	–	Coefficient
λ	–	Fitting factor; cladding reduction factor; buckling wave length
ν	–	Coefficient
£	–	Plasticity correction ratio for skin-stringer panel

Chapter 1.0

GENERAL OVERVIEW

1.1 OVERVIEW

A knowledge of structural design methods and an understanding of the load paths are essential to advancement in nearly all phases of airframe engineering. Approximately ten percent of the engineering hours spent on a new airframe design project are chargeable directly to:
- The obtaining of design loads in members
- The strength checking of drawings of the finished parts and assemblies

No aircraft engineer can expect to go far in advanced drafting, layout work, and design without a working knowledge of stress analysis. Thus, whether an engineer expects to go into the actual analysis work or to engage in some other type of aircraft engineering he will find a good working knowledge of stress analysis to be of great value in doing a better job.

The types of analysis used in airframe work are very different in many respects from those used in other types of engineering. The reason lies in the fact that the challenge of structural weight savings must be paramount if an efficient performance aircraft, i.e., one which carries a high percentage of pay load, is to be obtained.

(A) SIZING SCENARIO

In airframe structures, there are mostly redundant structures such as the typical wing box beam, which require the use of computer analysis. A three-stringer box beam, for instance, is a statically determinate structure but each additional stringer adds one more redundancy for each cross section. For airframe structures, the number of redundancies is of the order of thousands and the solution of such problems by conventional methods for solving highly indeterminate structures is extremely tedious and is, indeed, not feasible; computer analysis, such as the Finite Element Modeling (FEM) method and the method of successive approximation are the only reasonable methods to use in these cases.

If a simple structure with stringers of equal size and spacing is considered, analytical methods can greatly simplify the solutions to the problem for preliminary sizing work. If either the method or the structure analyzed is simplified when doing preliminary sizing, cost-effectiveness is increased, as shown below:

(a) Type of structures – For example, it may be possible to convert the wing box beam into a rectangular section with parallel shear webs and be symmetrical about a vertical plane.

(b) Lumping stringers – Reduce the complexity of large numbers of stringers by combining adjacent stringers into so-called 'lumped stringers'.

(c) Type of loading – Choose one of the critical or primary loads from the axial (from bending), shear flow (from vertical shear load or torsion) or lateral (from external aerodynamic or concentrated loads) loads to do the sizing.

(d) Limitation of stress distribution – Obtain the main or key stress distributions but not the detailed stresses.

Before going into detail sizing (also called production stress analysis), a rapid and reasonably accurate (i.e., approximate 0 – 10%) sizing of the structural dimensions is required for the preliminary aircraft design stage.

Another extremely important issue in every engineer's mind is that computer analysis will soon replace most, if not all, hand calculations in the aircraft industry. Computer analysis is a black-box operation in contrast to preliminary sizing which gives the engineer a feeling for the real world. With black-box operations, the engineer has no way of justifying the output results. When computer input is performed by the engineer, how can he verify that no mistakes were made in the input of numbers or decimals? Since Murphy's Law is always in operation, some mistakes are unavoidable in either design or analysis. All results or output should be reviewed by an experienced and/or knowledgeable engineer. If computer analysis is not performed by a knowledgeable engineer, it will turn out to be a "garbage in/garbage out" process which could result in very dangerous and questionable results. Preliminary sizing can be used in parallel to double check computer output to assure that the result is in the ballpark. All aerospace and aircraft engineers should learn preliminary sizing techniques to diagnose computer black-box problems. The difference between preliminary sizing and detail sizing (or detail stress analysis) can be summarized as follows:

(a) Preliminary sizing:
- Is a challenge and diagnosis job
- Most airframe structures are highly redundant structures, i.e., skin-stringer panels, cut-outs, tapered wing box structures, tapered fuselage cylindrical structures, etc.
- First simplify the structure into an equivalent, if it is a highly redundant one
- For simple members (and most airframe structures are not simple) there may be no difference between preliminary and detail sizing
- Sometimes requires the engineer to make assumptions and judgments based on previous experience
- Basically it is hand calculations plus help from a simple desk-top calculator
- Approximation methods may be used
- Apply simple methods of stress analysis approaches rather than detail analysis
- Prior to running airframe Finite Element Modeling (FEM), preliminary structural sizing input data are required for the first analytical cycles as shown in Fig. 1.1.1. and even the second or third cycles

(b) Detail sizing:
- A time-consuming and costly process
- Using the correct method and designing to the proper margin of safety (MS) is adequate as far as it goes.
- Gives more accurate results but usually requires help from a computer
- Involves more computer work including different software
- Most detail sizing is impossible by hand calculation within a reasonable time frame and cost because it involves too many equations (from a dozen to a few hundred):
  – load distribution in redundant structures.
  – cutout analysis
  – damage tolerance and fatigue analysis

Once the general features of an aircraft design have been decided, proceed as follows:

(a) First, lay out a structure which will accommodate those features and form a skeleton on which to hang the necessary installations.

(b) Next, determine the loading conditions which will cause the highest loads in the structures and make a preliminary sizing or analysis to find the effect of these loads. This preliminary sizing is necessary in order to determine approximate dimensions, since aircraft installations are as compact as possible and clearance allowances are low.

(c) After the final design features are decided upon and all installations, such as those for power plant, electrical system, control systems and furnishings, are placed it is possible to estimate the final dead weight and its distribution and to proceed with the final analysis (see Fig. 1.1.1).

The engineer should be not only thoroughly familiar with the requirements for routine strength checking but also have the knowledge necessary for:

- Detail design for fatigue considerations
- All structures must withstand hail and lightning strikes
- Must operate in and be protected against corrosive environments indigenous to all climates

The structure must have a serviceable life of 20 years or more with minimum maintenance and still be lighter than any vehicle built to date. Under stringent competition, the design must incorporate new materials such as advanced composite materials and processes that advance the state-of-the-art to improve aircraft performance.

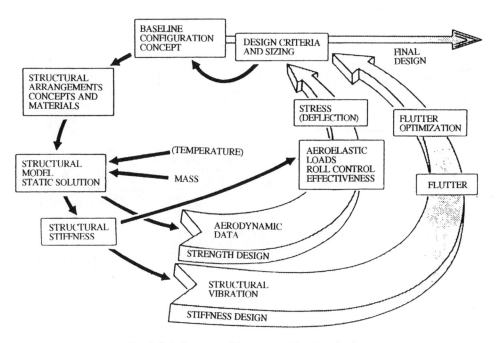

Fig. 1.1.1 Structural Analytical Design Cycles

(B) SPECIFICATIONS AND REQUIREMENTS

A good overall structural concept incorporating all these factors is initiated during preliminary design and sizing. At the very beginning of a preliminary design effort, a designer writes a set of specifications consistent with the needs. It should be clearly understood that during preliminary design it is not always possible for the designer to meet all the requirements of a given set of specifications such as shown in Fig. 1.1.2 and the U.S. FAA certification flowchart as shown in Fig. 1.1.3.

COMMERCIAL AIRCRAFT:

- FEDERAL AVIATION REGULATIONS (FAR), VOL. III, PART 23 – AIRWORTHINESS STANDARDS: NORMAL, UTILITY, AND AEROBATIC CATEGORY AIRPLANES
- FEDERAL AVIATION REGULATIONS (FAR), VOL. III, PART 25 – AIRWORTHINESS STANDARDS: TRANSPORT CATEGORY
- JAR (Joint Airworthiness Requirements) – EUROPEAN COUNTRIES

MILITARY AIRCRAFT:

- MIL-A-8860(ASG) – GENERAL SPECIFICATION FOR AIRPLANE STRENGTH AND RIGIDITY (U.S.A.)

Fig. 1.1.2 Regulations and Specifications

In fact, it is not at all uncommon to find certain minimum requirements unattainable. It is necessary to compromise as shown in Fig. 1.1.4 which indicates what might happen if each design or production group where allowed to take itself too seriously.

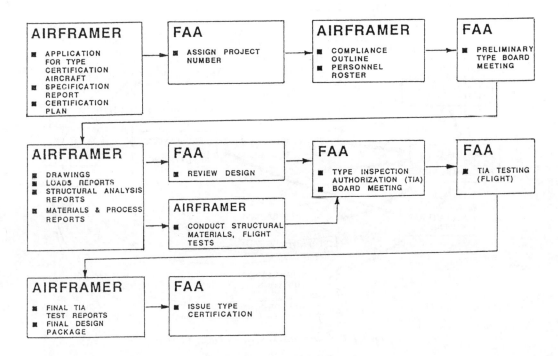

Fig. 1.1.3 U. S. FAA Certification Flowchart

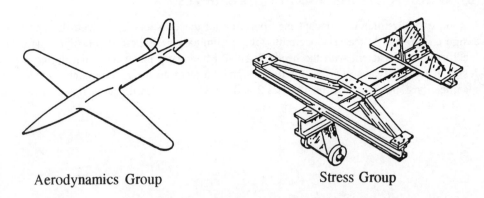

Aerodynamics Group   Stress Group

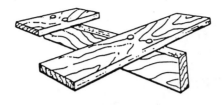

Production Group

Fig. 1.1.4 Results of Group Dominance in Aircraft Design

(C) DESIGNER'S FUNCTIONS

The extent to which compromises can be made must be left to the judgment of the designer based on the designer's interactions with other disciplines. The structural designer is merely a focal point for the collect of all necessary data and information from all disciplines prior to making the engineering drawing.

Designer's functions (as shown in Fig. 1.1.5):

- Provide focal point
- May need to depend heavily on the support of others for scientific and technical analysis
- Must really understand and apply basics of engineering in the thought process of design
- Should understand specialized engineering disciplines sufficiently to evaluate influence on design
- Think design and engineering basics and their interplay

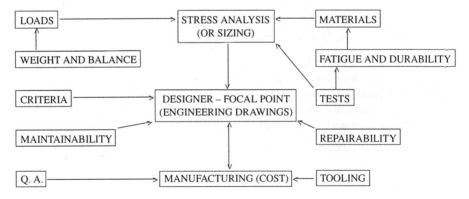

Fig. 1.1.5 Structural Designer's Function (make compromise with other groups)

(D) ENGINEER'S FUNCTIONS

- Many engineering functions do not require doing design, for example:
  - Analysis
  - Testing
  - Research
  - Metallurgy
  - Materials and Processes
  - Flight dynamics
  - Thermodynamics
  - Structural mechanics
  - Electromagnetic fields
- All engineers should have sufficient formal training in design and shop training subjects to understand the design process
- Some will not have an affinity or talent for design but the exposure will better equip them for productive work

However, it must be kept in mind that to achieve a design most adaptable to the specified purpose of the aircraft, sound judgment must be exercised in considering the value of the necessary modifications and compromises.

(E) DESIGN CONSIDERATIONS

Airframe design considerations and information are as follows:

- Designer should familiarize himself with all existing airframes or structures of a similar type to that proposed
- Use available design guides, design handbooks, design manuals, etc. that are generally the airframe company's proprietary information or data not ordinarily covered by classical textbooks or standard handbooks
- Provide for the smooth flow of stress from one section of a part to another and from one part to another
- For long-life structures, detail design is more important than the loading spectrum and operating stress level; it is possible to miss the desired structural life by a factor of 100 if the detail design is flawed
- Design for assembly – easier assembly or installation always provide better quality
- Structural tolerance – shims used where they could have been avoided are costly and wasteful; shims not used where needed or not used properly can be far more costly
- Damage tolerance design – built-in flaws:
  - Damage tolerance design (use lower stress level)
  - Fail-safe design (use higher stress level)
- Structural consideration for aircraft future growth
- Design for manufacturing to lower cost
- Design for maintainability and repairability to lower operating cost

1.2 DISCUSSION OF OPTIMUM DESIGN

The optimum design theory learned in college, just like the well-known classic beam theory, cannot be directly used on airframe structures because of its thin sheet and buckling characteristics. Instead, the empirical methods of analysis from NACA reports, which combine both theory and test results, should be the primary methods for airframe structures. One well-known method is the incomplete diagonal tension shear web analysis (NACA TN 2661).

However, the optimum design theory points the engineer in the right direction in pursuing lightweight structures.

The primary function of the structure is to transmit forces or loads. From the structural engineer's point of view, the objective is to do this with the minimum possible weight and minimal cost. In the general sense, the optimum structure is one that does the best over-all job of minimizing the undesirable qualities (weight, cost, repair, maintenance, etc.). If a structure could be found that would give minimum values on all counts, the problem would be solved. However, it may happen that the structure that has minimum weight will not be as low in cost as one only slightly heavier. In such cases the structural engineer must make a decision based on the function of the aircraft and other criteria that have to be met. Experience states that for every pound of structure which is ideally necessary to carry the basic loads in an aircraft, there is about 50% or more structural weight added onto it which consists of doublers for cutouts, brackets, splice fittings, tension and lug fittings, plus extra materials due to fabricating technique, avoiding corrosion, and other considerations.

(A) OPTIMUM AND NON-OPTIMUM EFFECTS

It is necessary to understand the distinction between the optimum weight and the extra weight caused by non-optimum weight which is represented by joints, cutouts, non-tapered sheet, fatigue, stiffness, cost, etc. The optimum and non-optimum effects are as follows:

(a) Optimum effects (minimum structure weight):
- Weight/strength ratio
- Material and proportions
- Structural configuration

(b) Non-optimum effects (practical design):
- Joints
- Cutouts
- Fatigue and fail-safe
- Stiffness
- Crippling
- Yield strength
- Minimum gages
- Corrosion
- Standard gages
- Non-tapered members
- Landing gear mounts
- Engine mounts
- Integral fuel tank
- Temperature

(c) Other effects:
- Cost
- Repairability
- Maintainability

(B) STRUCTURAL INDEXES

Since structural research is aimed primarily at finding the optimum structures for the design conditions encountered, the application of this principle must be considered as an essential part of such research. Likewise, it should be possible to reduce the design and analysis time considerably by taking advantage of all the structures that have previously been designed and tested.

The structural index is particularly useful in design work because it contains:

- The intensity of the loads
- Dimensions which limit the size of the structure.

The structural index offers the designer a guide to the optimum type of structure. Once the design data for a range of indexes has been worked out, the designer can tell at once which type of structure should be used. Furthermore, it is possible to determine the relative efficiencies of different types of structures as a function of the structural index. The structural index will indicate how much of a weight penalty must be accepted if for some reason the optimum structure is not selected.

For structural-engineering purposes, this optimum structural design can be done most conveniently by using the structural index because:

(a) The structural index is most useful in determining what constructions and/or material to fit a particular loading.

(b) The structural index can be used to:
- Quickly size structures
- Select the most efficient material
- Identify the lightest type of construction

One of the great advantages of this index approach is that design proportions that are optimum (minimum structural weight) for a particular structure are also optimum for structures of any size, provided that they all have the same structural index. This principle also permits the results of actual tests to be extended over a wide range of structures without the necessity of additional stress analysis.

As shown in the examples in Fig. 1.2.2, the structural index for an individual configuration is based on its own function:

- For shear beams, it is the shear load and the depth of the beam
- For columns, the axial load and the length of the column
- For circular shells or tubes, the bending moment and the diameter of the shell

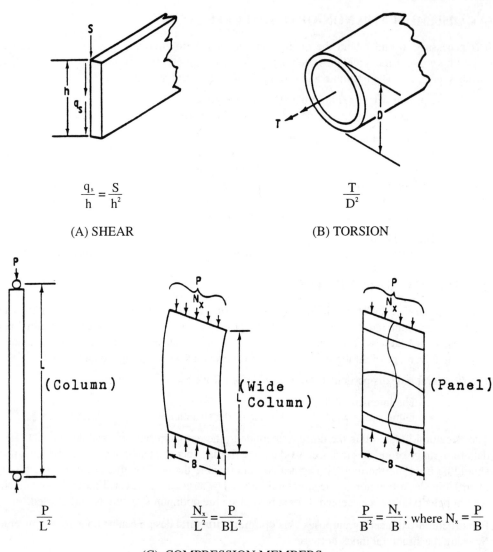

(A) SHEAR $\dfrac{q_s}{h} = \dfrac{S}{h^2}$

(B) TORSION $\dfrac{T}{D^2}$

(Column) $\dfrac{P}{L^2}$

(Wide Column) $\dfrac{N_x}{L^2} = \dfrac{P}{BL^2}$

(Panel) $\dfrac{P}{B^2} = \dfrac{N_x}{B}$; where $N_x = \dfrac{P}{B}$

(C) COMPRESSION MEMBERS

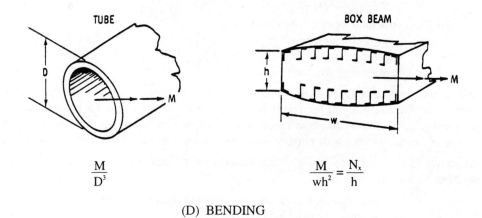

TUBE $\dfrac{M}{D^3}$

BOX BEAM $\dfrac{M}{wh^2} = \dfrac{N_x}{h}$

(D) BENDING

Fig. 1.2.2 Structural Index

For any given value of the structural index, it is possible to find the combination of material and cross-sectional shape that will give the lightest structure. In determining the optimum design on a weight/strength basis, the two major variables are the material and the proportions, or configuration, of the structure:

(a) For the simple case of pure tension, the optimum configuration is a straight line, and the proportions of the cross section have no effect on the weight. The weight/strength factor is then determined entirely by the properties of the material.

(b) In the case of pure compression, the optimum configuration is again a straight line, but the size and shape of the cross section now play an important part in the buckling strength of the member.

In choosing a material, the structural engineer is strongly influenced by many factors besides the weight/strength factor; e.g., by material cost, availability, simplicity of design, corrosion, formability, machinability, etc. It is therefore, most important to give the structural engineer complete information on the weight/strength comparisons of various materials, showing the weight advantages or penalties involved in selecting any particular material.

1.3 STRUCTURAL WEIGHT

The structural designers or stress engineers who determine structural sizes must be concerned with weight. Nevertheless, with today's specifications, there is a tendency to narrow one's viewpoint to the mechanics of the job and to forget the fundamental reasons for that job. It has been said, sometimes in jest and often in earnest, that the weight engineer is paid to worry about weight. However, unless aircraft engineers in a company are concerned about weight, that company may find it difficult to beat the competition or, to design a good performance airplane. On a typical aircraft one pound of excess empty weight will result in the addition of ten pounds to the aircraft gross weight to maintain performance.

Weight engineers can estimate or calculate the weight of an airplane and its component parts but actual weight savings are always made by stress engineers. Today, a very small margin of weight can determine the difference between the excellent and poor performance of an airplane. If the structure and equipment of a successful model are increased by only 5% of its gross weight, the consequent reduction in fuel or payload may well mean cancellation of a contract. In transport aircraft the gross-weight limit is definitely stipulated; thus, any increase in empty weight will be accompanied by a reduction in fuel or payload.

The weight breakdown of aircraft structures over the years shows a remarkable consistency in the values of the structural weight expressed as percentages of the take-off gross weight (or all-up weight) realized in service, irrespective of whether they were driven by propeller or by jet; see Fig. 1.3.1. At the project stage, if performance and strength are kept constant, a saving of structural weight is also accompanied by savings in fuel, the use of smaller engines, smaller wings to keep the same wing loading and so on, so that the saving in take-off weight of the aircraft to do the same job, is much greater than the weight saved on the structure alone.

The objective of structural design is to provide the structure that will permit the aircraft, whether military or civil, to do job most effectively, that is with the least total effort, spread over the whole life of the aircraft from initial design until the aircraft is thrown on the scrap heap. There is thus an all-embracing simple criterion by which the success of the structural design can be judged. It is not sufficient to believe that the percentage of structural weight is of itself an adequate measure of effective design, either of the complete airplane or of the structure itself: a well-known example is the effect of increased wing aspect ratio:

- Increased structural weight
- May give increased fuel economy while cruising

Nevertheless, the percentage of structural weight is a useful measure, provided its limitations are recognized.

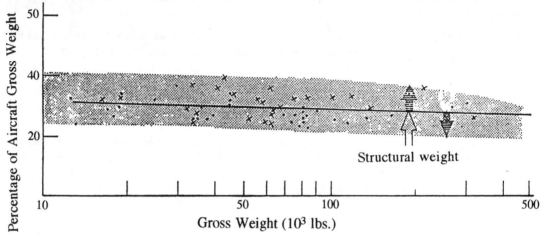

Fig. 1.3.1 Typical Structural Weight of an Aircraft

However, it is very difficult to control structural weight during the production stage because of the constraints of schedule and on-time delivery. It is not unusual for the first lots of an airframe to be slightly heavier than the predicted target weight. The engineer can reduce structural weight by the tail end of production by utilizing a "weight reduction program" illustrated by the form shown in Fig. 1.3.2.

Type of Change (Check one or more)

Structure Change			Systems Change	
A	Fuselage	_____	H Propulsion	_____
B	Wing Center Section	_____	J Controls	_____
C	Wing Pylon	_____	K Electrical/Avionic	_____
D	Outer Wing	_____	L Hydraulic	_____
E	Empennage	_____	M Environmental Control/	
F	Engine Pod	_____	Anti-Icing	_____
G	Landing Gear	_____	N Flight Station	_____
			O Interior	_____

Brief Description of Change (Use additional pages if necessary) _____

_____
_____
_____
_____
_____

Reference (If Applicable)

  Basic Spec Paragraph Number _____
  Layout/Dwg. No.   _____
  Part No.   _____
  Other   _____

  Rough estimate of weight reduction _____ lb/airplane

                    Signed _____

	Accepted for Review		Approval	Date
Date Received by Weight Department				
Weight Department	Yes	No		
Project Group Engineer	Yes	No		
Project Department Manager	Yes	No		
Assistant Division Engineer	Yes	No		
Division Engineer	Yes	No		
Chief Engineer	Yes	No		

If unacceptable — state reason and return to originator _____

_____

_____

Signature _____ Date _____

Fig. 1.3.2 Example of a Weight Reduction Proposal Form

1.4 DESIGN FOR MANUFACTURING

Since an airframe structure is one of the most expensive and complicated systems on an aircraft, a team effort (or concurrent engineering) to design for manufacturing is crucial in reducing overall cost. Cost savings has to be the bottom line to survive in today's highly competitive market.

The cost of developing a new transport aircraft can be divided into two major cost categories:

(a) Non-recurring costs
- Engineering
- Planning
- Testing

(b) Recurring costs
- Material
- Labor
- Quality Assurance (Q. A.)

Recurring costs (or manufacturing costs) make up approximately 80% to 90% of the above total cost and design for manufacturing is an important asset in lowering manufacturing costs. Methods of cost reduction include the following:

- Lower part counts
- Use standard parts, and commonality
- Simplify fabrication and assembly
- Use proven methods
- Allocate funds for research and development to pursue new methods
- Relax tolerances
- Use CAD/CAM
- Automate manufacturing processes

Design and manufacturing, as shown in Fig. 1.4.1, are successive phase of a single operation, the ultimate objective of which is the emergence of an acceptable final product. In an aerospace context, such acceptability has several components:

- Market viability
- Operational efficiency
- Capacity for further development
- Structural integrity

Less obvious but just as important, a structure must not be so complex or difficult in concept that its realization will create great difficulties, or increase unduly the cost of the manufacturing process. Production, emerging as a specialized branch of engineering, is sandwiched between the designer's drawings and the final product. Consequently, its achievement is less apparent, and frequently in the past it has not been accorded a like degree of consideration or credit. Yet, it is the production phase of the operation that translates the design into hardware with reasonable cost.

Chapter 1.0

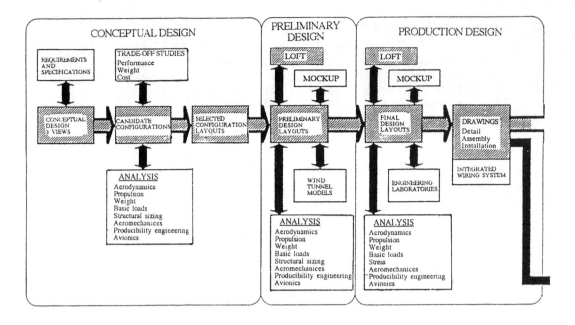

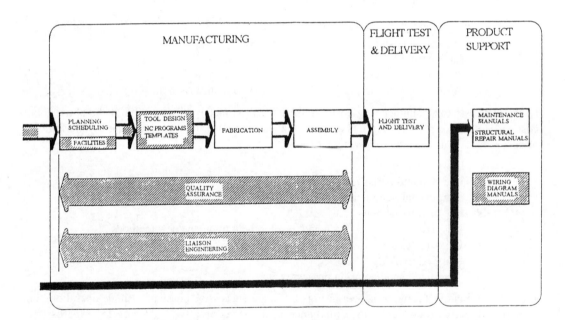

Fig. 1.4.1 How an Aircraft is Fabricated

An airframe is conceived as a complete structure, but for manufacturing purposes must be divided into sections, or main components, which are in turn split into sub-assemblies of decreasing size that are finally resolved into individual detail parts. Each main component is planned, tooled and built as a separate unit and joined with the others in the intermediate and final assembly stages as shown in Fig. 1.4.2.

General Overview

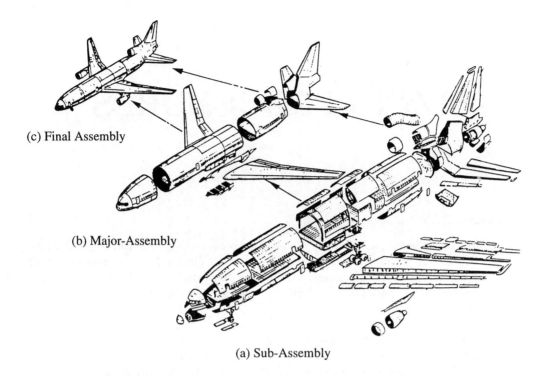

(c) Final Assembly

(b) Major-Assembly

(a) Sub-Assembly

Fig. 1.4.2 Lockheed L-1011's Sub-assemblies and Final Assembly

Tooling is required for each stage of the building of each component: detail tooling of individual parts of which there may be many thousands, followed by assembly tooling of increasing size and complexity for the stages of progressive assembly.

There is nothing new in attempting to design an aircraft to give trouble-free operation. This has, of course, always been one of the major parts of a design job. Today, there has been incorporated the requirements of both structural maintenance and repair in every engineering drawing before it is released and goes into production. If airframes have become too difficult to maintain and repair, it is not entirely due to a lack of appreciation of the problem at the drawing stage, but is mainly due to the very great increase in complexity of modern aircraft. A great deal of this complication is due to equipment and automatic gadgetry.

Between 25 to 40% of the total direct operating cost (DOC) of an aircraft is due to maintenance and repair, quite apart from the losses due to the aircraft not being serviceable when required.

References

1.1 Niu, C. Y., "AIRFRAME STRUCTURAL DESIGN", Conmilit Press Ltd., P.O. Box 23250, Wanchai Post Office, Hong Kong, (1988).
1.2 Bruhn, E. F., "ANALYSIS AND DESIGN OF FLIGHT VEHICLE STRUCTURES", Jacobs Publishers, 10585 N. Meridian St., Suite 220, Indianapolis, IN 46290, (1975).
1.3 NASA TM X-73306, "Astronautic Structures Manual, Vol. I, Vol. II, and Vol. III", George C. Marshall Space Flight Center, Marshall Space Flight Center, Alabama 35812, (August, 1975).

Chapter 2.0

SIZING PROCEDURES

2.1 OVERVIEW

This Chapter contains the requirements for stress analysis or sizing reports and emphasizes the importance of preparation and of writing concise and readable reports. It is the stress engineer's responsibility to determine that the design chosen for a particular item is the lightest possible compromise of the various considerations which affect the weight of the proper design.

(a) Design loads:

Design loads for stress analysis are the ultimate loads which are limit loads multiplied by the 1.5 safety factor. All loads coming from the Loads Department are limit loads (which are the maximum loads on aircraft structures, not ultimate loads) except those specified as ultimate loads for other purposes. The safety factor of 1.5 is intended to cover some or all the following:

- Uncertainties in loads
- Inaccuracies in structural analysis
- Variations in strength properties of materials
- Deterioration during service life
- Variations in fabrication between nominally identical components

(b) Coverage (Classification of parts is based on common sense or determined by authoritative engineers in the company):

- Critical – all primary structures are to be fully substantiated, e.g., wing, empennage, fuselage, etc.
- Near critical parts – Select number of points and components are analyzed, e.g., wing and empennage trailing edge panels, radomes, etc.
- Non-critical parts, e.g., fairings, cabin interior panels, etc.

(c) Begin with the overall picture and work down to the details:

- What is being analyzed?
- Why is it being analyzed?
- How is it being analyzed?
- Load conditions and references
- Calculation of stresses or loads and allowables
- Both structural and computer model elements should be identified when these are necessary for completeness in analysis
- State engineering assumptions
- Use company's Design Manuals or Handbooks (approved by certifying agency) if they are available
- Maintain consistency with the same calculated numbers throughout the calculations

- Arrange it so that data is developed in the sequence needed
- Show the final Margin of Safety (MS) and state the type of failure and load condition

(d) The sequence of stress work can be generally classified into the following three categories:

- Preliminary sizing – the project is just beginning and the aircraft is in the preliminary design stage
- Production (or project) stress analysis – continuation of the project
- Final formal stress report for aircraft certification purpose – end of the project

2.2 PRELIMINARY SIZING

Unlike formal stress reports, preliminary sizing calculations may be handwritten but should be legible enough to make good reproduction copies. Stress sheets are merely for project use but should be kept for future information and stored in a good filing system so that data can be efficiently retrieved whenever needed.

Preliminary sizing calculations:

- Consist of unsubmitted data and information
- Provide data to begin project
- Take 6 months to one year (depending on market situation)
- Require the most experienced engineers
- Based on given conditions and loads
- Require a knowledge of basic methods of stress analysis
- Use rough or approximation methods of calculation to obtain quick answers
- Use rough sizing within reasonable accuracy (conservatively not more than 10%)
- Must meet fit and function requirements
- Require knowledge of fatigue and damage tolerant design
- Stress check usually not required
- Require basic knowledge of:
  - material selection
  - manufacturing and fabrication costs
  - repairability, maintainability, and assembly procedures
  - future aircraft growth consideration
- May not be neat but must be legible
- Must be saved for follow-on production design use

In general, these steps should be followed whenever doing preliminary sizing:

Step 1 – Recognize the structural function and configuration of the component:
- Wing box, tail box, fuselage barrel
- Control surfaces
- Joint, splice or fittings
- Panel with cutout

Step 2 – Basic loads (static, fatigue, fail-safe, and crash-condition loads):
- Tension
- Compression
- Any combination of above loads
- Shear
- Normal surface pressure loading

Step 3 – Material selection:
- Static
- Fracture toughness
- Fatigue

Step 4 – Fastener and repairability:
- Fastener selection
- Adequate edge margin
- Joint or splice requirements

Step 5 – Compromise efficient structure:
- Configurations
- Installation and assembly
- Fabrication
- Design for stiffness requirements (flutter)

Step 6 – To meet low cost:
- Manufacturing and assembly
- To meet market competition
- Aircraft performance requirements

2.3 PRODUCTION STRESS ANALYSIS

Long before the formal stress report is written, the production (project) stress analysis will be done to support structural design (engineering drawings). All the stress analysis sheets are kept on file for use in the later formal stress report.

Production stress analysis is the most important factor in determining whether or not the airframe will meet the target weight (usually they are slightly overweight during the production design stage). Engineering manpower is at its peak during this stage with additional help provide by job shoppers (temporary workers).

Production stress work:
- Consists of unsubmitted data and information
- Takes about two years (commercial transports)
- Consists of detail stress analysis
- Based on given structure, materials, loads, etc.
- Requires knowledge of methods of stress analysis:
  – from airframe company's design and structural manuals, if available
  – from documents, books, reports, etc.
- Requires use of computers for more accurate stress analysis
- Requires Finite Element Modeling structural analysis to determine more accurate load distribution within redundant structures
- Defines Margin of Safety (MS) under static loads (ultimate load conditions)
  – keep MS close or near to 0 to save structural weight
  – allow high MS only for special requirements such as fatigue, stiffness, test result, etc.
- Stress checks are required only for the move important areas

2.4 FORMAL STRESS REPORTS

Unlike preliminary sizing and production stress analysis, the formal stress report is typed (or word processed) and requires a high standard of readability and neatness for submission to the certifying agency. The report should be written in the third person and the writer should place himself (or herself) in the position of the unfamiliar person who will be reading and checking the report.

Formal stress report:
- Consist of submitted data and information
- Usually takes a little over a year
- Consist of wrap-up and condensation of production stress analysis
- Requires more effort
- Must be neat and clean report with minimal page count
- Requires checking and approval
- Must use approved methods of analysis
- Must specify drawing references
- Must use company's stress sheet form

- Is sent to certifying agency with several copies kept in project office for reference
- May be revised to make corrections or substantiate a down stream derivative airframe

These reports are a means of satisfying the certifying agency that airframe strength requirements have been fulfilled before issuing the aircraft type certificate. They also serve as:

- An important basis for airframe modification and conversion
- The basis for resolving production salvage and retrofit issue

(A) COVERAGE:

(a) Primary structures:

- All primary structures are to be fully substantiated
- All significant internal loading shall be summarized for parts that are substantiated by ultimate strength testing in lieu of stress analysis
- These loadings are to be included in the submitted reports or unsubmitted reports kept in Project Department file systems

(b) Repetitive structures:

- Covered by analysis of the critical or typical one of each group
- Structures such as typical ribs, bulkheads, clips, etc.)

(c) Near critical parts will be guided by:

- Select number of points and components analyzed
- Fatigue life
- Stiffness

(d) Non-critical parts:

- Aircraft exterior fairing and most of the interior cabin parts are non-critical structures
- Formal stress reports for these parts are usually not required by the certifying agency
- Unsubmitted stress analysis is still required and kept on file for information

(B) CONSIDERATIONS AND REQUIREMENTS:

(a) A good stress report can be produced without excess bulk by:

- Use of tabular calculations
- Reduction or elimination of routine calculations
- Use of computer output results

(b) Begin with the overall picture and work down to the details:

[See Chapter 2.1(c)]

(c) Calculations not shown:

- All repetitious and simple calculations should be omitted
- One sample calculation may be shown and the remaining results summarized in tabular or graphical form
- Computer results of structural allowables, section properties, etc.
- Test details

(d) Unsubmitted data and information:

- Includes final computer output data
- Kept and maintained by each project group or department office
- Considered to be part of the basis for the submitted analysis
- Provides basis for the formal stress report

- A complete file of unsubmitted analysis shall be maintained for the licensing agency and for future reference until the end of the project
- Good filing systems are necessary to store and retrieve data efficiently

(e) Report should be reviewed by an independent checker to catch errors (signs of loading are especially important) and to check:
- Numerical calculations
- Correct method of analysis
- Completeness of coverage
- Readability

(f) Sketches:

Sketches should be kept to the absolute minimum that will fulfill the following requirements:
- Describe the parts
- Show location of parts on airframe
- Show type of loading and reaction involved
- Identify all parts by drawing numbers
- Show the dimensions that are to be used in the analyses

Also:
- Do not number figures except to identify them when more than one appear on a page
- Draw neatly and approximately to scale, but show no more than necessary
- Use should be made of reductions of production drawings

(g) Standard symbols and sign conventions:
- Define abbreviations, acronyms, and nomenclature
- Use your company's standards or those found in this book
- Standard symbols can be found in MIL-HDBK-5
- Use the common sign conventions found in this chapter

(h) References:

References used in formal stress analysis shall be approved by a certifying agency and/or the company's project office. Every figure, sketch, formula, and analysis should be referenced to its source except for the following exceptions:
- Items, formulas, etc., easily found on the same page
- Common equations or theorems of algebra and trigonometry
- Abbreviations, acronyms, and nomenclature

(C) PREPARATION

(a) Paper forms :
- A size paper (8.5" x 11"):
  - Title page
  - Revision pages
  - Margin of Safety pages
  - Analysis pages (basic pages used in the body of the report)
  - Distribution list pages
- B size paper (17" x 11"): Fold-out pages should be kept to a minimum

(b) Formal stress reports shall contain the following data and information:
- Title page
- List of pages to prevent pages are missing

Sizing Procedures

- Abstract
- Three-view drawing of the aircraft
- Table of Contents
- Engineering drawing list
- Symbols, acronyms, and nomenclature
- Load conditions with coding data
- Summary of test results
- Summary of final Margins of Safety (MS)
- References
- Detail analysis
- Appendixes as required
- Distribution list of controlled copies

2.5 SIGN CONVENTIONS

A significant problem and challenge for engineers doing the stress or load analysis is the selection of sign conventions. The following recommendations will prevent confusion:

- Engineers should use standard sign conventions so that calculations can be read and checked without confusion
- It is strongly recommended to assume that unknown loads are positive and to represent them as such in calculations
- If an unknown load was assumed to be positive but turns out to be negative, the load direction must be reversed
- Once the engineer has assigned a sign direction, whether positive or negative (not recommended), to an assumed load, it must be maintained throughout the entire calculations; otherwise confusion will occur and correct results will be impossible to obtain

(A) SIGN CONVENTIONS

Positive sign conventions representing axial load, shear load, bending moment, and torsion should be assigned, as shown in Fig. 2.5.1 and described below:

- Axial load is considered positive when the load puts the element in tension, or acts away from the element
- Positive shear load is defined as acting downward on the right side of the element and acting upward on the left side of the element; positive shear flow is defined in the same manner
- Positive bending moment is defined as acting counter-clockwise on the right face of the element and clockwise on the left side of the element, or that which produces compression in the top "fibers"
- Use a double-arrow symbol to represent torsion or moment in which the positive torque or moment is indicated by the right-hand rule with the thumb of right hand pointing in the positive moment direction, the curl of right-hand fingers indicating positive torque
- Positive angle rotation is in the counterclockwise direction and 0° points to the right-hand side

Chapter 2.0

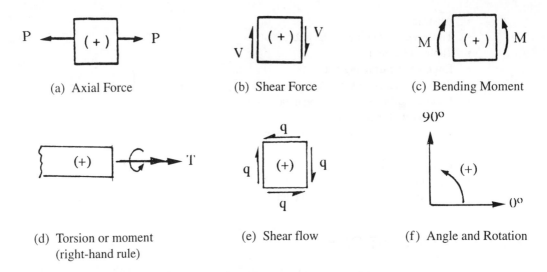

(a) Axial Force (b) Shear Force (c) Bending Moment

(d) Torsion or moment (right-hand rule) (e) Shear flow (f) Angle and Rotation

Fig. 2.5.1 Positive Sign Conventions

(B) REACTION LOADS

The signs used to represent reaction loads are shown in Fig. 2.5.2:
- If the direction of a reaction is unknown, assume it acts in the positive (+) direction
- If the result is positive, the assumption was correct. If not, change the sign
- The reaction loads occur generally at supports
- Moment and shear diagrams are generated by using applied loads in conjunction with the known reaction loads at supports

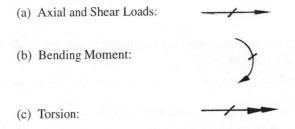

(a) Axial and Shear Loads:

(b) Bending Moment:

(c) Torsion:

Fig. 2.5.2 Symbols for Reaction Loads

(C) LOAD DIAGRAMS

Load diagram plots are shown in Fig. 2.5.3 and Fig. 2.5.4:
- The positive bending moment diagram is plotted on the compression stress side of the member
- The positive shear diagram is plotted:
  - on the upper side of the simple beam
  - on the outside of the frame, arch, etc.

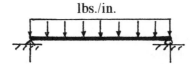

(a) Assume A Beam With Uniform Loading

(b) Bending Moments (c) Shear Loads

Fig. 2.5.3 Load Diagrams for a Simple Beam

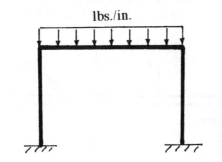

(a) Assume a Frame with Uniform Loading

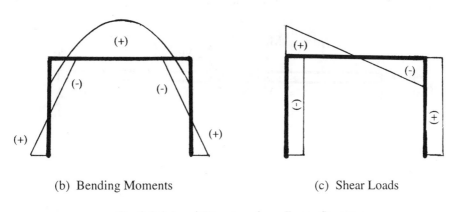

(b) Bending Moments (c) Shear Loads

Fig. 2.5.4 Load Diagrams for a Frame Structure

2.6 LOAD PATHS AND FREE-BODY DIAGRAMS

Whether the correct load path (or internal load distribution) in the analyzed structure is either determinate or indeterminate is one of the first things which must be decided before starting to perform structural analysis or sizing calculations. The correct load paths are determined by using Finite Element Modeling, a must to determine the correct load paths in highly redundant structures:

- Cut-out
- Shear-lag
- Fatigue load distribution around fasteners
- Multi-cell box beams
- Stringer run-out
- Skin-stringer panels
- Make sure the assumed boundary conditions are correct or very close to those of the real structure, otherwise the output will be garbage
- Never say, "the computer says" as if computer analysis is never mistaken since computer output is only as correct as what is inputted

In analyzing the action of forces on a given structural member body, it is absolutely necessary to isolate the body in question by removing all contacting and attached bodies and replacing them with vectors representing the forces that they exert on the isolated body. Such a representation is called a free-body diagram; see Fig. 2.6.1. The free-body diagram is the means by which a complete and accurate account of all forces acting on the body in question may be taken. Unless such a diagram is correctly drawn the effects of one or more forces will be easily omitted and error will result. The free-body diagram is a basic step in the solution of problems in mechanics and is preliminary to the application of the mathematical principles that govern the state of equilibrium or motion. Note that where an imaginary cut is made, the internal loads at that point become external loads on the free-body diagram.

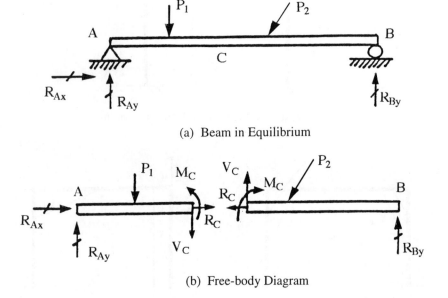

Fig. 2.6.1 Free Body of a Simple Beam

There are two essential parts to the procedure of constructing a free-body diagram:
- A clear decision must be made as to exactly what body (or group of bodies considered as a single body) is to be isolated and analyzed. An outline representing the external boundary of the body selected is then drawn.
- All forces, known and unknown, that are applied externally to the isolated body should be represented by vectors in their correct positions. Known forces should be labeled with their magnitudes, and unknown forces should be labeled with appropriate symbols.

In many instances the correct sense of an unknown force is not obvious at the outset. In this event the sense may be arbitrarily assumed and considered to be positive relative to the free-body diagram. A sign convention as shown in Fig. 2.5.1 and Fig. 2.5.2 must be established and noted at the beginning of each calculation. The sign convention selected must be consistently followed throughout each calculation. The correctness or error of the sense assumption will become apparent when the algebraic sign of the force is determined upon calculation. A plus sign indicates the force is in the direction assumed, and a minus sign indicates that the force is in the direction opposite to that assumed.

2.7 DRAWING TOLERANCES

The general procedure is to design to zero margin of safety on nominal dimensions and to show in the formal report. However, in cases where adverse tolerances could produce a significant strength reduction, their effect should be considered. The strength level below which tolerances cannot be neglected is that level at which there is an appreciable possibility that service difficulties will be encountered. The probabilities of the tolerances going bad, the material having minimum properties, the structure reacting to the loading condition as severely as predicted by strength analysis and the loading condition being achieved, should all enter into this decision.

Rule: Use mean dimension and total negative tolerance not exceeding the following values:

- 10% for single load path
- 20% for multiple load path

Example:

Given the following dimensional tolerances find the value of 'a' for stress analysis for a single load path:

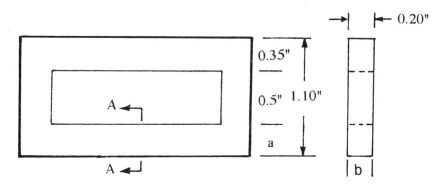

(All dimensional tolerances in this example are ± 0.030 in.)

The calculated dimensions or "a" and "b" are:

$a = 1.10 - 0.35 - 0.5 = 0.25$ in.;

$b = 0.20$ in. (Don't use this value)

Tolerances affecting the dimensions are:

$a \rightarrow \pm 0.03; \pm 0.03; \pm 0.03$

$b \rightarrow \pm 0.03$

Negative tolerances (interested values):

$$a = -(0.03 + 0.03) = -0.06 \text{ in.}$$
$$b = -0.03 \text{ in.}$$

If a required dimension is determined by using more than two other dimensions, only the two largest tolerances shall be added to obtain the total negative for the computed dimension. Therefore, the dimensions for this example should be:

$$a = 1.1 \, (0.25 - 0.06) = 0.209 \text{ in.}$$
$$b = 1.1 \, (0.20 - 0.03) = 0.187 \text{ in.}$$

(The above factor of 1.1 is for single load path; use 1.25 for a multiple load path)

The actual computed area at cross section A – A is

Area: $A = a \times b = 0.209 \times 0.187 = 0.039 \text{ in.}^2$ (Use this value for stress analysis)

If use nominal area: $A = a \times b = 0.25 \times 0.20 = 0.050 \text{ in}^2$ (Don't use this value)

Conclusion: The area difference is $\Delta A = \dfrac{0.05 - 0.039}{0.039} = 28.2\%$.

The following alternatives are recommended that to call out 'a' dimension on the engineering drawing, if the value 'a' is critical for strength:

$$a = \dfrac{0.260}{0.240} \quad \text{or} \quad a = 0.250 \pm 0.010 \quad \text{or} \quad a = 0.250 \, \begin{array}{l} -0.000 \\ +0.020 \end{array}$$

It should be noted that the nominal dimension is the same (0.250) for each of the three alternatives.

2.8 MARGIN OF SAFETY (MS)

Unless otherwise specified, a factor of safety of 1.5 must be applied to the limit load:

- Ultimate load = 1.5 × limit load
- The structure must be able to support ultimate loads without failure
- The structure must be able to support limit loads without detrimental permanent deformation

The general procedure is to design a structure to zero margin. The Margin of Safety (MS) for the stress analysis is equal to zero or greater, but is never a negative. Follow these to calculate the MS to meet airworthiness requirements (also refer to Chapter 4.4):

(a) First step – Under the ultimate load case:

$$MS_{(xxxx)} = \dfrac{F}{f} - 1 \geq 0$$

where: F – Use capital letters to represent allowable stress (F), moment (M), load (P), etc.
        f – Use small letters to represent applied ultimate stress (f), moment (m), load (p), etc.
        (xxx) – Option for showing the critical condition, i.e.:
        Tension
        Compression
        Shear
        Buckling
        Bearing
        Fastener
        Etc.

(b) Second step – Check material yield conditions:

$$MS = \frac{F_{yield}}{f_{limit}} - 1 \geq 0$$

where: F_{yield} – Use capital letters to represent allowable stress (F_{cy}, F_{bry}, etc.), moment (M), load (P), etc.

f_{limit} – Use small letters to represent applied limit stress (f), moment (m), load (p), etc.

(c) Third step – The final MS is the smallest MS from either (a) or (b).

The above mentioned MS which is based on ultimate static strength will be used unless it is overridden by the following criteria:

- Adequate fatigue life for the loading spectrums
- Sufficient rigidity for the aero-elastic or dynamic considerations
- Damage tolerance

2.9 STIFFNESS REQUIREMENTS

In many instances in the past aircraft manufacturers have experienced service failure and malfunction of various mechanical devices due to insufficient rigidity or improper allowables for possible and/or probable build-up of material and manufacturing tolerances, for instance:

(a) Loss of doors and/or improper seating of doors due to excessive deflections and adverse tolerance build-up.

(b) Failure of "over center" and "dead center" mechanisms due to adverse tolerances and unaccounted-for deflections.

(c) Loss of canopies due to adverse tolerances and excessive deflection.

Guarding against such failures and malfunctions is a joint responsibility of the Design and Stress Departments. Some of the reasons for these malfunctions should be more apparent to the stress engineer than the design engineer since during the computation of loads the effects of variations in the design can be qualitatively evaluated.

Specifications require that all devices maintain full strength and function properly under deflections and stresses at limit loads:

(a) The Stress Department is responsible for sufficient control system rigidity and the effect of tolerances, where applicable, should be considered when calculating the rigidity of control systems.

(b) Consideration should be given to the effects of limit deflections on the device, whether due to limit load on the device or due to limit load on other parts of the airplane.

(c) In addition to the specification requirements, the Structural Dynamics department often specifies rigidity requirements for various items:
  - Requires structural stiffness of EI and GJ values for wing empennage and engine pylon for divergence to prevent flutter in high speed flight
  - GJ values for control surface structures

The Stress department should take responsibility for coordination with the Structural Dynamics department.

It is recognized that using some of these criteria will result in designs that are considerably over strength. The design penalty may seem large in the preliminary stages but actually it is only a fraction of the cost, dollarwise and timewise, having to make a change at a later date. The weight penalty is more than justified in terms of practicality and reliability.

2.10 DISCUSSION

The practical stress analysis and sizing can be drawn as following:

- Practically, stress analysis is not an exact solution but is a good approximation and its derivation is covered by the safety factor of 1.5
- Conventional methods of stress (or structural) analysis obtained from college study, textbooks, handbooks of structural formulas or equations, etc. that mostly cannot be directly used on airframe sizing because there are many effects from various boundary conditions, elastic supports, loading conditions, etc.
- The input of engineering judgment and/or assumption is frequently to be used to modify the conventional method of analysis to fit the need
- Sometimes structural tests may be required to justify or modify the conventional method of analysis for critical components, e.g., cutouts, critical joint analysis, etc.
- The best computer is "you" not the computerized method which is merely a very good tool for your use to gain calculation speed and to solve highly redundant structural systems
- Engineers shall understand what is in the computer analysis rather than just perform the input and output process; otherwise it makes no difference between "you" and a robot (so called "engineering crisis" which will end up to be a Murphy's Law result!)
- Traditionally, when in "doubt", the engineer uses a conservative method or approach to size airframe structures and the level of conservatism is based upon engineering judgment

One of the most important things for a successful engineer to do is to maintain a very efficient filing system for each individual use (honestly speaking, this is the weakest point for almost every engineer) or engineering project purpose. Engineers need not remember every piece of data or information but must have quick and efficient means of locating whatever is required.

A good and efficient filing system in an engineering project should be set up in such manner that:

- Retrieval of information and data is quick and easy
- There is one central filing system in a group unit (e.g., a group of engineers under a lead engineer or supervisor could be a group unit) rather than many separate files maintained by each individual engineer
- Anyone can retrieve data in a group unit without help
- Storage space is reduced

Chapter 3.0

EXTERNAL LOADS

3.1 INTRODUCTION

This chapter provides a summary review of the external loads used in the structural design of transport aircraft. For consistency, the major design loads discussed herein will concentrate on commercial transports. The entire discussion is based on U.S. FAR 25 as a guideline, and for reference only. Engineers should refer to the updated regulations to which aircraft will be certified, i.e., FAR, JAR, etc. Military aircraft are basically similar to commercial transports, the main differences being their higher load factors, severe maneuvering capabilities and flight environments.

On a commercial transport jet, the number of primary airframe load conditions investigated could well be over 10,000. The number of actual critical design conditions is about 300 – 500, involving almost every type of maneuver required by FAR. Additionally, critical internal loads may occur at as many as near 20,000 locations on an average wide-body size jet transport, e.g., B747, B777, DC10, L-1011, A300, etc. Handling the numerical data generated would not be economically feasible without the capability of modern computers and the potential for organization inherent in matrix algebra techniques.

External loads, as shown in Fig. 3.1.1, are generally defined as those forces and loads applied to the aircraft structural components to establish the strength level of the complete aircraft. The strength level, or structural criteria, is defined by the aircraft speed, load factors, and maneuvers the aircraft must be capable of performing. The structural component loading may be caused by air pressure, inertia forces, or ground reactions during landing and taxiing. Some components may be designed for ground handling such as hoisting, packaging, and shipping.

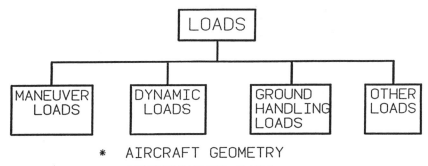

Fig. 3.1.1 Aircraft External Loads

The determination of design loads on the various structural components involves a study of the air pressures and inertia forces during certain prescribed maneuvers, either in the air or on the ground. The objective is to put sufficient strength into each component to obtain an aircraft with a satisfactory strength level compatible with structural weight. The methods by which this objective is obtained are sometimes relatively unrelated to the actual anticipated normal usage of the aircraft. Some of the prescribed maneuvers are, therefore, arbitrary and empirical, which is indicated by a careful examination of the structural criteria. The use of arbitrary criteria is justified since the usage and characteristics as well as the future development of any new aircraft are difficult to predict. Also, limitations of the "state-of-the-art" preclude the determination, in some cases, of exact loading for a given maneuver or configuration. It is then expedient to design a new aircraft to a strength level which is simply defined and compatible with the strength of an existing satisfactory aircraft. The specified conditions for structural design become simply a transfer system of experience from existing aircraft to new aircraft.

Fig. 3.1.2 shows several simple methods of load analyses used in the past (also see Ref. 3.2 and 3.3) and new methods and research are continually advancing the "state-of-the-art". This fact should not be overlooked and new designs cannot be based solely on past experience. Rational maneuvers and loading must be considered when the "state-of-the-art" permits their use. The availability of electronic computing equipment permits investigations of aircraft and loading behaviors as rationally as can be described mathematically. Care must be exercised, however, to ensure that the amount of "rationality" and computing time spent is compatible with the required starting, or input, data.

Lifting Line	**Lifting Surface**	**Distributed Approach**
• No streamwise chord bending	• Includes streamwise chord bending	• The lift consists of the sum of weighted function that can define the lift at any arbitrary point
• Remote pitch balance	• Mutual induction of entire planform is simulated	• Includes both spanwise and chordwise lift
• Remote fuselage effect	• Complex, time consuming and expensive analysis	• Exact solution for an elliptic wing
• Simplified, economical analysis	• Low aspect ratio, i.e., delta wings	• Aerodynamic influence coefficient (AIC) would not necessarily be symmetrical
• High aspect ratio	• More accurate method	• Accurate method
• Less accurate method		
	Reference documents	
• NACA TM 1120 (1949)	• NACA TM 1120 (1949)	• NASA TR R48 (1959)
• NACA TN 1862	• ARC REP. R&M 1910 (1943)	
• NACA TN 3030 (1953)	• ARC REP. R&M 2884 (1955)	
• NACA TN 1491 (1947)		
• NACA TN 1772 (1948)		

Fig. 3.1.2 Methods for Estimating Wing Lift (Subsonic Flow)

3.2 STRUCTURAL DESIGN CRITERIA

In order to apply structural design criteria to a particular aircraft it is necessary to have available the characteristics of the aircraft related to the determination of external and internal loads on the aircraft. The key characteristics are:

- Gross weight requirements - loadability
- Performance capabilities

- Stiffness
- Aerodynamic characteristics
- Landing gear features and characteristics
- Operating altitudes
- Loads analysis requirements

The structural design criteria defines the maneuvers, speeds, useful load, and aircraft design weights which are to be considered for structural design and sizing. These items may be grouped as those criteria which are under control of the aircraft operator. In addition, the structural criteria must consider such items as inadvertent maneuvers, effects of turbulent air, and severity of ground contact during landing and taxiing. The basic structural design criteria are based largely on the type of aircraft and its intended use.

(a) The first group of criteria which are under the control of the aircraft operator are based on conditions for which the pilot will expect the aircraft to be satisfactory. Loading for these conditions are calculated from statistical data on what pilots do with and expect from an aircraft. The strength provided in the aircraft structure to meet these conditions should be adequate for the aircraft to perform its intended mission as follows:

(i) Commercial aircraft – Must be capable of performing its mission in a profitable as well as safe manner, and it is assumed to be operated by qualified personnel under well-regulated conditions.

(ii) Military type aircraft – Do not usually lend themselves as readily to a definition of strength level based on a mission, because the mission may not be well-defined or a number of missions of various types may be contemplated. Military aircraft, also, are not always operated under well-regulated conditions and thus necessitate a wide range of design limits.

(b) The second group of criteria covering loading, over which the pilot has little or no control, includes the effects of turbulent air, sinking speeds during landing, taxiing over rough terrain, engine failure probability, and similar conditions. The strength necessary for these conditions is based almost entirely on statistical data and probability of occurrence. Statistics are constantly being gathered and structural design conditions modified or formulated from the interpretations of these data. Statistics can easily be used to design an airplane similar to the one on which the data were collected. Unfortunately, however, the data cannot always be successfully extrapolated to a new and different design.

When a new aircraft is designed in a category to which the criteria does not apply, deviations to or applicable interpretations of existing methods and requirements are necessary. Occasionally, a condition is experienced which is outside the statistics and was not foreseen. Then, a failure may occur. A one-hundred percent record of no failure during initial operation of a new aircraft is difficult, if not impossible, to obtain, particularly when production of an economical aircraft in a reasonable time span is essential. To attempt to account for these unforeseen conditions, a factor of safety is used and the concept of fail-safe was developed.

The amount of analysis used in the derivation of the aircraft basic loads is dependent on the size and complexity of the design and knowledge and data available. The time element is also important during design and the structural design is dependent on basic loads. Therefore, the loads must be determined early in the design process to preclude the possibility of delaying design work. The time available and quality of input data often governs the amount of load analysis that can be made. Fig. 3.2.1 shows typical wing loading along a wing span which is the basic load (bending M_x, shear S_z, and torsion M_y) requirements for preliminary sizing.

Chapter 3.0

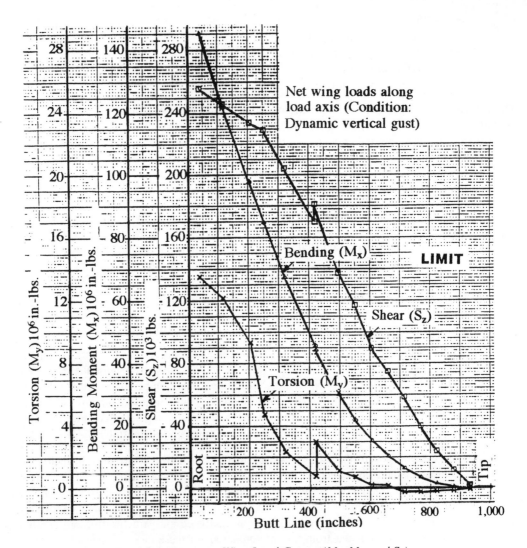

Fig. 3.2.1 Transport Wing Load Curves (M_x, M_y, and S_z)

Another consideration in determining the extent of the load analysis is the amount of structural weight involved. Since weight is always of great importance, a refinement of the methods used to compute loads may be justified. A fairly detailed analysis may be necessary when computing operating loads on such items as movable surfaces, doors, landing gear, etc. Proper operation of the system requires an accurate prediction of loads. In cases where loads are difficult to predict, recourse must be made to wind tunnel measurements. The final basic loads must be an acceptable compromise of all the considerations and Fig. 3.2.2 shows basic load organization interfaces and flow of data.

External Loads

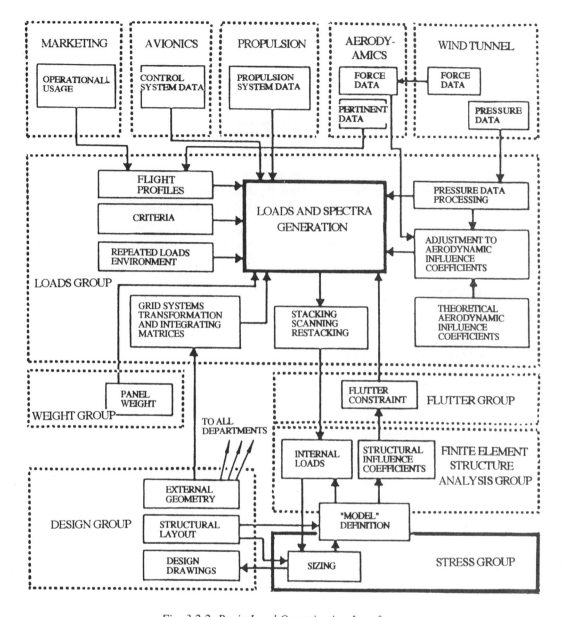

Fig. 3.2.2 Basic Load Organization Interfaces

(A) STRUCTURAL STIFFNESS

In high speed aircraft, structural stiffness requirements are important to provide adequate strength to prevent some high speed phenomena such as:

- Flutter
- Control surface reversal – whereby the control surface has lost its effectiveness due to weak torsional stiffness of the wing box
- Static divergence – whereby the wing structure becomes torsionally unstable as the angle of attack increases due to the applied loads

It is therefore necessary to have an estimate of structural stiffness before studies in these areas are started. As a design progresses it is sometimes necessary to update these stiffness estimates to agree with the actual design. When this happens, it is necessary to also update the dynamic work to find if the aircraft still complies with requirements.

(B) AERODYNAMIC CHARACTERISTICS

The overall aerodynamic characteristics of the aircraft are important to the determination of structural loads. Some of the necessary information that must be determined by analysis and wind tunnel tests are:

- Basic aircraft stability characteristics
- Pressure distributions over the wing, empennage, and fuselage
- Control surface hinge moments
- Pressure distributions over high lift devices, such as flaps and slates

These are usually obtained from wind tunnel tests of force and pressure models. These are conducted in subsonic and transonic wind tunnels to get the effects of compressibility. When wind tunnel tests are not available, estimates are made from wind tunnel and flight tests of past similar models.

Landing gear shock-absorbing characteristics are determined by analytical means and then verified by drop test of the gear.

(C) LOAD ANALYSIS

Design loads are determined for the various structural components using computers. Some of the analyses required are:

- Maneuver loads analysis
- Gust loads analysis
  - Static approaches
  - Dynamic approaches
- Landing loads analysis including both rigid airframe and dynamic analyses
- Ground handling loads including both rigid and elastic aircraft characteristics
- Control surface reversal characteristics and load distributions
- Dynamic analyses for control surface oscillatory conditions due to "black box" (autopilot or yaw damper) failure
- Flutter analysis and tests including both wind tunnel and flight testing to verify the adequacy of the damping characteristics of the aircraft throughout the speed range

3.3 WEIGHT AND BALANCE

Structural design loads affect the weight of the airframe structure, and the weight of the aircraft influences the magnitude of design loads. This interdependence suggests that a judicious selection of the preliminary design weight is mandatory to the economical design of an airframe.

(A) WEIGHT REQUIREMENTS

Structural weight requirements are established as part of the basic design of the aircraft to accomplish the mission desired. These weights vary from the minimum gross weight to the maximum gross weight for which the aircraft is designed. Other considerations concerning the weight distribution within the aircraft are as follows:

- Center of gravity (c.g.) limits – established to provide the desired loadability of passengers and cargo
- Weight distribution of fixed items such as engines, fuel tanks, etc., to allow maximum use of the aircraft cargo and passenger compartments
- Fuel requirements – affects c.g. requirements and gross weight capabilities

During flight the c.g. travel is limited by the stability and control capabilities of the aircraft. Forward c.g. is limited by the ability of the tail to maintain balance during flaps-down landing approaches. Aft c.g. is limited by the so-called "maneuver point" at which the aircraft reaches neutral stability. It tends to become unstable and to diverge in pitch and be difficult to control. Preliminary limits are determined from wind tunnel data and final limits are selected from flight demonstrations which allow some margin from the absolute limits.

(B) MAJOR AIRCRAFT WEIGHT

Aircraft gross weight and detailed distribution of weight both have a large influence on structural design loads. For example, realizing that the lifting forces by air on an aircraft wing must be as great as the aircraft gross weight in order to support the aircraft in flight, it can be seen that wing-up-bending will be almost a direct function of aircraft gross weight as shown in Fig. 3.3.1.

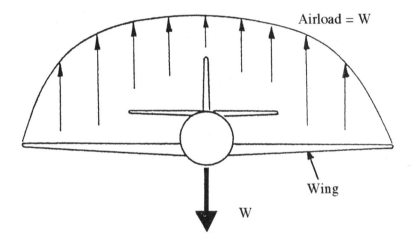

Fig. 3.3.1 Lifting Force vs. Gross Weight of the Aircraft

Another apparent example of the effect of the aircraft gross weight on design loads is the vertical load imposed on the landing gear when it contacts the ground at a given sinking speed. The gross weight to be used for structural design is determined from the mission requirements of the aircraft. Mission requirements and/or design criteria spell out maximum and minimum amounts of fuel and payload to be considered at various stages of aircraft operation. Typical aircraft major design weights are:

(a) Take-off gross weight – Maximum gross weight to perform the specified mission. The aircraft is generally considered as containing capacity fuel and maximum payload. This is varied in some instances to suit the particular needs of the design. Take-off gross weight is considered for taxiing conditions and flight conditions, but the aircraft need not be designed for landing at this weight.

(b) Design landing weight – Maximum weight for the landing operation. It is reasonable to expect that a predetermined amount of fuel is used up before the aircraft is expected to land. It would be unnecessarily penalizing the aircraft structure to design it to loads imposed during a landing at weight above the design landing weight. In many instances fuel dump provisions are installed in an aircraft so that aircraft weight may be reduced to the design weight in emergency landings. Critical for both wing and fuselage down-bending during landing.

(c) Zero fuel weight — Maximum weight with zero fuel aboard. This somewhat confusing title for a design weight is very popular in structures because it is descriptive. It defines the summation of the weight of empty aircraft, items necessary for operation (crew, engine oil, etc.), and cargo and/or passengers. This condition is usually critical for wing up-bending.

All design weights are negotiated during the formulation of design criteria. Using past experience, an attempt is made to establish the design weight at a level that will include the natural growth of aircraft gross weight between preliminary and final design as well as for future aircraft growth (this growth is sometimes as high as more than 30% of the original aircraft gross weight), and yet not be so excessive as to penalize the design.

(C) CENTER OF GRAVITY ENVELOPE

The combination of design weight and aircraft center of gravity location are of considerable importance. A plot is made of the variation of c. g. with the design weight. An envelope enclosing all extremes of the variation is determined for design. An attempt is made to set the c. g. limits such that they will include the extremes that can result from changes that occur as the design progresses. A typical example of a center of gravity envelope is shown in Fig. 3.3.2.

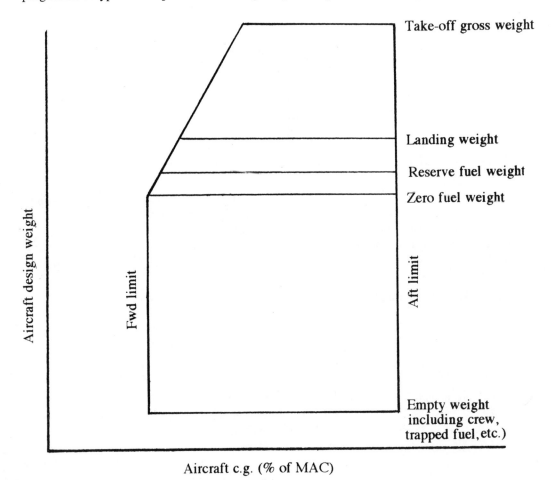

Fig. 3.3.2 A Typical Center of Gravity Envelope

(D) WEIGHT DISTRIBUTION

Once the critical loading is determined, a distribution of the dead weight is necessary. An accurate distribution of weight is important because the dead weight of fuselage, wings, cargo, etc., contributes to a large part of the loading. The effect of the distribution of weight can be realized by considering the distribution of mass items of cargo within a fuselage. Whether these large masses are condensed about the centerbody or placed at extreme forward or aft locations will influence greatly the magnitude of down-bending experienced by the fuselage forebody or aftbody during a hard landing.

The amount and disposition of fuel weight in the wing is particularly important in that it can be utilized to provide bending relief during flight. This is easily explained in terms of wing bending in an aircraft containing fuel in wing boxes. Fig. 3.3.3 illustrates how the fuel weight in the wing acts to relieve the bending caused from air load. The illustration points out that placing fuel as far outboard as possible and using fuel from the most inboard tanks first provides the optimum arrangement for wing bending during flight.

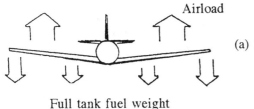

(a) Fuel weight provides relief to wing up-bending

Full tank fuel weight

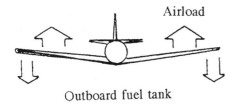

(b) Inboard fuel expended, outboard fuel providing more relief

Outboard fuel tank

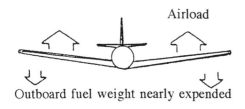

(c) Outboard fuel nearly expended, bending relief decaying faster than airload up-bending

Outboard fuel weight nearly expended

Fig. 3.3.3 Wing Up-bending Relief Due to Fuel Weight

Thinking in terms of the large inertia effects in a down-bending direction during a hard landing and taxiing point out that extreme arrangement of wing fuel is not necessarily the best solution. The illustration does make it apparent that fuel management (sequence of usage) should be given considerable attention in order that the final design be the best compromise.

3.4 FLIGHT LOADS

Aircraft operating variables for which critical values must be determined in showing compliance with the flight load requirements are:

- Flight altitude
- Operating weight
- Center of gravity (c.g.) travel

(A) MANEUVER LOADS

SYMMETRICAL:

Use the common maneuver V-n diagram ('V' represents aircraft speed and 'n' represent load factor), as shown in Fig. 3.4.1, to represent the maneuver load requirements for an aircraft. Because of the effects of compressibility on wing lift and the effects of varying air density on aircraft speed capability, V-n diagrams have different shapes at different attitudes. All critical combinations of altitude, weight, and c.g. travel position must be investigated.

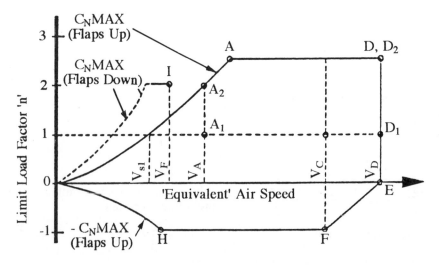

Fig. 3.4.1 Typical Maneuver V-n Diagram (Commercial Transport)

Within the limits of the diagrams the aircraft must be checked for both balanced and pitching maneuver conditions. In a balanced condition the aircraft is assumed to be in equilibrium with zero pitching acceleration. This can be achieved in a steady state pull-up or turn. All critical points around the periphery of the diagram must be checked for this condition.

(a) Flap-up case:
For airworthy transport category aircraft, design to a minimum positive limit load factor of 2.5 and a negative load factor of – 1.0

(b) Flap-down case:
Flap-down maneuver requirements are derived in the same way as those for flap-up except as affected by the following differences:

- Because greater wing lift coefficients are possible with flap-down, the aerodynamic lift curve is steeper
- A maximum limit load factor of 2.0 is required
- The V-n diagram is limited by a speed (V_F) which is equal to the stall speed multiplied by a factor of 1.6 or 1.8, depending on the aircraft weight condition and flap setting

External Loads

UNSYMMETRICAL:

(a) Rolling case:

Rolling occurs due to the application of roll control devices and it is an unbalanced moment about the aircraft c.g. that is to be reacted by aircraft inertia forces. The effect of wing torsional deflections on the rolling effectiveness of the ailerons and spoilers must be accounted for. Fig. 3.4.2 shows the load relationships and symmetrical load factor requirements for rolling. The aileron and spoiler deflection requirements at each of the required speeds are as follows:

- At V_A, sudden deflection of ailerons and spoilers to stops (limited by pilot effort)
- At V_C, ailerons and spoiler deflection to produce roll rate same as above at V_A
- At V_D, ailerons and spoiler deflection to produce roll rate $\frac{1}{3}$ of above at V_A

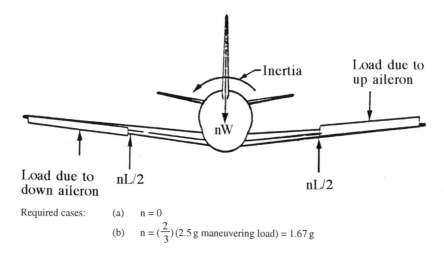

Required cases: (a) $n = 0$
(b) $n = (\frac{2}{3})(2.5 \text{ g maneuvering load}) = 1.67 \text{ g}$

Fig. 3.4.2 Rolling Maneuver Forces and Load Factors

(b) Yawing case:

The yawing maneuver occurs due to rudder deflection and produces three design requirements:

- With the aircraft in unaccelerated flight at a speed VA (the design maneuvering speed), and at zero yaw, the rudder is suddenly displaced to the maximum deflection. This causes a high yawing acceleration
- The aircraft then yaws to a maximum side slip angle. This is a momentary dynamic overyaw condition
- With the aircraft settled down to a static side slip angle, the rudder is then returned to neutral

(B) GUSTS

SYMMETRICAL:

(a) Flap-up case:

Gust criteria are based upon the fact that turbulent conditions of varying intensity occur in the air through which the aircraft flies. As with maneuver requirements, all critical combinations of altitude, weight, and c.g. position must be investigated for all critical points around the periphery of the gust V-n diagram as shown in Fig. 3.4.3. Special considerations apply when accelerations caused by gusts are computed for flexible swept wing configuration.

Chapter 3.0

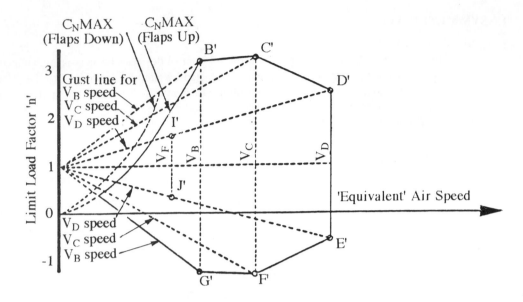

Fig. 3.4.3 Typical Gust V-n Diagram (Commercial Transport)

(b) Flap-down case:
   The requirements call for positive and negative 25 fps (feet per second) gusts for speeds up to V_F (flap speed).

UNSYMMETRICAL:

(a) Rolling case:
   The condition of unsymmetrical gusts must be considered by modifying the symmetrical gusts at B' or C' on the gust V-n diagram. 100% loads are applied to one side and 80% on the other.

(b) Yawing case:
   The aircraft is assumed to encounter lateral gusts while in unaccelerated flight. The derived gusts at the related aircraft speed of V_B (design speed for maximum gust intensity), V_C (design cruise speed), and V_D (design dive speed) must be investigated for flap-up and flap-down conditions.

3.5 GROUND LOADS

This discussion is based on aircraft configurations with conventional arrangements of main and nose landing gears. For configurations with tail gear, refer to U.S. FAR 23 or FAR 25 for further load requirements.

(A) LANDING CONDITIONS

The load factors (n) required for design for landing conditions are related to the shock absorbing characteristics of the main and nose gears. These characteristics are usually based on the result of drop tests. From these test results, the required design limit load factors are determined. For commercial transports:

- For landing weight, the required descent velocity at initial impact is 10 fps
- At the design takeoff weight it is 6 fps

External Loads

LEVEL LANDING (LL):

During level landing conditions, as shown in Fig. 3.5.1, the wings are supporting the aircraft weight, W, during landing impact. In the two point landing case (2 pt. landing or LL2), vertical loads (V) and horizontal loads (D), are applied at the wheel axles. These loads are reacted by aircraft inertia loads, nW and T. The moments created by this combination of loads is placed in equilibrium by the pitching inertia of the aircraft. In the three point landing case (3 pt. landing or LL3), the main and nose gears contact the ground simultaneously.

Two Point Landing:

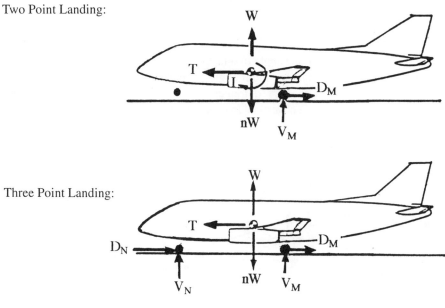

Three Point Landing:

Fig. 3.5.1 Level Landing

The aircraft must be good for the landing loads which are equivalent to those determined for the combinations of gross weight and sink speed. The aircraft speed ranges from standard day stall speed to 1.25 times hot day stall.

During the first part of landing impact, the wheel is spun up to aircraft rolling velocity by friction between the tire and the runway. This friction induces a short time, aft acting load at the axle which causes the gear to deflect aft as shown in Fig. 3.5.2.

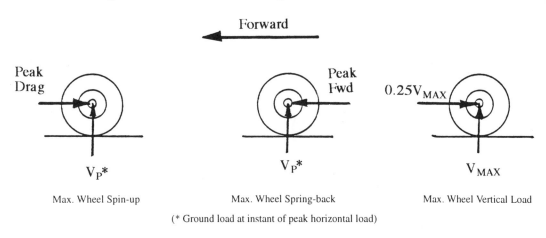

Max. Wheel Spin-up   Max. Wheel Spring-back   Max. Wheel Vertical Load

(* Ground load at instant of peak horizontal load)

Fig. 3.5.2 Level Landing – Gear Load Combinations

Aircraft are designed for the loads acting on the gear at the instant of peak drag. These loads are determined by rational analysis, taking into account the time history of the events that occur during landing impact; gear inertia loads are also included.

Following spin up, the gear springs back in a forward direction. A peak loading occurs at the point of maximum forward deflection. The aircraft must be designed for the vertical and horizontal loads occurring at this instant.

The three level landing situation which must be investigated is the one which occurs at the instant of maximum vertical ground reaction. An aft-acting horizontal load of $0.25V_{max}$ is applied at the axle in this situation.

TAIL-DOWN LANDING:

The tail-down landing is made at an extreme angle of attack as shown in Fig. 3.5.3. The angle is to be limited either by the stall angle or the maximum angle for clearance with the ground at main wheel impact, whichever occurs first. Loading requirements are identical to those of level landing except that the condition of maximum wheel vertical load is not required.

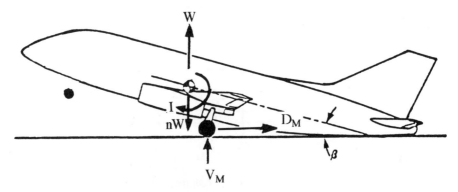

(β – Angle for main gear and tail structure contacting ground unless exceeds stall angle.)

Fig. 3.5.3 Tail-down Landing

ONE-WHEEL LANDING:

The ground reactions are equal to those on one side for level landing, as shown in Fig. 3.5.4. The unbalanced loads are reacted by aircraft rolling inertia.

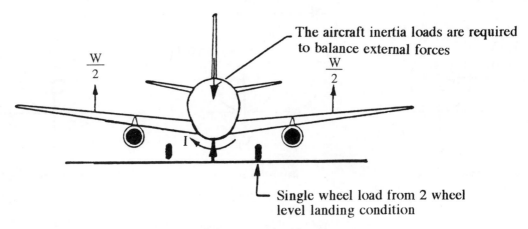

Fig. 3.5.4 One-wheel Landing

LATERAL DRIFT LANDING (LD):

The aircraft is in the level landing attitude with only the main wheels on the ground, as shown in Fig. 3.5.5. The main gear loads are equal to one half the level landing maximum vertical main gear loads. Side loads of 0.8 of the vertical ground load on one side acting inward and 0.6 of the vertical ground load on the other side outward are applied at the ground contact point. These loads are resisted by aircraft transitional and rotation inertia.

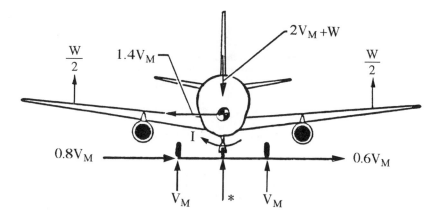

V_M – One-half the max. vertical ground reaction obtained at each main gear in the level landing conditions

* Nose gear ground reactions equal to zero.

Fig. 3.5.5 Lateral Drift Landing

REBOUND LANDING:

Since an aircraft may rebound into the air after a hard landing, the gear and its supporting structure must be checked out for this condition. With the landing gear fully extended and not in contact with the ground, a load factor of 20 must be applied to the unsprung weights of the gear. The load factor must act in the direction of Oleo movement.

(B) GROUND HANDLING

TAXIING:

Taxiing conditions are investigated at the take-off weight and no wing lift is considered. Shock absorbers and tires are placed in their static position.

TAKE-OFF RUN:

This loading occurs when the aircraft is taxied over the roughest ground and it is design practice to select a limit factor depending on the type of main gear:

- 2.0 for single axle gear
- 1.67 for truck type gear

BRAKE AND ROLL (BR):

The aircraft is taxiing in a level attitude at either 1.2 times the aircraft design landing weight, or at 1.0, the design ramp weight (for abort take-off condition), without using thrust reversal power. The brakes are applied and a friction drag reaction (coefficient of friction is 0.8) is applied at the ground contact point of each wheel with brakes. The drag reaction is equal to 0.8 times the vertical reaction load unless it can be shown that lesser loads cannot be exceeded because of brake capacity. Two attitudes must be considered (see Fig. 3.5.6):

- Only the main gear are in contact with the ground with the pitching moment resisted by angular acceleration of the aircraft (2 pt. braking or BR2)
- All wheels are on the ground and equilibrium is obtained by ground reactions (3 pt. braking or BR3)

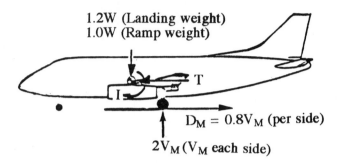

(A) BR2 Condition – Main gear only

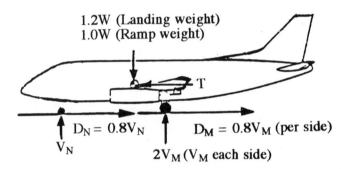

(B) BR3 Condition – Main and nose gear

Fig. 3.5.6 Brake and Roll (BR)

GROUND TURNING:

Assume the aircraft is in a steady turn with loads as shown in Fig. 3.5.7. The combination of turn radius and velocity is such that a total side inertia factor of 0.5 is generated by side loads on the wheels. The vertical load factor is 1.0. The side ground reaction on each wheel is 0.5 of the vertical reaction.

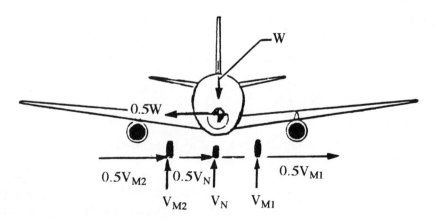

Fig. 3.5.7 Ground Turning

NOSE GEAR YAWING

There are two independent nose wheel yawing conditions, as shown in Fig. 3.5.8, and they are:

(a) Nose gear side load:
   - The aircraft is standing on the runway at a vertical load factor of 1.0
   - A side load of 0.8 of the nose gear vertical load is applied at the nose wheel ground contact point

(b) Unsymmetrical braking (a moving situation in which brakes are applied to one main gear):
   - A drag load is applied to this gear at the ground with a value equal to the maximum brake capacity but limited to 0.8 of the vertical load on the gear
   - Equilibrium is obtained by balancing side and vertical loads on the nose and main gear
   - Any moment which would cause a side load on the nose gear greater than $0.8V_N$ is balanced by aircraft inertia

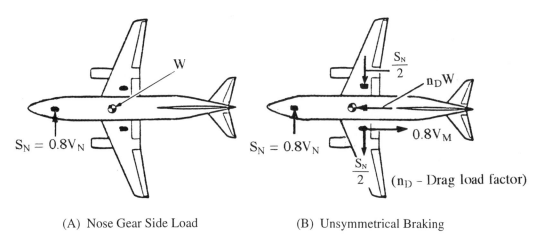

(A) Nose Gear Side Load (B) Unsymmetrical Braking

Fig. 3.5.8 Nose Gear Yawing

PIVOTING:

The aircraft pivots about one main gear which has its brakes locked as shown in Fig. 3.5.9. The limit vertical load factor is 1.0 and the coefficient of friction is 0.8. The aircraft is placed in static equilibrium by loads at the ground contact points. The aircraft is in the three point attitude and pivoting is assumed to take place about one main landing gear unit.

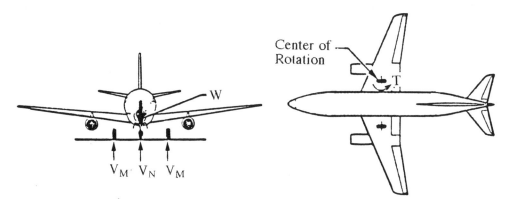

V_N (nose gear) and V_M (main gear) are static ground reactions.

Fig. 3.5.9 Pivoting

Chapter 3.0

REVERSED BRAKING:

In this condition, the aircraft is in the three point attitude at a limit load factor of 1.0 and moving aft when the brakes are applied as shown in Fig. 3.5.10:

- Horizontal loads are applied at the ground contact point of all main wheels with brakes
- These loads must equal 1.2 times the load generated at nominal maximum static brake torque except that they need not exceed 0.55 times the vertical load at each wheel
- The pitching moment is to be balanced by rotational inertia

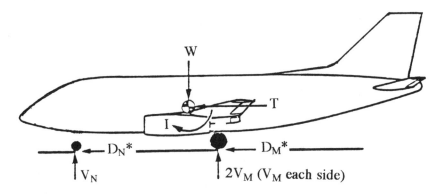

* 1.2 load for nominal max. static brake torque (limited to 0.55V)

Fig. 3.5.10 Reverse Braking

TOWING:

Towing requirements consist of many (approximately twelve) separate conditions which provide for main or nose gear towing fittings and fittings at other locations. The tow loads are obtained by multiplying the take-off weight by varying factors. As shown in Fig. 3.5.11, for example, a forward-acting load of 0.15W (take-off gross weight) occurs at the nose gear towing fitting. The requirements call for a reaction equal to the vertical gear load to be applied at the axle of the wheel to which the tow load is applied. In this condition, it would be V_N at the nose gear and the unbalanced load is reacted by aircraft inertia.

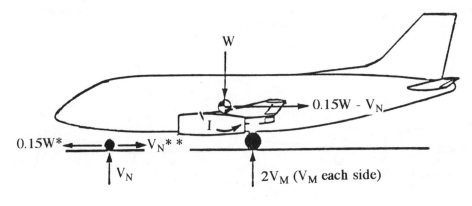

* Applied at tow fitting
** Applied at axle

Fig. 3.5.11 Towing at Nose Gear

(C) UNSYMMETRICAL LOADS ON MULTIPLE WHEEL UNITS

The requirements for unsymmetrical loads on multiple wheel units (i.e., bogie, truck, etc.) also involve all the ground load conditions that have been covered previously. For conditions in which all tires are inflated, provision must be made for such variables as:

- Wheel arrangement
- Tire diameter
- Tire inflation
- Runway crown
- Aircraft attitude
- Structural deflections

The case of one deflated tire on multiple wheel units or two deflated tires on gear with four or more wheels must be investigated. There is usually a reduction in gear load requirements for deflated tire conditions but this is not always true. For instance, in the drift landing condition, 100% of the vertical load for the undeflated tire condition must be applied.

3.6 DYNAMIC LOADS

In actual circumstances aircraft structures will deform and vibrate under certain rates of loading. At critical aircraft speed conditions the flexible structure can be self-destructive. Structural motions caused by these flexible characteristics require the evaluation of dynamic effects on flight and ground loads as well as investigation of the dynamic stability margins for the overall aircraft structure.

(A) DYNAMIC LOADS

It must be shown by analysis that the aircraft can meet the symmetrical and unsymmetrical gust load conditions in flight when various rates of gust load application are considered. These dynamic flight loads must then be compared to the corresponding static gust loads and the increment of dynamic amplification must be considered in the ultimate design loads. Similarly, it must be shown by analysis that the aircraft can withstand the rate of loading applied to the structure by the landing gear during design landing conditions.

(B) FLUTTER AND DIVERGENCE

Flutter is a self-excited structural oscillation of the aircraft which derives its energy from the air mass in which it is flying. It must be shown by analysis, and tests where applicable, that the aircraft will be free from destructive flutter at speeds up to 1.2 times the design dive speed. Divergence, unlike flutter, is dependent only on the flexibility of the structure rather than on the characteristic of the structural oscillations. The external air loads will cause a redistribution to occur which can reinforce the original loading. When a speed is reached in which the structure can no longer support this loading, divergence deflection will occur and the analysis must prevent this divergence.

(C) BUFFET AND VIBRATION

Buffeting is a dynamic structural loading caused by unstable aerodynamic flow. High speed buffet is associated with shock wave formations at high Mach number speeds when local flow separation occurs behind the shock. This can cause some vibrations in structure and it is required that there will be no buffeting conditions in normal flight severe enough to:

Chapter 3.0

- Interfere with the control of the aircraft
- To cause excessive fatigue to the crew
- To cause structure damage

Aircraft must be shown in flight to be free from excessive vibration under any speed up to design dive speed of the aircraft.

3.7 CONTROLLABLE SURFACES

Controllable surfaces are those moving components which make aircraft flyable in the air and they are:

- Control surfaces – ailerons, elevators, rudder
- Leading edge flaps and/or slats and their supports
- Trailing edge flaps and their supports

The following flight load conditions should be reviewed:

(a) Structural loads:
  - Due to air loads in flight
  - Consider stiffness and flutter
  - Due to ground gust

(b) Hinge loads:
  - Due to air loads in flight.
  - Due to ground gust.
  - Hinge loads (perpendicular to hinge line) on control surface due to wing or empennage box bending

(c) For miscellaneous loads on control surfaces, refer to Chapter 3.9.

3.8 FUSELAGE PRESSURIZATION LOADS

Fuselage pressurization is an important structural loading which induces hoop and longitudinal stresses in the fuselage. Pressurization must be combined with flight and ground loading conditions. The most important consideration for establishing the fuselage design pressure is the effect of altitude on cabin pressure and the fuselage must be designed to maintain cabin pressure.

When aircraft fly above 15,000 ft. altitude, it is usually required that the cabin be pressurized to insure comfortable conditions for passengers during flight.

Determination of pressure differential:

- Pressure differential can be obtained from a nomogram as shown in Fig. 3.8.1
- In general, most commercial transport cabins are designed for 8,000 ft. altitude pressure:
  - at 8,000 ft. altitude the atmospheric pressure is 10.92 psi
  - the constant cabin pressure is 10.92 psi from the altitude of 8,000 ft and up to the designed altitude
- Some executive aircraft may require at sea-level (14.7 psi) pressure differential; for example see Fig. 3.8.2:
  - if the aircraft is designed for a max. flight altitude of 45,000 ft, the pressure differential is 12.55 psi
  - pressure differential is 8.77 psi for 8,000 ft. altitude pressure
- Most military transport cabins are designed for 1,000 ft. altitude pressure
- Pressure differential is used to design the minimum thickness requirement for fuselage shell structure

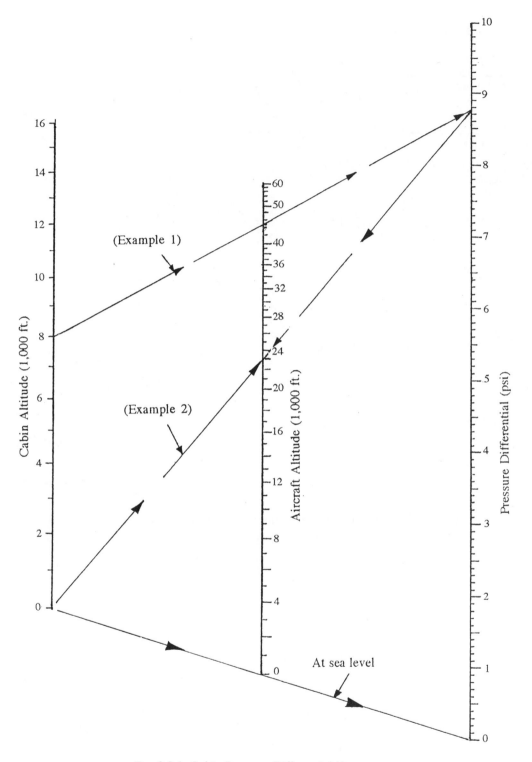

Fig. 3.8.1 Cabin Pressure Differential Nomogram

Chapter 3.0

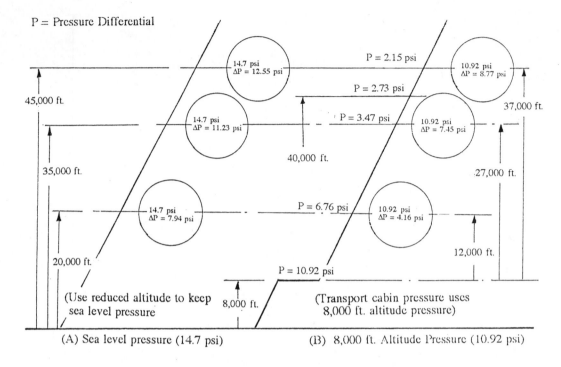

Fig. 3.8.2 Pressure Differential (Sea Level vs. 8,000 ft. Altitude)

Example 1:

Assume a commercial transport provided an 8,000 ft. pressure differential and the maximum flight altitude is at 45,000 ft.

(a) From nomogram in Fig. 3.8.1 with 8,000 ft. altitude, obtain a pressure differential of 8.77 psi.

(b) The pressure differential used with flight and ground conditions on the aircraft are:

Conditions	Max. Positive pressure (Burst) in psi	Max. Negative Pressure (collapsing) in psi
Flight (all)	9.02[1]	−0.5[2]
Ground	1.0	−0.5

(In addition, a pressure of 2.0 x 9.02 = 18.04 psi is considered to act alone)

(1) Upper limit of positive relief valve setting = 8.77 psi (cabin pressure differential, see Fig. 3.8.1) + 0.25 psi (assumed relief valve tolerance) = 9.02 psi.
(2) Upper limit of negative relief valve setting = 0.5 psi.

Example 2:

A commercial transport which normally has operational settings for 8,000 ft.-cabin-differential pressure is limited to 8.77 psi at the flight altitude of 45,000 ft. If this aircraft will maintain sea level cabin pressure (14.7 psi) in flight, based on Fig. 3.8.2, its flight altitude should be reduced to 23,000 ft.

External Loads

3.9 WING FUEL PRESSURE LOADS

Fuel pressure loads in the fuel tank can induce pressure which is normal to the tank wall structure (e.g., upper and lower covers, inb'd and outb'd ribs, and front and rear spars). The most critical location is near the most outboard tank. The factors that affect the fuel pressure loads (lbs./in.^{2}) are:

- Tank geometry, as shown in Fig. 3.9.1
- Load factors (i.e., n_x, n_y, and n_z) on the aircraft
- Fuel density: $\rho = 0.0307$ lbs./in.^{3}
- Aircraft maneuvers (i.e., $\dot{\phi}, \ddot{\phi}, \ddot{\psi}$, pitch, vertical gust, etc.)

  where: Angular velocity ($\dot{\phi}$) in radian/sec
  Angular acceleration ($\ddot{\phi}$ and $\ddot{\psi}$) in radian/sec^2
  Acceleration of gravity: $g = 386$ in./sec^2

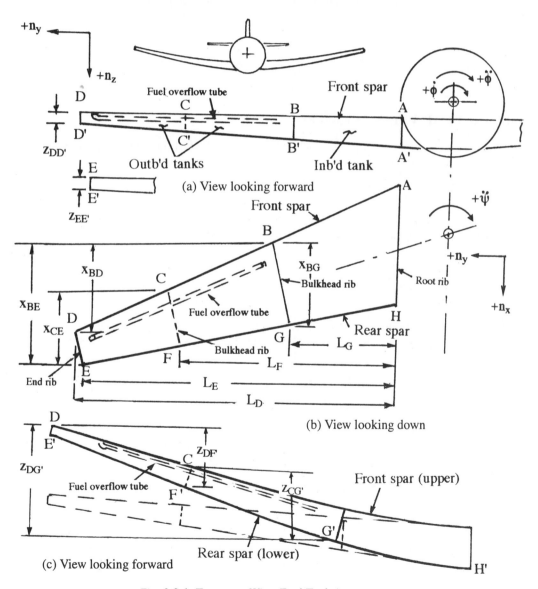

Fig. 3.9.1 Transport Wing Fuel Tank Arrangement

Chapter 3.0

(A) SYMMETRICAL MANEUVER (PITCH OR VERTICAL GUST)

An aircraft under symmetrical pitch maneuver ($n_y = 0$) is shown in Fig. 3.9.1(c) and the fuel pressure at point G':

$$p = 1.5[(n_z z_{DG'} + n_x x_{BG})\rho] \quad \text{Eq. 3.9.1(a)}$$

The equation is applicable when the two outboard fuel tanks are connected by a fuel overflow tube; if no tube exists, the following equation should be used:

$$p = 1.5[(n_z z_{CG'} + n_x x_{BG})\rho] \quad \text{Eq. 3.9.1(b)}$$

Example 1:

Determine the ultimate fuel pressure at point G' of the fuel tank geometry shown in Fig. 3.9.1(c) induced by a climb maneuver (assume the fuel overflow tube exists):

$n_x = 1.3; \quad n_y = 0; \quad n_z = 2.5$

$z_{DG'} = 150"; \quad x_{BG} = 100$

From Eq. 3.9.1(a), the ultimate fuel pressure at point G' is:

$p = 1.5[(2.5 \times 150 + 1.3 \times 100) \times 0.0307] = 23.3$ psi

(B) ANGULAR VELOCITY ($\dot{\phi}$) ROLL MANEUVER

The ultimate pressure at point E' in a fuel tank [see Fig. 3.9.1(a)] due to an angular velocity roll maneuver (centrifugal fuel pressure) and load factors of n_x and n_z is:

$$p = 1.5[(\frac{\dot{\phi}^2}{2g})(L_E^2 - L_G^2) + n_x \times x_{BE} + n_z \times z_{EE'}]\rho \quad \text{Eq. 3.9.2(a)}$$

This equation is applicable when the two outboard fuel tanks are connected by a fuel overflow tube; if no tube exists, the following equation should be used:

$$p = 1.5[(\frac{\dot{\phi}^2}{2g})(L_E^2 - L_F^2) + n_x \times x_{CE} + n_z \times z_{EE'}]\rho \quad \text{Eq. 3.9.2(b)}$$

Example 2:

Determine the ultimate fuel pressure at point D' of the fuel tank geometry shown in Fig. 3.9.1 (a) induced by an angular roll velocity maneuver (assume the fuel overflow tube is exists):

$\dot{\phi} = 45°/\text{sec} = 0.78$ radian/sec

$n_x = 0.363; \quad n_y = 0; \quad n_z = 1.33$

$L_D = 880"; \quad L_G = 340"; \quad z_{DD'} = 12.87"; \quad x_{BD'} = 370"$

From Eq. 3.9.2(a):

$$p = 1.5[(\frac{\dot{\phi}^2}{2g})(L_D^2 - L_G^2) + n_x \times x_{BD} + n_z \times z_{DD'}]\rho$$

$$= 1.5[(\frac{0.78^2}{2 \times 386})(880^2 - 340^2) + 0.363 \times 370 + 1.33 \times 12.87] \times 0.0307$$

$$= 1.5[519.2 + 134.3 + 17.1] \times 0.0307 = 30.88 \text{ psi}$$

The ultimate fuel pressure, $p_r = 30.93$ psi, is a normal load on the end rib D-E [see Fig. 3.9.1(b)], the local wing upper and lower surfaces, and the front and rear spars at the point D and E.

External Loads

(C) ANGULAR ACCELERATION ($\ddot{\phi}$) ROLL MANEUVER

The delta limit load factor, Δn_z, is due to an angular acceleration roll maneuver and the Δn_z acts on the lower surface (left-hand wing box) and on the upper surface (right-hand wing box) at the location of radius R.

$$\Delta n_z = (\frac{R\ddot{\phi}}{g}) \qquad \text{Eq. 3.9.3}$$

Example 3:

Determine the delta limit load factor Δn_z at point D' of the fuel tank shown in Fig. 3.9.1(a), induced by an angular acceleration roll maneuver:

$\ddot{\phi} = 58°/\text{sec.}^2 = 1.012 \text{ radian/sec.}^2$

$n_x = 0.363; \qquad n_y = 0; \qquad n_z = 1.33$

$R = L_D = 890"; \qquad z_{DD'} = 13"; \qquad x_{BD} = 100"$

From Eq. 3.9.3 to obtain the delta limit load factor of Δn_z:

$$\Delta n_z = (\frac{R\ddot{\phi}}{g}) = (\frac{890 \times 1.012}{386}) = 2.33$$

The ultimate fuel pressure at point D' (total limit load factor = $n_z + \Delta n_z$) is:

$p = 1.5[(n_z + \Delta n_z)z_{DD'} + n_x \times x_{BD}]\rho$
$= 1.5[(1.33 + 2.33) \times 13 + 0.363 \times 100] \times 0.0307 = 3.86 \text{ psi}$

(D) ANGULAR ACCELERATION ($\ddot{\psi}$) YAW MANEUVER

The delta limit load factor of Δn_x occurs when the aircraft performs an angular acceleration yaw maneuver and the Δn_x occurs at the location of radius R.

$$\Delta n_x = (\frac{R\ddot{\psi}}{g}) \qquad \text{Eq. 3.9.4}$$

Example 4:

Determine the ultimate fuel pressure load at point E' (lower surface) of the fuel tank shown in Fig. 3.9.1(a) and (b), induced by an angular acceleration yaw maneuver:

$\ddot{\psi} = 27.2°/\text{sec}^2 = 0.475 \text{ radian/sec}^2$

$n_x = 0.363; \qquad n_y = 0; \qquad n_z = 1.33$

$R = L_E = 880"; \qquad x_{BE} = 150"; \qquad z_{EE'} = 13"$

From Eq. 3.9.4 obtain the delta limit load factor Δn_x:

$$\Delta n_x = (\frac{R\ddot{\psi}}{g}) = (\frac{880 \times 0.475}{386}) = 1.08$$

The ultimate fuel pressure load at point E' is:

$p = 1.5[(n_x + \Delta n_x)x_{BE} + n_z \times z_{EE'}]\rho$
$= 1.5[(0.363 + 1.08) \times 150 + 1.33 \times 13] \times 0.0307 = 10.76 \text{ psi}$

3.10 MISCELLANEOUS LOADS

(A) CONTROL SURFACES (EXCLUDE INFLIGHT LOADS)

(a) Ground gust:
   - Forward moving air velocities of not over 88 fps load each of the control surfaces
   - Specified arbitrary hinge moment factors

(b) Pilot loads:
   - Specify limits on pilot forces on aileron, elevator, and rudder
   - Specify limits on pilot forces with or without mechanical assists, such as tabs or servos

(c) Hinge loads:
   - Due to air loads
   - Hinge loads (perpendicular to hinge line) on control surface due to wing or empennage box bending (called forced bending loads)

(d) Arbitrary inertia load factors, acting along the hinge line, must be applied to the control surface weight, i.e.,
   - 24.0 for vertical surfaces
   - 12.0 for horizontal surfaces

(e) Arbitrary inertia load factors, acting perpendicular to the hinge line, must be applied to the control surface balance weight. Refer to certifying regulations and/or company's requirements for these load factors.
   - Ailerons, elevators – load factors perpendicular to the hinge line
   - Rudder(s) – load factors on both perpendicular to and along the hinge line

(B) EMERGENCY CRASH LOADS

Provision must be made for the protection of the occupants in wheels-up crash landing on land or water. The aircraft structure involved must be designed for the following ultimate load factors (g):

	FAR 25	MIL-A-8860(ASG)
Forward	9.0 g	16.0 g
Up	3.0 g	7.5 g
Down	6.0 g	16.0 g
Side	± 3.0 g	± 5.0 g
Aft	1.5 g	6.0 g

In addition, all support structures and cargo restraints for items of mass that could injure an occupant if they came loose must be designed for the same factors. The design loads for seat belt attachment fittings and seat attachment fittings must include an added factor of 1.33.

(C) DITCHING

When aircraft operate over water, the aircraft are equipped for ditching (refer to FAR 25.801). From a structural standpoint, the aircraft must remain intact to the extent that:
   - There will be no immediate injury to the occupants
   - The aircraft will stay afloat long enough to allow the occupants to leave the aircraft and enter life rafts

External Loads

(D) BREAKAWAY DESIGN

Breakaway design conditions need to be considered for such parts as engine pylons (see Fig. 3.10.1), landing gear (see Fig. 3.10.2), flap track supports (on low wing design), etc., which must break away cleanly when striking an obstacle, ditching or any other emergencies occur.

- Inflight case
- On ground crash landing case

(E) HAND LOADS

(a) All stiffeners, whether mechanically fastened or cocured or bonded, are to be designed to one of the following three categories:

- Horizontal lower framing stiffeners around an access cutout – apply 300 lbs. ultimate load over 4 inches
- Vertical and horizontal upper framing stiffeners around an access cutout – apply 150 lbs. ultimate over 4 inches
- All stiffeners not in either of the above categories – 75 lbs. ultimate over 4 inches

(b) External doors in opened position.

(c) Interior equipment.

(F) GROUND GUST LOADS

External doors (such as large cargo doors) in an opened position.

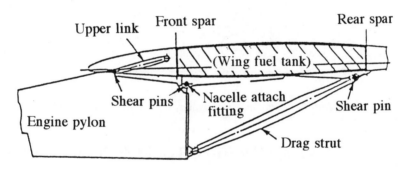

(a) Break shear pins

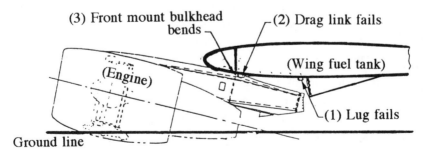

(b) Break lug or shear pins

53

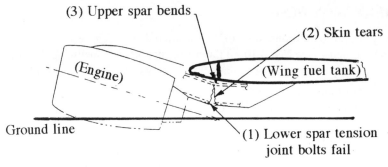

(c) Break tension bolt

Fig. 3.10.1 Wing-mounted Engine Breakaway Cases

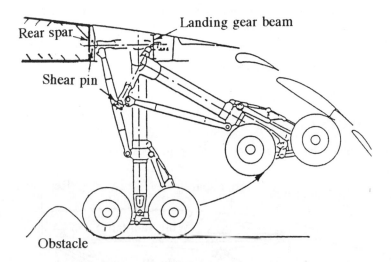

Fig. 3.10.2 Main Landing Gear Breakaway Case (Break Shear Pins)

3.11 LOAD CONDITIONS SUMMARY FOR COMMERCIAL TRANSPORT:
(REFERENCE ONLY)

(a) Flight: Pilot-induced maneuvers and aircraft system malfunction

- Combination stabilizer-elevator maneuvers
- Aileron and/or asymmetric spoiler maneuvers
- Rudder maneuvers
- Combinations of above

(b) Flight: Atmospheric turbulence

- Power spectral approach
  - 'Mission Analysis' based on operational spectra
  - 'Design Envelope Analysis' based on extremes of operation envelopes
- Discrete gust approach
  - FAR gust load factor equation
  - Tuned discrete gusts (i.e., one-minus-cosine shaped gusts with varying wave lengths)

(c) Landing
- FAR landing condition (i.e., LL2, LL3, DL)
- Rational landing time histories (based on FAR-designated sinking speeds, wing lifts, pitch attitudes, and speeds)

(d) Ground handling
- FAR ground handling (i.e., BR2, BR3, turning, pivoting, unsymmetrical braking, towing)
- Rational taxiing (i.e., bump encounter during take-off runs, rejected take-off, landing run-out, taxiing)
- Rational ground maneuvering
- Jacking

(e) Fail-safe and breakaway design requirements

(A) WING:

(a) General
- Positive high angle of attack
- Negative high angle of attack
- Positive low angle of attack
- Negative low angle of attack
- Flaps down maneuver – take-off configuration
- Flaps down maneuver – landing configuration
- Maneuver with certain wing fuel tanks empty
- Control surface reversal
- Unsymmetrical spanwise lift distribution
- Fuel vapor or refueling pressure
- Dive maneuvers
- Taxi
- Jacking
- Flutter
- Roll initiation
- Fatigue
- Fail-safe
- Thermal gradients
- Lightning strike

(b) Spar conditions
- Fuel slosh
- Fuel head – crash conditions
- Concentrated shear loads (i.e., supports for engine mounts, landing gear, etc.)
- Fuselage pressure in wing center section

(c) Rib conditions
- Concentrated load redistribution
- Wing surface panel stabilization
- Sonic fatigue (near jet engine exhaust path)
- Rib crushing loads
- Fuel slosh
- Fuel head

(d) Leading edge conditions
- Hail strike
- Sonic fatigue – jet engine reverse thrust
- Thermal anti-icing
- Duct rupture

(e) Trailing edge and fairing load conditions
- Sonic fatigue – near jet engine exhaust path
- Slosh and gravel impact
- Positive and negative normal air pressure
- Buffet
- Minimum gage

(B) TAIL:

(a) General

- Control surface reversal
- Control surface effectivity
- Rib crushing
- Control surface support
- Concentrated load redistribution
- Stabilizing surface panel structure
- Fatigue
- Sonic fatigue
- Flutter
- Hoisting
- Minimum skin gage
- Actuator support

(b) Leading edge

- Hail strike
- Thermal anti-icing (if required)

(c) Horizontal stabilizer

- Instantaneous elevator deflection
- Positive maneuver
- Unsymmetrical spanwise load distribution
- Fin gust (on T-tail)
- Negative maneuver

(d) Vertical stabilizer

- Instantaneous rudder – yaw initiation
- Dynamic overyaw
- Check maneuver
- Engine out
- Fin lateral gust

(e) Aft fuselage

- Redistribute fin and stabilizer concentrated loads
- Tail skid loads (for emergency ditching condition)

(C) FUSELAGE:

(a) General

- Internal pressure
- External pressure
- Dynamic landing
- Braked roll
- Ground turn
- Taxiing
- Unsymmetrical braking
- Vertical gust
- Fin lateral gust
- Instantaneous rudder deflection
- Rudder deflection for engine out
- Instantaneous elevator deflection
- Positive maneuvers
- Negative maneuvers
- Crash loading
- Ditching
- Jacking
- Fail-safe
- Fatigue
- Fuel slosh
- Fuel head crash landing
- Fuel vapor and refueling pressure

(b) Frame and bulkhead

- Stiffness to maintain monocoque stability
- Concentrated load redistribution

(c) Skin

- Cabin pressure and fuselage shear load

(d) Door

- Cabin pressure, wind and hand loads

(e) Floor
- Cargo and passenger loads
- High heels

(f) Other conditions
- Decompression of one compartment
- Hail strike
- Bird strike

(D) ENGINE MOUNT:

(a) Mount
- Yawing gyroscopic (alone or combined with vertical and thrust)
- Pitching gyroscopic (alone or combined with vertical and thrust)
- Landing impact (–4.0 to + 8.0 load factors on engine weight)
- 3.0 load factors on engine thrust (alone or combined with vertical load)
- 3.0 load factors on reverse thrust (alone or combined with vertical load)
- 9.0 load factors forward on engine weight – used as ultimate load factors for crash condition
- Roll side loads or 2.5 load factors side on engine weight
- Combinations of thrust with vertical and side loads
- Aerodynamic vertical and side loads
- Dynamic unbalance and shock mounting
- Fatigue design
- Breakaway condition

(b) Engine Inlet and Duct
- Engine surge or hammer shock
- Various pressures for normal and abnormal operation

(c) Engine Cowling
- Internal pressure (if applicable)
- Aerodynamic pressure
- Elevated temperature
- Latching
- Duct burst
- Sonic fatigue
- Handling

References

3.1 "14 Code of Federal Regulations (14 CFR)", U.S. Government Printing Office, Washington, 1992.
3.2 Niu, C. Y., "AIRFRAME STRUCTURAL DESIGN", Conmilit Press Ltd., P.O. Box 23250, Wanchai Post Office, Hong Kong, 1988.
3.3 Peery, D. J., 'AIRCRAFT STRUCTURES', McGraw-Hill Book Company, Inc., New York, NY. 1950.
3.4 F. M. Hoblit, N. Paul, J. D. Shelton, and F. E. Ashford, "Development of a Power – Spectral Gust Design Procedure for Civil Aircraft", FAA-ADS-53, January 1966.
3.5 J. E. Wignot, "Structural Design of Transport Airplanes for Transient Environments", Lockheed-California Company paper, January 1966.
3.6 Lomax, T.L., "STRUCTURAL LOADS ANALYSIS FOR COMMERCIAL TRANSPORT AIRCRAFT", AIAA Education series, 1801 Alexander Bell Drive, Reston, VA22091, (1966).
3.7 Taylor, J., "MANUAL ON AIRCRAFT LOADS", Pergamon Press Inc., New York, (1965).
3.8 Hoblit, E.M., "GUST LOADS ON AIRCRAFT: CONCEPTS AND APPLICATIONS", AIAA Education series, 1801 Alexander Bell Drive, Reston, VA 22091, (1988).

Chapter 4.0

MATERIAL PROPERTIES

4.1 INTRODUCTION

Airframe material allowables are equally as important as aircraft loads, and allowable properties directly affect structural weight. Besides the strength allowables, other characteristics of materials also should be considered during the design stage to ensure airframe safety and long-life structures. Material selection criteria are as follows:

- Static strength efficiency
- Fatigue strength
- Fracture toughness and crack growth rate
- Availability and cost
- Fabrication characteristics
- Corrosion and embrittlement phenomena
- Compatibility with other materials
- Environmental stability

In recent decades a most notable event has been the development of a branch of applied mathematics, known as fracture mechanics, by engineers concerned with airframe and, to a lesser extent, turbine engines. The designer no longer chooses a material solely on the basis of its textbook qualities, but on its proven ability to withstand minor damage in service without endangering the safety of the airframe. The residual strength after damage, described as the toughness, is now uppermost in the engineer's mind when he chooses alloys for airframe use. Damage caused by fatigue is the main worry because it is difficult to detect and can disastrously weaken the strength of critical components.

So, whereas aluminum alloys looked as if they had reached a technical plateau, airframe engineers have been able to clarify their needs as a result of the work done on fracture mechanics, and metallurgists have changed their compositions and treatment techniques to meet the new toughness requirements. In the years ahead, there will be the introduction of new materials with greater resistance to damage, as well as the development of more refined techniques for damage detection.

(a) For pressurized fuselage cabins and lower wing skins – two areas particularly prone to fatigue through the long-continued application and relaxation of tension stresses – the standard material is an aluminum alloy designated 2024-T3.

(b) For upper wing skins, which have to withstand mainly compression stresses as the wing flexes upwards during flight, 7075-T6 is used. This alloy is also used extensively for military fighter aircraft structures, which generally have stiffer wings and – except for the cockpit area – an unpressurized fuselage. The 7075-T6 alloy is almost twice as strong as 2024-T3, and therefore the weight can be reduced correspondingly in suitable applications.

Stress corrosion cracking is a defect which usually occurs after some years of service. It often, however, has its origins during the manufacturing process, when temperature differences caused by the unequal rates of cooling in various parts of a component give rise to stresses because the cooler parts contract more than the hotter areas. The very minute and virtually undetectable cracks which occur, frequently buried in the body of the component, gradually spread and reduce its strength.

A brief description of the major materials used in airframe structures can be found in military handbook MIL-HDBK-5 (use the latest version) which is the authoritative and official source of design data for both military and commercial airframe materials.

(A) WROUGHT FORMS

The wrought material forms commonly used on airframe structures are the following:

(a) Sheet and plate:
- Sheet – A rolled flat metal product below 0.250 inch in thickness
- Plate – A rolled flat metal product 0.250 inch in thickness or greater
- Clad – A thin coating of metal bonded to the base alloy by co-rolling (The cladding alloy is chosen so as to give maximum corrosion protection to the base metal. Nominal thickness of clad is 2.5 – 5% of composite thickness per side depending on alloy)

(b) Extrusions: Conversion of a billet into a length of uniform cross section by forcing metal at an elevated temperature through a die of desired cross sectional outline.
- Stepped extrusion – A single product with one or more abrupt cross section changes obtained by using multiple die segments.
- Impact extrusion – Resultant product of the process (cold extrusion) of a punch striking an unheated slug in a confining die (The metal flow may be either between the punch and die or through another opening)
- Hot impact extrusion – Similar to the above cold impact extrusion except that a preheated slug is used and the pressure application is slower

(c) Forging: Plastically deforming metal (usually hot) formed into shapes with compressive forces, with or without shaped dies. Forging provides the highest possible properties in predetermined locations and the most efficient material utilization.
- Hand forging – Requires no shaped dies. Obtainable in two categories:
  - Forged block – results in unidirectional grain only.
  - Shaped hand forging – results in longitudinal grain in two or more directions.
- Precision forging – A conventional forging that approximates the machined part as closely as possible thereby resulting in minimum machining or no machining at all.
- Blocked type die forging – Has a cross section at least 50% larger than the final machined dimensions.

(d) Casting: A part formed by the solidification of a material in a mold of desired configuration.
- Sand casting – Uses compacted sand for the mold material
- Die casting – Casting formed by injection of melted metal under pressure into metal dies
- Investment casting – Uses a refractory type of mold (normally gypsum plaster) and an expendable wax, plastic or other material

General wrought form applications on airframe structures are as follows:
- Sheet and plate:
  - Sheets are used on fuselage skin, control surfaces, rib webs, pressure bulkhead domes, structural repairs, general aviation airframes, etc., which generally are low-cost parts
  - Plates are machined to varying thicknesses to save weight on airframe components such as wing box skins, tail box skins, wing and tail spar webs, bulkhead rib webs, thicker fuselage skin near cutouts and areas of high shear loads, etc., which generally are high-cost parts
- Extrusions:
  - For higher strength applications
  - Used as is, and is usually a standard industry part or one of a company's own standard parts which are all low-cost parts. Used on vertical stiffeners of spars, ribs, deep beams, etc.
  - Usually the approximate shape desired is extruded and then machined to various shapes and thicknesses along the entire length
- Forging:
  - Has good grain flow direction
  - Used on wing and tail bulkhead ribs and spars, fuselage bulkhead frames, etc. which basically are machined parts for weight saving purpose
  - Precision forgings are used on clips, window cutout frames, etc., since machining is not required (cost savings)
- Castings:
  - The purpose to use castings is cost saving
  - Use them with caution! Most airframe companies do not allow the use of castings on any primary structures and may allow their use on secondary structures only under certain conditions

(B) ALUMINUM ALLOY GROUPS

The major aluminum alloys are divided into groups by an index system in which the first digit of the identifying number indicates the alloy used, as shown below:

- Group 1000   contains 99% elemental aluminum
- Group 2000   Copper as the major alloying element
- Group 3000   Manganese as the major alloying element
- Group 4000   Silicon as the major alloying element
- Group 5000   Magnesium as the major alloying element
- Group 6000   Magnesium and Silicon as the major alloying element
- Group 7000   Zinc as the major alloying element

There are four basic tempers used for aluminum and its alloys. They are:
- O – Annealed
- F – As fabricated
- H – Strain hardened
- T – Heat treated (All aluminum alloys used in primary airframe applications are the heat treated tempers)

A typical heat treat designation for a extrusion is listed below:

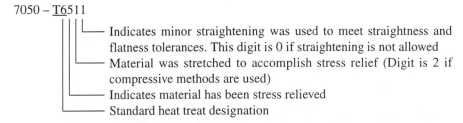

The aluminum alloy groups 2000 and 7000 are backbone materials for airframe structures and they are primarily used are as follows:

- Group 2000 – Primarily used in tension applications where fatigue and damage tolerant design is critical, e.g., lower wing surfaces, pressurized fuselage skin, etc.
- Group 7000 – Primarily used in compression applications where fatigue and damage tolerant design is not critical, e.g., upper wing surfaces, wing ribs, floor beams, etc.

4.2 STRESS-STRAIN CURVES

Before performing stress analysis, the stress engineer must understand the characteristics of stress-strain curves as this is not only an important aspect but the very foundation of stress analysis. The stress (F) vs. strain (ε) curve is basically a load (tension or compression axial load) vs. deformation curve plotted from a tensile test specimen, as shown in Fig. 4.2.1. A better name for the stress-strain curve would be "axial stress-strain curve" as this is a better description of what is tested and plotted.

It is interesting to note that not much information is available dealing with "shear stress-strain curve" which would have the same general shape as the material's stress-strain curves. In thicker webs or skins under a shear load or an enclosed section under a torsion load, the ultimate shear stress may reach into the plastic range and, therefore, a similar curve is needed for calculations.

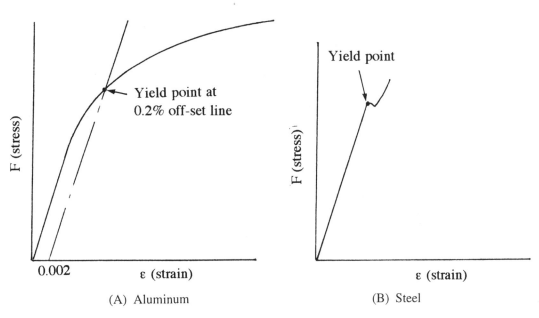

Fig. 4.2.1 Typical Stress-strain Curves for Aluminum and Steel

Chapter 4.0

For airframe structures, the most important stress-strain curves are obviously those for aluminum alloys which are quite different from other materials because they have longer elongation, as shown in Fig. 4.2.2. This curve shows a typical aluminum stress-strain relationship that every stress engineer shall understand its characteristics. This is the first step prior to performing stress analysis or sizing work correctly.

The "flat top" of the stress-strain curve indicates that for a large proportion of the total allowable strain, the corresponding stress is at, or very near, the allowable ultimate or "breaking" stress of the material. This means that any fiber of the material will support ultimate stress while undergoing considerable strain. It is this effect which permits other fibers, operating at lower stress engineer shall understand its characteristics. This is the first step prior to performing stress analysis or sizing work correctly.

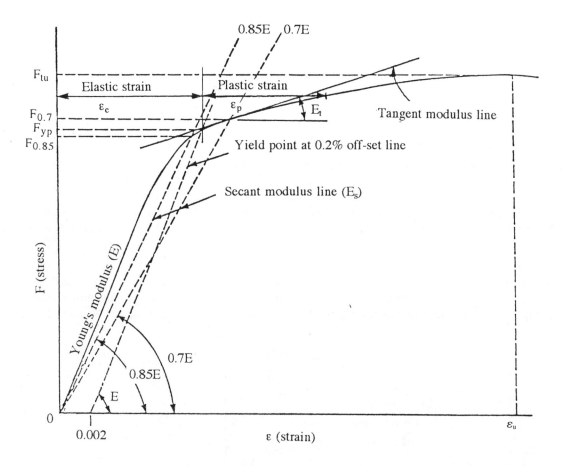

Fig. 4.2.2 Typical Aluminum Stress-Strain Curve – Definition of Terms

The stress-strain curve shape can be expressed by the Ramberg-Osgood formula with E, $F_{0.7}$, and n parameters (these values can be obtained from material data) in non-dimensional form:

$$\frac{E\varepsilon}{F_{0.7}} = \frac{F}{F_{0.7}} + \frac{3}{7}\left(\frac{F}{F_{0.7}}\right)^n \qquad \text{Eq. 4.2.1}$$

'n' values range between 8 and 35 and:
- Higher values are good for column strength and buckling
- Lower values are good in redundant structures which have better redistribution of loads

(A) MODULUS

Up to the elastic limit the slope of the f (stress) vs. ε (strain) curve is constant and above the elastic limit E is replaced by E_t (tangent modulus) which varies with stress level. This explains why the material stress-strain curve is divided into two main sections, namely:

- Elastic limit (proportional range)
- Inelastic limit (plastic range)

A group of stress-strain curves for several aluminum alloys (including the 2000 alloy series and the 7000 alloy series) are plotted in Fig. 4.2.3 and it is possible to see how strength characteristics are affected by the use of the different alloying elements.

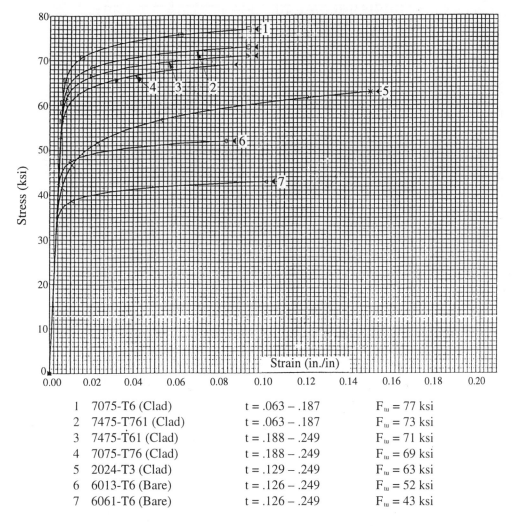

1	7075-T6 (Clad)	t = .063 – .187	F_{tu} = 77 ksi
2	7475-T761 (Clad)	t = .063 – .187	F_{tu} = 73 ksi
3	7475-T61 (Clad)	t = .188 – .249	F_{tu} = 71 ksi
4	7075-T76 (Clad)	t = .188 – .249	F_{tu} = 69 ksi
5	2024-T3 (Clad)	t = .129 – .249	F_{tu} = 63 ksi
6	6013-T6 (Bare)	t = .126 – .249	F_{tu} = 52 ksi
7	6061-T6 (Bare)	t = .126 – .249	F_{tu} = 43 ksi

Fig. 4.2.3 Stress-Strain Curves of Selected Aluminum Alloys (Sheets)

(a) Young's modulus (E):

Young's modulus is actually the first linear portion of the tangent modulus:

$$E = \frac{f}{\varepsilon} \qquad \text{(proportional or elastic range)} \qquad \text{Eq. 4.2.2}$$

(b) Tangent modulus (E_t):

$$E_t = \frac{f}{\varepsilon} \quad \text{(plastic or inelastic range)} \qquad \text{Eq. 4.2.3}$$

Typical aluminum alloy compressive stress-strain and compressive tangent-modulus curves for 7075-T651 plate and 7075-T651 extrusion are shown in Figs. 4.2.4 and 4.2.5, respectively.

(c) Modulus of rigidity or shear modulus (G):

$$G = \frac{E}{2(1+\mu)} \quad \text{(in elastic range)} \qquad \text{Eq. 4.2.4}$$

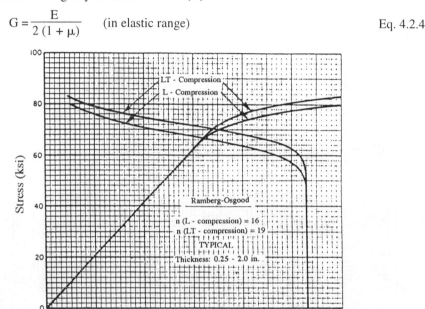

Fig. 4.2.4 *Compressive Stress-Strain and Tangent-Modulus (E_t) Curves for 7075-T651 Plate*

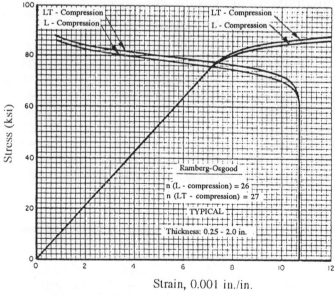

Fig. 4.2.5 *Compressive Stress-Strain and Tangent-Modulus (E_t) Curves for 7075-T651 Extrusion*

Material Properties

(B) STRESS AND STRAIN

(a) Stress:

When external forces are applied to a structural member, the member deforms and external forces, P, are transmitted throughout the member per unit cross-sectional area, A. The term stress is expressed as:

$$f = \frac{P}{A} \text{ (psi)} \qquad \text{Eq. 4.2.5}$$

(a) Strain:

Deformation, more commonly referred to as strain, is the inevitable result of stress. All materials are deformable under force and strain is defined as the forced change in the dimensions of a member. Materials are elastic within a certain range of stress and are able to recover their original dimensions upon removal of stress in the elastic range (not in the plastic range).

A member subjected to axial compression or tension will deflect an amount, δ. If the original length of the member is L, then axial strain, ε, may be mathematically written as follows:

$$\varepsilon = \frac{\delta}{L} \text{ (in./in.)} \qquad \text{Eq. 4.2.6}$$

(C) POISSON'S RATIO

As a material undergoes axial deformation, a certain amount of lateral contraction or expansion takes place, as shown in Fig. 4.2.6.

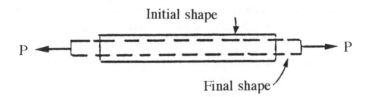

Fig. 4.2.6 Poisson's Effect

The ratio of the absolute value of the strain in the lateral direction to the strain in the axial direction is the Poisson's ratio:

$$\mu = -\frac{\varepsilon_y \text{ (lateral strain)}}{\varepsilon \text{ or } \varepsilon_x \text{ (axial strain)}} \qquad \text{Eq. 4.2.7}$$

The Poisson's ratio for aluminum alloys is approximately 0.33, and it ranges from 0.25 to 0.35 for other airframe metallic materials.

Chapter 4.0

Example:

For a round coupon, L, specimen is being loaded by P = 1,500 lbs. and the measured elongation across the 2.0 inch gage length is ΔL = 0.00259 inch. The change in the diameter dimension was measured to be 0.00016 inch.

Find the elastic modulus (E) and the Poisson's ratio (μ) for the material.

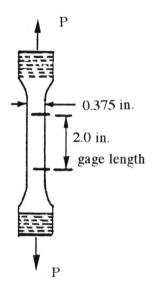

$$\varepsilon_x = \frac{\Delta L}{L} = \frac{0.00259}{2} = 0.0013 \text{ (axial strain)}$$

$$\varepsilon_y = -\frac{0.00016}{0.375} = -0.00043 \text{ (lateral strain)}$$

From Eq. 4.2.7, Poisson's ratio, $\mu = -\frac{-0.00043}{0.0013} = 0.33$

$$\text{Modulus, E} = \frac{f}{\varepsilon_x} = \frac{\frac{P}{A}}{\varepsilon_x}$$

$$= \frac{\frac{1,500}{(\frac{\pi \times 0.375^2}{4})}}{0.0013}$$

$$= 10.5 \times 10^6 \text{ psi}$$

(D) ELONGATION EFFECT

Elongation, e, is presented by %, a measure of the ductility of a material based on a tension test which can been seen from stress-strain curve on the horizontal scale and the bigger the elongation the more ductility for certain materials. For the most common aluminum materials the ultimate elongation values, e_u, range approximately from 0.08 (or 8%) to 0.15 (15%) in./in.

When a panel consisting of skin and stringers which have different material properties is used in tension or compression applications. Account must be taken of the fact that both the stringers and skin must elongate equally or the combination will fail (e.g., tension) when the less ductile material reaches its ultimate elongation.

The method of analysis for dissimilar material is:

$$N = \frac{P_{sk} + P_{st}}{b_s} = \frac{F_{sk} \times A_{sk} + F_{st} \times A_{st}}{b_s} \text{(lbs./in.)} \qquad \text{Eq. 4.2.8}$$

where: b_s – Stringer spacing (in.)
sk – Skin
st – Stringer or stringer
A – Cross-sectional area (in.^{2})
F_{tu} – Material ultimate allowable tensile stress
$e_{u,sk}$ – Skin strain obtained from the actual material stress-strain curve
$e_{u,st}$ – Stringer strain obtained from the actual material stress-strain curve
e_u' – The smaller of $e_{u,sk}$ and $e_{u,st}$
F' – The allowable stress in each material corresponding to e_u', to be obtained from the design stress-strain curve for the material
N – Allowable tensile load/inch of the combined materials

Example:

Given a fuselage skin-stiffener panel with following materials and determine its tensile allowable load.

Skin: 2024-T3 clad; $b_s = 5.0$ in.; $A_{sk} = 0.288$ in.^{2}

Stringer: 7475-T761 clad; $A_{st} = 0.144$ in.^{2}

Solution by tabular form is as follows:

	Skin 2024-T3 clad (curve 5 from Fig. 4.2.3)	Stringer 7475-T761 clad (curve 3 from Fig. 4.2.3)
e_u (Elongation)	$e_{u,sk} = 0.15$	$e_{u,st} = 0.093$
e_u' (Elongation)	0.093 (use smaller of the two values of $e_{u,sk}$ and $e_{u,st}$)	
F' (Stress)	60 ksi	71 ksi
A (Area)	0.288 in.^{2}	0.144 in.^{2}
P (Load)	$P_{sk} = 0.288 \times 60 = 17.28$ kips	$P_{st} = 0.144 \times 71 = 10.22$ kips

$$N = \frac{P_{sk} + P_{st}}{b_s} = \frac{17.28 + 10.22}{5.0} = 5.5 \text{ kips/in.} \quad \text{(From Eq. 4.2.8)}$$

4.3 ALLOWABLES

A standardized document containing metallic materials design data which is acceptable for government certification is required as a source of design allowables. This document should also contain information and data on the following properties and characteristics:

- Fracture toughness
- Fatigue
- Creep
- Rupture
- Crack propagation rate
- Stress corrosion cracking

The following sources provide mechanical material properties for airframe grade metallic materials commonly used by airframe manufacturers:

- MIL-HDBK-5 (Military Handbook – Metallic Materials and Elements for Aerospace Vehicle Structures) is used by both U. S. military and commercial airframe manufacturers
- Additional information pertaining to material specifications and physical properties is available in company manuals

(A) ALLOWABLE BASES

For some materials more than one set of allowables exists under the basic headings of A, B, or S, which represents the basis upon which primary strength allowable properties were established.

- A basis – The allowable value above which at least 99% of the population values will fall with a 95% confidence level
- B basis – The allowable value above which at least 90% of the population values will fall with a 95% confidence level
- S basis – The minimum guaranteed value from the governing material specification and its statistical assurance level is not known

Chapter 4.0

To design a single member whose loading is such that its failure would result in loss of structural integrity, the minimum guaranteed design properties, or A basis or S basis values, must be used. On redundant or fail-safe structure design, where if one member experiences failure the loads are safely redistributed to adjacent members, the B values may be used.

(B) GRAIN DIRECTION EFFECT

The mechanical and physical properties of the material grain, as shown in Fig. 4.3.1, are not always equal in all three directions as described below:

- longitudinal (L): Parallel to the working direction or the main direction of grain flow in a worked metal (This is the strongest strength)
- Long transverse (LT): Across or perpendicular to the direction of longitudinal (L) grain (The second strongest strength and sometimes equal to that of longitudinal direction)
- Short transverse (ST): Shortest dimension of the part perpendicular to the direction of working. This is the weakest strength and the tendency toward stress corrosion cracking is usually greatest in this direction

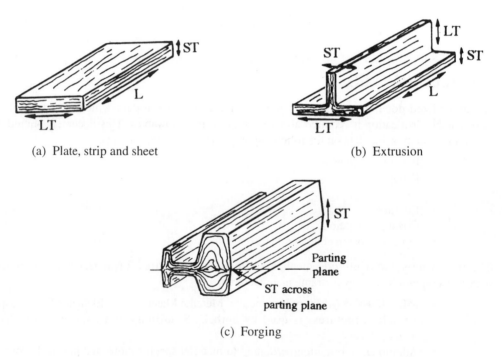

(a) Plate, strip and sheet

(b) Extrusion

(c) Forging

Fig. 4.3.1 Material Grain Direction

(C) DESIGN MECHANICAL PROPERTIES

Design mechanical properties for metallic materials can be found in MIL-HDBK-5. Properties for 2024 and 7075 alloys are shown in Figures 4.3.2 through 4.3.6 for easy reference since this data is used in many of the calculations in this book. The property data shown could be used for preliminary sizing but should not be used in formal stress analysis which is submitted to a certifying agency.

Specification	QQ-A-250/5									
Form	Flat sheet and plate									
Temper	T42									
Thickness, in.	0.008–0.009		0.010–0.062		0.063–0.249		0.250–0.499	0.500–1.000	1.001–2.000	2.001–3.000
Basis	A	B	A	B	A	B	S	S	S	S
Mechanical properties: F_{tu}, ksi:										
L	55	57	57	59	60	62	60	59	58	56
LT	55	57	57	59	60	62	60	59	58	56
F_{ty}, ksi:										
L	34	35	34	35	36	38	36	36	36	36
LT	34	35	34	35	36	38	36	36	36	36
F_{cy}, ksi:										
L	34	35	34	35	36	38	36	36	36	36
LT	34	35	34	35	36	38	36	36	36	36
F_{su}, ksi	33	34	34	35	36	37	36	35	35	34
F_{bru}^{c}, ksi:										
($\frac{e}{D} = 1.5$)	83	86	86	89	90	93	90	89	87	83
($\frac{e}{D} = 2.0$)	104	108	108	112	114	118	114	112	110	106
F_{bry}^{c}, ksi:										
($\frac{e}{D} = 1.5$)	48	49	48	49	50	53	50	50	50	50
($\frac{e}{D} = 2.0$)	54	56	54	56	58	61	58	58	58	58
e, percent (S-basis): LT	10	–	d	–	15	–	12	8	d	4
E, 10^3 ksi: Primary	10.5				10.5			10.7		
Secondary	9.5				10.0			10.2		
E_c, 10^3 ksi: Primary	10.7				10.7			10.9		
Secondary	9.7				10.2			10.4		
G, 10^3 ksi μ	0.33									
Physical properties: ω, lb/in.³	0.101									

(Bearing allowables are 'dry pin' values)

Fig. 4.3.2 Mechanical and Physical Properties of Clad 2024-T42 Alloy Sheet and Plate (Ref. 4.1)

Specification	QQ-A-250/4															
Form	Sheet					Plate										
Temper	T3					T351										
Thickness, in.	0.008-0.009	0.010-0.128		0.129-0.249		0.250-0.499		0.500-1.000		1.001-1.500		1.501-2.000		2.001-3.000		3.000-4.000
Basis	S	A	B	A	B	A	B	A	B	A	B	A	B	A	B	A	B
Mechanical properties:																	
F_{tu}, ksi:																	
L	64	64	65	65	66	64	66	63	65	62	64	62	64	60	62	57	59
LT	63	63	64	64	65	64	66	63	65	62	64	62	64	60	62	57	59
ST	–	–	–	–	–	–	–	–	–	–	–	–	–	52	54	49	51
F_{ty}, ksi:																	
L	47	47	48	47	48	48	50	48	50	47	50	47	49	46	48	43	46
LT	42	42	43	42	43	42	44	42	44	42	44	42	44	42	44	41	43
ST	–	–	–	–	–	–	–	–	–	–	–	–	–	38	40	38	39
F_{cy}, ksi:																	
L	39	39	40	39	40	39	41	39	41	39	40	38	40	37	39	35	37
LT	45	45	46	45	46	45	47	45	47	44	46	44	46	43	45	41	43
ST	–	–	–	–	–	–	–	–	–	–	–	–	–	46	48	44	47
F_{su}, ksi	39	39	40	40	41	38	39	37	38	37	38	37	38	35	37	34	35
F_{bru}, ksi:																	
($\frac{e}{D}$ = 1.5)	104	104	106	106	107	97	100	95	98	94	97	94	97	91	94	86	89
($\frac{e}{D}$ = 2.0)	129	129	131	131	133	119	122	117	120	115	119	115	119	111	115	106	109
F_{bry}, ksi:																	
($\frac{e}{D}$ = 1.5)	73	73	75	73	75	72	76	72	76	72	76	72	76	72	76	70	74
($\frac{e}{D}$ = 2.0)	88	88	90	88	90	86	90	86	90	86	90	86	90	86	90	84	88
e, percent (S-basis):																	
LT	10	–	–	–	–	12	–	8	–	7	–	6	–	4	–	4	–
E, 10^3 ksi	10.5					10.7											
E_c, 10^3 ksi	10.7					10.9											
G, 10^3 ksi	4.0					4.0											
µ	0.33					0.33											
Physical properties:																	
ω, lb/in.^{3}	0.101																

(Bearing allowables are 'dry pin' values)

Fig. 4.3.3 Mechanical and Physical Properties of 2024-T3 Alloy Sheet and Plate (Ref. 4.1)

Material Properties

Specification	QQ-A-250/12															
Form	Sheet					Plate										
Temper	T6 and T62					T651										
Thickness, in.	0.008-0.011	0.012-0.039		0.040-0.125		0.126-0.249		0.250-0.499		0.500-1.000		1.001-2.000		2.001-2.500		2.500-3.000
Basis	S	A	B	A	B	A	B	A	B	A	B	A	B	A	B	A	B
Mechanical properties:																	
F_{tu}, ksi:																	
L	–	76	78	78	80	78	80	77	79	77	79	76	78	75	77	71	73
LT	74	76	78	78	80	78	80	78	80	78	80	77	79	76	78	72	74
ST	–	–	–	–	–	–	–	–	–	–	–	–	–	70	71	66	68
F_{ty}, ksi:																	
L	–	69	72	70	72	71	73	69	71	70	72	69	71	66	68	63	65
LT	61	67	70	68	70	69	71	67	69	68	70	67	69	64	66	61	61
ST	–	–	–	–	–	–	–	–	–	–	–	–	–	59	61	56	58
F_{cy}, ksi:																	
L	–	68	71	69	71	70	72	67	69	68	70	66	68	62	64	58	60
LT	–	71	74	72	74	73	75	71	73	72	74	71	73	68	70	65	67
ST	–	–	–	–	–	–	–	–	–	–	–	–	–	67	70	64	66
F_{su}, ksi	–	46	47	47	48	47	48	41	44	44	45	44	45	44	45	42	43
F_{bru}, ksi:																	
($\frac{e}{D} = 1.5$)	–	118	121	121	124	121	124	117	120	117	120	116	119	114	117	108	111
($\frac{e}{D} = 2.0$)	–	152	156	156	160	156	160	145	148	145	148	143	147	141	145	134	137
F_{bry}, ksi:																	
($\frac{e}{D} = 1.5$)	–	100	105	102	105	103	106	97	100	100	103	100	103	98	101	94	97
($\frac{e}{D} = 2.0$)	–	117	122	119	122	121	124	114	118	117	120	117	120	113	117	109	112
e, percent (S-basis):																	
LT	5	7	–	8	–	8	–	9	–	7	–	6	–	5	–	5	–
E, 10^3 ksi	10.3					10.3											
E_c, 10^3 ksi	10.5					10.6											
G, 10^3 ksi	3.9					3.9											
μ	0.33					0.33											
Physical properties: ω, lb/in.^{3}	0.101																

(Bearing allowables are 'dry pin' values)

Fig. 4.3.4 Mechanical and Physical Properties of 7075 Alloy Sheet and Plate (Ref. 4.1)

Specification	QQ-A-200/11												
Form	Extrusion (rod, bar, and shapes)												
Temper	T6, T6510, T6511, and T62												
Cross-sectional area, in^2	≤ 20												>20, ≤ 32
Thickness, in.	Up to 0.249		0.250-0.499		0.500-0.749		0.750-1.499		1.500-2.999		3.000-4.499		
Basis	A	B	A	B	A	B	A	B	A	B	A	B	S
Mechanical properties: F_{tu}, ksi:													
L	78	82	81	85	81	85	81	85	81	85	81	84	78
LT	76	80	78	81	76	80	74	78	70	74	67	70	65
F_{ty}, ksi:													
L	70	74	73	77	72	76	72	76	72	76	71	74	70
LT	66	70	68	72	66	70	65	68	61	65	56	58	55
F_{cy}, ksi:													
L	70	74	73	77	72	76	72	76	72	76	71	74	70
LT	72	76	74	78	72	76	71	74	67	71	61	64	60
F_{su}, ksi	42	44	43	45	43	45	42	44	41	43	40	41	38
F_{bru}, ksi:													
($\frac{e}{D}=1.5$)	112	118	117	122	117	122	116	122	115	120	109	113	105
($\frac{e}{D}=2.0$)	141	148	146	153	146	153	145	152	144	151	142	147	136
F_{bry}, ksi:													
($\frac{e}{D}=1.5$)	94	99	97	103	96	101	95	100	93	98	89	92	87
($\frac{e}{D}=2.0$)	110	117	115	121	113	119	112	118	110	116	105	110	104
e, percent (S-basis): LT	7	–	7	–	7	–	7	–	7	–	7		6
E, 10^3 ksi: E_c, 10^3 ksi: G, 10^3 ksi μ	10.4 10.7 4.0 0.33												
Physical properties: ω, lb/in.^{3}	0.101												

(Bearing allowables are 'dry pin' values)

Fig. 4.3.5 Mechanical and Physical Properties of 7075 Alloy Extrusion (Ref. 4.1)

Material Properties

Alloy					Hy-Tuf 4330V	D6AC 4335V	AISI 4340 D6AC	AISI 4340	0.40C 300M	0.42C 300M
Form	All wrought forms				All wrought forms			Bar, forging, tubing		
Condition	Quenched and tempered				Quenched and tempered			Quenched and tempered		
Basis	S	S	S	S	S	S	S	S	S	S
Mechanical properties:										
F_{tu}, ksi	125	160	180	200	220	220	260	260	270	280
F_{ty}, ksi	100	142	163	176	185	190	215	215	220	230
F_{cy}, ksi	109	154	173	181	193	198	240	240	236	247
F_{su}, ksi	75	96	108	120	132	132	156	156	162	168
F_{bru}, ksi:										
($\frac{e}{D}$ = 1.5)	194	230	250	272	297	297	347	347	414	430
($\frac{e}{D}$ = 2.0)	251	300	326	355	385	385	440	440	506	525
F_{bry}, ksi:										
($\frac{e}{D}$ = 1.5)	146	202	230	255	267	274	309	309	344	360
($\frac{e}{D}$ = 2.0)	175	231	256	280	294	302	343	343	379	396
e, percent:										
L					10	–	–	10	8	7
T					5	–	–	–	–	–
E, 10^3 ksi:	29.0									
E_c, 10^3 ksi:	29.0									
G, 10^3 ksi	11.0									
μ	0.32									
Physical properties: ω, lb/in.^{3}	0.283									

(Bearing allowables are 'dry pin' values)

Fig. 4.3.6 Mechanical and Physical Properties of Low-Alloy Steel (Ref. 4.1)

4.4 AIRWORTHINESS REQUIREMENTS

The term plasticity of a material implies yielding under steady load, and it is defined as the property of sustaining visible permanent deformation without rupture. In aluminum alloys, there are two basic properties (ultimate and yield strength) in stress-strain curves which directly affect the value of the margin of safety (MS) and the ability of the material to support limit loads without detrimental permanent deformation; both must be considered to meet airworthiness requirements.

   (a)   Curve A: f_{ty} or $F_{cy} > (0.67) F_{tu}$ property (see Fig. 4.4.1):

   In this case, the material's allowable stresses of ultimate tensile (f_{tu}) or bearing (f_{bru}) govern the MS which meets airworthiness requirements (by ultimate loads).

Chapter 4.0

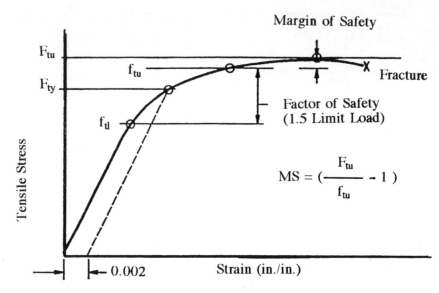

where: f_{tl} – Actual stress at limit load
(Ultimate Design Allowable Stress = F_{tu})

Fig. 4.4.1 Stress-Strain Curves for Aluminum Alloy ($F_{ty} > 0.67 F_{tu}$)

(b) Curve B: F_{ty} or F_{cy} < (0.67) F_{tu} property (see Fig. 4.4.2):

In this case, the material's allowable stresses of yield (f_{cy}) or bearing (f_{bry}) govern the MS needed to meet airworthiness requirements and prevent detrimental permanent deformation (by limit loads).

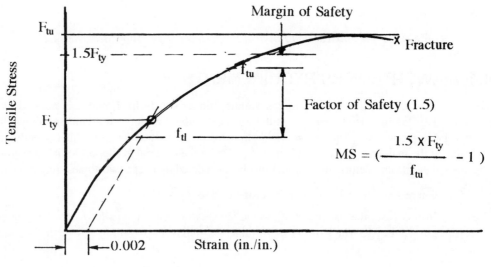

where: f_{tl} – Actual stress at limit load
(Ultimate Design Allowable Design Stress = 1.5 × F_{ty})

Fig. 4.4.2 Stress-Strain Curves for Aluminum Alloy ($F_{ty} < 0.67 F_{tu}$)

4.5 TOUGHNESS AND CRACK GROWTH RATE

The toughness (K = stress intensity factor) of a material may be defined as the ability of a component with a crack or defect to sustain a load without catastrophic failure as shown in Fig. 4.5.1. This toughness data become increasingly important in the evaluation of a design for a fail-safe structure. Material toughness is also known as the quality of a material to resist failure in the presence of a fatigue crack, as shown in Fig. 4.5.2. Fracture toughness and fatigue resistance must be designed to handle tension structures in an airframe.

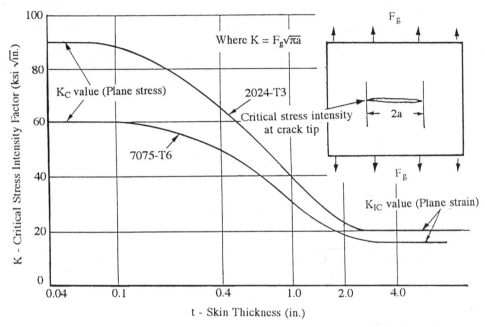

(After a certain thickness is reached, the critical stress intensity factor K remains constant as K_{IC})

Fig. 4.5.1 Aluminum Alloy Stress Intensity Factor (Toughness) vs. Thickness

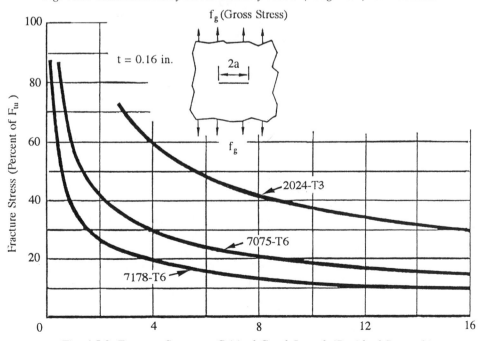

Fig. 4.5.2 Fracture Stress vs. Critical Crack Length (Residual Strength)

Application of cyclic loads less than the critical static residual strength of a cracked part will result in a crack propagation rate ($\frac{da}{dN}$). Crack growth characteristics of a material under cyclic stresses are a measure of its ability to contain a crack and prevent it from rapidly attaining critical length, as shown in Fig. 4.5.3. For most good, ductile, structural materials this is an orderly, stable process having the same general characteristics. Continued cycling eventually raises the crack size to be critical for that stress level and catastrophic failure occurs.

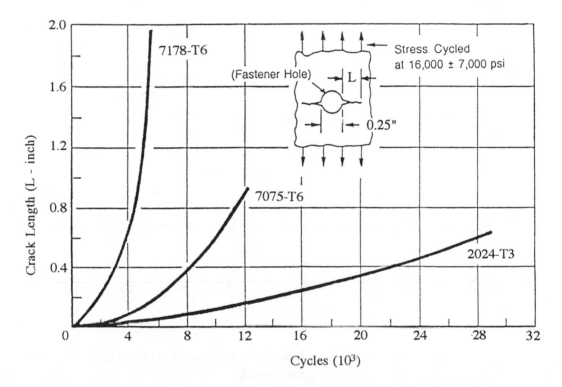

Fig. 4.5.3 Crack Length vs. Cycling Load

4.6 MATERIAL USAGE

General aircraft applications are listed below for reference (Note: Structural efficiency, SE = ratio of material ultimate stress/density and the higher the SE the ligher the structure):

(a) Aluminum Alloys (density = 0.101 lbs/in.^{3}; SE = 752):
- The most widely used materials in airframe structures
- Inexpensive, easily formed and machined
- 2000 series alloys provide better medium-to-high strength than other aluminum alloys
- 2019 alloy has comparatively superior strength at cryogenic temperatures
- 2024 alloy materials are widely used in fatigue-critical areas; slow crack propagation
- 5000 and 6000 series alloys can be readily brazed and welded
- 7075 alloy is a high strength material used primarily in high compression construction (This alloy is not usually welded, and has poorer ductility, notch toughness, and formability than other aluminum alloys)
- Al-Lithium alloy is 10% lighter and stiffer than other aluminum alloys, but costs more

(b) Magnesium Alloys (density = 0.064 lbs/in.^{3}; SE = 610):

- This material is seldom used for primary structures and is usually prohibited from use on airframes
- Also has compatibility problems with other materials, especially in humid or salty atmospheres

(c) Titanium Alloys (density = 0.16 lbs/in.^{3}; SE = 981):

- Good structural efficiencies and good cryogenic temperature endurance
- More expensive than aluminum alloys
- Lightweight and corrosion resistant
- Can be formed, machined, and welded
- Good toughness
- Ti-6Al-4V alloys are widely used in annealed and heat-treated forms
- Ti-5Al-2.5Sn alloy is extremely weldable

(d) Steel Alloys (density = 0.283 lbs/in.^{3}; SE = 884):

- Steel alloys are not widely used on airframe structures with the exception of landing gear
- Do not use any PH corrosion resistant steel such as 17-7PH, 17-4PH, or AM350 in the annealed or solution treated forms because their corrosion resistance is appreciably lowered
- For low strength requirements, use 300 series stainless steel alloys
- For high strength, use type 301-1/4H, 301-1/2H or 301-H
- Non-magnetic temperature resistant material required:
  - A286 for low strength
  - Inconel 718 for high strength
  - Titanium can be used to save weight
- Use only types 321 or 347 stainless steel when fusion welding or when a 300 series stainless steel is required
- AISI 4130 steel – thickness less than 0.5 inch and H.T. $\leq$ 180 ksi
  AISI 4340 steel – thickness greater than 0.5 inch and H.T. $\geq$ 180 ksi

(e) Nickel Alloys (density = 0.3 lbs/in.^{3}; SE = 517):

- Exhibit good properties from the cryogenic range to 1800 – 2000 °F
- Corrosion-resistant, readily welded (solution-treated)
- Typical materials are Inconel X-750, X-600, X-625, and X-718

(f) Beryllium Alloys (density = 0.066 lbs/in.^{3}; SE = 1152):

- Has very high structural efficiency (the ratio of material ultimate stress/density)
- Expensive material
- Has limited formability; difficult to join, drill and machine; notch sensitive and not weldable
- Produces toxic gas

(g) Uranium material (density = 0.68 lbs/in.^{3}):

- This is a non-structural material and is used for balance weight because of its very high density

4.7 MATERIAL SELECTION PROCEDURES

Follow these steps to select the final material:

(1) MATERIAL APPLICATIONS

(a) Operational features – Description of principal loads and environment for the component

(b) Principal design requirements – Most important design properties for satisfying the operational features

(c) Material form – Sheet, plate, extrusion, forging, casting, bar, etc.

(2) RATING CATEGORIES

(a) Static strength properties:
- Property basis – A, B or S value
- F_{tu} – Ultimate tensile stress
- F_{ty} – Tensile yield stress
- F_{cy} – Compressive yield stress
- F_{su} – Ultimate shear stress
- F_{bru}, F_{bry} – Ultimate bearing stress; bearing yield stress
- E, E_t – Modulus of Elasticity and tangent Modulus

(b) Durability and damage tolerance properties:
- K_{IC}, K_C – Fracture toughness: plane strain and plane stress
- K_{ISCC} – Stress corrosion resistance crack
- SCC – threshold stress corrosion resistance
- $\Delta K(\frac{da}{dN})$ – Fatigue crack propagation rate
- S-N curves – Fatigue toughness
- Corrosion resistance – Resistance to exfoliation, pitting, galvanic corrosion, etc.

(c) Producibility:
- Cost – Basic raw material cost
- Availability – Lead time for material delivery
- Fabrication – Machinability, formability and heat treatment characteristics
- Current applications – Similar applications for new aircraft
- Specifications – Data and information available

(d) Serviceability:
- Reliability – Performs satisfactorily for at least a given period of time when used under stated conditions
- Resists cracking, corrosion, thermal degradation, wear and foreign object damage for a specified period of time
- Performs in accordance with prescribed procedures and conforms to specified conditions of utility within a given period of time
- Ease of inspection for damage

(e) Final selection:
- Rank by the total ratings listed above
- Select the highest rank

References

4.1 MIL-HDBK-5 (Military Handbook – Metallic Materials and Elements for Flight Vehicle Structures), U.S. Government Printing Office, Washington, D.C.
4.2 Niu, C. Y., "AIRFRAME STRUCTURAL DESIGN", Conmilit Press Ltd., P.O. Box 23250, Wanchai Post Office, Hong Kong, 1988.
4.3 Ramberg, W. and Osgood, W. R., "Description of Stress-strain Curves of Three Parameters", NACA TN 902, July, 1943.

Chapter 5.0

STRUCTURAL ANALYSIS

The purpose of this Chapter is merely to review some of the common methods which are practically useful for airframe structural analysis and sizing. All the derivation procedures for equations, which can be found in textbooks or in the references in the back of this Chapter, are generally not repeated herein. Sign conventions should be the same as those defined in Chapter 2.0.

5.1 REVIEW OF HOOKE'S LAW

All structural members which undergo stress are strained as described in the material stress-strain curve in Chapter 4.0. Hooke's law states that when an element undergoes stress it can be expressed simply as "stress is proportional to strain" which is valid until the material is stressed to its elastic limit.

(a) In an one-dimensional element of an isotropic material there is only one elastic constant of Modulus 'E'. The strain can be expressed as a linear function of the stress or vice versa:

$$\varepsilon = \frac{f}{E} \qquad \text{Eq. 5.1.1}$$

(b) In a two-dimensional element of an isotropic material there are only two independent elastic constants (namely modulus 'E' and Poisson's ratio 'μ'). Each of the three strain components can be expressed as a linear function of the three stress components or vice versa:

- Axial strains:

$$\varepsilon_x = \frac{f_x - \mu f_y}{E} \qquad \text{Eq. 5.1.2}$$

$$\varepsilon_y = \frac{f_y - \mu f_x}{E} \qquad \text{Eq. 5.1.3}$$

- Shear strains:

$$\gamma_{xy} = \frac{f_{xy}}{G} \qquad \text{Eq. 5.1.4}$$

(c) In a three-dimensional element of an isotropic material there are also two independent elastic constants (E and μ). Each of the six strain components can be expressed as a linear function of the six stress components or vice versa:

- Axial strains:

$$\varepsilon_x = \frac{f_x - \mu(f_y + f_z)}{E} \qquad \text{Eq. 5.1.5}$$

$$\varepsilon_y = \frac{f_y - \mu(f_z + f_x)}{E} \qquad \text{Eq. 5.1.6}$$

Structural Analysis

$$\varepsilon_z = \frac{f_z - \mu(f_x + f_y)}{E}$$
Eq. 5.1.7

- Shear strains:

$$\gamma_{xy} = \frac{f_{xy}}{G}$$
Eq. 5.1.8

$$\gamma_{yz} = \frac{f_{yz}}{G}$$
Eq. 5.1.9

$$\gamma_{zx} = \frac{f_{zx}}{G}$$
Eq. 5.1.10

(d) In a three-dimensional element of anisotropic materials there are a total of 36 elastic constants.

Refer to Chapter 4.0 for the three elastic constants E, G, and μ shown in above equations.

Each material has a proportionality constant E, shown in Eq. 5.1.1, and the expression for axial deflection can be obtained as shown:

$$E = \frac{f}{\varepsilon} = \frac{\frac{P}{A}}{\frac{\delta}{L}}$$

where: $f = \frac{P}{A}$ (stress)

$\varepsilon = \frac{\delta}{L}$ (strain)

Therefore, the member deformation (or elongation) by axial load is

$$\delta = \frac{PL}{AE}$$
Eq. 5.1.11

Example 1:

Determine the load applied to the angle A by plate B in the structure shown when the modulus of elasticity for the steel is $E = 30 \times 10^6$ psi and for the aluminum is $E = 10 \times 10^6$ psi.

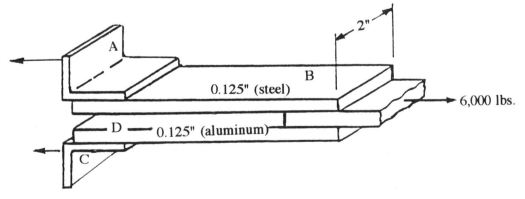

Assuming the deflection (δ) is the same for plates B and D, the following relationship may be derived. From Eq. 5.1.11:

$$\delta_B = \frac{P_B L_B}{A_B E_B}$$

Chapter 5.0

$$\delta_D = \frac{P_D L_D}{A_D E_D}$$

Assume δ_B equal to δ_D, then

$$\frac{P_B L_B}{A_B E_B} = \frac{P_D L_D}{A_D E_D} \quad \text{Eq. (A)}$$

A_B equals A_D and L_B equals L_D Therefore Eq. (A) reduces to :

$$\frac{P_B}{E_B} = \frac{P_D}{E_D} \quad \text{or} \quad P_B = \frac{P_D E_B}{E_D} \quad \text{Eq. (B)}$$

To balance the free body $P_B + P_D$ must equal 6,000 lbs, then

$$P_B = 6{,}000 - P_D \quad \text{Eq. (C)}$$

Combine Eq. (B) and Eq. (C) to obtain:

$$P_D \frac{E_B}{E_D} = 6{,}000 - P_D$$

$$P_D = \frac{6{,}000}{1 + \frac{30 \times 10^6}{10 \times 10^6}} = 1{,}500 \text{ lbs.}$$

$$P_B = 4{,}500 \text{ lbs.}$$

Example 2:

Consider the same problem as given in Example 1, except let plate B be aluminum and its thickness equal to 0.063 inch. From Eq. 5.1.11:

$$\delta_B = \frac{P_B L_B}{A_B E_B}$$

$$\text{and} \quad \delta_D = \frac{P_D L_D}{A_D E_D}$$

For this analysis, $\delta_B = \delta_D$, $L_B = L_D$, and $E_B = E_D$

Therefore, $\dfrac{P_B}{A_B} = \dfrac{P_D}{A_D}$ or $P_B = P_D \dfrac{A_B}{A_D}$ $\quad$ Eq. (D)

To balance the free body $P_B + P_D$ must equal 6,000 lbs and then,

$$P_B = 6{,}000 - P_D \quad \text{Eq. (E)}$$

Combine Eq. (D) and Eq. (E), to obtain:

$$P_D \frac{A_B}{A_D} = 6{,}000 - P_D$$

$$P_D = \frac{6{,}000}{1 + \frac{A_B}{A_D}}$$

$$= \frac{6{,}000}{1 + \frac{2 \times 0.063}{2 \times 0.125}}$$

$$= 4{,}000 \text{ lbs.}$$

$$\text{and} \quad P_B = 2{,}000 \text{ lbs.}$$

(**Note:** The previous examples show that the loads transferred through a multi-member structure are proportional to their areas and modulus of elasticity when the members' lengths are the same)

5.2 PRINCIPAL STRESSES

Rational design methods for structural members comprise essentially three operations:

- The determination of loads to which the structure may be subjected
- The stresses produced by the loads
- The proportioning of the component members of the structure to resist these stresses without rupture or undue deformation

In the analysis of stressed members, it is usually convenient to determine stresses only on specific planes at a given point under consideration. In some cases the load-carrying capacity of a member is controlled by its ability to resist bending stresses or longitudinal shear stresses, but often the critical stresses act on planes other than those on which the stresses have been determined. Hence a problem arises:

- Given the stresses on a certain plane at a point in a stressed member, find the maximum normal and shear stresses or a critical combination of normal and shear stresses at a point

Consider a point situated at some arbitrary position in a given stressed member, the stresses acting upon the horizontal and vertical planes passing through this point are assumed to be known. The stresses, shown in Fig. 5.2.1, that act upon some arbitrary chosen inclined plane (angle φ) and two stress components on this inclined plane are

- One perpendicular to the plane (f_ϕ normal stress)
- One parallel to the plane (f_ϕ shear stress)

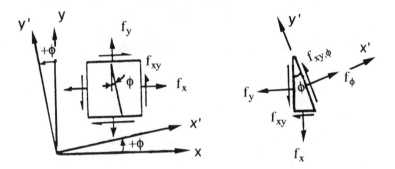

Axial stress: $f_\phi = \dfrac{f_x + f_y}{2} \pm (\dfrac{f_x - f_y}{2})\cos 2\phi + f_{xy}\sin 2\phi$

Shear stress: $f_{xy,\phi} = (\dfrac{f_x - f_y}{2})\sin 2\phi - f_{xy}\cos 2\phi$

Fig. 5.2.1 Principal Stresses

The principal stresses can be calculated by the following equations:

$$f_\phi = \dfrac{f_x + f_y}{2} \pm \sqrt{(\dfrac{f_x - f_y}{2})^2 + f_{xy}^2} \qquad \text{Eq. 5.2.1}$$

(a) The maximum principal stress:

$$f_{max} = \dfrac{f_x + f_y}{2} + \sqrt{(\dfrac{f_x - f_y}{2})^2 + f_{xy}^2} \qquad \text{Eq. 5.2.2}$$

(b) The minimum principal stress:

$$f_{min} = \frac{f_x + f_y}{2} - \sqrt{(\frac{f_x - f_y}{2})^2 + f_{xy}^2}$$ Eq. 5.2.3

These principal stresses act on inclined planes (principal planes), defined by the angle φ in Eq. 5.2.4.

$$\tan 2\phi = \frac{2f_{xy}}{f_x - f_y}$$ Eq. 5.2.4

The maximum and minimum shear stresses act on planes defined by Eq. 5.2.5; these planes are always 45° from the principal planes.

$$f_{xy, \phi} = \pm \sqrt{(\frac{f_x - f_y}{2})^2 + f_{xy}^2}$$ Eq. 5.2.5

(a) The maximum principal shear stress:

$$f_{xy, max} = + \sqrt{(\frac{f_x - f_y}{2})^2 + f_{xy}^2}$$ Eq. 5.2.6

(b) The minimum principal shear stress:

$$f_{xy, min} = - \sqrt{(\frac{f_x - f_y}{2})^2 + f_{xy}^2}$$ Eq. 5.2.7

MOHR'S CIRCLE GRAPHIC METHOD

Mohr's circle, as shown in Fig. 5.2.2, is a simple method of obtaining geometric properties of stresses which are plotted on f_y and $-f_{xy}$ at point A and on f_x, f_{xy} at point B. Draw a circle through points A and B and locate the center of the circle at point O on the f_x, f_y axis. The principal stresses of f_{max}, f_{min}, and 2ϕ are read from the graph. This is a good approximation method and was often used in the good old days when engineers used slide rules to do calculations.

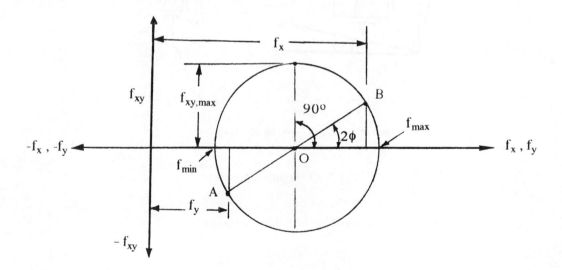

Fig. 5.2.2 Mohr's Circle Graph

5.3 EQUILIBRIUM AND COMPATIBILITY

These are the two important conditions which must be satisfied by any stable structure:

(a) Condition of equilibrium – For a stable structure the resultant of all external loads (both applied and reactive) must be zero and the external loads must be balanced by internal stresses.

(b) Condition of compatibility – Similarly in the deformed state the compatibility or the continuity of the structure as a whole and each individual member must be maintained along with the applied constraints.

Every member or particle in the universe is in static or dynamic equilibrium; therefore, all the external forces and inertial force acting on a member or particle are always in equilibrium or perfect balance. Consequently, the algebraic summation of all the forces or of all the moments acting on a member must equal zero.

A member must be shown to balance freely in space before valid analysis can proceed and it may not always be necessary to define all the forces numerically in order to establish the validity of the analysis. If two or more free bodies that are shown to be in balance are interconnected, the interfacing forces become internal forces and the remaining external forces still produce a perfect balance.

As shown in Fig. 5.3.1, a force in space is most conveniently defined by its components in three directions and its moments about three axes. The six equations of equilibrium are summarized as:

Forces: $\Sigma F_x = 0;\quad \Sigma F_y = 0;\quad \Sigma F_z = 0$ Eq. 5.3.1

Moments: $\Sigma M_x = 0;\quad \Sigma M_y = 0;\quad \Sigma M_z = 0$ Eq. 5.3.2

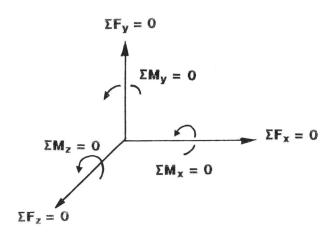

Fig. 5.3.1 Equilibrium of Forces (Right-hand Rule)

Rectangular axes (mutually perpendicular) are generally preferred. Six equations of equilibrium may be used to obtain the magnitude of balancing forces on a free body. If the forces or moments shown on a member being structurally analyzed are not in perfect balance, then one or more of the forces shown is in error. Therefore, any subsequent analysis of internal forces or stresses may be erroneous.

Stress analysis has to satisfy both equilibrium and compatibility conditions which enable us to write down as many equations as required to evaluate the unknown quantities. This can be done in

Chapter 5.0

two ways using either the force (flexibility) approach or displacement (stiffness) method:

(a) Force method – In this method, the redundant reactions or member forces are treated as unknowns and are obtained by satisfying compatibility of structural nodes.

(b) Displacement method – In this method the nodal displacements of the structure are regarded as unknowns and are obtained by satisfying equilibrium of structural nodes.

Example:

Assume the total load applied on this structure is P and obtain the load distribution in these two rods (areas A_1 and A_2).

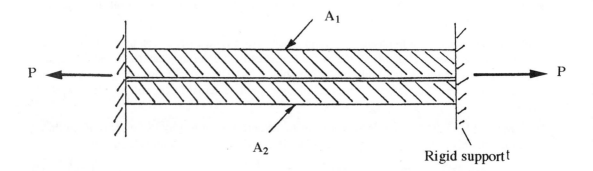

The structure is redundant to the first degree as one of the two rods can safely be removed.

(a) Force method:

Let P_1 be the load in rod A_1

From statics:

$$P_2 = P - P_1$$

$$\delta_1 = \frac{P_1 L}{A_1 E}$$

$$\delta_2 = \frac{P_2 L}{A_2 E}$$

As the two rods are rigid against rotation, $\delta_1 = \delta_2$ (compatibility condition)

$$\therefore \frac{P_1 L}{A_1 E} = \frac{(P - P_1) L}{A_2 E}$$

Therefore,

$$P_1 = \frac{P A_1}{A_1 + A_2}$$

and $P_2 = \dfrac{P A_2}{A_1 + A_2}$

(b) Displacement method:

Let δ be the elongation of the rods and

$$P_1 = \frac{\delta A_1 E}{L}$$

and $P_2 = \dfrac{\delta A_2 E}{L}$

But, $P = P_1 + P_2$ (Equilibrium condition)

$$P = \dfrac{\delta A_1 E}{L} + \dfrac{\delta A_2 E}{L} = \dfrac{\delta E(A_1 + A_2)}{L}$$

and $\delta = \dfrac{PL}{(A_1 + A_2)E}$

Therefore, $P_1 = \dfrac{PA_1}{A_1 + A_2}$ and $P_2 = \dfrac{PA_2}{A_1 + A_2}$

(Conclusion: Same result as obtained in force method)

5.4 DETERMINATE STRUCTURES

A statically determinate structure is one in which all external reactions and internal stresses for a given load system can be found by use of the equations of static equilibrium, and a statically indeterminate structure is one in which all reactions and internal stresses cannot be found by using only the equations of equilibrium.

For a planar structure such as beam or frame structures only three relationships (or three equations as shown in Eq. 5.4.1 through Eq. 5.4.3) must be satisfied:

Horizontal forces: $\Sigma F_x = 0$ Eq. 5.4.1

Vertical forces: $\Sigma F_y = 0$ Eq. 5.4.2

Moments: $\Sigma M_z = 0$ Eq. 5.4.3

For a statically determinate structure the reactions under given loading can be calculated using these relationships. For any structure in equilibrium, any part of it, isolated as a "free-body", must be in equilibrium under the applied loading.

(1) BEAMS

In aircraft design, a large portion of the beams are tapered in depth and section, and also carry distributed load. Thus to design, or check the various sections of such beams, it is necessary to have the values of the axial loads, vertical shear loads and bending moments along the beam. If these values are plotted as a function of the x-coordinate, the resulting curves are referred to as axial load (if any), shear loads, and bending moment diagrams. The principal of superposition can be used for beam analysis:

- This principle is used for linear structural analyses and enables engineers to determine separately the displacements due to different loads
- The final displacements are then the sum of the individual displacements

(A) THE METHOD OF SECTIONS

The internal loads acting on any cross section along a member can be determined by the method of sections using free-body diagrams. This method is generally applied after all external loads acting on the structural member are computed.

To compute the axial load, shear load, and bending moment as a function of the x-coordinate, the beam is sectioned off at some arbitrary distance x-coordinate. Applying the equations of equilibrium to the sectioned structure, namely $\Sigma F_x = 0$, $\Sigma F_y = 0$, and $\Sigma M_z = 0$, one obtains the axial load, shear load, and bending moment as a function of x-coordinate. They will be discontinuous at points where the beam loading abruptly changes, i.e., at points where concentrated loads or moments are applied. Therefore, the beam must be sectioned between any two points where the loading is discontinuous.

Example:

Use the method of sections to draw axial load, shear load, and bending moment diagrams for the beam as shown under uniform loading.

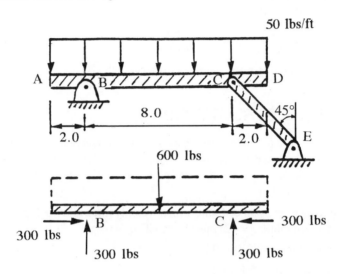

(A) Axial load (P), shear load (V), and bending moment (M) calculations:

(a) Beam section just to the left of B, $0 < x < 2$ (Section between point A and B)

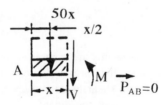

Axial loads: $\Sigma F_x = 0$, $\therefore P_{AB} = 0$

Shear loads: $\Sigma F_y = 0$, $-V_{AB} - 50x = 0$,

$\therefore V_{AB} = -50x$

Moment: $\Sigma M_z = 0$, $M_{AB} + (50x)(\frac{x}{2}) = 0$

$\therefore M_{AB} = -25x^2$

(b) Beam section just to the left of C, $2 < x < 10$ (Section between point B and C)

Axial loads: $\Sigma F_x = 0$, $P_{BC} + 300 = 0$

$\therefore P_{BC} = -300$ lbs.

Shear loads: $\Sigma F_y = 0$, $300 - V_{BC} - 50x = 0$,

$\therefore V_{BC} = 300 - 50x$

Structural Analysis

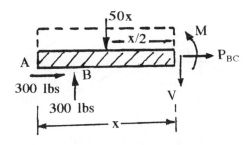

Moment: $\Sigma M_z = 0$, $(50x)(\frac{x}{2}) + M_{BC} - 300(x-2) = 0$

$\therefore M_{BC} = -25x^2 + 300x - 600$

(c) Beam section just to the left of D, $10 < x < 12$ (Section between C and D)

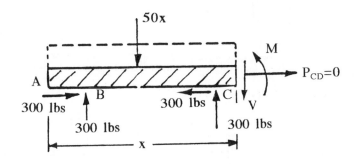

Axial loads: $\Sigma F_x = 0$, $P_{CD} + 300 - 300 = 0$ $\therefore P_{CD} = 0$
Shear loads: $\Sigma F_y = 0$, $300 + 300 - 50x - V_{CD} = 0$, $\therefore V_{CD} = 600 - 50x$
Moment: $\Sigma M_z = 0$, $(50x)(\frac{x}{2}) + M_{CD} - 300(x-2) - 300(x-10) = 0$

$\therefore M_{CD} = -25x^2 + 600x - 3{,}600$

(B) Axial load, shear load, and bending moment diagrams:

Beam Section	Axial Load	Shear Load	Bending Moment
$0 \leq x \leq 2$ (Section A – B)	$P_{AB} = 0$	$V_{AB} = -50x$	$M_{AB} = -25x^2$
$2 \leq x \leq 10$ (Section B – C)	$P_{BC} = -300$	$V_{BC} = -50x + 300$	$M_{BC} = -25x^2 + 300x - 600$
$10 \leq x \leq 12$ (Section C – D)	$P_{CD} = 0$	$V_{CD} = -50x + 600$	$M_{CD} = -25x^2 + 600x - 3{,}600$

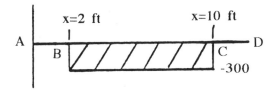

(a) Axial Load (lbs) Diagram

89

Chapter 5.0

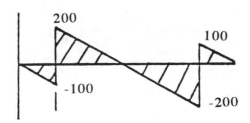

(b) Shear Load (lbs) Diagram

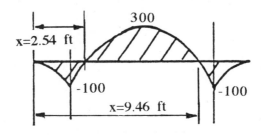

(c) Bending Moment (ft-lbs) Diagram

(B) SUMMATION METHOD

In cases where a beam is subjected to several concentrated loads and/or moments, using the method of sections to construct shear and bending moment diagrams can be quite laborious. The summation method which is based on the differential relations between applied load, shear load, and bending moment can then be useful. Without any derivation, the four basic relationships between load, shear, and bending moment can be expressed as follows:

(a) The slope of the shear diagram at any point is equal to the load intensity at that point:

$$\frac{dV}{dx} = w \qquad \text{Eq. 5.4.4}$$

(b) The slope of the moment diagram at any point is equal to the intensity of the shear load at that point:

$$\frac{dM}{dx} = V \qquad \text{Eq. 5.4.5}$$

(c) The change in shear between two points (a and b) is equal to the area of the loading curve between a and b:

$$dV_{ab} = \int_a^b w \, dx \qquad \text{Eq. 5.4.6}$$

(d) The change in moment between two points (a and b) is equal to the area under the shear diagram between a and b:

$$dM_{ab} = \int_a^b V \, dx \qquad \text{Eq. 5.4.7}$$

The above relations simplify the drawing of shear and moment diagrams for beams with concentrated loads and any distributed loads which have constant load intensity. For more complicated distributed loading, the method of sections is probably easier to use.

Example:

Use the summation method, draw a shear and bending moment diagram for the beam with the loading shown below.

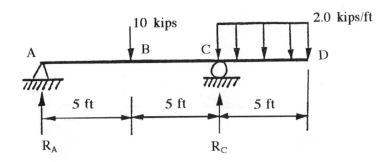

(a) First solve for the reactions:

$\Sigma M_C = 0$ $R_A \times 10 - 10 \times 5 + 10 \times 2.5 = 0$ $\therefore R_A = 2.5$ kips

$\Sigma M_A = 0$ $10 \times 5 - R_C \times 10 + 10 \times 12.5 = 0$ $\therefore R_C = 17.5$ kips

(b) Shear diagram:
- The shear load at the left-hand end of the beam (section A – B) is equal to the reaction R_A
- Since there is zero load between A and B, the change in shear between A and B is zero
- At B, the shear changes by 10 kips due to the 10 kips applied load applied between B and C
- At C, the shear changes by 17.5 kips due to the reaction, $R_C = 17.5$ kips, at C
- Between C. and D, the shear load changes by the area under uniform load = (2kips/ft.)(5 ft.) = 10 kips. Since the load is constant between C and D, the slope of the shear diagram is constant (or a straight line)
- The shear load is zero at D, since it is a free end

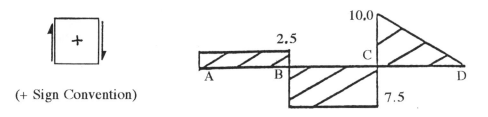

(+ Sign Convention)

(Shear Diagram in kips)

(c) Bending moment diagram:
- The bending moment at A must be zero, for this is a hinge support
- The moment between A and B increases by the area under the shear curve between A and B, area = (2.5 kips)(5 ft.) = 12.5 ft.–kips. [Since the shear load is constant between A and B, the slope of the bending moment diagram must be constant (or a straight line)]
- The change in moment between B and C must decrease by the area (note: negative area), area = (7.5 kips)(5 ft.) = 37.5 ft.–kips. (Since the shear load is constant, the slope of the moment diagram is constant)
- The change in moment between C and D must be the area under the shear diagram

Chapter 5.0

between C and D, area = (10 kips)(5 ft.)$(\frac{1}{2})$ = 25 ft.–kips. (Since the area is positive, the moment must increase. The shear diagram between C and D is always positive but must decrease. The moment diagram must have a decreasing positive slope)

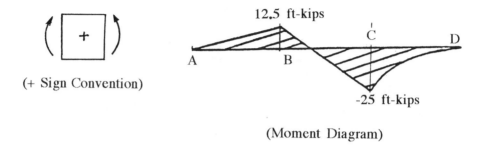

(+ Sign Convention)

(Moment Diagram)

(2) TRUSS ANALYSIS

This section introduces two methods of solving planar truss structures, namely the method of joints and the method of sections, as shown in Fig. 5.4.1.

(A) THE METHOD OF JOINTS

Each joint in the truss is isolated as an individual 'free-body' and placed in a state of equilibrium. Since all loads pass through a common point, only two equilibrium equations are required, $\Sigma F_x = 0$ and $\Sigma F_y = 0$, to solve for two unknown loads. For example, at point C solve for P_{BC} and P_{CD}, the forces in members of BC and CD respectively, as shown in Fig. 5.4.1(a). Then consider each other joint in sequence in a similar manner.

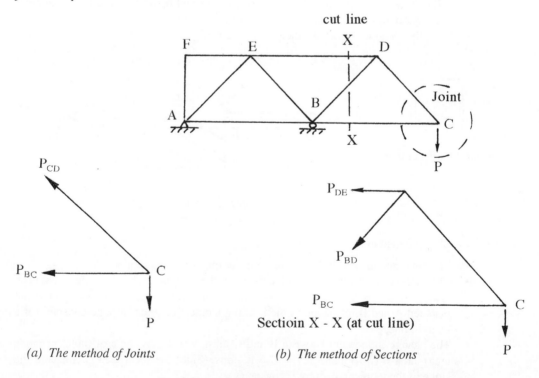

(a) The method of Joints   (b) The method of Sections

Fig. 5.4.1 A Typical Planar Truss Structure

(B) THE METHOD OF SECTIONS

A cross section is taken through the truss and the part of the truss to one side or the other of the section is treated as a "free-body", as shown in Fig. 5.4.1(b), and placed in equilibrium. In this case, three equations of equilibrium are required, $\Sigma F_x = 0$, $\Sigma F_y = 0$, and $M_z = 0$. This method is useful when the loads in only a few specific members are required. It should be noted that only three members whose loads are unknown may be cut at any one time. Consider the truss to the right-hand side of Section X–X, as a "free-body" as shown in Fig. 5.4.1(b), then solve for P_{BD}, P_{DE}, and P_{BC} using three equations of equilibrium.

Example 1:

Determine forces in all members of the truss structure as shown.

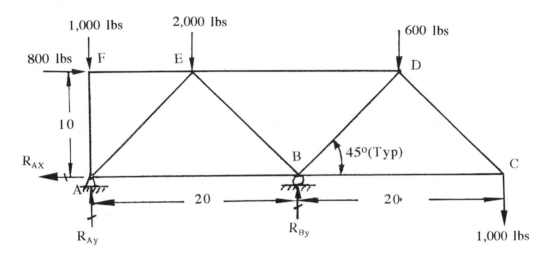

(a) Use the method of joints:

The first step is to place the structure in equilibrium, i.e., calculate reactions at A and B.

Take moments about A, then

$$\Sigma M_A = 10 \times 800 + 10 \times 2{,}000 + 30 \times 600 + 40 \times 1{,}000 - 20 \times R_{By} = 0$$

$$\therefore R_{By} = 4{,}300 \text{ lbs.}$$

$$\Sigma P_y = -1{,}000 - 2{,}000 - 600 - 1{,}000 + 4{,}300 + R_{Ay} = 0$$

$$\therefore R_{Ay} = 300 \text{ lbs.}$$

$$\Sigma P_x = 800 - R_{Ax} = 0$$

$$\therefore R_{Ax} = 800 \text{ lbs.}$$

Consider joint 'C',

$$\Sigma P_y = -1{,}000 + P_{CD} \sin 45° = 0$$

$$\therefore P_{CD} = \frac{1{,}000}{\sin 45°} = 1{,}414 \text{ lbs.}$$

$$\Sigma P_x = -P_{CD} \cos 45° + P_{BC} = 0$$

$$= -(\frac{1{,}000}{\sin 45°}) \cos 45° + P_{BC} = 0$$

$$\therefore P_{BC} = 1{,}000 \text{ lbs.}$$

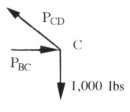

Consider joint 'D'

$$\Sigma P_y = -600 - (\frac{1,000}{\sin 45°})\cos 45° + P_{BD} \cos 45° = 0$$

$$\therefore P_{BD} = \frac{1,600}{\cos 45°} \text{lbs.} = 2,263 \text{ lbs.}$$

$$\Sigma P_x = P_{CD} \sin 45° + P_{BD} \sin 45° - P_{DE} = 0$$

$$= (\frac{1,000}{\sin 45°})\sin 45° + (\frac{1,600}{\cos 45°})\sin 45° - P_{DE} = 0$$

$$\therefore P_{DE} = 2,600 \text{ lbs.}$$

Continue in the same manner for joints 'A', 'B', 'E', and 'F'.

Example 2:

Use the same structure shown in Example 1 to determine forces by the method of sections in members CD, BD, and BC.

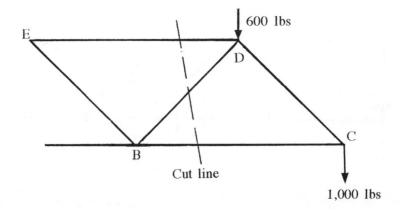

First, cut a cross section through the members to be solved (only three members whose loads are unknown may be cut at any one time). Treat the structure to the right side of the section line as a free body and place it in equilibrium. Take moment about 'B', then

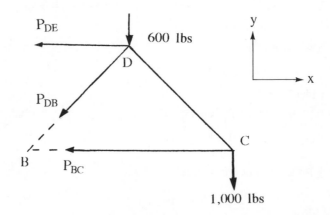

$\Sigma M_B = 1,000 \times 20 + 600 \times 10 - P_{DE} \times 10 = 0$

$\therefore P_{DE} = 2,600$ lbs.

$\Sigma P_y = -600 - 1,000 - P_{BD} \cos 45° = 0$

$\therefore P_{BD} = \dfrac{1,600}{\cos 45°}$ lbs.

$\Sigma P_x = -2,600 - P_{BD} \sin 45° + P_{CB} = 0$

$\quad = -2,600 - (-\dfrac{1,600}{\cos 45°}) \sin 45° + P_{BC} = 0$

$\therefore P_{BC} = 1,000$ lbs.

(3) SPACE TRUSS ANALYSIS

A space truss (or three dimensional structure) is one designed to react forces in three mutually perpendicular axes. This method is mainly used on determinate structures for landing gear and engine truss support analysis. The method of solving these structure is essentially the same as for the planar structure already covered except that six equations of equilibrium must be satisfied, see Eq. 5.3.1 and Eq. 5.3.2. As shown in Fig. 5.4.2, the components of force along three mutually perpendicular axes x, y, and z can be obtained as follows:

$P_x = P \cos \alpha$ Eq. 5.4.8

$P_y = P \cos \gamma$ Eq. 5.4.9

$P_z = P \cos \beta$ Eq. 5.4.10

If all three component forces are known, then the resultant force P can be obtained from:

$P = \sqrt{P_x^2 + P_y^2 + P_z^2}$ Eq. 5.4.11

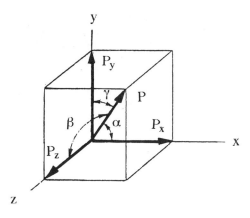

Fig. 5.4.2 Force Components Relationship of Space Truss Member

Example:

Determine forces on a aircraft nose landing gear as shown. Applied loads: $P_D = 20,000$ lbs and $P_V = 40,000$ lbs.

    Determine the true length of drag brace

$\quad\quad\quad$ Length of brace $= \sqrt{(70^2 + 30^2)} = 76$ in.

Component forces R_{Ax} and R_{Az}:

$$R_{Ax} = R_A \frac{70}{76.16} = 0.919 \, R_A$$

$$R_{Az} = R_A \frac{30}{76.16} = 0.394 \, R_A$$

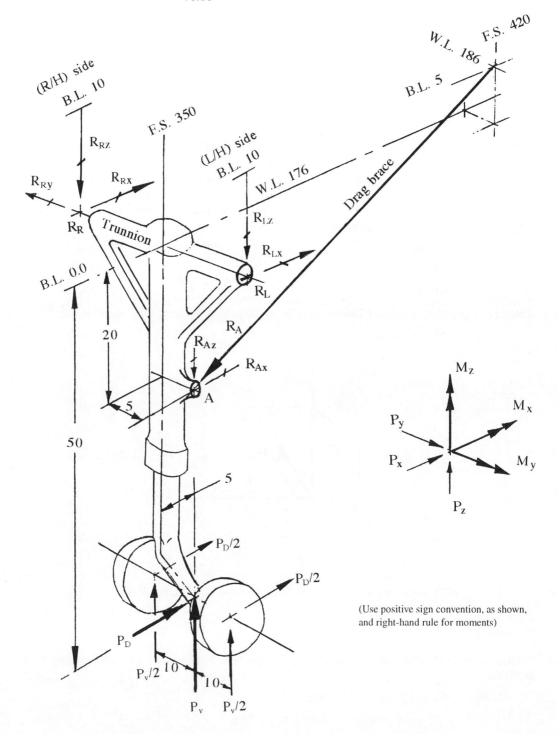

(Use positive sign convention, as shown, and right-hand rule for moments)

Take moments about trunnion center line,

$$\Sigma M_y = 5 \times 40{,}000 + 50 \times 20{,}000 - 20 \times R_{Ax} = 0$$

$$\therefore R_{Ax} = 60{,}000 \text{ lbs.}$$

$$\therefore R_{Az} = 25{,}723 \text{ lbs.}$$

$$\therefore R_A = 65{,}288 \text{ lbs.}$$

Take moments about z-axis at trunnion right-hand bearing,

$$\Sigma M_z = 20{,}000 \times 10 - 60{,}000 \times 15 + 20 \times R_{Lx} = 0$$

$$\therefore R_{Lx} = 35{,}000 \text{ lbs.}$$

$$\Sigma P_x = 20{,}000 - 60{,}000 + 35{,}000 + R_{Rx} = 0$$

$$\therefore R_{Rx} = 5{,}000 \text{ lbs.}$$

Take moments about x-axis at trunnion right-hand bearing:

$$\Sigma M_x = -40{,}000 \times 10 + 25{,}800 \times 15 + 20 \times R_{Lz} = 0$$

$$\therefore R_{Lz} = 650 \text{ lbs.}$$

$$\Sigma P_z = 40{,}000 - 25{,}800 - 650 - R_{Rz} = 0$$

$$\therefore R_{Rz} = 13{,}550 \text{ lbs.}$$

Resultant force on left-hand bearing: $R_L = \sqrt{35{,}000^2 + 650^2} = 35{,}006$ lbs.

Resultant force on right-hand bearing: $R_R = \sqrt{13{,}550^2 + 5{,}000^2} = 14{,}443$ lbs.

5.5 INDETERMINATE (REDUNDANT) STRUCTURES

There are many methods for calculating beam deflections and load distributions and each particular method has its own merits and drawbacks, To choose which one is most applicable to airframe structures is not a simple matter. For instance, the 'Moment Distribution Method' was the most favored means for structural engineers in the past (slide rule age) but now it is no longer needed since the advent of FEM (Finite Element Modeling) on computer.

The calculation of deflections of beam or element analysis is important in airframe structures for two main reasons:

- A knowledge of the load-deflection characteristics of the airframe is of primary importance in studies of the influence of structural flexibility, i.e., Structural Influence Coefficients (SIC), on airplane load performance
- Calculation of deflections are necessary in solving for the internal load distributions of complex indeterminate (or redundant) structures

For a planar structure such as a simple beam or frame structure only three force relationships (see the following three equations, Eq. 5.4.1 through Eq. 5.4.3) must be satisfied to solve three unknowns and they are:

Horizontal forces: $\Sigma F_x = 0$

Vertical forces: $\Sigma F_y = 0$

Moments: $\Sigma M_z = 0$

Indeterminate structures which have more unknowns than the three available equations are those for which the stress distribution cannot be determinated by these three equations alone.

A structure may be indeterminate in support reactions or may be indeterminate integrally and a physical interpretation of structural redundancy can be made using the concept of load paths. If there are more than the minimum number of load paths required to maintain stability, the structure is statically indeterminate or redundant.

The easiest way of determining the degree of redundancy is to find out the maximum number of reactive forces, as shown in Fig. 5.5.1(b), and/or members, as shown in Fig. 5.5.1(c), that can be removed without affecting the stability of the structure. As in Fig. 5.5.1(b), any of the supports, and in Fig. 5.5.1(c), the member BC can safely be removed, both the structures are therefore redundant to the first degree.

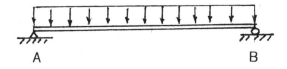

(a) Determinate Structures

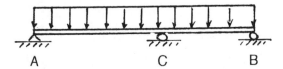

(b) Indeterminate Structures (First degree of redundancy)

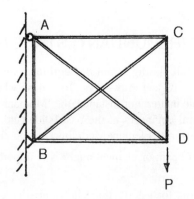

(c) Indeterminate Truss Structure (First degree of redundancy)

Fig. 5.5.1 Statically Determinate vs. Indeterminate Structures

(1) STRAIN ENERGY (OR WORK) METHOD

Work as defined in mechanics is the product of force times distances. If the force varies over the distance then the work is computed by integral calculus. The total elastic strain energy of a structure is

$$U = \int \frac{P^2 dx}{2AE} + \int \frac{M^2 dx}{2EI} + \int \frac{T^2 dx}{2GJ} + \int \frac{V^2 dx}{2AG} + \iint \frac{q^2 dx\, dy}{2Gt}$$ Eq. 5.5.1

P – Axial force (lbs.)
M – Moment (in.-lbs.)
T – Torsion (in.-lbs.)

V – Shear force (lbs.)
q – Shear flow (lbs./in.)
G – Modulus of rigidity (psi)
E – Modulus of elasticity (psi)
J – Torsional constant (in.^{4})
I – Moment of inertia (in.^{4})
A – Cross section area (in.^{2})
t – Web thickness (in.)

It is seldom that all the terms of the above equation need be employed in a calculation; some are of localized or secondary nature and their energy contribution may be ignored.

(A) VIRTUAL WORK (MINIMUM POTENTIAL ENERGY) METHOD

One of the most powerful deflection methods is the "Virtual Work method", which is also called the "Unit Load Method". The rate of change of strain energy with respect to deflection is equal to the associated load. In structural analysis the most important use of this theorem is made in problems concerning buckling instability and other non-linearity.

$$\frac{dU}{d\delta} = P \quad \text{(load)} \qquad \qquad \text{Eq. 5.5.2}$$

(B) CASTIGLIANO'S THEOREM METHOD

The rate of change of strain energy with respect to load is equal to the associated deflection. Castigiliano's theorem is quite useful in performing deflection calculations. Do not use this method for non-linear problems.

$$\frac{dU}{dP} = \delta \quad \text{(deflection)} \qquad \qquad \text{Eq. 5.5.3}$$

(C) LEAST WORK (CASTIGLIANO'S SECOND THEOREM) METHOD

In any loaded indeterminate structure the values of the loads must be such as to make the total elastic internal strain energy, resulting from the application of a given system of loads, a minimum.

Castigliano's second theorem provides a powerful method for analyzing indeterminate structures, and is very effective in the analysis of articulated indeterminate structures. In the case of continuous beams or frames, in the past, most analysts used some other method, usually that of moment distribution.

The basic equations for this method are:

$$1.0 \times \delta = \int_a^b (\frac{mM}{EI})dx \quad \text{(for deflection)} \qquad \qquad \text{Eq. 5.5.4}$$

$$1.0 \times \theta = \int_a^b (\frac{mM}{EI})dx \quad \text{(for rotation)} \qquad \qquad \text{Eq. 5.5.5}$$

where: m – bending moment caused by unit load or moment
M – bending moment caused by given loads
δ – beam deflection where unit load is applied
θ – beam rotation where unit moment is applied
L – beam length

These equations are valid for bending deformations only. The term "m" represents the bending moment due to the virtual load (unit load) or the virtual moment (unit moment), depending on whether bending deflections (δ) or rotations (θ) are desired.

Example 1:

Calculate the deflection and rotation at the free end of the beam due to the load P.

Bending moment, M = –Px

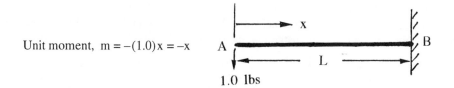

(a) Deflection at free end (at point 'A')

Apply a unit load where deflection is required and obtain unit moment "m",

Unit moment, $m = -(1.0)x = -x$

Set up virtual work equation from Eq. 5.5.4 and solve

$$1.0 \times \delta_A = \int_0^L \left(\frac{mM}{EI}\right) dx$$

$$= \int_0^L \left[\frac{(-x)(-Px)}{EI}\right] dx$$

$$= \frac{P}{EI} \int_0^L x^2 dx$$

or $\quad \delta_A = \dfrac{PL^3}{3EI}$

The deflection is positive in the direction of the unit load (downward).

(b) Rotation at free end (at 'A')

Apply a unit moment where rotation is required and obtain unit moment 'm',

Unit moment: $m = -1$

Set up virtual work equation from Eq. 5.5.5 and solve

$$1.0 \times \theta_A = \int_0^L \left(\frac{mM}{EI}\right) dx$$

$$= \int_0^L \left[\frac{(-1)(-Px)}{EI}\right] dx$$

$$= \frac{P}{EI}\int_0^L x\,dx$$

or $\theta_A = \dfrac{PL^2}{2EI}$

The rotation is positive (counter clockwise) in the direction of the unit moment.

Example 2:

Calculate the mid-span bending deflection of the beam shown due to the uniformly tapered load.

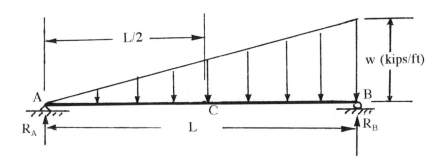

$$\Sigma M_B = 0,\ R_A \times L - \left(\frac{wL}{2}\right)\left(\frac{L}{3}\right) = 0$$

$$\therefore R_A = \frac{wL}{6}\ \text{and}\ R_B = \frac{wL}{3}$$

Determine the bending moment equation for the beam,

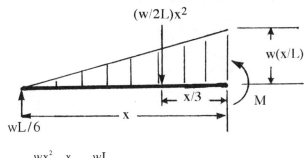

$$M + \left(\frac{wx^2}{2L}\right)\left(\frac{x}{3}\right) - \frac{wL}{6}x = 0$$

$$\therefore M = -\frac{w}{6L}x^3 + \frac{wL}{6}x$$

Apply a unit load at mid-span and solve for bending moment,

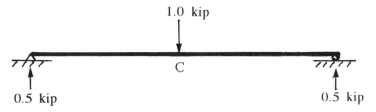

For first half of beam $0 < x < \dfrac{L}{2}$

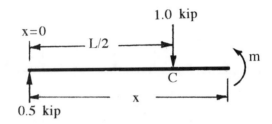

$$m = \dfrac{x}{2}$$

For second half of beam $\dfrac{L}{2} < x < L$

$$m + (1.0)(x - \dfrac{L}{2}) - \dfrac{x}{2} = 0$$

$$\therefore m = \dfrac{L}{2} - \dfrac{x}{2}$$

Set up virtual work equation from Eq. 5.5.4 and integrate,

$$1.0 \times \delta_c EI = \int_0^{\frac{L}{2}} \dfrac{x}{2}(\dfrac{wLx}{6} - \dfrac{wx^3}{6L})dx + \int_{\frac{L}{2}}^{L} (\dfrac{L}{2} - \dfrac{x}{2})(\dfrac{wLx}{6} - \dfrac{wx^3}{6L})dx$$

Solving the above equation yields,

$$EI\delta_c = (\dfrac{wL^4}{288} - \dfrac{wL^4}{1,920}) + (\dfrac{wL^4}{32} - \dfrac{15wL^4}{768} - \dfrac{7wL^4}{288} + \dfrac{31wL^4}{1,920}) = \dfrac{5wL^4}{768}$$

or $\delta_c = \dfrac{5wL^4}{768EI}$ (Deflection at mid-span at 'C')

(2) ELASTIC-WEIGHTS METHOD

The deflection at point A on the elastic curve of a beam is equal to the bending moment at A due to the M/EI diagram acting as a distributed beam load.

Likewise, the angular change at any section of a simply supported beam is equal to the shear at that section due to the $\dfrac{M}{EI}$ diagram acting as a beam load.

(A) MOMENT-AREA METHOD

In many designs this method proves to be a simple and quick solution to determine beam slope and deflection. Their relationship is shown in Fig. 5.5.2.

Structural Analysis

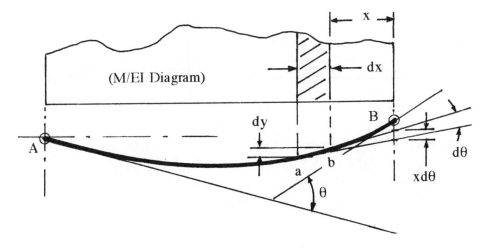

Fig. 5.5.2 Beam Slope and Deflection

The moment-area method is based upon the following propositions:

(a) 1st proposition – The change in slope of the elastic line of a beam between any two points A and B is numerically equal to the area of the $\frac{M}{EI}$ diagram between these two points.

$$d\theta_A = \int_A^B (\frac{M}{EI}) dx \qquad \text{Eq. 5.5.6}$$

(b) 2nd proposition – The deflection of a point A on the elastic line of a beam in bending normal to the tangent of the elastic line at a point B is equal numerically to the applied moment of the $\frac{M}{EI}$ area between points A and B about point A.

$$\delta = x d\theta_A = x \int_A^B (\frac{M}{EI}) dx \qquad \text{Eq. 5.5.7}$$

Example:

Determine the maximum deflection in the beam under a load of 3 kips (kip = 1,000 lbs.) as shown.

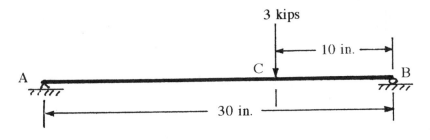

(Assume EI = 10×10^3 kips.–in.^{2})

103

Chapter 5.0

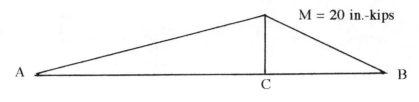

(Moment diagram)

Assume x = 0 at point 'A'

$R_A = 1.0$ kips and $R_B = 2$ kips

Change of slope between A and B:

$$\frac{M}{EI} = \frac{1}{EI}(20 \times \frac{20}{2} + 20 \times \frac{10}{2}) = \frac{300}{EI}$$

Vertical distance between B and tangent at A will be given by the moment of the $\frac{M}{EI}$ diagram about B,

$$\frac{M}{EI} = \frac{1}{EI}[(20 \times \frac{20}{2})(10 + \frac{20}{3}) + (20 \times \frac{10}{2})(2 \times \frac{10}{3})]$$

$$= \frac{1}{EI}(4,000)$$

$$= \frac{4,000}{10 \times 10^3}$$

$$= 0.4 \text{ in. as measured from B}$$

Slope at A is $\frac{0.4}{30} = 0.0133$ radian or $0.0133 \times (\frac{180}{\pi}) = 0.76°$

Maximum deflection will occur where slope is equal to zero, therefore the change in slope ($\frac{M}{EI}$ diagram) between A and x should be equal to zero.

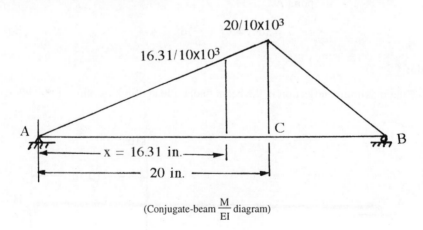

(Conjugate-beam $\frac{M}{EI}$ diagram)

$$0.0133 - \frac{20(\frac{x}{20})(\frac{x}{2})}{10 \times 10^3} = 0$$

$x^2 = 266$ ∴ x = 16.31 in. from A

From $\frac{M}{EI}$ diagram, taking moment about A and the vertical distance between A and tangent at x = 16.31 in. which is the max. deflection,

$$(\frac{16.31}{10 \times 10^3})(\frac{16.31}{2})(2 \times \frac{16.31}{3}) = 0.145 \text{ in. deflection at x = 16.31 in.}$$

(B) CONJUGATE-BEAM METHOD

The conjugate structure is an extension of the method of elastic weights into two dimensions. In the evaluation of deflections of single-story rigid frames, either single or multi-span, the method is extremely useful.

In general, the conjugate-beam method is of much greater practical importance than the moment-area method.

- The conjugate structure is positioned in a horizontal plane
- The slope and deflection:

Real beam support		Conjugate beam
Slope	→	Shear
Deflection	→	Moment

- The end of the conjugate structure corresponding to the end of the real structure that deflects always has a fixed support
- The conjugate structure, for a given real structure, is the members and their relative position as shown in Fig. 4.5.3

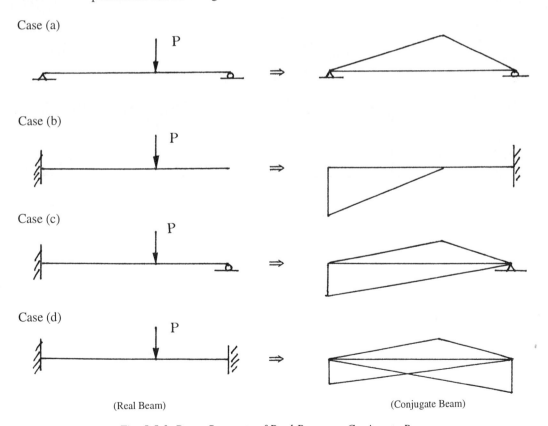

(Real Beam)   (Conjugate Beam)

Fig. 5.5.3 Beam Supports of Real Beam vs. Conjugate Beam

Chapter 5.0

Before deciding to use the conjugate beam method, it is important to note that:

- The $\frac{M}{EI}$ diagrams have been drawn on the side of the conjugate beam corresponding to the compression side of the real beam
- The $\frac{M}{EI}$ load is always considered to push against the conjugate beam, whether it be up or down
- Moments causing tension at the bottom side of the beam are considered to be positive
- The real beam at any location will deflect toward the tension side of the conjugate beam at the corresponding location

Example:

Find the maximum deflection at the middle of the beam at point 'C' by using the conjugate beam method and E = 10,700 ksi

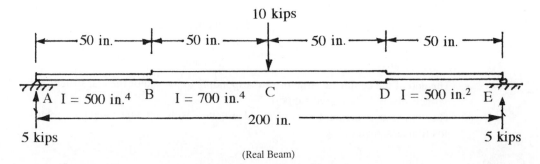

(Real Beam)

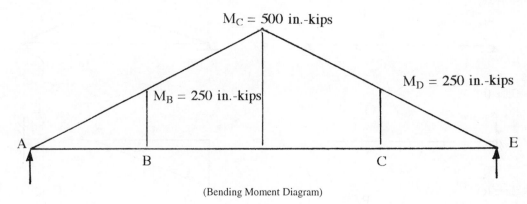

(Bending Moment Diagram)

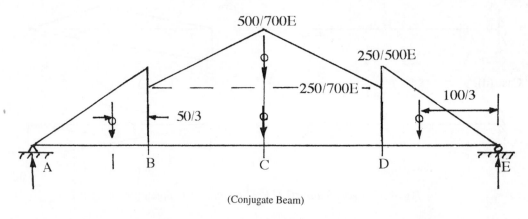

(Conjugate Beam)

106

$$200R_A - (\frac{250}{500E})(\frac{50}{2})[(150+\frac{50}{3})+\frac{100}{3}] - (\frac{250}{700E})(100)(100) - (\frac{500}{700E} - \frac{250}{700E})(\frac{100}{2})(100)$$
$$= 0$$

$$200R_A - \frac{2,500}{E} - \frac{3571.43}{E} - \frac{1785.72}{E} = 0$$

$$\therefore R_A = \frac{39.286}{E} \quad \text{or} \quad R_A = 0.00376 \text{ radian at support 'A'}$$

The maximum deflection at 'C',

$$\delta_C = \frac{39.286}{E}(100) - (\frac{250}{500E})(\frac{50}{2})(50+\frac{50}{3}) - (\frac{250}{700E})(50)(25) - (\frac{500}{700E} - \frac{250}{700E})(\frac{50}{2})(\frac{50}{3})$$

$$\delta_C = \frac{3,928.6}{E} - \frac{833.33}{E} - \frac{446.43}{E} - \frac{148.81}{E}$$

$$\delta_C = \frac{3,928.6}{E} - \frac{1,428.57}{E}$$

$$\delta_C = \frac{2,500}{E} \quad \text{or} \quad \delta_C = 0.234 \text{ in.}$$

(3) ELASTIC CENTER AND COLUMN ANALOGY METHOD

Inside an aircraft fuselage body there are numerous numbers of rings or closed frames to maintain shape and provide stabilizing supports for the longitudinal shell stringers. At some point relatively heavy frames or bulkheads are needed to transfer concentrated loads between the fuselage and tail, landing gear, power plant, etc. In general, these frames undergo bending forces in transferring the applied loads to the other resisting portions of the fuselage. The Elastic Center method (also similar is the Column Analogy Method which will not be discussed here) can give a reasonably close approximation and it is recommended for use in obtaining frame and closed ring bending loading for preliminary sizing purposes.

The Elastic Center Method is:

- Particular to the analysis of a fixed-end arch and airframe fuselage ring applications
- Both the Elastic Center and Column Analogy methods are limited to a maximum of three redundant reaction components for preliminary sizing

As applied to single-span rigid fixed-end frames and arches, the Column Analogy method is very similar to that of the Elastic Center method and selection of one in place of the other is a matter of personal preference.

In this text, it has been demonstrated that the analysis of these types of structures, i.e., arches, frames, rings, etc., can be considerably simplified if the redundant reaction components are considered to act at the centroid of the elastic areas. This point 'o' is known as the elastic center or, some times, as the neutral point.

By applying the Elastic Center method:

- For the frame example, shown in Fig. 5.5.4, remove either one of the end supports from the frame structure (in this case support at A was removed)
- The redundant forces are assumed to be acting at the elastic center 'o' of the frame which gives resulting equations (independent of each other)
- Support 'A' and centroid 'o' are connected by a rigid bar or bracket (a non-deformed member)
- In general, moment diagrams are plotted on the compression stress side of the members. However, in this example the positive moment is plotted on the outside of

Chapter 5.0

the member and the negative moment is plotted on the inside of the member

It is good practice to try a condition such that the $\frac{M}{EI}$ diagram is symmetrical about one axis or, if possible, about both x and y axes through the elastic center. It makes one or both of the redundant X_o and Y_o equal to zero, thus reducing considerably the amount of numerical calculation.

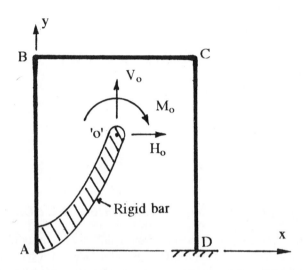

(The point of 'o' is the point of the centroidal axes for the values of $\frac{ds}{EI}$ of the structure and the value of M_o, H_o, and V_o are the forces going through this point)

Fig. 5.5.4 Remove End Supports of a Frame Structure

The redundant loads designated as H_o, V_o, and M_o which act on the centroid of the elastic areas and the general terms of the equations are:

$$M_o = -\frac{\Sigma \Psi_s}{\Sigma \frac{ds}{EI}} \qquad \text{Eq. 5.5.8}$$

$$H_o = \frac{\Sigma \Psi_s y - \Sigma \Psi_s x (\frac{I_{xy}}{I_y})}{I_x (1 - \frac{I_{xy}^2}{I_x I_y})} \qquad \text{Eq. 5.5.9}$$

$$V_o = \frac{\Sigma \Psi_s x - \Sigma \Psi_s y (\frac{I_{xy}}{I_x})}{I_y (1 - \frac{I_{xy}^2}{I_x I_y})} \qquad \text{Eq. 5.5.10}$$

M_o – Moment (in.-lbs.)
H_o – Horizontal forces in x-axis (lbs.)
V_o – Vertical forces in y-axis (lbs.)

$\Psi_s = M \frac{ds}{EI}$ is the value of area of the $\frac{M}{EI}$ diagram

Structural Analysis

For symmetry about one axis (the y-centroidal axis is often an axis of symmetry in airframe structures) through the Elastic Center eliminate E (Modulus of elasticity) since, in practical design, they are constant over the entire structure:

$$M_o = \frac{\text{Area of } \frac{M}{I} \text{ diagram}}{\text{Total elastic weight of structure}}$$

$$= -\frac{\Sigma \Psi_s}{\Sigma \frac{ds}{I}} \qquad \text{Eq. 5.5.11}$$

$$H_o = \frac{\text{Moment of } \frac{M}{I} \text{ diagram about x-axis}}{\text{Elastic moment of inertia about x-axis}}$$

$$= \frac{\Sigma \Psi_s y}{I_x} \qquad \text{Eq. 5.5.12}$$

$$V_o = \frac{\text{Moment of } \frac{M}{I} \text{ diagram about y-axis}}{\text{Elastic moment of inertia about y-axis}}$$

$$= -\frac{\Sigma \Psi_s x}{I_y} \qquad \text{Eq. 5.5.13}$$

Before solving the three equations, the frame must become a determinate structure by the removal of supports as shown in Fig. 5.5.5

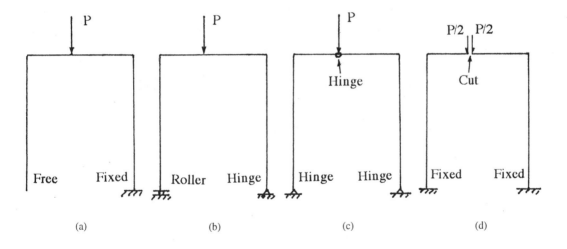

Fig. 5.5.5 Frame Configurations After Removal of Support

Example 1:

(This is an one-axis symmetrical case).
Given a frame structure with fixed supports at point A and D and reacting a concentrated load of 20 kips (kip = 1,000 lbs.), find the bending moment diagram along entire frame structure.

109

Chapter 5.0

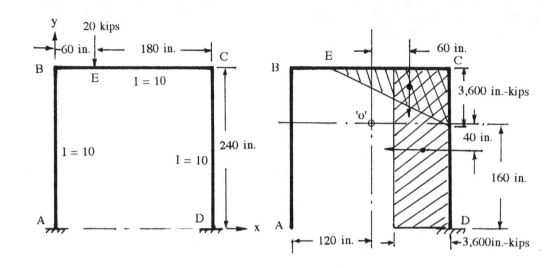

$M_{m,A} = M_{m,B} = M_{m,E} = 0$
$M_{m,C} = -3,600$ in.-kips
$M_{m,D} = -3,600$ in.-kips

Mm, moment Diagram
[Moments (in.-kips) are plotted on compression stress sides of the frame members]

Mem.	$\Omega = \dfrac{ds}{I}$	x	y	Ωx	Ωy	Ωx^2	Ωy^2
A-B	$\dfrac{240}{10} = 24$	0	120	0	24 × 120 = 2,880	0	24 × 120² = 345,600
B-C	$\dfrac{240}{10} = 24$	120	240	24 × 120 = 2,880	24 × 240 = 5,760	24 × 120² = 345,600	24 × 240² = 1,382,400
C-D	$\dfrac{240}{10} = 24$	240	120	24 × 240 = 5,760	24 × 120 = 2,880	24 × 240² = 1,382,400	24 × 120² = 345,600
	$\Sigma \dfrac{ds}{I} = 72$			Σ 8,640	Σ 11,520	Σ 2,304,0000	Σ 1,843,200

I'$_x$	I'$_y$	$I_{xx} = \Omega y^2 + I'_x$	$I_{yy} = \Omega x^2 + I'_y$
115,200	0	115,200 + 345,600 = 460,800	0
0	115,200	1,382,400 + 0 = 1,382,400	115,200 + 345,600 = 460,800
115,200	0	345,600 + 115,200 = 460,800	1,382,400 + 0 = 1,382,400
		$\Sigma I_{xx} = 2,304,000$	$\Sigma I_{yy} = 1,843,200$

Calculate the terms of I'$_x$ and I'$_y$ which are the elastic moments of inertia of each member about its own centroidal x and y axes:

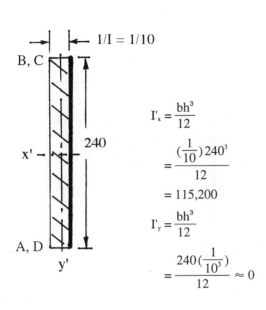
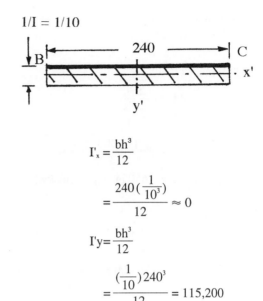

$$I'_x = \frac{bh^3}{12}$$
$$= \frac{(\frac{1}{10})240^3}{12}$$
$$= 115,200$$
$$I'_y = \frac{bh^3}{12}$$
$$= \frac{240(\frac{1}{10^3})}{12} \approx 0$$

$$I'_x = \frac{bh^3}{12}$$
$$= \frac{240(\frac{1}{10^3})}{12} \approx 0$$
$$I'_y = \frac{bh^3}{12}$$
$$= \frac{(\frac{1}{10})240^3}{12} = 115,200$$

Location of elastic center 'o':

$$X_o = \frac{\Sigma \Omega x}{\Sigma \Omega} = \frac{8,640}{72} = 120''$$

$$Y_o = \frac{\Sigma \Omega y}{\Sigma \Omega} = \frac{11,520}{72} = 160''$$

Find moments of inertia at elastic center 'o':

$$I_{xo} = I_{xx} - \Sigma \Omega Y_o^2 \qquad I_{yo} = I_{yy} - \Sigma \Omega X_o^2$$
$$= 2,304,000 - 72 \times 160^2 \qquad = 1,843,200 - 72 \times 120^2$$
$$= 2,304,000 - 1,843,200 \qquad = 1,843,200 - 1,036,800$$
$$= 460,800 \qquad = 806,400$$

Find the value of $\psi_s = M(\frac{ds}{EI})$, the area of the $\frac{M}{EI}$ diagram:

$M(\frac{ds}{EI})$ on member of B – C:

$$\psi_{s,B-C} = (\frac{-3,600}{10})(\frac{180}{2}) = -32,400$$

$M(\frac{ds}{EI})$ on member of C – D:

$$\psi_{s,C-D} = (\frac{-3,600}{10})240 = -86,400$$

Total $M(\frac{ds}{EI})$ on entire frame:

$$\Sigma \psi_s = (\psi_{s,B-C}) + (\psi_{s,C-D}) = -32,400 - 86,400 = -118,800$$

Chapter 5.0

For symmetry about y-axis through Elastic Center 'o' and eliminate E (Modulus is constant over the entire structure):

From Eq. 5.5.11:

$$M_o = -\frac{\Sigma \psi_s}{\Sigma \frac{ds}{I}}$$

$$= -\frac{-118,800}{72}$$

$$= 1,650 \text{ in.-kips (clockwise)}$$

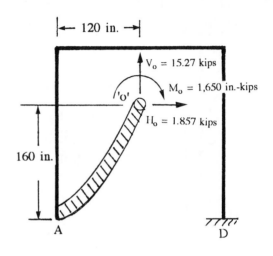

From Eq. 5.5.12:

$$H_o = \frac{\Sigma \psi_s y}{I_x}$$

$$= \frac{-32,400 \times 80 + 86,400 \times 40}{460,800}$$

$$= 1.875 \text{ kips} \rightarrow \text{(to the right)}$$

From Eq. 5.5.13:

$$V_o = \frac{\Sigma \psi_s x}{I_y}$$

$$= \frac{-32,400 \times 60 - 86,400 \times 120}{806,400}$$

$$= 15.27 \text{ kips} \uparrow \text{(up)}$$

Moments due to M_o, H_o and V_o at elastic center 'o':

$M_{o,A} = M_o + H_o \times 160 - V_o \times 120$
$\quad = 1,650 - 1.875 \times 160 - 115.27 \times 120$
$\quad = 117.8$ in.-kips

$M_{o,B} = M_o - H_o \times 80 - V_o \times 120$
$\quad = 1,650 - 1.875 \times 80 - 15.27 \times 120$
$\quad = -332$ in.-kips

$M_{o,C} = M_o - H_o \times 80 + V_o \times 120$
$\quad = 1,650 - 1.875 \times 80 + 15.27 \times 120$
$\quad = 3,332$ in.-kips

$M_{o,D} = M_o + H_o \times 160 + V_o \times 120$
$\quad \doteq 1,650 + 1.875 \times 160 + 15.27 \times 120$
$\quad = 3,782$ in.-kips

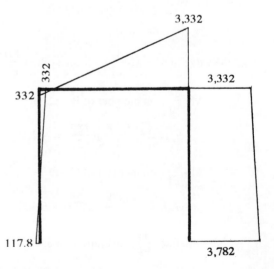

Structural Analysis

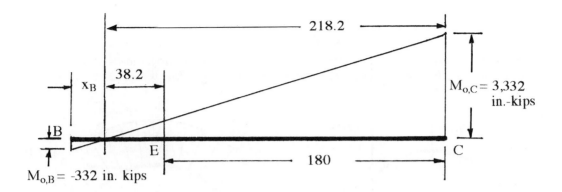

$$\frac{240 - x_B}{3{,}332} = \frac{x_B}{332} \qquad \therefore \ x_B = 21.8 \text{ in.}$$

$$M_{o,E} = 332 \frac{38.2}{21.8} = 581.8 \text{ in.-kips}$$

Final moments due to applied load of 20 kips:

$$M_A = M_{m,A} + M_{o,A} = 0 + 117.8 = 117.8 \text{ in.-kips}$$
$$M_B = M_{m,B} + M_{o,B} = 0 - 332 = -332 \text{ in.-kips}$$
$$M_C = M_{m,C} + M_{o,C} = -3{,}600 + 3{,}332 = -268 \text{ in.-kips}$$
$$M_D = M_{m,D} + M_{o,D} = -3{,}600 + 3{,}782 = 182 \text{ in.-kips}$$
$$M_E = M_{m,E} + M_{o,E} = 0 + 581.8 = 581.8 \text{ in.-kips}$$

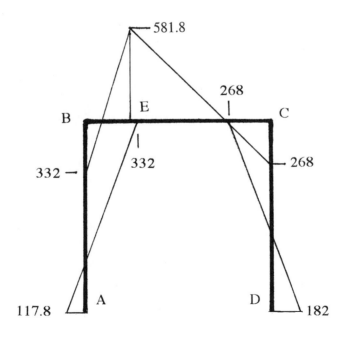

[Positive moments (in.-kips) are plotted on outside of frame members]

Chapter 5.0

Example 2:

(This is a two-axes symmetrical case analysis).

Given a ring structure of constant cross-section subjected to symmetrical loads, P = 5 kips, as shown determine bending moment diagram along the entire ring.

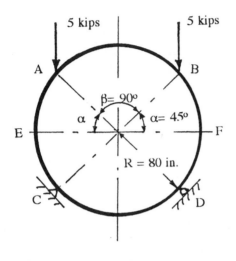
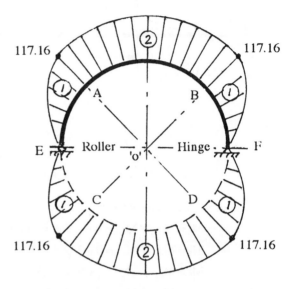

(Given ring and loads)　　　　　　　　Mm, moment diagram [Positive moments (in.-kips) are plotted on outside of ring member]

The elastic center coincides with the center of the ring due to its symmetry about two axes.

The value of area of the $\frac{M}{EI}$ diagram:

$$\Sigma \frac{ds}{EI} = \pi \times \frac{160}{I} = \frac{502.65}{I}$$

The elastic moment of inertia about x and y axes through center of the ring,

$$I_x = I_y = \pi \times R^3 = \pi \times 80^3 = 1,608,495 \quad \text{(assume ring width = 1.0 in.)}$$

The bending moments at A, B, C, and D due to two concentrated loads of 5 kips:

$$M_{m,A} = M_{m,B} = M_{m,C} = M_{m,D} = 5(80 - 80 \cos 45°) = 117.16 \text{ in.-kips}$$

The area of portion ①,

$$\psi_{s,1} = PR^2(\alpha - \sin \alpha) \quad \text{where } \alpha = 45°$$

$$= 5 \times 80^2 (45 \times \frac{\pi}{180} - 0.707)$$

$$= 5 \times 80^2 (0.7854 - 0.707) = 2,508.8$$

The area of portion ②,

$$\psi_{s,2} = PR^2 \beta (1 - \cos \alpha) \quad \text{where } \alpha = 45° \text{ and } \beta = 90° = \frac{\pi}{2}$$

$$= 5 \times 80^2 (\frac{\pi}{2})(1 - 0.707) = 14,727.79$$

The areas of portion ① and ②:

$$\Sigma\psi_s = 4 \times \psi_{s,1} + 2 \times \psi_{s,2} = 4 \times 2{,}508.8 + 2 \times 14{,}727.79$$
$$= 10{,}015.2 + 29{,}455.58$$
$$= 39{,}490.78$$

Because of two axis symmetry,

$$\Sigma\psi_s y = 0 \qquad \therefore H_o = 0$$
$$\Sigma\psi_s x = 0 \qquad \therefore V_o = 0$$

The bending moment due to M_o,

$$M_o = M_{o,A} = M_{o,B} = M_{o,C} = M_{o,D}$$
$$= -\frac{\Sigma\psi_s}{\Sigma\dfrac{ds}{I}}$$
$$= -\frac{39{,}490.78}{502.65}$$
$$= -78.57 \text{ in.-kips}$$

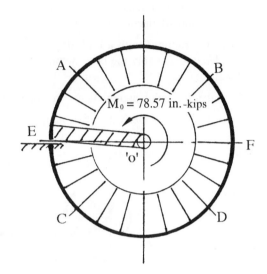

The final moments:

$$M_A = M_{m,A} + M_{o,A}$$
$$= 117.16 - 78.57 = 38.6 \text{ in.-kips}$$
$$M_A = M_B = M_C = M_D = 38.6 \text{ in.-kips}$$
$$M_E = M_F = M_{m,E} + M_{o,E}$$
$$= 0 + (-78.57)$$
$$= -78.57 \text{ in-kips}$$

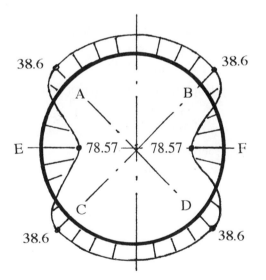

[Positive moments (in.-kips) are plotted on outside of ring member]

(4) SLOPE-DEFLECTION METHOD

The slope-deflection method is a displacement method which is another widely used method for analyzing all types of statically indeterminate beams and frames. In this method:

- All joints are considered rigid such that all angles between members meeting at a joint are assumed not to change in magnitude as loads are applied to the structure

- For a member bounded by two end joints, the end moments can be expressed in terms of the end rotations
- For static equilibrium the sum of the end moments on the members meeting at a joint must be equal to zero
- The rotations of the joints are treated as the unknowns

The equations of static equilibrium provide the necessary conditions to handle the unknown joint rotations and when these unknown joint rotations are found, the end moments can be computed from the Slope-deflection equations.

Thus for structures with a high degree of redundancy, the Slope-deflection method should be considered as possibly the best method of solution which can be programmed by Finite Element Modeling.

The end deformation and moments are obtained by using the following equations:

(a) Case 1 – Derivation by rotation:

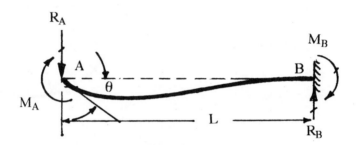

Using 1st moment-area proposition,

$$\frac{(M_A + M_B)L}{2EI} = \theta_A$$

Using 2nd moment-area proposition and taking moment about 'A',

$$(\frac{M_A L}{2EI})(\frac{L}{3}) + (\frac{M_B L}{2EI})(\frac{2L}{3}) = 0$$

Solving,

$$M_A = \frac{4EI\theta_A}{L}$$ Eq. 5.5.14

$$M_B = -\frac{2EI\theta_A}{L}$$ Eq. 5.5.15

$$R_A = -R_B = \frac{6EI\theta_A}{L^2}$$

(b) Case 2 – Derivation by deflection:

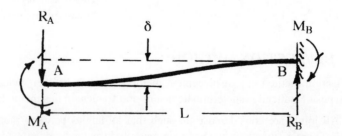

Using 1st proposition from moment-area method,

$$\frac{(M_A + M_B)L}{2EI} = 0$$

Using 2nd proposition from moment-area method and taking moment about 'A',

$$\left(\frac{M_A L}{2EI}\right)\left(\frac{L}{3}\right) + \left(\frac{M_B L}{2EI}\right)\left(\frac{2L}{3}\right) = \delta$$

Solving,

$$M_A = \frac{6EI\delta}{L^2} \qquad \text{Eq. 5.5.16}$$

$$M_B = -\frac{2EI\delta}{L^2} \qquad \text{Eq. 5.5.17}$$

$$R_A = -R_B = \frac{12EI\delta}{L^3}$$

Using above relationship, the moments M_{AB} and M_{BA} for the beam A-B can be expressed as follows:

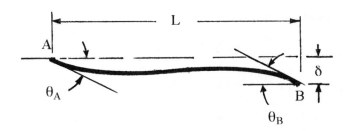

$$M_{AB} = \frac{4EI\theta_A}{L} + \frac{2EI\theta_B}{L} - \frac{6EI\delta}{L^2}$$
$$= 2k(2\theta_A + \theta_B - 3\phi) \qquad \text{Eq. 5.5.18}$$

$$M_{BA} = \frac{4EI\theta_B}{L} + \frac{2EI\theta_A}{L} - \frac{6EI\delta}{L^2}$$
$$= 2k(2\theta_B + \theta_A - 3\phi) \qquad \text{Eq. 5.5.19}$$

where $k = \frac{EI}{L}$ and $\phi = \frac{\delta}{L}$ (assumes clockwise moments as positive).

The fixed end moments (FEM) of a beam are calculated when the beam is held fixed at beam ends of 'A' and 'B' under the applied loading. The rotation of θ_A and θ_B are then calculated by using equilibrium equations for the beam end of 'A' and 'B' as follows:

$$M_{AB} = FEM_A \quad \text{(at 'A' end of the beam)} \qquad \text{Eq. 5.5.20}$$

$$M_{BA} = FEM_B \quad \text{(at 'B' end of the beam)} \qquad \text{Eq. 5.5.21}$$

Symmetrical frameworks can be solved by a similar procedure by treating the rotations of joints (in place of the supports as in a continuous beam) as unknowns.

Example:

Given the following continuous beam with a 30 lbs./in. uniformly distributed load and one concentrated load of 800 lbs. as shown below. Determine moments at supports.

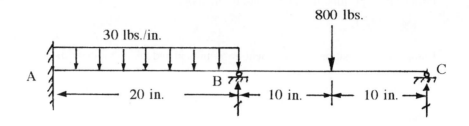

$(EI_{AB} = 20 \times 10^7 ; \ EI_{BC} = 10 \times 10^7)$

Calculate fixed-end moments (FEM) at A and B,

$$FEM_{AB} = -30 \times \frac{20^2}{12} = -1,000 \text{ in.-lbs.}$$

$$FEM_{BA} = 1,000 \text{ in.-lbs.}$$

$$FEM_{BC} = -800 \times \frac{20}{8} = -2,000 \text{ in.-lbs.}$$

$$FEM_{CB} = 2,000 \text{ in.-lbs.}$$

$$k_{AB} = 20 \times \frac{10^7}{20} = 10^7 \text{ in.-lbs.}$$

$$k_{BC} = 10 \times \frac{10^7}{20} = 0.5 \times 10^7 \text{ in.-lbs.}$$

$$\phi = \frac{\delta}{L} = 0 \text{ for all supports as there is no translation}$$

Slope-deflection Equations:

$$M_{AB} = FEM_{AB} + 2k_{AB}(2\theta_A + \theta_B) = -1,000 + 2(10^7) \times \theta_B$$

$$M_{BA} = FEM_{BA} + 2k_{AB}(2\theta_B + \theta_A) = 1,000 + 2(10^7) \times 2\theta_B$$

$$M_{BC} = FEM_{BC} + 2k_{BC}(2\theta_B + \theta_C) = -2,000 + 2(0.5 \times 10^7)(2\theta_B + \theta_C)$$

$$M_{CB} = FEM_{CB} + 2k_{BC}(2\theta_C + \theta_B) = 2,000 + 2(0.5 \times 10^7)(2\theta_C + \theta_B)$$

For equilibrium, $\Sigma M_B = 0$ and $\Sigma M_C = 0$:

$$1,000 + 4 \times 10^7 \theta_B - 2,000 + 2 \times 10^7 \theta_B + 10^7 \theta_C = 0$$

$$2,000 + 2 \times 10^7 \theta_C + 10^7 \theta_B = 0$$

$$\therefore \ 6\theta_B + \theta_C = 1,000 \times 10^{-7}$$

$$\therefore \ \theta_B + 2\theta_C = -2,000 \times 10^{-7}$$

Solving:

$$\theta_B = 363.6 \times 10^{-7}$$

$$\theta_C = -1,181.8 \times 10^{-7}$$

$$M_A = M_{AB} = -1,000 + 2 \times 10^7 \times 363.6 \times 10^{-7}$$

$$= -272.8 \text{ in.-lbs.}$$

$$M_B = M_{BA} = 1,000 + 4 \times 10^7 \times 363.6 \times 10^{-7}$$

$$= 1,000 + 1,454.4$$

$$= 2,454.4 \text{ in.-lbs.}$$

(5) MOMENT-DISTRIBUTION METHOD

The moment-distribution method is:

- An approximation method
- A practical and convenient way of analyzing such structures as continuous beams, stiff-jointed frames etc.
- A displacement approach using what is mathematically called a relaxation technique which does away with the necessity of direct solution of a large number of simultaneous equations
- The method is simple, rapid and particularly adapted to the solution of continuous structures of a high degree of redundancy
- It avoids the usual tedious algebraic manipulations of numerous equations
- Many who use moment distribution do not actually understand the why of it; they merely know how to perform the balancing operation
- This whole operation can be largely automatic and can require very little thought

The moment-distribution method considers:

- Any moment considered at the end of a member will always be the moment the member exerts on the joint or support
- In order to balance any joint, it is necessary only to determine the magnitude and sign of the unbalanced internal moment that must be distributed to the various members intersecting at that joint to make ΣM for the internal moments equal zero
- This unbalanced internal moment, with sign reversed, is distributed to the members intersecting at the joint in accordance with the distribution factors
- The carry-over operation then performed will usually upset the previous balance of one or more joints, and these must be rebalanced
- The carry-over from these joints will upset other joints, and so on, the corrections becoming smaller and smaller until they are no longer significant

Prior to the advent of computer calculation, this method was a proven means whereby many types of continuous frames, which were formerly designed only by approximation methods could be analyzed with accuracy and comparative ease.

The moment-distribution method uses the following displacement results in its derivations:

(a) The far end is fixed:

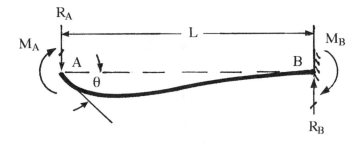

$$M_A = \frac{4EI\theta}{L}$$ Eq. 5.5.22

$$R_A = \frac{6EI\theta}{L^2}$$ Eq. 5.5.23

$$M_B = \frac{2EI\theta}{L} \qquad \text{Eq. 5.5.24}$$

$$R_B = -\frac{6EI\theta}{L^2} \qquad \text{Eq. 5.5.25}$$

(b) The far end is pinned:

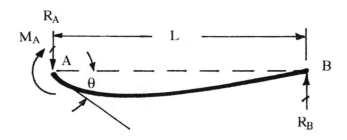

$$M_A = \frac{3EI\theta}{L} \qquad \text{Eq. 5.5.26}$$

$$R_A = \frac{3EI\theta}{L^2} \qquad \text{Eq. 5.5.27}$$

$$M_B = 0 \qquad \text{Eq. 5.5.28}$$

$$R_B = -\frac{3EI\theta}{L^2} \qquad \text{Eq. 5.5.29}$$

Consider a case, shown in Fig. 5.5.6, where three members AB, AC and AD meet in a stiff joint 'A' and B and D are fixed while C is pinned. Let joint 'A' be restrained against translation.

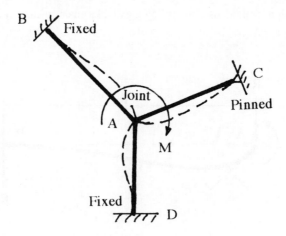

Fig. 5.5.6 A Moment Applied on a Joint with Three Members

Apply a moment M to joint 'A' and let θ be the rotation of the joint due to the applied moment of M,

$$M_{AB} = \frac{4EI_{AB}\theta}{L_{AB}} ; \qquad M_{AC} = \frac{3EI_{AC}\theta}{L_{AC}} ; \qquad M_{AD} = \frac{4EI_{AD}\theta}{L_{AD}}$$

For equilibrium:

$$M = M_{AB} + M_{AC} + M_{AD}$$

$$M = \theta \left(\frac{4EI_{AB}}{L_{AB}} + \frac{3EI_{AC}}{L_{AC}} + \frac{4EI_{AD}}{L_{AD}} \right)$$

$$M = \theta (K_{AB} + K_{AC} + K_{AD}) \qquad \text{Eq. 5.5.30}$$

- $K_{AB} = \dfrac{4EI_{AB}}{L_{AB}}$ can be considered as the stiffness of AB

- $K_{AC} = \dfrac{3EI_{AC}}{L_{AC}}$ can be considered as the stiffness of AC

- $K_{AD} = \dfrac{4EI_{AD}}{L_{AD}}$ can be considered as the stiffness of AD

- Total stiffness at joint 'A' is $K_A = K_{AB} + K_{AC} + K_{AD}$

Therefore, the applied moment, M, carried by each member is proportional to its stiffness 'K' and, thus, the distribution factor (DF) for a member at a joint is the ratio of the member stiffness to joint stiffness.

$$M_{AB} = M \frac{K_{AB}}{K_A} \qquad \text{Eq. 5.5.31}$$

$$M_{AC} = M \frac{K_{AC}}{K_A} \qquad \text{Eq. 5.5.32}$$

$$M_{AD} = M \frac{K_{AD}}{K_A} \qquad \text{Eq. 5.5.33}$$

The carry-over factors (COF) are obtained from the following equations:

$$M_{BA} = \frac{M_{AB}}{2} \qquad \text{(fixed-end at 'B')}$$

$$M_{CA} = 0 \qquad \text{(Pinned-end at 'C')}$$

$$M_{DA} = \frac{M_{AD}}{2} \qquad \text{(fixed-end at 'D')}$$

Carry-over factors are $\frac{1}{2}$ for fixed-end and 0 for pinned-end.

The moment-distribution method uses these distribution factors and carry-over factors to solve indeterminate structures by the following approach:

- Clamp all joints and calculate fixed end moments due to applied loads
- Satisfy equilibrium at all joints by applying additional moments equal and opposite to the out-of-balance moments; distribute and carry over the applied moments by using the appropriate carry-over factors

Chapter 5.0

- Check each joint for equilibrium and if it is not satisfied, continue the process of balancing, distributing and carrying over until equilibrium is satisfied within reasonable limits

Example:

Given the following continuous beam with two uniformly distributed loads and two concentrated loads, as shown, find the moments at the supports.

(a) Distribution factors (DF):

$$K_{BA} = 4E \times \frac{3I}{20} = \frac{3EI}{5}; \quad K_{BC} = 4E \times \frac{2I}{15} = \frac{8EI}{15}; \quad K_B = K_{BA} + K_{BC} = (\frac{3}{5} + \frac{8}{15})EI = \frac{17EI}{15}$$

$$\therefore DF_{BA} = \frac{\frac{3EI}{5}}{\frac{17EI}{15}} = \frac{9}{17} = 0.53$$

$$\therefore DF_{BC} = \frac{\frac{8EI}{15}}{\frac{17EI}{15}} = \frac{8}{17} = 0.47$$

$$K_{CB} = 4E \times \frac{2I}{15} = \frac{8EI}{15}; \quad K_{CD} = \frac{3EI}{10}; \quad K_C = (\frac{8}{15} + \frac{3}{10})EI = \frac{5EI}{6}$$

$$\therefore DF_{CB} = \frac{\frac{8EI}{15}}{\frac{5EI}{6}} = \frac{16}{25} = 0.64$$

$$\therefore DF_{CD} = \frac{\frac{3EI}{10}}{\frac{5EI}{6}} = \frac{9}{25} = 0.36$$

(b) Fixed-end moments (FEM):

$$FEM_{AB} = FEM_{BA} = 60 \times \frac{20^2}{12} = 2,000 \text{ in.-lbs.}$$

$$FEM_{BC} = 300 \times 9^2 \times \frac{6}{15^2} + 600 \times 9 \times \frac{6^2}{15^2}$$
$$= 648 + 864 = 1,512 \text{ in.-lbs.}$$

$$FEM_{CB} = 300 \times 6^2 \times \frac{9}{15^2} + 600 \times 9^2 \times \frac{6}{15^2}$$
$$= 432 + 1,296 = 1,728 \text{ in.-lbs.}$$

$$FEM_{CD} = FEM_{DC} = 30 \times \frac{10^2}{12} = 250 \text{ in.-lbs.}$$

The calculation can be accomplished in a tabular form as shown below:

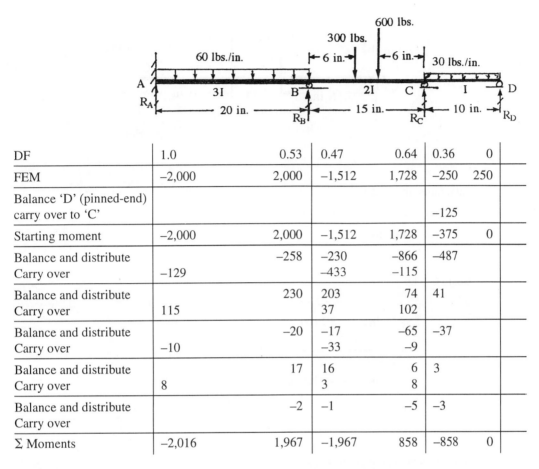

DF	1.0		0.53	0.47	0.64	0.36	0
FEM	−2,000		2,000	−1,512	1,728	−250	250
Balance 'D' (pinned-end) carry over to 'C'						−125	
Starting moment	−2,000		2,000	−1,512	1,728	−375	0
Balance and distribute			−258	−230	−866	−487	
Carry over	−129			−433	−115		
Balance and distribute			230	203	74	41	
Carry over	115			37	102		
Balance and distribute			−20	−17	−65	−37	
Carry over	−10			−33	−9		
Balance and distribute			17	16	6	3	
Carry over	8			3	8		
Balance and distribute			−2	−1	−5	−3	
Carry over							
Σ Moments	−2,016		1,967	−1,967	858	−858	0

Beam support reactions can be obtained from free-body diagram method:

(a) Beam A – B:

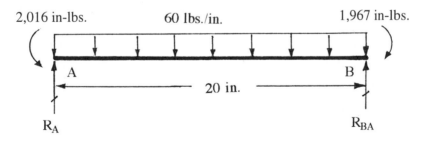

Take moment about point 'B',

$$R_A \times 20 - 2,016 - \frac{60 \times 20^2}{2} + 1,967 = 0$$

$R_A = 602.5$ lbs.

$R_{BA} = 1,200 - 602.5 = 597.5$ lbs.

(b) Beam B – C:

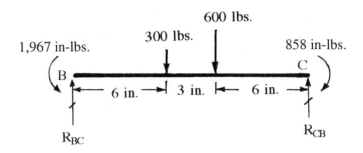

Take moment about point 'C',

$R_{BC} \times 15 - 1{,}967 - 300 \times 9 - 600 \times 6 + 858 = 0$

$R_{BC} = 494$ lbs.

$R_{CB} = 900 - 494 = 406$ lbs.

(c) Beam C – D:

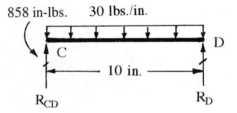

Take moment about point 'D',

$$R_{CD} \times 10 - 858 - \frac{30 \times 10^2}{2} = 0$$

$R_{CD} = 235.7$ lbs.

$R_D = 300 - 235.7 = 64.3$ lbs.

(d) Summation of beam support reactions at A, B, C and D are as follows:

$R_A = 602.5$ lbs.

$R_B = R_{BA} + R_{BC} = 597.5 + 494 = 1{,}091.5$ lbs.

$R_C = R_{CB} + R_{CD} = 406 + 235.7 = 641.7$ lbs.

$R_D = 64.3$ lbs.

5.6 FINITE ELEMENT MODELING (FEM)

Finite Element Modeling (FEM) is a powerful computer tool for determining stresses and deflections in a given structure which is too complex for classical analysis. Material properties such as Young's modulus (E) and Poisson's ratio (μ) are entered along with boundary conditions such as displacements (δ), applied loads (P), etc.

The FEM method has these characteristics:

- Arrays of large matrix equations that appear complicated to the novice
- Fundamentally simple concepts involving basic stiffness and deflection equations
- The first step is the construction of a structural model that breaks a structure into simple shapes or elements located in space by a common coordinate grid system
- The coordinate points, or nodes, are locations in the model where output data are provided
- Essentially, FEM geometrically divides a structure into small elements with easily defined stress and deflection characteristics

The method appears complex because a model of an airframe structure can have thousands of elements or members, each with its own set of equations. Because of the very large number of equations and corresponding data involved, the finite-element method is only possible when performed by computer.

With FEM, modeling is critical because it establishes the structural locations where stresses are evaluated, thus:

- If a component is modeled inadequately, the resulting computer analysis could be quite misleading in its predictions of areas of maximum strain and deflection
- Modeling inadequacies include the incorrect placement of elements and attempting to define a structure with an insufficient number of elements
- Such errors can be avoided by anticipating areas of maximum strain, but doing so requires engineering experience
- In most cases, the finer the grid, the more accurate the results
- However, the computer capacity and time required and cost of analysis increase with the number of elements used in the model
- The efficiency can be increased by concentrating elements in the interested areas of high stress while minimizing the number of elements in low stress areas

It is not uncommon to develop FEM for prototype design for which experimental data can be obtained. Strain gaging is probably the most common method of obtaining experimental data in structural tests. Once FEM results and experimental data have been correlated, design modifications can be made, and these subsequent changes are often tested through FEM before being implemented on the actual prototype.

- FEM is useful in design work, such as structural repair or modification, where a structural beef-up or change is contemplated
- A FEM baseline model can be made for an existing structure for which stress and deflection data are known
- A comparison is made between FEM results and known experimental data to calibrate the FEM results
- Proposed design modifications can then be evaluated to the baseline modeling knowing that the new FEM results will have the same accuracy and requires the same calibration as the baseline case

(A) ANALYTICAL MOTIVATION

The requirements of airframe structural analysis for both new design and repair are undergoing change due to different requirements and a broadened analytical concept as shown below:

- Different environment (aircraft flight altitude, speed and etc.)
- Different construction
- Refined detail
- Expanded analysis coverage

- Broadened analytical concept
- Expanded use of computers

Motivated by these changes in analytical requirements, the effective use of the computer is of paramount necessity and importance. Environment change has precipitated the search for improved structures to meet the requirements of high performance aircraft. The changes in altitude and speed have not only motivated a search for lighter structures but have necessitated the considerations of heated structure.

As shown in Fig. 5.6.1, a major structural discontinuity occurs at the juncture of components such as the wing and fuselage section. At such structural junctions, a major redistribution of stresses must occur and the flexural dissimilarities of the wing and fuselage must be accounted for in design. Regardless of the construction details at this juncture, the major components affect each other. In those example cases where the proportions of a component are such that beam theory can be employed, it is common practice to assume the behavior of one of the components and correspondingly analyze the other component. In the example case of the wing/ fuselage juncture just mentioned, such a procedure could assume the fuselage to provide a cantilevered support boundary for the wing and then analyze the wing by beam theory.

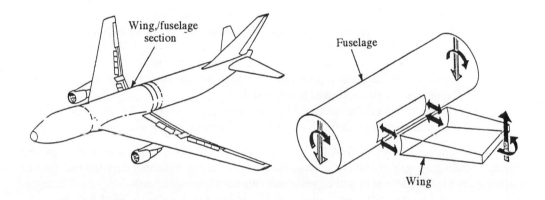

Fig. 5.6.1 Equilibrium and Compatibility Analysis of a Wing / Fuselage Junction

Such conventional procedures will essentially ensure that the analytical forces that occur between the wing/ fuselage will be in equilibrium. Unfortunately, the actual elastic structural compatibility usually enforces a different distribution of forces between these two major structures (wing and fuselage). Thus a change in analytical requirements is required and analytical technology must be powerful enough to ensure both stress (force equilibrium) and deflection (compatibility) at the structural discontinuity.

(B) AIRFRAME INTERNAL LOADS

Of most interest to stress engineers are the structural internal loads which are used to size the structures or to obtain a minimum Margin of Safety (MS). The accuracy of internal loads not only affects aircraft safety but structural weight issues as well. Airframe requirements must be satisfied by both:

- Force equilibrium
- The elastic deflection compatibility

REDUNDANT STRUCTURES:

Redundant structures represent a broadening of the analytical concepts to include both equilibrium and compatibility requirements which gives the actual structural load or stress distribution. Furthermore, most airframe structures are very redundant (or indeterminate) which means, in other words, the analysis of redundant structures leads to the need to solve sets of more than thousands of simultaneous linear algebraic equations. If the actual redundant structure is large, the set of simultaneous equations will be very large. In the good old days, when faced with redundant structures the structural engineer was tempted to make assumptions on the relative force-displacement behavior of connecting structural members, and thereby use equilibrium methods and avoid the problem of solving large ordered sets of simultaneous equations.

It is important to indicate the structural characteristics that should be present to justify the use of equilibrium methods. Fig. 5.6.2 is the first example of a case which examines a one-bay uniform stiffened cylinder in bending:

- The twelve stiffeners are equal in area
- Stiffener spacing is equal
- Skin thickness is constant

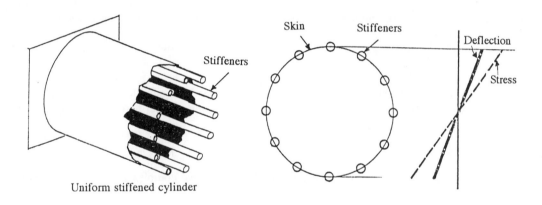

Fig. 5.6.2 Stress and Deflection Distribution of Uniform Stiffened Cylinder

If it is assumed that the skins are carrying only pure shear load, the problem has nine unknown stiffener forces. The strain energy solution of this problem, which requires solution of nine simultaneous equations, results not only in linear displacements along the cross section but also linear stresses as shown in Fig. 5.6.2. However, if the beam bending theory ($\frac{My}{I}$ – an equilibrium method) had been used, an engineer would get the same results without solving any simultaneous equations.

For the second example of this case, use the same circular cross section, same stiffener spacing, and same shell thickness, but use irregular stiffener cross sectional areas, as shown in Fig. 5.6.3. The stiffener cross sectional areas are unequal, ranging from one to four ratio. The application of the Finite Element Modeling method yields very nonlinear stress and deflection distributions. This solution requires the solution of nine simultaneous equations to obtain the stresses that are in equilibrium and corresponding compatible deflection.

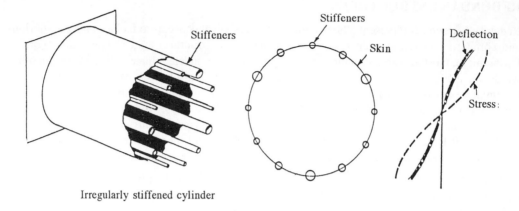

Irregularly stiffened cylinder

Fig. 5.6.3 Stress and Deflection Distribution of an Irregularly Stiffened Cylinder

It is clear that in the second example of this case, the equilibrium method alone (i.e., simple beam theory) would be an inadequate analysis. Thus, there are very restricted conditions that will justify the use of the equilibrium method only. It is readily seen that the analysis of the practical structures will lead to a large computation problem which has three major components:

- Speed
- Reliability
- Accuracy

Obviously, when the analysis involves the solution of thousands of simultaneous equations, the Finite Element Modeling method, which uses a computer, is an essential part of the analysis and computation task.

(C) ANALYSIS METHODS

Matrix analysis is the basis of Finite Element Modeling to solve complex redundant structures. There are basically two approaches:

(a) In the matrix force approach the internal element loads and external reactions are considered to be the unknowns, the correct system of loads being that which satisfies the compatibility equations.

(b) In the matrix displacement approach the displacements are considered to be the unknowns and the equations of equilibrium are enforced to give the correct displacement system.

In order to compute the aerodynamic loading and/or structural internal loads for an aircraft and its deflected shape owing to this loading, and to analyze such parameters as dynamic response and aeroelastic characteristics, the deflection and slope flexibility matrices are required. The technique used in the forming of these flexibility matrices is a complex analysis which is beyond the scope of this book. There are many books and reports within public domain for the interested engineer.

(D) EQUIVALENT STRUCTURES

Modeling experience is invaluable to simplify certain areas or segments of structures to reduce the cost of Finite Element Modeling:

- Use equivalent structural elements to reduce a large number of unnecessary grids and nodes
- Accuracy is not sacrificed because the retained segments have stiffness properties such that their response is consistent with the original structures

Structural Analysis

EQUIVALENT SHEAR WEB THICKNESS:

In finite element analysis, the web or skin with cutouts may be temporarily analyzed by first filling in the cutouts by introducing "fictitious" elements with equivalent thickness (T_e).

- The equivalent thickness defined by equal shear deflection is calculated based on the same energy under shear
- The shear stiffness (Gt) of a solid web or skin not subjected to buckling

The following formulas calculate the shear stiffness of various types of shear webs in terms of the thickness of an equivalent solid plate.

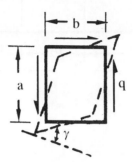

$$\gamma = \frac{f_s}{G} = \frac{q}{tG} = \frac{q}{T_e G_e} \qquad \text{Eq. 5.6.1}$$

$$\text{Energy} = \frac{q^2(ab)}{2tG} \qquad \text{Eq. 5.6.2}$$

where: γ – Shear strain
f_s – Shear stress
t – Web thickness
T_e – Equivalent web thickness
q – Shear flow
G – Modulus of rigidity (or shear modulus)
G_e – Equivalent modulus of rigidity of equivalent thickness (T_e) material

(a) Web with mixed material and thickness

- Web with two different thicknesses:

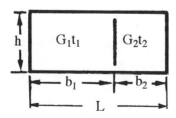

$$T_e = \frac{L}{G_e \left(\frac{b_1}{t_1 G_1} + \frac{b_2}{t_2 G_2} \right)} \qquad \text{Eq. 5.6.3}$$

where: G_e – modulus of rigidity of equivalent thickness (T_e) material
G_1 – modulus of rigidity of t_1 material
G_2 – modulus of rigidity of t_2 material

If all use the same material ($G_e = G_1 = G_2$):

$$T_e = \frac{L}{(\frac{b_1}{t_1} + \frac{b_2}{t_2})}$$ Eq. 5.6.4

- Web with three different thicknesses:

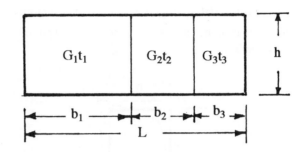

$$T_e = \frac{L}{(\frac{b_1}{t_1} + \frac{b_2}{t_2} + \frac{b_3}{t_2})}$$ Eq. 5.6.5

(b) Picture Frame (without web):

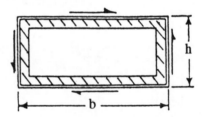

$$T_e = \frac{24E}{G_e b h (\frac{b}{I_b} + \frac{h}{I_h})}$$ Eq. 5.6.6

where: I_b – moment of inertia of top or bottom member for bending within plane of panel.
I_h – moment of inertia of side members

(c) Truss beam:

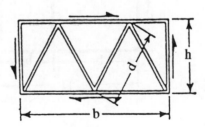

$$T_e = (\frac{bh}{G_e})(\frac{EA}{\Sigma d^3})$$ Eq. 5.6.7

where: A – cross-sectional area of one diagonal member and summation extends over all web members.

(d) Shear web with circular holes:

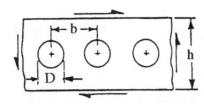

$$T_e = (\frac{Gt}{G_e})(1 - \frac{D}{b})[1 - (\frac{D}{h})^3]$$ Eq. 5.6.8

The above equation is empirical and valid only below buckling stress. Within accuracy limits of the formula, stiffness is unaffected by flanges around holes. (Ref. NACA Wartime report L-323)

(e) Shear web with deformable fasteners:

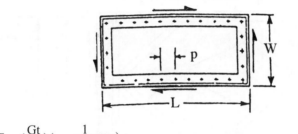

$$T_e = (\frac{Gt}{G_e})(\frac{1}{1 + \frac{np^2 Gt}{cWL}})$$ Eq. 5.6.9

where: n – total number of fasteners
p – pitch of fasteners
c – spring constant of fastener (shear force necessary to produce unit relative displacement of two strips jointed by one fastener)

(f) Heavy web with rectangular cutout:

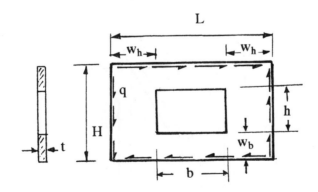

$$T_e = \frac{1}{\frac{1}{t_e} + (\frac{1}{tL})(\frac{bH}{2W_b} + 2W_h)}$$ Eq. 5.6.10

where: $t_c = \dfrac{24E}{Gbh}[\dfrac{1}{(\dfrac{b}{I_b}+\dfrac{h}{I_h})}]$ (see Eq. 5.6.6)

$I_b = t\dfrac{w_b^3}{12}$

$I_h = t\dfrac{w_h^3}{12}$

$L = 2w_h + b$

$H = 2w_b + h$

(g) Machined web with diamond-shaped cutout:

A_d – area of diagonal member

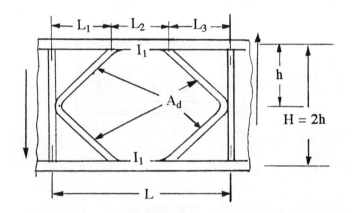

$T_e = \dfrac{L}{(\dfrac{G}{2E})[(\dfrac{HL_1^3}{12I_1}) + \dfrac{(\sqrt{H^2+4L_2^2})^3}{A_d H}]}$ Eq. 5.6.11

(h) Diagonal truss web:

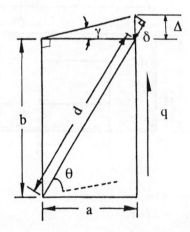

Load on diagonal member (d): $p_d = qd$

Diagonal member displacement: $\delta = \dfrac{qd^2}{AE}$

$$\gamma = \dfrac{q}{t_e G} = \dfrac{D}{a} = \delta(\dfrac{d}{b})(\dfrac{1}{a}) = q\dfrac{d^3}{AE(ab)}$$

$$T_e = (\dfrac{E}{G})\dfrac{(A\,a\,b)}{d^3} \qquad\qquad \text{Eq. 5.6.12}$$

(**Note:** shear deflection is the angular distortion of the field, while the edges remain without elongation)

EQUIVALENT BEAM AXIAL AREA:

The following formulas are given an equivalent axial area (A_e):

(a) The unbuckled web of a beam in bending will add to the beam stiffness and this addition is represented by adding the equivalent area (A_e) to the area of the beam caps (or chord).

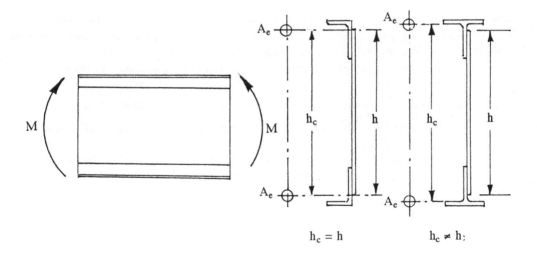

$h_c = h$ \qquad $h_c \neq h$:

The cap centroidal height is equal to the web height (h).

$$\dfrac{th^3}{12} = \dfrac{A_e h^2}{2}$$

The equivalent web area:

$$A_e = \dfrac{th}{6} \qquad\qquad \text{Eq. 5.6.13}$$

If the cap centroidal height (h_c) is different than the web height (h), then the equivalent web area is

$$A_e = \dfrac{th}{6}(\dfrac{h}{h_c})^2 \qquad\qquad \text{Eq. 5.6.14}$$

(b) Stepped member in axial load

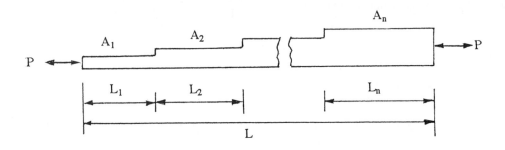

$$A_e = (\frac{E}{E_e})[\frac{L}{(\frac{L_1}{A_1} + \frac{L_2}{A_2} + ... + \frac{L_n}{A_n})}]$$ Eq. 5.6.15

where: E_e – Equivalent modulus of elasticity of equivalent area (A_e)

LUMPING STRINGER AREAS:

If using skin-stringer panels or integrally-stiffened panels to construct a wing box, obviously for analysis there are too many stringers and they should be evenly lumped (except spar caps) into a number of "lumped members" (substitute stringers, bundled stringers, etc.), as shown in Fig. 5.6.4, to simplify the Finite Element Modeling work. This lumping system has been proven by experience to give acceptable answers and to yield reasonably accurate values.

The following considerations should be borne in mind before lumping structural members:

- Use skin and web center line as box perimeter lines on which all lumped areas should be placed
- The lumped stringer areas placed on the perimeter lines should be proportionally reduced to be consistent with the same moment of inertia of the original box properties under beam bending, as shown in Fig. 5.6.5
- In compression panels, use the effective width of the buckled skin (use 15t on each side of the stringer as a ball-park number for preliminary sizing)
- If it is non-buckled skin, use full skin width (use engineering judgment to make this choice)
- For skin-stringer type boxes, the use of the spar cap plus the local skin as one individual lumped member is recommended, as shown in Fig. 5.6.5
- For multi-cell boxes, combine the spar cap plus the local skin as a single lumped member (see Fig. 5.6.5)
- All the results from calculated axial loads and shear flows should be de-lumped back to the original structural configuration
- Use internal loads, i.e., axial tension or compression loads and shear flows, rather than stresses from output of Finite Element Modeling analysis because this could easily be misleading

Structural Analysis

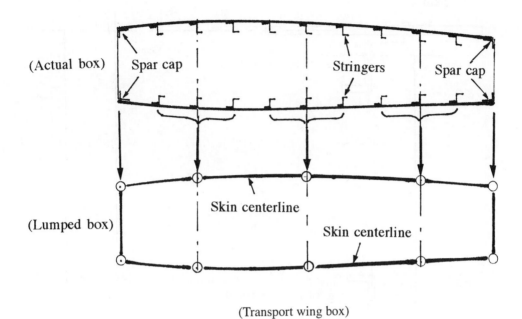

(Transport wing box)

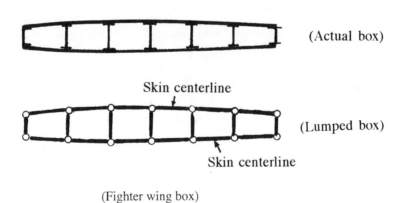

(Fighter wing box)

Fig. 5.6.4 Lumping Skin-Stringer Axial Areas into One Lumped Member

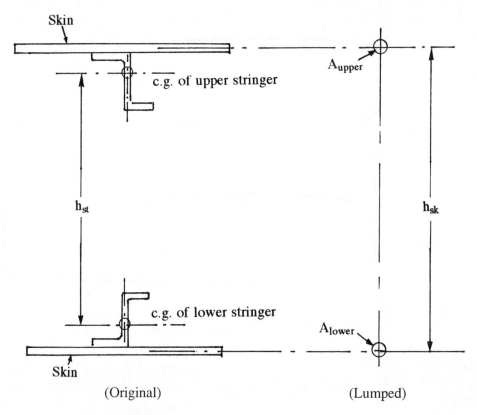

Lumped area (A) of skin and stringers:

A_{upper} or $A_{lower} = A_{sk} + A_{st,e}$

$A_{st,e} = A_{st}(\dfrac{h_{st}}{h_{sk}})^2$ (Equivalent Stringer area)

A_{sk} = Skin area and A_{st} = Stringer area

Fig. 5.6.5 Lumped Stringer Area

References

5.1 Timoshenko, S. and Young, D. H., "Elements of Strength of Materials", D. Van Nostrand Company, Inc., New York, NY, (1962).

5.2 Roark, R. J., "Formulas for Stress and Strain", McGraw-Hill Book Company, (1975).

5.3 Kinney, J. S., "Indeterminate Structural Analysis", Addison-Wesley Publishing Company, Inc., (1957).

5.4 Peery, D., "AIRCRAFT STRUCTURES", McGraw-Hill Book Company, Inc., New York, NY, (1950).

5.5 Rieger, N. F. and Steele, J. M., "Basic Course in Finite-element Analysis", Machine Design, June 25, 1981.

Airframe Stress Analysis and Sizing

Chapter 6.0

BEAM STRESS

6.1 BEAM THEORY

Beam theory, also called classic beam theory, considers an initially straight beam and applies a moment, M, so that the beam is bent with a radius of curvature, R, as shown in Fig. 6.1.1.

- Assume that plane sections remain plane after bending and that the strain along the neutral axis (NA) is zero because the length a-b is unaltered
- Another assumption is that the load acts in a plane of symmetry so that the beam does not twist

For the triangle b-c'-c'', the increase in the length of the beam at a distance, y, from the NA is δ and the strain at $y = \frac{\delta}{L_o} = \varepsilon$. It can be seen that the triangle O-b'-a' is geometrically similar to that of triangle b-c'-c''.

$$\therefore \frac{L_o}{R} = \frac{\delta}{y}$$

$$\text{or } \frac{y}{R} = \frac{\delta}{L_o} = \varepsilon \qquad \qquad \text{Eq. 6.1.1}$$

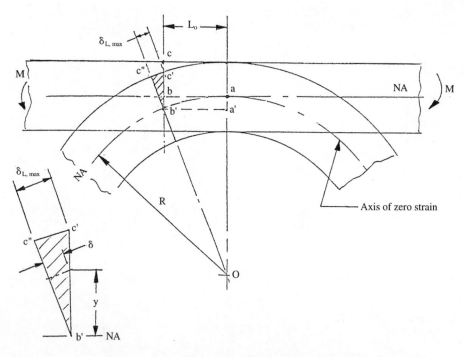

Fig. 6.1.1 Beam Bending within the Elastic Limit

The Young's Modulus ($E = \frac{f}{\varepsilon}$, see Eq. 4.2.2), which pertains to the elastic limit of the beam material, can be rewritten as follows:

$$\varepsilon = \frac{f}{E} \quad \text{also from Eq. 6.1.1,} \quad \varepsilon = \frac{y}{R}$$

$$\frac{f}{y} = \frac{E}{R} \quad \text{or} \quad f = \frac{Ey}{R} \qquad \text{Eq. 6.1.2}$$

where: f – stress
ε – strain
y – distance from Neutral Axis (NA)

For equilibrium, the internal moment due to the stress distribution across the beam cross section must be equal to the applied moment M, as shown in Fig. 6.1.2.

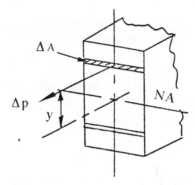

Fig. 6.1.2. Internal Balance Moment in a Beam Cross Section

$$\Delta p = \Delta A \times f$$

Substitute for f, see Eq. 6.1.2, then

$$\Delta p = \Delta A \frac{Ey}{R}$$

$$\Delta M = (\Delta A \frac{Ey}{R}) y$$

$$\Sigma M = (\frac{E}{R})(\Sigma y^2 dA)$$

But $\Sigma y^2 dA$ is the moment of inertia (I) or the second moment of area, then

$$\frac{M}{I} = \frac{E}{R} = \frac{f}{y}$$

From which, $f = \frac{My}{I}$ \qquad Eq. 6.1.3

This is one of the most commonly used equations in stress analysis and is known as the "beam bending theory", "bending formula", or 'flexure'.

6.2 SECTION PROPERTIES

When the internal load has been determined, each component or member must be analyzed for its strength and rigidity. In stress analysis the fundamental shape of features and other related properties of the member are expressed by equations that show the relationships of loads and shape to strength and rigidity. The shape characteristics which are associated with the cross section of the member are known as section properties. The section properties which are commonly used in design calculations are the following:

- Area (A, in.^{2})
- Centroid or neutral axis (NA, in.)
- Moment of inertia (I, in.^{4})
- Radius of gyration (ρ, in.)

(A) MOMENT OF INERTIA OF AREAS

To determine certain bending stresses in beams, and also in the design of columns, a quantity must be determined that is called the moment of inertia and denoted by the symbol 'I'. The moment of inertia of an area with respect to any axis not through its centroid is equal to the moment of inertia of that area with respect to its own parallel centroid axis added to the product of that area multiplied by the square of the distance from the reference axis to its centroid.

When the number of areas is finite, $I_x = \Sigma Ay^2$, by which equation the moment of inertia may be obtained. When each area is infinitesimal, or dA, and the number of areas is finite, $I_x = \int y^2 dA$, by which the exact moment of inertia may be obtained, as shown in Fig. 6.2.1.

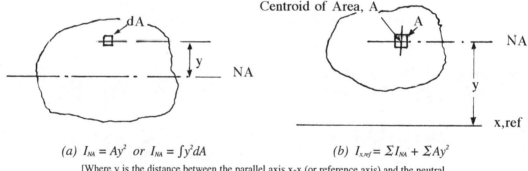

(a) $I_{NA} = Ay^2$ or $I_{NA} = \int y^2 dA$ (b) $I_{x,ref} = \Sigma I_{NA} + \Sigma Ay^2$

[Where y is the distance between the parallel axis x-x (or reference axis) and the neutral axis (NA) of the element areas]

Fig. 6.2.1 Moment of Inertia with Respect to a Reference Axis

In practical engineering language, the moment of inertia of the cross-sectional area of a beam or column of a given material with respect to its neutral axis may be said to represent the relative capacity of the section to resist bending or buckling in a direction perpendicular to the neutral axis.

A useful value for structural column problems is the radius of gyration of an area. For an area, having a moment of inertia about the x axis, the radius of gyration is defined as

$$\rho_x = \sqrt{\frac{I_x}{A}}$$
Eq. 6.2.1

Similarly, the radius of gyration about the y axis is

$$\rho_y = \sqrt{\frac{I_y}{A}}$$
Eq. 6.2.2

Example:

Find the moment of inertia 'I' and radius of gyration 'ρ' of the following T-section.

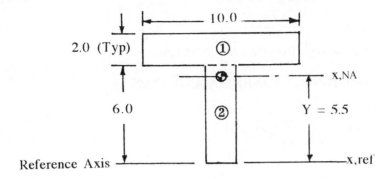

Segment	Area, A (in.²)	y (in)	Ay (in.³)	I_{NA} (in.⁴)	Ay^2 (in.⁴)
①	2 × 10 = 20.0	7	20 × 7 = 140.0	$10 \times \frac{2^3}{12} = 6.67$	20 × 7² = 980
②	2 × 6 = 12.0	3	12 × 3 = 36.0	$2 \times \frac{6^3}{12} = 36.0$	12 × 3² = 108
Σ	32.0		176.0	42.67	1,088

$$Y = \frac{\Sigma Ay}{A} = \frac{176.0}{32.0} = 5.5 \text{ in.}$$

$$I_{x,ref} = \Sigma I_{N.A.} + \Sigma Ay^2 = 42.67 + 1088 = 1130.67 \text{ in.}^4$$

$$I_{x,NA} = I_{x,ref} - AY^2 = 1130.67 - 32.0 \times 5.5^2 = 162.67 \text{ in.}^4$$

$$\rho_{x,NA} = \sqrt{\frac{I_{x,NA}}{\Sigma A}}$$

$$= \sqrt{\frac{162.67}{32}} = 2.255 \text{ in.}$$

Another section property used in structural analysis is the product of inertia for an area. This term gives a general indication as to how the area is distributed with respect to the x and y axes. The product of inertia, I_{xy}, is defined by

$$I_{xy} = \int xy \, dA \qquad \text{Eq. 6.2.3}$$

For a section which is symmetric with respect to the x and/or y axis, the product of inertia (I_{xy}) is zero. Unlike moment of inertia, the product of inertia can be a positive or negative value. The product of inertia about an inclined axes is shown in Fig. 6.2.2

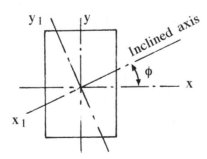

Fig. 6.2.2 Moment of Inertias about Inclined Axes

Chapter 6.0

The moments of inertia shown in Fig. 6.2.2. are:

$$I_{x1} = I_x \cos^2\phi - I_{xy} \sin 2\phi + I_y \sin^2\phi \qquad \text{Eq. 6.2.4}$$

$$I_{y1} = I_x \sin^2\phi + I_{xy} \sin 2\phi + I_y \cos^2\phi \qquad \text{Eq. 6.2.5}$$

$$I_{x1y1} = I_{xy} \cos 2\phi + \frac{(I_x - I_y)\sin 2\phi}{2} \qquad \text{Eq. 6.2.6}$$

where ϕ is the angle from the x axis to the x_1 axis

(B) PRINCIPAL MOMENTS OF INERTIA AND AXES

Principal moments of inertia,

$$I_{max} = \frac{I_x + I_y}{2} + \sqrt{I_{xy}^2 + (\frac{I_x - I_y}{2})^2} \quad \text{(Max. value)} \qquad \text{Eq. 6.2.7}$$

$$I_{min} = \frac{I_x + I_y}{2} - \sqrt{I_{xy}^2 + (\frac{I_x - I_y}{2})^2} \quad \text{(Min. value)} \qquad \text{Eq. 6.2.8}$$

The angle of principal axes is

$$\phi = \frac{1}{2} \tan^{-1}(\frac{2I_{xy}}{I_y - I_x}) \qquad \text{Eq. 6.2.9}$$

These special axes are called principal axes where I_x and I_y are maximum and minimum principal moment of inertias when $I_{xy} = 0$.

The above values of I_{max}, I_{min} and I_{xy} can be obtained by the Mohr's Circle graphic method as shown in Fig. 5.2.2 by replacing the values of f_{max}, f_{min} and f_{xy} with I_{max}, I_{min} and I_{xy} respectively in the graphic.

Example:

Calculate the moments of inertia about the x and y axes and the product of inertia for the following Z-cross section.

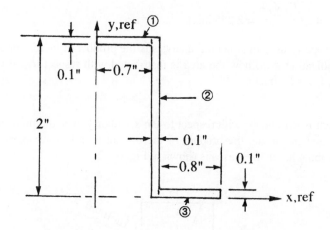

Segment	A(in.^{2})	x(in.)	y(in.)	Ax(in.^{3})	Ay(in.^{3})	Ax^2(in.^{4})	Ay^2(in.^{4})	Axy(in.^{4})	$I_{NA,x}$(in^4)	$I_{NA,y}$(in^4)
1	0.07	0.35	1.95	0.0245	0.1365	0.0086	0.2662	0.0478	0.0001	0.0029
2	0.20	0.75	1.0	0.150	0.20	0.1125	0.200	0.150	0.0667	0.0002
3	0.08	1.2	0.05	0.096	0.004	0.1152	0.0002	0.0048	0.0001	0.0043
Σ	0.35			0.2705	0.3405	0.2363	0.4664	0.2026	0.0669	0.0074

$$X = \frac{0.2705}{0.35} = 0.7729 \text{ in.}; \quad Y = \frac{0.3405}{0.35} = 0.9729 \text{ in.}$$

$I_{x,ref} = 0.0669 + 0.4664 = 0.5333 \text{ in.}^4$

$$\therefore I_x = I_{x,ref} - A Y^2 = 0.5333 - 0.35 (0.9729)^2 = 0.202 \text{ in.}^4$$

$I_{y,ref} = 0.0074 + 0.2363 = 0.2437 \text{ in.}^4$

$$\therefore I_y = I_{y,ref} - A X^2 = 0.2437 - 0.35 (0.7729)^2 = 0.0346 \text{ in.}^4$$

$I_{xy,ref} = 0.2026 \text{ in}^4$

$$\therefore I_{xy} = I_{xy,ref} - A XY = 0.2026 - 0.35 (0.7729 \times 0.9729) = -0.0606 \text{ in.}^4$$

Find principal moments of inertia,

Maximum and minimum moments of inertia from Eq. 6.2.7 and Eq. 6.2.8, respectively, are

$$I_{max} = \frac{I_x + I_y}{2} + \sqrt{I_{xy}^2 + (\frac{I_x - I_y}{2})^2}$$

$$= \frac{0.202 + 0.0346}{2} + \sqrt{(-0.0606)^2 + (\frac{0.202 - 0.0346}{2})^2}$$

$$= 0.222 \text{ in.}^4$$

$$I_{min} = \frac{I_x + I_y}{2} - \sqrt{I_{xy}^2 + (\frac{I_x - I_y}{2})^2}$$

$$= 0.015 \text{ in.}^4$$

The angle of the principal axis (see Eq. 6.2.9) is

$$\phi = \frac{1}{2} \tan^{-1}(\frac{2 I_{xy}}{I_x - I_y})$$

$$= \frac{1}{2} \tan^{-1}[2(\frac{-0.0606}{0.202 - 0.0346})] = 18°$$

6.3 BENDING OF SYMMETRICAL AND UNSYMMETRICAL SECTIONS

(A) SYMMETRICAL SECTIONS

If one of the axes is an axis of symmetry, the bending stress can be determined by:

$$f_b = \frac{-(M_y) x}{I_y} + \frac{(M_x) y}{I_x} \quad \text{Eq. 6.3.1}$$

If the bending stress calculated above is negative (–), this means the stress is in compression. The stress distribution for a section with two moments M_x and M_y applied along the x and y axes, respectively, is shown in Fig. 6.3.1.

Chapter 6.0

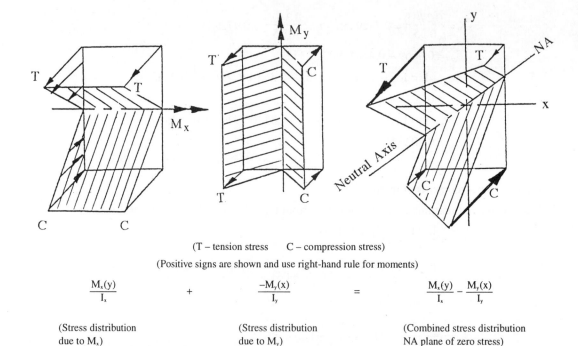

(T – tension stress   C – compression stress)
(Positive signs are shown and use right-hand rule for moments)

$$\frac{M_x(y)}{I_x} \quad + \quad \frac{-M_y(x)}{I_y} \quad = \quad \frac{M_x(y)}{I_x} - \frac{M_y(x)}{I_y}$$

(Stress distribution due to M_x)   (Stress distribution due to M_y)   (Combined stress distribution NA plane of zero stress)

Fig. 6.3.1 Bending Stresses on Symmetric Section

Example:

Symmetrical rectangular section with biaxial bending moments. Find stress at point A.

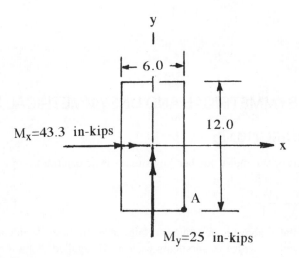

The moment of inertia:

$$I_x = \frac{bh^3}{12} = 6 \times \frac{12^3}{12} = 864 \text{ in.}^4$$

$$I_y = \frac{hb^3}{12} = 12 \times \frac{6^3}{12} = 216 \text{ in.}^4$$

From Eq. 6.3.1, the bending stress at point 'A' is:

$$f_{b,A} = \frac{-25 \times 3}{216} + \frac{43.3 \times (-6)}{864} = -0.647 \text{ ksi}$$

(B) UNSYMMETRICAL SECTIONS

The bending stress on an unsymmetrical section is shown in Fig. 6.3.2.

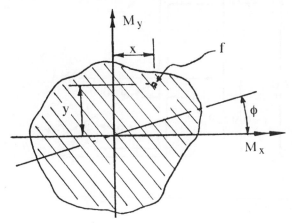

(Positive signs as shown, right-hand rule used for moments)

Fig. 6.3.2 Unsymmetric Section Geometry

The bending stress at any given point on an unsymmetrical section is determined by:

$$f_b = \frac{-(M_y I_x + M_x I_{xy})x + (M_x I_y + M_y I_{xy})y}{I_x I_y - (I_{xy})^2} \quad \text{Eq. 6.3.2}$$

The bending moments are assumed to be positive, as shown in Fig. 6.3.2 (right-hand rule), when they produce compression stress (−) in the first quadrant of the area, where x and y coordinates are positive.

From Eq. 6.2.9, the angle ϕ of the principal axis is:

$$\tan 2\phi = \frac{2 I_{xy}}{I_x - I_y}$$

where: I_x, I_y – moments of inertia about the origin of x and y coordinates, respectively
I_{xy} – product of inertia
x, y – distance to any point from the origin of x and y coordinates, respectively

(The above section properties were introduced in Section 6.2 of this Chapter)

When the X and Y axes are principal axes of the section or either one of the X and Y axes is an axis of symmetry, the product of inertia is zero ($I_{xy} = 0$), and the equation used to determine the bending stress at any given point on the section is exactly the same one shown previously (see Eq. 6.3.1):

$$f_b = \frac{-(M_y)x}{I_y} + \frac{(M_x)y}{I_x}$$

Example:

Determine the bending stress at point 'A' for the given cross section and loading shown below. Given section properties are:

$I_x = 0.202$ in.^{4}

$I_y = 0.0346$ in.^{4}

$I_{xy} = -0.0606$ in.^{4}

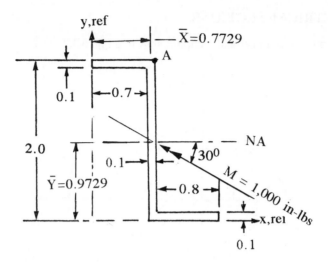

Applied moment components:

$$M_x = M \cos 30° = -1{,}000\,(0.866) = -866 \text{ in.-lbs.}$$
$$M_y = M \sin 30° = 1{,}000\,(0.5) = 500 \text{ in.-lbs.}$$

(Refer to Chapter 2.0 for use of the right-hand rule for moment sign convention)

From Eq. 6.3.2

$$f_b = \frac{-(M_y I_x + M_x I_{xy})x + (M_x I_y + M_y I_{xy})y}{I_x I_y - (I_{xy})^2}$$

$$f_b = \frac{-[(500 \times 0.202) + (-866)(-0.0606)]x + [(-866 \times 0.0346) + (500)(-0.0606)]y}{(0.202)(0.0346) - (-0.0606)^2}$$

$$f_b = \frac{-153.48x - 60.264y}{0.00332}$$

$$f_b = -46{,}229x - 18{,}152y$$

At point A, $x = 0.7 + 0.1 - 0.773 = 0.027$ in. and $y = 2.0 - 0.973 = 1.027$ in. and the bending stress at A is

$$f_{b,A} = -46{,}229\,(0.027) - 18{,}152\,(1.027)$$
$$= -1{,}248 - 18{,}642$$
$$= -19{,}890 \text{ psi}$$

6.4 PLASTIC BENDING

Few designs utilized in modern airframe structures are concerned with the effects of plasticity. The curve shown in Fig. 4.2.2 is the typical stress-strain relationship of common aluminum alloys. The "flat top" of the stress-strain curve for a particular material means that any fiber of that material will support ultimate stress while undergoing considerable strain and this characteristic is essential for plastic bending.

The term of plasticity implies yielding or flowing under steady load. It is defined as the property of sustaining visible permanent deformation without rupture. A beam of rectangular cross section subjected to bending is used to illustrate both the elastic and plastic ranges in Fig. 6.4.1.

Beam Stress

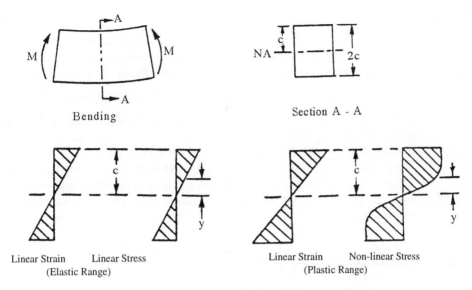

Fig. 6.4.1 Rectangular Cross Section Subjected to Bending

In the region below the elastic range (or proportional limit) on the stress-strain curve, the stress distribution follows the linear apparent stress $\frac{My}{I}$ shape. As the outer fiber stresses approach the ultimate stress, they undergo continuing strain until inner fibers also approach ultimate stress under proportional strain. The stress at any point on the cross section may be found by simply entering the stress-strain curve at the proper strain and reading the corresponding stress. The linear shape of the strain distribution is assumed to hold throughout the plastic range. By plotting the stress distribution and then computing the internal moment, the ultimate allowable external moment that can be applied to the section can be obtained.

(A) SECTION FACTOR

The ultimate bending moment of a section is a function of the cross-sectional shape. The shape can be represented by the section factor 'k' which is obtained by the following formula:

$$k = \frac{2Q}{\frac{I}{c}}$$
    Eq. 6.4.1

where: Q – The first moment about the principal axis of the area between the principal axis and the extreme fiber (in.^{3})
  I – Moment of Inertia of the whole section about the principal axis (in.^{4})
  c – Distance from the principal axis to the extreme fiber (in.)

Fig. 6.4.2 shows the range of values of 'k' for some common shapes. Fig. 6.4.3 through Fig. 6.4.6 give the values of 'k' for channels, I-beams, round tubes and T-sections.

k	1.0	1.5	1.0 - 1.5	1.0 - 1.5	1.27 - 1.7	1.5 - 2.0①
Section	⊥I⊤	▨	[	I	⊙	L ⊥

(① Sections with 'k' greater than 2.0 should not be used)

Fig. 6.4.2 Range of Values of Section Factors 'k'

Chapter 6.0

Before using plastic bending analysis, the engineer should understand the following:
- The section factor 'k' which is 1.5 for rectangular sections, varies with the shape of the section and can easily be computed for any symmetrical section
- For a rectangular section (i.e., end pad of a tension fitting, see Chapter 9.9), the total allowable moment becomes 1.5 times the moment produced by $F_{tu} = \dfrac{My}{I}$
- Round tube or solid pin sections are frequently used in pin bending for shear lug fitting (see Chapter 9.8)
- Compression crippling or lateral bowing cutoffs are not critical
- The application of plastic bending on airframe structures is not often used
- The most effective sections are the rectangular section or bulkier sections which are used only to react the very high bending moment under structural space limitation or restriction on shallow integrally-machined beams such as fighter wing spars, landing gear support beams, flap track supports, wing-mounted pylons, etc.
- It is preferable to design cross sections that are symmetrical about both axes (e.g., I-section, circular bar, tube, etc.), or at least symmetrical about one axis (such as a channel section) to avoid complex calculations
- Check the structural permanent deformation of material yield criteria

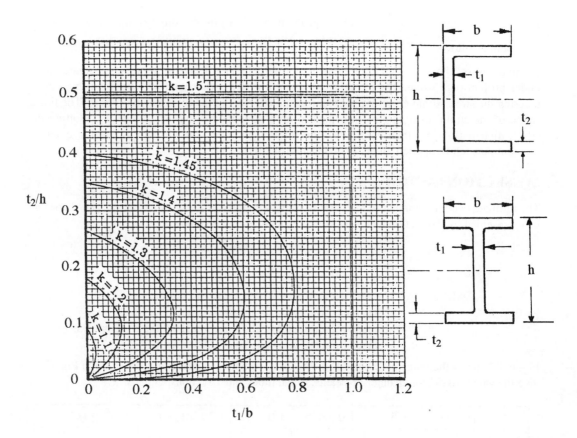

Fig. 6.4.3 Section Factor 'k' for Channels and I-beams

Beam Stress

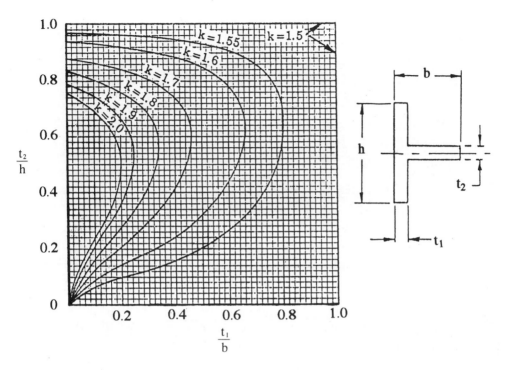

Fig. 6.4.4 Section Factor, k, for T-sections

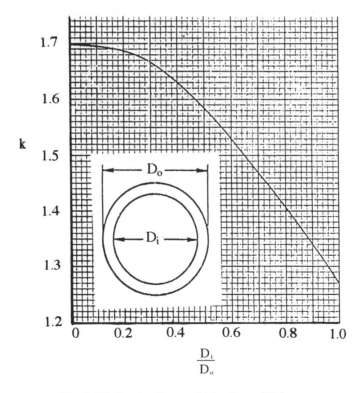

Fig. 6.4.5 Section Factor 'k' for Round Tubes

Chapter 6.0

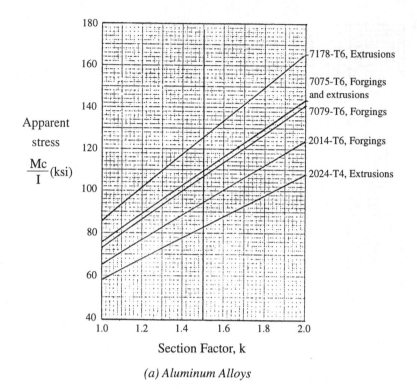

Fig. 6.4.6 Plastic Bending in Aluminum and Steel Alloys

(B) CALCULATION PROCEDURE

(a) Calculate the section factor 'k' or obtain from Figs. 6.4.3 through 6.4.5.

(b) Determine the "apparent stress", $\frac{Mc}{I}$, from Fig. 6.4.6:

- $\frac{Mc}{I}$ is the stress corresponding to a linear stress-strain distribution caused by the ultimate moment.
- It is not the actual stress in the member, but it is convenient for use in calculating the ultimate moment

(c) Multiply the "apparent stress" by $\frac{I}{c}$ to obtain the ultimate allowable moment (M_u).

(d) Calculate the MS_u:

$$MS_u = \frac{M_u}{m_u} - 1$$

where: MS_u – Ultimate margin of safety
M_u – Ultimate allowable moment
m_u – Ultimate applied moment

It is important to note that the yield margin of safety is usually critical for sections designed by plastic bending effects. The yield margin of safety must always be checked and in many cases will govern the design of the part.

Example 1:

For the rectangular section shown, calculate the Margin of Safety (MS) for ultimate strength. The ultimate applied moment is $M_u = 480,000$ in-lbs and the material is a 2014-T6 forging.

$Q = (\frac{bh}{2})(\frac{h}{4})$

$= (\frac{2 \times 4}{2})(\frac{4}{4}) = 4$ in^3

$I = \frac{bh^3}{12}$

$= \frac{2 \times 64}{12} = \frac{32}{3}$ in^4

$\frac{I}{c} = \frac{\frac{32}{3}}{2} = \frac{16}{3}$

$k = \frac{2Q}{\frac{I}{c}}$

$= \frac{2 \times 4}{\frac{16}{3}} = 1.5$

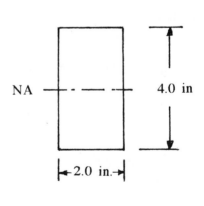

Refer to Fig. 6.4.6 with 2014-T6 and k = 1.5, obtain $\frac{Mc}{I} = 95,500$ psi

$M_u = 95,500 \times (\frac{16}{3}) = 509,000$ in-lbs

$MS_u = \frac{M_u}{m_u} - 1 = \frac{509,000}{480,000} - 1 = 0.06$

Chapter 6.0

Example 2:

The channel section shown is subjected to an ultimate bending moment of 35,000 in.-lbs. Determine the margin of safety using the plastic bending curves provided. Assume section is stable when extreme fibers in compression are stressed to F_{tu} (ultimate tensile stress of the material).

Material is 7075-T6 extrusion

Moment of inertia:

$I = 0.3937$ in.^{4}

$c = 1.0$ in.

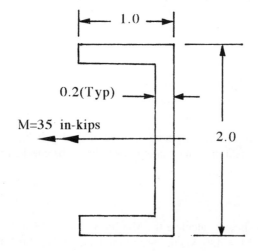

From Fig. 6.4.3, when $\frac{t_2}{h} = 0.1$ and $\frac{t_1}{h} = 0.2$, obtain $k = 1.24$

From plastic bending curves of Fig. 6.4.6:

$F_b = 92,000$ psi

$M_u = F_b(\frac{I}{c}) = 92,000(\frac{0.3937}{1.0}) = 36,220$ in.-lbs.

$MS_u = (\frac{M_u}{m_u}) - 1 = \frac{36,220}{35,000} - 1 = \underline{0.03}$

6.5 SHEAR STRESS IN BEAMS

Consider a rectangular cross section in which a vertical shear load acts on the cross section and produces shear stresses. It is assumed that:

(a) The vertical shear load produces vertical shear stresses over the exposed surface.

(b) The distribution of the shear stresses is uniform across the width of the beam.

Consider the element shown in Fig. 6.5.1: since one side of the element contains vertical shear stresses, equal shear stresses must be on the perpendicular faces to maintain equilibrium. At any point in the beam the vertical and horizontal shear stresses must be equal. If this element is moved to the bottom or top of the beam, it is apparent that the horizontal shear stresses vanish because there are no surface stresses on the outer surfaces of the beam. Therefore, the vertical shear stresses must always be zero at the top or bottom of the beam.

Beam Stress

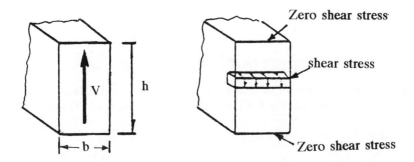

Fig. 6.5.1 The Shear Load on the Vertical Surface of an Element

(A) SYMMETRICAL SECTIONS

When a beam is bent, shearing occurs between the plane sections. The shear produces what is called the horizontal shear stress, as shown in Fig. 6.5.2

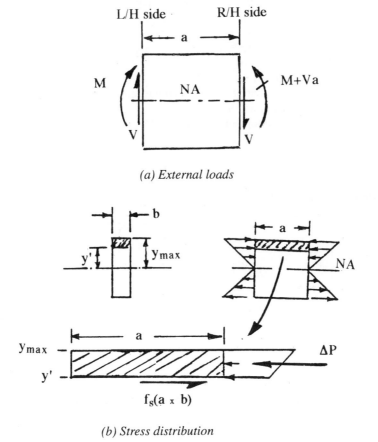

(a) External loads

(b) Stress distribution

Fig. 6.5.2 Shear Stress Distribution in a Beam Cross Section

For a portion of segment 'a' of a beam, the value of the shear, V, at the right-hand (R/H) side must be equal to the value of shear at the left-hand (L/H) side, since no external loads are applied; therefore the vertical components of the horizontal shear must be equal to the V value. From the flexure formula (see Eq. 6.1.3), the bending stress, f, due to M is

153

Chapter 6.0

L/H side: $f' = \dfrac{My}{I}$

R/H side: $f'' = \dfrac{My}{I} + V \times a \times \dfrac{y}{I}$

Take a section through the element at distance y' from the N.A. and balance the shaded portion above the section line as shown in Fig. 6.5.2(b). The out-of-balance load due to the moment is

$$\Delta P = (V \times a \times \dfrac{y}{I}) \Delta A$$

The load on the shear plane y' = f_s(a x b); for the load equilibrium, then

$$f_s (a \times b) - \int_{y'}^{y_{max}} (V \times a \times \dfrac{y}{I}) dA = 0$$

which gives

$$f_s = (\dfrac{V}{b} I) \int_{y'}^{y_{max}} y dA \qquad \text{Eq. 6.5.1}$$

The integral term of $\int_{y'}^{y_{max}} y dA$ is often noted as Q which is the first moment of area of the cross section above y' with the moment arm measured from the NA. Then the basic shear stress formula is

$$f_s = \dfrac{VQ}{Ib} \qquad \text{Eq. 6.5.2}$$

where V – applied shear load
I – Moment of inertia about the beam NA
b – Beam width
Q – first moment of area $\int_{y'}^{y_{max}} y dA$
f_s – shear stress

Example 1:

Find shear stress for a rectangular section as shown:

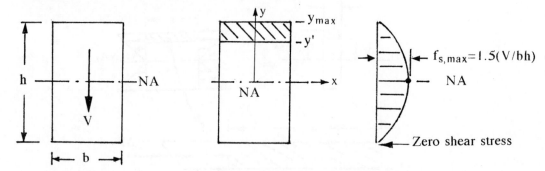

Calculate the shear stress at level y' for the rectangular section as shown,

$$Q = \int_{y'}^{y_{max}} y dA = b(\dfrac{h}{2} - y')[y' + \dfrac{(\dfrac{h}{2} - y')}{2}]$$

$$= \dfrac{b}{2}[(\dfrac{h^2}{4} - (y')^2] \quad \text{(The first moment of area of segment about the N.A.)}$$

$$f_s = \dfrac{VQ}{Ib} = \dfrac{V}{2I}[\dfrac{h^2}{4} - (y')^2]$$

The max. shear stress occur at y' = 0 which is at the N.A. (substituting $I = \frac{bh^3}{12}$):

$$f_{s,max} = 1.5\left(\frac{V}{bh}\right)$$

The peak stress is 1.5 times the average shear stress ($f_s = \frac{V}{bh}$) for the section and the stress is zero at the top and bottom of the section of the beam

Example 2:

Calculate the maximum shear stress in the section shown if the total shear load applied is 5,000 lbs. The properties for the upper and lower spar caps are given below and assume the web to be effective in resisting bending stress.

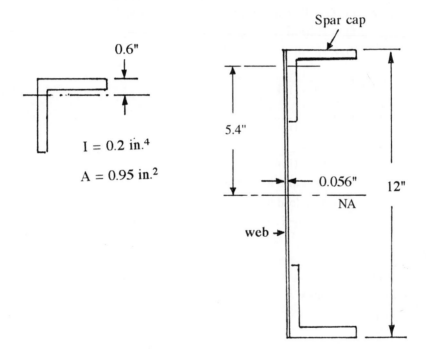

$$I = I_{web} + I_{caps}$$
$$= \left[\frac{0.056(12)^3}{12}\right] + 2\left[0.2 + 0.95\left(\frac{12}{2} - 0.6\right)^2\right]$$
$$= 63.87 \text{ in.}^4$$

The first moment of area :

$$Q = 0.95(5.4) + 6(0.056)(3) = 6.14 \text{ in.}^3$$

Maximum shear stress occurs at mid-height of the section:

$$f_s = \frac{VQ}{Ib}$$
$$= \frac{5,000(6.14)}{63.87(0.056)}$$
$$= 8,580 \text{ psi}$$

Chapter 6.0

(B) UNSYMMETRICAL SECTIONS

The basic shear stress formula of Eq. 6.5.2 does not apply for unsymmetric sections; It is based on $f = \dfrac{My}{I}$ (see Eq. 6.1.3) and is valid only for sections which are bent about their principal axes. For complex bending sections, a generalized bending formula was developed to be used to calculate bending stresses for unsymmetric sections. This generalized formula can also be rewritten in terms of external shear loads instead of external bending moments. The shear flow in an unsymmetric section can be calculated by the following equations:

$$q = -(K_3 V_x - K_1 V_y) Q_y - (K_2 V_y - K_1 V_x) Q_x \qquad \text{Eq. 6.5.3}$$

or shear stress :

$$f_s = \frac{q}{t} = \frac{-[(K_3 V_x - K_1 V_y) Q_y - (K_2 V_y - K_1 V_x) Q_x]}{t} \qquad \text{Eq. 6.5.4}$$

where: Q_x, Q_y – first moment of areas about the x and y axes, respectively

$Q_x = \int_0^{y_{max}} y \, dA$

$Q_y = \int_0^{x_{max}} x \, dA$

$K_1 = \dfrac{I_{xy}}{I_x I_y - (I_{xy})^2}$

$K_2 = \dfrac{I_y}{I_x I_y - (I_{xy})^2}$

$K_3 = \dfrac{I_x}{I_x I_y - (I_{xy})^2}$

V_x – shear load in x direction

V_y – shear load in y direction

t – thickness or width of the cross section

Example:

The loads, $V_x = 5{,}000$ lbs. and $V_y = 8{,}667$ lbs. are applied at the shear center for the Z-section: calculate the shear stress at points 'a' and 'b'.

Given: $I_x = 0.6035$ in.^{4}

$I_y = 0.0574$ in.^{4}

$I_{xy} = -0.1305$ in.^{4}

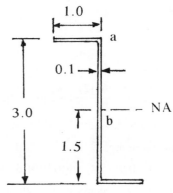

Therefore:

$K_1 = \dfrac{I_{xy}}{I_x I_y - (I_{xy})^2} = -5.88$

$K_2 = \dfrac{I_y}{I_x I_y - (I_{xy})^2} = 3.259$

$K_3 = \dfrac{I_x}{I_x I_y - (I_{xy})^2} = 34.269$

Shear flow:

$q = -[34.269(5{,}000) - (-5.88)(8{,}667)] Q_y - [3.259(8{,}667) - (-5.88)(5{,}000)] Q_x$

$= -(171{,}345 + 50{,}962) Q_y - (28{,}246 + 29{,}400) Q_x$

$= -222{,}307 Q_y - 57{,}646 Q_x$

(a) At point 'a':

$Q_x = 0.1 \times 1.0 \times 1.45 = 0.145$

$Q_y = 0.1 \times (-0.45) = -0.045$

$q_a = -222,307 Q_y - 57,646 Q_x$

$ = -222,307 \times (-0.045) - 57,646 \times 0.145 = 10,004 - 8,359$

$ = 1,645$ lbs./in.

or shear stress:

$f_s = \dfrac{1,645}{0.1} = 16,450$ psi

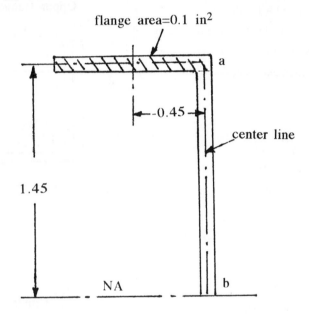

(b) At point 'b':

$Q_x = 0.145 + 1.4 \times 0.1 \times 0.7 = 0.243$

$Q_y = -0.045$

Shear flow:

$q_b = -222,307(-0.045) - 57,646(0.243) = 10,004 - 14,008$

$ = -4,004$ lbs./in.

or shear stress:

$f_s = \dfrac{-4,004}{0.1} = -40,040$ psi

(C) SHEAR STRESS IN THIN OPEN SECTIONS

A modification of the basic shear stress formula is applied to open thin sections. The following assumptions apply:

- The cross sectional thickness is thin, so that all section dimensions will be measured between the centerlines of the elements
- All shear stresses are zero which act perpendicular to the thickness or the section
- The cross section is open

Chapter 6.0

To calculate the shear stress in an open thin section, the following equation is used. Note that this is fundamentally the same as the basic shear equation of Eq. 6.5.2.

$$f_s = \frac{VQ}{Ib}$$

Example:

Load V is applied to the thin symmetric channel section shown such that it does not twist the cross section. Draw the shear stress distribution for the web and flanges.

The general shear equation is $f_s = VQ/It$ (see Eq. 6.5.2)

where, $Q = \int_0^s y \, dA$

s – length along flange or web

(a) Flange between 'a' and 'b' ($0 < s < w$)

The first moment of area:

$$Q = t_f s \left(\frac{h}{2}\right) = \frac{t_f h s}{2}$$

The shear stress:

$$f_s = \frac{VQ}{It} = \left(\frac{Vh}{2I_x}\right) s$$

Flange shear stress at point 'a':

$$f_{s,a} = 0$$

Flange shear stress at point 'b', when $s = w$:

$$f_{s,b} = \frac{Vhw}{2I_x}$$

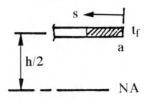

Upper flange

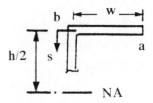

Upper flange

(b) Web segment between 'b' and 'c' ($w < s < \frac{h}{2}$)

The first moment of area:

$$Q = Q_{flange} + Q_{web}$$

$$Q = \underbrace{t_f w \left(\frac{h}{2}\right)}_{\text{(first moment of area of flange)}} + \underbrace{t_w s}_{\text{(web area)}} \times \underbrace{\left(\frac{h}{2} - \frac{s}{2}\right)}_{\text{(web moment arm)}}$$

At top of web at point 'b':

$$s = 0 \text{ and } Q = \frac{t_f w h}{2}$$

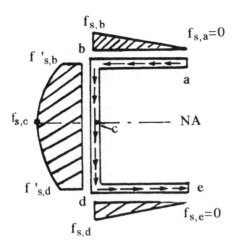

$$f'_{s,b} = (\frac{V}{I_x t_w})(\frac{t_f w h}{2})$$

At neutral axis (N.A.) of web, when $s = \frac{h}{2}$:

$$s = \frac{h}{2} \text{ and } Q = \frac{t_f w h}{2} + \frac{t_w h^2}{8}$$

$$f'_{s,c} = (\frac{V}{I_x t_w})(\frac{t_f w h}{2} + \frac{t_w h^2}{8})$$

Since this is a symmetric cross section, therefore, $f_{s,a} = f_{s,e} = 0$, $f_{s,b} = f_{s,d}$, and $f'_{s,b} = f'_{s,d}$.

(D) SHEAR FLOW

The activity of off-loading a panel boundary member by shearing the load into a thin sheet and transferring that load some distance to another boundary member while simultaneously developing the required balancing "couple forces" in other members has been given the descriptive name of "shear flow".

For beams consisting of flange members and shear panel webs, the panels are subjected to shear flow due to bending moment on the beam. The shear flow can be determined by considering "free body" diagrams for the flanges and the webs. The value of shear flow is constant along the length of a panel.

The basic assumptions are:

- The thin sheet carries all shear loads
- Longitudinal members such as beam caps, stringers, and longerons carry all axial loads. The cross-sectional areas of these members (plus their effective width of sheet) are assumed concentrated at their respective centroids
- A shear force acting on a member causes a change in axial load, or a change in axial load will produce a shear force
- Shear flow in the web of any rectangular panel is constant. The only way the shear flow can be changed is by changing the loads in the bounding members or by applying components of loads in skewed or slanting boundary members

The shear flow unit is lbs./in. which is a constant value along the four edges on a thin web or skin of a beam or closed-box structure which comes from:

- Shear forces on a simple beam ($q = \frac{V}{h}$)

Chapter 6.0

- Shear flows also can be found from torsion on a thin-walled closed section ($q = \frac{T}{2A}$, refer to Eq. 6.8.5 in this Chapter and to Chapter 8.0).

On a thin-web beam, axial loads are carried by the beam caps and leaving the web to carry shear load only, as shown in Fig. 6.5.3. Actually, the shear flow can be derived directly from the basic shear stress formula of Eq. 6.5.2 which can be rewritten and becomes:

$$f_s = \frac{VQ}{It} \quad \text{(replace beam width 'b' with web thickness 't')}$$

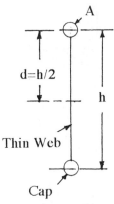

(Assume areas of upper and lower cap are equal)

Fig. 6.5.3 Idealized Thin-Web Beam

The first moment of area is

$$Q = A \times d = \frac{Ah}{2}$$

Moment of inertia (assume $d = \frac{h}{2}$):

$$I = 2Ad^2 = \frac{Ah^2}{2}$$

Then the shear flow on this thin web is

$$q = f_s \times t = (\frac{VQ}{It}) \times t = \frac{2VAh}{2Ah^2} = \frac{V}{h} \qquad \text{Eq. 6.5.5}$$

The shear flows are constant around four edges of a rectangular panel which always are two pairs of shear flows but opposite as shown in Fig. 6.5.4.

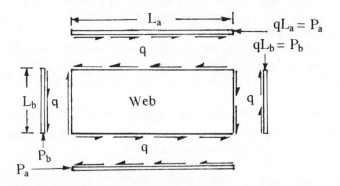

Fig. 6.5.4 Free-body of Shear Flows on a Thin Web

Beam Stress

Example:

Assume a cantilever beam carrying two 1,000 lbs. loads as shown below. Determine the shear flows and beam cap axial load distribution.

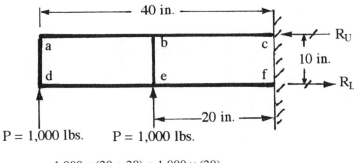

$$R_U = R_L = \frac{1,000 \times (20 + 20) + 1,000 \times (20)}{10} = 6,000 \text{ lbs.}$$

Shear flow around the beam web as shown below:

$$q_{abed} = \frac{1,000}{10} = 100 \text{ lbs./in.}$$

$$q_{bcfe} = \frac{2,000}{10} = 200 \text{ lbs./in.}$$

Load distribution along the length of 40" on upper cap, a-b-c:

(Axial compression load distribution on upper beam cap)

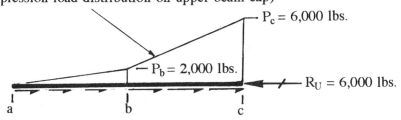

6.6 SHEAR CENTER

One of the assumptions of the basic flexure formula is that the load acts in a plane of symmetry so that the beam does not twist. If vertical loads are applied on a cross section that are not in a plane of symmetry, the resultant of the shear stresses produced by the loads will be a force parallel to the plane of loading, but not necessarily in that plane. Consequently, the member will twist. Bending without twisting is possible provided that the loads are in the same plane that the resultant shear stresses act, or in more general terms, bending moment without twisting will occur when the loads are applied through the "shear center". The shear center is defined as the point on the cross section of a member through which the resultant of the shear stresses must pass for any orientation of loading so that the member will bend but not twist.

Chapter 6.0

Example 1:

Find the shear center of the following channel section so that when load V_y passes through the shear center it produces no twist.

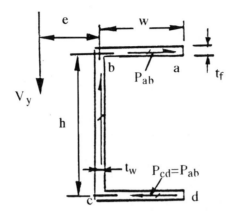

The resisting resultant shear loads in the flanges (P_{ab}) are shown. The upper and lower flange resultant must be equal so that $\Sigma F_x = 0$. The resultant P_{ab} can be calculated by integrating the area under the shear stress distribution.

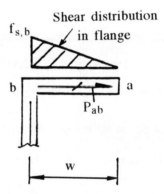

The first moment of area at 'b': $Q = t_f w(\frac{h}{2}) = \frac{t_f h w}{2}$

The shear stress at 'b': $f_{s,b} = \frac{V_y Q}{I_x t_f} = \frac{V_y h w}{2 I_x}$

Total shear load along flange a - b:

$$P_{ab} = \frac{(f_{s,b})(t_f w)}{2}$$

$$= \frac{(\frac{V_y h w}{2 I_x})(t_f w)}{2}$$

$$= (\frac{w^2 h t_f}{4 I_x}) V_y$$

The sum of the moments about the flange/web junction, point c, yields,

$$V \times e = P_{ab} h = (\frac{w^2 h^2 t_f}{4 I_x}) V_y$$

162

$$e = \frac{w^2 h^2 t_f}{4 I_x}$$

where: $I_x = \frac{t_w h^3}{12} + \frac{w t_f h^2}{2}$

so that the shear center distance

$$e = \frac{w^2 t_f}{(\frac{h t_w}{3} + 2 w t_f)}$$

Example 2:

Determine the position of the shear center 'e' and web shear flows between stringers under the given load of 10,000 lbs as shown.

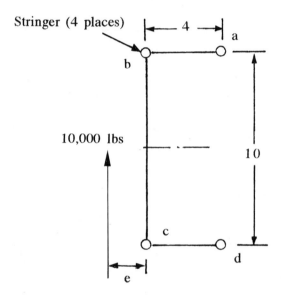

Stringer area = 1.0 in.²

$$I_{N.A.} = 2 \times 2 (1.0 \times 5^2) = 100 \text{ in.}^4$$

Load on stringers (assume the stringer length, L = 1.0 in.):

$$P = (\frac{M}{I_{N.A.}}) y A$$

$$= (\frac{10{,}000 \times 1.0}{100})(y)(1.0)$$

$$= 100 \, y$$

Stringer	y (in.)	P (lbs.)	L (in.)*	$\frac{P}{L}$ (lbs./in.)	$q_{web} = \frac{P}{L}$ (lbs./in.)
a	5.0	500	1.0	500	
b	5.0	500	1.0	500	500
c	−5.0	−500	1.0	−500	1,000
d	−5.0	−500	1.0	−500	500

(* Here assume the stringer length L = 1.0 in.)

Chapter 6.0

For zero twist on the section the applied shear must act through the shear center. Take the moment center at stringer 'c',

$\Sigma T_c = 10{,}000 \times e - 500 \times 4 \times 10 = 0$

Therefore, the shear center distance, e = 2.0 in.

6.7 TAPERED BEAMS

If the beam is constant in height the shear flow is constant throughout the web where a constant shear force is applied. In airframe structures, there are many applications where the beams are tapered in depth. Consider the beam shown in Fig. 6.7.1:

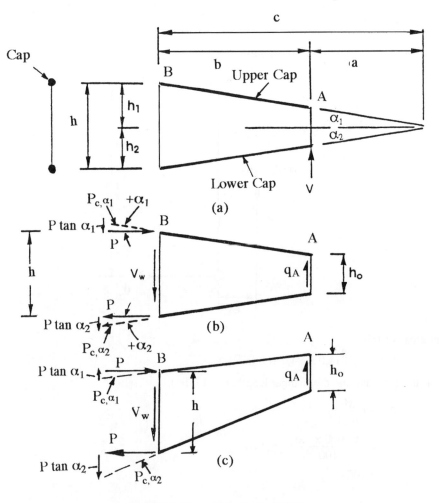

Fig. 6.7.1 Beam Tapered in Height

Fig. 6.7.1 shows a beam consisting of two concentrated sloping cap (flanges or chord) areas, separated by a thin web. An applied load, V, is applied at the free end of the beam at A, putting a total shear force of V throughout the length of the beam A-B. It is assumed that:

- The cap can take axial load only
- The web is ineffective in bending
- Sloping caps are inclined at angles α_1 and α_2 to the horizontal

164

Beam Stress

(A) RESISTANCE BY BEAM WEB AND CAPS:

Since the caps are inclined and can resist axial load only, the resultant cap loads can be separated into horizontal and vertical components. The couple loads, P, of the horizontal component of the caps resist the applied bending moment, M, and the vertical components of the caps resist part of the shear force resulting from V. Unlike a constant height beam where the web resists the total shear force, a tapered web beam's shear force is resisted by both the web (V_w) and the caps (V_c) as shown in Fig. 6.7.1(a).

$$V_c = P \times \tan \alpha_1 + P \times \tan \alpha_2 = P(\tan \alpha_1 + \tan \alpha_2) \qquad \text{Eq. 6.7.1}$$

For the loading shown in the Fig. 6.7.1(b),

$$P = V \times \frac{b}{h}$$

so

$$V_c = V\left(\frac{b}{c}\right)$$

$$V_w = V - V_c = V\left(\frac{a}{c}\right)$$

$$V_w = V - P(\tan \alpha_1 + \tan \alpha_2) \qquad \text{Eq. 6.7.2}$$

By similar triangles from Fig. 6.7.1(a), $\frac{a}{c} = \frac{h_o}{h}$, which leads to the final expressions for total shear force resisted by the web at B:

$$V_w = V\left(\frac{h_o}{h}\right) \qquad \text{Eq. 6.7.3}$$

At any section along the beam A-B, the total shear force:

$$V = V_c + V_w \qquad \text{Eq. 6.7.4}$$

(B) RESISTANCE BY BEAM WEB:

The shear flow in the web at A;

$$q_A = \frac{V}{h_o} \qquad \text{Eq. 6.7.5}$$

The shear flow in the web at B:

$$q_B = \frac{V_w}{h} = \frac{V\left(\frac{h_o}{h}\right)}{h} = \frac{V h_o}{h^2} \qquad \text{Eq. 6.7.6}$$

or

$$q_B = q_A \left(\frac{h_o}{h}\right)^2 \qquad \text{Eq. 6.7.7}$$

(C) BEAM CAPS:

The horizontal load component of the caps, denoted P, must be equal to the total bending moment, M, divided by the beam height, h. The vertical components of the cap loads are equal to P times the tangent of the angle of cap inclination (α). The total shear force resisted by the caps at B (watch positive or negative sign of angles of α) and from Eq. 6.7.1

$$V_c = P(\tan \alpha_1 + \tan \alpha_2) \quad \text{[see Fig. 6.7.1(b)]}$$

$$V_c = P(-\tan \alpha_1 + \tan \alpha_2) \quad \text{[see Fig. 6.7.1(c)]}$$

Chapter 6.0

By similar triangles from Fig. 6.7.1(a), $\frac{a}{c} = \frac{h_o}{h}$:

$$\tan \alpha_1 + \tan \alpha_2 = \frac{h}{c}$$

or
$$V_c = P \times (\frac{h}{c})$$

or
$$V_c = V(\frac{h - h_o}{h})$$
Eq. 6.7.8

(a) Upper sloping cap (α_1):

- Vertical component of shear load in the cap:

$$V_{c,\alpha_1} = \frac{V(\tan \alpha_1)(h - h_o)}{h(\tan \alpha_1 + \tan \alpha_2)}$$
Eq. 6.7.9

- Axial load in the cap:

$$P_{c,\alpha_1} = \frac{\frac{V(h - h_o)}{h(\tan \alpha_1 + \tan \alpha_2)}}{\cos \alpha_1}$$
Eq. 6.7.10

(b) Lower sloping cap (α_2):

- Vertical component of shear load in the cap:

$$V_{c,\alpha_2} = \frac{V(\tan \alpha_2)(h - h_o)}{h(\tan \alpha_1 + \tan \alpha_2)}$$
Eq. 6.7.11

- Axial load in the cap:

$$P_{c,\alpha_2} = \frac{\frac{V(h - h_o)}{h(\tan \alpha_1 + \tan \alpha_2)}}{\cos \alpha_2}$$
Eq. 6.7.12

It is seen that this secondary shear caused by the sloping beam caps can have an appreciable effect on the shear flows in the web and hence must not be neglected. Since the web shear flows are computed by the difference in stringer and cap loads at successive stations divided by the distance between stations, it is apparent that the shear flows as computed represent the average values between stations.

Another effect is that wherever the taper changes, such as the spar at the side of the fuselage (dihedral or anhedral angle) the component of load in each longitudinal due to the break must be transferred by a rib to the shear beams. This can impose sizable shear and bending loads on the rib.

Fig. 6.7.2 shows three cases of shear flows which should be kept in mind when evaluating shear flow in tapered beams.

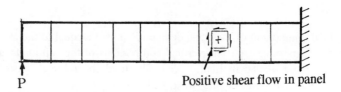

(a) Positive shear flow

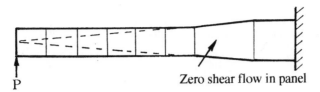

Zero shear flow in panel

(b) Zero shear flow

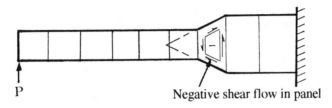

Negative shear flow in panel

(c) Negative shear flow

Fig. 6.7.2 Three Cases of Shear Flows

Example 1:

Find the shear stress on the 0.04 thickness beam web at section B-B:

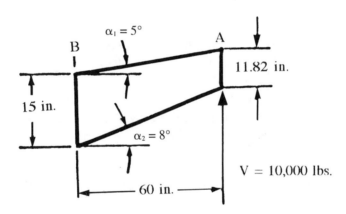

(a) Use Eq. 6.7.2, $V_w = V - P(\tan \alpha_1 + \tan \alpha_2)$:

Calculate the coupled load of $P = 10{,}000 \times \dfrac{60}{15} = 40{,}000$ lbs.

$V_w = 10{,}000 - 40{,}000(-\tan 5° + \tan 8°) = 10{,}000 - 2{,}121 = 7{,}879$ lbs.

$f_s = \dfrac{7{,}879}{15} \times 0.04 = 13{,}132$ psi

(b) Use Eq. 6.7.6, $q_B = \dfrac{V h_o}{h^2}$:

$q_B = \dfrac{V h_o}{h^2} = 10{,}000 \times \dfrac{11.82}{15^2} = 525.3$ lbs./in.

$f_s = \dfrac{525.3}{0.04} = 13{,}132$ psi (which agrees with f_s above)

167

Chapter 6.0

Example 2:

Calculate the shear flows in the web from x = 0 to x = 100 in. as shown below.

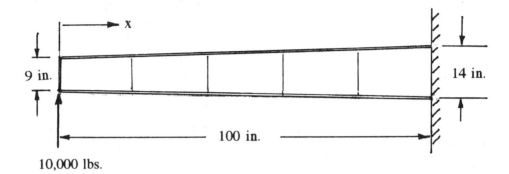

The solution lends itself to making a table by using Eq. 6.7.3 $V_w = V(\frac{h_o}{h})$:

x (in.)	h (in.)	$\frac{h_o}{h}$	V_w (lbs.)	$q = \frac{V_w}{h}$ (lbs./in.)
0	9	1	10,000	1,111.1
20	10	0.9	9,000	900
40	11	0.818	8,180	743.8
60	12	0.75	7,500	625
80	13	0.692	6,923	532.5
100	14	0.643	6,429	459.2

Example 3:

Find the shear flow in each web and draw a free-body diagram of elements. Assume caps resist axial load only and web is ineffective in bending.

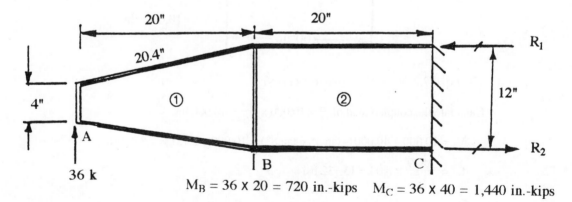

$M_B = 36 \times 20 = 720$ in.-kips $M_C = 36 \times 40 = 1,440$ in.-kips

(Given Beam and Load)

$R_1 = 36 \times \frac{40}{12} = 120$ kips and $R_2 = 120$ kips

168

Beam Stress

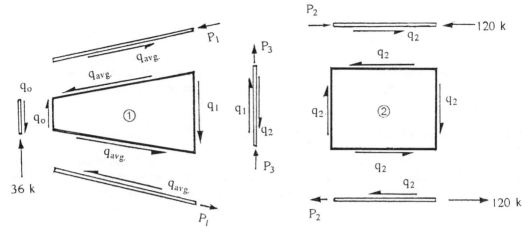

(Free-body Diagram)

$$q_o = \frac{36}{4} = 9 \text{ kips/in.} \quad \text{(From Eq. 6.7.5)}$$

$$q_1 = \frac{36(\frac{4}{12})}{12} = 1.0 \text{ kips/in.} \quad \text{(From Eq. 6.7.6)}$$

$$q_2 = \frac{36}{12} = 3 \text{ kips/in.}$$

or $\quad q_2 = \frac{2P_3}{12} + q_1 = 2 \times \frac{12}{12} + 1.0 = 3 \text{ kips/in.}$

$$P_1 = (\frac{36}{2})(\frac{12-4}{12})(\frac{20.4}{4}) = 61.2 \text{ kips}$$

$$q_{avg} = \frac{61.2}{20.4} = 3 \text{ kips/in.}$$

$$P_2 = 61.2(\frac{20}{20.4}) = 60 \text{ kips}$$

$$P_3 = 61.2(\frac{4}{20.4}) = 12 \text{ kips}$$

6.8 TORSION

For all uniform straight members in pure torque for unrestrained torsion in the elastic range, the applied torque is balanced by the section itself. Twisting of any section along the member is about a longitudinal axis of rotation passing through the shear center.

   (a)   The rate of change of the angle of twist, θ, is constant along the member and given in radians,

$$\theta = \frac{TL}{GJ} \qquad \text{Eq. 6.8.1}$$

        where: T – Applied torsional moment
                     L – Length of member
                     G – Modulus of rigidity
                     J – Torsional constant (see Fig. 6.8.3) dependent on the form of the cross section; use polar moment of inertia for circular sections

(b) The torsional shear stress, f_{ts}, at any section along the member is

$$f_{ts} = \frac{T}{J}$$ Eq. 6.8.2

For the torsional equations given above it is assumed that the material is stressed in the elastic range and Hooke's law applies. Stress in the plastic range will not be discussed because applications in airframe structure are rare.

(A) SOLID SECTIONS

When a solid circular section is twisted about its own axis, all sections remain plane. No warping occurs on the member. The torsional shear stress distribution is linear along any radial line from the axis of rotation and the maximum shear stress occurs at the outer surface, see Fig. 6.8.1.

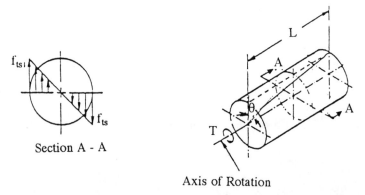

Fig. 6.8.1 Torsional Shear Stress Distribution - Circular Section

(B) NONCIRCULAR SECTIONS

When a noncircular solid section is twisted about its own axis, the sections warp with respect to their original planes, see Fig. 6.8.2. The torsional shear stress distribution is nonlinear with the following exceptions:

- At any point where the outer surface is normal to an axis of symmetry, the distribution is linear
- The maximum shear stress occurs at a point on the surface closest to the axis of rotation

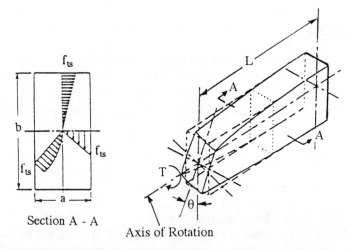

Fig. 6.8.2 Torsional Shear Stress Distribution - Noncircular Section

Beam Stress

The torsional shear stress, f_{ts}, of a solid section is

$$f_{ts} = \frac{nT}{J}$$
Eq.6.8.3

where: n – A coefficient related to the cross-sectional shape and the point at which the stress is to be determined, see Fig. 6.8.3.

Expressions for the torsional constant, J, and the torsional shear stress, f_{ts}, of solid sections are given in Fig. 6.8.3. The expression for $f_{ts,max}$ includes the specific expressions for value of n that are required to produce the maximum shear stress for each section.

Solid sections are included as thick-walled closed sections since their torsional shear stress distribution follows that of solid sections.

CROSS SECTION	TORSIONAL CONSTANT J	TORSIONAL SHEAR STRESS $f_{ts} = \frac{T}{J} n$
(circle, R, c, R_c)	$J = \frac{\pi R^4}{2}$	$f_{ts,max} = \frac{T}{J} R_c$ at the radius R_c $f_{ts,max} = \frac{T}{J} R$ at the outer surface
(hollow circle, R, R_c, R_1)	$J = \frac{\pi (R^4 - R_1^4)}{2}$	$f_{ts,max} = \frac{T}{J} R_c$ at the radius R_c $f_{ts,max} = \frac{T}{J} R$ at the outer surface
(ellipse, a, b, c)	$J = \frac{\pi a^3 b^3}{a^2 + b^2}$	$f_{ts,max} = \frac{2T}{\pi ab^2}$ at point c
(rectangle, a, b, b/2, c)	$J = k_1 a^3 b$ k_1 and k_2 are given below.	$f_{ts,max} = \frac{T}{k_2 a^2 b}$ at point c

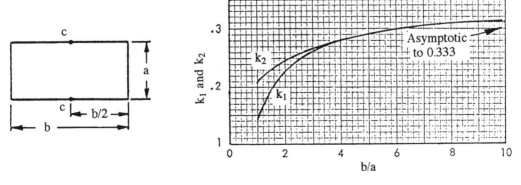

Coefficients k_1 and k_2 for Rectangular Sections

Fig. 6.8.3 Torsional Constant (J) and Torsional Shear Stress (f_{ts}) – Solid Sections

(B) THIN-WALLED CLOSED SECTIONS

Torsion of thin-walled closed section members is balanced at any section by a uniform shear flow 'q' around the median line of the section, see Fig. 6.8.4. The stress is assumed to be constant across the thickness at any point on the section.

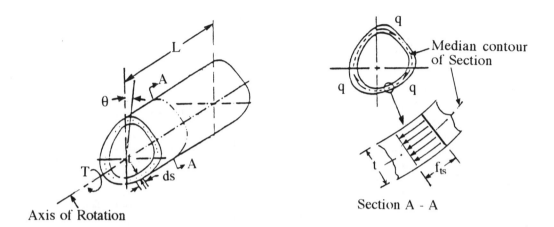

Fig. 6.8.4 Torsion of Thin-Walled Closed Sections

Maximum stress occurs at the thinnest point on the section. The torsional constant, J, for all thin-walled closed sections subjected to a torque, T, is given by:

$$J = \frac{4A^2}{\int \frac{ds}{t}} = \frac{4A^2}{\sum_{i=1}^{n} \frac{b_i}{t_i}}$$
Eq.6.8.4

where: A – The area enclosed by the median contour
         b – The segment length of the median contour
         ds – Length of a differential element of the median contour
         t – Wall thickness

The shear flow, q, and the torsional shear stress, f_{ts}, are

$$q = \frac{T}{2A}$$
Eq.6.8.5

and

$$f_{ts} = \frac{T}{2At}$$
Eq.6.8.6

Expressions for the torsional constant J and the torsional shear stress f_{ts} of thin-walled closed sections are given in Fig. 6.8.5. Thin-walled closed sections with stiffeners, or with more than one cell, are covered in Chapter 8.0.

Beam Stress

CROSS SECTION	TORSIONAL CONSTANT J	TORSIONAL SHEAR STRESS $f_{ts} = \dfrac{T}{2At}$
Circular tube, R to midpoint of t, t is constant	$J = 2\pi R^3 t$	$f_{ts} = \dfrac{T}{2\pi R^2 t}$
Elliptical tube, a and b to midpoint of t, t is constant	$J = \dfrac{4\pi a^2 b^2 t}{\sqrt{2(a^2+b^2)}}$	$f_{ts} = \dfrac{T}{2\pi abt}$
Rectangular tube with t_1, t_2, a, b	$J = \dfrac{2t_1 t_2 a^2 b^2}{at_2 + bt_1}$	$f_{ts1} = \dfrac{T}{2abt_1} \qquad f_{ts2} = \dfrac{T}{2abt_2}$
If $t_1 = t_2 = t$:		
	$J = \dfrac{2ta^2 b^2}{a+b}$	$f_{ts} = \dfrac{T}{2abt}$
Arbitrary closed contour	$J = \dfrac{4A^2}{\int \dfrac{ds}{t}} = \dfrac{4A^2}{\sum_{i=1}^{n} \dfrac{b_i}{t_i}}$	$f_{tsp} = \dfrac{T}{2At_p}$ at point c
	A = Area enclosed by the median contour b = Length of median contour	
	If t is constant:	
	$J = 4A^2 \dfrac{t}{b}$	$f_{ts} = \dfrac{T}{2At}$ at any point

1. All dimensions defining the shape of a section are taken at the median contour.
2. It is assumed that sufficient stiffness exists at the ends of the member so that a section will maintain its geometric shape during torsion.

Fig. 6.8.5 Torsional Constant (J) and Torsional Shear Stress (f_{ts}) – Thin Walled Sections

Example 1:

A thin-walled extruded aluminum section shown below is subjected to a torque of T = 8,000 in.-lbs., G = 4 x 10^6 psi. and member length L = 12 in.

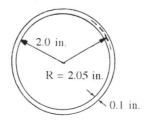

173

Chapter 6.0

(a) Find the max. torsional shear stress using equation of $f_{ts} = \dfrac{Tr}{J}$

$$J = \dfrac{\pi}{2}(R_o^4 - R_i^4) = \dfrac{\pi}{2}(2.1^4 - 2.0^4) = 5.42 \text{ in.}^4$$

$$f_{ts} = \dfrac{8,000 \times 2.1}{5.42} = 3,100 \text{ psi}$$

(b) Find the max. torsional shear stress using equation of $f_{ts} = \dfrac{T}{2At}$

$$f_{ts} = \dfrac{8,000}{2 \times \pi \times 2.05 \times 0.1} = 3,030 \text{ psi}$$

(c) The angle of twist using equation of $\theta = \dfrac{TL}{GJ}$

$$\theta = \dfrac{8,000 \times 12}{(4 \times 10^6)(5.42)} = 0.0044 \text{ rad.}$$

(d) The angle of twist using equation of $\theta = \dfrac{TLS}{4A^2Gt}$ (S = circumference)

$$\theta = \dfrac{(8,000 \times 12)(2 \times 2.05 \times \pi)}{4(2.05^2 \times \pi)^2(4 \times 10^6)(0.1)} = 0.0044 \text{ rad.}$$

Example 2:

The tubular rectangular section shown is subjected to an external torsional moment of T = 80,000 in.-lbs. and assume G = 4 × 10⁶ psi.

Calculate the torsional shear stress in each side and the angle of twist per unit length.

The solution lends itself to making a table :

Segment	ds (in.)	t (in.)	$\dfrac{ds}{t}$
①	25.25	0.04	631
②	15.7	0.05	314
③	25.3	0.032	791
④	13.4	0.04	335

$$\Sigma \dfrac{ds}{t} = 2,071$$

ds = 25.25"
① t = .050
② t = .040
④ t = .040
③ t = .032
ds = 13.40"
ds = 15.70"
ds = 25.30"

$$J = \dfrac{4A^2}{\Sigma \dfrac{ds}{t}} = \dfrac{4 \times 387.4^2}{2.071} = 290 \text{ in.}^4$$

$$q = \dfrac{T}{2A} = \dfrac{80,000}{2 \times 387.4} = 103 \text{ lbs./in.}$$

The shear flow is equal on all sides,

Segment	q (lbs./in.)	t (in.)	$f_{ts} = \dfrac{q}{t}$ (psi)
①	103	0.04	2,575
②	103	0.05	2,060
③	103	0.032	3,220
④	103	0.04	2,575

The angle of twist per unit length (assume L = 1.0 in.):

$$\theta/in. = \frac{TL}{GJ} = \frac{80,000 \times 1.0}{(4 \times 10^6)(290)} = 0.000069 \text{ rad./in.}$$

(C) THIN-WALLED OPEN SECTIONS

Torsion of a thin-walled open section is balanced at any section by a shear flow 'q' about the periphery of the section, see Fig. 6.8.6. The stress distribution is assumed to be linear across the thickness of the section. The stress at one edge is equal and opposite in direction to the stress at the other edge.

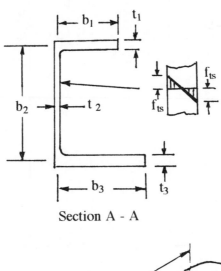

Section A - A

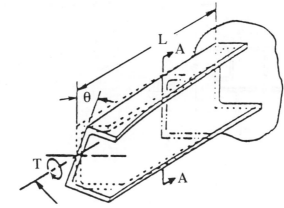

Axis of Rotation

Fig. 6.8.6 Torsion of Thin-Walled Open Section

The torsional shear stress, f_{ts}, for thin-walled open section is obtained by:

$$f_{ts} = \frac{Tt}{J} \qquad \text{Eq.6.8.7}$$

$$J = \frac{1}{3}(\sum_{i=1}^{n} b_i t_i^3) \qquad \text{Eq.6.8.8}$$

where: b_i – Length of ith element
t_i – Thickness of ith element

Expressions for the torsional constant, J, and the torsional shear stress, f_{ts}, of specific configurations of thin-walled open sections are given in Fig. 6.8.7.

Chapter 6.0

CROSS SECTION	TORSIONAL CONSTANT J	TORSIONAL SHEAR STRESS $f_{ts} = \dfrac{T}{J} t$
Thin Rectangular 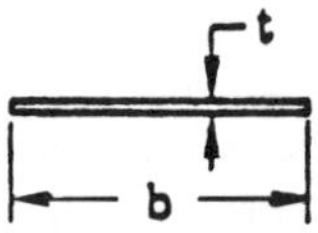	$J = \dfrac{1}{3} b t^3$	$f_{ts} = \dfrac{3T}{b t^2}$
Thin Circular 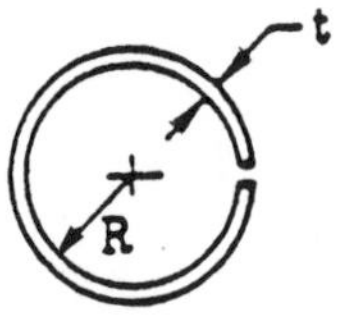 R = Mean radius	$J = \dfrac{2}{3} \pi R t^3$	$f_{ts} = \dfrac{3T}{2 \pi R t^2}$
Any thin formed section 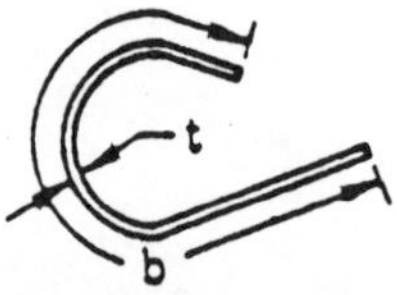	$J = \dfrac{1}{3} b t^3$	$f_{ts} = \dfrac{3T}{b t^2}$
Sections composed of 2 or more thin rectangular elements 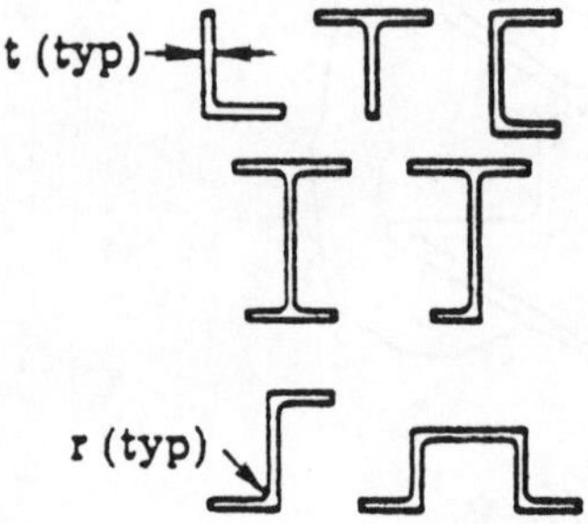	$J = \dfrac{1}{3} \sum_{i=1}^{n} b_i t_i^3$	At any element $i = j$ $f_{ts,j} = \dfrac{T}{J} t_j$ At $t_i = t_{max}$ $f_{ts,max} = \dfrac{T}{J} t_{max}$

Fig. 6.8.7 Torsional Constant (J) and Torsional Shear Stress (f_{ts}) – Thin-Walled Open Sections

Example 1:

The section shown below is subjected to a torque of T = 1,000 in.-lbs. and it is assumed that G = 4×10^6 psi. Calculate the max. torsional shear stress and the angle of twist per unit length.

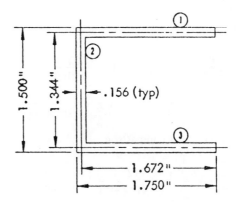

The solution lends itself to making a table :

Segment	b (in.)	t (in.)	bt³
①	1.672	0.156	0.00635
②	1.344	0.156	0.0051
③	1.672	0.156	0.00635
			$\Sigma bt^3 = 0.0178$

$$J = \frac{\Sigma bt^3}{3} = \frac{0.0178}{3} = 0.00593$$

The max. torsional shear stress:

$$f_{ts,max} = \frac{Tt}{J} = \frac{1,000 \times 0.156}{0.00593} = 26,307 \text{ psi}$$

The angle of twist per unit length (L = 1.0 in.):

$$\theta/\text{in.} = \frac{TL}{GJ} = \frac{1,000 \times 1.0}{(4 \times 10^6)(0.00593)} = 0.042 \text{ rad./in.}$$

Example 2:

The round tube below has an average radius of (R) and wall thickness (t). Assume the ratio of $\frac{R}{t} = 20$ and tube length L = 1.0.

Compare the torsional strength and the stiffness of this tube with a tube with identical dimensions except for a small slit as shown. Both tubes are thin-walled, so work with the centerline dimensions.

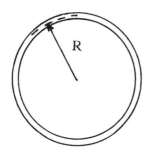
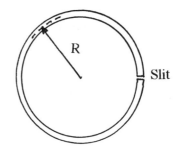

Chapter 6.0

Non-Slit Tube	Tube with Slit
$J = 2\pi R^3 t$	$J = \dfrac{2\pi R t^3}{3}$
$f_{ts} = \dfrac{TR}{J} = \dfrac{T}{2\pi R^2 t}$	$f_{ts} = \dfrac{3T}{(2\pi R)t^2}$
$\theta/\text{in.} = \dfrac{T}{GJ} = \dfrac{T}{(2\pi R^3 t)G}$	$\theta/\text{in.} = \dfrac{3T}{(2\pi R t^3)G}$

(a) The torsional shear stress comparison:

$$\frac{f_{ts}\,(\text{non-slit})}{f_{ts}\,(\text{with slit})} = \frac{\dfrac{T}{2\pi R^2 t}}{\dfrac{3T}{2\pi R t^2}} = \frac{T(2\pi R t^2)}{3T(2\pi R^2 t)} = \frac{t}{3R} = \frac{1}{60}$$

Note: The torsional shear stress in the slit tube is 60 times higher than the closed tube.

(b) The torsional stiffness comparison:

$$\frac{\theta\,(\text{non-slit})}{\theta\,(\text{with slit})} = \frac{\dfrac{TL}{(2\pi R^3 t)G}}{\dfrac{3TL}{(2\pi R t^3)G}} = \frac{t^2}{3R^2} = \frac{1}{3(20)^2} = \frac{1}{1,200}$$

Note: For a given torque, the open tube rotates 1,200 times as much as the closed tube.

6.9 CRUSHING LOAD ON BOX BEAMS

The crushing load occurs on a beam bending due to its curvature, as shown in Fig. 6.9.1. This load can be ignored on solid beams but not on box beams.

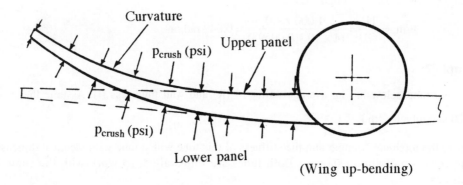

Fig. 6.9.1 Crushing Load

When an up-bending moment acts on a box beam, the upper panel is in compression which produces downward crushing load normal to the upper surfaces; the opposite occurs on the lower surface. In a box beam, these crushing loads are reacted by intermediate ribs (vertical stiffeners).

$$p_{crush} = \frac{2N^2}{E t_e h_c} \quad (\text{psi}) \qquad \text{Eq. 6.9.1}$$

$$\text{or} \quad p_{crush} = \frac{2f^2 t_e}{E h_c} \quad (\text{psi}) \qquad \text{Eq. 6.9.2}$$

$$\text{or} \quad p_{crush} = \frac{t_e h_c M^2}{2 E I^2} \quad (\text{psi}) \qquad \text{Eq. 6.9.3}$$

where: N – Load intensity in pounds per inch width (lbs./in.)
E – Panel material of modulus of elasticity (psi)
t_e – Equivalent panel thickness including skin and stringer per inch width (in.)
h_c – Beam depth between the centroids of the upper and lower panels (in.)
f – Applied stress in panel (psi)
M – Applied moment (in.-lbs.) at the beam cross section
I – Moment of inertia of the total beam cross section (in.⁴)

Actually the crushing load (psi) along the chordwise length of a wing box is not constant because the beam axial stress (f) or the load intensity (N) varies (high at rear spar and low in front spar); therefore, Eq. 6.9.1 and Eq. 6.9.2 should be used. Eq. 6.9.3 is used at the preliminary design stage due to the lack of chordwise variation of f or N.

Example:

Assume an ultimate load intensity N = 20,000 lbs./in. at the front spar and N = 28,000 lbs./in. at the rear spar. Determine the wing crushing load (p_{crush}) and the running load (w) at the rib.

$E = 10.7 \times 10^6$ psi

$t_e = 0.3"$

$h_c = 22"$ (at front spar)

$h_c = 20"$ (at rear spar)

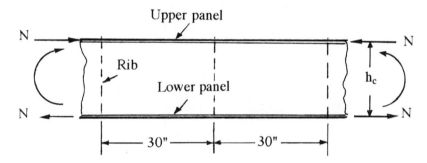

From Eq. 6.9.1:

$$p_{crush} = \frac{2N^2}{E t_e h_c} = \frac{2(20,000)^2}{10.7 \times 10^6 \times 0.3 \times 22} = 11.33 \text{ psi} \quad \text{(at front spar)}$$

$$p_{crush} = \frac{2N^2}{E t_e h_c} = \frac{2(25,000)^2}{10.7 \times 10^6 \times 0.3 \times 20} = 19.47 \text{ psi} \quad \text{(at rear spar)}$$

Assume the rib spacing is 30" and the running load (w – lbs./in.) along the wing chord length of the rib is:

$w_{front} = 11.33 \times 30 = 340$ lbs./in.

$w_{rear} = 19.47 \times 30 = 584$ lbs./in.

Chapter 6.0

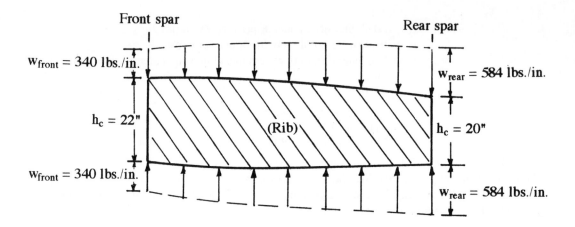

References

6.1 Timoshenko, S. and Young, D. H., "Elements of Strength of Materials", D. Van Nostrand Company, Inc., New York, NY, (1962).

6.2 Roark, R. J. and Young, W. C., "FORMULAS FOR STRESS AND STRAIN", McGraw-Hill Book Company, New York NY, (1975, 5th edition).

6.3 Bruhn, E. F., "ANALYSIS AND DESIGN OF FLIGHT VEHICLE STRUCTURES", Jacobs Publishers, 10585 N. Meridian St., Suite 220, Indianapolis, IN 46290, (1975).

6.4 Cozzone, F. P., "Bending Strength in the Plastic Range", Journal of the Aeronautical Science, (May, 1943).

Chapter 7.0

PLATES AND SHELLS

7.1 INTRODUCTION

The plate shown in Fig. 7.1.1 is a two-dimensional counterpart of the beam, in which transverse loads are resisted by flexural and shear stresses, with no direct stresses in its middle plane. The thin flat sheet, by deflecting enough to provide both the necessary curvature and stretch, may develop membrane stresses to support lateral loads. In the analysis of these situations no bending stress is presumed in the sheet.

The skin may also be classified as either a plate or a membrane depending upon the magnitude of transverse deflections under loads. Transverse deflections of plates are small in comparison with the plates' thickness, i.e., on the order of a tenth to half of the plate thickness. On the other hand, the transverse deflections of a membrane will be on the order of ten times of its thickness. However, most airframe skins fall between the above two extremes and hence behave as plates having some membrane stresses.

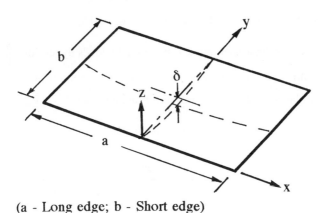

(a - Long edge; b - Short edge)

Fig. 7.1.1 Plate Subjected Normal Loading

The pressurized cabin of a modern aircraft, such as cylindrical fuselage and pressure dome, is a pressure vessel. Shells that have the form of surfaces of revolution find extensive application in various kinds of containers, tanks and domes. A surface of revolution is obtained by rotation of a plane curve about an axis lying in the plane of the curve. This curve is called the meridian and its plane is a meridian plane. The position of a meridian is defined by an angle θ and the position of a parallel circle is defined by the angle φ as shown in Fig. 7.1.2.

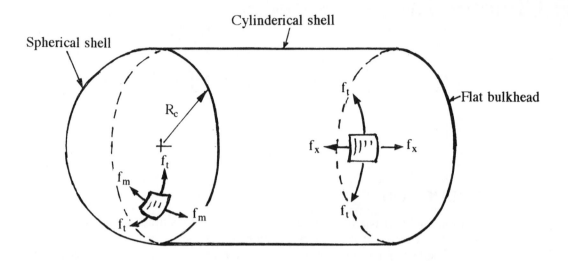

Fig. 7.1.2 Shells and Flat Panel Subjected to Internal Pressure

7.2 PLATES

Approximate criteria for classifying a plate subjected to normal loads may be classified in three groups based on relation of maximum deflection (δ) vs. thickness (t) :

- Thick plate – supports the load by bending ($\frac{\delta}{t} < 0.5$).

- Membrane (or diaphragm) plate – supports the loads by tension ($\frac{\delta}{t} > 5$).

- Thin plate – supports the load by both bending and tension and this plate is the primary class encountered in airframe structures ($0.5 < \frac{\delta}{t} < 5$).

Corresponding characteristics from engineering beam theory and plate analysis are compared below:

	Simple Beam	Plate
Dimensions	x-axis	x and y-axes
Poisson's ratio (μ)	No affect	Affect
Stiffness	EI	$D = \frac{Et^3}{12(1-\mu^2)}$
Moments	M_x	M_x and M_y
Shears	V_x	V_x and V_y

Boundary conditions for thin plates in airframe structures usually fall between hinged edges (simply supported) and fixed edges (built-in edges). It is recommended that in most cases an average of the two conditions be taken.

(A) THICK PLATE

In deriving the thick plate bending equations shown in Fig. 7.2.1, it was assumed that no stresses acted in the middle (neutral) plane of the plate (no membrane stresses). However, practically all loaded plates deform into surfaces which induce some middle surface stresses. It is the necessity to hold down the magnitude of these very powerful middle surface stretching forces that results in the severe rule-of-thumb restriction that plate bending formulae apply accurately only to problems in which deflections are a few tenths of the plate thickness.

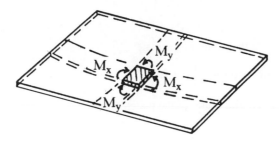

Fig. 7.2.1 Thick Plate Subjected Normal Loading

A common case is rectangular plates supporting uniform loads, p (psi). The major engineering results are the values of the max. deflections (for stiffness) and the max. stresses (for strength). These may be put in the form shown below (value "b" is the length of the short edge and assume Poisson's ratio $\mu = 0.3$):

Case 1 – Plate with 4 hinged edges under uniform load (p):

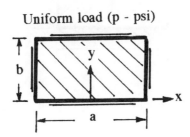

Max. bending stress at the center of the plate:

$$f = \beta (p \frac{b^2}{t^2}) \text{ (psi)} \qquad \text{Eq. 7.2.1}$$

Max. deflection at the center of the plate:

$$\delta = \alpha (p \frac{b^4}{Et^3}) \text{ (in.)} \qquad \text{Eq. 7.2.2}$$

Max. shear force at center of the long edge 'a':

$$V = \kappa\, p\, b \text{ (lbs./in.)} \qquad \text{Eq. 7.2.3}$$

Where the coefficients α, β and κ are given in Fig. 7.2.2

Case 2 – Plate with 4 hinged edges under concentrated load (P_z) at center (uniform over small concentric circle of radius r_o which is greater than 0.5t):

Chapter 7.0

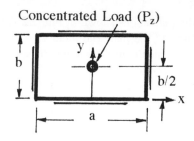

Bending stress at the center of the plate:

$$f = \left(\frac{3P_z}{2\pi t^2}\right)\left[(1+\mu)\ln\left(\frac{2b}{\pi r_o}\right) + \beta'\right] \qquad \text{Eq. 7.2.4}$$

Max. deflection at the center of the plate:

$$\delta = \alpha'\left(\frac{P_z b^2}{E t^3}\right) \text{ (in.)} \qquad \text{Eq. 7.2.5}$$

Where the coefficients α' and β' are given in Fig. 7.2.2

Case 3 – Plate with 4 fixed edges under uniform load (p):

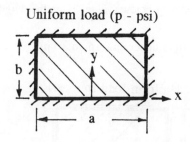

Max. bending stress at the center of the long edge 'a':

$$f = -\beta_1\left(p\frac{b^2}{t^2}\right) \text{ (psi)} \qquad \text{Eq. 7.2.6}$$

Bending stress at the center of the plate:

$$f = \beta_2\left(p\frac{b^2}{t^2}\right) \text{ (psi)} \qquad \text{Eq. 7.2.7}$$

Max. deflection at the center of the plate:

$$\delta = \alpha\left(p\frac{b^4}{E t^3}\right) \text{ (in.)} \qquad \text{Eq. 7.2.8}$$

Where the coefficients α, β_1 and β_2 are given in Fig. 7.2.3

Case 4 – Plate with 4 fixed edges under concentrated load (P_z) at center (uniform over small concentric circle of radius r_o which is greater than 0.5t):

Plate and Shell Stress

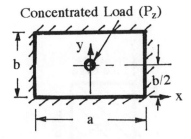

Bending stress at the center of the plate:

$$f = \left(\frac{3P_z}{2\pi t^2}\right)\left[(1+\mu)\ln\left(\frac{2b}{\pi r_o}\right) + \beta_1'\right]$$
Eq. 7.2.9

Bending stress at the center of the long edge 'a':

$$f = -\beta_2'\left(\frac{P_z}{t_2}\right) \text{ (psi)}$$
Eq. 7.2.10

Max. deflection at the center of the plate:

$$\delta = \alpha'\left(\frac{P_z b^2}{E t^3}\right) \text{ (in.)}$$
Eq. 7.2.11

Where the coefficients α', β_1' and β_2' are given in Fig. 7.2.3

Case 5 – Plate with 3 fixed edges and 1 free edge under uniform load (p):

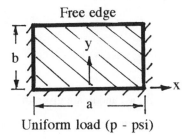

Max. bending stress at x = 0 and y = 0:

$$f = -\beta_1\left(p\frac{b^2}{t^2}\right) \text{ (psi)}$$
Eq. 7.2.12

Shear force at x = 0 and y = 0:

$$V = \kappa_1 p b \text{ (lbs./in.)}$$
Eq. 7.2.13

Bending stress at free edges, x = 0 and y = b:

$$f = \beta_2\left(p\frac{b^2}{t^2}\right) \text{ (psi)}$$
Eq. 7.2.14

Bending stress at free edges, $x = \pm\frac{a}{2}$ and y = b:

$$f = -\beta_3\left(p\frac{b^2}{t^2}\right) \text{ (psi)}$$
Eq. 7.2.15

Max. shear force at free edges, $x = \pm\frac{a}{2}$ and y = b:

$$V_a = \kappa_2 p b \text{ (lbs./in.)}$$
Eq. 7.2.16

Where the coefficients β_1, β_2, β_3, κ_1, and κ_2 are given in Fig. 7.2.4

Case 6 — Plate with 2 fixed edges and 2 free edges under uniform load (p):

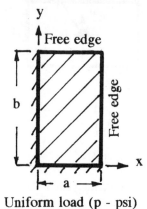

Uniform load (p - psi)

Max. bending stress at $x = a$ and $y = 0$:

$$f = -\beta_1 (p \frac{b^2}{t^2}) \text{ (psi)} \qquad \text{Eq. 7.2.17}$$

Shear force at $x = a$ and $y = 0$:

$$V = \kappa_1 \, p \, b \text{ (lbs./in.)} \qquad \text{Eq. 7.2.18}$$

Bending stress at free edges, $x = 0$ and $y = b$:

$$f = -\beta_2 (p \frac{b^2}{t^2}) \text{ (psi)} \qquad \text{Eq. 7.2.19}$$

Max. shear force at free edges, $x = 0$ and $y = b$:

$$V = \kappa_2 \, p \, b \text{ (lbs./in.)} \qquad \text{Eq. 7.2.20}$$

Where the coefficients β_1, β_2, κ_1, and κ_2 are given in Fig. 7.2.5

Case 7 — Cantilever plate of infinite length with a concentrated load (P_z) at $y = c$ and $x = 0$:

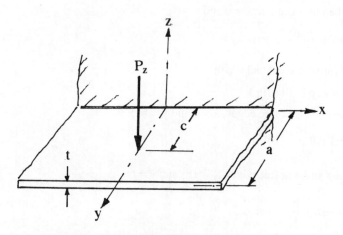

Bending stress at $x = 0$ and $y = y$:

$$f = k_m (\frac{6P_z}{t^2}) \text{ (psi)} \qquad \text{Eq. 7.2.21}$$

Deflection along the free edge, y = a and x = x (assume $\mu = 0.3$):

$$\delta = 3.476k_z\left(\frac{P_z a^2}{Et^3}\right) \text{ (in.)} \qquad \text{Eq. 7.2.22}$$

Where the coefficients k_m and k_z are given in Fig. 7.2.6

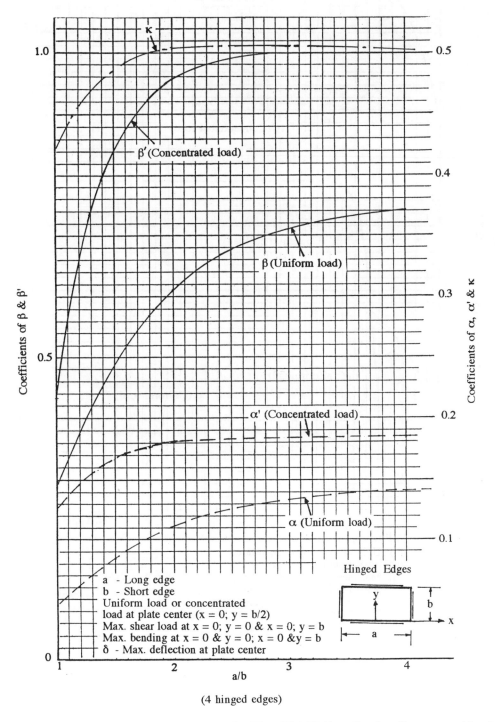

Fig. 7.2.2 *Coefficients α, β and κ for Rectangular Plate With Uniform Load or Concentrated load*

Chapter 7.0

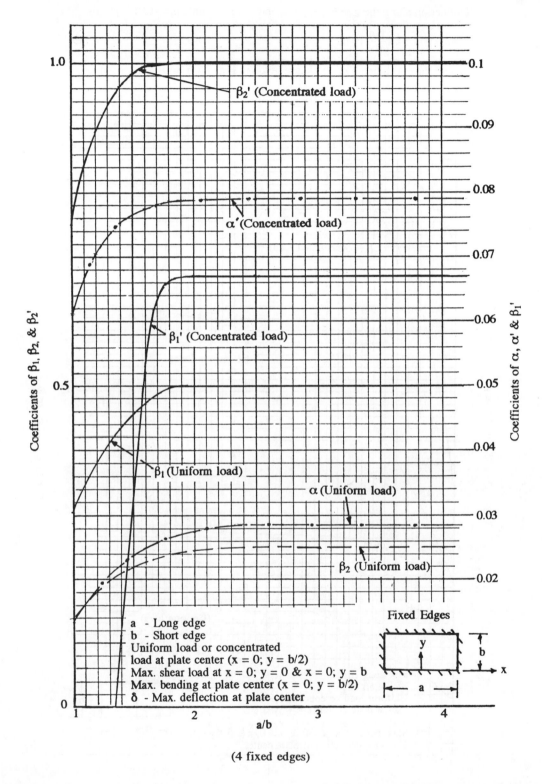

Fig. 7.2.3 Coefficients α, β and κ for Rectangular Plate With Uniform Load and Concentrated load

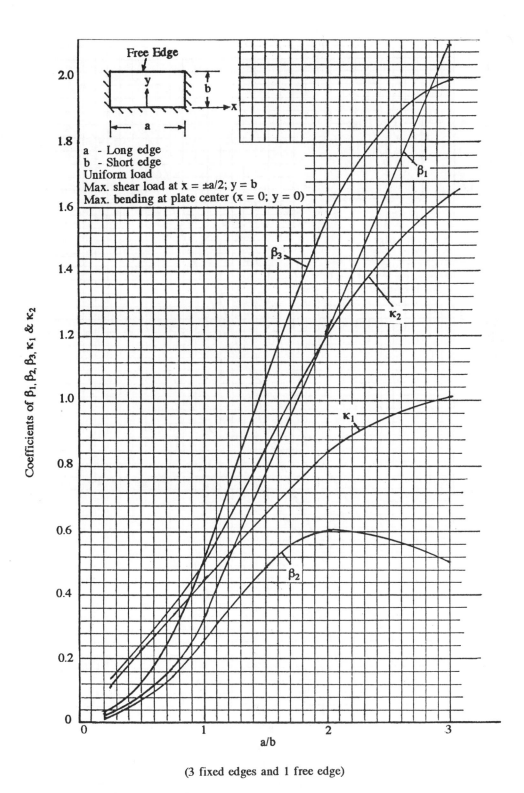

(3 fixed edges and 1 free edge)

Fig. 7.2.4 Coefficients α, β and κ for Rectangular Plate With Uniform Load

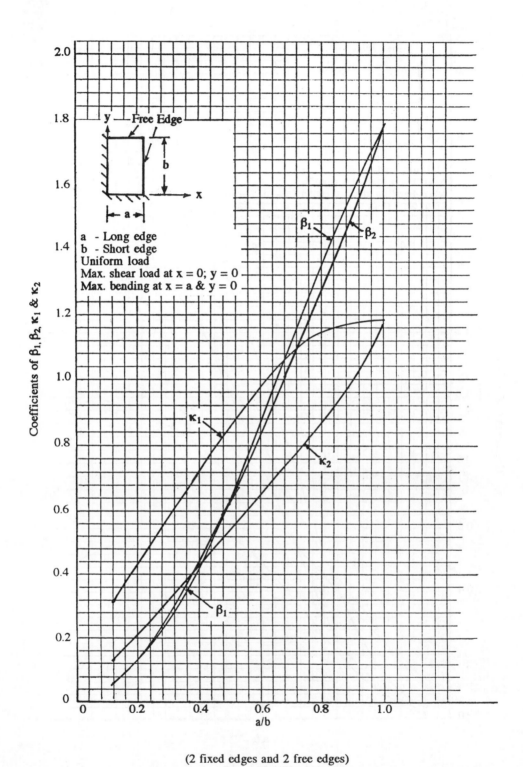

(2 fixed edges and 2 free edges)

Fig. 7.2.5 Coefficients α, β and κ for Rectangular Plate With Uniform Load

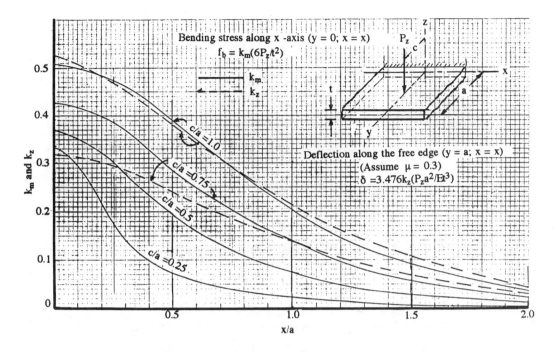

Fig. 7.2.6 Coefficients k_m and k_z for a Cantilever Plate of Infinite Length

Example:

Assume an aluminum plate ($E = 10 \times 10^6$ psi and $\mu = 0.3$), $a = 7.5''$ and $b = 5''$ with thickness $t = 0.1''$ under a uniform load $p = 10$ psi or concentrated load $P_z = 1,000$ lbs. Determine its max. stresses and deflections with following edge conditions:

(a) The max. bending stresses under uniform load of 10 psi:

4 hinged edges (Eq. 7.2.1 from Case 1 and β from Fig. 7.2.2):

$$f = \beta(p\frac{b^2}{t^2}) \text{ at the center of the plate}$$

$$= 0.485(10 \times \frac{5^2}{0.1^2}) = 12,125 \text{ psi}$$

4 fixed edges (Eq. 7.2.6 from Case 3 and β_1 from Fig. 7.2.3):

$$f = -\beta_1(p\frac{b^2}{t^2}) \text{ at the center of the long edge 'a'}$$

$$= -0.455(10 \times \frac{5^2}{0.1^2}) = -11,375 \text{ psi}$$

3 fixed edges and 1 free edge (Eq. 7.2.12 from Case 5 and β_1 from Fig. 7.2.4):

$$f = -\beta_1(p\frac{b^2}{t^2}) \text{ at } x = 0 \text{ and } y = 0$$

$$= -0.77(10 \times \frac{5^2}{0.1^2}) = -19,250 \text{ psi}$$

2 fixed edges and 2 free edges (Eq. 7.2.17 from Case 6 and β_1 from Fig. 7.2.5):

Chapter 7.0

$$f = -\beta_1(p\frac{b^2}{t^2}) \text{ at } x = a \text{ and } y = 0$$

$$= -1.17(10 \times \frac{5^2}{0.1^2}) = -29,250 \text{ psi}$$

(b) The max. deflections under uniform load of 10 psi:

4 hinged edges (Eq. 7.2.2 from Case 1 and α from Fig. 7.2.2):

$$\delta = \alpha(p\frac{b^4}{Et^3}) \text{ at the center of the plate}$$

$$= 0.085(10 \times \frac{5^4}{10 \times 10^6 \times 0.1^3}) = 0.053 \text{ in.}$$

4 fixed edges (Eq. 7.2.8 from Case 3 and α from Fig. 7.2.3):

$$\delta = \alpha(p\frac{b^4}{Et^3}) \text{ at the center of the plate}$$

$$= 0.024(10 \times \frac{5^4}{10 \times 10^6 \times 0.1^3}) = 0.015 \text{ in.}$$

(c) The max. bending stresses under concentrated load of $P_z = 375$ lbs (assume $r_o = 0.25$):

4 hinged edges (Eq. 7.2.4 from Case 2 and β' from Fig. 7.2.2):

$$f = (\frac{3P_z}{2\pi t^2})[(1+\mu)\ln(\frac{2b}{\pi r_o}) + \beta_1'] \text{ at the center of the plate}$$

$$= (\frac{3 \times 375}{2\pi 0.1^2})[(1+0.3)\ln(\frac{2 \times 5}{\pi \times 0.25}) + 0.84]$$

$$= (17,905)[(1.3)\ln(12.733) + 0.84] = 74,260 \text{ psi}$$

4 fixed edges (Eq. 7.2.9 from Case 4 and β_1' from Fig. 7.2.3):

$$f = (\frac{3P_z}{2\pi t^2})[(1+\mu)\ln(\frac{2b}{\pi r_o}) + \beta_1'] \text{ at the center of the plate}$$

$$= (\frac{3 \times 375}{2\pi 0.1^2})[(1+0.3)\ln(\frac{2 \times 5}{\pi 0.25}) + 0.032]$$

$$= (17,905)[(1.3)\ln(12.733) + 0.032] = 59,793 \text{ psi}$$

(d) The max. deflection under concentrated load of $P_z = 375$ lbs:

4 hinged edges (Eq. 7.2.5 from Case 2 and α' from Fig. 7.2.2):

$$\delta = \alpha'(\frac{P_z b^2}{Et^3}) \text{ at the center of the plate}$$

$$= 0.166(375 \times \frac{5^2}{10 \times 10^6 \times 0.1^3}) = 0.156 \text{ in.}$$

4 fixed edges (Eq. 7.2.11 from Case 4 and α' from Fig. 7.2.3):

$$\delta = \alpha'(\frac{P_z b^2}{Et^3}) \text{ at the center of the plate}$$

$$= 0.0765(375 \times \frac{5^2}{10 \times 10^6 \times 0.1^3}) = 0.053 \text{ in.}$$

(B) MEMBRANE (DIAPHRAGM) PLATE

It is useful to consider the limiting case of the flat membrane (diaphragm) plate which cannot support any of the lateral load by bending stresses and hence has to deflect and stretch to develop both the necessary curvatures and membrane stresses (tension stress) see Fig. 7.2.7. If the plate is made so thin as to necessitate supporting all pressure loading:

- It stretches and developes membrane stresses
- Permanent plate deformation results in quilting which is not acceptable for airframe design
- Strong framework of stringers and/or bulkheads is required
- Localized pad-up where the plate is attached to the framework by fasteners is required

Structurally the most efficient form to handle membrane stress is the spherical and cylindrical shell in which the lateral pressure loads are supported by tensile stress alone in the curved walls. Since the membrane plate is seldom used on airframe applications it will not be discussed.

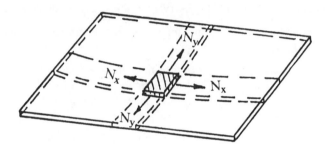

Fig. 7.2.7 Membrane Plate Subjected to Normal Loading

(C) THIN PLATE

The most efficient plate designs generally fall between the thick and membrane plates. The exact analysis of the two-dimensional plate which undergoes large deflections and thereby supports lateral loading partly by its bending resistance and partly by membrane action is very involved; see Fig. 7.2.8.

- An approximate solution of the large deflection plate can be obtained by combining the flat thick and membrane plate solutions

- No interaction between these stresses is assumed and since the analysis is non-linear, the result can be an approximation only

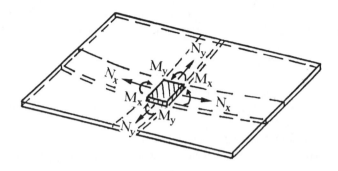

Fig. 7.2.8 Thin Plate Subjected to Normal Loading

The approximation equation is somewhat conservative. The pressure load coefficient ($\frac{pb^4}{Et^4}$) versus plate deflection (δ), membrane stress (f_d) and total stress (f = membrane + bending) are plotted in Fig. 7.2.9 and Fig. 7.2.10.

It is, in fact, useful to consider the analysis of the thin plate which cannot support any lateral loads by bending stresses alone. It has to deflect and stretch to develop the necessary curvatures to produce membrane stresses or a combination of partial bending and membrane (diaphragm) stresses to support the lateral load.

However, these presumed conditions are seldom encountered in practice for the following reasons:

(a) Since thin plate and membrane analyses are non-linear, it follows that the exact stresses cannot found by a superposition.

(b) If the skin is made so thin as to necessitate supporting all normal pressure loads by stretching and developing membrane stress, this results in permanent deformation which is not permitted in airframe design becasue it requires that structures have no permanent deformation under design limit loads.

(c) The heavy membrane tensile forces developed during large deflections will cause the surrounding supports to deflect towards each other thereby increasing the thin sheet skin deflection and relieving some of the stresses. In an actual design having the boundary conditions of elastic edge structural supports, such as a framework of stringers, a frame or bulkhead, errors on the order of 30% are likely if the framework elasticity is neglected in analysis.

(d) It is seldom that the design of a panel is such that the skin handles only normal pressure loads – most often the skin panel on the framework of stringers and frames or bulkheads must simultaneously transmit in-plane loading as well as normal pressure loads. Since thin plate analysis is a non-linear and cannot be calculated by a superposition method, the magnitude of the error is very difficult to estimate in the absence of an exact analysis. The alternative is expensive structural testing.

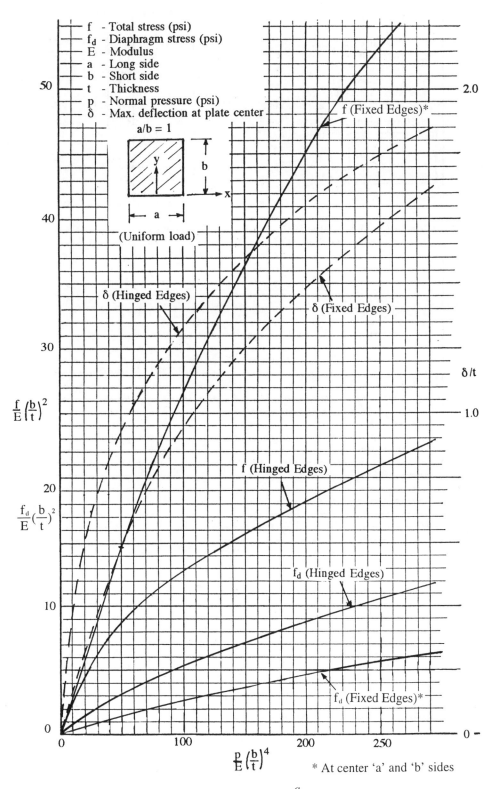

Fig. 7.2.9 Large Deflection of Square Plates ($\frac{a}{b} = 1.0$) under Uniform Load

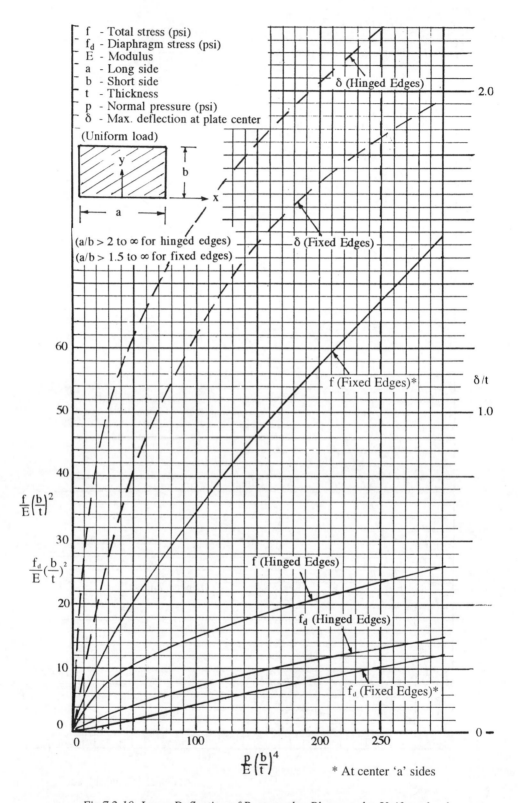

Fig.7.2.10 Large Deflection of Rectangular Plates under Uniform load

Example:

Given: Panel size: a = 20" long and b = 10" width, and thickness, t = 0.08".

Normal (lateral) pressure loading: p = 10.55 psi.

Material: Aluminum alloy, $E = 10.3 \times 10^6$ psi.

Find the max. deflection (δ), membrane or diaphragm stress (f_d) and total stress (f) at center of the plate.

(1) The panel aspect ratio: $\frac{a}{b} = \frac{20}{10} = 2$ and assume four edges are hinged supports.

(2) The pressure load coefficient, $\frac{pb^4}{Et^4} = \frac{(10.55)(10)^4}{(10.3 \times 10^6)(0.08)^4} = 250$.

(3) From Fig. 7.2.10, find:

The max. deflection: $\delta = 2.2\,t = 2.2 \times 0.08 = 0.176$ inch.

The membrane stress: $f_d = \frac{13.2\,(10.3 \times 10^6)(0.08)^2}{10^2} = 8,701$ psi

The total stress: $f = \frac{23.6\,(10.3 \times 10^6)(0.08)^2}{10^2} = 15,557$ psi

7.3 CYLINDRICAL SHELLS

Structurally, the most efficient form of pressure vessel is one in which the lateral pressures are supported by tensile stresses alone in the curved skins of the body, as shown in Fig. 7.3.1. The skins of these bodies have zero bending stiffness and hence have the properties of a membrane or diaphragm stress. The stresses developed, lying wholly in tangential directions at each point, are called membrane stresses.

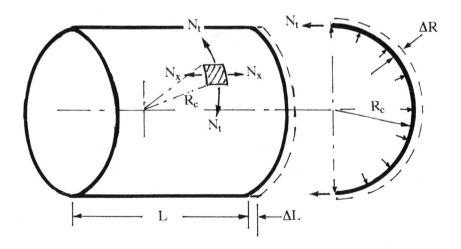

Fig. 7.3.1 Cylindrical Shell

Chapter 7.0

(A) BASIC EQUATIONS:

The hoop tension load: $\quad N_t = pR_c$ (lbs./in.) $\hspace{2cm}$ Eq. 7.3.1

The hoop tension stress: $\quad f_t = \dfrac{pR_c}{t}$ (psi) $\hspace{2cm}$ Eq. 7.3.2

Radial displacement: $\quad \Delta R = (\dfrac{pR_c^2}{Et})(1 - \dfrac{\mu}{2})$ (in.) $\hspace{2cm}$ Eq. 7.3.3

The longitudinal load: $\quad N_x = \dfrac{pR_c}{2}$ (lbs./in.) $\hspace{2cm}$ Eq. 7.3.4

The longitudinal stress: $\quad f_x = \dfrac{pR_c}{2t}$ (psi) $\hspace{2cm}$ Eq. 7.3.5

The longitudinal elongation: $\quad \Delta L = \dfrac{pR_c L}{Et}(0.5 - \mu)$ (in.) $\hspace{2cm}$ Eq. 7.3.6

where: $\;$ p – Internal pressure (psi) or pressure differential
$\hspace{1.5cm}$ E – Modulus of elasticity (psi)
$\hspace{1.5cm}$ μ – Poisson's ratio
$\hspace{1.5cm}$ R_c – Radius of the cylindrical shell
$\hspace{1.5cm}$ t – Thickness of the cylindrical skin

(B) DOUBLE-CYLINDER SECTION:

Double-cylinder (double-lobe or double-bubble) fuselage sections are frequently used on narrow body fuselages to give more lower deck cargo volume as shown in Fig. 7.3.2. The upper and lower cylinders still act as individual cylindrical shells but the floor member will pick up tension load (N_F):

$\quad N_F = p(R_{C,U} \cos \theta_U + R_{C,L} \cos \theta_L)$ (lbs./in.) $\hspace{2cm}$ Eq. 7.3.7

where: $\;$ p $\;$ – Internal pressure (psi) or pressure differential
$\hspace{1.5cm}$ $R_{C,U}$ – Radius of the upper cylindrical shell
$\hspace{1.5cm}$ $R_{C,L}$ – Radius of the lower cylindrical shell
$\hspace{1.5cm}$ θ_U $\;$ – Tangential angle of the upper cylindrical shell
$\hspace{1.5cm}$ θ_L $\;$ – Tangential angle of the lower cylindrical shell

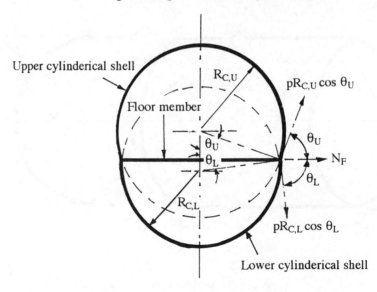

Fig. 7.3.2 Double-cylindrical Section

7.4 HEMISPHERICAL SHELLS

A hemispherical surface of a low curvature is commonly used as a pressure dome, as shown in Fig. 7.4.1, in aircraft fuselage design; this dome supports the pressure loading by membrane stresses. Generally, a reinforcing ring, placed at the seam, resists the radial component of these stresses.

The meridian and tangential (perpendicular to meridian) membrane stress (perpendicular to meridian) are identical which is equal to the smaller longitudinal stress (f_x) of cylinderical shell:

$$f_t = f_\phi = \frac{pR}{2t} \quad (psi) \qquad \text{Eq. 7.4.1}$$

where: f_t – Skin tangential stress (psi)
f_ϕ – Skin meridian stress (psi)
p – Internal pressure (psi)
R – Radius of the hemispherical shell (in.)
t – Thickness of the hemispherical shell skin (in.)

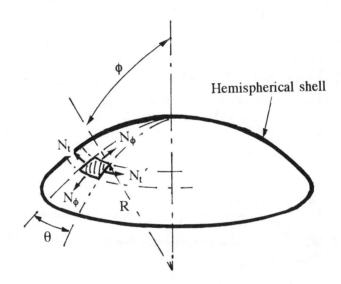

θ – Tangential angle
φ – Meridian angle

Fig. 7.4.1 Hemispherical Shell

The selection of end bulkhead configurations for pressurized cabin fuselage design are discussed below:

(a) Hemispherical bulkheads:

This type of bulkhead is highly desirable from a structural standpoint, but because of space limitations, as shown in Fig. 7.4.2, it is physically impossible to assemble the complex joint.

Chapter 7.0

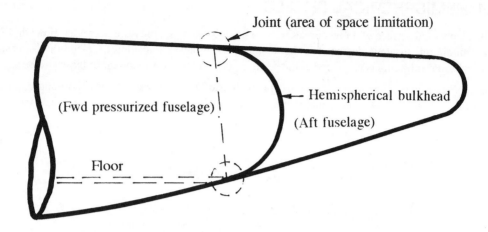

Fig. 7.4.2 Fuselage Bulkhead with Full Hemispherical Shell

(b) Flat bulkheads:

This type, as shown in Fig. 7.4.3, is structurally inefficient and heavier but it
* Provides more useful volume as shown in Fig. 7.4.3 in fwd fuselage radome area
* Is used when a passageway in aft bulkhead is requireed (e.g., DC9, B727, BAC111, etc.)

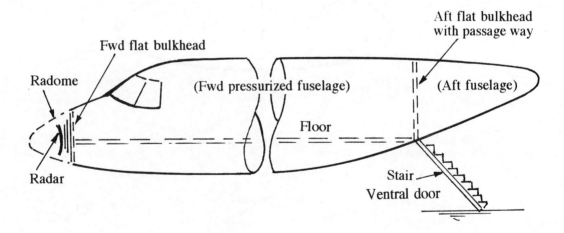

Fig. 7.4.3 Fuselage with Flat Bulkhead Applications

(c) Semi-spherical bulkheads (dashed head):

This type of bulkhead, shown in Fig. 7.4.4, is a compromise design which supports the pressure loading by membrane stresses in the same manner as a hemispherical bulkhead. A bulkhead ring, as shown in Fig. 7.4.5, is required to assemble this bulkhead to the fuselage and also to resist the compression loading which results from internal pressure on the semi-spherical shape.

$$N_{BR} = N_\phi \cos\phi = (\frac{pR}{2})\cos\phi$$

$$R = \frac{R_c}{\sin\phi}$$

$$N_{BR} = (\frac{p\cos\phi}{2})(\frac{R_c}{\sin\phi})$$

The compressive loads acting on the bulkhead ring are:

$$F_{BR} = (\frac{pR_c^2}{2})\cot\phi \qquad \text{Eq. 7.4.2}$$

where: ϕ – Meridian angle

p – Internal pressure (psi)

R – Radius of the semi-spherical shell (in.)

R_c – Radius of the cylindrical shell (in.)

N_{BR} – Uniform radial load acting inward on the bulkhead ring (lbs./in.)

F_{BR} – compression load acting on the bulkhead ring (lbs.)

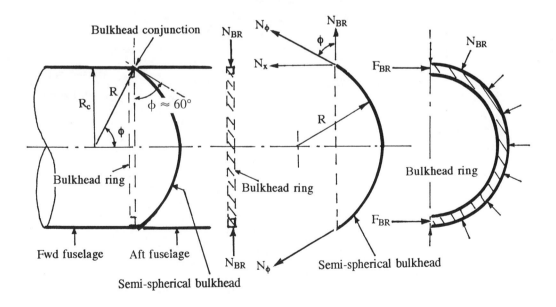

Fig. 7.4.4 Semi-spherical Bulkhead and Fuselage Body Conjunction

Chapter 7.0

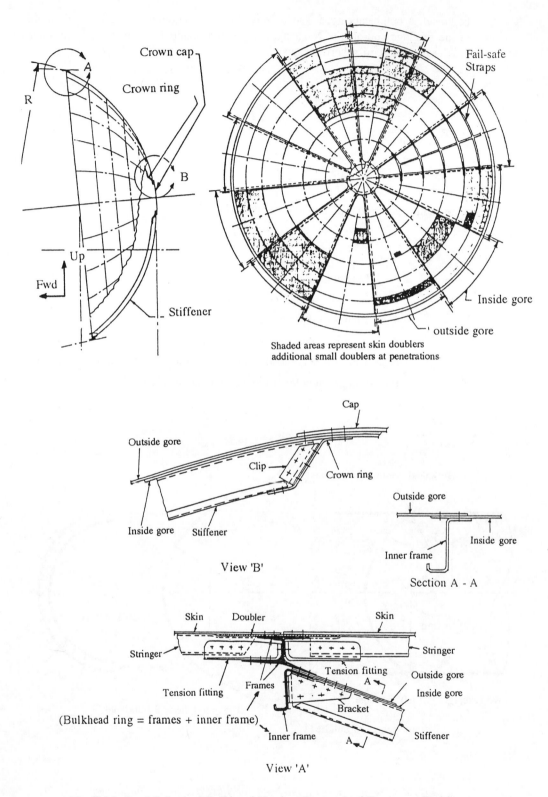

Fig. 7.4.5 Semi-Spherical Bulkhead Used on Pressurized Transport Fuselage

Example:

Assume a fuselage, as shown in Fig. 7.4.4, having cabin-differential pressure of 8.77 psi and a diameter of 200" at the pressurized semi-spherical bulkhead with a meridian angle $\phi = 60°$. Determine the bulkhead ring compression load.

From Eq. 7.4.2, the bulkhead ring compression load is:

$$F_{BR} = (\frac{pR_c^2}{2})\cot\phi = (\frac{8.77 \times 100^2}{2})\cot 60° = 25,317 \text{ lbs}.$$

7.5 HONEYCOMB PANELS

Applications for honeycomb materials which are used in sandwich construction are many and varied. The most desirable honeycomb structures is that made of composite materials because it has the lowest structural weight and is easily fabricated. Nearly every aircraft or missile that flies today, or that will fly in the future, utilizes composite honeycomb sandwich construction in the structures of the airborne unit (e.g., see Fig. 7.5.1).

The reasons for the aircraft designer's choice of a honeycomb panel rather than more conventional construction (e.g., skin-stringer type) usually include:
- Its high strength-to-weight ratio
- Vastly improved sonic damage resistance; impact resistance; resistance to battle damage
- High rigidity per unit weight
- Superior surface smoothness
- Insulation
- Economy (use of composite materials and precludes corrosion problems)

Structural panels (Exterior panels):
  Wing and empennage fixed leading and trailing edge panels, control surfaces, fairing, radome, tail cone, landing gear doors, fuselage floor panels, etc.
Non-structural panels (Interior panels):
  Fuselage side-wall panels, ceiling, partition, lavatory, decoration panels, etc.

Fig. 7.5.1 Honeycomb Structure Applications

The function of the honeycomb or sandwich panel, shown in Fig. 7.5.2, may best be described by making an analogy to an I-beam. The high density facing of the honeycomb panel corresponds to the flanges of an I-beam, the object being to place a high density, high strength material as far from the neutral axis as possible, thus increasing the section modulus. The honeycomb in the sandwich structure is comparable to the web of an I-beam, which supports the flanges and allows them to act as a unit. The honeycomb in a sandwich panel differs from the web of an I-beam in that it maintains a continuous support for the facings, allowing the facings to be worked up to or above their yield strength without crimping or buckling. The adhesive which bonds the honeycomb core to its facings must be capable of transmitting shear loads between these two components thus making the entire structure an integral unit. When a sandwich panel is loaded as a beam, the honeycomb and the bond resist the shear loads while the facings resist the forces due to bending moment, and hence carry the panel bending or tensile and compressive loads. The results from bending have been found to be influenced by several parameters:

- Facing thickness and material
- Core material and thickness.

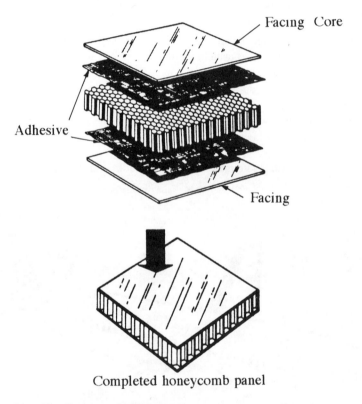

Completed honeycomb panel

(Note: The adhesive may be eliminated if prepreg material is used as facings)

Fig. 7.5.2 Honeycomb Structure

The facing thickness alone will cause variations because the skins are able to carry shear loads in addition to what the core carries and, furthermore, they are able to take on additional shear loads after the core has reached yield strength. Thus far the primary application for honeycomb panels on airframe structures is for normal pressure loading (aerodynamic pressure) cases. Whether honeycomb panels could ever be used as columns to carry primary compression loads remains to been seen due to lack of reliability.

(A) DESIGN REQUIREMENTS

Honeycomb structures should be designed to meet the basic structural criteria listed below when these criteria pertain to the type of loading under consideration:

- The facings shall be thick enough to withstand the tensile or compressive stresses imposed by the design load
- The core shall have sufficient strength to withstand the shear stresses induced by the load
- Honeycomb structures shall have sufficient flexural and shear rigidity to prevent excessive deflections under load
- The core shall have sufficient compressive strength to resist crushing by design loads acting normal to the panel facings
- The core shall be thick, strong and stiff enough to prevent overall buckling, shear crimping, face wrinkling and inter-cell dimpling when the honeycomb panel is loaded in edgewise compression
- The adhesive bond shall be strong enough to withstand any flatwise tensile and shear stresses induced by the loading
- The honeycomb panel close-outs and attachment points shall have sufficient strength and tie in with the core and facings to transmit load to the rest of the structure

Modes of failure in honeycomb structures are shown in Fig. 7.5.3.

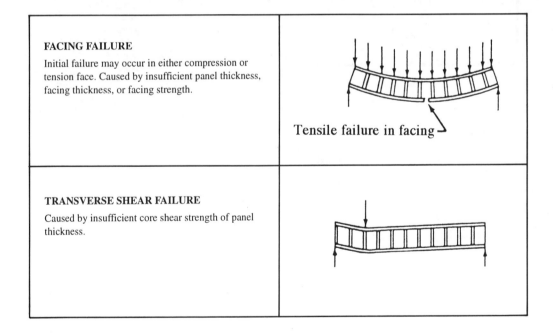

Fig. 7.5.3 Honeycomb Failure Modes under Normal Pressure

When using composite materials, both the facing fiber and core directions should be called out on the drawings to insure that the strength required is obtained; see Fig. 7.5.4.

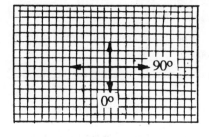
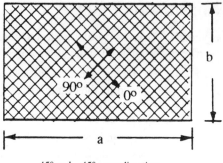

0° and 90° warp direction
(Strong in bending)

45° and – 45° warp direction
(Strong in in-plane shear which is
not used in plate bending)

(a) Facing Material (warp direction)

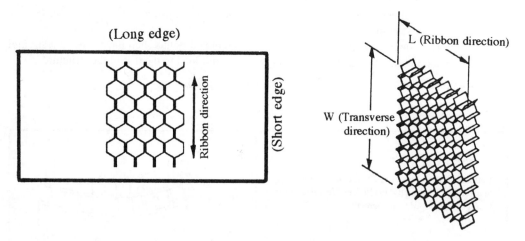

(Usually the ribbon direction (L) is placed along the short edge of the panel to carry high shear load)

(b) Core Material (Ribbon direction)

Fig. 7.5.4 Orientations

(B) FACING MATERIALS

After familiarity is gained in honeycomb construction as to where, why, and how it is used, the proper selection of material becomes quite obvious. The following points should be considered when selecting materials:

(a) Care should be taken in selection of different materials which would be in contact with each other and the effects of thermal expansion must be taken into consideration.

(b) It is very difficult to completely seal a honeycomb panel against outside environmental conditions. Even though it may be assumed that moisture or other contaminants would not come in contact with the core materials, in actual practice this is generally not true. The core material, as well as the facing material, must be able to maintain its properties when thus subjected to the outside elements.

(c) Care should be taken to select the most economical of the various material combinations suitable.

In a panel it is generally desirable to use the same facing material on each side of the honeycomb structure. In cases where dissimilar skins are required (not desirable), caution must be used to eliminate skin distortion due to unequal thermal expansion coefficients. This may be particularly troublesome during the honeycomb panel curing cycle.

Where parts are of a complex contoured shape, reinforced composite material faces should be considered. These may frequently be laid up prepreg plies of composite materials and cured as an integral assembly of skin and core. Even where pre-cured skins are a requirement, tooling costs may be less for fabrication of these materials than for formed metal facings, and the tolerance problem less severe.

Composite materials, i.e., fiberglass, Kevlar, graphite, etc., have been widely used in airframe secondary structures. Fig. 7.5.5 shows compressive and tensile stress allowables for fiberglass epoxy laminate materials.

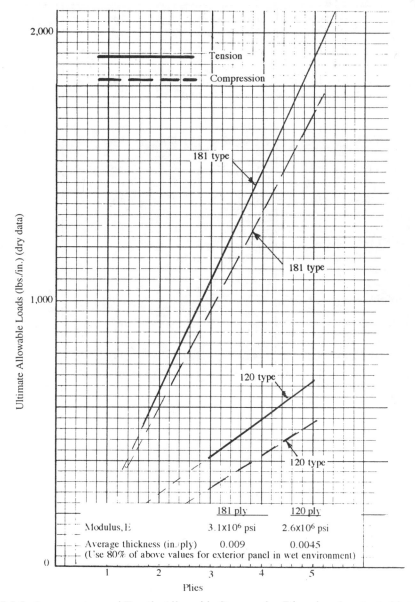

Fig. 7.5.5 Compressive and Tensile Allowable Stresses for Fiberglass Laminate Materials

(C) HONEYCOMB CORE

Honeycomb cores are produced in a variety of materials including paper, fiberglass, composite materials, aluminum of various alloys, stainless steel, and many super alloy materials. Fig. 7.5.6 shows the shear strength properties of non-metallic Nomex (HRH10) and it is apparent that normally expanded honeycomb is stronger in the 'L' direction than it is in the 'W' direction. Whenever possible, a structural engineer should take advantage of this fact.

Honeycomb designation	Plate Shear					
	"L" Direction			"W" Direction		
	Strength psi		Modulus ksi	Strength psi		Modulus ksi
Material – Cell – Density	Typ.	Min.	Typical	Typ.	Min.	Typical
HRH 10 – $\frac{1}{8}$ – 1.8	90	65	3.7	50	36	2.0
HRH 10 – $\frac{1}{8}$ – 3.0	190	165	7.0	100	85	3.5
HRH 10 – $\frac{1}{8}$ – 4.0	270	225	9.2	140	110	4.7
HRH 10 – $\frac{1}{8}$ – 5.0	325	235	—	175	120	—
HRH 10 – $\frac{1}{8}$ – 6.0	370	260	13.0	200	135	6.0
HRH 10 – $\frac{1}{8}$ – 8.0	490	355	16.0	275	210	7.8
HRH 10 – $\frac{1}{8}$ – 9.0	505	405	17.0	310	260	9.0
HRH 10 – $\frac{3}{16}$ – 2.0	110	80	4.5	60	45	2.2
HRH 10 – $\frac{3}{16}$ – 3.0	160	130	5.8	90	70	3.5
HRH 10 – $\frac{3}{16}$ – 4.0	245	215	7.8	140	110	4.7
HRH 10 – $\frac{3}{16}$ – 4.5	290	225	9.5	145	110	4.0
HRH 10 – $\frac{3}{16}$ – 6.0	390	330	14.5	185	150	6.0
HRH 10 – $\frac{1}{4}$ – 1.5	75	45	3.0	35	23	1.5
HRH 10 – $\frac{1}{4}$ – 2.0	110	72	4.2	55	36	2.8
HRH 10 – $\frac{1}{4}$ – 3.1	170	135	7.0	85	60	3.0
HRH 10 – $\frac{1}{4}$ – 4.0	240	200	7.5	125	95	3.5
HRH 10 – $\frac{3}{8}$ – 1.5	75	45	3.0	35	23	1.5
HRH 10 – $\frac{3}{8}$ – 2.0	110	72	4.2	55	36	2.2

(Test data obtained at 0.5 inch thickness)

Fig. 7.5.6 Mechanical Properties of HRH10 (Nomex) Honeycomb

Honeycomb core shear strength will vary with core thickness such that thicker cores have lower shear strength and rigidity than thinner ones. In view of typical core thickness values in actual usage, as well as aircraft company and military specifications, honeycomb core is generally tested at the thickness of:

- 0.625 inch for aluminum materials.
- 0.500 inch for non-metallic materials.

Honeycomb manufacturers are often asked to qualify core materials to other thickness values and the engineer should use the correction factor for each material being considered. Fig. 7.5.7 gives approximate correction factors for both aluminum and non-metallic honeycomb core for preliminary sizing use.

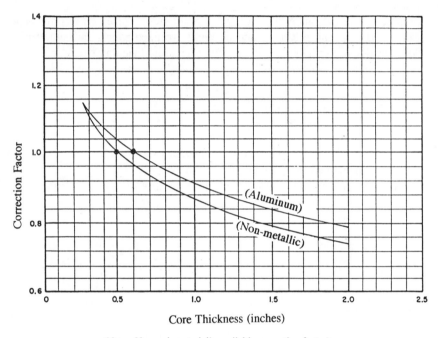

(Note: Use each material's available correction factor)

Fig. 7.5.7 Approximate Correction Factors for Honeycomb Core

(D) PANEL CLOSE-OUTS

It is generally good practice to close off the edges in honeycomb panel construction. The edge closure will protect the core from accidental damage, serve as a moisture seal, and provide edge reinforcement to facilitate transfer and distribution of edge loads; see Fig. 7.5.8.

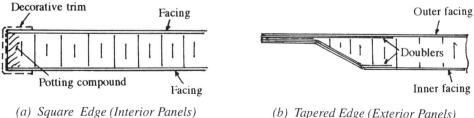

(a) Square Edge (Interior Panels)   (b) Tapered Edge (Exterior Panels)

Fig. 7.5.8 Panel Close-outs

Chapter 7.0

Good design always involves choosing the best approach from several alternatives. Certainly selecting ways to close-out a honeycomb panel and assure proper load transfer from the facings and core to the attachment points is probably the most critical aspect of honeycomb panel design. To do this with the least amount of cost and weight premium involves knowing how the part will be fabricated in shop. The most common configurations are:

(a) Square Close-outs:

Square close-outs, as shown in Fig. 7.5.8(a), are used for interior panels such as floor panels, side-wall panels, ceiling panels, partitions, etc., inside of the fuselage. Most of these panels, except the floor panels, are for decoration purposes and are non-structural panels which are easily made, and are light structures.

(b) Tapered Close-outs:

Tapered close-outs, as shown in Fig. 7.5.8(b), must be used on exterior panels such as wing and empennage leading and trailing edge fixed panels and removable doors and fairings due to their higher surface normal pressure load and moisture intrusion. Due to aerodynamic smoothness and low drag requirements, the close-out used must:

- To meet the shear strength requirements (see Fig. 7.5.9) in the tapered area caused by static, fatigue or sonic loads
- Form a continuous aerodynamic surface with adjacent structure
- Maintain minimum edge thickness as shown in Fig. 7.5.10
- Contain the fastener countersink (countersunk head height) within the edge thickness to avoid knife edge, as shown in Fig. 7.5.11
- Meet tension allowables for edge fastener pull-through, as shown in Fig. 7.5.12

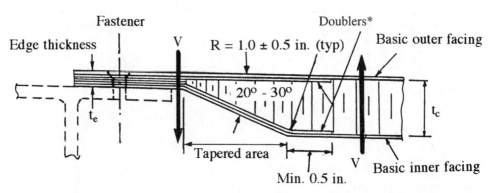

* Use same type and number of plies in the facing laminates on each side

Edge load (lbs./in.)	Facing laminate at tapered area (Basic panel facing plus additional doubler fabric plies below)	
	Inner facing	Outer facing
50	3 plies of 120	3 plies of 120
80	4 plies of 120	4 plies of 120
100	2 plies of 181	2 plies of 181
150	3 plies of 181	3 plies of 181

Fig. 7.5.9 Allowable Edge Load in Tapered Area of Panel Close-out Transition

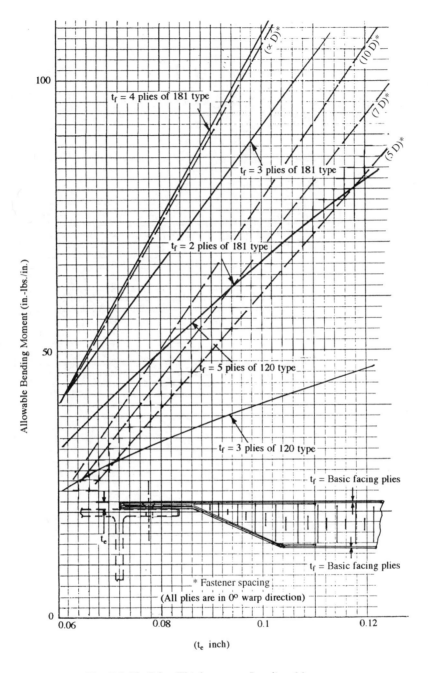

Fig. 7.5.10 Edge Thickness vs. Bending Moment

	Aluminum Rivets (standard)			Aluminum Rivets (shear head)			Steel Fasteners (shear head)		
D, Fastener dia. (in.)	$\frac{5}{32}$	$\frac{3}{16}$	$\frac{1}{4}$	$\frac{5}{32}$	$\frac{3}{16}$	$\frac{1}{4}$	$\frac{3}{16}$	$\frac{1}{4}$	$\frac{5}{16}$
Tension allowable (lbs.)	490	700	1,240	360	520	930	1,220	2,280	3,670
Head height – t_d (in.)	0.055	0.070	0.095	0.036	0.047	0.063	0.048	0.063	0.07

Note: Edge thickness: $t_e \approx 1.5 t_d$ or $t_e = t_{skin}$ whichever is greatest.

Chapter 7.0

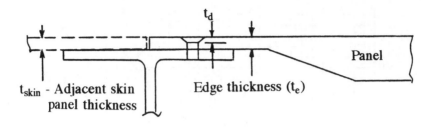

Fig. 7.5.11 Typical Countersunk Fastener Tension Allowables and Head Heights

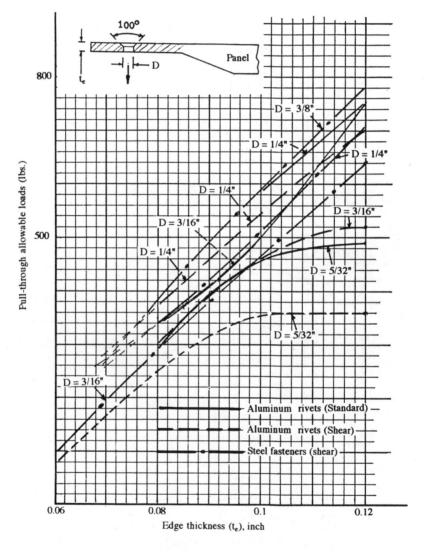

Note: For 120 type, use 85% of above values
100° countersunk fasteners
Edge distance $\frac{e}{D} \geq 3.0$
All plies in 0° Warp direction

Fig. 7.5.12 Fastener Tension Pull-through Allowables for 181 Type Fiberglass Fabric Laminates

Plate and Shell Stress

The tapered transition shown in Fig. 7.5.13, which is commonly used as the wedge of a wing flap, aileron, elevator, rudder, etc., is occasionally accomplished by bonding one facing to a constant thickness core; machining the core to the desired taper and then bonding a second facing to the core. This multi-stage bonding may reduce the strength of the adhesives which are subjected to thermal cycling, but in some cases multi-stage bonding is the most practical assembly technique.

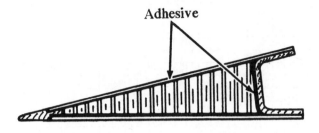

Fig. 7.5.13 Tapered Transition

(E) CORE-INSTALLED INSERTS

The load imposed on an insert is one of those shown in Fig. 7.5.14

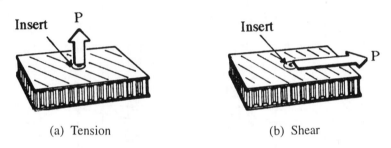

Fig. 7.5.14 Types of Load Imposed on a Insert

Since honeycomb panels generally have thin facings, the shear and tension loads in particular should be transmitted to the entire honeycomb core, as shown in Fig. 7.5.15, whenever possible. For this reason, most inserts are bonded to the core by means of a potting adhesive. Selection of such inserts depends on the following factors:

- Panel thickness
- Specific strength requirements
- Environmental exposure

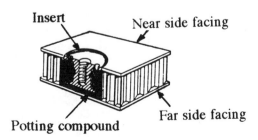

Fig. 7.5.15 Potted Inserts Used in Honeycomb Panels

The installation of a potted inserts, shown in Fig. 7.5.15, is used in non-metallic panels in which the insert is potted into the core and it provides the best structural strength since the adhesive bonds the fastener to both facings and the core. The near side facing is the most effective to react shear loads. Figs. 7.5.16 and 7.5.17 provide tension (P_t) and shear (P_s) allowable loads in panels of fiberglass facings and Nomex core.

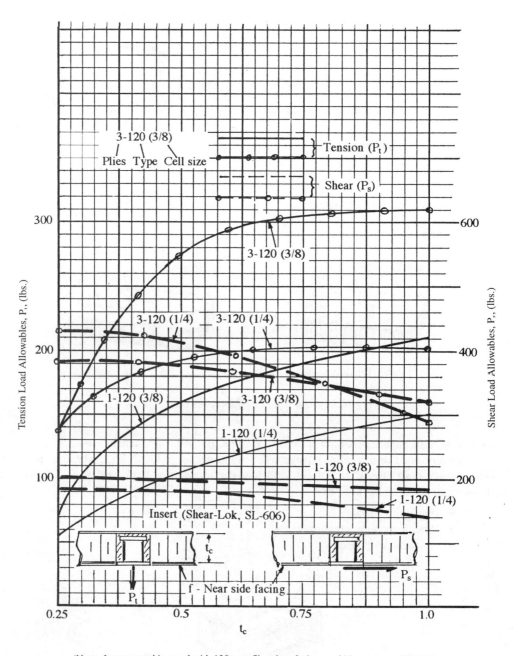

(Note: Insert potted in panel with 120 type fiberglass facings and Nomex core (HRH10))

Fig. 7.5.16 Tension and Shear Load Allowables through Insert Potted Honeycomb Panels

Plate and Shell Stress

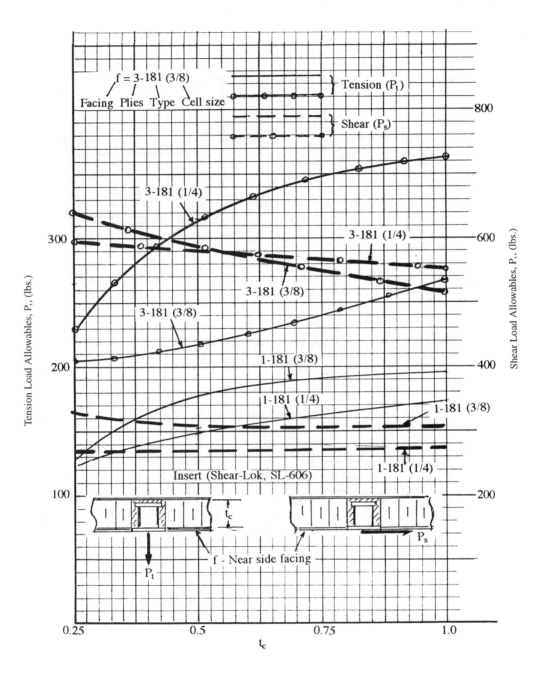

(Note: Insert plotted in panel with 181 type fiberglass facings and Nomex core (HRH10))

Fig. 7.5.17 Tension and Shear Load Allowables through Insert Potted in Honeycomb Panels

Chapter 7.0

Example:

Given: Panel size: a = 30 in. and b = 20 in.

Panel aspect ratio: $\frac{a}{b} = \frac{30}{20} = 1.5$

Normal uniform pressure load: p = 5 psi
(Assume a plate with 4 hinged edges)

Material: Facings – Fiberglass plies
(type 120 and 181 fabric)
Nomex core (HRH10)

Edge fasteners: Dia. = $\frac{3}{16}$" (100° steel countersunk standard head)

Requirement: The max. panel deflection (δ) under given uniform load should not be greater than 0.5".

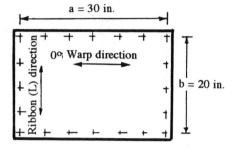

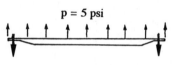

Assume a plate with 4 hinged edges under uniform load (Case 1 in Cahpter 7.2)

(1) Design and sizing practices are as follows:
- Use thicker facing on compression side, if possible
- Assume centroid depth (h) between two facings and core thickness, $t_c = H - \frac{t_{outer}}{2} - \frac{t_{inner}}{2}$
- This is a process of trial and error method
- First assume a honeycomb panel configuration as shown below:

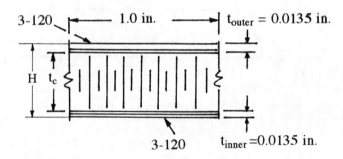

(2) Find max. bending stress at the center of the plate (use Eq. 7.2.1 and β from Fig. 7.2.2):

$$f = \beta(p\frac{b^2}{t^2}) \text{ at the center of the plate}$$

$$= 0.485 \, (5 \times \frac{20^2}{t^2})$$

$$t^2 = \frac{970}{f} \text{ (solid plate thickness)} \hspace{2cm} \text{Eq. (A)}$$

216

Plate and Shell Stress

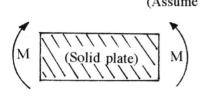
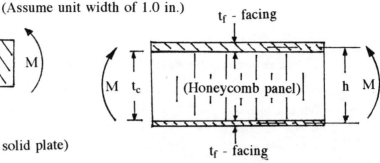

(Assume unit width of 1.0 in.)

$I_s = t^3/12$
(Moment of inertia of solid plate)

$I_h = t_f h^2 / 2$
(Moment of inertia of honeycomb panel)

Moment of inertia for solid plate: $I_s = \dfrac{t^3}{12}$

Moment of inertia for honeycomb panel: $I_h = \dfrac{t_f h^2}{2}$ where t_f = facing plies

Let $I_s = I_h$ and then,

$$\frac{t^3}{12} = \frac{t_f h^2}{2}$$

$$t^3 = 6 t_f h^2 \qquad \text{Eq. (B)}$$

$$t^2 = (6 t_f h^2)^{\frac{2}{3}} \qquad \text{Eq. (C)}$$

From Fig. 7.5.5, use 3 plies of 120 type ($t_f = 3 \times 0.0045 = 0.0135$ in.),

$$F_{120} = \left(\frac{300 \text{ lbs./in.}}{0.0135}\right) = 22{,}222 \text{ psi}$$

Combining Eq. (A) and Eq. (C) and let $f = F_{120} = 22{,}222$ psi to obtain,

$$t^2 = (6 t_f h^2)^{\frac{2}{3}} = \frac{970}{22{,}222}$$

$h = 0.34$ in. (Required honeycomb core thickness for bending moment strength)

(3) Max. deflection at the center of the plate (use Eq. 7.2.2 and α from Fig. 7.2.2):
(use $E = 2.6 \times 10^6$ psi, see Fig. 7.5.5)

$$\delta = \alpha \left(\frac{p b^4}{E t^3}\right)$$

$$= 0.085 \left(5 \times \frac{20^4}{2.6 \times 10^6 \times t^3}\right)$$

Substitute $t^3 = 6 t_f h^2$ [from Eq. (B)] into the above equation and let them equal to $\delta = 0.5$ in. (requirement):

$$\delta = 0.085 \left(5 \times \frac{20^4}{2.6 \times 10^6 \times 6 t_f h^2}\right) = 0.5$$

$h = 0.804$ in. (Required honeycomb core thickness for stiffness) > 0.34 in.

Therefore, the final core thickness: $t_c = h - t_f = 0.804 - 0.0135 = 0.79$ in. for this panel.

217

Chapter 7.0

(4) Find max. shear load at the center of the long edge of 'a' (Eq. 7.2.3 and $\kappa = 0.485$ from Fig. 7.2.2):

$V = \kappa\, p\, b = 0.485 \times 5 \times 20 = 48.5$ lbs./in.

(5) Determine the core material with $t_c = 0.79$ in.:

Use Nomex honeycomb core (HRH10 $- \frac{3}{16} - 2.0$); from Fig. 7.5.6 obtain the shear allowable 80 psi (L) and 45 psi (W).

With core thickness $t_c = 0.79$ in., from Fig. 7.5.7 find the correction factor = 0.91 and the shear allowable in the 'L' direction is

$V_{all} = 80 \times 0.91 \times 0.79 = 57.5$ lbs./in. $> V = 48.5$ lbs./in.   O.K.

(6) Close-out at tapered area :

From step (4), the max. shear load along the edge of a = 30" is V = 48.5 lbs./in.

Doublers required in tapered area; refer to Fig. 7.5.9,

Inner facing: Requires 3 plies of 120 type

Outer facing: Requires 3 piles of 120 type

(Discussion: Besides the basic outer facing of 3 plies of 120, add a set of doublers: 3 piles of 120.

(7) Close-out - edge thickness:

From step (4), the max. shear load along the edge of a = 30" is V = 48.5 lbs./in.

The moment distribution is shown below:

$M = 1.0 \times 48.5 = 48.5$ in.-lbs./in.

Assume: $M_t = 60\%\, M = 48.5 \times 60\% = 29.1$ in.-lbs./in. (at taper end)

$M_f = 40\%\, M = 48.5 \times 40\% = 19.4$ in.-lbs./in. (at fasteners)

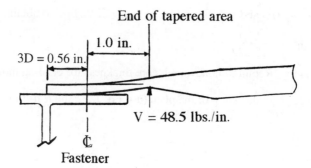

At end of taper, the moment is $M_t = 29.1$ in.-lbs./in.

From Fig. 7.5.10, the edge thickness, $t_{e,t} = 0.068$ in.

From Fig. 7.5.11, the min. edge thickness due to countersunk head height $t_d = 0.048$ (steel $D = \frac{3}{16}$ in.),

$t_{e,f} = 1.5 \times 0.048 = 0.072$ in.

The final edge thickness, $t_e = 0.072$ in.

Plate and Shell Stress

(8) Fasteners:

From step (7), the moment at fasteners is $M_f = 19.4$ in.-lbs./in. Assume fastener's pitch is 2.0 in. and use fastener edge distance of $e = 3D = 3(\frac{3}{16}) = 0.56$ in. The max. tension load on each fastener is:

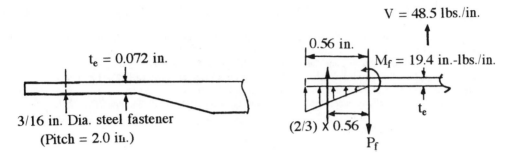

$$P_f = 2 \times (V + \frac{M_f}{0.56 \times \frac{2}{3}})$$

$$P_f = 2 \times (48.5 + \frac{19.4}{0.56 \times \frac{2}{3}})$$

= 201 lbs. tension load per fastener

From Fig. 7.5.12 with $t_e = 0.072$ in. and steel $D = \frac{3}{16}$ in. fasteners, obtain the pull-through allowable load of 208 lbs./fastener which is greater than the required $P_f = 201$ lbs. O.K.

(9) Final configuration as sketched below:

(Note: This is one of many alternatives for design of this panel; final selection depends on structural weight, fabrication cost, service experience, etc.)

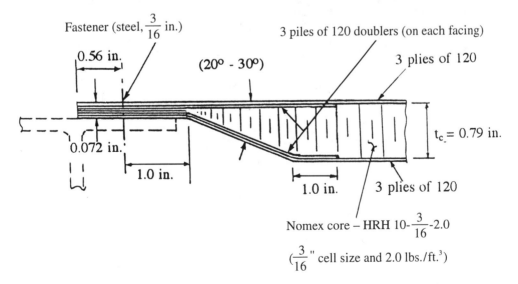

219

References

7.1 Sechler, E. E. and Dunn, L. G., "AIRPLANE STRUCTURAL ANALYSIS AND DESIGN", Dover Publications, Inc., New York, NY, (1942).
7.2 Roark, R. J. and Young, W. C., "FORMULAS FOR STRESS AND STRAIN", McGraw-Hill Book, Company, New York NY, (1975, 5th edition).
7.3 Timoshenko, S. and Woinowsky-Krieger, S., "THEORY OF PLATES AND SHELLS", McGraw-Hills Book 7.4 Flugge, W., "Stress Problems in Pressurized Cabins", NACA TN 2612, (1952).
7.5 Wang, C. T., "Bending of Rectangular Plates With Large Deflections", NACE TN 1462, (1948).
7.6 Anon., "Design Handbook for Honeycomb Sandwich Structures", Hexcel Corp., TSB 123, (1967).
7.7 Anon., "Structures Manual, Section 6, Thin Shells", Aerojet Inc., (1960).
7.8 Anon.: "Shur-Lok Specialty Fasteners for Industry", Shur-Lok Corp. P.O. box 19584, Irvine, CA. 92713.

Chapter 8.0

BOX BEAMS

8.1 INTRODUCTION

This chapter is primarily concerned with the box beam (Ref. 8.3) which is a construction of thin skins or webs and stringers (flanges, caps, chords, etc.). The design of such a box beam involves the determination of the distribution of shear, bending and torsion loads at any section of a wing or fuselage into axial load in the longitudinal stringers and shear in the skins (upper and lower skin-stringer panels) and webs (vertical spar webs). This chapter gives the engineer practical methods of analysis to perform the preliminary sizing of box beams.

Since the individual caps and shear web of a simple beam alone are incapable of resisting torsion (approximately true in most cases) it follows that a single-web beam can resist a shear load only when the load is applied to the shear center. The single-web beam cannot usually be used by itself as a structural member due to its inability to resist torsion (because of the very small torsional constant value of 'J', as discussed in Chapter 6.0).

It is therefore customary to use a closed box section, a cylindrical beam such as a circular fuselage section, a two-spar wing box to form a closed section, or by combining one or more webs to form a multi-cell box, as shown in Fig. 8.1.1. The shear flow distribution in a box beam cannot be determined by the bending stresses alone, as it is impossible to find a point of zero shear flow that will be independent of the external loads.

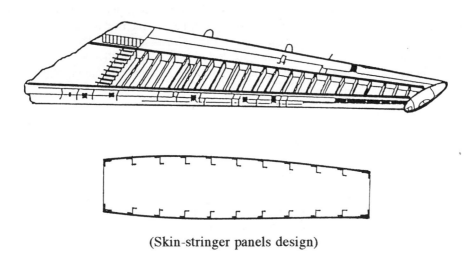

(Skin-stringer panels design)

(a) Transport Wing (Two-cell Box)

Fig. 8.1.1 Wing Box Beams

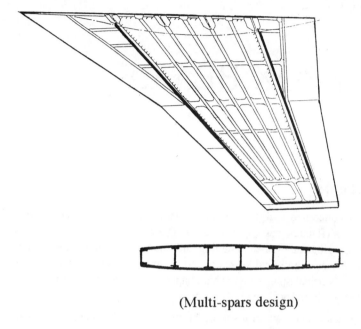

(Multi-spars design)

(b) Fighter Wing (Multi- cell Box)

Fig. 8.1.1 Wing Box Beams (cont'd)

Most box beams used in transport aircraft (in thicker areas of wings or in the tail) use single-cell boxes although in a very few cases two-cell box construction is used. The thinner wing boxes used on fighters and supersonic aircraft utilize multi-cell boxes which provide a much more efficient structure. Due to broad and comprehensive analysis of box beams, it is obviously impossible to cover all aspects in this chapter. However, the most important information and design data for practical design requirements and preliminary sizing will be covered in this chapter. The analyses in this chapter cover only one and two-cell boxes; refer to the references at the end of the chapter for information on other multi-cell box designs. References 8.9 and 8.10 describe the torsional shear flows in a multi-cell box beam by the method of successive approximation, which provides a simple and rapid method for determining the shear flows.

A practical description of a box beam is a rectangular shape with skin-stringer panels on top and bottom connected by two vertical beams (a single-cell box) or by more vertical beams (a multi-cells box). Box beams have the following characteristics:

- They consist of wide skin-stringer panels which are subjected to shear flow due to bending moment and torsion
- The thin skins take little of the axial compression load
- The thin skins carry shear loads very well and even though they may buckle under the loads they do not fail but continue to carry the buckling loads plus significant additional shear loads
- Stringers carry axial loads, and shear flow induces a change in axial load in the stringer
- The usual cross-section of a wing box is unsymmetrical both in shape and in distribution of material

- The resultant bending moment and shear at a section are usually not normal to the neutral axis
- Since conventional wing boxes taper in planform and section as well as in area, the effects from taper are accounted for directly
- It is necessary to compute the principal axes of the beam before the axial stresses can be determined

Once a good grasp of the fundamental analysis is obtained, it facilitates the design of actual box beam structures. The effects of "shear lag" due to cutouts and discontinuities, tapers in depth and planform, and building-in at the root which restrains warping, are quite important and may have significant effect on the design.

Not only is the box beam an important construction method, but applications are found in the primary structures of all current airframe designs. Much past research and testing indicates that box beam applications are generally recommended for airframe designs as follows:

- Single-cell box beams are used for thicker wing airfoils such as transport wings, tail sections, control surfaces, fuselage shells, etc.
- Two-cell box beams are used for single-spar airfoil sections in general aviation or for three-spar wing boxes designed for stiffness or other purposes
- Three or more cell box beams are basically for thin airfoil applications (approximate wing thickness ratio $\frac{t}{c} \approx 3 - 8\%$) such as fighter aircraft and supersonic aircraft

The "Unit Method of Beam Analysis" described in Reference 8.8 is a simple hand calculation method which covers:

- A single-cell box
- Shear lag effect
- Effect of taper in depth and width of the beam
- Principal axes
- Calculation done by hand using a simple desk-top calculator

It is customary to designate the reference axes of the box beam cross-section as x (horizontal), z (vertical) and y (perpendicular to the cross-section or the box beam) axes. The common coordinates used in box beam analysis are shown in Fig. 8.1.2.

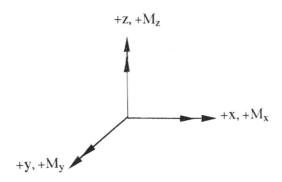

Fig. 8.1.2 Positive Sign Conventions

Use of positive signs to represent unknown loads is recommeded, as shown in Fig. 8.1.2. If calculations show the unknown load to be negative, the direction of the assumed load is then the opposite.

The usual cross-section of a wing box is unsymmetrical with respect to both vertical and horizontal axes, and the loads that act on the box are:

M_x – Beam bending moments
P_z – Beam shear
M_y – Torsion
P_x – Chord shear
M_z – Chord moment

M_y, P_x and M_z listed above are ignored in preliminary sizing because they are not the primary design loads except under special design conditions such as 9.0 g crash landings or ditching, etc. (see Chapter 3.0).

In this chapter it is appropriate to review the term "flexural center" which has been a source of much confusion to engineers since there are so many similar terms, i.e., center of shear, center of flexure, elastic center, elastic axis, torsional center, center of twist, center of least strain, etc., which are used in technical reports, books, papers, etc. The following observations relate to preliminary sizing:

- "Shear center" is used in open section beams and is defined as a point through which the line of action of a load must pass to not cause twist. This problem can usually be ignored in box beam design because the nature of the box resists torque and the loads not always pass through the "shear center"
- In box beam design use of the term "flexural center" refers to the neutral axes of the cross-section
- "Torsional center" is when torsional moment torques through the shear center and produces only rotational movement but no lateral translations
- The "Flexural center" does not quite coincide with the "torsional center" but approximates it for preliminary sizing

(A) SHORTCOMINGS OF BEAM THEORY

There is no doubt that the shell type of box beam is widely used in airframe structures, particularly for wing, tail, and fuselage applications. However, for determining beam axial stresses due to bending, the classic "beam theory" has serious shortcomings:

(a) Under pure bending (no shear) the assumptions of classic "beam theory" are valid

(b) When the beam bending is the result of the shear load which is being transmitted by the structure it can no longer be assumed that the plane cross section remains plane

- As shown in Fig. 8.1.3, the axial loads in the top and bottom panels of the box beam which increase from wing tip to root, are applied along the lines of attachment of the vertical shear webs and must travel across the horizontal panel material
- This relatively long path permits appreciable shear deflection which causes the cross section to distort from the assumed plane section
- The result is an increase in axial stress in the panel near the vertical shear webs and a decrease in that portion farthest away from the webs, as compared with the results of the classic beam theory
- This phenomenon, which is called "shear lag effect", has received much attention in box beam analysis because of the nature of a deformed thin panel box construction under shear loads

(c) Other fundamental errors involved in applying the classic beam theory for shear stress to the box beam occurs when the beam has taper in width and depth.

- Shear stresses derived from beam theory assumes a beam of constant cross section and does not give a true picture for beams with a tapered cross sections
- The effect of taper in depth for simple beams is discussed in Chapter 6.0 and box beams in Chapter 8.6

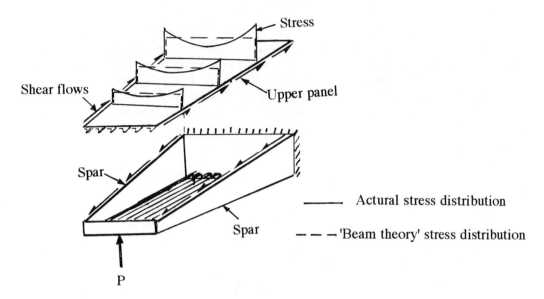

Fig. 8.1.3 Shear Lag Effect in a Wing Box Beam

(B) STIFFENED PANELS WITH THIN SHEETS

Variations in the configurations of stringer and end restraints under torsional moment or torque have the following effects on load distribution in the box:

(a) The effect of the torsional rigidity of open stringers such as Z, I, J, open hat-section, etc. on the design of skin-stringer panels is as follows:

- The torsional rigidity of each individual stringer as compared to the box beam section is so small that it can be ignored, at least in preliminary sizing
- The rigidity of the closed stringer is greater than that of the open section of the same size
- For detail sizing, the contribution from closed type stringers could be considered and a simplified procedure is given in Reference. 8.6
- In practice, the open stinger sections are used on the current wing, tail or fuselage box beam designs due to the need to inspect for corrosion problems, except in special cases such as using a few closed hat-sections as an air vent on a wing upper panel

(b) The effect of end restraint is not taken into consideration in preliminary design:

- For the equations used in this chapter it is assumed that the torsional boxes are free to warp out of their plane and that no stresses are produced normal to their cross-sections

- End restraint analysis should be done by computer due to its redundant nature
- The end restraint should be considered in detail analysis where:
  - the cantilever wing box attaches to the rigid fuselage side.
  - the landing gear support attaches to the wing, necessitating a heavy bulkhead
  - the pressure dome and fuselage meet, necessitating a heavy bulkhead

8.2 SHEAR FLOW DUE TO TORSION

The basic rules and assumptions governing the computation of shear flow in flat panels do not change if the panels are curved around a closed structure as in a section of a fuselage or wing box with many longitudinal stringers. The major differences in analysis come about as a result of differences in the manner of loading of the multiplicity of load paths in the structure. There are also some subtle complications such as out-of-plane load components on stringers. The engineer should realize that the thickness of the skin or web does not influence the shear flow distribution.

Since the single cell is a statically determinate structure it is usually simpler to determine the load distribution in the section by simple equilibrium. Various statically determinate sections are shown in Fig. 8.2.1.

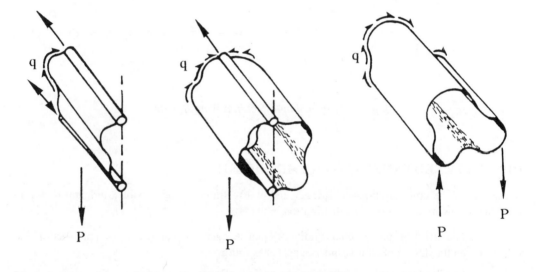

(a) Two-Stringer Open Shell   (b) Two-Stringer Closed Shell   (c) Closed Shell Box Beam

Fig. 8.2.1 Statically Determinate Sections

(A) TORSION AND SHEAR FLOWS

Assume a general closed section, as shown in Fig. 8.2.2, with sufficient support elsewhere to effectively maintain the shape even when loaded. In order to satisfy the equilibrium requirement that the applied torque, M_y, must not develop any net load unbalance, the shear flow, q, due to the torque must be constant around the perimeter and the sum of all the moment due to the shear flow must equal the applied torque. Then, for any incremental distance 'ds' along the perimeter,

$$dM_y = q \times ds \times h \qquad \text{Eq. 8.2.1}$$

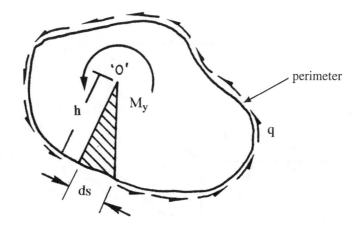

Fig. 8.2.2 Torque around a Closed Section

The moment of the force qxds about any arbitary selected point (reference point 'o') is the product of the force and the moment arm, h. Since the term of ds x h (segment area) is twice the area of the shaded triangle, and summing all torque and triangles around the section,

Torque: $M_y = q \times 2A$ Eq. 8.2.2

or Shear flow: $q = \dfrac{M_y}{2A}$ Eq. 8.2.3

where: A – Enclosed area along the entire perimeter

Such a uniform torsional shear flow applied to the end of a closed section structure develops constant reacting shear flows in the full structure identical to those developed in single flat panels by in-plane loads as discussed in Chapter 6.0.

(B) TORSION OF CELLULAR SECTIONS

Analysis of single-cell structures subjected to torsion has already been discussed. They could be analyzed in a straight forward manner by consideration of static load (e.g., equilibrium) alone. With multi-cell sections this is not possible and as in the indeterminate (redundant) structures already discussed, both equilibrium and compatibility have to be considered to develop enough relationships to evaluate all the unknowns (see References 8.6 and 8.7 for further information).

Fig. 8.2.3 shows a cellular section subjected to a torque, M_y. Consider the element a-b-c-d subjected to a shear flow, q, due to the torque. Let δ be the deflection due to the shear flow, and the strain energy stored in the element is given as:

From the strain energy equation:

$$dU = \frac{\text{Force} \times \text{displacement}}{2} = \frac{q \, ds \, \delta}{2}$$

Shear strain in element $= \gamma = \dfrac{\delta}{1.0} = \dfrac{f_{ts}}{G}$ per unit length

where f_{ts} = shear stress due to torsion $= \dfrac{q}{t}$ (q is shear flow and t is thickness)

Chapter 8.0

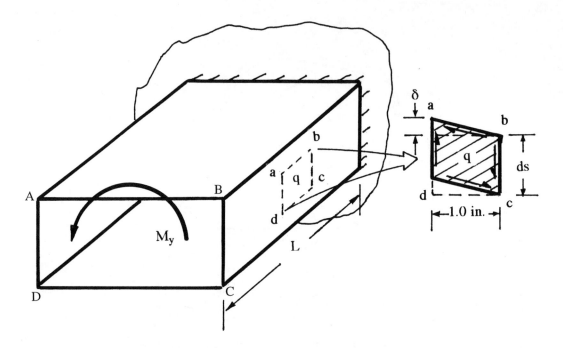

Fig. 8.2.3 Twist of Single Cell

Also $\quad q = \dfrac{M_y}{2A}$

$$\delta = \dfrac{f_{ts}}{G} = \dfrac{q}{Gt} = \dfrac{M_y}{2AGt}$$

where: A = Enclosed area of the box A-B-C-D

Strain energy:

$$dU = \dfrac{q\,\delta\,ds}{2}$$

$$= (\dfrac{M_y}{2A})(\dfrac{M_y}{2AGt})(\dfrac{ds}{2})$$

$$= (\dfrac{M_y^2}{8A^2Gt})ds$$

Total strain energy:

$$U = \oint (\dfrac{M_y^2}{8A^2Gt})ds$$

$$= (\dfrac{M_y^2}{8A^2G})\oint \dfrac{ds}{t}$$

where $\oint \dfrac{ds}{t}$ is the line integral and is the integral of the expression $\dfrac{ds}{t}$ around the cell. If web thickness is constant for each wall, $\oint \dfrac{ds}{t}$ is the summation of the ratio of wall length to wall thickness for the cell.

Using Castigliano's Theorem, the twist angle 'θ' will be obtained by:

$$\theta = \frac{\partial U}{\partial M_y} = \frac{\partial(\frac{M_y^2}{8A^2G})\oint \frac{ds}{t}}{\partial M_y}$$

$$= (\frac{M_y}{4A^2G})\oint \frac{ds}{t}$$

$$= (\frac{q\,2A}{4A^2G})\oint \frac{ds}{t}$$

$$= (\frac{1}{2AG})\oint q\frac{ds}{t}$$

The above expression has been derived from a unit length of the box. The twist over the entire length, L, of the box will therefore be obtained by:

$$\theta = (\frac{L}{2AG})\oint q\frac{ds}{t}$$

The above equation can be rewritten for a single section as below:

$$\theta = (\frac{L}{2AG})(\frac{M_y}{2A})(\frac{4A^2}{J}) = \frac{M_yL}{GJ}$$

where: $\oint \frac{ds}{t} = \frac{4A^2}{J}$ (From Eq. 6.8.4)

$q = \frac{M_y}{2A}$ (From Eq. 8.2.3)

Consider the two cell structure shown in Fig. 8.2.4 and let q_1 be the shear flow on the free walls of cell 1 and q_2 the shear flow for the free walls of cell 2.

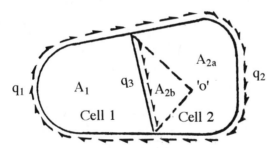

Fig. 8.2.4 Shear Flow Distribution of Two-Cell Box

Torsional moment in terms of internal shear flow:

Taking moment of shear flow about a joining point 'o',

$$M_{y,o} = 2q_1(A_1 + A_{2b}) + 2q_2A_{2a} - 2q_3A_{2b}$$
$$= 2q_1A_1 + 2q_1A_{2b} + 2q_2A_{2a} - 2q_3A_{2b}$$
$$= 2q_1A_1 + 2q_1A_{2b} + 2q_2A_{2a} - 2(q_1 - q_2)A_{2b}$$
$$= 2q_1A_1 + 2q_2A_2$$

When a given torque is applied to the multi-cell box beam, consideration of equilibrium requires that

$$M_y = \Sigma_{i=1}^{n}(2q_iA_i) \qquad \text{Eq. 8.2.4}$$

The moment of internal shears for a multi-cell box structure is equal to the sum of twice the area of each cell multiplied by the shear flow in its free walls.

If q_3 is the shear flow in the common wall, then for equilibrium:

$$q_3 = q_1 - q_2$$

Also condition of compatibility requires that the twist of all the cells must be the same, i.e.

$$\theta = (\frac{q_i}{2A_iG})\oint \frac{ds}{t} \qquad \text{Eq. 8.2.5}$$

Combining of Eq. 8.2.4 and Eq. 8.2.5 to obtain the values q_{in} and q_i ...... q_n and θ in a multi-cell box structure.

Example:

Determine shear flows in all walls as shown below:

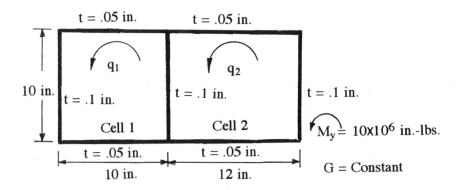

Assume the shear flow in cell 1 is q_1 and in cell 2 is q_2. Their angles of rotation are:

$$\theta_1 = (\frac{1}{2A_1G})\oint q \times \frac{ds}{t}$$

$$2G\theta_1 = (\frac{1}{10 \times 10})[q_1(\frac{10}{0.1} + \frac{10}{0.05} + \frac{10}{0.1} + \frac{10}{0.05}) - q_2(\frac{10}{0.1})]$$

$$2G\theta_1 = 6q_1 - q_2$$

$$\theta_2 = (\frac{1}{2A_2G})\oint q \times \frac{ds}{t}$$

$$2G\theta_2 = (\frac{1}{10 \times 12})[q_2(\frac{12}{0.05} + \frac{10}{0.1} + \frac{12}{0.05} + \frac{10}{0.1}) - q_1(\frac{10}{0.1})]$$

$$2G\theta_2 = (\frac{17}{3})q_2 - (\frac{5}{6})q_1$$

But $\theta_1 = \theta_2$

$$6q_1 - q_2 = (\frac{17}{3})q_2 - (\frac{5}{6})q_1$$

$$q_2 = (\frac{41}{40})q_1 \qquad \text{Eq. (A)}$$

Also for equilibrium:

$$M_y = 2A_1q_1 + 2A_2q_2$$

$$10^6 = 200q_1 + 240q_2 \qquad \text{Eq. (B)}$$

Solve Eq. (A) and Eq. (B), obtain:

$q_1 = 2{,}242$ lbs./in.; $q_2 = 2{,}298$ lbs./in.

$q_{12} = q_1 - q_2 = -56$ lbs./in.

Final loading:

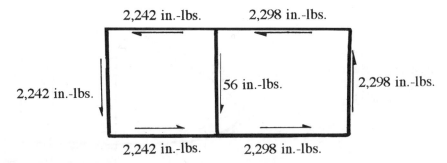

(C) SHEAR CENTER

The shear center for any beam bending becomes a main reference point for both shear and torsion loading. Since the torsional stiffness and strength of an open section is quite small, it is recommended to keep the applied load at or close to the shear center as shown in Fig. 8.2.5 and for any segments of shell structures, the following rules should be kept in mind:

- The applied load acts parallel to a straight line connecting the centroids of the two adjacent flanges (or stringers)
- Passes through the shear center a distance $e = \dfrac{2A_o}{h}$ from this straight line.
- Has the magnitude $= q\,h$

The shear center for open cells is located by maintaining rotational equilibrium between the applied load and reactive web shear flows.

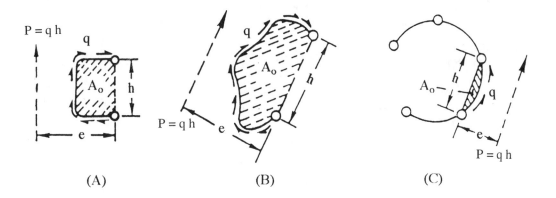

Fig. 8.2.5 Shear Center of Three Open Shell Sections

The familiar D section, shown in Fig. 8.2.6, which is a part of a nose section of an airfoil, is used for ailerons, elevators, rudder, flaps, etc., on general aviation and small aircraft airframes. The D-section may be considered a combination of a beam and a torsion box. Since this closed cell is stable under bending and torsion loads, a different procedure is required:

Chapter 8.0

P_n – Resultant of shear flow on nose skin
P_w – Resultant of shear flow on vertical web

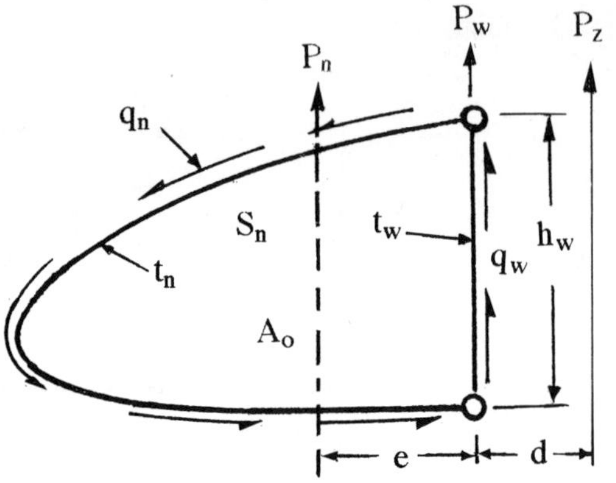

Fig. 8.2.6 Shear Center of a D-Cell Section

D-CELL WITH TWO-STRINGERS SECTION:

Assumptions

(a) Bending moment is taken by beam stringers only:
  - The vertical load, P_z, must act parallel to the vertical web of the D-cell beam
  - There can be no component of load in any other direction since there is no stringer to resist it

(b) Torsion moment is taken by the cell (the shear flows are carried by both the nose skin and vertical web).

(c) The total shear load acting on the cell may be resolved into two components:
  - The shear load P_w acting in the web plane
  - The shear load P_n in the forward or aft skin acting at the shear center of the section

From static:
$$\Sigma F_z = 0: \quad P_z = -P_w - P_n$$

$$\Sigma M_w = 0: \quad P_z d = P_n e_n \qquad \therefore P_n = \frac{P_z d}{e_n}$$

Since $e_n = \dfrac{2A_o}{h}$ then, $P_n = \dfrac{P_z dh}{2A_o}$

Therefore, $P_w = -P_z - \dfrac{P_z dh}{2A_o} = -P_z(1 + \dfrac{dh}{2A_o})$

Also, $q_n = \dfrac{P_n}{h} = -\dfrac{P_z d}{2A_o}$ \hfill Eq. 8.2.6

$q_w = \dfrac{P_w}{h} = -P_z(\dfrac{1}{h} + \dfrac{d}{2A_o})$ \hfill Eq. 8.2.7

With the skin and web thickness known, the shear stresses are:

Shear stress in nose skin: $\quad f_{ts,n} = \dfrac{q_n}{t_n}$

Shear stress in vertical web: $\quad f_{ts,w} = \dfrac{q_w}{t_w}$

The stringer (flange) stress can be found from the beam bending formula of $f = \dfrac{M_y}{I}$, which in this case becomes $f = \dfrac{M}{hA_f}$, where h is the beam height and A_f is the beam stringer area.

The angle of twist per unit length of this D-cell can be determined as follows:

$$\theta = \frac{T}{GJ}$$

Since $T = 2qA_o$ and $J = \dfrac{4A_o^2}{\oint \dfrac{ds}{t}}$

$$\theta = \frac{\Sigma q \dfrac{ds}{t}}{2A_o G} \qquad \text{Eq. 8.2.8}$$

Substituting Eq. 8.2.6 and Eq. 8.2.7 into Eq. 8.2.8, the angle of twist per unit length is:

$$\theta = \left(\frac{1}{2GA_o}\right)\left(\frac{q_n S_n}{t_n} + \frac{q_w h}{t_w}\right) \qquad \text{Eq. 8.2.9}$$

where: S_n – Perimeter of nose skin contour
t_n – Nose skin thickness
t_w – Web thickness
G – Modulus of rigidity (shear modulus)

Note that in Eq. 8.2.6 and Eq. 8.2.7:

- If P_z is applied at the web, q_n becomes zero
- If P_z is applied at the shear center for the nose skin, then q_w becomes zero
- If P_z is applied at the shear center for the cell, then the angle of twist, θ, is zero

The location of the shear center can be obtained from Eq. 8.2.9. Thus, assuming the shear modulus are equal, $G_n = G_w$ (common practice in airframe design).

$$\frac{S_n}{2A_o t_n} = -\frac{1}{dt_w} - \frac{h}{2A_o t_w}$$

From which, the distance from web to shear center, d = e

$$e = \frac{-2A_o t_n}{S_n t_w + h t_n} \qquad \text{Eq. 8.2.10}$$

Example:

Given section with geometry and load as shown below, determine the shear stress in the skin and web, and the location of the shear center for the cell.

$P_z = 3,000$ lbs. and $M_y = -24,000$ in.-lbs.

$t_n = 0.032$ in.

$S_n = 38$ in. (nose perimeter)

$t_w = 0.04$ in.

$h_w = 8.6$ in.

$A_o = 144$ in.^{2}

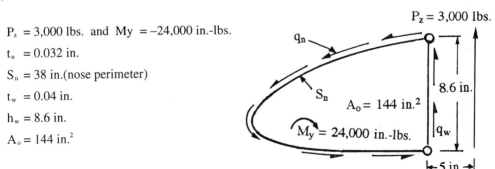

Chapter 8.0

From $\Sigma M_w = 0$ about lower point on the web:

$$-24,000 + 3,000 \times 5 = q_n \times 2 \times 144 \quad \therefore q_n = -31.3 \text{ lbs./in.}$$

From $\Sigma P_z = 0$:

$$3,000 = -31.3 \times 8.6 + q_w \times 8.6 \quad \therefore q_w = 318 \text{ lbs./in.}$$

The shear center can be determined from Eq. 8.2.10 (assuming that $G_n = G_w$):

$$e = \frac{-2 \times 144 \times 0.032}{38 \times 0.04 + 8.6 \times 0.032} = -5.13 \text{ in.}$$

The minus sign, of course, indicates the distance to the left of the web, see Fig. 8.2.6.

The shear stresses are:

In the nose skin: $f_{ts,n} = \dfrac{31.3}{0.032} = 980$ psi

In the vertical web: $f_{ts,w} = \dfrac{318}{0.04} = 7,950$ psi

D-CELL WITH THREE-STRINGERS SECTION:

The procedure for three-stringers section is:

(a) Solve for the closed cell web shear flows. If the applied loads are not applied at the shear center, the cell will undergo some angle of twist.

(b) Calculate the shear flow which would twist the cell back into its original position.

(c) The sum of shear flows from step (a) and (b) represent the shear flow in the cell when the load is applied at the shear center (zero twist).

(d) Locate the line of action of the shear center by maintaining rotation equilibrium between the shear flows in step (c) and the applied load.

Example:

Calculate the line of action of the shear center for the three-stringer closed cell as shown below.

$t_{ab} = 0.08"$ $\quad S_{ab} = 10"$
$t_{bc} = 0.06"$ $\quad S_{bc} = 10"$
$t_{ac} = 0.04"$ $\quad S_{ac} = 25"$
Cell enclosed area, $A_o = 100$ in.^{2}
Cell circumference = 45 in.

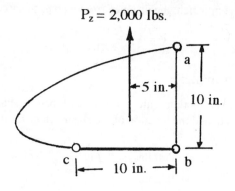

The web shear flows can be found by static:

$\Sigma M_b = 0$:

$$-2,000 \times 5 = 2 q_{ac} \times 100$$

$$\therefore q_{ac} = -50 \text{ lbs./in.}$$

$\Sigma F_x = 0$:

$q_{ac} \times 10 = q_{bc} \times 10$

$\therefore q_{bc} = -50$ lbs./in.

$\Sigma F_z = 0$:

$q_{ab} \times 10 + q_{ac} \times 10 = 2,000$

$\therefore q_{ab} = 150$ lbs./in.

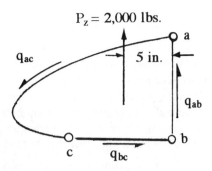

The cell with the 2,000 lbs. load applied at 5 inches away from a-b will undergo bending and rotation. A balance shear flow, q_t, is then applied to this cell to twist it back to its original position.

From Eq. 8.2.8, the angle of twist is $\theta = \dfrac{\Sigma q(\frac{ds}{t})}{2 A_o G}$. Therefore, $\Sigma q(\dfrac{ds}{t}) = 0$ and solve for q_t (counterclockwise is positive):

$$\dfrac{-50(25)}{0.04} + \dfrac{q_t(25)}{0.04} - \dfrac{50(10)}{0.06} + \dfrac{q_t(10)}{0.06} + \dfrac{150(10)}{0.08} + \dfrac{q_t(10)}{0.08} = 0$$

$\therefore q_t = 22.73$ lbs./in.

The resulting shear flows represent the shear flows in the cell when the 2,000 lbs. load is applied at the shear center. The final shear flows are shown below:

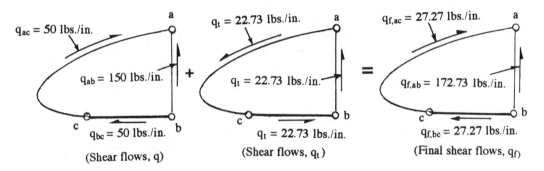

The shear center, e, is calculated below:

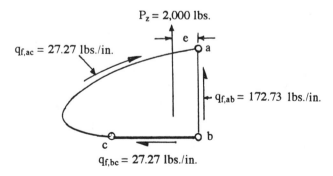

The distance 'e' can be then calculated from $\Sigma M_b = 0$:

$2,000 \times e = 2(27.27)(100)$

$\therefore e = 2.73$ in. (the distance to shear center)

Chapter 8.0

8.3 SINGLE-CELL BOX (TWO-STRINGER SECTION)

The method of analysis for a single-cell box was discussed previously under Shear Center; now a "cut" web method to determine shear flows will be introduced. The determination of shear flows in the webs of a single-cell section due to a concentrated load which is not applied at the shear center and which produces bending and torsion in the cell section is slightly more complicated than the determination of shear flows in an open-cell section. The complication arises because each end-load carrying element has two reaction load paths and only one equation of equilibrium, that is, $\Sigma F_z = 0$. To overcome the difficulty the following procedure can be adopted:

- Cut one of the webs, thus making the cell an open cell, and therefore a statically determinate structure
- Solve for the shear flows in the open cell, the so-called qo system
- Replace the cut web and apply a closing torsional shear flow qt system
- The sum of the q_o and q_t shear flow systems represents the final shear flows in the cell webs

Consider a typical nose D-cell of a wing section in which the vertical web represents the shear beam and the nose skin has no reinforced stringers. Obviously classic beam theory does not apply directly to this cell structure because it has two shear flow paths namely in the nose skin and the vertical web. The following example will illustrate the simple method of solution:

Example 1:

For a typical nose cell of a wing section, calculate the shear flows in the webs.

$P_z = 10,000$ lbs.

Area of upper and lower stringer:

$A = 1.0$ in.^{2}

Enclosed area of the D-cell

$A_o = 100$ in.^{2}

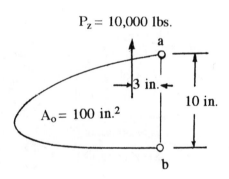

Cut the nose skin as shown to form an open cell:

Balance the open cell:

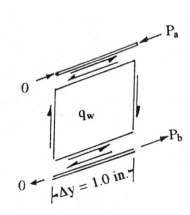

236

Moment of inertia about the neutral axis:
$$I_x = 2(1.0 \times 5) = 50 \text{ in.}^4$$

Upper and lower stringer axial loads:
$$P_b = -P_a = (P_z \times \frac{\Delta y}{I_x}) \times z_{stringer} \times A_{stringer}$$
$$= (10,000 \times \frac{1.0}{5}) \times 5 \times 1.0 = 1,000 \text{ lbs.}$$
$$q_w = \frac{\Delta P}{\Delta y} = \frac{1,000 - 0}{1.0} = 1,000 \text{ lbs./in.}$$

Close cell:

Determine the torque unbalance, $M_{y,t}$, which must be applied to create internal shear flows, q_t, (positive shear flow as shown) to establish equilibrium. Take the moment about point 'b' and, then,

$$\Sigma M_{y,t} = 0 = -10,000 \times 3 + q_t \times 2 \times 100$$
$$\therefore q_t = 150 \text{ lbs/in.}$$

Find the sum of the shear flows of q_o and q_t as shown below:

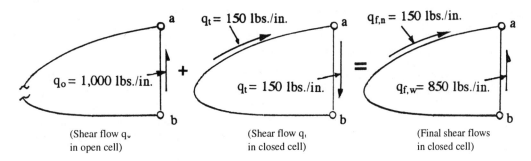

(Shear flow q_w in open cell)   (Shear flow q_t in closed cell)   (Final shear flows in closed cell)

Example 2:

Given a cylindrical shell (D = 10") with upper and lower stringers as shown below, determine the shear flows in the skins.

Enclosed area of the shell, $A_o = \frac{\pi 10^2}{4}$

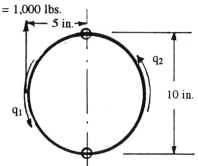

$\Sigma F_z = 0$:

$$-q_{z,1} = q_{z,2} = \frac{1,000}{20} = 50 \text{ lbs./in.}$$

$\Sigma M_y = 0$:

$$q_{t,1} = q_{t,2} = \frac{T}{2A_o} = \frac{-5,000}{2}(\pi \times \frac{10^2}{4}) = -32 \text{ lbs./in.}$$

Therefore, $q_1 = q_{z,1} + q_{t,1} = (-50) + (-32) = -82$ lbs./in.

$q_2 = q_{z,2} + q_{2,t} = 50 + (-32) = 18$ lbs./in.

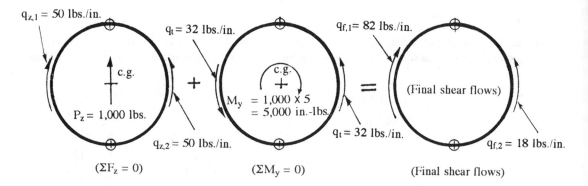

8.4 SINGLE-CELL BOX (MULTI-STRINGER SECTION)

As mentioned in simple beam analysis, it is assumed that the bending moments are resisted by the stringers (flanges) in proportion to the $\frac{Mz}{I}$ distribution. The shear flows must be such as to produce these stringer loads or vice versa. Since the distribution of stringer loads to resist bending is assumed to be the same for the open as for the closed sections:

- The closed cell could be simplified by temporarily assuming one of the skins or webs is cut
- The shear forces act at the shear center of this open section
- Having satisfied the bending conditions, the remaining unresisted loading on this section is the torsional moment caused by transferring the shear forces to the shear center
- This unbalanced moment will act on the torsion cell as a uniform shear flow $q = \frac{M_y}{2A}$ which will add directly into the shear flows for the temporarily cut section without affecting the stringer axial loads

It will be recalled that in an open section the shear center is located so as to produce shear flows in the webs that maintain the $\frac{Mz}{I}$ distribution in the stringers. In a cellular structure, however, the $\frac{Mz}{I}$ distribution will be maintained regardless of the location of the external shear force:

- The shear center for a cell is therefore thought of as the point at which an external shear force must be applied so as to obtain no torsional rotation or twist of the section
- It should be noted that the shear flows determined by equilibrium conditions are independent of the web thicknesses while the shear center of the closed cell will be greatly affected by the shear deflection and therefore the web thicknesses

The shear flows distributed to the webs or skins of the cell:
- Must be in static equilibrium with the externally applied forces at any cut section
- It is more convenient not to work with the shear center of the temporarily cut section
- Any reference point can be selected (usually select the c.g. of the section)
- The difference in torsional moment, M_y, from the externally applied loads and that of the cut section shear flows must simply be corrected by a uniform shear flow q = $\frac{M_y}{2A}$ acting all around the cell.

There are other considerations should be kept in mind when doing calculations:
- Creation of a tabulation with help from a desk top calculator facilitates calculation
- The basic data supplied to the computing group must be accurate
- The sign conventions must be consistant through out the entire calcualtion, see Chapter 2.0. for more information
- If there are too many stringers in a box beam, they should be evenly lumped (except spar caps) into a number of groups to simplify the calculation (refer to LUMPING STRINGER AREAS in Chapter 5.6)
- Results of calculations must be carefully inspected and spot-checks made
- An inadvertent reversal of the sign of a torsional moment or torque results in completely misleading shear flows

A simple multi-stringer single cell box is used to illustrate the procedure for determining shear flow in the webs or skin between stringers. Refer to Reference 8.8 for the method of detail analysis which should be consulted. This reference explains the procedure used by most of the aircraft industry for the analysis of box beam structures. It is essential that the stress engineer shall be familiar with this method.

Since the true neutral axis would vary with each condition investigated, the standard procedure is to work with reference axes parallel to the box axes through the c.g. of the section, as shown in Fig. 8.4.1. The box is usually unsymmetrical and bending stresses on this section should be considered; this phenomenon was discussed in Chapter 6.3 (refer to Eq. 6.3.2 and change the y to z) The true axial stresses in the longitudinals are then obtained from the following equation:

$$f_b = \frac{-(M_z I_x + M_x I_{xz})x + (M_x I_z + M_z I_{xz})z}{I_x I_z - (I_{xz})^2}$$ Eq. 8.4.1

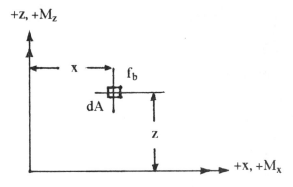

(Use right-hand rule for moments)

Fig. 8.4.1 The Principal Stresses

Chapter 8.0

(A) SHEAR FLOWS

From simple beam theory and Eq. 6.5.2 the shear stress is,

$$f_s = \frac{VQ}{Ib}$$

or the shear flow is $q = \frac{VQ}{I}$

$$q = (\frac{M}{y})[\frac{\int z(dA)}{I_x}]$$

$$= \frac{(\frac{M_{x,2} \times z}{I_x})(dA) - (\frac{M_{x,1} \times z}{I_x})(dA)}{\Delta y}$$

$$= \frac{(f_2)(dA) - (f_1)(dA)}{\Delta y}$$

$$= \frac{P_2 - P_1}{\Delta y}$$

Where P_2 and P_1 represent longitudinal axial load in a given stringer at successive stations apart, and, q, is the shear flow inducing this change, as shown in Fig. 8.4.2.

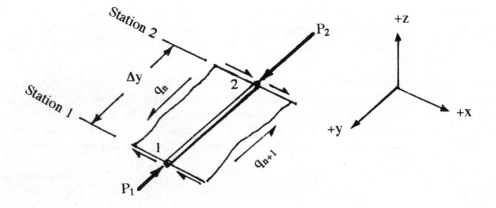

Fig. 8.4.2 *Axial Load and Shear Flows on a Member*

The longitudinal axial stresses are determined from $\frac{M_x z}{I_x}$. When calculating these stresses the principal axes must be considered. These stresses are multiplied by the appropriate longitudinal area to get load, P.

Once the axial stress is computed for adjacent sections, the procedure for computing shear flow in the skins or webs connecting the stringers is the same as previously described.

The procedure is as follows:

(a) Cut any one web or skin between stringers in the box (it is good practice to place the "cut" where the shear flow is near zero).

(b) Cut redundant webs and calculate "cut-web" shear flows by using:

- If not symmetrical (from Eq. 8.4.1):

$$f_b = \frac{-(M_zI_x + M_xI_{xz})x + (M_xI_z + M_zI_{xz})z}{I_xI_z - (I_{xz})^2}$$

- Let $M_z = 0$ because the contribution of stress on f_b is small since the M_z value is generally relatively small and the I_z value is quite large

$$f_b = \frac{-(M_zI_{xz})x + (M_xI_z)z}{I_xI_z - (I_{xz})^2} \qquad \text{Eq. 8.4.2}$$

- If symmetrical about x - axis ($I_{xz} = 0$) and let $M_z = 0$.

  Using $f_b = \dfrac{M_x z}{I_x}$ \qquad Eq. 8.4.3

- $\Delta P = f_{bi} A_i$

- "Cut-web" shear flows, $q_o = \dfrac{\Delta P}{\Delta y}$

(c) Impose:
  - $\Sigma M_y = 0$
  - Check balancing shear flows: q_t

(d) Add q_t to "cut-web" shear flows, q_o, to final shear flows for the box beam.

(B) SYMMETRICAL BOX

The procedural steps and assumptions are as follows:

(a) Cut any one web or skin between stringers in the box

(b) Determine "Cut-web" shear flows by using:

- Symmetrical about x - axis ($I_{xz} = 0$) and $M_z = 0$.

  Using $f_b = \dfrac{M_x z}{I_x}$

- $\Delta P = f_{bi} A_i$

- "Cut-web" shear flows, $q_o = \dfrac{\Delta P}{\Delta y}$

(c) Impose:
  - $\Sigma M_y = 0$
  - Check balancing shear flows: q_t

(d) Add q_t to "cut-web" shear flows, q_o, to obtain the final shear flows for the box beam.

Example 1A:

(Note: This is the first of four example calculations which use four different box beams as well as different methods of analysis to illustrate the solutions for these types of structures).

Given a constant symmetrical wing box section (symmetrical about the x-axis) and load as shown below, determine the shear flow in the various webs. Assume there are two adjacent stations, namely station 1 and station 2, and the distance between them is y = 10 in. (wing spanwise).

Assume a vertical load of $P_z = 5,000$ lbs. applied at station 1 and assume moment

- At station ① is $M_{x,1} = -500,000$ in.-lbs.
- At station ② is $M_{x,2} = M_{x,1} + (-P_z \times 10) = -550,000$ in.-lbs.

Stringer areas:

$A_1 = 0.5$ in.^{2}
$A_2 = 0.5$ in.^{2}
$A_3 = 0.5$ in.^{2}
$A_4 = 1.0$ in.^{2}
$A_5 = 1.0$ in.^{2}
$A_6 = 0.5$ in.^{2}

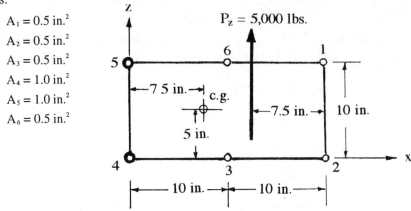

(1) Set up a table to determine section properties:

Stringer No.	A	x	z	A x	A z	A x^2	A z^2
1	0.5	20	10	10	5	200	50
2	0.5	20	0	10	0	200	0
3	0.5	10	0	5	0	50	0
4	1.0	0	0	0	0	0	0
5	1.0	0	10	0	10	0	100
6	0.5	10	10	5	5	50	50
Σ	4.0			30	20	500	200

Location of c.g.: $X = \frac{30}{4} = 7.5$ in. and $Z = \frac{20}{4} = 5$ in.

Moment of inertia:

$I_x = \Sigma Az^2 - \Sigma A \times Z^2 = 200 - 4(5)^2 = 100$ in.^{4}
$I_z = \Sigma Ax^2 - \Sigma A \times X^2 = 500 - 4(7.5)^2 = 275$ in.^{4}

(2) Determine stringer load P_1 (at station 1) and P_2 (at station 2) values:

(Note: Since P_z is up, the induced load in stringers No. 1, 5, and 6, which are above the neutral axis, must be compression).

Stringer No.	A (in.^{2})	(z - Z) (psi)	$f_1 = \frac{M_{x,1}(z-Z)}{I_x}$ (psi)	$f_2 = \frac{M_{x,2}(z-Z)}{I_x}$ (lbs.)	$P_1 = f_1 \times A$ (lbs.)	$P_2 = f_2 \times A$ (lbs.)
1	0.5	+5	−25,000	−27,500	−12,500	−13,750
2	0.5	−5	25,000	27,500	12,500	13,750
3	0.5	−5	25,000	27,500	12,500	13,750
4	1.0	−5	25,000	27,500	25,000	27,500
5	1.0	+5	−25,000	−27,500	−25,000	−27,500
6	0.5	+5	−25,000	−27,500	−12,500	−13,750

(3) With the station distance $\Delta y = 10.0$ in. and skin 6-1 temporarily cut and the 'cut' shear flows, $q_{o,}$:

Box Beams

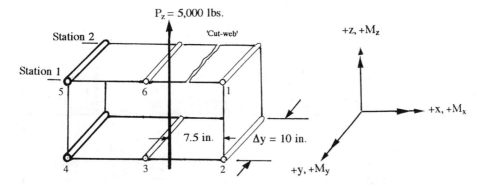

(Note: Counterclockwise shear flow is positive torque moment, M_y, (right-hand rule) and the following "cut" shear flows are independent of torque ($P_z \times d$) and web thickness).

Stringer No.	Web No.	$\Delta P = (P_2 - P_1)$	$\Delta q = \dfrac{\Delta P}{\Delta y}$	$q_o = \Sigma \Delta q$
1		−1,250	−125	
	1-2			−125
2		1,250	125	
	2-3			0
3		1,250	125	
	3-4			125
4		2,500	250	
	4-5			375
5		−2,500	−250	
	5-6			125
6		−1,250	−125	
	6-1(cut)			0

The following calculations are for the above "cut" shear flow, $q_o = \Sigma \Delta q$:

$q_{o, 1\text{-}2} = \Delta q_{1\text{-}2} = -125$ lbs./in.

$q_{o, 2\text{-}3} = q_{o, 1\text{-}2} + \Delta q_{2\text{-}3} = -125 + (125) = 0$

$q_{o, 3\text{-}4} = q_{o, 2\text{-}3} + \Delta q_{3\text{-}4} = 0 + (125) = 125$ lbs./in.

$q_{o, 4\text{-}5} = q_{o, 3\text{-}4} + \Delta q_{4\text{-}5} = 125 + (250) = 375$ lbs./in.

$q_{o, 5\text{-}6} = q_{o, 4\text{-}5} + \Delta q_{5\text{-}6} = 375 + (-250) = 125$ lbs./in.

$q_{o, 6\text{-}1} = q_{o, 5\text{-}6} + \Delta q_{6\text{-}1} = 125 + (-125) = 0$ (Shear flow in this web should be zero because it was cut)

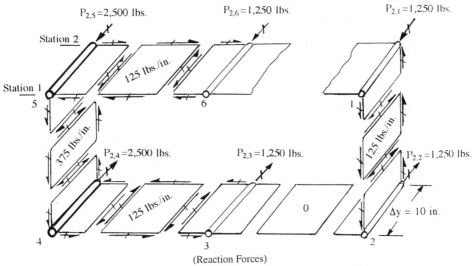

(Reaction Forces)
(Free-body Diagrams – counterclockwise shear flow is positive ⟲+)

Chapter 8.0

(4) Take the moments about the c.g. as the reference point "o" (any arbitrarily selected point could be used, but in this case the c.g. of the section was used) to calculate each "segment area" as shown:

$A_{o,1-2} = 10 \times (\frac{20 - 7.5}{2}) = 62.5$ in.^{2}

$A_{o,2-3} = \frac{10 \times 5}{2} = 25$ in.^{2}

$A_{o,3-4} = \frac{10 \times 5}{2} = 25$ in.^{2}

$A_{o,4-5} = \frac{10 \times 7.5}{2} = 37.5$ in.^{2}

$A_{o,5-6} = \frac{10 \times 5}{2} = 25$ in.^{2}

$A_{o,6-1} = \frac{10 \times 5}{2} = 25$ in.^{2}

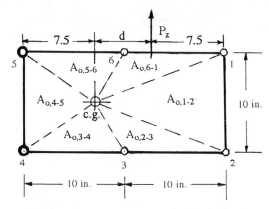

(Define the segment areas, A_o, in the box)

(5) The torsion moment, $M_{y,o}$, due to the "cut" shear flows, q_o, is computed below:

Web No.	q_o	2A	$M_{y,o} = q_o \times 2A$
1-2	−125	125	−15,625
2-3	0	50	0
3-4	125	50	6,250
4-5	375	75	28,125
5-6	125	50	6,250
6-1	0	50	0
		$\Sigma 2A = 400$ in.^{2}	$\Sigma M_o = 25,000$ in.-lbs.

(Above moment, $\Sigma M_{y,o} = 25,000$ in.-lbs. is due to a "cut" shear flow.)

(6) A moment summation is then made about the same point (at c.g.) as before in order to determine the torque unbalance, $M_{y,t}$, which must be applied to create internal shear flows to establish equilibrium.

$\Sigma M_{c.g.} = P_z \times d + M_{y,o} + M_{y,t} = 0$

$= 5,000 \times 5 + 25,000 + M_{y,t} = 0$

$M_{y,t} = -50,000$ in-lbs. $= 2 q_t A$

From which: $q_t = \frac{M_{y,t}}{2A} = \frac{-50,000}{400} = -125$ lbs/in. (clockwise shear flow)

(7) The final shear flows become:

Web No.	q_o (due to "cut")	q_t (due to balanced torque, $M_{y,t}$)	q (final shear flows)
1-2	−125	−125	−250
2-3	0	−125	−125
3-4	125	−125	0
4-5	375	−125	250
5-6	125	−125	0
6-1	0	−125	−125

(8) The correct final shear flows are shown on the sketch below and the arrows indicate the shear flow direction.

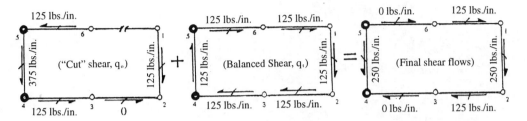

(Final Shear Flow Diagram – reaction forces)

Example 1B:

Use the same given beam as in the previous Example 1A, except assume a vertical load of $P_z = 5,000$ lbs. still applied at station 1 and assume moment at station 1 is $M_{x,1} = 0$ and then at station 2 is $M_{x,2} = -P_z \times 10 = -50,000$ in.-lbs. The total load application is exactly the same as example 1A. and the results are also the same as shown by the following calculation.

Determine stringer load P_1 and P_2 values:

Stringer No.	A (in.²)	(z – Z) (in.)	$f_1 = \dfrac{M_{x,1}(z-Z)}{I_x}$ (psi)	$f_2 = \dfrac{M_{x,2}(z-Z)}{I_x}$ (psi)	$P_1 = f_1 \times A$ (lbs.)	$P_2 = f_2 \times A$ (lbs.)
1	0.5	+5	0	−2,500	0	−1,250
2	0.5	−5	0	2,500	0	1,250
3	0.5	−5	0	2,500	0	1,250
4	1.0	−5	0	2,500	0	2,500
5	1.0	+5	0	−2,500	0	−2,500
6	0.5	+5	0	−2,500	0	−1,250

The "cut" shear flows, q_o:

Web No.	$\Delta P = (P_2 - P_1)$	$\Delta q = \dfrac{\Delta P}{\Delta y}$	$q_o = \Sigma \Delta q$
1-2	−1,250 − 0 = −1,250	−125	−125
2-3	1,250 − 0 = 1,250	125	0
3-4	1,250 − 0 = 1,250	125	125
4-5	2,500 − 0 = 2,500	250	375
5-6	−2,500 − 0 = −2,500	−250	125
6-1	−1,250 − 0 = −1,250	−125	0

The above values of 'q_o' are identical to that of the previous calculation (step (3) of example 1A) and final shear flows are therefore a repetition of those shown previously.

(C) UNSYMMETRICAL SECTIONS

The procedure steps are:

(a) Cut any one web or skin between stringers in the box.

(b) "Cut-web" shear flows by using:

Chapter 8.0

- let $M_z = 0$
- From Eq. 8.4.2:

$$f_b = \frac{-(M_x I_{xz})x + (M_x I_z)z}{I_x I_z - (I_{xz})^2}$$

- $\Delta P = f_{bi} A_i$
- "Cut-web" shear flows, $q_o = \dfrac{\Delta P}{\Delta y}$

(c) Impose:
- $\Sigma M = 0$
- Check balancing shear flows: q_t

(d) Add q_t to "cut-web" shear flows, q_o, to obtain the final shear flows for the box beam.

Example 2:

Use the same given loading and box beam used in Example 1A, except rearrange locations of stringers No. 5 and 6 to make this box become an unsymmetrical section as shown below:

- At station 1 is $M_{x,1} = 0$
- At station 2 is $M_{x,2} = 0 + (-P_z \times 10) = -50,000$ in.-lbs.

Stringer areas:

$A_1 = 0.5$ in.^{2}
$A_2 = 0.5$ in.^{2}
$A_3 = 0.5$ in.^{2}
$A_4 = 0.5$ in.^{2}
$A_5 = 1.0$ in.^{2}
$A_6 = 1.0$ in.^{2}

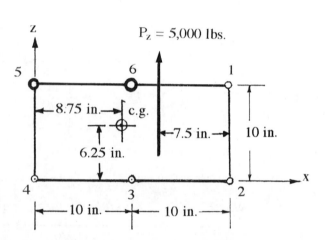

(1) Set up a table to determine section properties:

Segment	A(in.^{2})	x(in.)	z(in.)	Ax(in.^{3})	Az(in.^{3})	Ax^2(in.^{4})	Az^2(in.^{4})	Axz(in.^{4})
1	0.5	20	10	10	5	200	50	100
2	0.5	20	0	10	0	200	0	0
3	0.5	10	0	5	0	50	0	0
4	0.5	0	0	0	0	0	0	0
5	1.0	0	10	0	10	0	100	0
6	1.0	10	10	10	10	100	100	100
Σ	4.0			35	25	550	250	200

$X = \dfrac{35}{4.0} = 8.75$ in.; $Z = \dfrac{25}{4.0} = 6.25$ in.

$I_x = \Sigma Az^2 - A \times Z^2 = 250 - 4.0(6.25)^2 = 93.75$ in.^{4}
$I_z = \Sigma Ax^2 - A \times X^2 = 550 - 4.0(8.75)^2 = 243.75$ in.^{4}

$$I_{xz} = \Sigma Axz - A \times X \times Z = 200 - 4.0(8.75 \times 6.25) = -18.75 \text{ in.}^4$$

Let $\ni_1 = (I_xI_z) - (I_{xz})^2 = 93.75 \times 243.75 - (-18.75)^2 = 0.0225 \times 10^6 \text{ in.}^8$

(2) With the station distance $\Delta y = 10.0$ in. and skin 6-1 temporarily cut and the "cut" shear flows ($f_1 = 0$ and $P_1 = 0$ at station 1):

$$f_1 = \frac{-(M_{x,1}I_{xz})(x-X) + (M_{x,1}I_z)(z-Z)}{I_xI_z - (I_{xz})^2} = 0$$

$$f_2 = \frac{-(M_{x,2}I_{xz})(x-X) + (M_{x,2}I_z)(z-Z)}{I_xI_z - (I_{xz})^2}$$

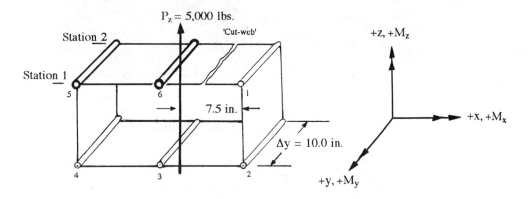

Stringer No.	A	(x – X)	(z – Z)	$\ni_2 = -(M_{x,2})(I_{xz})(x-X)$	$\ni_3 = (M_{x,2})(I_z)(z-Z)$	$\ni_2 + \ni_3$
1	0.5	11.25	3.75	-10.55×10^6	-45.7×10^6	-56.25×10^6
2	0.5	11.25	–6.25	-10.55×10^6	76.17×10^6	65.62×10^6
3	0.5	1.25	–6.25	-1.17×10^6	76.17×10^6	75×10^6
4	0.5	–8.75	–6.25	8.2×10^6	76.17×10^6	84.37×10^6
5	1.0	–8.75	3.75	8.2×10^6	-45.7×10^6	-37.5×10^6
6	1.0	1.25	3.75	-1.17×10^6	-45.7×10^6	-46.87×10^6

Stringer No.	Web No.	$f_2 = \dfrac{\ni_2 + \ni_3}{\ni_1}$	$P_2 = f_2 \times A$	$\Delta q = \dfrac{P_2 - P_1}{\Delta y}$	$q_o = \Sigma \Delta q$
1	1-2	–2,500	–1,250	–125	–125
2	2-3	2,916	1,458	146	21
3	3-4	3,333	1,667	167	188
4	4-5	3,750	1,875	187	375
5	5-6	–1,667	–1,667	–167	208
6	6-1	–2,083	–2,083	–208	0

The following calculations are for the above "cut" shear flow, $q_o = \Sigma \Delta q$:

$q_{o,1\text{-}2} = \Delta q_1 = -125$ lbs./in.

$q_{o,2\text{-}3} = q_{o,1\text{-}2} + \Delta q_2 = -125 + (146) = 21$ lbs./in.

$q_{o,3\text{-}4} = q_{o,2\text{-}3} + \Delta q_3 = 21 + (167) = 188$ lbs./in.

$q_{o,4\text{-}5} = q_{o,3\text{-}4} + \Delta q_4 = 188 + (187) = 375$ lbs./in.

$q_{o,5\text{-}6} = q_{o,4\text{-}5} + \Delta q_5 = 375 + (-167) = 208$ lbs./in.

$q_{o,6\text{-}1} = q_{o,5\text{-}6} + \Delta q_6 = 208 + (-208) = 0$ lbs./in. (Shear flow in this web should be zero because it was cut)

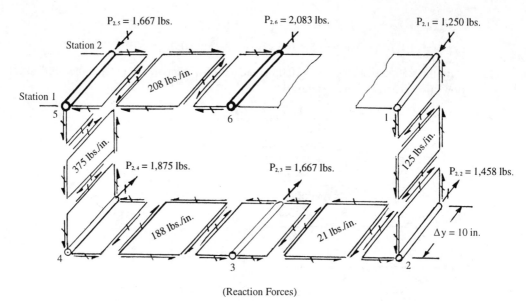

(Reaction Forces)
(Free-body Diagrams – counterclockwise shear flow is positive ↶(+))

(3) Taking moments about c.g. as the reference point 'o' and each segment area ($2A_o$):

$2A_{o,1-2} = 10 \times 11.25 = 112.5$ in.^{2}
$2A_{o,2-3} = 10 \times 6.25 = 62.5$ in.^{2}
$2A_{o,3-4} = 10 \times 6.25 = 62.5$ in.^{2}
$2A_{o,4-5} = 10 \times 8.75 = 87.5$ in.^{2}
$2A_{o,5-6} = 10 \times 3.75 = 37.5$ in.^{2}
$2A_{o,6-1} = 10 \times 3.75 = 37.5$ in.^{2}

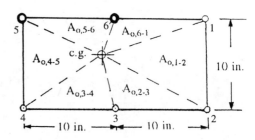

Web No.	q_o	$2A_o$	$M_{y,o} = q_o \times 2A_o$
1-2	−125	112.5	−14,063
2-3	21	62.5	1,313
3-4	188	62.5	11,750
4-5	375	87.5	32,813
5-6	208	37.5	7,800
6-1	0	37.5	0
		$\Sigma 2A = 400$ in.^{2}	$\Sigma M_{y,o} = 39,613$ in.-lbs.

(Above moment, $\Sigma M_{y,o} = 39,613$ in.-lbs. is due to a 'cut' shear flow)

(4) A moment summation is then made about the same point as before in order to determine the torque unbalance, $M_{y,t}$, which must be applied to create internal shear flows to establish equilibrium.

$M_{c.g.} = P_z \times d + \Sigma M_{y,o} + M_{y,t} = 0$
$= 5,000 \times (20 - 7.5 - 8.75) + 39,613 + M_{y,t} = 0$
$M_{y,t} = -18,750 - 39,613 = -58,363$ in-lbs. $= 2q_t A$

From which: $q_t = \dfrac{M_{y,t}}{2A} = \dfrac{-58,363}{400} = -146$ lbs./in.

(5) The final shear flows become:

Web No.	q_o (due to 'cut')	q_t (due to balanced torque, $M_{y,t}$)	q (final shear flows)
1-2	−125	−146	−271
2-3	21	−146	−125
3-4	188	−146	42
4-5	375	−146	129
5-6	208	−146	62
6-1	0	−146	−146

(6) The final shear flows are shown on the sketch below and the arrows indicate the shear flow direction.

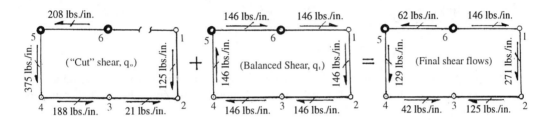

(Final Shear Flow Diagram – reaction forces)

8.5 TWO-CELL

(A) TWO-STRINGER BOXES (SINGLE SPAR)

Frequent practical application is made of this type of structure which is usually applied on aircraft control surfaces (ailerons, elevators and rudders), flaps, slats, single-spar wings or tail sections. If another cell is added to the section, as shown in Fig. 8.5.1, then the section becomes statically indeterminate.

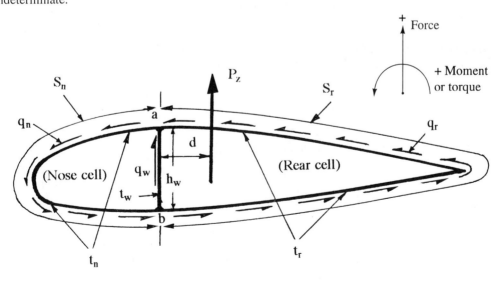

Fig. 8.5.1 Single-Spar Airfoil Section With Two Cells

Since there are three unknowns (q_n, q_w, and q_r), three equations relating to these quantities are needed.

The first is from $\Sigma F_z = 0$:

$$P_z = (q_n + q_w - q_r)h \qquad \text{Eq. 8.5.1}$$

The second is from $\Sigma M_b = 0$:

$$P_z \times d = q_n \times 2A_n + q_r \times 2A_r \qquad \text{Eq. 8.5.2}$$

The third equation is obtained from the condition that elastic continuity must be maintained, namely, the twist angle of nose cell 'n' must equal to that of tail cell 'r':

$$\theta_n = \theta_r$$

From Eq. 8.4.3, the twist angle per unit length is,

$$\theta_n = (\frac{1}{2A_n})(\oint \frac{q_n}{G_e t_n})(ds)$$

$$\theta_r = (\frac{1}{2A_r})(\oint \frac{q_r}{G_e t_r})(ds)$$

In evaluating the line integrals in the above equations, it should be noted that the integration must be complete for each cell and that the web, w, is common to both cells and must therefore be included with each integration.

$$G_e \theta_n = (\frac{1}{2A_n})(\frac{S_n q_n}{t_n} + \frac{h_w q_w}{t_w})$$

$$G_e \theta_r = (\frac{1}{2A_r})(\frac{-h_w q_w}{t_w} + \frac{S_r q_r}{t_r})$$

Therefore,

$$\frac{\frac{S_n q_n}{t_n} + \frac{h_w q_w}{t_w}}{A_n} = \frac{\frac{-h_w q_w}{t_w} + \frac{S_r q_r}{t_r}}{A_r} \qquad \text{Eq. 8.5.3}$$

The simultaneous equations of Eq. 8.5.1, Eq. 8.5.2 and Eq. 8.5.3 will determine the values of q_n, q_r, and q_w.

Example:

For the section shown in Fig. 8.5.1, given the following data, determine the shear stress in the skin of the nose, tail and vertical web.

$P_z = 2,000$ lbs. $d = 3.0$ in.
$A_n = 100$ in. $t_n = 0.025$ in. $S_n = 45$ in.
$A_r = 200$ in. $t_r = 0.025$ in. $S_r = 60$ in.
    $t_w = 0.04$ in. $h_w = 10.0$ in.

From Eq. 8.5.1:

$2,000 = (q_n + q_w - q_r) \times 10$

$q_n + q_w - q_r = 200$ \qquad Eq. (A)

From Eq. 8.5.2:

$2,000 \times 3.0 = q_n \times 2 \times 100 + q_r \times 2 \times 200$

$200 q_n + 400 q_r = 6,000$

$q_n = 30 - 2 q_r$ \qquad Eq. (B)

Combining Eq. (A) and Eq. (B) obtain,

$$q_w = 170 + 3q_r \quad \text{Eq. (C)}$$

From Eq. 8.5.3:

$$\frac{\dfrac{S_n q_n}{t_n} + \dfrac{h_w q_w}{t_w}}{A_n} = \dfrac{\dfrac{-h_w q_w}{t_w} + \dfrac{S_r q_r}{t_r}}{A_r}$$

$$\frac{\dfrac{45 \times q_n}{0.025} + \dfrac{10 \times q_w}{0.04}}{100} = \dfrac{\dfrac{-10 \times q_w}{0.04} + \dfrac{60 \times q_r}{0.025}}{200}$$

$$18q_n + 2.5q_w = -1.25q_w + 12q_r$$

$$18q_n + 3.75q_w - 12q_r = 0 \quad \text{Eq. (D)}$$

Solve above Eq. (B), Eq. (C) and Eq. (D), obtain,

$$18(30 - 2q_r) + 3.75(170 + 3q_r) - 12q_r = 0$$

$$q_r = 32 \text{ lbs./in.}$$

Shear stress, $f_{r,s} = \dfrac{32}{0.025} = 1{,}280$ psi

From Eq. (B): $q_n = 30 - 2(32) = -34$ lbs./in.

Shear stress, $f_{n,s} = \dfrac{34}{0.025} = 1{,}360$ psi

From Eq. (C): $q_w = 170 + 3(32) = 266$ lbs./in.

Shear stress, $f_{w,s} = \dfrac{266}{0.04} = 6{,}650$ psi

The final shear flows are shown below:

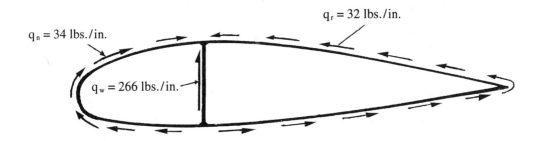

(Above are action shear flows)

(B) MULTI-STRINGER BOX (THREE SPARS)

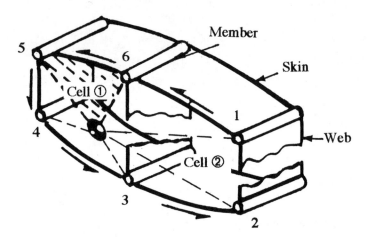

Fig. 8.5.2 Multi-Stringer Box Beam with Three Spars

The procedural steps to analyze the two-cell box shown in Fig. 8.5.2, are similar to those shown in Section 8.4 except as shown below:

(a) Cut any two webs or skins between stringers in the box.

(b) Calculate "Cut-web" shear flows by using:
  - If not symmetrical (from Eq. 8.4.1):
  $$f_b = \frac{-(M_z I_x + M_x I_{xz})x + (M_x I_z + M_z I_{xz})z}{I_x I_z - (I_{xz})^2}$$
  - Let $M_z = 0$ because the contribution of stress on f_b is small since the M_z value is generally relatively small and the I_z value quite large. From Eq. 8.4.2,
  $$f_b = \frac{-(M_x I_{xz})x + (M_x I_z)z}{I_x I_z - (I_{xz})^2}$$
  - Using $\frac{M_x z}{I_x}$, if symmetrical about x - axis ($I_{xz} = 0$) and $M_z = 0$.
  - $\Delta P = f_{bi} A_i$
  - "Cut-web" shear flows, $q_o = \frac{\Delta P}{L}$

(c) Impose $\theta_1 = \theta_2$
$$\text{Calculate } \theta i = \Sigma \frac{q \times ds \times L}{2A \times G \times t}$$
(Be sure to include both "cut-web" shear flows in all twist calculations)

(d) Impose:
  - $\Sigma M = 0$
  - $\theta_1 = \theta_2$
  - Check balancing shear flows: $q_{t,1}$ in cell ① and $q_{t,2}$ in cell ②

(e) Add $q_{t,1}$ and $q_{t,2}$ to "cut-web" shear flows, q_o, to obtain the final shear flows for the box beam.

Example 3:

Use the same given loading and box beam used in Example 1A except add a middle spar to make the box become a two-cell section (three spars) as shown below.

- At station 1 is $M_{x,1} = 0$
- At station 2 is $M_{x,2} = 0 + (-P_z \times 10) = -50,000$ in.-lbs.

Stringer areas are the same as those used in Example 1A:

Stringer areas:

$A_1 = 0.5$ in.^{2}
$A_2 = 0.5$ in.^{2}
$A_3 = 0.5$ in.^{2}
$A_4 = 1.0$ in.^{2}
$A_5 = 1.0$ in.^{2}
$A_6 = 0.5$ in.^{2}

Web thickness, t = 0.1 in. (typ)

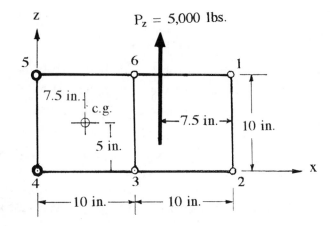

(1) The section properties (from Example 1A):

Location of c.g.: $X = \dfrac{30}{4} = 7.5$ in. and $Z = \dfrac{20}{4} = 5$ in.

Moments of inertia:
$I_x = 100$ in.^{4} $I_z = \Sigma A x^2 - \Sigma A \times X^2 = 500 - 4 \times 7.5^2 = 275$ in.^{4}

(2) With the station distance $\Delta y = 10.0$ in. and web 1-2 and 3-6 temporarily cut and the "cut" shear flows, $q_o,$: (Both ΔP and Δq are from Example 1A)

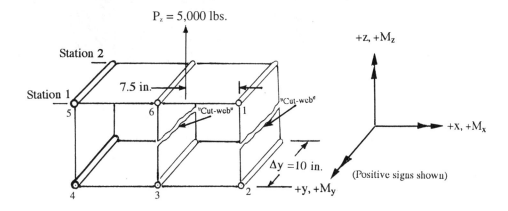

Chapter 8.0

Stringer No.	Web No.	$\Delta P = (P_2 - P_1)$	$\Delta q = \dfrac{\Delta P}{\Delta y}$	$q_o = \Sigma \Delta q$
		(from Example 1A)		
1	1-2 (cut)	−1,250	−125	0
2	2-3	1,250	125	125
3	3-4	1,250	125	250
4	4-5	2,500	250	500
5	5-6	−2,500	−250	250
6	6-1	−1,250	−125	125
	3-6 (cut)			0

$q_{o,2-3} = \Delta q_2 = 125$ lbs./in.

$q_{o,3-4} = q_{o,2-3} + \Delta q_3 = 125 + (125) = 250$ lbs./in.

$q_{o,4-5} = q_{o,3-4} + \Delta q_4 = 250 + (250) = 500$ lbs./in.

$q_{o,5-6} = q_{o,4-5} + \Delta q_5 = 500 + (-250) = 250$ lbs./in.

$q_{o,6-1} = q_{o,5-6} + \Delta q_6 = 250 + (-125) = 125$ lbs./in.

$q_{o,1-2} = q_{o,6-1} + \Delta q_1 = 125 + (-125) = 0$ (web 1-2 was cut)

$q_{o,3-6} = 0$ (web 3-6 was cut)

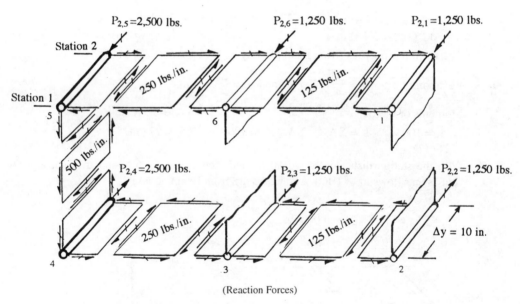

(Reaction Forces)

(Free-body Diagrams – counterclockwise shear flow is positive ↶)

(3) Angle of rotation θ_1 and θ_2:

From Eq. 8.2.5, assume L = 1.0 (unit length):

$\theta_1 = \dfrac{\Sigma q_{t,1} ds}{2 A_1 G_1 t}$ or $2 A_1 G_1 \theta_1 = q_{t,1} \Sigma \dfrac{ds}{t}$

$\theta_2 = \dfrac{\Sigma q_{t,2} ds}{2 A_2 G_2 t}$ or $2 A_2 G_2 \theta_2 = q_{t,2} \Sigma \dfrac{ds}{t}$

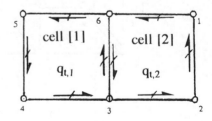

(The above equations include shear flow in "cut-web")

Box Beams

In cell [1], the torsional angle, θ_1:

$$2A_1G_1\theta_1 = \frac{250(10)}{0.1} + \frac{500(10)}{0.1} + \frac{250(10)}{0.1} - \frac{q_{t,1}(10)}{0.1} - \frac{q_{t,1}(10)}{0.1} - \frac{q_{t,1}(10)}{0.1} - \frac{q_{t,1}(10)}{0.1} + \frac{q_{t,2}(10)}{0.1}$$

$$2A_1G_1\theta_1 = 25,000 + 50,000 + 25,000 - 100q_{t,1} - 100q_{t,1} - 100q_{t,1} - 100q_{t,1} + 100q_{t,2}$$

$$2A_1G_1\theta_1 = 100,000 - 400q_{t,1} + 100q_{t,2} \qquad \text{Eq. (A)}$$

In cell [2], the torsional angle, θ_2:

$$2A_2G_2\theta_2 = \frac{125(10)}{0.1} + \frac{125(10)}{0.1} - \frac{q_{t,2}(10)}{0.1} - \frac{q_{t,2}(10)}{0.1} - \frac{q_{t,2}(10)}{0.1} - \frac{q_{t,2}(10)}{0.1} + \frac{q_{t,1}(10)}{0.1}$$

$$2A_2G_2\theta_2 = 25,000 - 400q_{t,2} + 100q_{t,1} \qquad \text{Eq. (B)}$$

In this case, $A_1 = A_2$ and $G_1 = G_2$. These two cells will rotate at the same angle, namely $\theta_1 = \theta_2$.

$$100,000 - 400q_{t,1} + 100q_{t,2} = 25,000 - 400q_{t,2} + 100q_{t,1}$$

$$500q_{t,1} - 500q_{t,2} = 75,000 \quad \text{or} \quad q_{t,1} - q_{t,2} = 150 \qquad \text{Eq. (C)}$$

(4) Shear flows required to balance the unbalanced torque $\Sigma M_{y,t} = 0$:

Define the segment areas, Ao, in the cross-section:

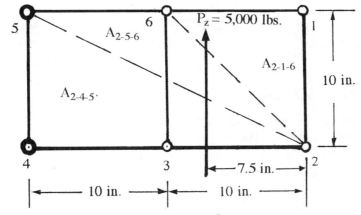

Enclosed area of cell [1]; $2A_1 = 200$ in.^{2}

Enclosed area of cell [2]; $2A_2 = 200$ in.^{2}

$$2A_{2-1-6} = 2(\frac{10 \times 10}{2}) = 100 \text{ in.}^2$$

$$A_{2-5-6} = 2(\frac{10 \times 10}{2}) = 100 \text{ in.}^2$$

$$2A_{2-4-5} = 2[\frac{(10+10) \times 10}{2}] = 200 \text{ in.}^2$$

$\Sigma M_{y,t} = 0$ at stringer No. 2:

$$-5,000(7.5) + 125(2A_{2-1-6}) + 250(2A_{2-5-6}) + 500(2A_{2-4-6}) - q_{t,1}(2A_1) - q_{t,2}(2A_2) = 0$$

$$-5,000(7.5) + 125(100) + 250(100) + 500(200) - 200q_{t,2} - 200q_{t,1} = 0$$

$$200q_{t,2} + 200q_{t,1} = 100,000 \quad \text{or} \quad q_{t,1} + q_{t,2} = 500 \qquad \text{Eq. (D)}$$

Solving Eq. (C) and Eq. (D) obtain:

$q_{t,1} = 325$ lbs./in. and $q_{t,2} = 175$ lbs./in.

(5) Final shear flows are obtained by adding above $q_{t,1}$ and $q_{t,2}$ to the 'cut-web' shear flows given below:

Web No.	q_o (due to "cut")	$q_{t,1}$ (due to balanced torque, $M_{y,t}$)	$q_{t,2}$	q (final shear flows)
1-2 (web)	0	0	175	175
2-3	125	0	175	300
3-4	250	325	0	575
4-5 (web)	500	325	0	825
5-6	250	325	0	575
6-1	125	0	175	300
3-6 (web)	0	325	175	$q_{t,1} - q_{t,2} = 150$

(6) The final shear flows are shown below and the arrows indicate the shear flow direction (reaction forces):

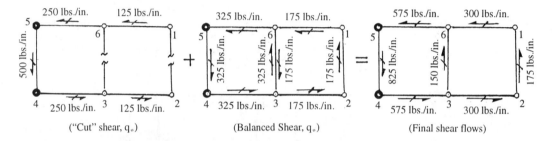

8.6 TAPERED CROSS-SECTIONS

For most current aircraft the wing sections and tail sections are tapered in planform as well as in vertical depth (spar web) as the wing root is approached. The effect on a simple beam was discussed in Chapter 6.0.

Since the box beam shear webs carry load only in the vertical web and the longitudinal stringers can only carry axial load from bending, any sudden change in taper can impose large loads on the rib at the "kink" station as shown in Fig. 8.6.1. A typical case where this occurs is where the wing is attached to the fuselage (root joint locations).

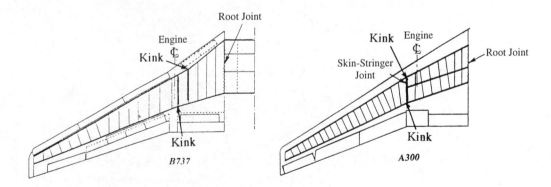

Fig. 8.6.1 Wing Box Planform Taper with Kink

(A) TAPER IN DEPTH

The case of a shear load applied on a beam tapered in depth, where the shear flow varies throughout the vertical spar web, was discussed in Chapter 6.0. Fig. 8.6.2 is a typical commercial transport wing box depth distribution showing maximum depth of the wing box as well as its front and rear spars. The following example uses a rectangular box beam with constant width but a taper in depth to illustrate the nature of the error involved in using classic beam theory to calculate the shear distribution on a tapered beam.

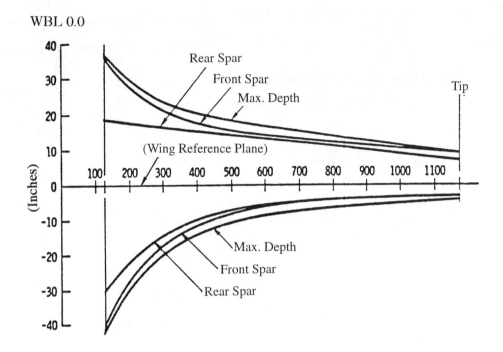

Fig. 8.6.2 Typical Transport Wing Box Depth Distribution

Example 4:

Use the same given box beam used in Example 1A in Chapter 8.4 except this is a tapered box beam with different loading arrangements as shown below.

At cross-section A - A (station 1), the moment is $M_{x,1} = -P_z \times 50 = -5,000 \times 50 = -50,000$ in.-lbs.

Stringer areas:

$A_1 = 0.5$ in.^{2}
$A_2 = 0.5$ in.^{2}
$A_3 = 0.5$ in.^{2}
$A_4 = 1.0$ in.^{2}
$A_5 = 1.0$ in.^{2}
$A_6 = 0.5$ in.^{2}

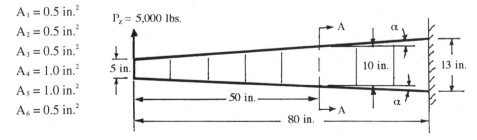

(1) The section properties (from Example 1A):

Location of c.g.: $X = \frac{30}{4} = 7.5$ in.

and $Z = \frac{20}{4} = 5$ in.

Moment of inertia:

$I_x = 100$ in.^{4}

$I_z = \Sigma Ax^2 - \Sigma A \times X^2 = 500 - 4(7.5)^2$

$= 275$ in.^{4}

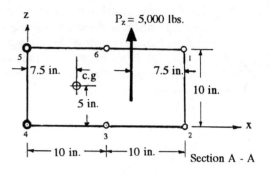

Section A - A

(2) Find the net shear load at cross-section A - A:

$$f_b = \frac{(M_{x,1})(z-Z)}{I_x} = \frac{(250,000)(z-Z)}{100} = 12,500 \text{ psi}$$

$$\tan \alpha = \frac{\frac{13-5}{2}}{80} = 0.05$$

Stringer No.	A (in.²)	(z – Z)	P = f_b × A (lbs.)	V_f = (P × tan α) (lbs.)
1	0.5	5	–6,250	–312.5
2	0.5	–5	6,250	312.5
3	0.5	–5	6,250	312.5
4	1.0	–5	12,500	625
5	1.0	5	–12,500	–625
6	0.5	5	–6,250	–312.5
				$\Sigma V_f = 2,500$

$P_z = V_w + \Sigma V_f = 5,000$ lbs.

V_w = Shear taken by web = $V_z(\frac{h_o}{h}) = 5,000(\frac{5}{10}) = 2,500$ lbs.

ΣV_f = Shear taken by stringers = $V_z(\frac{h-h_o}{h}) = 5,000(\frac{5}{10}) = 2,500$ lbs.

(3) Determine "cut" shear flows, q_o, by using the vertical total shear load taken by web, V_w = 2,500 lbs.:

In this case, use the station distance $\Delta y = 1.0$ in. (unit distance) and web 4-5 temporarily cut (Note: If $\Delta y = 10$ in. is used instead of $\Delta y = 1.0$, the result should be the same).

Assume at Station 1 the $M_{x,1} = 0$ and therefore the values of $P_1 = 0$ and $f_1 = 0$

The bending moment at Station 2 is:

$M_{x,2} = 2,500 \times 1.0 = 2,500$ in.-lbs. (use this $M_{x,2}$ to calculate P_2 and f_2)

$\Delta f = \frac{M_{x,2}(z-Z)}{I_x} = \frac{2,500 \times 5.0}{100} = 125$ psi

$\Delta P = \Delta f \times A$

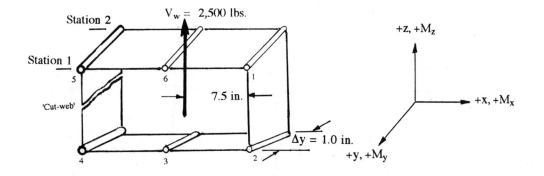

Str. No.	Web No.	A (in.²)	(z – Z) (in.)	$\Delta f = \dfrac{M_{x,2}(z-Z)}{I_x}$ (psi)	$\Delta q = \Delta P = \Delta f \times A$ (lbs./in.)	$q_o = \Sigma \Delta q$ (lbs./in.)
1	1-2	0.5	+5	−125	−62.5	−250
2	2-3	0.5	−5	125	62.5	−187.5
3	3-4	0.5	−5	125	62.5	−125
4	4-5 (cut)	1.0	−5	250	125	0
5	5-6	1.0	+5	−250	−125	−125
6	6-1	0.5	+5	−125	−62.5	−187.5

$q_{o,5\text{-}6} = \Delta q_5 = -125$ lbs./in.

$q_{o,6\text{-}1} = q_{o,5\text{-}6} + \Delta q_6 = -125 + (-62.5) = -187.5$ lbs./in.

$q_{o,1\text{-}2} = q_{o,6\text{-}1} + \Delta q_1 = -187.5 + (-62.5) = -250$ lbs./in.

$q_{o,2\text{-}3} = q_{o,1\text{-}2} + \Delta q_2 = -250 + (62.5) = -187.5$ lbs./in.

$q_{o,3\text{-}4} = q_{o,2\text{-}3} + \Delta q_3 = -187.5 + (62.5) = -125$ lbs./in.

$q_{o,4\text{-}5} = q_{o,3\text{-}4} + \Delta q_4 = -125 + (125) = 0$ (cut web)

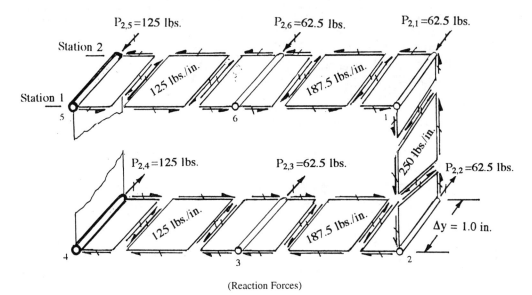

(Reaction Forces)

(Free-body Diagrams – counterclockwise shear flow is positive ↶+)

Chapter 8.0

(4) Assume a shear flow q_t (positive shear flow as shown below) required to balance the unbalanced torque $\Sigma M_{y,t} = 0$ and taking a reference point at stringer No. 2:

Box enclosed area:

$2A_o = 2(10 \times 20)$
$= 400 \text{ in.}^2$

Segment areas:

$2A_{o,2-1-6} = 2(\dfrac{10 \times 10}{2})$
$= 100 \text{ in.}^2$

$2A_{o,2-5-6} = 2(\dfrac{10 \times 10}{2})$
$= 100 \text{ in.}^2$

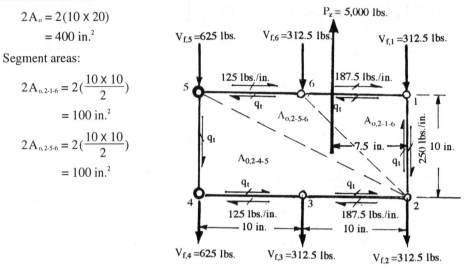

$\Sigma M_{y,t} = -5,000(7.5) + (V_{f,4} + V_{f,5})(20) + (V_{f,3} + V_{f,6})(10) - (q_{o,6-1})(2A_{o,2-1-6}) - (q_{o,5-6})(2A_{o,2-5-6}) + q_t(2A_o) = 0$

$\Sigma M_{y,t} = -5,000(7.5) + 2 \times 625(20) + 2 \times 312.5(10) - (187.5)(100) - (125)(100) + q_t(400) = 0$

$q_t = 93.8$ lbs./in.

(5) The final shear flows become:

Web No.	q_o (due to 'cut' web)	q_t (due to balanced torque, $M_{y,t}$)	q (final shear flows)
1-2	−250	93.8	−156.2
2-3	−187.5	93.8	−93.7
3-4	−125	93.8	−31.2
4-5	0 (cut web)	93.8	93.8
5-6	−125	93.8	31.2
6-1	−187.5	93.8	−93.7

(6) Final shear flows are shown in the sketch below and the arrows indicate the shear direction

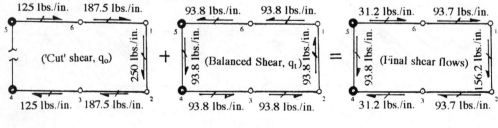

(Reaction Forces)

Box Beams

(B) TAPER IN WIDTH

A box beam which is tapered in planform is another sizing consideration for wing box structures. Important practical knowledge on how the stringers should be arranged in a tapered box beam, including interface with ribs and spar and cost effective fabrication is illustrated below:

(a) It will be noted that the wing surface panel as shown in Fig. 8.6.3 is poorly designed as the increase in bending material occurs along the center of the panel. This requires the shear flow to be carried across the entire panel. This not only causes very high panel shear stresses but also, as a secondary effect, causes undesirable shear lag effects which result in higher stringer stresses along the beams than is shown by the $f = \dfrac{My}{I}$ beam theory.

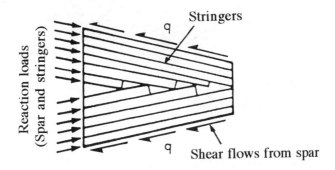

Fig. 8.6.3 Wing Stringers Convergence to Center (Poor Design)

(b) In the case where the stringers are all parallel each other, the shear web shear flow causes a chordwise running load which must be resisted by the skin or ribs all along the wing span, as shown in Fig. 8.6.4.

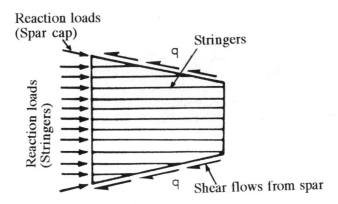

Fig. 8.6.4 Wing Stringers in Parallel

The factors which control the lay out of the stringer orientation are as follows:

- Locate a series of spanwise access manholes (a high priority on lower wing panels) as close to the center as possible
- Make all stringers parallel to the rear spar cap to avoid or minimize stringer run-outs which are areas of very high stress concentration
- Fig. 8.6.5 shows an optimum arrangement of stringer run-outs at the front spar where the axial stress is generally lower

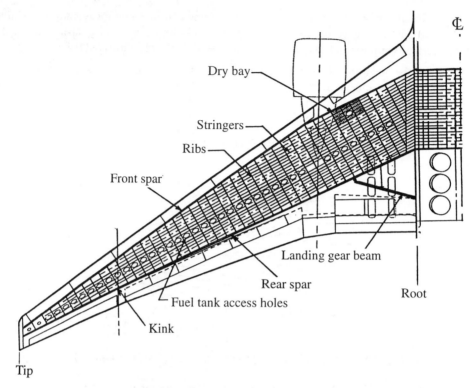

Fig. 8.6.5 Stringer Layout and Access Holes on Wing Box

8.7 SHEAR LAG

"Shear lag" (Reference. 8.3) can be described as the effect of stresses in idealized shells under loading conditions that produce characteristically large and non-uniform axial (stringer) stresses. Axial loading by concentrated forces and bending produced by transverse loading in wide beam flanges (skin-stringer panels) diminishes with the distance from the beam web and this stress diminution is called "shear lag" as shown in Fig. 8.7.1.

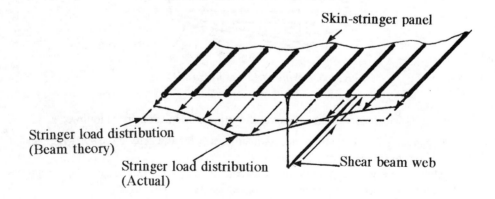

Fig. 8.7.1 Shear Lag Effect on Wide-flange Beam

(A) SWEPT WING BOX

It is common engineering knowledge that in a box beam the bending stresses do not always conform very closely to the predictions of classic beam theory.

- The deviations from the theory are caused primarily by the shear deformations in the cover skin-stringer panels of the box that constitute the flanges of the beam
- The effect of skin shear strains in the skin-stringer panel causes some stringers to resist less axial load than those calculated by beam theory ($\frac{My}{I}$) which assumes that plane sections remain plane after bending

The problem of analyzing these deviations (shear lag) is not taken into consideration in classic beam theory. Shear lag is more pronounced in shells of shallow section (e.g., wing and empennage box structures) than in shells of deep section (e.g., circular fuselage shell structures). Consideration of shear lag effects is therefore much more important for wings than for the fuselage (if the basic method of construction is similar).

In a wing box, the cover panel skin is loaded along the edges by shear flows from the beam spar webs. These shear flows are resisted by axial forces developed in the longitudinal members of spars and stringers. According to the beam theory, the stringer stresses should be uniform chordwise at any given beam station. However, the central stringers tend to lag behind the others in picking up the load because the intermediate skin, which transfers the loads in from the edges, is not perfectly rigid in shear. In such a case the inside stringers would be out of action almost entirely. With the inside stringer stresses lagging, the outside stringers and spar caps must carry an over-stress to maintain equilibrium.

In general, the "shear lag" effect in sheet-stringer box structures is not appreciable except for the following situations:

- Cutouts which cause one or more stringers to be discontinued
- Large abrupt changes in external load applications
- Abrupt changes in stringer areas

A wing root of a swept wing configuration, which produces a triangular stress re-distribution area (see shape triangular area), higher stress occurs toward the rear spar at point 'B' because of shear lag effect, as shown in Fig. 8.7.2. More material should be added in the area of 'A-C' to build-up axial stiffness to relieve the stress or loads in the point 'B' area.

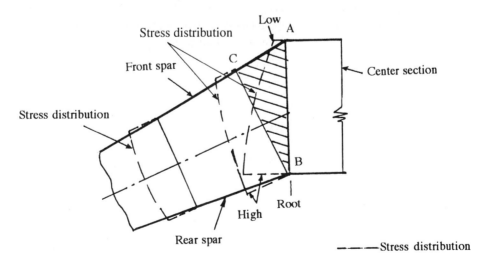

Fig. 8.7.2 Stress Distribution Along Wing Box Due an Up-Bending Moment

(B) TWO-STRINGER PANELS

The methods shown in Fig. 8.7.3 (see Reference 8.3) are useful to give a quick answers when analyzing certain redundant structures such as joints, doubler reinforcements, etc.

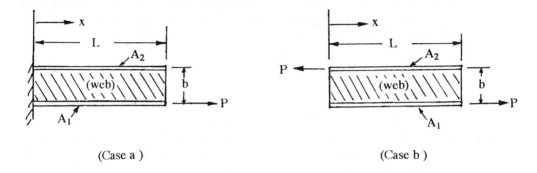

Fig. 8.7.3 Loading Cases on a Two Skin-stringer Panel

The shear lag parameter 'K':

$$K = \sqrt{(\frac{Gt}{Eb})(\frac{1}{A_1} + \frac{1}{A_2})}$$ Eq. 8.7.1

where   K – Shear-lag parameter (for panels of constant cross-section)
    $A_T = A_1 + A_2$
    A_1 – Cross-section area of member 1
    A_2 – Cross-section area of member 2
    b – The distance between the centroids of members 1 and 2
    G – Modulus of rigidity
    E – Modulus of elasticity
    t   – Thickness of the web between members

The following comments relate to these formulas:

- These calculations relate to the elastic range
- These equations are only valid for conditions of constant web thickness (t) and constant modulis (E and G).
- In practical design, the finite length of 'L' should always be used.
- Cases 'a' and 'b' become identical if the panel is very long.

The following formulas are used for panel analysis:

(a) Case 'a' (Local doubler reinforcement), see Fig. 8.7.3(a):

Axial stress in member 1:

$$f_1 = (\frac{P}{A_T})(1 + \frac{A_2 \cosh Kx}{A_1 \cosh KL})$$ Eq. 8.7.2

Axial stress in member 2:

$$f_2 = (\frac{P}{A_T})(1 - \frac{\cosh Kx}{\cosh KL})$$ Eq. 8.7.3

Shear stress in web:

$$f_s = (\frac{PK}{t})(\frac{A_2 \sinh Kx}{A_T \cosh KL})$$ Eq. 8.7.4

Case 'a' can be applied in the sizing of a doubler which is either bonded or fastened to a skin panel, as shown in Fig. 8.7.4.

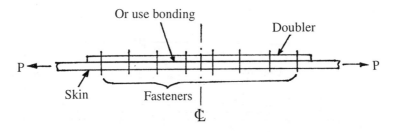

Fig. 8.7.4 Doubler Reinforcement

(b) Case 'b' (load transferring), see Fig. 8.7.3(b):
Axial stress in member 1:

$$f_1 = (\frac{P}{A_T A_1 \sinh KL})[A_1 \sinh KL + A_2 \sinh Kx - A_1 \sinh K(L-x)] \qquad \text{Eq. 8.7.5}$$

Axial stress in member 2:

$$f_2 = (\frac{P}{A_T A_2 \sinh KL})[A_2 \sinh KL - A_2 \sinh Kx + A_1 \sinh K(L-x)] \qquad \text{Eq. 8.7.6}$$

Shear stress in web:

$$f_s = (\frac{PK}{tA_T \sinh KL})[A_1 \cosh K(L-x) - A_2 \cosh Kx] \qquad \text{Eq. 8.7.7}$$

This case may be used to determine load transfer in a fastened or bonded joint, as shown in Fig. 8.7.5 below.

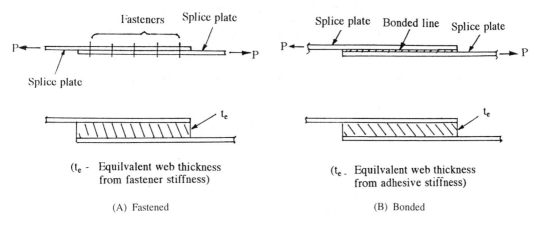

Fig. 8.7.5 A Fastened or Bonded Splice

Example 1:

Use Case 'a' formulas to determine the end fastener shear load on a reinforced doubler as shown below.
- This is an approximate method to calculate the end fastener load on a constant cross-section doubler

Chapter 8.0

- Assume a stiffener (member 1) reinforced by a doubler (member 2) with fasteners and symmetrical about the centerline

(1) Given doubler reinforced joint as shown below:

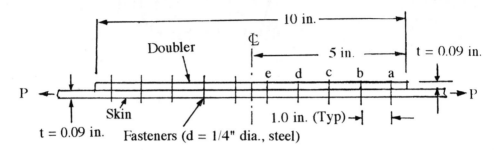

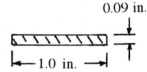

(Aluminum plates; $E = 10 \times 10^6$ psi and $G = 3.9 \times 10^6$ psi)

$A_1 = A_2 = 1.0" \times 0.09 = 0.09$ in.^{2}

b = distance between centerlines of plates

$= \dfrac{0.09}{2} + \dfrac{0.09}{2} = 0.09$ in.

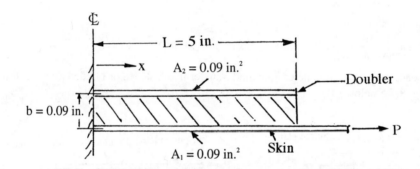

(2) Convert fastener spring constant 'k' to an equivalent web thickness (t_e):

Shear strain, $\gamma = \dfrac{q}{G \times t_e}$

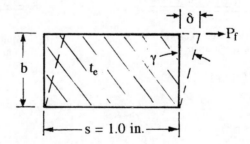

$\delta = \gamma \times b = \dfrac{q \times b}{G \times t_e} = \left(\dfrac{P_f}{s}\right)\left(\dfrac{b}{G \times t_e}\right)$

P_f = applied force of fastener

s = fastener's spacing or pitch

266

Fastener spring constant, $k = \dfrac{P_f}{\delta} = P_f \left(\dfrac{s \times G \times t_e}{P_f \times b} \right)$

$t_e = \dfrac{b \times k}{s \times G}$

$= \dfrac{0.09 \times 0.16 \times 10^6}{1.0 \times 3.9 \times 10^6} = 0.00316$ in.

Where $k = 0.16 \times 10^6$ lbs/in. (spring constant for dia. $= \dfrac{1}{4}$ in., steel fastener from Fig. B.6.2)

(3) Stress distribution in member 2 (doubler):

The shear lag parameter (from Eq. 8.7.1):

$K = \sqrt{\left(\dfrac{Gt}{Eb}\right)\left(\dfrac{1}{A_1} + \dfrac{1}{A_2}\right)}$

$= \sqrt{\left(\dfrac{3.9 \times 10^6 \times 0.00369}{10 \times 10^6 \times 0.09}\right)\left(\dfrac{1}{0.09} + \dfrac{1}{0.09}\right)} = 0.596$

(4) Axial stress in doubler (member 2) from Eq. 8.7.3:

$f_2 = \left(\dfrac{P}{A_t}\right)\left(1 - \dfrac{\cosh Kx}{\cosh KL}\right)$

$= \left(\dfrac{P}{0.18}\right)\left(1 - \dfrac{\cosh 0.596 x}{\cosh 0.596 \times 5.0}\right)$

$= (5.555 P)\left(1 - \dfrac{\cosh 0.596\, x}{9.869}\right)$

(5) End fastener shear load (P_a) at 'a' ($x = 4$ in.):

Doubler stress: $f_{2,\,x=4} = (5.555 P)\left(1 - \dfrac{\cosh 0.596 \times 4}{9.869}\right)$

$= (5.555 P)\left(1 - \dfrac{5.47}{9.869}\right) = 2.476 P$

Doubler axial load $P_{2,\,x=4} = f_{2,\,x=4} \times A_2$

$= 2.476 P \times 0.09 = 0.223 P$

or fastener shear load at 'a': $P_a = P_{2,\,x=4} = 0.223 P$

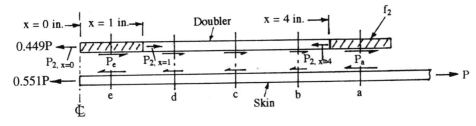

(6) Fastener shear load (P_b) at 'b' ($x = 3$ in.):

Doubler stress: $f_{2,\,x=3} = (5.555 P)\left(1 - \dfrac{\cosh 0.596 \times 3}{9.869}\right)$

$= (5.555 P)\left(1 - \dfrac{3.072}{9.869}\right) = 3.83 P$

Doubler axial load $P_{2,\,x=3} = f_{2,\,x=3} \times A_2$

$= 3.83 P \times 0.09 = 0.345 P$

Then, the fastener shear load at 'b': $P_b = P_{2, x=3} - P_{2, x=4}$
$$= 0.345P - 0.223P = 0.122P$$

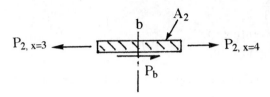

(7) Fastener shear (P_e) at 'e' (x = 0 in.):

At x = 0 in.:

Doubler stress: $f_{2, x=0} = (5.555P)(1 - \dfrac{\cosh 0.596 \times 0}{9.869})$

$$= (5.555P)(1 - \dfrac{1}{9.869}) = 4.99P$$

Doubler axial load $P_{2, x=0} = f_{2, x=0} \times A_2$
$$= 4.99P \times 0.09 = 0.449P$$

At x = 1.0 in.:

Doubler stress: $f_{2, x=1} = (5.555P)(1 - \dfrac{\cosh 0.596(1.0)}{9.869})$

$$= (5.555P)(1 - \dfrac{1.183}{9.869}) = 4.889P$$

Doubler axial load $P_{2, x=1} = f_{2, x=1} \times A_2$
$$= 4.889P \times 0.09 = 0.44P$$

Then, the fastener shear load at 'e': $P_e = P_{2, x=0} - P_{2, x=1}$
$$= 0.449P - 0.44P = 0.009P$$

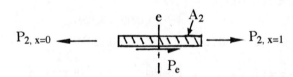

(8) Discussion: This calculation relates to the elastic range and the fastener load distribution as follows:

There are 5 fasteners on each side of the centerline, and the doubler should take 50% of the ultimate load of 'P', and each fastener should carry an **equal amount of load as:**

$$P_{ult.} = \dfrac{\dfrac{P}{2}}{5} = 0.1P$$

(a) Fastener shear load (P_a) at 'a':

$$\text{Load ratio} = \dfrac{P_a}{P_{ult.}} = \dfrac{0.223P}{0.1P} = 2.23$$

The fastener shear load (P_a) in the elastic range is 2.23 times higher than that of ultimate fastener shear load ($P_{ult.}$) and P_a load is very important in estimating structural fatigue life. Therefore, try to reduce fastener load, P_a, by any means to meet the structural fatigue life requirements (refer to Chapter 9.0)

(b) Fastener shear load (P_e) at 'e':

$$\text{Load ratio} = \frac{P_a}{P_{ult.}} = \frac{0.009P}{0.1P} = 0.09$$

Since the fastener shear load of P_e in the elastic range is only 9% of the ultimate fastener load, which is obviously very low, structural fatigue life will not be affected.

Example 2:

Use the method from Case 'b' to determine the end fastener load on a spliced joint held together by fasteners (dia. = $\frac{1}{4}$ in., steel) as shown below.

- This is an approximate method to calculate the end fastener load on a constant cross-section doubler
- Assume two plates (member 1 and member 2, identical to those in Example 1) are fastened
- Determine the end (first) fastener shear load

(1) Given the splice joint shown below:

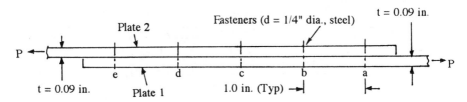

(Aluminum sheets)

$A_1 = A_2 = 1.0 \times 0.09 = 0.09$ in.^{2}

Use b = distance between centerlines of plates

$$= \frac{0.09}{2} + \frac{0.09}{2} = 0.09 \text{ in.}$$

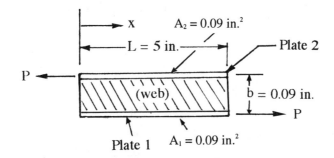

(2) From Example 1, the equivalent web thickness:

$t_e = 0.00369$ in.

(3) From Example 1, shear lag parameter:

K = 0.596

(4) Axial stress in plate 2 from Eq. 8.7.6:

$$f_2 = (\frac{P}{A_T A_2 \sinh KL})[A_2 \sinh KL - A_2 \sinh Kx + A_1 \sinh K(L-x)]$$

$$f_2 = [\frac{P}{(0.18)(0.09)\sinh(0.596 \times 5)}][(0.09)\sinh(0.596 \times 5) - (0.09)\sinh 0.596x + (0.09)\sinh 0.596(5-x)]$$

$$f_2 = [\frac{P}{(0.18)(0.09)(9.819)}][0.884 - (0.09)\sinh 0.596x + (0.09)\sinh 0.596(5-x)]$$

$$f_2 = [6.287P][0.887 - (0.09)\sinh 0.596x + (0.09)\sinh 0.596(5-x)]$$

(5) End fastener shear load (P_a) at 'a' (x = 4 in.):

$$f_2 = [6.287P][0.884 - (0.09)\sinh 0.596(4) + (0.09)\sinh 0.596(5-4)]$$

$$f_2 = [6.287P][0.884 - 0.484 + 0.057] = 2.873P \text{ (psi)}$$

Plate 2 axial load $P_{2,x=4} = f_{2,x=4} \times A_2$

$$= 2.873P \times 0.09 = 0.259P$$

or fastener shear load at 'a' : $P_a = P_{2,x=4} = 0.259P$

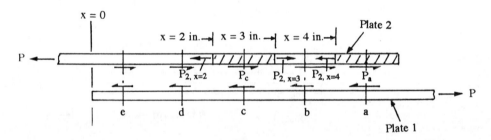

(6) Fastener shear (P_c) load at 'c' :

At x = 3 in.:

$$f_2 = [6.287P][0.884 - (0.09)\sinh 0.596 \times 3 + (0.09)\sinh 0.596(5-3)]$$

$$f_2 = [6.287P][0.884 - 0.261 + 0.135] = 4.766P$$

Plate 2 axial load $P_{2,x=3} = f_{2,x=3} \times A_2$

$$= 4.766P \times 0.09 = 0.429P$$

At x = 2 in.:

$$f_2 = [6.287P][0.884 - (0.09)\sinh 0.596 \times 2 + (0.09)\sinh 0.596(5-2)]$$

$$f_2 = [6.287P][0.884 - 0.135 + 0.261] = 6.35P$$

Plate 2 axial load $P_{2,x=2} = f_{2,x=2} \times A_2$

$$= 6.35P \times 0.09 = 0.572P$$

Then, the fastener load at 'c' : $P_c = P_{2,x=2} - P_{2,x=3}$

$$= 0.572P - 0.429P = 0.143P$$

(7) Discussion: There are 5 fasteners on this joint and each fastener, under ultimate load design, should carry on equal amount of loads as :

$$P_{ult.} = (\frac{P}{5}) = 0.2P$$

(a) Fastener load at 'a' (P_a should be much higher than that of ultimate load):

$$\text{Load ratio} = \frac{P_a}{P_{ult.}} = \frac{0.259P}{0.2P} = 1.3$$

Since the fastener load in the elastic range is 30% higher than that of the ultimate fastener load, this value is very important in estimating structural fatigue life. Therefore, try to reduce fastener load, P_a, by any means (see Chapter 9.12) to meet or increase the fatigue life.

(b) Fastener load at 'c' (P_c should be lower than that of ultimate load):

$$\text{Load ratio} = \frac{P_c}{P_{ult.}} = \frac{0.143P}{0.2P} = 0.72$$

Since the fastener shear load in the elastic range is only 72% of the ultimate shear load, the structural fatigue life is obviously not affected (compared with $P_a = 0.259$).

References

8.1 Shanley, F. R., "Aircraft Structural Research", Aircraft Engineering, (July, 1943), p.200.

8.2 Williams, D., "AN INTRODUCTION TO THE THEORY OF AIRCRAFT STRUCTURES", Edward Arnold Ltd., London, (1960).

8.3 Kuhn, Paul., "STRESSES IN AIRCRAFT AND SHELL STRUCTURES", McGraw-Hill Book Co., New York, NY, (1956).

8.4 Duncan, W. J., "Technical Notes. The Flexural Centre or Centre of Shear", Journal of The Royal Aeronautical Society, Vol. 57, (Sept. 1953).

8.5 Kuhn, Paul, "Remarks on The Elastic Axis of Shell Wings", NACA TN 562, (1936).

8.6 Kuhn, Paul, "The Initial Torsional Stiffness of Shells With Interior Webs", NACA TN 542, (1935).

8.7 Kuhn, Paul, "Some Elementary Principles of Shell Stress Analysis With Notes on The Use of The Shear Center", NACA TN 691, (1939).

8.8 Shanley, F. R. and Cozzone, F. P., "Unit Method of Beam Analysis", Lcokheed Paper No. 46, 1940. and also published in Journal of the Aeronautical Sciences, Vol. 8, (April, 1941), p. 246.

8.9 Benscoter, "Numerical Transformation Procedures for Shear Flow Calculations", Journal of Institute of Aeronautical Sciences, (Aug. 1946).

8.10 Samson, D. R., "The Analysis of Shear Distribution for Multi-Cell Beams in Flexure by Means of Successive Numerical Approximations", Journal of The Royal Aeronautical Society, (February, 1954), p. 122.

8.11 Bruhn, E. F., "ANALYSIS AND DESIGN OF FLIGHT VEHICLE STRUCTURES", Jacobs Publishers, 10585 N. Meridian St., Suite 220, Indianapplis, IN 46290, (1975).

Chapter 9.0

JOINTS AND FITTINGS

9.1 INTRODUCTION

The ideal airframe would be a single complete unit of the same material involving one manufacturing operation (and this may occur in future composite material airframe construction). However, the majority of the present aluminum airframe structures consist of built-up construction. Also, the requirements of repair and maintenance dictate a structure of several main units held together by fastened joints utilizing many rivets, bolts, bonding, lugs, fittings, etc.

For economy of manufacturing, structural and stress engineers should have a thorough knowledge of shop processes and operations. The cost of fitting fabrication and assembly varies greatly with the type of fitting, shape and the required tolerance. Poor layout of a major fitting arrangement may require very expensive tools and jigs for shop fabrication and assembly.

The philosophy of airframe stress analysis differs from the usual approach in that it is generally based on a statistical as well as an analytical basis. Either by governmental specification or by company regulation, the stress engineer is told what load he can put on a fastening element.

- Such policy is the result of some unfortunate past service experience in which riveted aircraft joints became loose, fastened joint failed in fatigue, and other fasteners failed to function after what was considered a short service life
- The replacement of even a small percentage of fasteners is a serious problem, since rivets and rivet installation are very expensive
- A good joint design is one for which fasteners can be installed under normal shop conditions, without producing a structural weakness in the service life of the aircraft

Practically all aircraft are constructed by attaching many small parts together with rivets, screws, and bolts. Since holes must be drilled in the members to make the attachments, it is not possible to attain ultimate allowable stresses in the gross areas of tension members.

- This reduction in area has very little effect on stiffness because only a very small percentage of the total material is removed
- A study of the stress variation around holes (in the commonly used aluminum alloys) shows that in the elastic range the stress at the edge of holes is as much as three times the average stress
- In the plastic range (ultimate load design case), however, the load tends to distribute uniformly over the remaining material

Three reasons why the effect of holes and stress concentrations are not more serious at ultimate loads are:

- The material develops permanent deformation at local points of high stress concentration and shifts excess loads to adjacent material
- Holes are filled by rivets or screws which help prevent inward deflection

Chapter 9.0

- The biaxial tensile stresses which develop around the holes help "streamline" the flow of load

Joints are the most common source of failure for airframes as well as other structures, and failures may occur because of many factors, all of which are difficult to evaluate to an exact degree. These factors not only affect static strength and stiffness but have a great influence on the fatigue life of the joint and the adjacent structures. Splice joints generally fail in one of the modes shown in Fig. 9.1.1.

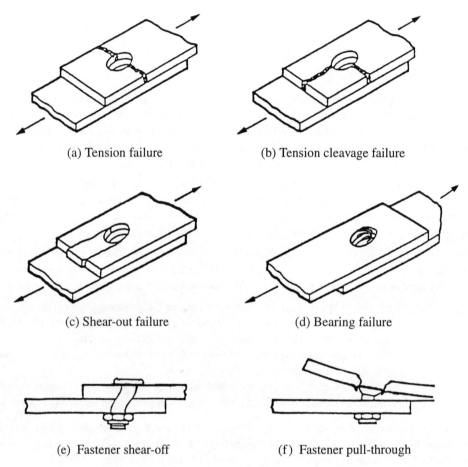

(a) Tension failure   (b) Tension cleavage failure

(c) Shear-out failure   (d) Bearing failure

(e) Fastener shear-off   (f) Fastener pull-through

Fig. 9.1.1 Typical Failure Modes of a Splice Joint

The following are the general design requirements for joints:

(a) Fitting factor:
 An ultimate fitting factor of 1.15 (per FAR 25.625) shall be used in the joint analysis:
 - This factor of 1.15 shall apply to all portions of the fitting including the fastening and bearing on the joined member
 - For each integral fitting, the part must be treated as a fitting up to the point at which the section properties become typical of the member

 No fitting factor need be used:
 - For joints made under approved practices and based on comprehensive test data
 - With respect to any other design factors for which a larger special factor is used

(b) Overall joint efficiency:
It is a primary consideration that the efficiency of the joint will be equal to or greater than that of the parent structure. One side of the joint should not be designed for maximum efficiency at the expense of a weight and fabrication cost penalty on the other. The joint should be located at support structures such as stringers, stiffeners, bulkheads, etc. to improve joint efficiency as shown in Fig. 9.1.3.

(c) Eccentricities (unsupported joint):
Moment produced by eccentricities in a joint, especially the lap joint as shown in Fig. 9.1.2, will induce excessive loading and secondary tension loads on end fasteners. The stress on the plate is no longer uniform stress, and the maximum tensile stress in the elastic range (critical for fatigue consideration) is,

$$f = \frac{P}{t} + (\frac{Pt}{2})(\frac{6}{t^2}) = \frac{4P}{t} \quad \text{(1.0 inch-wide strap)} \quad \text{Eq. 9.1.1}$$

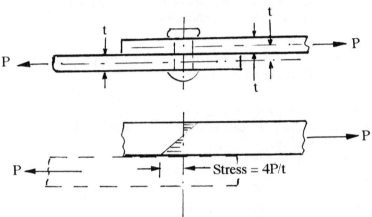

Fig. 9.1.2 Excessive Stress Due to Eccentricity

The excessive stress induced by eccentricity on the members of a lap joint is reduced in the plastic range for ultimate strength design.

(d) Supported joint:
All airframe structural joints are supported joint designs, as shown in Fig. 9.1.3, which provide structural integrity to reduce significant local high local stress due to eccentricity. The supported joint is located at a stringer, stiffener, bulkhead, etc. and it always uses double or staggered rows (in fuselage hoop tension applications triple rows may be used) of fasteners for wing fuel tank design.

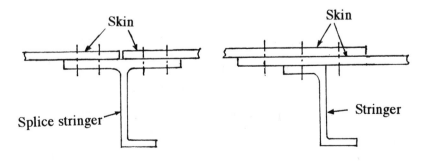

Fig. 9.1.3 Supported Joints

Chapter 9.0

(e) Joint rigidity:
Insufficient or excessive rigidity of the surrounding structure may cause excessive deflections under highly concentrated loads and consequently changes in direction and magnitude of the loads on a joint, e.g., in structural support areas of engine pylons, landing gear, flaps, etc.

(f) Mixed fasteners and fits:
Generally it is not good practice to use mixed fasteners and different fits on the same joint since the tighter fit holes will start to pick up load early and the looser holes will not pick up the load until the joint starts to deflect. This will cause premature failure of structural life. Every fastener should be installed wet (it reduces bearing allowables compared to dry) per the specifications of the government, industry, or the individual company.

(g) Mixed splice materials:
Joints should be strain analyzed based on the stress-strain data of the splice and parent materials to avoid the problem of insufficient or excessive rigidity.

(f) Fastened and bonded joint:
It is difficult to determine the load distribution for a fastened and bonded joint but the bonded area will pick up most of the load:

- The use of bonding on a fastened joint will provide extra fatigue life and increased corrosion resistance at the bondline
- Provide sufficient fasteners to carry all of the static ultimate loads, as no load is carried by the bonding

(h) Permanent set:
Under the limit load no permanent set is allowed to occur as is the requirement for any airframe structure. Use the material bearing yield allowable, F_{bry}, if it is lower than $\frac{F_{bru}}{1.5}$ (see Fig. 4.4.2) and for the similar situation of F_{ty}.

(i) Splices adjacent to continuous members:
This joint should be designed to be as rigid as geometry will permit using ample material and interference or tight fit fasteners to minimize slippage which might overload the continuous members and cause premature failure. An example of this type of joint, as shown in Fig. 9.1.4, is splicing a stringer which is attached to a continuous fuselage skin (frequently used in repair of broken stringer).

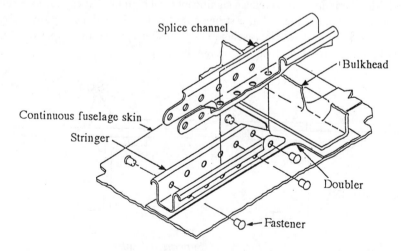

Fig. 9.1.4 Fuselage Stringer Splice (Continuous Skin)

(j) Fastener spacing and edge distance ($\frac{e}{D}$):

In normal metallic sizing, the minimum fastener spacing (pitch) is 4D and edge distance in the direction of load is $\frac{e}{D} = 2.0$ (D is the diameter of the fastener and e is the distance from the center of the fastener to edge of the part plus an additional margin of 0.03 inch for tolerance or misdrill) as shown in Fig. 9.12.4.

Minimum edge distance, $\frac{e}{D} = 1.5$ may be used, provided the following criteria are met:

- Low load transfer such as spar or rib vertical stiffener attached to web
- Assume non-buckled skin

(k) Countersunk fastener:
Since the knife or feather edge, as shown in Fig. 9.1.5, is not allowed in airframe design because of fatigue requirements, the minimum plate thickness (t) is:

$$t \approx 1.5 \times t_d \qquad \text{Eq. 9.1.2}$$

where: t – The plate thickness
        t_d – Countersunk fastener head height

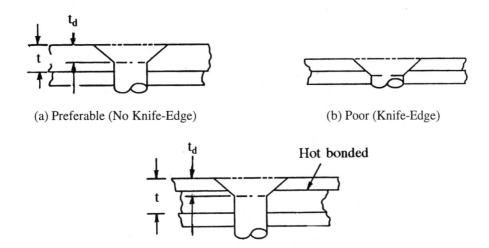

(a) Preferable (No Knife-Edge) (b) Poor (Knife-Edge)

(c) Acceptable (Bonded to Eliminate Effect of Knife-Edge)

Fig. 9.1.5 Countersunk Fastener Knife Edge

(l) Adjacent skin buckling:
Buckling of the skin adjacent to the splice may change the local in-plane load distribution and introduce a prying and tension load on a critical fastener. This buckling not only affects the static strength and stiffness of the joint but also has a great influence on the fatigue life of both the joint and the adjacent structures as well.

(m) Fastener symbol code:
A fastener symbol system, based on the NAS 523 standard, is used on engineering drawings. The symbol, as shown in Fig. 9.1.6, consists of a single cross with code letters or numbers in the quadrants identifying fastener features.

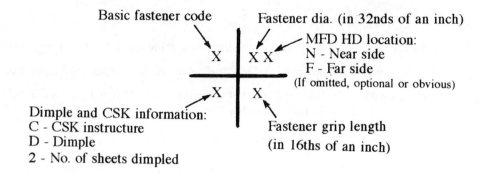

(a) Symbol code

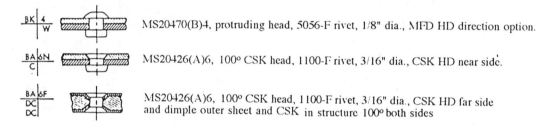

(b) Examples

XZK-6
(Protruding head, dia. = 3/16")

XTJ-8
(Protruding shear head, dia. = 1/4")

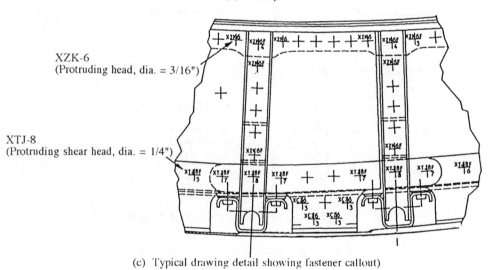

(c) Typical drawing detail showing fastener callout)

Fig. 9.1.6 Fastener Symbol Code

Joints and Fittings

9.2 FASTENER INFORMATION

There are basically four groups of fastener systems, namely:

- Permanent fasteners (rivets), shown in Fig. 9.2.1. and Fig. 9.2.2
- Removable fasteners (screws bolts), shown in Fig. 9.2.3
- Nuts/nut-plates, shown in Fig. 9.2.4
- Washers, shown in Fig. 9.2.4

In making fastener selection, the engineer must consider all the conditions to be encountered by the overall design as well as the allowable strength required. Blind fasteners which are part of the permanent fastener group shown in Fig. 9.2.1 are only used in blind areas where conventional installation or assembly is impossible.

All fasteners are installed wet (with sealant to prevent corrosion) per standard industry or company specification.

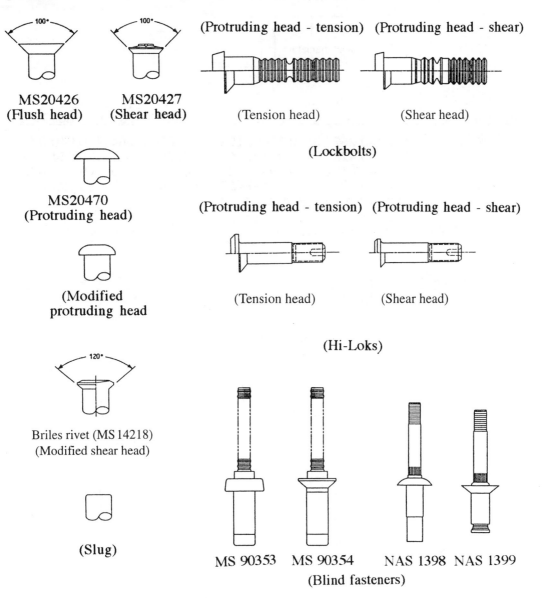

Fig. 9.2.1 Permanent Fasteners

279

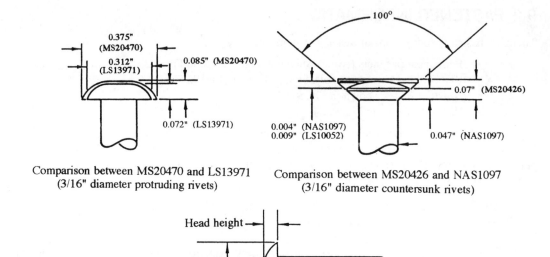

Comparison between MS20470 and LS13971
(3/16" diameter protruding rivets)

Comparison between MS20426 and NAS1097
(3/16" diameter countersunk rivets)

Rivet diameter (in.)	Head diameter MS20470	LS13971	Head height MS20470	LS13971
1/8	0.238-0.262	0.178-0.196	0.054-0.064	0.040-0.050
5/32	0.296-0.328	0.238-0.262	0.067-0.077	0.054-0.064
3/16	0.356-0.0394	0.296-0.328	0.080-0.090	0.067-0.077
1/4	0.475-0.525	0.415-0.459	0.107-0.117	0.093-0.103

Note: For countersunk rivets, see Appendix B.

Fig. 9.2.2 Permanent Fasteners – Comparison of Head Configurations

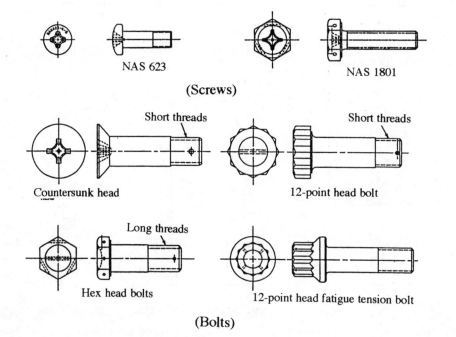

Fig. 9.2.3 Removable Fasteners

Tension Nut   Shear Nut   Nut-Plates

(a) Nut and Nut-plates

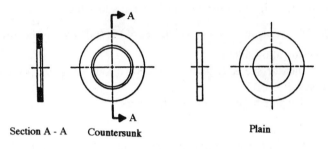

Countersunk (C) and Plain (P) Washers

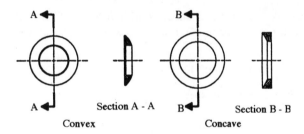

Self-Aligning Washer

(b) Washers

Fig. 9.2.4 Nuts and Washers

(A) BASIC CRITERIA FOR FASTENER STRENGTH ALLOWABLES

The allowable loads are based on the lowest values of the following criteria:

(a) Fastener Shear-off load:

$$P_{s,all} = F_{su} \left(\frac{\pi D^2}{4} \right) \quad\quad \text{Eq. 9.2.1}$$

where: F_{su} – Allowable ultimate shear stress of the fastener material from Ref. 9.1.
          D – Nominal fastener shank diameter.

Fig. 9.2.5 and Fig. 9.2.7 show some design allowables.

(b) Sheet bearing load (protruding head only):

$$P_{b,all} = F_{bru} D t \quad\quad \text{Eq. 9.2.2}$$

Chapter 9.0

where: F_{bru} – Allowable ultimate bearing stress of the sheet material frome Ref. 9.1.
D – Nominal fastener shank diameter.
t – Nominal sheet thickness.

Use "wet pin" bearing allowable data (the data from Ref. 9.1 is mostly "dry pin" bearing allowables and it should be noted that these are higher than "wet pin" allowables. Use 80% of the "dry pin" bearing values until data is established by test results).

(c) Countersunk fastener (flush head) and sheet combinations – the allowable ultimate and yield loads are established from actual test data only.

Fig. 9.2.8 through Fig. 9.2.13 show some design allowables.

(d) The tension allowable in a joint is whichever of the following is the lowest:

- Rivet tension allowable:
Use test data or see Ref. 9.1; Fig. 9.2.10 through Fig. 9.2.13 show some design allowables.
- Tensile allowable for threaded steel fasteners:

$$P_{t,all} = F_{tu} A_m$$
Eq. 9.2.3

where: F_{tu} – Allowable ultimate tensile stress of the fastener material.
A_m – Minor area of "the first thread" (not the shank) of fastener (see Fig. 9.2.6).

Use data from Ref. 9.1 or design allowables in Fig. 9.2.6 and Fig. 9.2.7.

- Countersunk fastener and sheet combinations – Allowable ultimate loads are established from test data only (failure from fastener head breakage or fastener pull-through).

Fig. 9.2.10 through Fig. 9.2.13 give a few design allowables

(e) Yield strength (fasteners and sheet combinations) to satisfy permanent set requirements at limit load.

Fastener Diameter		$\frac{3}{32}$	$\frac{1}{8}$	$\frac{5}{32}$	$\frac{3}{16}$	$\frac{1}{4}$	$\frac{5}{16}$	$\frac{3}{8}$	$\frac{7}{16}$	$\frac{1}{2}$	$\frac{9}{16}$
Fastener Material	F_{su} (Ksi)	\multicolumn{10}{c}{Ultimate Single Shear Load (lbs./fastener)}
1100F (A)	9										
5056 (B)	28	203	363	556	802	1450	2295	3275			
2117-T3 (AD)	30	217	388	596	862	1555	2460	3510			
2017-T31 (D)	34	247	442	675	977	1765	2785	3980			
2017-T3 (D)	38	275	494	755	1090	1970	3115	4445			
2024-T31 (DD)	41	296	531	814	1175	2125	3360	4800			
7075-T73 (E)	41										
7075-T731 (E)	43	311	558	854	1230	2230	3525	5030			
A-286 CRES	90	—	—	1726	2485	4415	6906	9938			
Ti-6Al-4V & Alloy Steel	95	—	—	1822	2623	4660	7290	10490	14280	18650	23610
Alloy Steel	108	—	—	2071	2982	5300	8280	11930	16240	21210	26840
	125	—	—	2397	3452	6140	9590	13810	18790	24540	31060
	132	—	—	2531	3645	6480	10120	14580	19840	25920	32800

(Ref. 9.1)

Fig. 9.2.5 Rivet Shear-off Strength Allowables

Tensile strength of fastener, ksi			Ultimate tensile strength, lbs				
Fastener diameter		Maximum minor area	160	180	220	260	300
In.	D		MIL-S-8879				
0.112	4-40	0.0054367	869	979	1,196	1,414	1,631
0.138	6-32	0.0081553	1,305	1,468	1,794	2,120	2,447
0.164	8-32	0.012848	2,055	2,313	2,827	3,340	3,854
0.190	10-32	0.018602	2,976	3,348	4,090	4,840	5,580
0.250	$\frac{1}{4}$-28	0.034241	5,480	6,160	7,530	8,900	10,270
0.312	$\frac{5}{16}$-24	0.054905	8,780	9,880	12,080	14,280	16,470
0.375	$\frac{3}{8}$-24	0.083879	13,420	15,100	18,450	21,810	25,160
0.438	$\frac{7}{16}$-20	0.11323	18,120	20,380	24,910	29,440	33,970
0.500	$\frac{1}{2}$-20	0.15358	24,570	27,640	33,790	39,930	46,100
0.562	$\frac{9}{16}$-18	0.19502	31,200	35,100	42,900	50,700	58,500
0.625	$\frac{5}{8}$-18	0.24700	39,520	44,500	54,300	64,200	74,100
0.750	$\frac{3}{4}$-16	0.36082	57,700	64,900	79,400	93,800	108,200
0.875	$\frac{7}{8}$-14	0.49327	78,900	88,800	105,500	128,300	148,000
1.000	1-12	0.64156	102,600	115,500	141,100	166,800	192,500
1.125	$1\frac{1}{8}$-12	0.83129	133,000	149,600	182,900	216,100	249,400
1.250	$1\frac{1}{4}$-12	1.0456	167,300	188,200	230,000	271,900	313,700
1.375	$1\frac{3}{8}$-12	1.2844	205,500	231,200	282,600	333,900	385,300
1.500	$1\frac{1}{2}$-12	1.5477	247,600	278,600	340,500	402,400	464,300

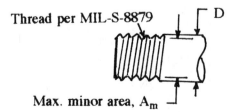

(Ref. 9.1)

All values are for 3A threads per MIL-S-8879.
Nuts designed to develop the ultimate tensile strength of the fastener are required.

Fig. 9.2.6 Ultimate Tensile Strength Allowables for Threaded Steel Fasteners

Bolt Dia. (in)	Bolt Area (in^2)	Moment of Inertia (in^4)	Single Shear (lbs.)	Tension (lbs.)	Bending (in-lbs.)
$\frac{1}{4}$	0.0491	0.000192	3680	4080	276
$\frac{5}{16}$	0.0767	0.000468	5750	6500	540
$\frac{3}{8}$	0.1105	0.00097	8280	10100	932
$\frac{7}{16}$	0.1503	0.001797	11250	13600	1480
$\frac{1}{2}$	0.1963	0.00307	14700	18500	2210
$\frac{9}{16}$	0.2485	0.0049	18700	23600	3140
$\frac{5}{8}$	0.3068	0.00749	23000	30100	4320
$\frac{3}{4}$	0.4418	0.01553	33150	44000	7450
$\frac{7}{8}$	0.6013	0.02878	45000	60000	11850
1.0	0.7854	0.0491	58900	80700	17670

Note: AN steel bolts, $F_{tu} = 125$ ksi, $F_{su} = 75$ ksi

Fig. 9.2.7 AN Steel Bolt Strength Allowables ($F_{tu} = 125$ ksi and $F_{su} = 75$ ksi)

Rivet Type	MS 20426 AD (2117-T3) ($F_{su} = 30$ ksi)		MS 20426 D (2017-T3) ($F_{su} = 38$ ksi)		MS 20426 DD (2024-T3) ($F_{su} = 41$ ksi)				
Sheet Material	Clad 2024-T42 and higher strength aluminum alloys								
Rivet Diameter, in. (Nominal Hole Diameter, in.)	$\frac{3}{32}$ (0.096)	$\frac{1}{8}$ (0.1285)	$\frac{5}{32}$ (0.159)	$\frac{3}{16}$ (0.191)	$\frac{5}{32}$ (0.159)	$\frac{3}{16}$ (0.191)	$\frac{1}{4}$ (0.257)	$\frac{3}{16}$ (0.191)	$\frac{1}{4}$ (0.257)
	Ultimate Strength, lbs.								
Sheet thickness, in.:									
0.020	132	163	...	...	...	...	...	...	...
0.025	156	221	250	...	...	...	...	...	...
0.032	√178	272	348	...	...	...	...	324	...
0.040	193	√309	418	525	476	...	...	555	...
0.050	206	340	√479	628	√580	726	...	758	975
0.063	216	363	523	√705	657	√859	1200	√886	1290
0.071	...	373	542	739	690	917	1338	942	1424
0.080	...	...	560	769	720	969	1452	992	1543
0.090	...	...	575	795	746	1015	√1552	1035	√1647
0.100	...	...	...	818	...	1054	1640	1073	1738
0.125	...	...	...	853	...	1090	1773	1131	1877
0.160	...	...	...	...	...	...	1891	...	2000
0.190	...	...	...	...	...	...	1970	...	2084
Rivet shear strength	217	388	596	862	755	1090	1970	1180	2120

Rivet Type	MS 20426 AD (2117-T3) (F_{su} = 30 ksi)			MS 20426 D (2017-T3) (F_{su} = 38 ksi)			MS 20426 DD (2024-T3) (F_{su} = 41 ksi)		
Sheet Material	Clad 2024-T42 and higher strength aluminum alloys								
Rivet Diameter, in. (Nominal Hole Diameter, in.)	$\frac{3}{32}$ (0.096)	$\frac{1}{8}$ (0.1285)	$\frac{5}{32}$ (0.159)	$\frac{3}{16}$ (0.191)	$\frac{5}{32}$ (0.159)	$\frac{3}{16}$ (0.191)	$\frac{1}{4}$ (0.257)	$\frac{3}{16}$ (0.191)	$\frac{1}{4}$ (0.257)
	Yield Strength, lbs.								
Sheet thickness, in.:									
0.020	91	98	...	...	...	...	...	...	...
0.025	113	150	110	...	...	...	...	...	...
0.032	132	198	200	...	...	...	...	204	...
0.040	153	231	265	273	270	...	...	362	...
0.050	188	261	321	389	345	419	...	538	594
0.063	213	321	402	471	401	515	610	614	811
0.071	...	348	453	538	481	557	706	669	902
0.080	...	...	498	616	562	623	788	761	982
0.090	...	...	537	685	633	746	861	842	1053
0.100	...	...	...	745	...	854	1017	913	1115
0.125	...	...	...	836	...	1018	1313	1021	1357
0.160	...	...	...	...	...	...	1574	...	1694
0.190	...	...	...	...	...	...	1753	...	1925

(Ref. 9.1)

Use 80% of above bearing values for "wet pin" until data is established by testing.
√ Values above line are for knife-edge condition.

Fig. 9.2.8 Bearing Allowables – Aluminum Countersunk 100° Rivet and 2024-T42 Clad Sheet

Fastener Type	HL 11 Pin (F_{su} = 95 ksi), HL 70 Collar			
Sheet Material	Clad 7075-T6			
Fastener Diameter, in. (Nominal Shank Diameter, in.)	$\frac{5}{32}$ (0.164)	$\frac{3}{16}$ (0.190)	$\frac{1}{4}$ (0.250)	$\frac{5}{16}$ (0.312)
	Ultimate Strength, lbs.			
Sheet thickness, in.:				
0.040	734 √	837 √	...	...
0.050	941	1083	1343	...
0.063	1207	1393	1762	2170
0.071	1385	1588	2012	2463
0.080	1557	1779	2281	2823
0.090	1775	2050	2594	3193
0.100	1876	2263	2919	3631
0.125	1950	2542	3765	4594
0.160	2007	2660	3970	5890
0.190	...	2694	4165	6105
0.250	...	...	4530	6580
0.312	...	...	4660	7050
0.375	...	...	...	7290
Fastener shear strength	2007	2694	4660	7290

Fastener Type	HL 11 Pin (Fsu = 95 ksi), HL 70 Collar			
Sheet Material	Clad 7075-T6			
Fastener Diameter, in. (Nominal Shank Diameter, in.)	$\frac{5}{32}$ (0.164)	$\frac{3}{16}$ (0.190)	$\frac{1}{4}$ (0.250)	$\frac{5}{16}$ (0.312)
	Yield Strength, lbs.			
Sheet thickness, in.:				
0.040	674	794	...	...
0.050	835	982	1325	...
0.063	1038	1230	1655	2141
0.071	1130	1355	1813	2338
0.080	1230	1480	2062	2620
0.090	1342	1625	2250	2880
0.100	1440	1750	2470	3420
0.125	1670	2020	2930	3860
0.160	1891	2360	3480	4620
0.190	...	2560	3840	5150
0.250	...	...	4440	6170
0.312	...	...	4660	6900
0.375	...	...	...	7290
Head height (nom.), in.	0.040	0.046	0.060	0.067

(Ref. 9.1)

Use 80% of above bearing values for "wet pin" until data is established by testing.
√ Values above line are for knife-edge condition.

Fig. 9.2.9 Bearing Allowables – Ti-6Al-4V Countersunk 100° Fastener and 7075-T6 Clad Sheet

SIZE	$\frac{3}{32}$	$\frac{1}{8}$	$\frac{5}{32}$	$\frac{3}{32}$	$\frac{1}{8}$	$\frac{5}{32}$	$\frac{3}{32}$	$\frac{1}{8}$	$\frac{5}{32}$
STRENGTH MS20470AD	217	388	596	261	467	715	261	467	715
LS13971AD	—	388	596	—	327	501	—	327	501
SHEET	6013-T6 THRU 7075-T6			6013-T6 THRU 2024-T62			7075-T73 & HARDER		
THICKNESS	SHEAR			TENSION			TENSION		
0.020	163			80			145		
0.025	188	278		118	125		185	240	
0.032	211	329	437	172	197	205	230	317	382
0.040	217	368	507	228	279	307	261	390	491
0.050		388	566	256	382	434		458	603
0.063			596	261	455	599		467	710
0.071					467	672			715
0.080						715			

Yield values are not shown because yield strength is greater than 66.7% of the ultimate value.

Fig. 9.2.10 Protruding Head Aluminum Rivets (MS20470AD and LS13971AD - 30 ksi shear) with Aluminum Sheet

SIZE	$\frac{3}{32}$	$\frac{1}{8}$	$\frac{5}{32}$	$\frac{3}{16}$	$\frac{1}{4}$	$\frac{3}{32}$	$\frac{1}{8}$	$\frac{5}{32}$	$\frac{3}{16}$	$\frac{1}{4}$
STRENGTH LS15906E MS20470DD	296	531	814	1175	2125	356	638	977	1410	2552
LS13971E	—	—	—	1175	2125	—	—	—	987	1787
THICKNESS			SHEAR					TENSION		
0.025	228					151				
0.032	271	394				221	252			
0.040	292	465	609			296	359	392		
0.050	296	516	718	905		340	493	558	604	
0.063		531	796	1067		356	599	774	863	
0.071			808	1119	1703		619	862	1023	1194
0.080			814	1158	1860		638	924	1177	1435
0.090				1175	1977			955	1304	1703
0.100					2063			977	1348	1972
0.125					2127				1410	2393

Yield values are not shown because yield strength is greater than 66.7% of the ultimate value.

Fig. 9.2.11 Protruding Head Aluminum Rivets (MS20470DD and LS13971E - 41 ksi shear) with Aluminum 6013-T6, 2024-T42 and 2024-T3 Sheet

SIZE	$\frac{1}{8}$	$\frac{5}{32}$	$\frac{3}{16}$	$\frac{1}{4}$	$\frac{1}{8}$	$\frac{5}{32}$	$\frac{3}{16}$	$\frac{1}{4}$
STRENGTH	531	814	1175	2125	471	721	1040	1883
CSK head height	.028	.036	.047	.063	.028	.036	.047	.063
THICKNESS		SHEAR				TENSION		
0.032	263				154			
0.040	352	408			229	241		
0.050	443	546	621		324	356	374	
0.063	507	691	833		415	509	555	
0.071	521	743	937	1207	444	583	668	747
0.080	531	781	1033	1408	467	645	784	915
0.090		802	1108	1599	471	689	894	1104
0.100		814	1137	1774	594	721	951	1294
0.125			1175	2025			1040	1653
0.160				2127				1867

Yield values are not shown because yield strength is greater than 66.7% of the ultimate value.

Fig. 9.2.12 Countersunk 100° Shear Head Aluminum Rivets (LS10052DD – 41 ksi shear) with Aluminum 6013-T6 or 2024-T3 Sheet

SIZE	$\frac{3}{32}$	$\frac{1}{8}$	$\frac{5}{32}$	$\frac{3}{16}$	$\frac{1}{4}$	$\frac{3}{32}$	$\frac{1}{8}$	$\frac{5}{32}$	$\frac{3}{16}$	$\frac{1}{4}$
STRENGTH	296	531	814	1175	2125	332	594	909	1312	2376
CSK head height	.036	.042	.055	.070	.095	.036	.042	.055	.070	.095
THICKNESS	\multicolumn{5}{c}{SHEAR}						
	SHEAR					TENSION				
0.025	125					118				
0.032	160√	214				172√	197			
0.040	198	267√	331			232	279√	307		
0.050	238	334	413√	497		284	382	434√	472	
0.063	283	406	521	626√		317	491	599	671√	
0.071	296	446	576	705	949	328	527	689	793	931
0.080		489	635	788	1069	332	560	763	923	1116
0.090		531	696	870	1203√		580	818	1056	1322√
0.100			754	945	1336		594	859	1127	1528
0.125			814	1119	1615			909	1255	1955
0.160				1175	1956				1312	2238

Yield values are not shown because yield strength is greater than 66.7% of the ultimate value.
√ Values above line are for knife-edge condition.

Fig. 9.2.13 Countersunk 100° Tension Head Aluminum Rivets (MS20426DD – 41 ksi shear) with Aluminum 7075-T73 Sheet

(B) COMMENTS

(a) Permanent fasteners:

- Solid aluminum rivets such as 2017, 2024, 2117 and 7050 are the most commonly used
- Use 7050-T73(E) rivet to replace the 2024-T32(DD) "ice box" rivet
- Tension type – carries a higher tension load due to its greater head depth
- Shear type – Use of shallow countersunk head fasteners allows thinner sheet to be used
- Hi-Loks and lockbolts are considered to be permanent fasteners
- Blind fasteners are generally used in blind areas where assembly is impossible

(b) Removable fasteners:

- Use 12-point protruding heads for ultra-high concentrated load applications
- Standard aircraft bolts have rolled threads
- Screw threads must not be placed in bearing
- There are tension and shear bolts and screws
- Screw identifications:
  – AN – Airforce/Navy Standard.
  – NAS – National aircraft Standard.

(c) Nut/nut-plates:

- Select nut-plates which have a tensile strength equal to or greater than the mating bolt or screw

- Strength of tension nuts are 180-220 ksi
- Strength of shear nuts are 125 ksi
- Nut-plates are attached with two small rivets
- Use a nut whose strength is compatible to that of the bolt
- There are two groups of nuts, namely tension nuts and shear nuts

(d) Washer applications:
- Plain washers are used under nuts
- Make sure a countersunk washer is used under bolt (manufactured) heads
- Self-aligning washers are used with fasteners in non-parallel joints
- Washers are used under high tension pre-loaded bolts

9.3 SPLICES

Splices are necessary for the following reasons:
- Limitations on sheet width and length (manufacturing consideration)
- To obtain desired spanwise taper of section area (Cost consideration)
- For fail-safe design (safety consideration)

Spanwise (i.e., wing) splices are designed for shear flows. In some instances there is sufficient chordwise loading to consider it in the splice analysis, but in the majority of cases chordwise loads are small and can be ignored. Fuselage longitudinal skin splices are primarily designed for hoop tension loading due to cabin pressurization but sometimes the local shear load must also be considered. The analysis must consider both the skin and attachment. If the skin is designed to a low margin of safety, as it should be, fastener hole-out efficiency must be maintained at approximately 75 to 80 percent.

The desired object in the design of splices is to obtain the required strength at lowest possible weight and cost. However, it is considered good practice to "balance" the design:

- If the sheet or skin margin is 10%, then the rivet margin should be approximately the same
- Experienced designers frequently design splices to the strength of the sheet and, usually, the cost of adding a few more rivets is minor compared to the gain in capacity
- However, reducing the number of rivets is also important to lower the overall cost

In modern airframe design the most important joint is the wing root chordwise joint which is usually located at the side of the fuselage; some designs may require a secondary chordwise joint between the root and tip due to manufacturing limitations which do not allow the fabrication of a one piece wing skin panel or stringer. Practically, however, it is not desirable to have such a secondary joint.

Example:

The skin is lap-spliced at a given panel. One skin thickness is 0.055 inch and the other skin thickness is 0.071 inch with countersunk head fasteners (F_{su} = 41,000 psi) of diameter $D = \frac{3}{16}$ inch. Find the margin of safety for both skins and rivets.

Chapter 9.0

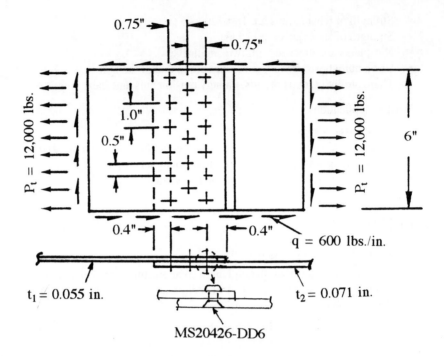

Given material (2024-T42 clad sheet, use "B" values):

$F_{tu} = 59,000$ psi; $F_{ty} = 35,000$; $F_{su} = 35,000$ psi;

$F_{bru} = 112,000$ psi and $F_{bry} = 56,000$ psi ($\frac{e}{D} = 2.0$)

Check: $\frac{F_{tu}}{F_{ty}} = \frac{59,000}{35,000} = 1.68 > 1.5$

Therefore, use $F_{tu,y} = 1.5 \times 35,000 = 52,500$ psi

$F_{su,y} = F_{su}(\frac{F_{tu,y}}{F_{tu}}) = 35,000(\frac{52,500}{59,000}) = 31,100$ psi

Check: $\frac{F_{bru}}{F_{bry}} = \frac{112,000}{56,000} = 2 > 1.5$ O.K.

Therefore, use $F_{bru,y} = 1.5 \times 56,000 = 84,000$ psi

(1) The load on each rivet is the resultant of loads on the skin,

$P = [(\frac{12,000}{17})^2 + (6 \times \frac{600}{17})^2]^{\frac{1}{2}}$

$= 737$ lbs. per rivet

(2) Use DD6 rivets (MS 20426; $D = \frac{3}{16}$ inch, 2024-T31, $F_{su} = 41,000$ psi):

Rivet shear allowable: $P_{all} = 1,175$ lbs. (see Fig. 9.2.5)

$MS = \frac{P_{all}}{P} = \frac{1,175}{737} - 1 = \underline{0.59}$ O.K.

(3) MS of rivets in the skin of $t_1 = 0.055$ in.:

Bearing allowable load:

$$F_{bru,y} \times D \times t_1 = 84,000 \times \frac{3}{16} \times 0.055 = 866 \text{ lbs.}$$

$$MS = \frac{866}{737} - 1 = \underline{0.18} \qquad \text{O.K.}$$

(4) Allowable load for $\frac{3}{16}"$ countersunk head bearing in plate of $t_2 = 0.071$ inch is 942 lbs. and its bearing yield strength is 669 lbs.:

$$\frac{942}{669} = 1.41 < 1.5 \text{ (not critical)}$$

$$MS = \frac{942}{737} - 1 = \underline{0.28} \qquad \text{O.K.}$$

(5) MS of the net section of skin, $t_1 = 0.055$ in.:

$$f_{t,1} = \frac{12,000}{6 \times 0.055}$$

$$= 36,364 \text{ psi (gross area tensile stress)}$$

$$f_{s,1} = \frac{600}{0.055}$$

$$= 10,900 \text{ psi (gross area shear stress)}$$

Skin section reduction due to rivet holes ($D = \frac{3}{16}"$):

$$(\frac{3}{16}) \times 0.055 = 0.0103 \text{ in.}^2 \text{ per rivet hole}$$

Gross area: $A_g = 6 \times 0.055 = 0.33$ in.^{2}

Reduced area: $A_r = 6 \times 0.0103 = 0.0619$ in.^{2} (total holes in first row)

Therefore, $\frac{A_r}{A_g} = 0.188$ and rivet hole efficiency: $\eta = 1 - 0.188 = 0.812$

$$R_t = \frac{f_{t,1}}{\eta F_{tu,y}}$$

$$= \frac{36,364}{0.812 \times 52,500} = 0.853$$

$$R_s = \frac{f_{s,1}}{\eta F_{su,y}}$$

$$= \frac{10,900}{0.812 \times 31,100} = 0.432$$

$$MS = \frac{1}{[(R_t)^2 + (R_s)^2]^{\frac{1}{2}}} - 1$$

$$= \frac{1}{[(0.853)^2 + (0.432)^2]^{\frac{1}{2}}} - 1 = \underline{0.046} \qquad \text{O.K.}$$

(6) MS of the net section of skin, $t_2 = 0.071$ inch:

(This skin is not critical compared to that of $t_1 = 0.055$ in.)

(7) Therefore, the min. MS = $\underline{0.046}$ (Net section critical of skin $t_1 = 0.055$ in.)

9.4 ECCENTRIC JOINTS

Concentric riveted connections which carry no moment are assumed to be loaded evenly, i.e., the load is distributed equally to the rivets. This is approximately true, even when the rivets are in a single line the end fasteners (or fisrt fasteners) are not overloaded as might be expected. All fastened joints must be checked for the

- Shear value of the fasteners
- Bearing value of the fastener in the attached sheets

Where fastener clusters must carry moment load (M) as well as shear force (P) in members they must be investigated for combined loads on the fastener clusters, as shown in Fig. 9.4.1. The solution of forces for a group fasteners subjected to moment is rather simple if the engineer assumes that the force on each fastener is proportional to its distance from the centroid. This is true of course only if the fasteners are all of the same size (also assume the same material).

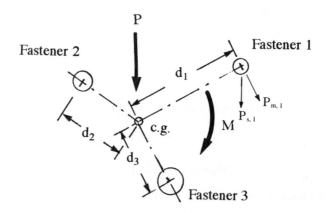

Fig. 9.4.1 Fastener Clusters

Shear load on given fastener 1, due to a concentrated load, P

$$P_{s,1} = P\left(\frac{A_1}{\Sigma A}\right) \quad \text{Eq. 9.4.1}$$

Shear load on given fastener 1, due to moment, M,:

$$P_{m,1} = M\left(\frac{A_1 d_1}{\Sigma A d^2}\right) \quad \text{Eq. 9.4.2}$$

where: A – Fastener area
d – Distance from centroid of the fastener clusters to given fastener

The above equations are based on the following:

- Fastener materials are the same
- Fastener bearing on the same material and thickness
- Fastener shear load assumes straight line distribution

Procedure for determining fastener shear loads is as follows:

(1) Find out fastener areas and their shear allowables.

(2) Determine centroid (c.g.) of the fastener clusters from a designated fastener:

X of c.g. – x-axis from a designated fastener
Y of c.g. – y-axis from a designated fastener

(3) Fastener shear load (see Eq. 9.4.1 and assume at fastener 1) due to a concentrated load, P:

$$P_{s,1} = P(\frac{A_1}{\Sigma A})$$

(Assume load, P, goes through the centroid of fastener clusters)

(4) Fastener shear load (see Eq. 9.4.2 and assume at fastener 1) due to moment, M,:

$$P_{m,1} = M(\frac{A_1 d_1}{\Sigma A d^2})$$

If areas of fasteners are equal and the fastener shear load (assume at fastener 1) due to moment , M,:

$$P_1 = M(\frac{d_1}{\Sigma d^2})$$

(5) Construct the vector diagrams for the loads on the assumed fastener 1 ($P_{s,1}$ and $P_{m,1}$) and measure the resultant vector ($P_{Sm,1}$).

(6) Margin of safety:

$$MS_{Shear} = \frac{\text{Fastener 1 shear allowable}}{P_{Sm,1}} - 1$$

$$\text{or } MS_{bearing} = \frac{\text{Sheet bearing allowable on fastener 1}}{P_{Sm,1}} - 1$$

Example 1 (Symmetrical Joint With Same Size Rivets):

Assume a symmetrical rivet pattern of five rivets of equal size ($D = \frac{3}{8}"$), carrying a shear load of $P_v = 10,000$ lbs. upward and a moment load of $M = 12,000$ in-lbs. acting as shown in Fig. A.

- For this case the centroid of the rivet joint about which moment load will take place is by inspection at the No. 5 rivet
- The center rivet cannot take any moment load because it has no lever arm, being at the centroid
- The four outer rivets are equal distance from the centroid and take equal moment loads
- The forces on the rivets are shown and the resultant forces on all rivets are different, being the resultants of the forces due to the vertical concentrated load of (P_V) and moment load (M)
- For the condition shown rivets No. 1 and No. 4 carry the greatest load and would be critical

Chapter 9.0

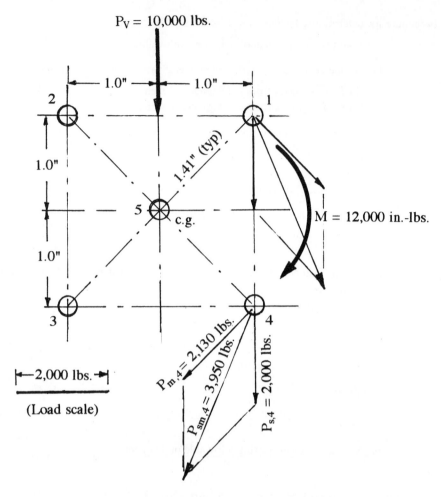

Fig. A Five Equal Sizes of Rivets ($D = \frac{3}{8}''$) with Symmetrical Rivet Pattern

(Centroid of rivet group is at rivet No. 5 and therefore rivet No. 5 has no load from moment)

(1) Use D-rivet (2017-T31, F_{su} = 34 ksi) and rivet shear strength of dia. = $\frac{3}{8}''$ is

P_{all} = 3,980 lbs. (Assume bearing is not critical)

(2) Centroid of rivet group is at rivet No. 5 and radial distance to each rivet effective in moment load is 1.41".

(3) Rivet shear load due to P_V = 10,000 lbs.:

$$P_s = \frac{10,000}{5} = 2,000 \text{ lbs./rivet}$$

(Assume P_V = 10,000 lbs. load goes through the centroid of the group of rivets)

(4) Rivet shear loads due to M = 12,000 in.-lbs.:

$$P_m = \frac{12,000}{4 \times 1.41} = 2,130 \text{ lbs./rivet}$$

(5) The final loads on the respective rivets are the vector resultants of the P_s = 2,000 lbs. and P_m = 2,130 lbs. loads; the max. resultant load is at rivets No. 1 and No. 4 and is $P_{sm,4}$ = 3,950 lbs. (by measuring the vector diagram):

(6) Margin of safety:

$$MS = \frac{P_{all}}{P_{sm,4}} - 1 = \frac{3,980}{3,950} - 1 = \underline{0.01}$$ O.K.

Example 2 (Unsymmetrical Joint With Same Size Rivets):

Design the unsymmetrical rivet pattern shown in Fig. B to carry the external shear P_V = 10,000 lbs and moment M = 12,000 in-lbs. with equal size rivets (D = $\frac{3}{8}$").

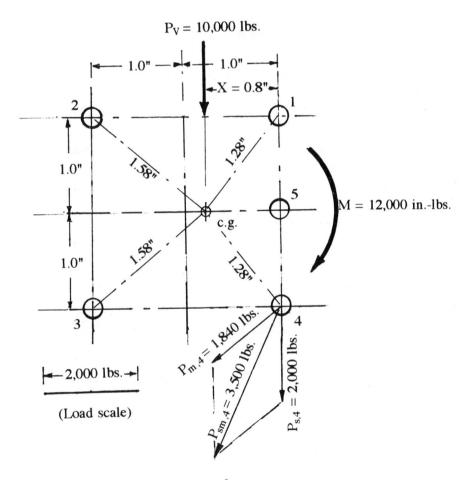

Fig. B Five Equal Sizes of Rivets (D = $\frac{3}{8}$") with Unsymmetrical Rivet Pattern

(1) Use D-rivet (2017-T31, F_{su} =34 ksi) and rivet shear strength of dia. = $\frac{3}{8}$" is

P_{all} = 3,980 lbs. (assume bearing strength is ample)

(2) Rivet shear load due to P_V = 10,000 lbs.:

$$P_s = \frac{10,000}{5} = 2,000 \text{ lbs./rivet}$$

Chapter 9.0

(3) Assume each $\frac{3}{8}"$ rivet equal to 1.0 unit:

$$X = \frac{2(2 \times 1.0)}{5} \times 1.0 = 0.8"$$

(4) Rivet loads due to M = 12,000 in.-lbs. are:

$$P_m = \frac{12,000}{2 \times 1.58 + 2 \times 1.28 + 0.8} = 1,840 \text{ lbs./rivet}$$

(5) The final loads on the respective rivets are the vector resultants of the P_s = 2,000 lbs. and P_m = 1,840 lbs. loads; the max. resultant load is at rivets No.1 and No. 4 and is $P_{sm, 4}$ = 3,500 lbs. (by measuring the vector diagram):

(6) Margin of safety:

$$MS = \frac{P_{all}}{P_{sm, 4}} - 1 = \frac{3,980}{3,500} - 1 = \underline{0.14} \qquad \text{O.K.}$$

Example 3 (Unsymmetrical Joint With Different Size Rivets):

Design a joint of unsymmetrical rivet pattern with unequal size rivets (D = $\frac{3}{8}"$ and D = $\frac{1}{4}"$) as shown in Fig. C to carry the vertical shear force P_V = 10,000 lbs and moment M = 12,000 in-lbs. (**Note:** This type of joint design using different size of rivets is not recommended in new design except as rework or repair work)

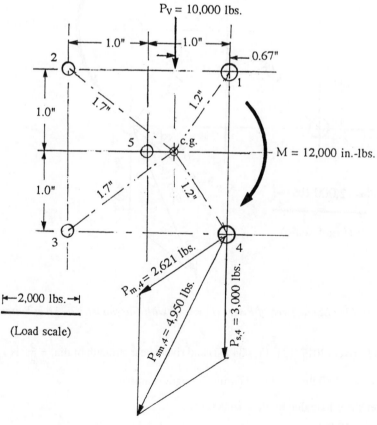

Fig. C Unequal Sizes of Rivets, $D = \frac{1}{4}"$ and $D = \frac{3}{8}"$, with Unsymmetrical Rivet Pattern

(1) Use E-rivet (material = 7075-T731, F_{su} = 43,000 psi) and assume bearing strength is ample:

Shear strength of rivets No. 2, 3, and 5 ($D = \frac{1}{4}"$ and rivet area = 0.04908 in.^{2}) is

P_{all} = 2,230 lbs/rivet

Shear strength of rivets No. 1 and 4 ($D = \frac{3}{8}"$ and rivet area = 0.1105 in.^{2}) is

P_{all} = 5,030 lbs/rivet

Rivet areas:

Area for $D = \frac{1}{4}"$: $A_{\frac{1}{4}}$ = 0.0491 in.^{2}

Area for $D = \frac{3}{8}"$: $A_{\frac{3}{8}}$ = 0.1105 in.^{2}

(2) Determine centroid (c.g.) of the fastener clusters from a designated fastener 1 and 4 (Y = 0 because it is symmetrical about x-axis):

$$X = \frac{2 \times (2 \times 0.0491) + 1 \times (1 \times 0.0491)}{2 \times 0.1105 + 3 \times 0.0491} = 0.67"$$

(3) Rivet shear load due to P_V = 10,000 lbs.:

On $D = \frac{1}{4}"$ rivet:

$$P_{s,2} = P_{s,3} = P_{s,5} = \frac{10,000 \times 0.0491}{3 \times 0.0491 + 2 \times 0.1105} = 1,333 \text{ lbs.}$$

On $D = \frac{3}{8}"$ rivet:

$$P_{s,1} = P_{s,4} = \frac{10,000 \times 0.1105}{3 \times 0.0491 + 2 \times 0.1105} = 3,000 \text{ lbs.}$$

(4) Rivet shear load due to M = 12,000 in.-lbs.:

The resisting moment for this rivet group which has a centroid at X = 0.67" and ΣAd^2 would be (put the solution into tabular form as follows):

Rivet No.	D (in.)	A (in.^{2})	d (in.)	Ad^2
1	$\frac{3}{8}$	0.1105	1.2	0.159
2	$\frac{1}{4}$	0.0491	1.7	0.142
3	$\frac{1}{4}$	0.0491	1.7	0.142
4	$\frac{3}{8}$	0.1105	1.2	0.159
5	$\frac{1}{4}$	0.0491	0.33	0.005
				$\Sigma Ad^2 = 0.607$

Then the actual load on any rivet is equal to:

$$\frac{Ad}{\Sigma Ad^2} M$$

The load on rivets No.1 and 4 ($D = \frac{3}{8}"$):

$$P_{m,1} = P_{m,4} = \frac{A_1 d_1}{\Sigma A d^2} M = \frac{0.1105 \times 1.2}{0.607} 12,000 = 2,621 \text{ lbs.}$$

The load on rivets No.2 and 3 ($D = \frac{1}{4}"$):

$$P_{m,2} = P_{m,3} = \frac{A_2 d_2}{\Sigma A d^2} M = \frac{0.0491 \times 1.7}{0.607} 12,000 = 1,650 \text{ lbs.}$$

(6) The final loads on the respective rivets are the vector resultants of the $P_{s,4} = 3,000$ lbs. and $P_{m,4} = 2,621$ lbs. loads; the max. resultant load is at rivets No.1 and No. 4 and is $P_{sm,4} = 4,950$ lbs (by measuring the vector diagram):

$$MS = \frac{P_{all}}{P_{s,m}} - 1 = \frac{5,030}{4,950} - 1 = 0.02 \qquad \text{O.K.}$$

9.5 GUSSET JOINTS

Gusset joints are generally used on truss beams, e.g., spars, ribs, floor beams, etc. However, these types of structures are very seldom used in airframe primary construction today due to the problems of weight, cost, and repair difficulty except very special applications.

- The stress or load distribution in gusset joints is usually uncertain and complicated and there are no rules that will fit all cases
- Engineers must rely upon their judgment of the appearance of the gusset and the possibilities of peculiarities of stress distribution as well as upon computations

A single gusset plate (on one side) is generally used on airframe structures and the eccentricity between gusset and attached member will induce additional excessive stress (bending stress) on the gusset plate. The stress is obviously not "$\frac{P}{t}$" but it is not as bad as "$\frac{4P}{t}$" (see Fig. 9.1.2) and it is somewhere between them. Use following aluminum material design allowables (until the verification test data is available) are:

- In tension or compression – Use 40% F_{tu} of the material allowable as shown in Fig. 9.5.1
- In shear net section – Use 70% F_{tu} for thickness $t = 0.125$ in. and 50% F_{tu} for $t = 0.04$ in. as shown in Fig. 9.5.2

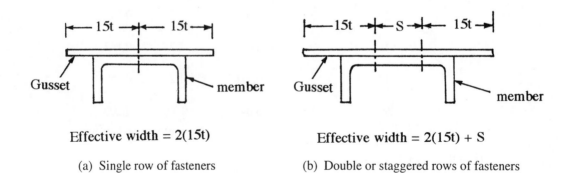

(a) Single row of fasteners

(b) Double or staggered rows of fasteners

Fig. 9.5.1 Effective Width

Joints and Fittings

Fig. 9.5.2 Shear Allowable vs. Plate Thickness

Design considerations are as follows:

- All load paths of members should pass through the "load center" to prevent moment occuring due to eccentricity effect
- A minimum of two fasteners is required for each member attached to the gusset
- Use of tension-head aluminum rivets or shear-head Hi-Lok fasteners is acceptable
- In general, gussets should be equal to or thicker than the members to which they are attached
- Since compression and combined shear will tend to cause free flange buckling, gusset must be designed with greater thickness to keep the compression and shear stresses low (see Fig. 9.5.1 and Fig. 9.5.2)
- Since gusset plates cannot be allowed to buckle, reduce the amount of free flange, use a thicker plate or use a stiffening method (bend-up flange, beaded plate, etc.) to prevent buckling (see Fig. 9.5.3)
- When the angle between two members is more than 45°, bend-up flanges are required to prevent the gusset from buckling under compression
- Do not use a single long row of fasteners on a diagonal member
- Check shear-out along every line of fasteners
- The necessity of using proper fasteners to connect the gusset to the cap is important and it is not good design practice to use the minimum number of fasteners necessary to carry the design load; the fastener pattern must be spaced so that the free flange of the gusset is tied to the members as securely as possible.

The following gusset types are those generally used in the joining of truss structures:

(a) Type I – Diagonal member in tension (see Fig. 9.5.3):

- Use cut-back as shown for weight savings
- Check the fasteners and net shear strength (use lower shear stress, see Fig. 9.5.2, for free flange design) along section A-A from the horizontal component load of member b-e
- Check Section B-B with the vertical component load from diagonal member b-e
- Check Section C-C and fasteners with the load from member b-e
- Unsupported or free flange at section D-D may fail in compression
- Check distance 'L' for compression buckling

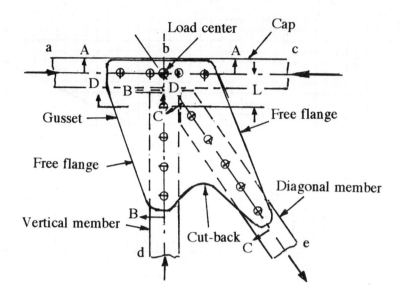

Fig. 9.5.3 Diagonal Member in Tension

(b) Type II – Diagonal member in compression (see Fig. 9.5.4):
- The most critical location is any free flange area where the gusset is joined to the diagonal member
- The greater gusset material thickness than that of type I is required due to higher compression on the diagonal member
- A cut-back between the vertical and diagonal members is generally not used
- Check all sections at fastener lines same as instructed for type I gusset

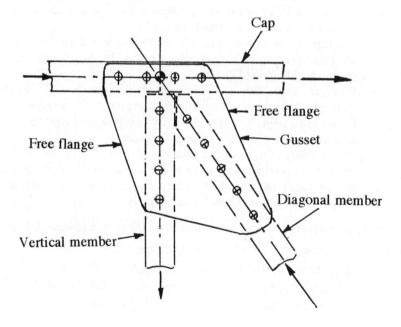

Fig. 9.5.4 Diagonal Member in Compression

(c) Type III – A-shaped gusset (Warren truss with two equal diagonal members) as shown in Fig. 9.5.5:

- One diagonal member takes tension and the other will take compression with equal load
- Tendency toward compression failure at section A-A
- Generally use a thicker gusset
- Use of cut-back is generally not recommended
- Use ample fasteners in the cap to help to support the free flange
- Bend-up flange is required between the two diagonal members

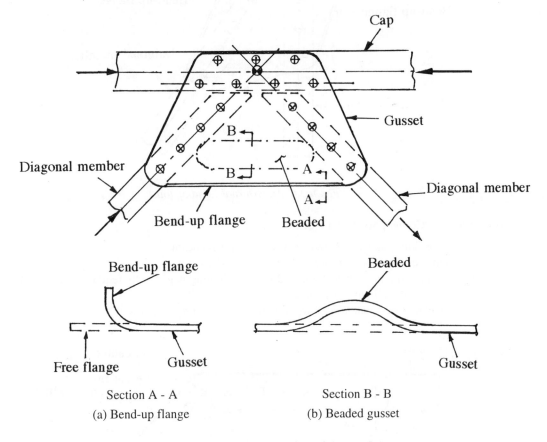

Fig. 9.5.5 A-shaped Gusset Joint

(d) Type IV – Gusset joint subjected to externally applied load (see Fig. 9.5.6):

- Applied load could be either applied at cap member or directly applied on gusset plate
- If external concentrated load is applied at cap, check section A-A which must be designed to withstand the externally applied compression load (or tension load in net section)
- Generally use a much thicker gusset
- Sometimes a wider gusset plate is required to transfer external concentrated load from cap to gusset and thus a greater free flange is required

Chapter 9.0

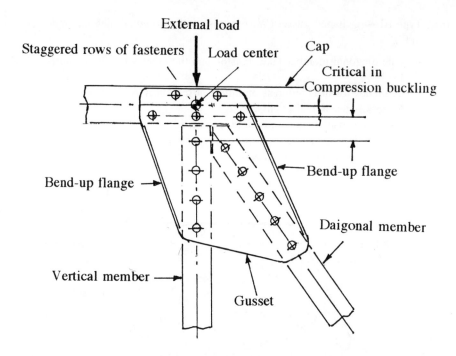

Fig. 9.5.6 An External Load at Joint

(e) Type V – Corner gusset joint (see Fig. 9.5.7):
- Corner joints usually are required at ends of truss beam areas
- Joints consist of two or three members joined by a gusset
- Lateral stiffness is critical and additional stiffening is generally required because of the lack of continuous cap material
- Use extra thick plate for gusset or additional stiffening methods

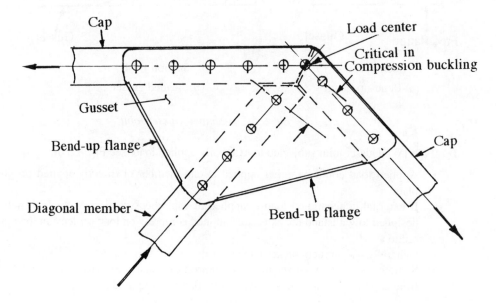

Fig. 9.5.7 Corner Joint

Joints and Fittings

Example:

Assume a corner gusset joint as shown below subjected internal loads at joint 'D' given in previous Example 1 of Chapter 5.0 (Section 5.4).

Given materials: All members (Channel sections as shown below) – 7075-T6 Extrusion with the following material allowables (see Fig. 4.3.5):

$F_{tu} = 82,000$ psi $F_{su} = 44,000$ psi $F_{bru} = 148,000$ psi

$E = 10.5 \times 10^6$ psi

Gusset plate – 7075-T6 with the following allowables (see Fig. 4.3.4):

$F_{tu} = 80,000$ psi $F_{su} = 48,000$ psi $F_{bru} = 160,000$ psi

Tension head rivets: MS20470(E)5, $D = \frac{5}{32}"$, $F_{su} = 41,000$ psi (see Fig. 9.2.5):

$P_{fastener, shear} = 814$ lbs.

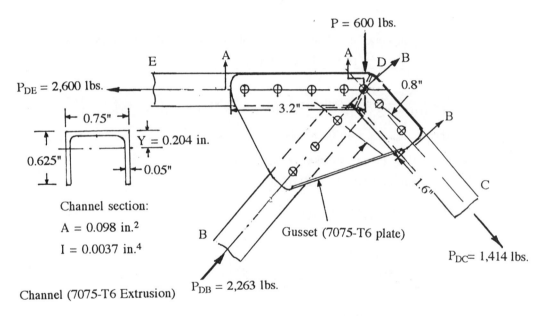

Fig. A A Gusset Joint Design

(1) Member D-B and its $P_{DB} = 2,263$ lbs. in compression:

Use single row of 3 MS20470(E) rivets (pitch = 5D = 0.78")

Gusset effective width (7075-T6):

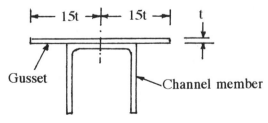

Single row of fasteners

Fig. B Gusset Effective Width

303

$2(15t) \times t \times 80,000 \times 40\% = 2,263$

$t = 0.0486"$ use $t = 0.05"$

Fastener bearing allowable:

$P_{fastener, bru} = F_{bru} Dt = (160,000 \times 80\%) \times \frac{5}{32} \times 0.05 = 1,000$ lbs. (Gusset)

$P'_{fastener, bru} = F_{bru} Dt = (148,000 \times 80\%) \times \frac{5}{32} \times 0.05 = 925$ lbs. (Channel members)

$P_{fastener, shear} = 814$ lbs. (Fastener shear) – critical

$MS_{fastener, shear} = \dfrac{814}{\frac{2,263}{3}} - 1 = \underline{0.08}$ 	O.K.

Gusset plate buckling check:

The effective width (single row of fasteners): $30t = 30 \times 0.05 = 1.5"$

The moment of inertia: $I = 1.5 \times \dfrac{0.05^3}{12} = 15.6 \times 10^{-6}$ in.^{4}

Use 'simply supported end fixity' ($c = 1.0$) column equation (Eq. 10.2.1, from Chapter 10):

$P_{cr} = \dfrac{\pi^2 EI}{L^2} = \dfrac{\pi^2 \times 10.5 \times 10^6 \times 15.6 \times 10^{-6}}{0.8^2} = 2,526$ lbs.

$F_{cr} = \dfrac{2,526}{0.05 \times 1.5} = 33,680$ psi $> F_{compression} = 32,000$ psi 	O.K.

where: $F_{compression} = 40\% \times 80,000 = 32,000$ psi

(2) Member D-E and its $P_{DE} = 2,600$ lbs. in tension:

Use single row of 4 MS20470(E) rivets

$MS_{fastener, shear} = \dfrac{814}{\frac{2,600}{4}} - 1 = \underline{0.25}$ 	O.K.

Shear at net section A-A:

$P_{gusset, shear} = (\dfrac{5D - D}{5D})(3.2)(t)(52.5\% \, F_{su})$

$= (\dfrac{4}{5})(3.2)(0.05)(48,000 \times 52.5\%)$

$= 3,226$

$MS_{gusset, shear} = \dfrac{3,226}{2,600} - 1 = \underline{0.24}$ 	O.K.

(3) Member D-C and its $P_{DC} = 1,414$ lbs. in tension:

Use single row of 2 MS20470(E) rivets

$MS_{fastener, shear} = \dfrac{814}{\frac{1,414}{2}} - 1 = \underline{0.15}$ 	O.K.

Shear at net section B-B:

$$P_{gusset, shear} = (\frac{5D - D}{5D})(1.6)(t)(52.5\% \ F_{su})$$

$$= (\frac{4}{5})(1.6)(0.05)(48,000 \times 52.5\%)$$

$$= 1,613$$

$$MS_{gusset, shear} = \frac{1,613}{1,414} - 1 = \underline{0.14} \qquad \text{O.K.}$$

9.6 WELDED JOINTS

The strength of welded joints depends greatly on the skill of the welder. The stress conditions are usually uncertain, and it is customary to design weld joints with liberal margins of safety. It is preferable to design joints so that the weld is in shear or compression rather than in tension (except for electronic beam – EB welding):

- Tubes in tension are usually spliced by "fish-mouth" joints which are designed so that most of the weld is in shear and so that the local heating of the tube at the weld is not confined to one cross section
- It is therefore necessary to reduce the allowable unit stresses in the welds and in the adjacent structure by using a weld efficiency factor or reduction factor

Typical welded connections are shown in Fig. 9.6.1(a).

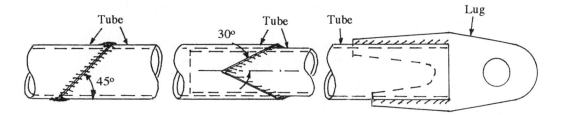

(a) Tube Connections

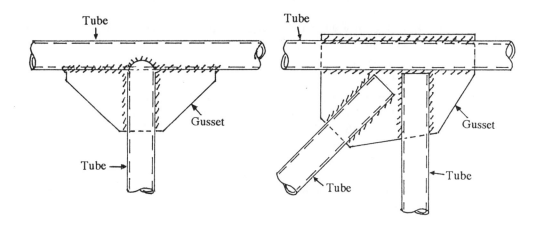

(b) Welded Joints of Gusset Plate and Tubes

Fig. 9.6.1 Typical Welding Joints

When members are joined by welding, the cross section through the welds themselves are greater than those through the adjacent members. This accounts, in part, for the fact that on test loads, weld assemblies rarely break in the welds themselves, but fail in the partially annealed structure just outside the weld.

It is generally necessary to use gussets to assist in transferring the loads in a joint as shown in Fig. 9.6.1(b). Tension members in particular are apt to have insufficient welds to develop the strengths necessary. Where tubes are joined in a cluster fitting which includes an assembly of "fish-mouthed" sleeves joined by gussets it is not necessary to calculate the strength of the weld between tubes and sleeves if standard fish-mouth angles of 30° to 37° are maintained. In this case the weld will always develop at least 80% of the strength of the original tube before it was welded.

Spotwelding should be used in non-structural applications where the primary loads on the spotwelds are shear. Fasteners pull the skins being fastened together and apply a clamping action. Spotwelds, however, force the skins apart a minimum of 0.001 inches. A structure of multiple material skins spotwelded together, loaded in compression, will be weaker than the same structure riveted together. Since spotwelds are detrimental to the fatigue life of the structure, spotwelding will not be discussed further in this section. More data and information can be found in Ref. 9.1 of MIL-HDBK-5, Section 8.2.2.3.

In the aircraft industry, welding is seldom used on airframe primary structures because of its uncertainty of strength but a few applications on which it is used include truss members for mounting propeller engines, landing gear lugs, pylon lugs, etc. Because of its simplicity and low cost, fusion and resistance welding have been used on the following steel structures (i.e., steel tubing):

- Supporting framework used in the testing of airframe structures
- Steel assembly jigs (stiffness is an important requirement)
- Backup framework for supporting mold tooling (generally used for fabricating of composite parts).

For certain materials and types of structural parts, welding plays an important role in joining since it will improve the structural fatigue life and leak-free character of the structure.

(A) DESIGN CONSIDERATIONS

(a) Classification of welds:

Class I – Vital Welds:
Class IA – Class IA weld is a vital weld in single load path structure and proof tests are required.
Class IB – Class IB weld is a vital weld which can be shown by analysis or test to be fail-safe. With the exception of proof testing, Class IB welds require the same type of inspection as Class IA welds. Proof tests of Class IB weld other than flash and pressure welds are not required.

Class II – Non-vital Welds which are further classified as follows:
Class IIA – Class IIA weld is a Class II weld which cannot be classified IIB.
Class IIB – Class IIB weld is one of secondary importance with minimal service requirements.

(b) Inspection:

The following data shall be incorporated on engineering drawings for welds in accordance with classification requirements:

Class of Weld	Required Inspection Drawing Notes*
IA	100% proof test
	100% X-ray inspection
	100% magnetic particle inspection
	100% ultrasonic inspection (EB welds only)
IB	Same as for Class IA, except that proof testing of welds, other than flashwelds and pressure welds, is not required.
IIA	X-ray sampling inspection
	100% magnetic particle inspection
IIB	Magnetic particle sampling inspection

(* Weldment drawings shall be reviewed for possible reduced inspection, commensurate with the degree of reliability required)

(c) Welding methods

- Machine welding methods:
  Welds produced by automatically mechanized and machine welding methods can be controlled better and kept more uniform than welds that are produced manually. All vital and important structural welds (Class I) shall be accomplished by mechanized or machine welding and shall be specified on the engineering drawing
- Manual welding methods:
  Manual welds are permissible for Class II applications
- Typical methods of fusion welding are:
  - TIG – Inert-gas shielded-arc welding with tungsten electrode
  - MIG – Inert-gas shielded metal arc welding using covered electrodes
  - EB – Electron beam (this is an expensive method which is used for special tension joining applications, class I)

(d) Stress Concentrations

No stress concentration (bend radius, formed flanges, joggles, tube bends, sharp radii, drilled holes, etc.) shall be allowed to occur in or adjacent to the weld, including the heat affected zone.

(e) Low Ductility Weld Materials

The ductility of the weld material is usually lower than that of the parent metal, even though the static strength is not reduced. Avoid placing welds in areas where unusual flexure is likely to occur or areas where significant load redistribution occurs.

(f) Fatigue Consideration

The fatigue life of simple butt welds shall be based on the assumption that the stress concentration factor at the weld is at least equal to $K_t = 3$. This does not permit the inclusion of a hole in the weld-affected-areas of the part. Greater stress concentration factors may be appropriate for configurations other than simple butt welds (see Fig. 9.6.2).

The non-geometric influences upon stress concentration in welds are so dominant that labeling any geometry with a unique K_t value is unrealistic. This is particularly true of fusion-butt joints, which are characteristically non-homogeneous; they may suffer from:

- poor penetration
- porosity
- or undue surface roughness

The mechanical properties of the material and its heat treatment or contour machining (if any) after welding will influence the importance of these defects on stress concentration.

For the 100 percent efficient V-butt welded joint, K_t of the perfect weld is 1.0. However, perfect welds are never attained and seldom approached. Fig. 9.6.2 shows how the fatigue quality of fusion butt-welded joints can be altered by slight changes in geometry alone.

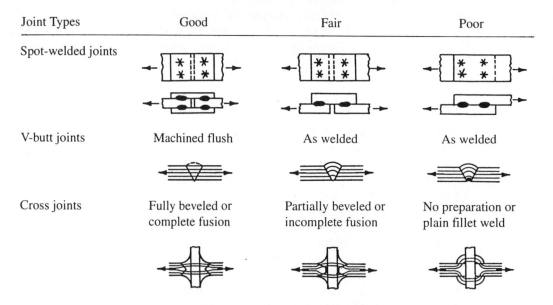

Joint Types	Good	Fair	Poor
Spot-welded joints			
V-butt joints	Machined flush	As welded	As welded
Cross joints	Fully beveled or complete fusion	Partially beveled or incomplete fusion	No preparation or plain fillet weld

Fig. 9.6.2 Comparison of Welded Joints

(g) Fail-safe Design

Fail-safe design of welds requires that the structure will support fail-safe loads in the event of complete weld failure or weld failure between effective crack stoppers in the weld.

(B) BUTT WELDS

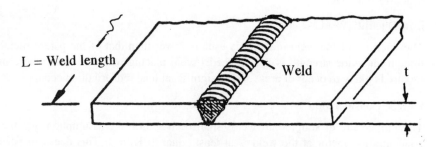

Fig. 9.6.3 Typical Butt Welded Joint

When computing the strength, use the material thickness (t) as the weld thickness,

    Joint strength = A × (allowable weld stress)

    where: A = t × (weld length)

Allowable weld stress – given in Fig. 9.6.4 or use the weld efficiency multiplied by the design property stress, both which are given in Fig. 9.6.5 and Fig. 9.6.6)

Allowable weld stress = Weld Efficiency × Design Property Stress

Materials	Heat Treatment Subsequent to Welding	Welding Rod or Electrode	
		F_{su} (ksi)	F_{tu} (ksi)
Carbon and Alloy Steels	None	32	50
Alloy Steels	None	43	72
Alloy Steel	Stress relieved	50	85
		60	100
Steels:	Quench and temper:		
4130	125 ksi	62	105
4140	150 ksi	75	125
4340	180 ksi	90	150

(The above data from Ref. 9.1 gives the allowable weld material strengths for the various steels. These weld metal allowables are based on 85% of the respective minimum tensile ultimate test values)

Fig. 9.6.4 Strength of Fusion Welded Joints of Alloy Steels

Material and Condition	Weld Efficiency [1]	Base Metal Property Condition after Welding
Alloys in T4 condition [3] before welding and left as welded	0.7	T4 [4]
Alloys in T6 condition [3] before welding and left as welded	0.6	T6
Alloys head treated to T4 [2] or T6 condition after welding	0.85	T4 [4] or T6
Work hardenable alloys [3] left as welded	1.0	Annealed
All alloys, annealed and left as welded	1.0	Annealed

(1) Allowable weld stress = Weld efficiency × base metal design allowable stress.
(2) For alloys heat treated to T4 or T6 condition after welding, the noted weld efficiency shall be reduced by multiplying by an additional 0.9 when the weld bead of butt welds is dressed flush on both sides.
(3) For aluminum alloys hardened before welding, the heat affected zone extends 1.0 inch or 15t on each side of the weld, whichever is less.
(4) Alloys in the T4 condition are yield critical and parts which may be loaded should be aged to T6 after welding.

Fig. 9.6.5 Weld Efficiency and Design Property Stress of Aluminum Alloys

Material and Condition	Weld Efficiency [1]	Base Metal Property Condition after Welding
Non-heat treatable 300 series [2] (321, 347, 304L)	1.0	Annealed
Heat treatable steels and solution heat treated and aged after welding	1.0	Solution heat treated and aged
Heat treatable steels and solution heat treated before welding.[2] Aged after welding or cold work Aged after welding	0.9	Solution heat treated and aged
Heat treatable steels not heat treated after welding	1.0	Annealed

(1) Allowable weld stress = Weld efficiency x base metal design allowable stress.
(2) For alloys cold work or solution heat treated before welding, the heat affected zone extends approximately one inch or 15t on each side of the weld, whichever is less.

Fig. 9.6.6 Weld Efficiency and Design Property Stress of Corrosion Resistant Steels

Example 1

A butt joint is formed by welding two plates as shown in Fig. 9.6.3. The materials are alloy steel with no heat treatment subsequent to welding and $F_{tu} = 72$ ksi (see Fig. 9.6.4). Find the margin of safety of this welded joint under the load of 5.0 kips.

Assume weld length, $L = 1.0$ inches; $t = 0.071$ inches and $F_{tu} = 72$ ksi

$P_{all} = F_{tu} \times L \times t = 72 \times 1.0 \times 0.071 = 5.112$ kips

$MS = \dfrac{5.112}{5.0} - 1 = \underline{0.02}$ O. K.

(C) FILLET WELDS

The fillet weld, as shown in Fig. 9.6.7, should be able to develop the ultimate tensile strength of its base material. The allowable stress in the weld metal and the heat affected zone is equal to the base metal design property stress multiplied by the weld efficiency. Therefore, the weld efficiency is the ratio of the ultimate property stress of the welded material divided by the base design property stress.

SIMPLE LAP WELDED JOINT:

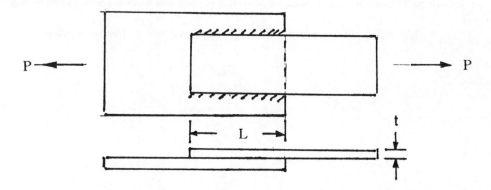

Fig. 9.6.7 Simple Lap Joint

The allowable load can be calculated by the following equation:

$$P_{all} = (F_{su} \text{ or } F_{tu}) \times (L) \times (t)$$

where: F_{su} or F_{tu} (ksi) – Allowable weld stress from Fig. 9.6.4 or use weld efficiency multiplied by design allowable stress from Fig. 9.6.5 and Fig. 9.6.6

L – Length of welded seams in inches

t – thickness (in inches) of the thinnest plate joined by the weld in the case of lap welds between two plates, or between plates and tubes, or the average thickness in inches of the weld material in the case of tube assemblies; but not to be greater than 1.25 times the thickness of the welded stock in any of the cases

Example 2:

A lap joint is formed by welding two plates (t = 0.065 in.) as shown in Fig. 9.6.7. The materials are alloy steel with no heat treatment subsequent to welding and F_{su} = 43 ksi (see Fig. 9.6.4). Find the margin of safety of this welded joint under the load of 5.0 kips.

$$P_{all} = (F_{su}) \times (L) \times (t) = 43 \times (1.0 \times 2) \times 0.065$$
$$= 5.59 \text{ kips}$$

$$MS = \frac{5.59}{5.0} - 1 = \underline{0.12} \qquad \text{O. K.}$$

GUSSET OR FITTING PLATE WELDED JOINT:

When computing the joint strength of a fillet welded joint, as shown in Fig. 9.6.8, assume the fillet dimension (S) to be 0.25 inches or twice the thickness of the thinnest adjacent material, whichever is less. If a larger fillet is required, specify fillet size on the drawing.

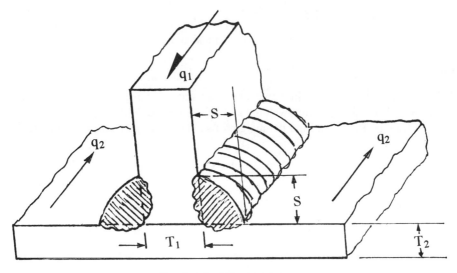

Fig. 9.6.8 Fillet Welded Joint

The joint is critical across T_1 or T_2:

$$q_{all, 1} = F_{su} \times (\text{Weld efficiency}) \times t_1$$

where across the double fillets

$$t_1 = \text{Minimum of } 2 \times 0.707 \times S \text{ or } T_1$$

$$q_{all, 2} = F_{su} \times (\text{Weld efficiency}) \times t_2$$

where across the two fillets

t_2 = Minimum of $0.707 \times S$ or T_2

$$MS = \text{Minimum of } \left(\frac{q_{all,1}}{q_1} \text{ or } \frac{q_{all,2}}{q_2}\right) - 1$$

Example 3:

A welded joint is formed by welding two plates as shown in Fig. 9.6.8. The materials are alloy steel with no heat treatment subsequent to welding and F_{su} = 43 ksi (see Fig. 9.6.4). Find the margin of safety of this welded joint under the load of 5.0 kips/in. at T_1 plate.

Assume T_1 = 0.125 In.; T_2 = 0.065 in. and S = 0.25 in.

(1) At plate T_1:

T_1 = 0.125 in.

or $t_1 = 2 \times 0.707 \times 0.25 = 0.354''$

$q_{all,1} = F_{su} \times$ (Weld efficiency) $\times T_1 = 43 \times 0.125 = 5.375$ kips/in.

$$MS = \frac{q_{all,1}}{q_1} - 1 = \frac{5.375}{5.0} - 1 = \underline{0.08} \qquad \text{O.K.}$$

(2) At plate T_2:

$$q_2 = \frac{q_1}{2} = \frac{5.0}{2} = 2.5 \text{ kips/in.}$$

T_2 = 0.065 in.

or $t_2 = 0.707 \times 0.25 = 0.177$ in.

$q_{all,2} = F_{su} \times$ (Weld efficiency) $\times T_2 = 43 \times 0.065 = 2.795$ kips/in.

$$MS = \frac{q_{all,2}}{q_2} - 1 = \frac{2.795}{2.5} - 1 = \underline{0.12} \qquad \text{O.K.}$$

(3) Therefore, the min. MS = $\underline{0.08}$ which is critical at T_1.

Example 4:

A welded joint is formed by welding two plates as shown in Fig. 9.6.8. The materials are aluminum alloy in T_4 condition before welding and left as welded. Weld efficiency = 0.7 (see Fig. 9.6.5) F_{su} = 36 ksi (see Ref. 9.1). Find the margin of safety of this welded joint under the load of 3.0 kips/in. at T_1 plate.

Assume T_1 = 0.125 In.; T_2 = 0.065 in. and S = 0.25 in.

(1) At plate T_1:

q_1 = 3.0 kips/in.

T_1 = 0.125 in.

or $t_1 = 2 \times 0.707 \times 0.25 = 0.354''$

$q_{all,1} = F_{su} \times$ (Weld efficiency) $\times T_1 = 36 \times 0.7 \times 0.125 = 3.15$ kips/in.

$$MS = \frac{q_{all,1}}{q_1} - 1 = \frac{3.15}{3.0} - 1 = \underline{0.05} \qquad \text{O.K.}$$

(2) At plate T_2:

$$q_2 = \frac{q_1}{2} = \frac{3.0}{2} = 1.5 \text{ kips/in.}$$

$T_2 = 0.065$ in.

or $t_2 = 0.707 \times 0.25 = 0.177$ in.

$q_{all,2} = F_{su} \times$ (Weld efficiency) $\times T_2 = 36 \times 0.7 \times 0.065 = 1.64$ kips/in.

$$\text{MS} = \frac{q_{all,2}}{q_2} - 1 = \frac{1.64}{1.5} - 1 = \underline{0.09} \qquad \text{O.K.}$$

(3) Therefore, the min. MS = $\underline{0.05}$ which is critical at T_1.

9.7 BONDED JOINTS

Most of the bonding applications on airframe structures are secondary bonding using adhesive such as the joining of skins together or bonding stringers to skins. The main purposes are:

- To improve fatigue life (no fasteners)
- Use the bonding of multiple thicknesses to replace expensive machined skin panels

The secondary bonded joints in primary joints are seldom used due to their poor reliability, except under stringent control conditions. But in composite structures, the secondary bonding or co-cured joining plays a very important rule to obtain the structural integrity and lower manufacturing cost. Detailed design information and data can be found in Ref. 9.4, 9,5 and 9.6.

Since the use of bonding on metallic airframes is not being considered for primary applications, this chapter will give a simple overview of basic design principles. There are a number of general precautions that should be observed by the engineer for any type of adhesive or bonded structure:

- Adhesives that are pressure sensitive should never be used in an application where a known magnitude of positive pressure cannot be assured
- No component should incorporate a design that results in a peeling (weak area in bonded joint) condition when subjected to external loads
- Peel stresses cannot be accurately analyzed
- Bonding dissimilar metals or materials with wide variations in thermal coefficients of linear expansion should be avoided whenever possible
- Residual stresses induced in the bond line by variation in thermal coefficients and bonding (or cure) temperature should be included with the stresses resulting from external loads
- Fig. 9.7.1 illustrates the graphical explanation of shearing behavior in adhesives

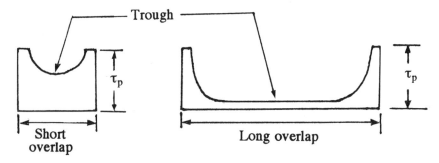

Fig. 9.7.1 Shear Stress Behavior in Bonded Joints

(a) Design considerations:
- Each adhesive has a characteristic cohesive strength
- The method of surface preparation also affects bond strength
- The load carrying ability increases and the average shear stress decreases with an increase in the ratio of adherend overlap to adherend thickness
- Adhesive bond strength decreases with increasing temperature. Further degradation may be experienced after long-time exposure to heat and moisture
- Long-time exposure of a bond, with or without load, to salt atmosphere may cause serious degradation for specific adhesives

(b) Failure modes:
- Adhesive stress and strain is far from uniform
- Adhesive strains are maximum at the ends and minimum at the middle
- At high loads adhesive stresses have plastic plateaus at the ends
- Load transfer depends on the area under the shear stress/overlap curve, as shown in Fig. 9.7.1
- Adhesive failure occurs when strain reaches a critical value at overlap edges
- Analysis confirms that dividing failure load by bonded area to get average stress "allowable" is entirely invalid

(c) Concerns for thermal mismatches:
- If two different materials are bonded, it is possible for the joint to break during cool-down without the application of any mechanical loading
- Analysis provides an upper bound on the strength reduction factor as viscoelastic effects tend to reduce thermal stresses relative to elastic estimates
- There is no thermal mismatch problem, if adherends of the same material are used

(d) Classification and inspection:

Bonded joints shall be classified for structural applications (similar to that of welded joints) as follows:

- Designating levels
- Types of inspection
- Peel strength

Class I – Vital bonded joints (e.g., aircraft bonded control surface panel):
: This class includes all bonded joints in primary structures and important secondary structures. It includes bonds of which a single failure would cause loss of a major component.

Class II – Non-vital bond joints:
: This class includes all bonded joints not falling within the vital classification.

Class IIA – Failure of bonded joint will degrade performance:
: e.g., fuselage bonded tail cone.

Class IIB – Bonded joints which are essentially non-structural, with minimal service requirements:
: e.g., fuselage interior cabin furnishings.

(f) Inspection:

To insure the integrity of adhesive bonds, rigorous process control, inspection and tests are necessary.

- Class I and Class II bonds require testing of coupons accompanying each batch or testing of a lot of coupons that can be cut from the parts after bonding and will be representative of the bonds obtained in the parts
- Class I bonds also require destructive testing of production parts

Joints and Fittings

(A) BALANCED DOUBLE LAP JOINTS

A joint is balanced (or the stiffness is balanced in a joint) if the extension stiffness of the adherends carrying the load in the two directions is equal. Fig. 9.7.2 shows a double-lap joint in which the two outer adherends transmitting the load to the left have a combined membrane stiffness equal to that of the single center adherend carrying the load to the right.

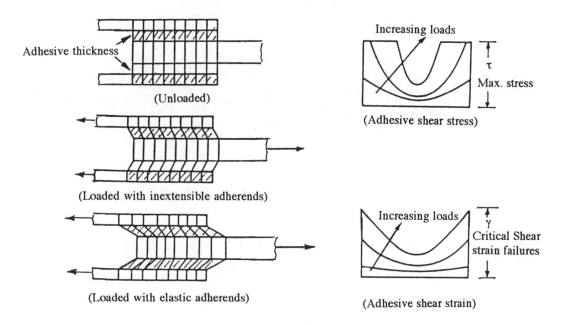

Fig. 9.7.2 Shear Stress/Strain Behavior in a Balanced Bonded Joint

Design considerations:

- Strain gradients in adherends depend on adherend stiffness
- Treating strain gradients as linearly elastic is satisfactory
- Max. load transfer occurs when all adhesive is fully plastic, i.e., no trough in stress/overlap curves (see Fig. 9.7.1). Adhesive thickness lines are indicated by hatched lines in Fig. 9.7.2
- From Fig. 9.7.3, each horizontal line is a ratio of $(\frac{\gamma_p}{\gamma_e})$, and $\gamma_e + \gamma_p$ is the total failure strain
- The highest benefit is obtained by increasing adhesive ductility by increasing the $\frac{\gamma_p}{\gamma_e}$ ratio value; Example 1 illustrates this benefit

Chapter 9.0

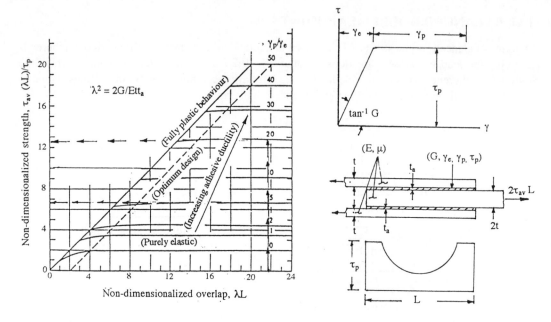

Fig. 9.7.3 Shear Stress/Overlap Curve of a Bonded Lap Joint

Example 1:

Consider a double lap bonded joint of 2024-T42 aluminum alloy (F_{tu} = 62 ksi) as shown below:

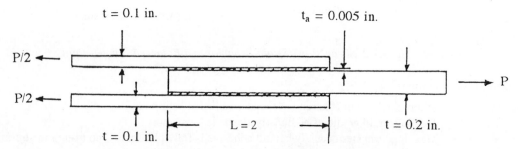

(Balanced Double Lap Joint of 2024-T3 Aluminum Alloy)

(1) The adherend's (2024-T3 sheet) strength data:

  F_{tu} = 62 ksi; E = 10.1 × 10^6 psi

  Adherend strength, P = F_{tu} × t = 62 × 0.2 = 12.4 kips/in.

(2) To develop adhesive strength to carry the above load P = 12.4 kips/in.:

  Assume given adhesive data:

  G = 0.3 × 10^6 psi; t_a = 0.005 in.; $\frac{\gamma_p}{\gamma_e}$ = 5; τ_p = 6 ksi

  Compute:

  $$\lambda = (\frac{2G}{Et\,t_a})^{\frac{1}{2}}$$

  $$= (\frac{2 \times 0.3 \times 10^6}{10.1 \times 10^6 \times 0.1 \times 0.005})^{\frac{1}{2}}$$

  $$= 10.9$$

  and λL = 10.9 × 2.0 = 21.8

316

Use above $\frac{\gamma_p}{\gamma_e} = 5$ and $\lambda L = 21.8$ and from Fig. 9.7.3 gives $\frac{\tau_{av}}{\tau_p}\lambda L = 6.7$ and

$\tau_{av} = 6.7 \times \frac{6}{21.8} = 1.84$ ksi

$P_{all} = 2 \times \tau_{av} \times L = 2 \times 1.84 \times 2.0 = 7.36$ kips/in.

$MS = \frac{P_{all}}{P} - 1 = \frac{7.36}{12.4} - 1 = \underline{-0.41}$ NO GOOD

(3) Try using $\frac{\gamma_p}{\gamma_e} = 20$ (ductile adhesive) and $\lambda L = 21.8$ [$\lambda = 10.9$ which is the same as in step (2)] and enter it on the curve shown in Fig. 9.7.3, it gives $\frac{\tau_{av}}{\tau_p}\lambda L = 12.5$

$\tau_{av} = \frac{12.5 \times 6}{21.8} = 3.44$ ksi

$P_{all} = 2 \times \tau_{av} \times L = 2 \times 3.44 \times 2.0 = 13.76$ kips/in.

$MS = \frac{13.76}{12.4} - 1 = \underline{0.11}$ O.K.

Therefore, use of ductile adhesive improves the joint strength by $\frac{13.76}{7.36} = 1.87$ (or 87% stronger)

(B) UNBALANCED DOUBLE LAP JOINTS

An unbalanced joint design (a joint for which stiffness is not balanced) is shown in Fig. 9.7.4 where all three adherends have a similar amount of extension stiffness so that the single one in the center carrying the load to the right has only about half the stiffness (i.e. the balanced doubler lap joints are equal stiffness) of the outer pair transmitting load to the left.

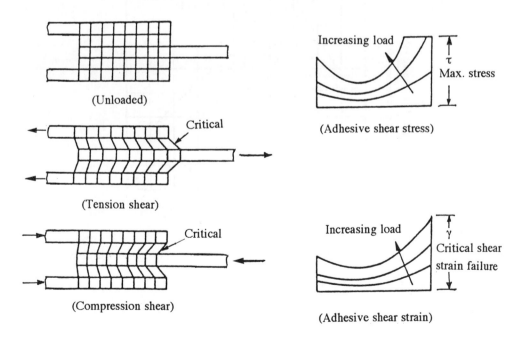

Fig. 9.7.4 Shear Stress/Strain Behavior in Unbalanced Bonded Joints

(a) Effects which occur in an unbalanced joint are summarized below:
- Strains are highest at the least stiff end
- Load transfer (area under stress/overlap curve) can be markedly less than for balanced joints
- Effects occurs regardless of load direction
- It is advisable to consider thickening the softer side unless the stiffer side is over-designed

(b) By comparing Fig. 9.7.2 and Fig. 9.7.4, it can be seen that a substantially smaller load is transferred when a joint is unbalanced.
- This highlights the importance of trying to design stiffness-balanced bonded joints
- The greater the unbalance, the less efficient the joint.
- Local thickening of thinner adherends should be considered to relieve this problem

(c) Unbalanced joints lead to strength loss as illustrated below:
- Start a joint with $E_i t_i = E_o t_o$
  where: Subscripts i – inner adherend and o – outer adherend
- $\dfrac{E_i t_i}{2E_o t_o} = 0.5$; enter this value into Fig. 9.7.5 and it gives a strength reduction ratio for adhesive failure of 0.62
- If $\dfrac{E_i t_i}{2E_o t_o} = 1.0$, it gives a strength reduction ratio 1.0
- Note that the adherend strength is obviously reduced by half

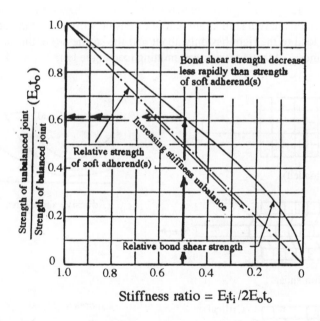

Fig. 9.7.5 Strength Reduction Ratio for Stiffness Unbalanced Joint

(C) SINGLE LAP JOINTS (UNSUPPORTED)

For an illustration of unsupported lap joints, see Fig. 9.1.2.

ADHEREND FAILURE MODE
- Max. adherend stress is the sum of membrane and bending
- Consider as a stress concentration due to bending

- Increasing lap length is beneficial as it enlarges moment arm of forces reacting eccentric moment
- Fig. 9.7.6 gives average adherend stress for a balanced joint

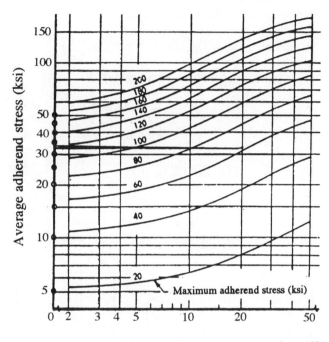

Fig. 9.7.6 *Adherend Failure of an Unsupported Single Lap Bonded Joint*

Example 2:

Find adherend strength (P_{all}) for a balanced single lap bonded joint as shown below.

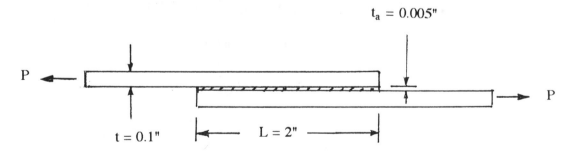

(1) Compute scaled overlap:
$$\frac{L}{t}[\frac{12(1-\mu^2) \times 10^6}{E}]^{\frac{1}{2}} = \frac{2}{0.1}[\frac{12(1-0.3^2) \times 10^6}{10.1 \times 10^6}]^{\frac{1}{2}}$$
$$= 20.8$$

(2) Select max. adherend (plate) stress allowable (2024-T3):

$F_{tu} = 62$ ksi

(3) Fig.9.7.6 gives average adherend stress $\tau_{av} = 33$ ksi

Load per unit width $P_{all} = \tau_{av} \times t = 33 \times 0.1 = 3.3$ kips/in.

Chapter 9.0

ADHESIVE FAILURE MODE

- The design curves shown in Fig. 9.7.7 are similar in concept to those presented for double lap joints
- Added complication due to eccentricity

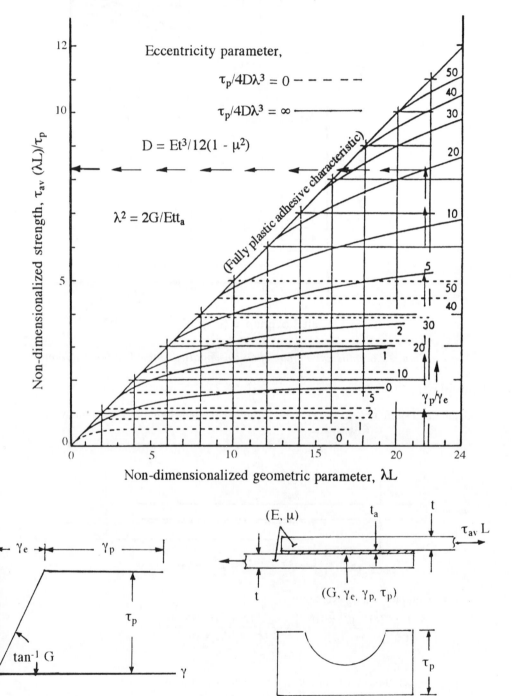

Fig. 9.7.7 *Adhesive Failure of an Unsupported Single Lap Bonded Joint*

Example 3:

Find adhesive strength (P_{all}) for a balanced single lap bonded joint as shown in Example 2. Determine the adhesive strength.

(1) Given data from Example 2:

The adherend's (2024-T42 sheet) data:

$F_{tu} = 62$ ksi; $E = 10.1 \times 10^6$ psi

Adherend strength, $P = 3.3$ kips/in.

The adhesive data:

$G = 0.3 \times 10^6$ psi; $t_a = 0.005$ in.; $\frac{\gamma_p}{\gamma_e} = 20$; $\tau_p = 6$ ksi

(2) Compute:

$$\lambda = (\frac{2G}{Et\,t_a})^{\frac{1}{2}}$$

$$= (\frac{2 \times 0.3 \times 10^6}{10.1 \times 10^6 \times 0.1 \times 0.005})^{\frac{1}{2}}$$

$$= 10.9$$

and $\lambda L = 10.9 \times 2.0 = 21.8$

(3) Use $\frac{\gamma_p}{\gamma_e} = 20$ and $\lambda L = 21.8$, from Fig. 9.7.7 gives $\frac{\tau_{av}}{\tau_p}\lambda L = 8.3$

$$\tau_{av} = \frac{8.3 \times 6}{21.8} = 2.28 \text{ ksi}$$

$P_{all} = \tau_{av} \times L = 2.28 \times 2.0 = 4.56$ kips/in.

$MS = \frac{4.56}{3.3} - 1 = \underline{0.38}$ O.K.

9.8 LUG ANALYSIS (BOLT IN SHEAR)

The lug analysis and sizing method presented in this chapter is from Ref. 9.7 and considers both the lug and pin acting together, since the strength of one can influence the strength of the other. Lugs should be sized conservatively, as their weight is usually small relative to their importance, and inaccuracies in manufacture are difficult to control. This method has a theoretical basis, and its validity has been verified by test. This method is applicable only to aluminum and steel alloy double shear lugs of uniform thickness, as shown in Fig. 9.8.1, with static load applied axially, transversely, or obliquely.

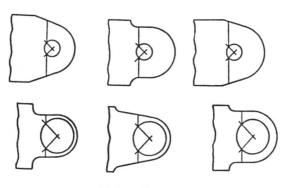

(a) Lug Shapes

Chapter 9.0

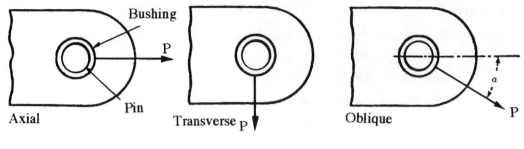

(b) Applied Load Cases

Fig. 9.8.1 Lug Configurations and Three Different Applied Load Cases

The typical applications require rotation movement and the transfer of very highly concentrated loads, e.g.,

- Trunnion joints of landing gear (includes fail-safe fuse)
- Engine pylon mount pin (includes fail-safe fuse)
- Hinges of control surfaces (aileron, elevator, rudder, spoiler, etc.)
- Horizontal tail pivot joints
- Removable joints for fighter wing root mounts
- Door hinges

A lug under axial load can fail in any one of the following cases which are outlined below:

(a) Axial load case ($\alpha = 0°$):

- Shear tear-out and bearing failure
- Under tension failure the net section cannot carry uniform $\frac{P}{A}$ stress due to the stress concentration effect
- Hoop tension failure occurs at the lug tip; no separate calculation is needed because shear and bearing has taken care of this type of failure
- Failure of shear-off pin
- Pin bending failure
- Excessive yielding of bushing (if used)

(b) Transverse load case ($\alpha = 90°$):

- This case can fail in the same manner as the failure types listed above
- Sizing is based on empirical curves

(c) Oblique load case (α between 0° and 90°):

- Sizing is based on interaction equations

(A) DESIGN CONSIDERATIONS

- Most applications for hinge designs use symmetrical double shear lugs or multiple shear lugs which are only used for fail-safe conditions (see Fig. 9.8.2)
- A fitting factor of $\lambda = 1.15$ should be used (both ultimate and yield strength)
- In any sizing, if both fitting and casting factors are involved, only the larger factor shall be used
- In addition to the factors mentioned above, lug sizing shall show a minimum MS of 20%
- All required factors and min. MS can be waived if the lug has been verified by test

- The ratio of lug thickness to hole diameter ($\frac{t}{D}$) should be greater than 0.3
- The grain directions (L, LT, and ST) of the lug material as shown in Fig. 9.8.3 must be carefully oriented as this affects the lug strength
- Use forging materials and pressed-fit bushings to improve fatigue life

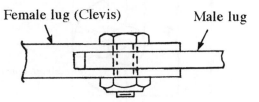

(a) Double Shear

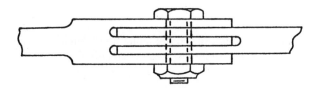

(b) Multiple Shear or Finger Type

Fig. 9.8.2 Lug Types

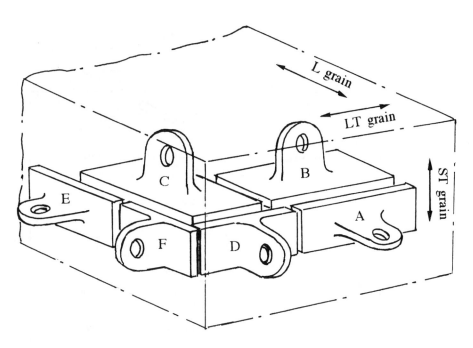

(Configurations B, C, D and F are not recommended for aluminum material except when using forging or xxxx-T73 condition materials which have better ST allowables)

Fig. 9.8.3 Material Grain Direction Effects

Chapter 9.0

(B) CASE I – AXIAL LOAD ($\alpha = 0°$)

The lug failure modes for this load case are shown in Fig. 9.8.4.

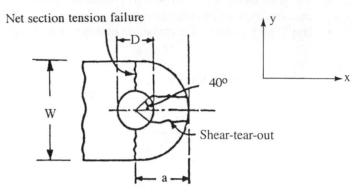

Fig. 9.8.4 Lug Tension and Shear-tear-out Failures

(a) Shear-bearing failure:

Failure consists of shear tear-out of the lug along a 40° angle on both sides of the pin (see Fig. 9.8.4) while bearing failure involves the crushing of the lug by the pin bearing. The ultimate load for this type of failure is given by the equation:

$$P_{bru} = k_{br} F_{tux} A_{br} \qquad \text{Eq. 9.8.1}$$

where: P_{bru} – Ultimate load for shear tear-out and bearing failure
k_{br} – Shear-bearing efficiency factor from Fig. 9.8.5
A_{br} – Projected bearing area ($A_{br} = Dt$)
D – Pin diameter or bushing outside diameter D_b (if bushing is used)
t – Lug thickness
F_{tux} – Ultimate tensile stress in x-direction

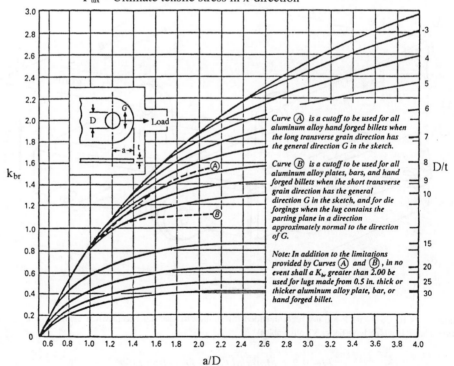

Fig. 9.8.5 Shear-Bearing Efficiency Factor, k_{br}

324

(b) Tension failure:

Tensile failure is given by:

$$P_{tu} = k_t F_{tux} A_t \qquad \text{Eq. 9.8.2}$$

where: P_{tu} – Ultimate load for tension failure
k_t – Net tension efficiency factor from Fig. 9.8.6
F_{tux} – Ultimate tensile stress of the lug material in x-direction
A_t – Minimum net section for tension [$A_t = (W - D)t$]
W – Width of the lug

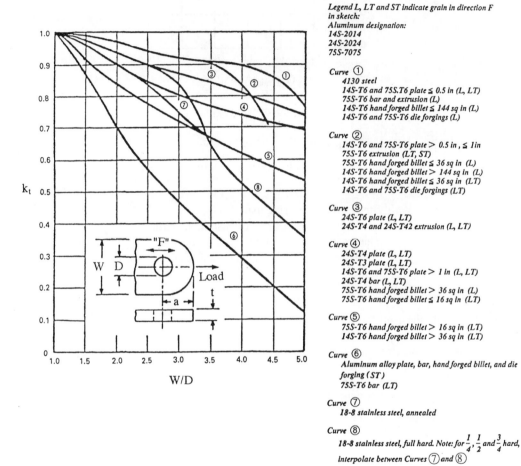

Fig. 9.8.6 Lug Efficiency Factor for Tension, k_t

(c) Yield failure – lug:

Lug yield load attributable to shear-bearing is given by:

$$P_y = C \left(\frac{F_{tyx}}{F_{tux}}\right) (P_u)_{min} \qquad \text{Eq. 9.8.3}$$

where: P_y – Yield load
C – Yield factor from Fig. 9.8.7
F_{tyx} – Tensile yield stress of lug material in load direction
F_{tux} – Ultimate tensile stress of lug material in load direction
$(P_u)_{min}$ – The smaller P_{bru} (Eq. 9.8.1) or P_{tu} (Eq. 9.8.2)

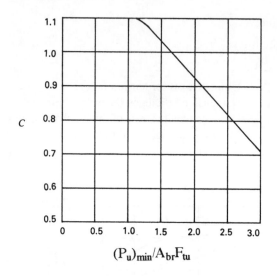

Fig. 9.8.7 Yield Factor, C

(d) Yield failure – bushing (if used):

Bushing yield bearing load attributable to shear-bearing is given by:

$$P_{bry} = 1.85 \, F_{cy} \, A_{brb} \qquad \text{Eq. 9.8.4}$$

where: P_{bry} – Bushing yield bearing load
F_{cy} – Compression yield stress of bushing material
A_{brb} – The smaller of the bearing areas of bushing on pin or bushing on lug (the latter may be the smaller as a result of external chamfer on the bushing).

(e) Pin shear-off failure:

Pin single shear-off failure is given by:

$$P_{p,s} = F_{su} \left(\frac{\pi D^2}{4}\right) \qquad \text{Eq. 9.8.5}$$

where: $P_{p,s}$ – Ultimate load for pin shear-off failure
F_{su} – Ultimate shear stress of the pin material

In most design cases, the lug design is in double shear:

$$P_{p,s} = 2 F_{su} \left(\frac{\pi D^2}{4}\right) \qquad \text{Eq. 9.8.6}$$

or $P_{p,s} = 2 P_{s,all}$ $\qquad$ Eq. 9.8.7

where: $P_{s,all}$ – Allowable ultimate single shear load from Fig. 9.8.8

(f) Pin bending failure:

If the pin used in the lug is too small, the pin can bend enough to precipitate failure in the lug because, as the pin bends, the stress distribution acting on the inner side of the lug tends to peak rather than form an even distribution as shown in Fig. 9.8.9.

Joints and Fittings

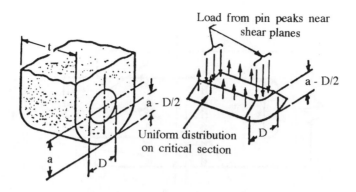

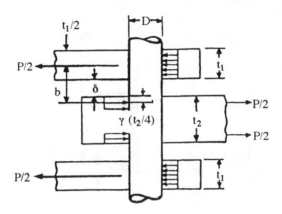

Fig. 9.8.8 Pin Moment Arm for Determination of Bending Moment

Since a weak or smaller pin can cause an inner lug (t_2) to fail at a smaller load, larger pins (ample MS) are always recommended. The moment arm is given by:

$$b = \frac{t_1}{2} + \delta + \gamma(\frac{t_2}{4})$$ Eq. 9.8.8

Compute the following two values:

$$\frac{(P_u)_{min}}{A_{br}F_{tux}}$$

$$r = \frac{a - \frac{D}{2}}{t_2}$$

where: t_1 – Outer lug thickness
  t_2 – Inner lug thickness
  $(P_u)_{min}$ – The smaller of P_{tux} (Eq. 9.8.1) and P_{tu} (Eq. 9.8.2) for the inner lug
  P_{tux} – Lug material across grain "F" (see sketch in Fig. 9.8.5)
  D – Pin diameter or D_h (if bushing used)
  δ – Gap (lug chamfer or use flange bushings) as shown in Fig. 9.8.9
  γ – Reduction factor (only applies to the inner lug) as given in Fig. 9.8.10

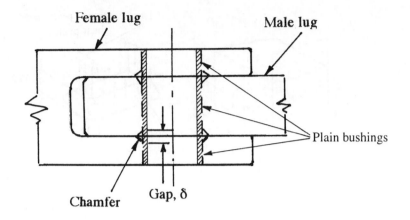

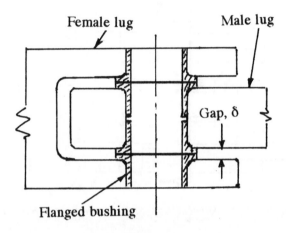

Fig. 9.8.9 Gap Definition, δ

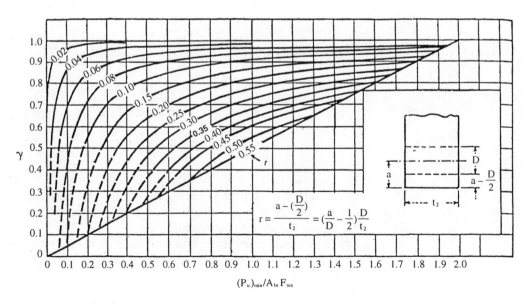

(Dashed lines indicate region where these theoretical curves are not substantiated by test data)

Fig. 9.8.10 Inner Lug Peaking Factors for Pin Bending, γ

The pin bending moment, $M = \frac{P}{2}b$ Eq. 9.8.9

$$MS = \frac{\text{Ultimate bending moment}}{\lambda M} - 1 \quad \text{where } \lambda - \text{Fitting factor}$$

The ultimate bending moment for $F_{tu} = 125$ ksi AN steel pin can be obtained from Fig. 9.2.7; use the plastic bending method to obtain the ultimate bending moment for other H.T. steel pin materials from Chapter 6.4.

Example 1:

Find the MS of a lug under an axial concentrated load of P = 15 kips (case I) with the following given data:

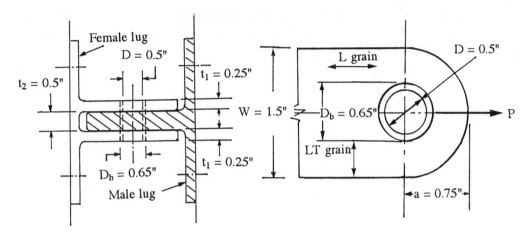

Lug: Aluminum 7075-T6 Extrusion (for material properties, see Fig. 4.3.5)

       W = 1.5"   a = 0.75" t_1 = 0.25" t_2 = 0.5"

       F_{tu} = 81 ksi; F_{ty} = 72 ksi; F_{cy} = 72 ksi   ('A' values of L-grain direction)

       F_{tu} = 74 ksi; F_{ty} = 65 ksi; F_{cy} = 71 ksi   ('A' values of LT-grain direction)

Pin: (H.T. = 125 ksi, material properties, see Fig. 4.3.6)

       F_{tu} = 125 ksi; F_{su} = 75 ksi; F_{cy} = 109 ksi   ('S' values)

       D = 0.5" (for pin shear allowable, see Fig. 9.2.5)

Bushing: (H.T. = 125 ksi; for material properties, see Fig. 4.3.6)

       D_b = 0.65" (outside diameter)

Compute:

$$\frac{a}{D_b} = \frac{0.75}{0.65} = 1.15; \quad \frac{W}{D_b} = \frac{1.5}{0.65} = 2.3$$

$$\frac{D_b}{t} = \frac{0.65}{0.5} = 1.3; \quad A_{br} = Dt = 0.65 \times 0.5 = 0.325$$

$$A_t = (W - D_b)t = (1.5 - 0.65) \times 0.5 = 0.425$$

Design requirements:

- Use fitting factor $\lambda = 1.15$
- Minimum MS = 0.2

(a) Shear-bearing failure (from Eq. 9.8.1):

$$P_{bru} = k_{br}F_{tux}A_{br}$$
$$= 0.97 \times 72 \times 0.325 = 22.7 \text{ kips}$$
$$MS = \frac{22.7}{1.15 \times 15} - 1 = 0.32 > 0.2 \qquad \text{O.K.}$$

The above 1.15 is fitting factor (λ)

(b) Tension failure (from Eq. 9.8.2):

$$P_{tu} = k_t F_{tu} A_t$$
$$= 0.95 \times 82 \times 0.425 = 33.1 \text{ kips}$$
$$MS = \frac{33.1}{1.15 \times 15} - 1 = 0.92 > 0.2 \text{ (min. MS requirement)} \qquad \text{O.K.}$$

(c) Yield failure – lug (from Eq. 9.8.3):

$$P_y = C\left(\frac{F_{tyx}}{F_{tux}}\right)(P_u)_{min}$$
$$= 1.1 \times \frac{72}{81} \times 22.7 = 22.2 \text{ kips}$$
$$MS = \frac{22.2}{1.15 \times \frac{15}{1.5}} - 1 = 0.93 > 0.2 \qquad \text{O.K.}$$

(d) Yield failure – bushing (from Eq. 9.8.4):

$$P_{bry} = 1.85 F_{cy} A_{brb}$$
$$= 1.85 \times 75 \times 0.25 = 34.7 \text{ kips}$$
$$MS = \frac{34.7}{1.15 \times \frac{15}{1.5}} - 1 = \text{high} > 0.2 \qquad \text{O.K.}$$

(e) Pin shear-off failure (from Eq. 9.8.7):

$$P_{p,s} = 2 P_{s,\text{all}}$$
$$= 2 \times 14.7 = 29.4 \text{ kips}$$

where: $P_{s,\text{all}} = 14.7$ kips from Fig. 9.2.7.

$$MS = \frac{29.4}{1.15 \times 15} - 1 = 0.7 > 0.2 \qquad \text{O.K.}$$

(f) Pin bending failure (from Eq. 9.8.8):

$$b = \frac{t_1}{2} + \delta + \gamma\left(\frac{t_2}{4}\right) \qquad \text{Eq. 9.8.8}$$

$$r = \frac{a - \frac{D_h}{2}}{t_2} = \frac{0.75 - \frac{0.65}{2}}{0.5} = 0.85$$

$$\frac{(P_u)_{min}}{A_{br}F_{tux}} = \frac{22.7}{0.325 \times 81} = 0.86$$

$\gamma = 0.43$ (from Fig. 9.8.10) and $\delta = 0.02"$ (gap from chamfers)

$$b = \frac{0.25}{2} + 0.02 + 0.43 \frac{0.5}{4} = 0.2$$

$$M = \frac{P}{2}b$$

$$= \frac{15}{2}0.2 = 1.5 \text{ in.-kips}$$

The allowable pin bending moment is 2.21 in.-kips (H.T. = 125 ksi) from Fig. 9.2.7.

$$\text{MS} = \frac{2.21}{1.15 \times 1.5} - 1 = \underline{0.28} > 0.2 \qquad \text{O.K.}$$

(g) The final minimum MS = 0.22 [pin bending is critical from Step (f)].

(C) CASE II – TRANSVERSE LOAD ($\alpha = 90°$)

The lug failure modes for this transverse load case is shown in Fig. 9.8.11.

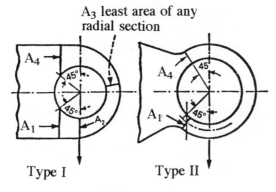

(a) Common shapes

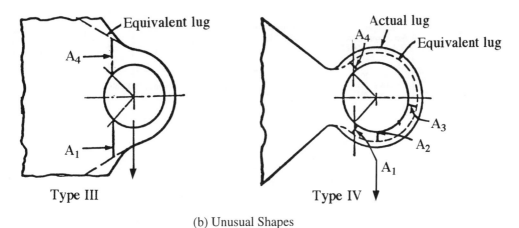

(b) Unusual Shapes

(Locations of Cross-Section Areas A_1, A_2, A_3, and A_4 are shown)

Fig. 9.8.11 Lugs Subjected to Transverse Load

Chapter 9.0

(a) Compute:

$A_{br} = D\,t$

$$A_{av} = \frac{6}{\dfrac{3}{A_1} + \dfrac{1}{A_2} + \dfrac{1}{A_3} + \dfrac{1}{A_4}} \qquad \text{Eq. 9.8.10}$$

$\dfrac{A_{av}}{A_{br}}$

(b) The ultimate load is obtained using:

$$P_{tru} = k_{tru} A_{br} F_{tuy} \qquad \text{Eq. 9.8.11}$$

where: P_{tru} – Ultimate transverse load
k_{tru} – Efficiency factor for transverse ultimate load from Fig. 9.8.12
A_{br} – Projected bearing area
F_{tuy} – Ultimate tensile stress of lug material in y-direction

The load that can be carried by cantilever beam action is indicated very approximately by curve A in Fig. 9.8.12; if the efficiency factor falls below curve A, a separate calculation as a cantilever beam is warranted, as shown in Fig. 9.8.13.

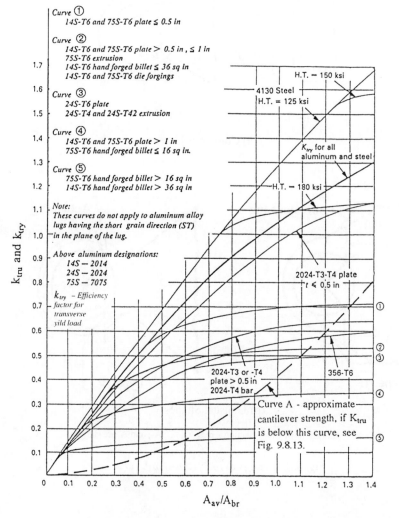

Fig. 9.8.12 Efficiency Factor for Transverse Load, k_{tru} and k_{try}

Joints and Fittings

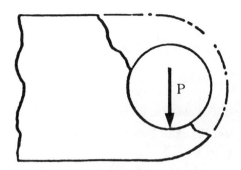

Fig. 9.8.13 Cantilever Beam Action of the Portion of the Lug Under Load

(c) The yield load is given by:

$$P_y = k_{try} A_{br} F_{tyy}$$ Eq. 9.8.12

where: P_y – Yield transverse load
k_{try} – Efficiency factor for transverse yield load from Fig. 9.8.12
A_{br} – Projected bearing area
F_{tyy} – Tensile yield stress of lug material in y-direction

(d) Determine the yield load using Eq. 9.8.4, namely $P_{bry} = 1.85 F_{cy} A_{brb}$

Example 2:

Assume use of the same lug given in Example 1 except use the Case II transverse load (P = 15 kips) application as shown below:

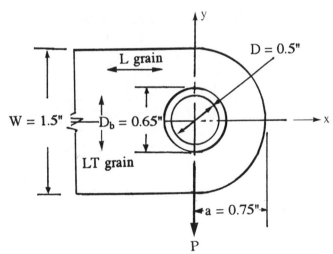

(a) Compute:

$$A_1 = A_4 = (0.75 - \frac{0.65}{2} \sin 45°) \times 0.5 = (0.75 - 0.23) \times 0.5 = 0.26$$

$$A_2 = A_3 = (0.75 - \frac{0.65}{2}) \times 0.5 = 0.21$$

$$A_{br} = 0.65 \times 0.5 = 0.325$$

From Eq. 9.8.10:

$$A_{av} = \frac{6}{\frac{3}{0.26} + \frac{1}{0.21} + \frac{1}{0.21} + \frac{1}{0.26}} = 0.24$$

$$\frac{A_{av}}{A_{br}} = \frac{0.24}{0.325} = 0.74$$

(b) The ultimate load is obtained using Eq. 9.8.11:

From curve ② of Fig. 9.8.12, obtain $k_{tru} = 0.51$:

$$P_{tru} = k_{tru} A_{br} F_{tuy}$$
$$= 0.51 \times 0.325 \times 74 = 12.3 \text{ kips}$$

(c) The yield load is obtained using Eq. 9.8.12:

From curve k_{try} of Fig. 9.8.12, obtain $k_{tru} = 0.86$

$$P_y = k_{try} A_{br} F_{tyy}$$
$$= 0.86 \times 0.325 \times 65 = 18.2 \text{ kips}$$

(d) The min. MS $= \dfrac{12.3}{1.15 \times 15} - 1 = -0.29$ (ultimate load)   NO GOOD

(e) Therefore the allowable ultimate load (considering fitting factor of 1.15 and 20% margin) is:

$$P = \frac{12.3}{1.15 \times 1.2} = 8.9 \text{ kips instead of the given load of 15 kips.}$$

(D) CASE III – OBLIQUE LOAD ($\alpha = 0°$ BETWEEN 90°)

Use the following interaction equation to size the oblique load case as shown in Fig. 9.8.14 below:

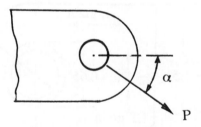

Fig. 9.8.14 Oblique Load Case

(a) Ultimate load:

$$MS = \frac{1}{(R_{a,u}^{1.6} + R_{tr,u}^{1.6})^{0.625}} - 1 \qquad \text{Eq. 9.8.13}$$

where: $R_{a,u}$ – Axial component ($\alpha = 0°$) of applied ultimate load divided by smaller of P_{bru} (from Eq. 9.8.1) or P_{tu} (from Eq. 9.8.2)

$R_{tr,u}$ – Transverse component ($\alpha = 90°$) of applied ultimate load divided by P_{tru} (from Eq. 9.8.11)

(b) Yield load:

$$MS = \frac{1}{(R_{a,y}^{1.6} + R_{tr,y}^{1.6})^{0.625}} - 1 \qquad \text{Eq. 9.8.14}$$

where: $R_{a,y}$ – Axial component ($\alpha = 0°$) of applied limit load divided by $P_{y,o}$ ($P_{y,o} = P_y$ from Eq. 9.8.3)

$R_{tr,y}$ – Transverse component ($\alpha = 90°$) of applied limit load divided by $P_{y,90}$
($P_{y,90} = P_y$ from Eq. 9.8.12)

Example 3:

Assume use of the same lug given in Example 1 except use the Case III oblique load (P = 15 kips) application as shown below:

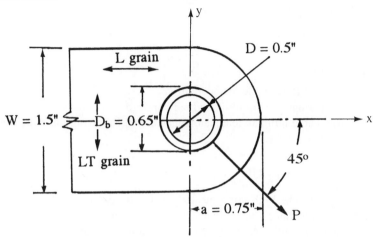

(a) Ultimate load:

  $P_{bru} = 22.7$ kips [see Step (a) of Example 1]

  $P_{tu} = 33.1$ kips [see Step (b) of Example 1]

  $P_{tru} = 12.3$ kips [see Step (b) of Example 2]

  $$R_{a,u} = \frac{P \cos 45°}{\text{Smaller of } P_{bru} \text{ or } P_{tu}}$$

  $$= \frac{15 \times 0.707}{22.7} = 0.47$$

  $$R_{tr,u} = \frac{P \sin 45°}{P_{tru}}$$

  $$= \frac{15 \times 0.707}{12.3} = 0.86$$

From Eq. 9.8.13:

$$MS = \frac{1}{1.15(0.47^{1.6} + 0.86^{1.6})^{0.625}} - 1 = -0.17 < 0.2 \qquad \text{N.G.}$$

Note: See discussion in Step (c).

(b) Yield load:

  $P_{y,o} = 22.2$ kips [see Step (c) of Example 1]

  $P_{y,90} = 18.2$ kips [see Step (c) of Example 2]

  $$R_{a,y} = \frac{\frac{P}{1.5} \cos 45°}{P_{y,o}}$$

  $$= \frac{\frac{15}{1.5} \times 0.707}{22.2} = 0.32$$

$$R_{tr,y} = \frac{\frac{P}{1.5}\sin 45°}{P_{y,90}}$$

$$= \frac{\frac{15}{1.5} \times 0.707}{18.2} = 0.39$$

From Eq. 9.8.14:

$$MS = \frac{1}{1.15(0.32^{1.6} + 0.39^{1.6})^{0.625}} - 1 = 0.59 > 0.2 \qquad \text{O.K.}$$

(c) Discussion
- Ample MS for yield load condition
- Not enough MS for ultimate load condition
- Allowable ultimate load can be calculated following the same procedures or re-size this lug

(E) TUBULAR PIN IN LUG

Under conditions of very high concentrated loads, the use of a tubular pin to encase a large diameter pin in the lug can provide greater bearing area and less pin bending deformation due to its greater moment of inertia. A nearly even distribution of bearing or compression stress on lug thickness can be thus obtained in addition to a reduction of wear and an improvement of the fatigue life of the lug. There are basically two types of design approaches:

- A tubular pin can be used as the shear fuse (breakaway design conditions for landing gear trunnion and engine pylon mount fittings)
- Double pin design (another smaller pin is used inside the tubular pin) provides fail-safe capability for vital hinge applications (e.g., horizontal tail pivot and certain hinges on control surfaces)

Design considerations:

- $\frac{D}{t}$ ratio $5 \leftrightarrow 10$

- $0.5 > \frac{t_1}{D} < 1.0$

  where: D – Tubular pin outside diameter
  t – Tubular wall thickness
  t_1 – Female lug wall thickness (see Fig. 9.8.8)

Use the following interaction equation to determine MS:

(a) Ultimate load:

$$MS = \frac{1}{R_{s,u}^2 + R_{b,u}} - 1 \qquad \text{Eq. 9.8.15}$$

where: $R_{s,u}$ – Ultimate shear stress divided by allowable shear stress (F_s from Fig. 9.8.15)

$R_{b,u}$ – Ultimate bending stress divided by modulus of rupture bending stress (F_b from Fig. 9.8.16) for the tubular pin

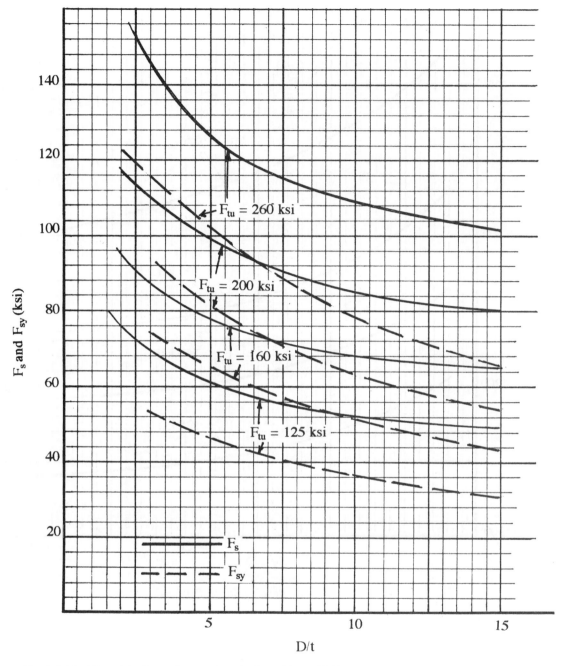

Fig. 9.8.15 Allowable Shear Stress, F_s and Allowable Yield Shear Stress, F_{sy} (Tubular Steel Pins)

Chapter 9.0

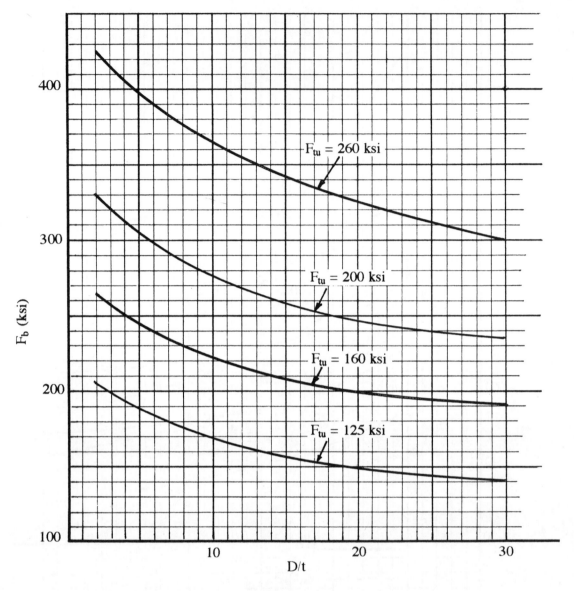

Fig. 9.8.16 Modulus of Rupture Bending Stress, F_b (Tubular Steel Pins)

(b) Yield load:

$$MS = \frac{1}{R_{s,y}^2 + R_{b,y}} - 1 \qquad \text{Eq. 9.8.16}$$

where: $R_{s,y}$ – Yield shear stress (f_{sy}) divided by allowable yield shear stress (F_{sy} from Fig. 9.8.15)

$R_{b,y}$ – Yield bending stress ($f_{by} = \frac{m_{limit} D}{2I}$) divided by allowable yield material stress (F_{ty}) for the tubular pin.

Example 4:

Assume use of the same lug given in Example 1 except for the use of a tubular pin with higher H.T. = 260 ksi as shown below:

Compute:

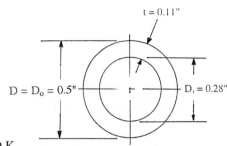

$\dfrac{D}{t} = 4.55$ O.K.

Cross-sectional area: $A_{tube} = \dfrac{\pi}{4}(0.5^2 - 0.28^2) = 0.135$ in.^{2}

Moment of Inertia: $I_{tube} = \dfrac{\pi}{4}(0.25^4 - 0.14^4) = 0.00276$ in.^{4}

Tube material properties: $F_{tu} = 260$ ksi $F_{ty} = 215$ ksi
   $F_{cy} = 240$ ksi $F_{su} = 156$ ksi

(a) Ultimate load (P = 15 kips):

Shear stress (gross area): $f_s = \dfrac{\frac{P}{2}}{A_{tube}} = \dfrac{\frac{15}{2}}{0.135} = 55.6$ ksi

From Fig. 9.8.15 obtain:

$F_s = 130$ ksi

$R_{s,u} = \dfrac{f_s}{F_s} = \dfrac{55.6}{130} = 0.43$

From Fig. 9.8.16 obtain the modulus of rupture bending stress:

$F_b = 400$ ksi

M = 1.5 in.-kips [from Step (f) of Example 1]

Bending stress: $f_b = \dfrac{M\frac{D}{2}}{I_{tube}} = \dfrac{1.5\frac{0.5}{2}}{0.00276} = 135.9$ ksi

$R_{b,u} = \dfrac{f_b}{F_b} = \dfrac{135.9}{400} = 0.34$

From Eq. 9.8.15:

$MS = \dfrac{1}{R_{s,u}^2 + R_{b,u}} - 1$

$= \dfrac{1}{1.15(0.43^2 + 0.34)} - 1 = 0.66 > 0.2$ O.K.

(b) Yield load ($P_{limit} = \dfrac{15}{1.5} = 10$ kips):

Shear stress (gross area): $f_{sy} = \dfrac{\frac{P_{limit}}{2}}{A_{tube}} = \dfrac{\frac{10}{2}}{0.135} = 37$ ksi

From Fig. 9.8.15 obtain:

$F_{sy} = 106$ ksi

$R_{s,y} = \dfrac{f_{sy}}{F_{sy}} = \dfrac{37}{106} = 0.35$

M = 1.5 in.-kips (use the same moment value from Step (f) of Example 1 divided by 1.5)

$$m_{lmit} = \frac{M}{1.5} = \frac{1.5}{1.5} = 1.0 \text{ in.-kips}$$

Bending stress: $f_{by} = \frac{m_{limit}\frac{D}{2}}{I_{tube}} = \frac{1.0\frac{0.5}{2}}{0.00276} = 90.6$ ksi

$$R_{b,u} = \frac{f_{by}}{F_{ty}} = \frac{90.6}{215} = 0.42$$

From Eq. 9.8.16:

$$MS = \frac{1}{R_{s,y}^2 + R_{b,y}} - 1$$

$$= \frac{1}{1.15(0.35^2 + 0.42)} - 1 = \underline{0.6} > 0.2 \qquad \text{O.K.}$$

Discussion: Practically, use lower H.T. material for new design and leave extra margin for future heavy growth aircraft use.

(F) COMBINED BOLT LOADS

The MS of the combined ultimate shear and tension load allowables for tension steel bolts is given in the interaction curve shown in Fig. 9.8.17.

$$R_t = \frac{P_t}{P_{t,all}} = \text{Applied tension load/Tension allowable strength of bolt}$$

$$R_s = \frac{P_s}{P_{s,all}} = \text{Applied shear load/Shear allowable strength of bolt}$$

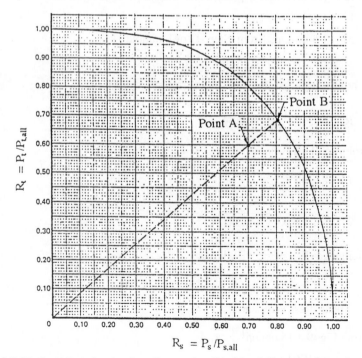

Fig. 9.8.17 Steel Bolt Interaction Curves for Combined Shear and Tension

The MS of the bolt in combined tension and shear may be determined from the above curve by the following procedures:

- Locate the point on the chart representing the applied values of R_s and R_t as illustrated by point A
- Draw a straight line through the point and the origin and extend it to intersect the curve as illustrated by point B
- Read the allowable on the straight line between the origin and point B divided by the straight line between the origin and point A

Example 5:

Use the lug and ultimate load of P = 15 kips as given in Example 1 with the lug installation shown below; find the MS of the bolts (F_{tu} = 220 ksi) fastened on the female lug.

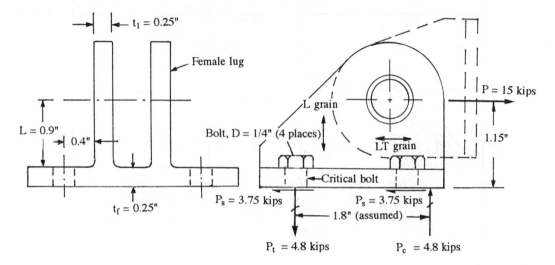

$$P_s = \frac{15}{4} = 3.75 \text{ kips (shear load on each bolt)}$$

$$P_t = 15 \times \frac{1.15}{2 \times 1.8} = 4.8 \text{ kips (fasteners with tension load)}$$

Use steel bolts with D = 0.25" and **shank area = 0.049 in.^{2}**

Steel material properties: H.T. = F_{tu} = 220 ksi and F_{su} = 132 ksi (see Fig. 4.3.6)

$P_{t,\text{all}}$ = 220 × 0.049 = 10.8 kips

(Note: Use bolt shank area to calculate the tension allowable load rather than using the max. minor bolt area as shown in Fig. 9.2.6)

$P_{s,\text{all}}$ = 132 × 0.049 = 6.5 kips

Compute:

$$R_s = \frac{\lambda P_s}{P_{s,\text{all}}} = \frac{1.15 \times 3.75}{6.5} = 0.66$$

$$R_t = \frac{\lambda P_t}{P_{t,\text{all}}} = \frac{1.15 \times 4.8}{10.8} = 0.51$$

The above 1.15 is the fitting **factor** (λ)

Obtain the margin of safety from the interaction curve in Fig. 9.8.17:

MS = 0.24 > 0.2 (min. MS requirement)

(G) LUG CLAMP-UP

Female lugs (clevis) should be checked for excessive residual stresses under clamp-up. Fig. 9.8.18 defines the maximum permissible gap which can be tolerated without requiring an interaction with the applied external loads.

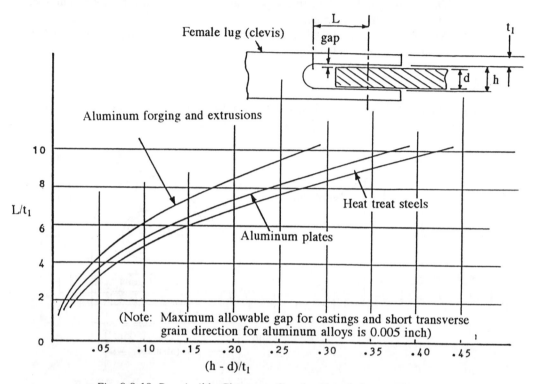

Fig. 9.8.18 Permissible Clamp-up Gap for Female Lugs (Clevis)

The lug installation shown in Fig. 9.8.18 is not practical on rotating hinge designs. The following design practices should be used:

- Use the spherical bearing shown in Fig. 9.8.19 to provide a small amount of unrestricted alignment that is very important for all control surface hinges
- Use a plain bushing, or the flanged bushing, shown in Fig. 9.8.9 and Fig. 9.8.19, to avoid clamp-up action
- The cross-sectional areas of the bushing should be equal to or greater than the steel bolt shank area (check the compression strength of the bushing due to the preload in bolt)

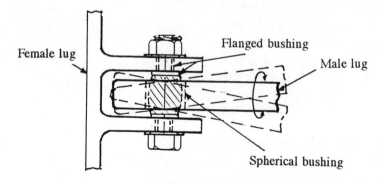

Fig. 9.8.19 Spherical Bushing Application

Example 6:

Using the same female lug given in Example 5, determine the maximum permissible clamp-up gap for this lug.

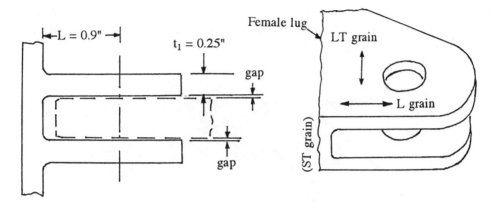

$\dfrac{L}{t_1} = \dfrac{0.9}{0.25} = 3.6$; from Fig. 9.8.18 obtain,

$\dfrac{h-d}{t_1} = \dfrac{\text{gaps}}{t_1} = 0.035"$

The max. permissible gaps = $0.035 \times 0.25 = 0.0088"$

9.9 TENSION FITTINGS (BOLT IN TENSION)

This section contains an introduction applicable to three common types of tension fittings, namely the angle, channel and double angle types, shown in Fig. 9.9.1. Tension fittings, like lugs, should be sized conservatively as their weight is usually small relative to their importance, and inaccuracies in manufacture are difficult to control.

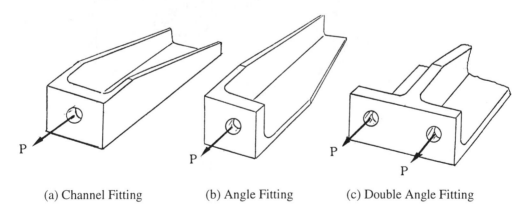

(a) Channel Fitting (b) Angle Fitting (c) Double Angle Fitting

Fig. 9.9.1 Typical Highly Loaded Tension Fittings

A tension fitting is one in which the bolt is loaded primarily in tension and they are used in the following applications:

- Removable wing root joints of fighter wings
- Wing root joints, especially at the four corners of the wing or tail box as shown in Fig. 9.9.2(a) and (b)
- Circumferential production joints on assemblies of fuselage sections

Chapter 9.0

- Axial load transfer of the fuselage stringer by using tension bolt to go through major bulkhead web rather than by cutting a large hole in the web as shown in Fig. 9.9.2(c)
- Back-up fittings for support pylons, flap track, etc., which have to go through wing box (critical for wing fuel tank box design case)

Since the state of stress in a tension fitting is very complex, conventional structural analysis is extremely difficult to obtain with reasonable results. This section will present a simplified method of sizing tension fittings (comparable to that of Ref. 9.9).

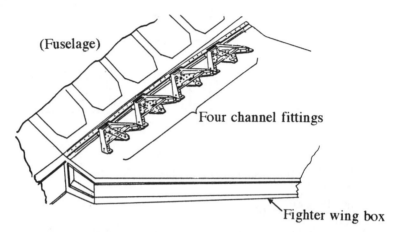

(a) Removable Fighter Wing Root Joint

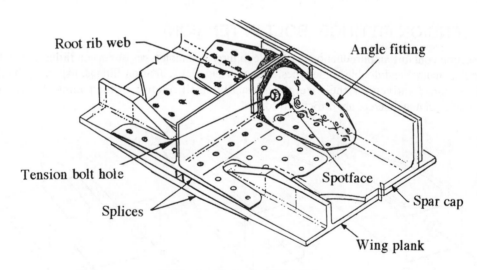

(b) Corner Fittings of Wing Box

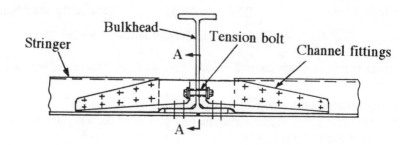

344

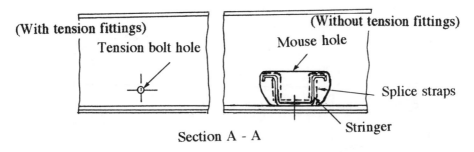

(c) Fuselage Stringer Joint at a Bulkhead

Fig. 9.9.2 Applications of Tension Fittings

In addition, the pertinent fitting and casting factors listed below should always be considered prior to sizing the fitting:

- Use fitting factor $\lambda = 1.15$
- Keep the $MS \geq 0.2$ for both ultimate and yield strength
- If in any application both a fitting and casting factor (used when casting material is used) are applicable, they must not be multiplied, but only the larger factor should be used
- Eccentricity in fitting tension produces prying action in which the tension bolt will pick up additional tension load; therefore, a higher MS on the bolt is recommended
- All these factors can be waived if the fitting has been verified by test data

The sizing of tension fittings in this chapter is based on ultimate design using the design rules below:

- The ultimate tensile stress (F_{tu}) of the fitting material should be less than 1.5 F_{ty}, with reasonable elongation, to take care of yield strength which is usually not easy to predict
- The use of forging is recommended since it gives better grain flow, see Fig. 9.9.3. If the fitting is machined from plate or forging, carefully orient the material grain direction (L, LT and ST)
- Bolts highly loaded in tension should be assembled with a washer under both the head and the nut
- Take care when installing washers to place the countersunk (or chamfered) side of the washer against the manufactured head of the bolt, as shown in Fig. 9.9.4
- Eccentricities in fitting tension bolts should be kept to a minimum by the use of spotface
- The back-up structure must be strong enough to carry the eccentric moment
- The use of Hi-Lok tension fasteners to install the fitting to back-up structure is recommended to resist kick load (see Fig. 9.9.4) and also to prevent premature tension failure in the fitting wall

Chapter 9.0

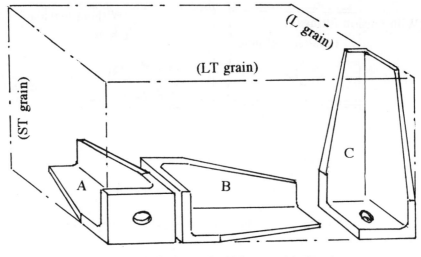

(Use forging or select higher strength in ST grain)

Fig. 9.9.3 Material Grain Direction Affects Fitting Strength

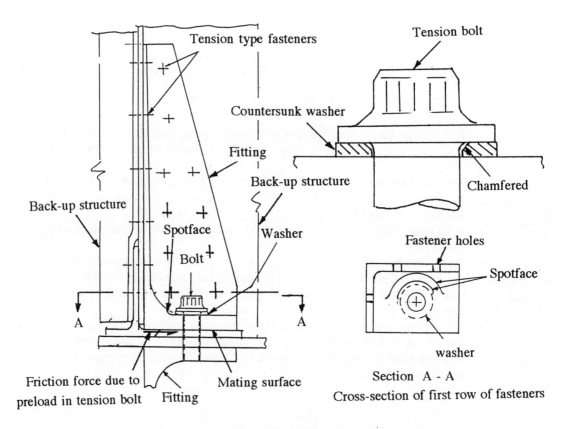

Fig. 9.9.4 Fitting Installation

346

Tension fittings are also fatigue critical structures and require the following procedures:
- Spotface should be used to minimize tension bolt eccentricity
- Clean sharp edges, especially in the spotface area, to minimize crack initiation
- Shot-peening of the entire surface of the fitting is required after machining to improve fatigue life
- Anodize all surfaces of the fitting
- Ream all fastener holes after drilling
- A preload applied to the tension bolt is necessary to improve the fatigue life of both the fitting and the tension bolt

(A) ANGLE FITTINGS

The angle fitting is the most popular tension fitting for use in airframe back-up support structures. The two walls of the fitting must be attached to the load transfer member by fasteners to obtain sound support, as shown in Fig. 9.9.5.

The following specific limitations should be noted prior to the sizing of angle fittings:
- Select a ductile material which has good elongation characteristics and reasonable ultimate tensile strength in the ST grain direction
- $\dfrac{F_{tu}}{F_{ty}} \leq 1.5$
- $\dfrac{b}{D} \approx 3$
- $\dfrac{c_y}{D} \approx 1.5$
- $\dfrac{a}{b} \approx 1.0 \leftrightarrow 1.5$
- $\dfrac{t_a}{t_b} \approx 1 \leftrightarrow 1.5$
- $\dfrac{t_a}{t_b} \approx 2 \leftrightarrow 3$

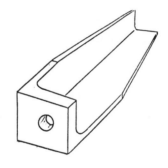

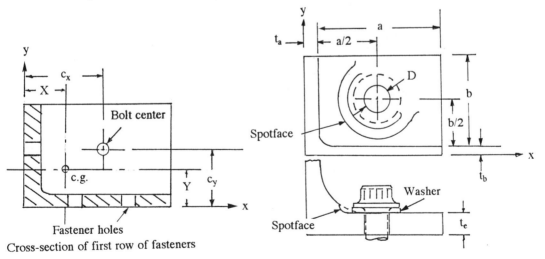

Fig. 9.9.5 Angle Fitting

Chapter 9.0

(a) Fitting wall:

(1) Tension:

$$f_{ta} = \frac{P}{A}$$

where: P – Applied ultimate axial load
A – Net cross-sectional area (less fastener holes)

(2) Bending:

Find the neutral axes (X and Y) of the cross-section and compute their moment of inertia:

I_x – Moment of inertia about the neutral axis x-x

I_y – Moment of inertia about the neutral axis y-y

Determine the bending moments:

$m_x = P(c_y - Y)$, moment along the neutral x-x axis

$m_y = P(c_x - X)$, moment along the neutral y-y axis

Calculate the max. tension stress (f_{bx} and f_{by}) due to the above moments.

Find the max. tension stress on the wall:

$$f_{tu} = f_{ta} + f_{bx} + f_{by}$$

(3) Margin of Safety for ultimate load:

$$MS = \frac{F_{tu}}{\lambda f_{tu}} - 1 \geq 0.2$$

where: λ – Fitting factor

(4) Check attachments between fitting and attached member.

(b) Fitting end:

(1) Bending:

Compute: $\frac{a}{b}$ ratio and from Fig. 9.9.6 obtain coefficient k_a

$$f_{bu} = \frac{k_a P}{t_e^2} \qquad \text{Eq. 9.9.1}$$

$$MS = \frac{F_{bu}}{\lambda(f_{bu})} - 1 \geq 0.2$$

where: F_{bu} – Assume a rectangular section and use a section factor of k = 1.5 to find the plastic bending stress of $F_{bu} = \frac{My}{I}$ (apparent stress) from Fig. 6.4.6 of Chapter 6.4

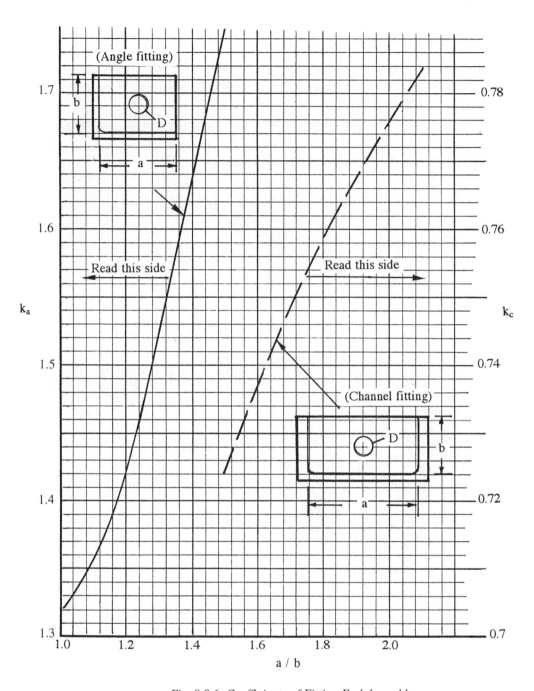

Fig. 9.9.6 *Coefficients of Fitting End, k_a and k_c*

(2) Shear through bolt hole:

$$f_{su} = \frac{P}{0.6 \pi D_w t_c}$$ Eq. 9.9.2

$$MS = \frac{F_{su}}{\lambda f_{su}} - 1 \geq 0.2$$

where: D_w – Outside diameter of the washer
 0.6 – Assume 60% effective
 F_{su} – Ultimate shear allowable of the fitting end materiel

Chapter 9.0

(B) CHANNEL FITTING

The channel fitting, as shown in Fig. 9.9.7, is much stronger than the angle fitting for highly concentrated loads. The channel fitting is used in airframe back-up support structures where the center fitting wall is attached to the load transfer member by fasteners and the top and bottom of the fitting is strong enough to carry the induced kick load.

The following specific limitations should be noted prior to the sizing of channel fittings:

- Select a ductile material which has good elongation characteristics and reasonable ultimate tensile strength in the ST grain direction
- $\dfrac{F_{tu}}{F_{ty}} \leq 1.5$
- $\dfrac{b}{D} \approx 3$
- $\dfrac{c_y}{D} \approx 1.5$
- $\dfrac{a}{b} \approx 1.5 \leftrightarrow 2.0$
- $\dfrac{t_a}{t_b} \approx 1 \leftrightarrow 1.5$
- $\dfrac{t_a}{t_b} \approx 2 \leftrightarrow 3$

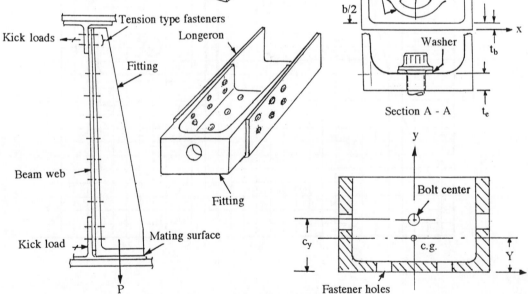

Fig. 9.9.7 Channel Fitting

(a) Fitting wall:

(1) Tension:

$$f_{ta} = \dfrac{P}{A}$$

where: P – Applied ultimate axial load
A – Net net cross-sectional area (less attachment holes)

(2) Bending:

Find the neutral axes (X and Y) of the cross-section and compute their moments of inertia:

I_x – Moment of inertia about the neutral axis x-x

I_y – Moment of inertia about the neutral axis y-y

Determine the bending moments:

$m_x = P(c_y - Y)$, moment along the neutral x-x axis

$m_y = P(c_x - X)$, moment along the neutral y-y axis

Calculate the max. tension stress (f_{bx} and f_{by}) due to above moments.

The total max. tension stress:

$f_{tu} = f_{ta} + f_{bx} + f_{by}$

(3) Margin of Safety for ultimate load:

$$MS = \frac{F_{tu}}{\lambda f_{tu}} - 1 \geq 0.2$$

where: λ – Fitting factor

(4) Check attachments between fitting and attached member.

(b) Fitting end:

(1) Bending:

Compute: $\frac{a}{b}$ ratio and from Fig. 9.9.6 obtain coefficient k_c

$$f_{bu} = \frac{k_c P}{t_e^2} \qquad \text{Eq. 9.9.3}$$

$$MS = \frac{F_{bu}}{\lambda f_{bu}} - 1 \geq 0.2$$

where: F_{bu} – Use a section factor of k = 1.5 to find the plastic bending stress of $F_{bu} = \frac{My}{I}$ (apparent stress) from Fig. 6.4.6 of Chapter 6.4

(2) Shear through bolt hole:

$$f_{su} = \frac{P}{0.6 \pi D_{w,0} t_e} \qquad \text{Eq. 9.9.4}$$

$$MS = \frac{F_{su}}{\lambda f_{su}} - 1 \geq 0.2$$

where: $D_{w,0}$ – Outside diameter of the washer
0.6 – Assume 60% effective.
F_{su} – Ultimate shear allowable of the fitting end material

(C) DOUBLE ANGLE FITTING

The sizing of a double angle fitting follows the same procedure as that of a single angle fitting. The long fitting walls must be attached to the back-up member or web by fasteners to obtain sound support, as shown in Fig. 9.9.8. This type of fitting is used as a back-up fitting to support a removable lug fitting, for example:

- Wing engine pylon mounting lug.
- External flap track mounting lug.

The double angle fitting has the same specific limitations as those for a single angle fitting.

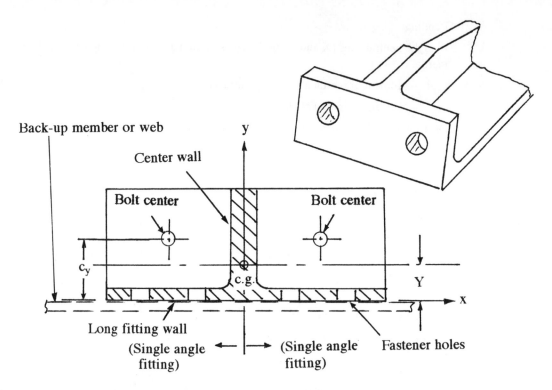

Fig. 9.9.8 Double Angle Fitting

(a) Fitting wall:

The sizing procedures are the same as those for a single angle fitting except that it is T-shaped instead of L-shaped (angle fitting) and the center wall is critical.

(b) Fitting end:

The sizing procedures are the same as those for a single angle fitting.

(D) BOLT PRELOAD

The preload on a tension bolt shall not produce a stress (based on the max. minor area see Fig. 9.2.6) greater than the bolt material yield strength (F_{ty}); generally use 75% of F_{ty}. Preloaded tension bolts are important for the following reasons:

- To improve the fatigue life of the fitting as well as the attached back-up structures
- To make the fitting stiffness comparable to that of adjacent structures
- The preload on the bolt produces friction between the fitting end and the attached bottom surface which balances the kick load caused by the eccentricity of the tension bolt

Under tension limit load the mating surfaces must always be in contact. The total load on a bolt that is subjected to both an initial tension preload and an external tension load is shown in Fig. 9.9.9.

Joints and Fittings

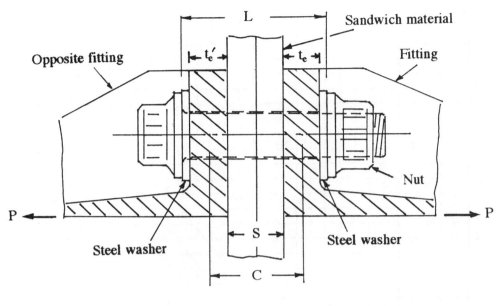

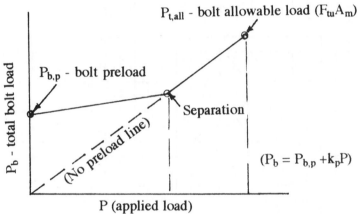

Fig. 9.9.9 Preloaded Tension Fitting

(a) Total load on tension bolt:

Preload on bolt due to tightening:

$$P_{b,p} = 75\% F_{ty} A_m \qquad \text{Eq. 9.9.5}$$

Total load on tension bolt:

$$P_b = P_{b,p} + k_p P \qquad \text{Eq. 9.9.6}$$

or $P_b = P$ (after separation of mating surfaces)

where: A_m – The max. minor area of the tension bolt (in.^{2}), see Fig. 9.2.6.
P – Applied ultimate tension load.
C – The distance C is measured between the centerlines of the fitting end thickness ($C = \frac{t_e}{2} + \frac{t_e'}{2} + S$)
L – Distance between bolt head and nut ($L = t_e + t_e' + S + 2t_w$)
k_p – Bolt tension preload coefficient, see Fig. 9.9.10.

353

Chapter 9.0

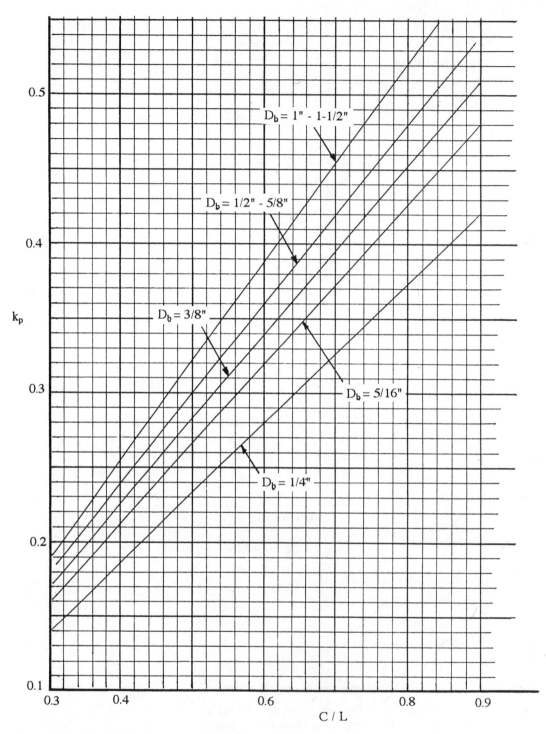

Above values are based on the following requirements:
1. D_b – Bolt diameter.
2. Steel bolt and aluminum fittings and sandwich.
3. Use steel washer outside diameter ($D_{w,o}$) data per Fig. 9.9.11.
4. Use of larger $D_{w,o}$ is conservative.
5. Use of titanium bolts in conservative.

Fig. 9.9.10 Preload Coefficient, k_p

(b) Torque:

The average wrench torque applied to the nut to produce a specified preload or stress is given by

$$T = \frac{\beta P_{b,p}}{A_m} \qquad \text{Eq. 9.9.7}$$

where: T – Wrench torque (±30% for dry nut and ±15% for specific lubricated nut)
β – Wrench torque ratio (from Fig. 9.9.11)

Bolt Diameter D_h (in.)	Steel Washer, D_w (in.)		β, Nut turning on steel washer	
	Outside ($D_{w,o}$)	Inside ($D_{w,i}$)	Lubricated per MIL-T-5544	Dry
$\frac{3}{16}$	0.365	0.193	0.000354	0.001088
$\frac{1}{4}$	0.53	0.253	0.000788	0.00245
$\frac{5}{16}$	0.59	0.316	0.001497	0.00509
$\frac{3}{8}$	0.69	0.378	0.002468	0.00829
$\frac{7}{16}$	0.78	0.441	0.004013	0.0127
$\frac{1}{2}$	0.88	0.503	0.006203	0.0203
$\frac{9}{16}$	0.97	0.566	0.008994	0.0294
$\frac{5}{8}$	1.06	0.629	0.01226	0.0421
$\frac{3}{4}$	1.25	0.754	0.02073	0.0713
$\frac{7}{8}$	1.44	0.879	0.03312	0.1136
1	1.63	1.004	0.04993	0.1742
$1\frac{1}{8}$	1.88	1.129	0.07129	0.246
$1\frac{1}{4}$	2.13	1.254	0.09859	0.333

Notes:
1. Above values are for turning nut.
2. Lubricant applied to threads and to bearing surfaces of nut and bolt head.
3. Both plain and countersunk washers are required.
4. Both bolt and washers are steel cadmium plated.
5. Use larger $D_{w,o}$, if H.T. of tension bolt is greater than 220 kis. (see Ref. 9.12 and 9.13).

Fig. 9.9.11 Tension Bolt Wrench Torque Coefficient, β

Be careful to select a compatible tension nut (shear nuts are not recommended) which matches the tension allowable bolt load.

Example:

Size a tension angle fitting which can carry an applied ultimate load of P = 11 kips using the following data:

Materials: Angle fitting – Aluminum 7050-T7452 hand forging ("A" values from Ref. 9.1):

$F_{tu} = 72$ ksi (L); $\quad F_{ty} = 62$ ksi (L); $\quad F_{su} = 41$ ksi

$F_{tu} = 70$ ksi (LT); $\quad F_{ty} = 60$ ksi (LT); $\quad F_{bru} = 129$ ksi (L) (dry pin allowable)

$F_{tu} = 67$ ksi (ST); $\quad F_{ty} = 55$ ksi (ST); $\quad E = 10.2 \times 10^3$ ksi

Bolt: H.T. = 220 – 240 ksi

$F_{tu} = 220$ ksi, $F_{ty} = 185$ ksi, $E = 29 \times 10^3$ ksi

Allowable tension load, $P_{all} = 42.9$ kips. (see Fig. 9.2.6)

Bolt diameter, $D_b = \dfrac{9}{16}$ in. (0.562")

Tension bolt shank area: $A_s = 0.2485$ in.^{2} (see Fig. 9.2.7)

Bolt max. minor area $A_m = 0.195$ in.^{2} (see Fig. 9.2.6)

Washer: Steel with a thickness of $t_w = 0.104"$ and outside diameter: $D_{w,o} = 0.97"$

Check angle fitting limitations:

- Select fitting configuration "A" as shown in Fig. 9.9.3. and use 'L' grain value ($F_{tu} = 70$ ksi) for fitting wall and ST grain value ($F_{tu} = 67$ ksi) for fitting end.

- $\dfrac{F_{tu}}{F_{ty}} = \dfrac{67}{55} = 1.2 < 1.5$ O.K.

- $\dfrac{b}{D} = \dfrac{1.9 - 0.2}{0.6} = 2.8 \approx 3$ O.K.

- $\dfrac{c_y}{D} = \dfrac{1.1}{0.6} = 1.83 \approx 1.5$ say O.K.

- $\dfrac{a}{b} = \dfrac{2.2 - 0.3}{1.9 - 0.2} = 1.1 \approx 1.0 \leftrightarrow 1.5$ O.K.

- $\dfrac{t_a}{t_b} = \dfrac{0.3}{0.2} = 1.5 \approx 1 \leftrightarrow 1.5$ O.K.

- $\dfrac{t_a}{t_b} = \dfrac{0.5}{0.2} = 2.5 \approx 2 \leftrightarrow 3$ O.K.

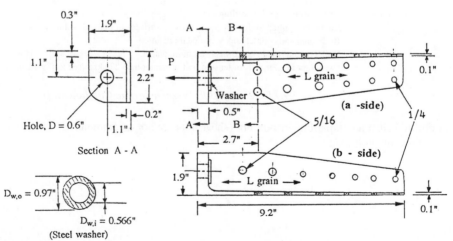

(a) Fitting wall:

Compute:

$I_x = 0.268$ in.^{4} $I_y = 0.248$ in.^{4}

$X = 0.622$ in. $Y = 0.533$ in.

$A_n = 0.63$ in.^{2} (Net cross-section area)

Net section tensile stress,

$$f_{ta} = \frac{P}{A_n} = \frac{11}{0.63} = 17.5 \text{ ksi}$$

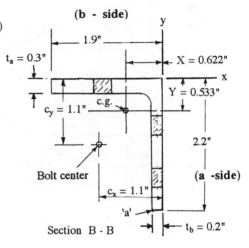

Moments:

$m_x = 11(1.1 - 0.533) = 6.2$ in.-kips

$m_y = -11(1.1 - 0.622) = -5.3$ in.-kips

Bending stress from m_x,

$$f_{bx} = \frac{m_x(2.2 - 0.533)}{I_x}$$

$$= \frac{6.2(2.2 - 0.533)}{0.268} = 38.6 \text{ ksi}$$

Bending stress from m_y,

$$f_{by} = \frac{m_y(0.622 - 0.2)}{I_y}$$

$$= \frac{-5.3(0.622 - 0.2)}{0.248} = -9 \text{ ksi}$$

The max. stress occurs at point 'a':

$f_t = f_{ta} + f_{bx} + f_{by}$

$= 17.5 + 38.6 - 9$

$= 47.1$ ksi

$$MS = \frac{F_{tu}}{\lambda f_t} - 1$$

$$= \frac{72}{1.15 \times 47.1} - 1$$

$= \underline{0.33 > 0.2}$ O.K.

Check fitting wall fasteners (Wet-installed titanium Hi-Lok fasteners):

Shear allowable (from Fig. 9.2.5):

$P_{s, all} = 7.29$ kips for $D = \frac{5}{16}"$

$P_{all, s} = 4.66$ kips for $D = \frac{1}{4}"$

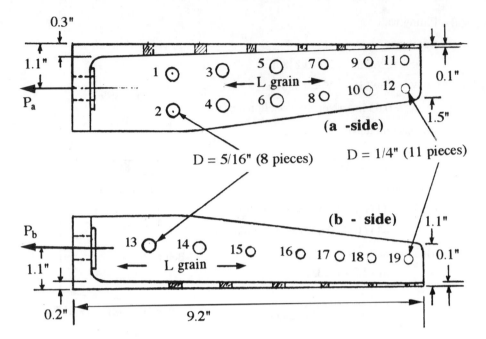

Bearing allowable [F_{bru} = 129 ksi (L) (dry pin allowable)]; use 80% reduction of dry pin bearing allowable:

$P_{b, all}$ = D × t × (80% F_{bru}) = D × t × 103.2 and $P_{b, all}$ as calculated below:

Bearing allowable on 'a' – side

Fast. No.	D	t	80% F_{bru}	$P_{b, all}$
1	.313"	.25"	103.2	8.1 (7.29)*
2	.313"	.25"	103.2	8.1 (7.29)*
3	.313"	.22"	103.2	7.1
4	.313"	.22"	103.2	7.1
5	.313"	.185"	103.2	6
6	.313"	.185"	103.2	6
7	.25"	.16"	103.2	4.1
8	.25"	.16"	103.2	4.1
9	.25"	.135"	103.2	3.5
10	.25"	.135"	103.2	3.5
11	.25"	.11"	103.2	2.8
12	.25"	.11"	103.2	2.8

* Shear critical
(Total allowable on a-side Σ = 61.6 kips)

Bearing allowable on 'b' – side

Fast. No.	D	t	80% F_{bru}	$P_{b, all}$
13	.313"	.185"	103.2	6
14	.313"	.17"	103.2	5.5
15	.25"	.155"	103.2	4
16	.25"	.14"	103.2	3.6
17	.25"	.12"	103.2	3.1
18	.25"	.11"	103.2	2.8
19	.25"	.1"	103.2	2.6

(Total allowable on b-side Σ = 27.6 kips)

Total allowable = 89.2 kips

Applied load at 'a'-side: $P_a = 11 \frac{61.6}{89.2}$ = 7.6 kips

Applied load at 'b'-side: $P_b = 11 \frac{27.6}{89.2}$ = 4.4 kips

Calculate the MS per the method for eccentric joints from Chapter 9.4 for the fastener patterns on each wall. The fitting factor (λ = 1.15) and minimum MS ≥ 0.2 are applicable here.

Since the allowable load (89.2 kips) is much highter than the applied load (P = 11 kips) the MS is not critical.

(b) Fitting end:

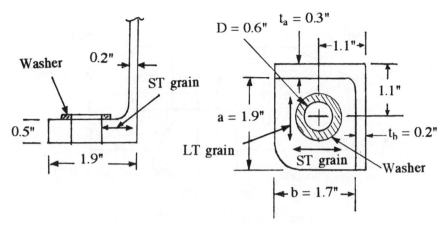

Section A - A

(1) Bending:

Use $\frac{a}{b} = 1.1$ ratio and from Fig. 9.9.6 obtain coefficient $k_c = 1.357$ and from Eq. 9.9.3:

$$f_{bu} = \frac{k_c P}{t_e^2}$$

$$= \frac{1.357 \times 11}{0.5^2} = 59.7 \text{ ksi}$$

Use section factor of $k = 1.5$ from Fig. 6.4.6(a) of Chapter 6.4. to find the plastic bending stress (apparent stress) for $F_{tu} = 67$ ksi (use smaller of grain values):

$F_{bu, 1.5} = 110$ ksi for $k = 1.5$ and $F_{bu, 1.0} = 76$ ksi for $k = 1.0$

[from Fig. 6.4.6(a) with 7075-T6 forging]

Use following ratio to determine "apparent stress" for $F_{tu} = 67$ ksi:

$$F'_{bu, 1.5} = \frac{F_{bu, 1.5}}{F_{bu, 1.0}} (F_{tu})$$

$$= \frac{110}{76}(67) = 97 \text{ ksi}$$

$$MS = \frac{F'_{bu, 1.5}}{\lambda (f_{bu})} - 1$$

$$= \frac{97}{1.15 \times 59.7} - 1 = \underline{0.41 > 0.2} \qquad \text{O.K.}$$

(2) Shear through bolt hole from Eq. 9.9.4:

$$f_{su} = \frac{P}{0.6 \pi D_{w, 0} t_e}$$

$$= \frac{11}{0.6 \pi 1.05 \times 0.5} = 12.03 \text{ ksi}$$

$$MS = \frac{F_{su}}{\lambda (f_{su})} - 1$$

$$= \frac{41}{1.15 \times 12.03} - 1 = \underline{\text{High}} \qquad \text{O.K.}$$

(c) Tension bolt:

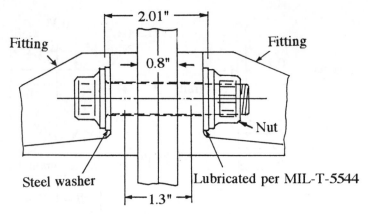

Steel bolt (H.T.= 220 - 240 ksi)

Compute:

Preload in bolt from Eq. 9.9.5:

$$P_{b,p} = 75\% F_{ty} A_m$$
$$= 75\% \times 185 \times 0.195$$
$$= 27 \text{ kips}$$
$$C = \frac{t_e}{2} + \frac{t_e'}{2} + S = \frac{0.5}{2} + \frac{0.5}{2} + 0.8 = 1.3$$
$$L = t_e + t_e' + S + 2t_w = 0.5 + 0.5 + 0.8 + 2 \times 0.104 = 2.01$$
$$\frac{C}{L} = \frac{1.3}{2.01} = 0.65 \text{ and from Fig. 9.9.10 obtaining } k_p = 0.38.$$

From Eq. 9.9.5

$$P_b = P_{b,p} + k_p P$$
$$= 27 + 0.38 \times 11 = 31.1 \text{ kips}$$
$$MS = \frac{P_{all}}{\lambda P_b} - 1$$
$$= \frac{42.9}{1.15 \times 31.1} - 1 = 0.2 \qquad \text{O.K.}$$

(4) Wrench torque applied to the nut:

From Fig. 9.9.11, $\beta = 0.008994$ (lubricated)

From Eq. 9.9.7:

$$\text{Torque,} \quad T = \frac{\beta \times P_{b,p}}{A_m}$$
$$= \frac{0.008994 \times 27}{0.195}$$
$$= 1.25 \text{ in.-kips}$$

On the engineering drawing the torque should be specified (use ±15% tolerance) as:

$$T = 1.25 + 1.25(0.15) = 1.06 \text{ in.-kips} \quad (\text{max.})$$
$$T = 1.25 - 1.25(0.15) = 1.44 \text{ in.-kips} \quad (\text{min.})$$

9.10 TENSION CLIPS

Tension clips are fairly common in airframe structures and are used to transmit relatively light tension loads from clip to clip or to another member. Clips are usually made from formed sheet or extrusions into various shape such as angles, back-to-back angles, T-shapes, double T-shapes and channels, as shown in Fig. 9.10.1. When a tension load is applied normal to the outstanding leg of an angle or T-shape, tension, bending, and shear stresses occur. If the outstanding leg is not rigidly supported, the allowable load is usually limited by considerations of permissible deflection and permanent set. If the fastener points through which the load is transmitted to the angle are not spaced sufficiently close, the full strength of the angle is not developed since local deformation at the fastener points becomes the limiting factor. Tests have revealed that:

- Clips formed from sheet exhibit good ductility and best tensile properties when the longitudinal grain direction (L) of the material runs around the bend
- Clips machined from extrusion may have the ST grain direction oriented around the bend and this results in less efficiency (see Fig. 9.10.2) and can be improved by shot-peening process
- Maintaining the original shape of the extrusion is recommended (non-machined part)

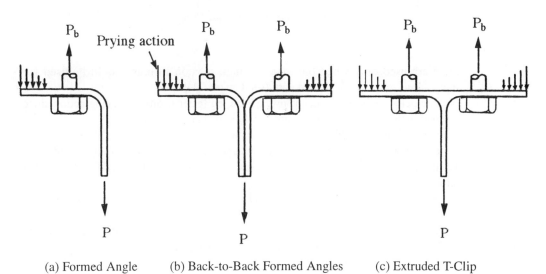

(a) Formed Angle   (b) Back-to-Back Formed Angles   (c) Extruded T-Clip

Fig. 9.10.1 Common Tension Clips

The applications of tension clips on airframe structures are frequently seen in the areas of:

- The frame bended flange which is fastened to fuselage skin
- The wing or empennage rib cap flange fastened to wing skin
- The rib flange which is fastened to the skin of control surfaces, e.g., flaps, slats, elevators, rudders, etc.

A clip is a complex structure and as was the case with tension fittings, conventional structural analysis is not applicable to this type of structure. Airframe companies take the results of numerous tests and compile them into a series of design curves in order to provide reliable data for engineer use. Both extruded angle clips and T-clips (ST grain if machined from extrusion) are relatively poor for sustaining repeated loading, with angle clips having the poorer performance of the two. This fatigue shortcoming is obviously due to bending stresses in the material's long transverse grain (LT) or short transverse grain dirction (ST), as shown in Fig. 9.10.2(b). For this reason all tension clip applications, which are subject to repeated loadings, require design for fatigue life rather than for static strength.

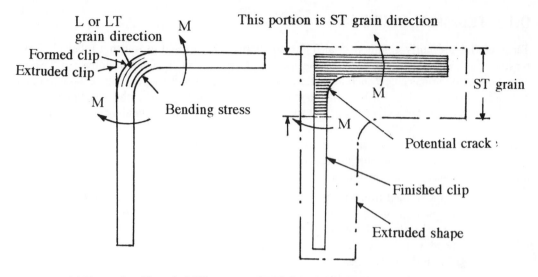

(a) Formed or Extruded Clips   (b) Finished Clip is Machined from Extrusion

Fig. 9.10.2 Formed vs. Extruded Clips

Design considerations:

- Clips are used only when the load is small; tension fittings should be used for applications in the primary load path (see Chapter 9.9)
- For thick clips failure may occur in the base leg fastener due to prying action since the load in the fastener is always greater than the applied load
- Tension fasteners should be used with tension clips
- Generous corner fillets on extruded angles and minimum bend radii on formed sheet metal angles should be used in all cases where applied loads tend to "open" or "close" the angle
- For maximum strength, the fastener head should be adjacent to the point of tangency of the fillet or bend radius
- High local stresses and large deflections in angle points prohibit the use of clips in applications subject to repeated or alternating loads
- Clips should not be used in continuous load transfer joints
- The method of fastener installation for clips is given in Fig. 9.10.3

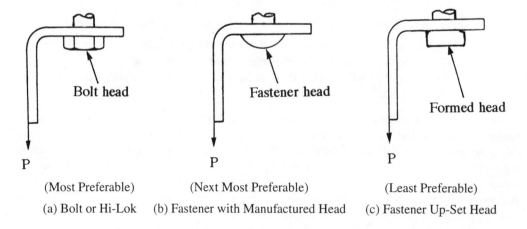

(Most Preferable)   (Next Most Preferable)   (Least Preferable)
(a) Bolt or Hi-Lok   (b) Fastener with Manufactured Head   (c) Fastener Up-Set Head

Fig. 9.10.3 Clip Installation

(A) ANGLE CLIPS

(a) Light angle clips:

This type of clip is generally formed from standard sheet stock and is not designed to carry the larger primary tension load. Use of tension-type aluminum rivets is acceptable and the design curves shown in Fig. 9.10.4, Fig. 9.10.5, and Fig. 9.10.6 are used to size light angle clips.

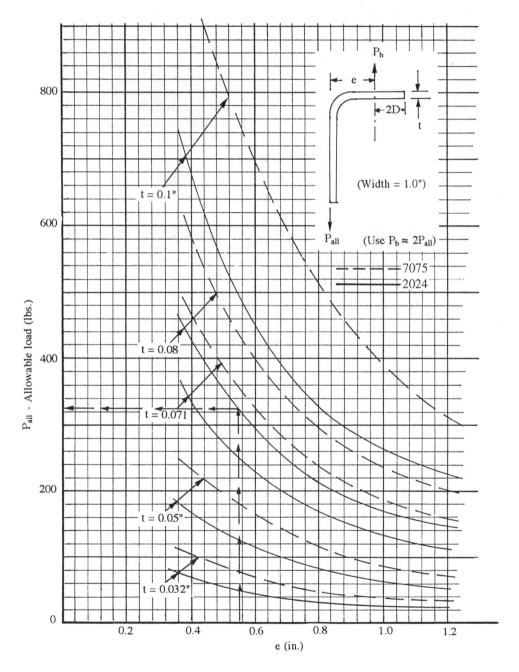

Fig. 9.10.4 Formed-Sheet Angle Clips (2024 and 7075)

Chapter 9.0

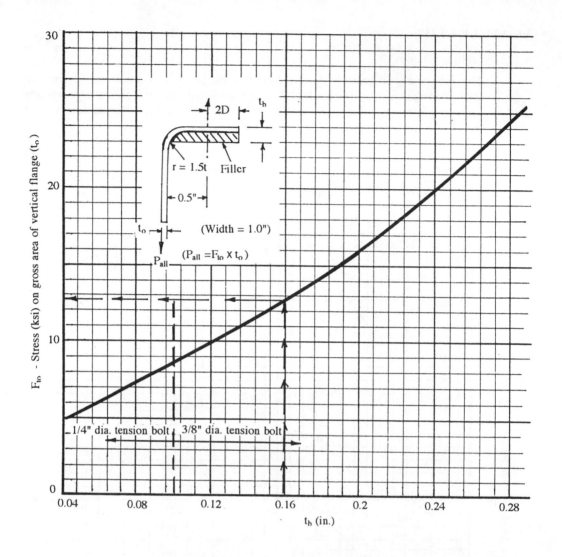

(It is improtant to ensure that radius fillers fit snugly in the clip bend radius)

Fig. 9.10.5 2024-T4 Formed-Sheet Angle Clips with Filler

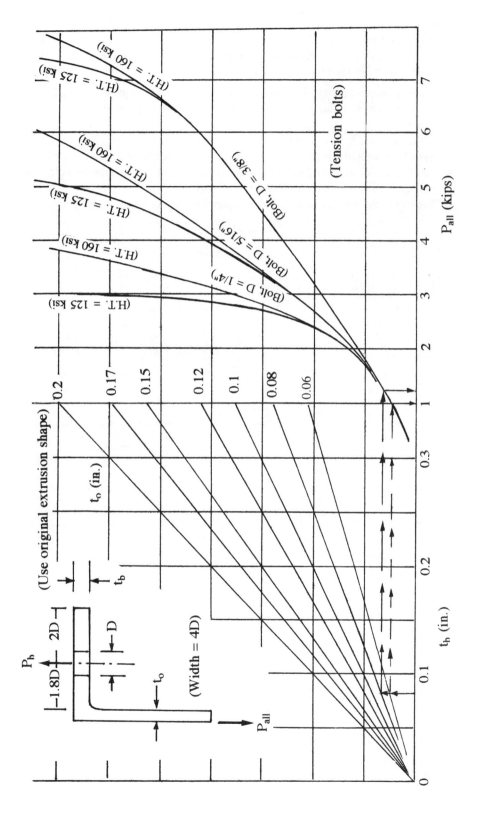

Fig. 9.10.6 Extruded Angle Clips (2024)

Chapter 9.0

(b) Heavy angle clips:

Tension angle clips which must carry large loads are usually made from either extrusions, or are machined from forging billets and joined together with tension bolts. On thicker clips the bolt becomes more critical due to prying action; Fig. 9.10.7 shows the relationship between the angle clip geometry and the tension load on the bolt.

- An increase in "e" will increase the bending moment on the clip and decrease the "a" value which in turn will increase the couple load acting on the bolt and the toe of the clip
- If the "a" value is increased to compensate, the thickness will also have to be increased in order to keep the increased bending stresses in the base within acceptable limits
- Small changes to any of these dimemsions can result in a significant bolt axial load increase causing the bolt head to fail in service
- Preload in the bolt is required (see Chapter 9.9) to avoid separation up to limit load

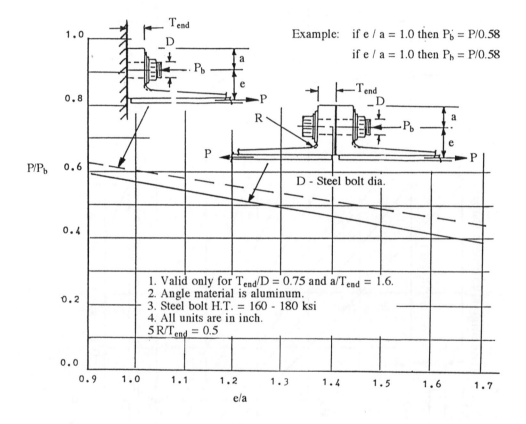

Fig. 9.10.7 Heavy Aluminum Angle Clips

(c) Fastener tension load (see Fig. 9.10.8):

The fastener tension load is expressed as below:

$$P_b = \frac{P(e + k \times c)}{k \times c} \quad \text{(No separation condition)} \qquad \text{Eq. 9.10.1}$$

($P_b = 1.5P - 3P$ approximate range)

After separation:

$P_b = P$ (No prying load exists)

Due to the complexity of clip mounting, as shown in Fig. 9.10.8, it is impossible to determine the value of k because it is a function of the following parameters:

- Geometry and material of the clip
- Type, material, and location of the fastener
- Preload in fastener
- Shape of prying load curve
- $\frac{t_b}{t_o}$ ratio effect

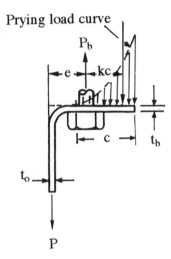

Fig. 9.10.8 Tension Load on Fastener of Angle Clip

Tests for the combination of both clip and fastener are frequently conducted by airframe manufacturers to assist the engineer in obtaining reliable data for clip sizing.

(B) T-CLIPS

T-Clips can be made from back-to-back formed angles or from extruded T-clips, as shown in Fig. 9.10.1.

(a) Extruded T-Clips:

If the finished T-clip configuration is obtained by machining an extruded shape, the grain direction of the outstanding flange may be in the ST grain direction and if so this is a matter of concern (requires shot-peening process). The design curves are shown in Fig. 9.10.9.

Chapter 9.0

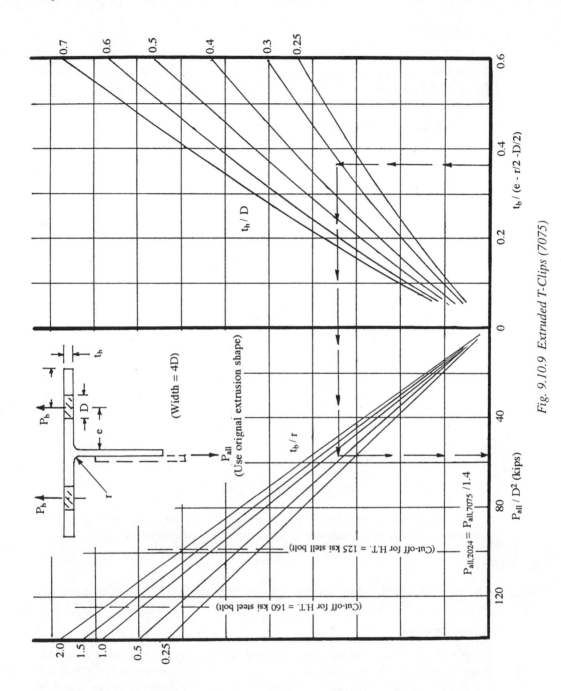

Fig. 9.10.9 Extruded T-Clips (7075)

(b) Back-to-back angles:

This type of clip consists of two single angle clips which are joined together and it is not as strong as a single extruded T-clip, but it provides better fatigue life. Use Fig. 9.10.6 to determine the allowable load (P) for a single formed angle, and then multipy it by 2.5 to arrive at the allowable load (P) for back-to-back formed angles.

Joints and Fittings

Example:

Determine the tension allowable loads (P_{all}) for each of the following given clips (2024 material) and fasteners:

(a) Formed angle clip:

Use $e = 0.55"$

$t = 0.08"$

From Fig. 9.10.4 obtain:

$P_{all} = 325$ lbs.

$P_b = 2 \times 325 = 650$ lbs.

(use $D = \frac{5}{32}"$ of MS20470 DD rivet

and $P_{t, all} = 977$ lbs from Fig. 9.2.11)

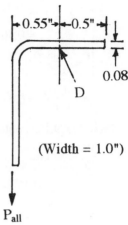

(b) Formed angle clip with filler:

Use $t_b = 0.16"$

From Fig. 9.10.5 obtain:

$F_{to} = 12.7$ ksi

$P_{all} = F_{to} \times t_o = 12.7 \times 0.08 = 1.02$ kips

Use steel bolt, $D = \frac{3}{8}"$ with H.T. = 125 ksi

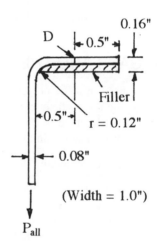

(c) Extruded angle clip:

Determine P_{all} and fastener:

Use $t_b = t_o = 0.08"$

From Fig. 9.10.6 obtain:

$P_{all} = 1.25$ kips

Use steel bolt, $D = .25"$ with H.T. = 125 ksi

If using $t_o = 0.06"$

$P_{all} = 1.0$ kips. < above 1.25 kips value

(which means that t_o also makes an important contribution)

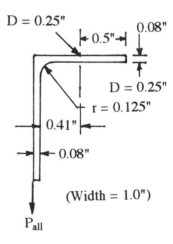

369

(d) Extruded T-clip:

$$\frac{t_b}{e - \frac{r}{2} - \frac{D}{2}} = \frac{0.08}{0.41 - \frac{0.125}{2} - \frac{0.25}{2}}$$

$$= 0.36$$

$$\frac{t_b}{D} = \frac{0.08}{0.25} = 0.32$$

$$\frac{t_b}{r} = \frac{0.08}{0.125} = 0.64$$

From Fig. 9.10.9, obtain

$$\frac{P_{all}}{D^2} = 57 \text{ kips}$$

$$P_{all} = 57 \times 0.25^2 = 3.6 \text{ kips} \quad \text{(for 7075 Al.)}$$

$$P_{all} = \frac{3.6}{1.4} = 2.6 \text{ kips} \quad \text{(for 2024 Al.)}$$

Use steel bolts, D = .25" with H.T. = 125 ksi

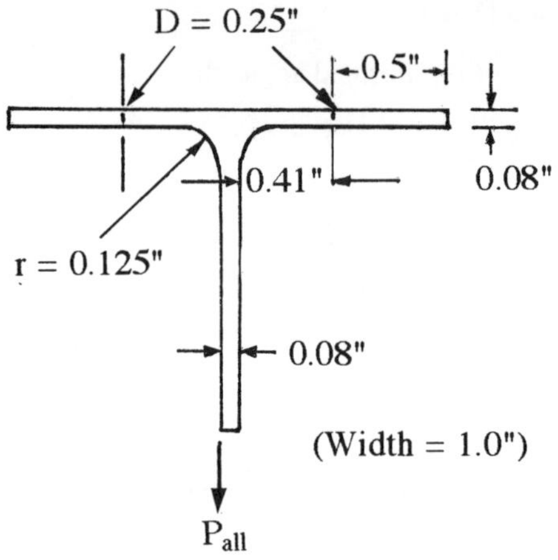

9.11 GAPS AND THE USE OF SHIMS

Shims are used in airframe production to control structural fit-up gap, to maintain contour or alignment, and for aesthetic purposes but they are costly. The engineer should decide whether the gap should be shimmed or pulled up by fasteners without shim as shown in Fig. 9.11.1. which gives the approximate maximum gap allowable without shim required.

Type of Structure	Maximum Allowable Gap (in.)
Heavy structure – members jointed are ≥ 0.3" thickness	0.003
Heavy structure – members jointed are < 0.3" thickness	0.006
Light structure members are short and rigid	0.01
Light 'spring' structure – members such as skin stringer	0.02

Fig. 9.11.1 Maximum Allowable Gap without Shim

Design considerations to control fit-up (pull-up) stresses:
- Design to eliminate shims where possible
- Specify where shims are required
- Specify maximum thickness of shims
- Define permissible gap
- Define type of shim
- Determine fit-up sequences
- Elicit feed-back from liaison and manufacturing

The stress corrosion pattern is governed by the material sustained tensile allowable:
- Parts machined from thick sections are more susceptible to stress corrosion than those made from thin sections
- The material grain strength in the center of thicker extrusions and forgings will be larger and more susceptible to stress corrosions than those that are thinner
- The ST grain direction is the most susceptible to stress corrosion

- The cracks always start on the surface of a part at the edge of fastener holes
- Other surface cracks start at a machined fillet radius due to assembly pull-up

Each case is unique and requires a study of deflections which induce bending stresses, as shown in Fig. 9.11.2.

(Clamping stress using cross section as ▨ with 1 inch width)

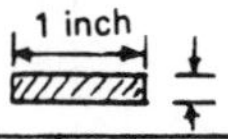

Case	Loading condition	Stress
(1) 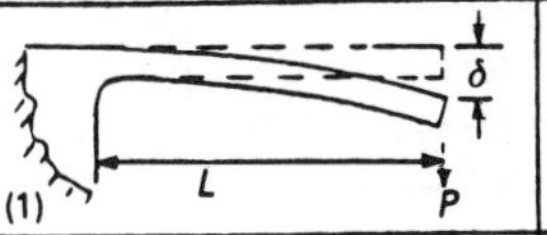	$P = \dfrac{E\delta}{4}\left(\dfrac{t}{L}\right)^3$	$f_b = \dfrac{1.5\,E\,\delta\,t}{L^2}$
(2) 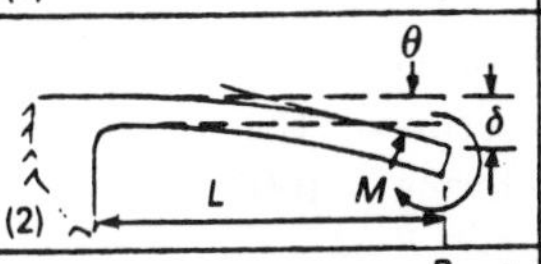	$M = \dfrac{E\,\delta\,t^3}{6L^2} = \dfrac{E\,\theta\,t^3}{12L}$	$f_b = \dfrac{E\,\delta\,t}{L^2} = \dfrac{E\,\theta\,t}{2L}$
(3) 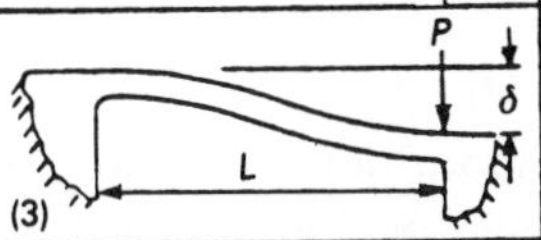	$P = E\,\delta\left(\dfrac{t}{L}\right)^3$	$f_b = \dfrac{3E\,\delta\,t}{L^2}$
(4) 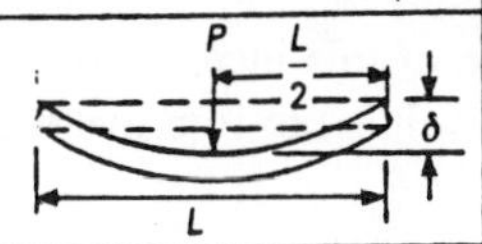	$P = 4E\,\delta\left(\dfrac{t}{L}\right)^3$	$f_b = \dfrac{6E\,\delta\,t}{L^2}$
(5) 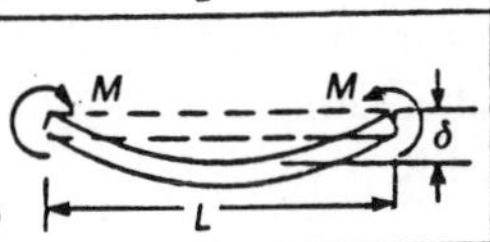	$M = \dfrac{E\,\delta\,t^3}{1.5L^2}$	$f_b = \dfrac{4E\,\delta\,t}{L^2}$
(6) 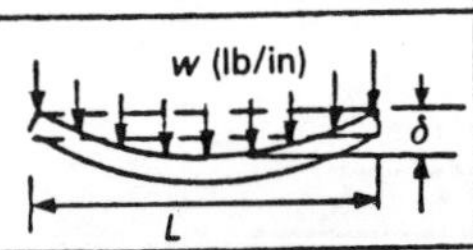	$w = \dfrac{6.4\,E\,\delta\,t^3}{L^4}$	$f_b = \dfrac{4.8\,E\,\delta\,t}{L^2}$
(7) 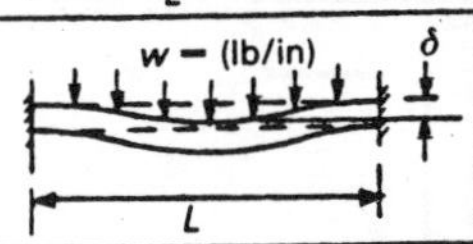	$w = \dfrac{32\,E\,\delta\,t^3}{L^4}$	$f_b = \dfrac{16\,E\,\delta\,t}{L^2}$
(8) 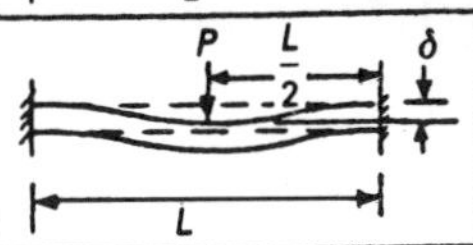	$P = 16E\,\delta\left(\dfrac{t}{L}\right)^3$	$f_b = \dfrac{12\,E\,\delta\,t}{L^2}$
(9) 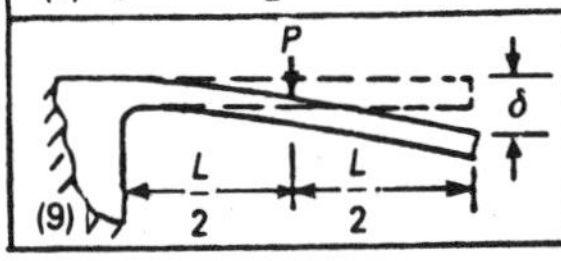	$P = \dfrac{4E\,\delta}{5}\left(\dfrac{t}{L}\right)^3$	$f_b = \dfrac{2.4\,E\,\delta\,t}{L^2}$

Fig. 9.11.2 Pull-up (or clamp-up) Stress

The bending stress from Fig. 9.11.2 should not exceed the sustained tensile stress allowable, shown in Fig. 9.11.3, under which stress corrosion will not occur.

Alloy	Grain Dir.	Plate	Rod and bar	Extruded shapes > 1.0"	Extruded shapes 1.0 – 2.0"	Forging
2024-T3 or T4	L	35	30	> 50	> 50	—
	LT	20	—	37	18	—
	ST	7	10	—	18	—
7075-T3 or T4	L	50	50	60	60	35
	LT	45	—	50	32	25
	ST	7	15	—	7	7
7075-T73	L	—	> 50	—	—	>50
	LT, ST	—	> 47	—	—	48
Alloy steel (Under 200 ksi)	L		(All forms = 115)			
	LT		(All forms = 95)			
	ST		(All forms = 95)			
4330M (220-240 ksi)	L		(All forms = 150)			
	LT		(All forms = 115)			
	ST		(All forms = 100)			

(**Note:** See Ref. 9.1 for additional design data)

Fig. 9.11.3 Sustained Tensile Stress Allowables (ksi)

Pull-up stress increases the mean stress and the significance of this increase is illustrated in Fig. 9.11.4. Designer may use smaller values (i.e. 50% of these values) to improve the fatigue life of the structure in regards to pull-up stress.

- Residual stresses usually increase the mean stress rather than the alternating stress
- Although fatigue life primarily depends on alternating stress, a large increase in mean stress also can significantly affect fatigue life

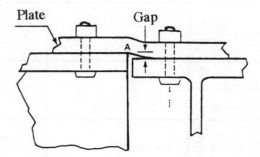

(a) Pull-up stress due to the structural mismatch

Joints and Fittings

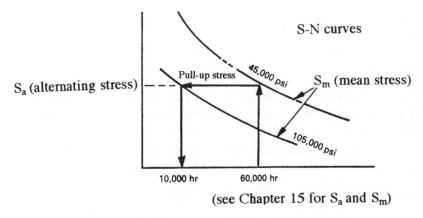

(b) Reduction of fatigue life from 60,000 to 10,000 flight hours.

Fig. 9.11.4 Pull-Up Stress Reduces Fatigue Life

Shimming limits (gap vs. fastener diameter):
- Shims are classified as structural and non-structural shims
- Solid or laminated shims may be used (stock may be used for gaps up to 0.05 in.)
- Gaps greater than 0.05 in. should be filled with combination of solid and laminated shims

Fig. 9.11.5 gives the guidelines of the maximum allowable gap for non-structural shim thickness:

Fastener Type	Ratio of Shim Thickness to Fastener Diameter	Non-structural Shim	Structural Shim
Aluminum Rivets	up to 20%	√	
	21% and greater		√
Hi-Loks (shear)	up to 30%	√	
	31% and greater		√
Hi-Loks (tension)	up to 40%	√	
	41% and greater		√

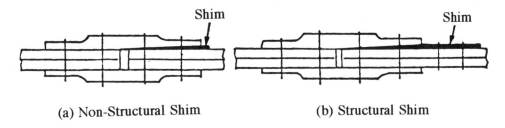

(a) Non-Structural Shim   (b) Structural Shim

Fig. 9.11.5 Shim Thickness vs. Fastener Diameter

9.12 FATIGUE CONSIDERATIONS

To meet today's requirement of long life structural joints, airframe structural splices and joints are primarily designed by fatigue requirements in fatigue critical areas, but must also be capable of carrying ultimate loads without failure. Most fatigue damage ocurs at much lower loading than the limit loading which is within the material elastic or proportional range. Therefore, when the load distribution for a group of fasteners in a splice joint is based on Young's modulus, it is no longer true that each fastener will carry equal load as they would under ultimate loading.

This chapter will introduce the necessary information and a simple method of analysis to assist the structural engineer in designing a successful joint which meets fatigue life requirements.

(A) STRESS CONCENTRATIONS

Stress concentration (also called a stress riser) is the primary factor which affects structural fatigue life and, therefore, good detail design is a major factor in improving fatigue performance. The most typical stress concentrations are caused by:

- Fasteners
- Eccentricity
- Abrupt cross-sectional change
- Loose fastener fit
- Open fastener hole
- Notches
- Sharp edges
- Around corners of rectangular cutouts

The fastener hole is one of the common stress concentration factor (K_t) areas and potential crack initiation areas. Fig. 9.12.1 shows two typical stress concentration factors at fastener holes (Ref. 9.14).

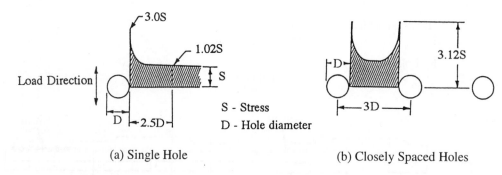

Fig. 9.12.1 Stress Distribution at Fastener Open Hole(s) – Elastic Range

Fastener hole conditions and riveting installation methods also affect the stress concentration factor, as shown in Fig. 9.12.2.

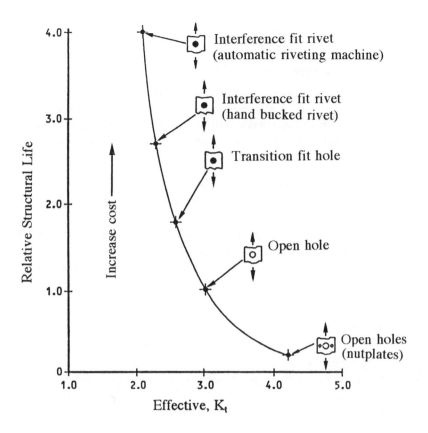

Fig. 9.12.2 Effects of Hole Conditions vs. K_t

(B) FASTENER PATTERN GUIDELINES

Fastener patterns directly affect joint static strength; they have even more effect on fatigue performance. Therefore, the following general design guidelines should be followed during sizing to prevent fatigue problems.

(a) General guidelines:
- Edge distance must be selected carefully to meet static strength and fatigue quality requirements such as minimum 2D
- The minimum spacing is 4D which is net section critical for both tension (stress concentration will increase rapidly if less than 4D) and shear (hole-out) efficiency
- The maximum spacing is approximately 6D – 8D to prevent failure due to inter-rivet compression buckling from occuring (see Chapter 14.0)
- Alloy steel screws less than $D = \frac{1}{4}$ inch diameter should not be used in structural applications because they are easily over-torqued and may yield in tension
- For transport airframe structural joint applications, Hi-Lok fasteners should be $D = \frac{3}{16}$ inch minimum diameter, and aluminum rivets should be $D = \frac{5}{32}$ inch minimum diameter

(b) Single-row pattern:
- Generally, this pattern is used on skin-stringer panels which joins the stringer to skin as an integral panel
- Do not use on the spanwise splice joints of transport wing fuel tank boxes

Chapter 9.0

- For fail-safe design, carefully check the fastener shear strength for the broken part (either the stringer and/or the skin) to ensure that fasteners are strong enough to transfer load onto the adjacent part to complete the load path

(c) Double-row pattern (see Fig. 9.12.3):
- Double-row fastener patterns have the most efficient load transfer
- Generally, if a minimum of 4D pitch is maintained, no tension failure will take place
- Watch that shear-out does not occur if fasteners are spaced too closely together
- Use on transport wing root-joints, empennage root-joints, wing box spanwise splices, etc.

(d) Staggered-row (zigzag) pattern [see Fig. 9.12.3(a)]:
- Staggered patterns are used where special requirements such as pressure tight splice joints or space limitations dictate
- Staggered fastener spacing causes reduction in the effective net area in the zigzag direction of the sheet cross section
- Use on transport wing box spanwise splices which have fuel seal requirements; double-row patterns can also be used for this application but generally not single-row patterns

(e) Triple-row pattern [see Fig. 9.12.3(b)]:
- This pattern is seldom used except for case with special requirements because of assembly cost
- Generally seen on transport fuselage longitudinal lap splice joints (thin skin applications) which carry fuselage circumferential hoop tension load
- This pattern reduces the eccentric effect and has superior fatigue life

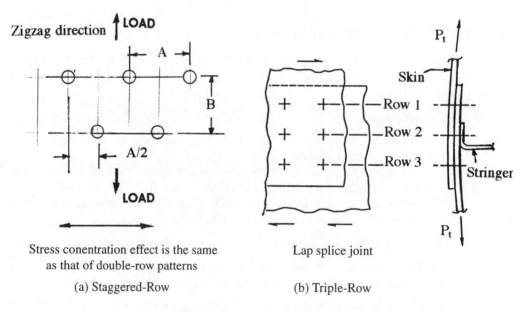

Fig. 9.12.3 Fastener Patterns

Fig. 9.12.4 gives minimum fastener spacing patterns for use in sizing.

Fastener	Pattern	A	B	C	e = 2D + 0.03
$D = \frac{5}{32}$ (0.156")	Single-row	0.63	—	—	0.34
	Staggered-row	1.0	0.39	0.63	0.34
	Double-row	0.63	0.55	—	0.34
$D = \frac{3}{16}$ (0.188")	Single-row	0.75	—	—	0.41
	Staggered-row	1.18	0.47	0.75	0.41
	Double-row	0.75	0.66	—	0.41
$D = \frac{1}{4}$ (0.25")	Single-row	1.0	—	—	0.53
	Staggered-row	1.56	0.63	1.0	0.53
	Double-row	1.0	0.9	—	0.53
$D = \frac{5}{16}$ (0.312")	Single-row	1.25	—	—	0.66
	Staggered-row	1.8	0.78	1.19	0.66
	Double-row	1.25	1.1	—	0.66
$D = \frac{3}{8}$ (0.375")	Single-row	1.5	—	—	0.78
	Staggered-row	2.35	0.94	1.5	0.78
	Double-row	1.5	1.3	—	0.78

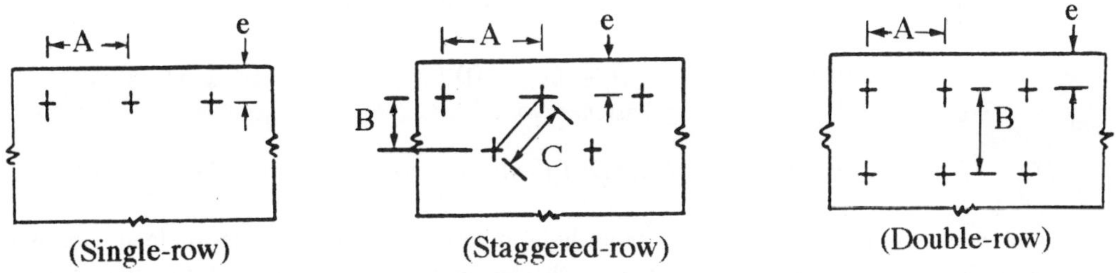

(Single-row)   (Staggered-row)   (Double-row)

Note: Typical minimum fastener spacing and all values are in inches.
Stress requirement may supersede any of the above values.

Fig. 9.12.4 Minimum Fastener Spacing Requirements

(C) FASTENER LOAD DISTRIBUTION

Joint material undergoes plastic deformation and the resultant yielding causes all the fasteners to load up. But tests and theory show that at operating load levels (mostly under 1.0 – 1.5g conditions) it is not equally distributed and the end fasteners will carry most of the load. At operating load levels the material is being stressed within the elastic limits, so load distribution will depend upon relative strains. Variation of load distribution with fastener and plate stiffness (see Fig. 9.12.5) is dependent on:

- $\frac{D}{t}$ ratio
- Fastener flexibility
- Fastener fit
- Plate tapering
- Materials of fastener and plates
- Fastener preload (clamp-up)
- A nearly uniform distribution is obtained from tapering, flexible fasteners, etc.

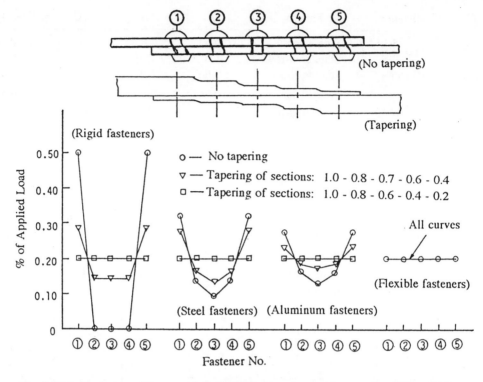

Fig. 9.12.5 Variation of Fastener Load Distribution due to Fastener and Plate Stiffness

In the real world, splice eccentricities, fastener hole imperfections, and fastener preload, etc., will cause some load transfer variation to occur in the fasteners. Fig. 9.12.6 illustrates that if a lap splice plate is used, the highest load will be carried by the end fasteners. To minimize this variation, every fastener installation should be tightly controlled per a specification on the engineering drawing.

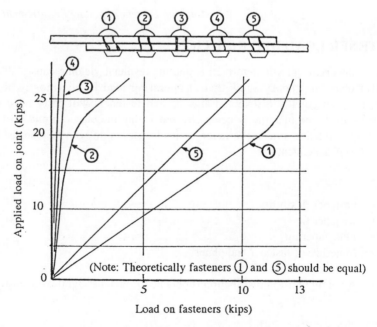

Fig. 9.12.6 Test Results for Fastener Load Distribution on a Lap Joint

The best tool to determine the fastener load is finite element modeling (FEM, see Chapter 5.6). The most important load is that of the end fastener since it takes the highest load in a group of fasteners in a joint splice. An accurate analysis of load distribution is very difficult to obtain because there are so many variables, but FEM will give the best result. An approximation method, which works only for constant splice plate thicknesses, to calculate the end fastener load for preliminary sizing purposes was explained in Chapter 8.7. Fastener load distribution for several types of joint design can be found in Chapter 7.7 of Reference 9.2. A sample of structural finite element modeling for a lap joint with four fasteners is illustrated in Fig. 9.12.7.

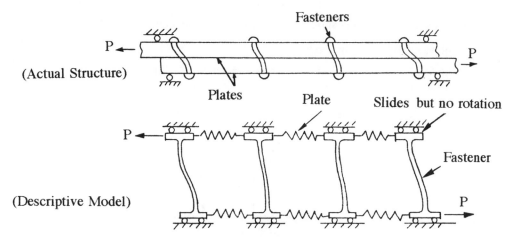

Fig. 9.12.7 Four Fastener Lap Joint

When there are several fasteners in a row (normal to the loading direction), certain fasteners can be simply lumped (added) together and be considered as one large fastener. Since the end fastener is the most highly loaded it is best to leave the end fastener alone, do the least lumping next to the end fastener and do the most lumping in the middle, as shown in Fig. 9.12.8.

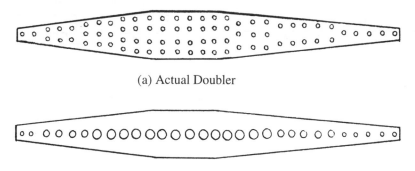

Fig. 9.12.8 Lumping of Fasteners

Another important consideration is the fastener spring constant (k) which is a function of:

- Type and material of the fastener
- Fastener hole condition
- Fastener fit and preload
- Geometry and material of the splice plates

The preliminary estimation value of the fastener spring constant (k) can be obtained from Ref. 9.2, 9.15 and 9.16, but the test results from a simple coupon test (see Appendix B) should be used on aircraft programs.

Determine the equivalent fastener spring constant for finite element modeling:

(a) Equivalent beam (I_e) method:

The fixed-end beam deflection equation is given in Fig. 9.12.9.

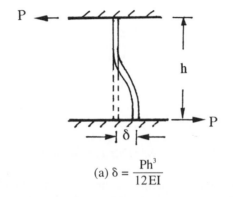

(a) $\delta = \dfrac{Ph^3}{12EI}$

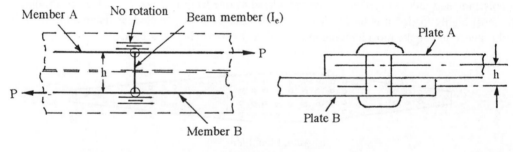

(b) FEM set-up

Fig. 9.12.9 Fixed-End Beam Deflection

Therefore, the equivalent moment of inertia (I_e) per single shear fastener used in FEM:

$$k = \frac{P}{\delta} = \frac{12EI}{h^3}$$

$$I_e = \frac{kh^3}{12E} \qquad \text{Eq. 9.12.1}$$

where: k – Fastener spring constant (lbs./in.)
E – Modulus of the beam material in FEM
h – The length of the beam in FEM (usually use the centerlines of the two adjacent plates

(b) Equivalent axial (A_e) member method:

The fixed-end beam deflection equation is given in Fig. 9.12.10.

Therefore, the equivalent area (A_e) per single shear fastener used in FEM:

$$k = \frac{P}{\delta} = \frac{AE}{L}$$

$$A_e = \frac{kL}{E}$$

Eq. 9.12.2

where: L – The length of the axial member in FEM.

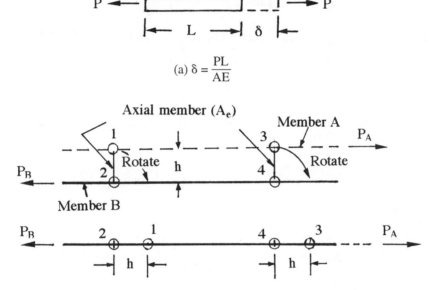

Fig. 9.12.10 Axial Member Elongation

Notes: 1. Member A and member B are overlapped
2. The load transfer are through the tension axial members of 1-2 and 3-4.

(c) Equivalent web thickness (t_e) method:

This method was discussed in Chapter 8.7.

Example:

Use the axial member method to set up FEM of a J-stringer is joined as shown below

(1) Actual structure:

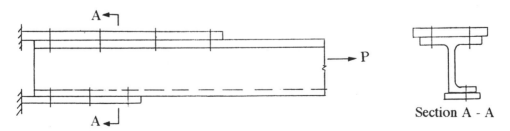

(2) FEM set-up:

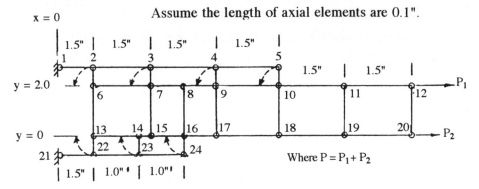

Grid No.	x	y	Grid No.	x	y	Grid No.	x	y
1	0.0	2.0	9	4.6	2.0	17	4.6	0.0
2	1.5	2.0	10	6.1	2.0	18	6.1	0.0
3	3.0	2.0	11	7.5	2.0	19	7.5	0.0
4	4.5	2.0	12	9.0	2.0	20	9.0	0.0
5	6.0	2.0	13	1.6	0.0	21	0.0	0.0
6	1.6	2.0	14	2.6	0.0	22	1.5	0.0
7	3.1	2.0	15	3.1	0.0	23	2.5	0.0
8	3.6	2.0	16	3.6	0.0	24	3.5	0.0

Grid coordinates

(D) FATIGUE ANALYSIS

The fatigue quality index (K) must be determined prior to the start of fatigue analysis to estimate joint fatigue life. The severity factor (SF) is the local peak stress caused by load transfer and bypass load, as shown in Fig. 9.12.11 and given by the following equation, is necessary to determine the fatigue quality index (K):

$$SF = (\frac{\alpha\beta}{\sigma})[\sigma_1 + \sigma_2]$$

$$SF = (\frac{\alpha\beta}{\sigma})[(\frac{K_{th}\Delta P}{Dt})\theta + (\frac{K_{tg}P}{Wt})] \quad \text{Eq. 9.12.3}$$

where: α – Fastener hole condition factor:
   Standard drilled hole   1.0
   Broached or reamed   0.9
   Cold worked hole   0.7 – 0.8
β – Hole filling factor:
   Open hole   1.0
   Steel lock bolt   0.75
   Rivets   0.75
   Threaded bolts   0.75 – 0.9
   Hi-Lok   0.75
   Taper-Lok   0.5
σ – Reference stress (i.e., σ_{ref}) in the structure
σ_1 – Local stress caused by load transfer, ΔP
σ_2 – Local stress caused by bypass load, P
P – Bypass load
ΔP – Load transfer through the fastener
D – Fastener diameter
t – Splice plate thickness

P   – Bypass load
W   – Width of the splice plate
K_{tb} – Bearing stress concentration factor, see Fig. 9.12.12
θ   – Bearing distribution factor, see Fig. 9.12.13
K_{tg} – Stress concentration factor, see Fig. 9.12.14

The SF may not yield the same results as the fatigue quality index (K) and the factor for the range of discrepancy is given in Fig. 9.12.15. This factor can be determined by testing coupons or components from selected critical areas in the airframe.

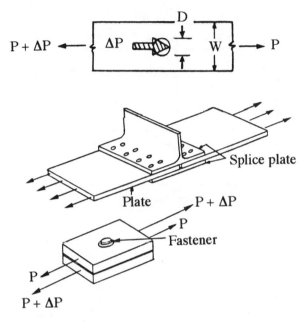

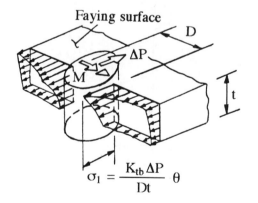

(a) Loads in a Fastener

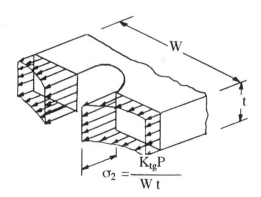

(b) Local Stress Caused by Load Transfer, ΔP   (c) Local Stress Caused by Bypass Load, P

Fig. 9.12.11 Stresses at a Loaded Fastener Hole

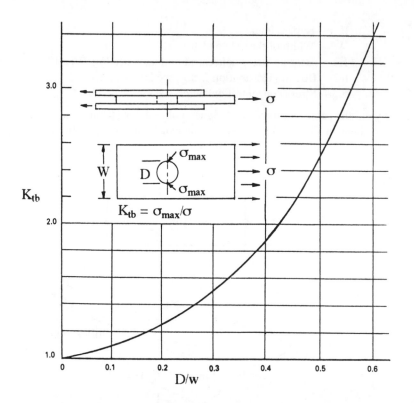

Fig. 9.12.12 K_{tb} – Bearing Stress Concentration Factor

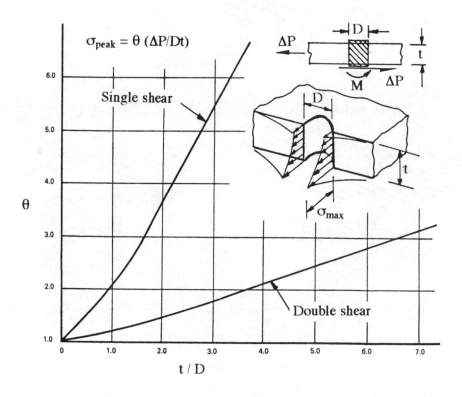

Fig. 9.12.13 θ – Bearing Distribution Factor

Joints and Fittings

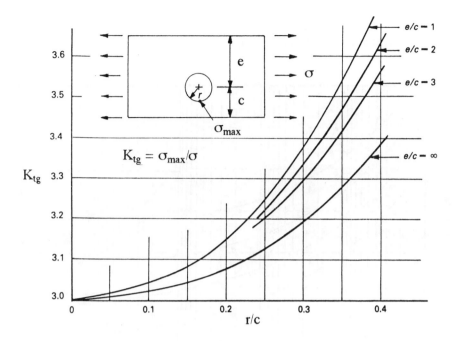

Fig. 9.12.14 K_{tg} – Stress Concentration Factor

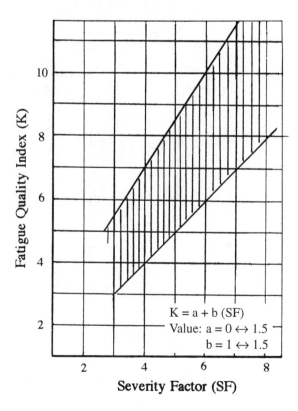

Fig. 9.12.15 Discrepancy Factor (Severity Factor – SF vs. Fatigue Quality Index – K)

Chapter 9.0

After determining the stress concentration factor value of K_t select the appropriate S-N curve (e.g., see Fig. 9.12.16) to use in calculating joint fatigue life.

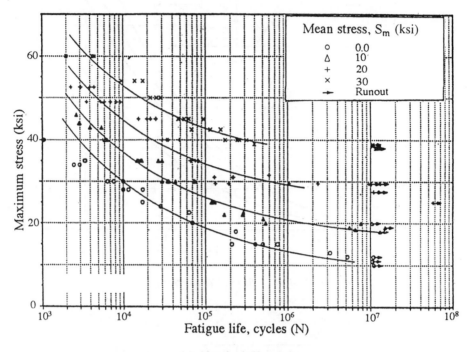

(a) Notched, $K_t = 2.0$

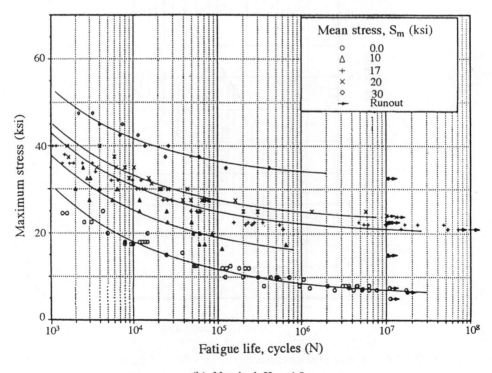

(b) Notched, $K_t = 4.0$

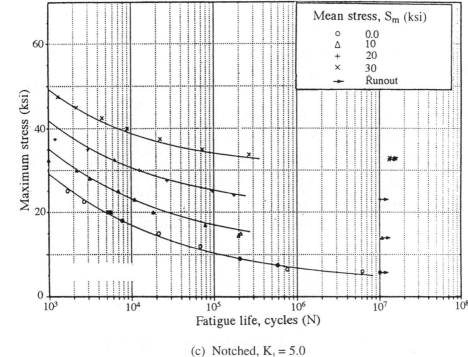

(c) Notched, $K_t = 5.0$

Fig. 9.12.16 S-N Curves for 2024-T3 Sheets (Ref. 9.1)

Example:

An aluminum doubler (t = 0.05 in. of 2024-T3) is fastened onto a fuselage skin (t = 0.05 in. of 2024-T3) which is subjected a hoop tension stress (reference stress) of $\sigma = 15,000$ psi as given below; determine the fatigue life (or cabin pressurization cycles) for both without and with the tapered end on the doubler.

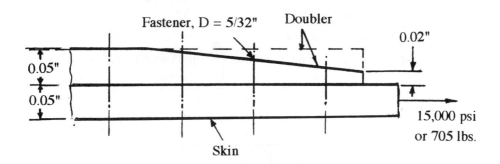

Fig. A Skin with a Tapered Doubler

Skin width: W = 0.94 in.

Fastener diameter: $D = \frac{5}{32}$ in. (Rivets)

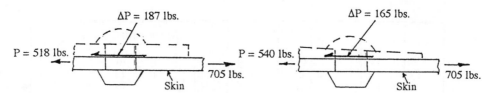

Fig. B Doubler without Tapered End   Fig. C Doubler with Tapered End

(1) Doubler without tapered end (see Fig. B)

    σ = 15,000 psi (Reference stress of the basic fuselage skin panel)

    ΔP = 187 lbs. (Fastener load)

    P = 15,000(W × t) − ΔP = 15,000(0.94 × 0.05) − 187 = 518 lbs.

    (Bypass load on splice plate)

Use α = 1.0 (Standard drilled hole); β = 0.75 (Rivets)

Use $\dfrac{D}{W} = \dfrac{0.156}{0.94} = 0.17$ and from Fig. 9.12.12 obtain K_{tb} = 1.2

Use $\dfrac{t}{D} = \dfrac{0.05}{0.156} = 0.32$ and from Fig. 9.12.13 obtain θ = 1.4 (single shear)

Use $\dfrac{r}{c} = \dfrac{\frac{0.156}{2}}{\frac{0.94}{2}} = 0.17$ and $\dfrac{e}{c} = 1$ and from Fig. 9.12.14 obtain K_{tg} = 3.1

From Eq. 9.12.3:

$$SF = \left(\dfrac{\alpha\beta}{\sigma}\right)\left[\left(\dfrac{K_{tb}\Delta P}{Dt}\right)\theta + \left(\dfrac{K_{tg}P}{Wt}\right)\right]$$

$$= \left(\dfrac{1.0 \times 0.75}{15,000}\right)\left[\left(\dfrac{1.2 \times 187}{0.156 \times 0.05}\right) \times 1.4 + \left(\dfrac{3.1 \times 518}{0.94 \times 0.05}\right)\right]$$

$$= (5 \times 10^{-5})(40,277 + 34,166) = 3.72$$

Assume discrepancy factor = 1.2 between SF and the fatigue quality index (K):

    K = 1.2 × 3.72 ≈ 4.5

For this case, assume K = K_t = 4.5

The stresses of S_{max} (maximum stress) and S_m (mean stress) are as follows (see Fig. D):

$S_{max} = \sigma$ = 15,000 psi and $S_m = \dfrac{S_{max} + S_{min}}{2} = \dfrac{15,000}{2} = 7,500$ psi

where: S_{min} = 0

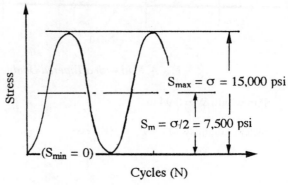

Fig. D The S_{max} and S_m vs. Cycles (N)

From Fig. 9.12.16(b) with $K_t = 4.0$ obtain $N_4 = 3 \times 10^5$ cycles

From Fig. 9.12.16(c) with $K_t = 5.0$ obtain $N_5 = 10^5$ cycles.

$$N_{4.5} = N_4 - (N_4 - N_5)(\frac{N_{4.5}}{K_{t.5} - K_{t.4}})$$

$$= 3 \times 10^5 - (3 \times 10^5 - 10^5)(\frac{4.5}{5 - 4})$$

$$= 1.5 \times 10^5 \text{ cycles}$$

Use fatigue scatter factor = 3.0 and the estimated fatigue life for doubler without tapered end:

$$N = \frac{1.5 \times 10^5}{3.0} = 50{,}000 \text{ cycles.}$$

(2) Doubler with tapered end (see Fig. C)

$\sigma = 15{,}000$ psi (Reference stress of the skin panel)

$\Delta P = 165$ lbs. (Fastener load)

$P = 15{,}000 (W \times t) - \Delta P = 15{,}000 (0.94 \times 0.05) - 165 = 540$ lbs.

Use the same values of α, β, K_{th}, θ, and K_{tg} as shown in calculation (1) and from Eq. 9.12.3:

$$SF = (\frac{\alpha\beta}{\sigma})[(\frac{K_{th}\Delta P}{Dt})\theta + (\frac{K_{tg}P}{Wt})]$$

$$= (\frac{1.0 \times 0.75}{15{,}000})[(\frac{1.2 \times 165}{0.156 \times 0.05}) \times 1.4 + (\frac{3.1 \times 540}{0.94 \times 0.05})]$$

$$= (5 \times 10^{-5})(33{,}538 + 35{,}617) = 3.6$$

Use discrepancy factor = 1.2 and the fatigue quality index: $K = 1.2 \times 3.6 \approx 4.3 = K_t$

Use the same stresses of S_{max} and S_m from calculation (1):

From Fig. 9.12.16(b) with $K_t = 4.0$ obtain $N_4 = 3 \times 10^5$ cycles.

From Fig. 9.12.16(c) with $K_t = 5.0$ obtain $N_5 = 10^5$ cycles.

$$N_{4.3} = N_4 - (N_4 - N_5)(\frac{N_{4.3}}{K_{t.5} - K_{t.4}})$$

$$= 3 \times 10^5 - (3 \times 10^5 - 10^5)(\frac{4.3}{5 - 4})$$

$$= 2.1 \times 10^5 \text{ cycles}$$

Use fatigue scatter factor = 3.0 and the estimated fatigue life for doubler with tapered end:

$$N' = \frac{2.1 \times 10^5}{3.0} = 70{,}000 \text{ cycles.}$$

(3) Fatigue life comparison between with and without tapered end:

$$\frac{N'}{N} = \frac{70{,}000}{50{,}000} = 1.4 \text{ or } 40\% \text{ improvement by using tapered end.}$$

(E) IMPROVEMENT OF FATIGUE LIFE

Good detail design is the most important means to decrease the stress concentration factors which will significantly increase the fatigue life of the joint. The trade-off between increasing fatigue life and cost depends on how critical an area is for fatigue. For example, the wing root-joint area requires careful detail design and the conducting of a series of tests from coupon up to full scale. The following methods can improve the fatigue life of a joint:

(a) Reduce stress concentration:

This can be controlled by detail design. Also one of the most effective methods is to design to a lower stress level to improve fatigue life. Ref. 9.18 is one of the best handbooks for techniques to reduce stress concentration.

(b) Interference-fit fastener hole conditions:

A careful trade-off must be made in the selection of proper interference-fit fastener hole conditions, as shown in Fig. 9.12.17, to increase the fatigue life but it also increases cost. The degree of interference is restricted by splice material stress corrosion (especially in the ST grain direction), as shown in Fig. 9.11.3, and the following examples are close tolerance fit fastener hole sizes for 7075-T76 materials:

- Interference fit between $-0.0035"$ and $+0.0000"$: used at fatigue critical areas
- Interference and clearance fit is between $-0.0015"$ and $+0.0025"$: used where fatigue is a major consideration but where installation problems preclude the use of an interference-fit
- Clearance fit is between $-0.0000"$ and $+0.0040"$: used where stress corrosion may be encountered or when the fastener grip exceeds 4 times the fastener diameter

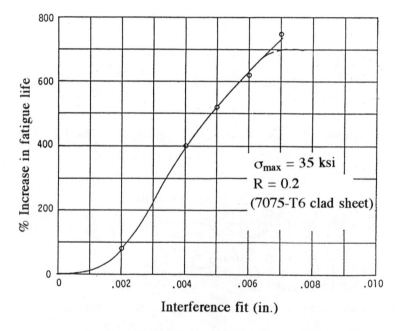

Fig. 9.12.17 *Interference-Fit Fastener vs. Fatigue Life*

(c) Reduce end-fastener load:

If the ends of straps, splices and doublers are tapered, the end fastener load transfer will be lessened, and as a result, joint fatigue life is greatly increased or improved:

- The ends should be tapered at a slope of 20:1 to a minimum of 0.04 inch or $\frac{1}{3}$ the thickness of the plate being spliced, as shown in Fig. 9.12.18
- The tapered ends can be accomplished by machined steps, machine tapering, or by stacking doublers
- The taper need not extend over more than four fastener locations
- All fasteners in the tapered portion should be capable of carrying static strength

Joints and Fittings

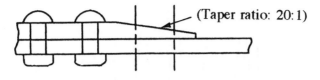

(a) Machined taper

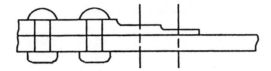

(b) Machined steps

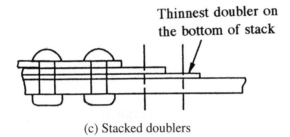

(c) Stacked doublers

Fig. 9.12.18 Tapered Ends

(d) Cold work fastener hole:

The use of interference fit fasteners alone has been proven to increase fatigue life significantly but additional increase can be obtained by the cold work method (It is sometimes argued that the combination cannot be so, see Ref. 9.19) shown in Fig. 9.12.19. These processes reduce both alternative stress (S_a) and the maximum stress (S_{max}), which improves fatigue life, as shown in Fig. 9.12.20. The selection of interference fit and/or cold work process depends upon the areas of criticality as this is a very tight control process and very costly.

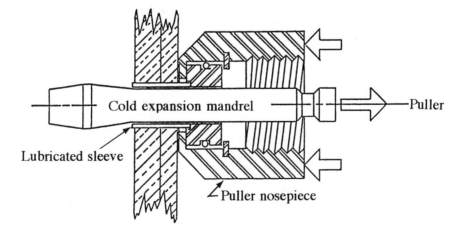

Fig. 9.12.19 Split Sleeve Cold-Expansion Process

391

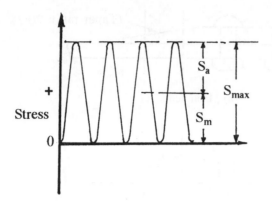

(a) Applied stress

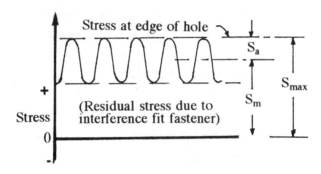

(b) Interference-fit

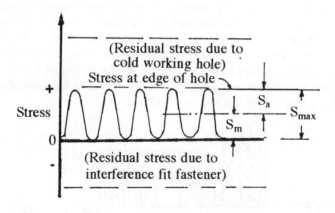

(c) Interference-fit and cold work

(**Note:** Increase fatigue life by reducing S_{max} and S_a.)

Fig. 9.12.20 Fatigue Effect of Fastener Interference Fit and Cold Work

(e) Fastener preload (clamp-up) effect on fatigue:

Preloading fasteners will greatly increase fatigue life. Tension bolts or Hi-Loks must be used to obtain the required preload on fasteners.

References

9.1 MIL-HDBK-5, "Metallic Materials and Elements for Flight Vehicle Structures", U.S. Government Printing Office, Washington, D.C.
9.2 Niu, C. Y., "AIRFRAME STRUCTURAL DESIGN", Conmilit Press Ltd., P.O. Box 23250, Wanchai Post Office, Hong Kong, (1988).
9.3 Smith, C. R., "Effective Stress Concentrations for Fillets in Landed Structures", Experimental Mechanics, (April, 1971).
9.4 Hart-Smith, L. J., "Adhesive Bonded Double Lap Joints", NASA CR 112235, (1973).
9.5 Hart-Smith, L. J., "Adhesive Bonded Single Lap Joints", NASA CR 112236, (1973).
9.6 Hart-Smith, L. J., "Adhesive Bonded Scarf and Step Joints", NASA CR 112237, (1973).
9.7 Melcon, M. A. and Hoblit, F. M., "Dvelopments in the Analysis of Lugs and Shear Pins", Product Engineering, (June, 1953).
9.8 Ekvall, J. C., "Static Strength Analysis of Pin-Loaded Lugs", J. Aircraft, Vol. 23, No. 5, (1986).
9.9 Anon., "Engineering Stress Memo Manual - No. 88", Lockheed Aircraft Corp.
9.10 Blake, Alexander, "Flanges That Won't Fail", Machine Design, (Dec. 26, 1974).
9.11 Blake, Alexander, "Structural Pin Design", Design News, (Jan. 21, 1974).
9.12 Anon., "SPS Threaded Fasteners – Section I, Reference Gudie to Bolts ans Screws", Standard Pressed Steel Co., Santa Ana, CA. 92702.
9.13 Anon., "SPS Threaded Fasteners – Section II, Reference Guide to Self-Locking Nuts", Standard Pressed Steel Co., Santa Ana, CA. 92702.
9.14 Peterson, R. E., "STRESS CONCENTRATION DESIGN FACTORS", John Wiley and Sons, (1953).
9.15 Tate, M. B. and Rosenfeld, S. J., "Preliminary Investigation of the Loads Carried by Individual Bolts in Bolted Joints", NACA TN 1051, (May, 1946).
9.16 Swift, T., "Repairs to Damage Tolerant Aircraft", Presented to International Symposium on Structural Integrity of Aging Airplanes. Atanta, Georgia. (March 20-22, 1990).
9.17 Jarfall, L. E., "Optimum Design of Joints: The Stress Severity Factor Concept", The Aeronautical Research Institute of Sweden, (1967).
9.18 Anon., "Fatigue and Stress Corrosion Manual for Designers", Lockheed-California Co. (April, 1976).
9.19 Broek, David, "THE PRACTICAL USE OF FRACTURE MECHANICS", Kluwer Academic Publishers, (1988).

Chapter 10.0

COLUMN BUCKLING

10.1 INTRODUCTION

A column is a structural member subjected to a uniaxial compressive stress. Its normal failure mode is some form of instability.

(a) Primary column failure is defined as any type of failure in which the cross section is
   - Translated
   - Rotated
   - Translated and rotated in its own plane.

(b) Secondary failure involves local distortion in the plane of the cross section of the column (crippling).

A perfectly straight, untwisted, and centrally loaded column will theoretically support an increasing load up to the critical load, P_{cr}, without translating or rotating. However, when P_{cr} is attained, the column experiences large deflections immediately with no corresponding increase in load as shown in Fig. 10.1.1 below.

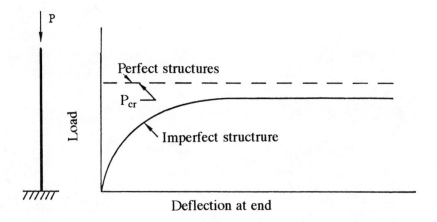

Fig. 10.1.1 Perfect vs. Imperfect Sructures of an Column

Actually the geometrically perfect, centrally loaded column never occurs in practice. As a result, actual columns are not capable of sustaining the P_{cr} associated with a perfect column. The difference between perfect and imperfect compression members must be recognized and the structural limitations imposed by these imperfections understood. Eccentricially loaded columns must be capable of carrying the compression load and the moment induced by the eccentricity and these members must analyzed as beam columns, as will be discussed later.

A column may fail in one of the following modes and therefore the strength of a column is the lowest strength associated with any of the four failure modes:

(a) Flexural instability – This mode is concerned with general buckling of the column and is dependent on the end fixity, cross section, and material.

(b) Crippling stress – Crippling (discussed in Section 11.0), or local buckling of the cross sectional shape, occurs mainly in extrusions, formed sheetmetal shapes, and thin walled tubes. It is dependent upon the material and cross section geometry and dimensions.

(c) Interaction between local crippling and flexural instability – When the critical buckling stress reaches a value of one half local crippling stress, some interaction between flexural instability and crippling may occur.

(d) Torsional instability – Torsional failure (Ref. 10.1 and 10.6) is relatively rare in columns, but it is to be expected in the following sections:
- I – sections of short length and very wide flanges or webs
- T – sections of short and intermediate length
- L – sections with equal and unequal legs of all lengths
- In general, torsional buckling is critical in sections having wide flanges and short column lengths

The theoretical buckling load for various types of columns frequently encountered in airframe practical design and sizing can be obtained from the formulas, curves, and tabular forms given in this chapter. Furthermore, the methods given apply only to perfectly straight columns without side load, whereas in most practical cases eccentricities due to manufacturing tolerances introduce such severe bending moments as to reduce the strength of the member. In such cases the member would have to be treated as a beam-column.

10.2 EULER EQUATION (LONG COLUMN)

Column failures by lateral translation are well known. Below the proportional limit of the material, the critical buckling load, P_E, is given by the Euler formula and still remains the basis of all buckling phenomena including the buckling of thin sheets (Chapter 11). In terms of load, the Euler equation for the buckling load of a simple pin-ended column can be written as follows:

$$P_E = \frac{\pi^2 EI}{L^2} \qquad \text{Eq. 10.2.1}$$

where: P_E – Euler buckling load.
E – Modulus of elasticity (in elastic range).
I – Smallest moment of inertia for the column cross section.
L – Length of the column.

(A) EFFECTIVE COLUMN LENGTH

By inspecting this equation it is readily seen that it actually describes the bending stiffness of the column. The quantities affecting the bending stiffness of a column members are:
- Material modulus of elasticity
- Moment of inertia
- Length of the column

Eq. 10.2.1 can be rewritten for any other end-fixity by:

$$P_E = \frac{\pi^2 EI}{(L')^2} \qquad \text{Eq. 10.2.2}$$

where: c – Column end-fixity, c. (values of 'c' for various column end-fixity and loading conditions are shown in Fig. 10.2.1 and Fig. 10.2.2 and Fig. 10.2.3)

$L' = \dfrac{L}{\sqrt{c}}$ – Effective column length, use smallest moment of inertia for the column cross section

Rewrite the Euler equation in terms of stress by:
- Dividing the Euler load, P_E by the column area (A)
- Introduce the slenderness ratio term ($\dfrac{L'}{\rho}$)

$$F_{cr} = \dfrac{\pi^2 E}{(\dfrac{L'}{\rho})^2}$$ Eq. 10.2.3

where: $\rho = \sqrt{\dfrac{I}{A}}$ – Least radius of gyration of column cross-section

When the critical stress in a column is above the proportional limit, it is necessary to substitute the tangent modulus of elasticity (E_t) for the modulus of elasticity (E). The resultant formula is known as the Euler-Engesser equation:

$$F_{cr} = \dfrac{\pi^2 E_t}{(\dfrac{L'}{\rho})^2}$$ Eq. 10.2.4

where: E_t – Tangent modulus of elasticity of the column material

In practice, pure column cases in airframe structures seldom occur; the beam-column is of most importance and interest to structural engineers when sizing airframe parts. However, engineers must fully understand the Euler equation because most beam-column design equations are functions or magnitudes of the Euler equation

Column shape and end fixity	End fixity coefficient	Column shape and end fixity	End fixity coefficient
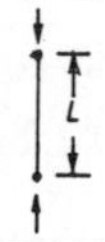 Uniform column, axially loaded, pinned ends	$c = 1$ $\dfrac{1}{\sqrt{c}} = 1$	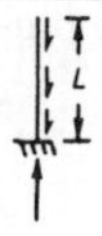 Uniform column, distributed axis load, one end fixed, one end free	$c = 0.794$ $\dfrac{1}{\sqrt{c}} = 1.12$
 Uniform column, axially loaded, fixed ends	$c = 4$ $\dfrac{1}{\sqrt{c}} = 0.5$	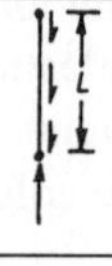 Uniform column, distributed axis load, pinned ends	$c = 1.87$ $\dfrac{1}{\sqrt{c}} = 0.732$
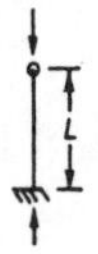 Uniform column, axially loaded, one end fixed, one pinned end	$c = 2.05$ $\dfrac{1}{\sqrt{c}} = 0.7$	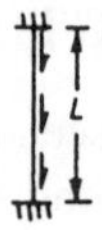 Uniform column, distributed axis load, fixed ends	$c = 7.5$ (Approx.) $\dfrac{1}{\sqrt{c}} = 0.365$
 Uniform column, axially loaded, one end fixed, one end free	$c = 0.25$ $\dfrac{1}{\sqrt{c}} = 2$	 Uniform column, distributed axis load, one end fixed, one end pinned	$c = 6.08$ $\dfrac{1}{\sqrt{c}} = 0.406$

$$P_{cr} = \dfrac{c \pi^2 EI}{L^2}$$

Fig. 10.2.1 Column End-Fixity Coefficients

Column Buckling

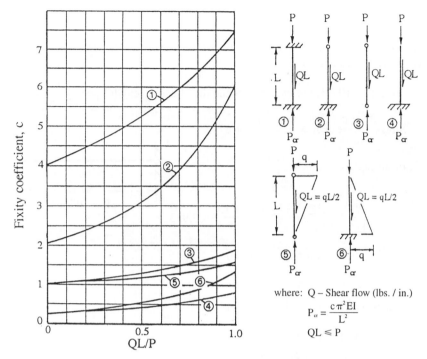

Fig. 10.2.2 Column End-Fixity Coefficients – Axial and Distributed Shear Load

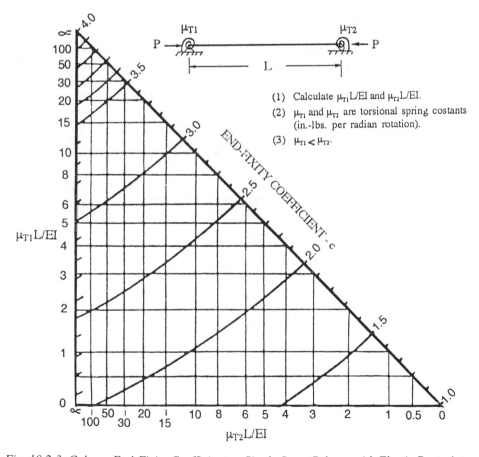

Fig. 10.2.3 Column End-Fixity Coefficients – Single Span Column with Elastic Restraints

Example:

Given an I-section column ($E = 10.7 \times 10^3$ ksi) as shown below. Size this column so that the compressive stress does not exceed 36 ksi.

From Fig. 10.2.1, obtain c = 2.05 and the effective length $L' = \dfrac{50}{\sqrt{2.05}} = 34.9$ in.

Stress, $f = \dfrac{P}{A}$ or $A = \dfrac{P}{f} = \dfrac{60}{36} = 1.67$ in.^{2}

From Eq. 10.2.3, $F_{cr} = \dfrac{\pi^2 E}{(\dfrac{L'}{\rho})^2}$

$\therefore (\dfrac{L'}{\rho})^2 = \dfrac{\pi^2 E}{F_{cr}} = \dfrac{\pi^2 \times 10.7 \times 10^3}{36} = 2{,}933$

$\dfrac{L'}{\rho} = 54.2$ and $\rho = \dfrac{34.9}{54.2} = 0.64$

(P = 60 kips), 50 in.

Therefore, the required moment of inertia:

$I = A\rho^2 = 1.67(0.64)^2 = 0.68$ in.^{4}

(B) COLUMN WITH ELASTIC SUPPORTS

Continuous compression members on airframe structures require many inermediate supports to shorten the column length, and these supports are actually elastic supports (e.g., wing or fuselage stringers supported by wing ribs or fuselage frames). Here are a few cases which will help engineers to size columns.

(a) Column with elastic supports (Fig. 10.2.4):

Case	Loading	Critical load
1	(span a, a with spring ω)	$P_{cr} = (\dfrac{\pi^2 EI}{4a^2}) + 0.375\overline{\omega}a$
2	(span a, a with spring ω)	$P_{cr} = (\dfrac{\pi^2 EI}{a^2}) + 0.4\overline{\omega}a$
3	More than two spans (a, na, a)	$P_{cr} = (\dfrac{\pi}{2})\sqrt{\dfrac{\overline{\omega}EI}{a}}$
4	Uniform elastic support	$P_{cr} = 2\sqrt{\dfrac{\overline{\omega}EI}{a}}$

Column Buckling

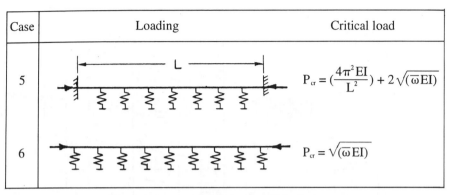

Case	Loading	Critical load
5		$P_{cr} = (\dfrac{4\pi^2 EI}{L^2}) + 2\sqrt{(\overline{\omega}EI)}$
6		$P_{cr} = \sqrt{(\overline{\omega}EI)}$

($\overline{\omega}$ – Spring constant (lbs./in.) and use E_t in plastic range)

Fig. 10.2.4 Column with Elastic Supports

(b) A column on elastic supports with unequal spacing (Fig. 10.2.5 and Fig. 10.2.6):

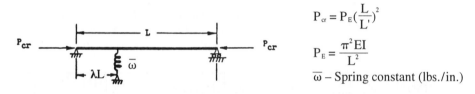

$P_{cr} = P_E (\dfrac{L}{L'})^2$

$P_E = \dfrac{\pi^2 EI}{L^2}$

$\overline{\omega}$ – Spring constant (lbs./in.)

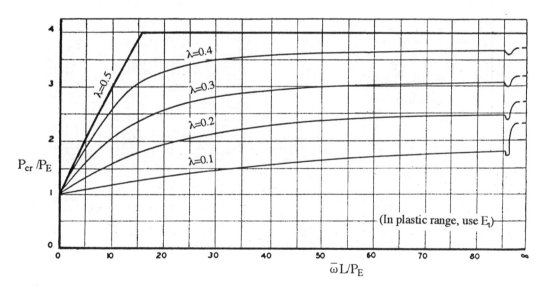

Fig. 10.2.5 Column on Elastic Supports with 2 Unequal Spacings

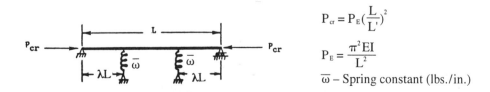

$P_{cr} = P_E (\dfrac{L}{L'})^2$

$P_E = \dfrac{\pi^2 EI}{L^2}$

$\overline{\omega}$ – Spring constant (lbs./in.)

Fig. 10.2.6 Column on Elastic Supports with 3 Unequal Spacings

Chapter 10.0

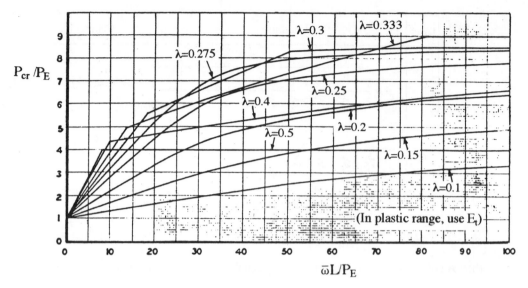

Fig. 10.2.6 Column on Elastic Supports with 3 Unequal Spacings (cont'd)

(c) A column on elastic supports with rotational restraints (Fig. 10.2.7):

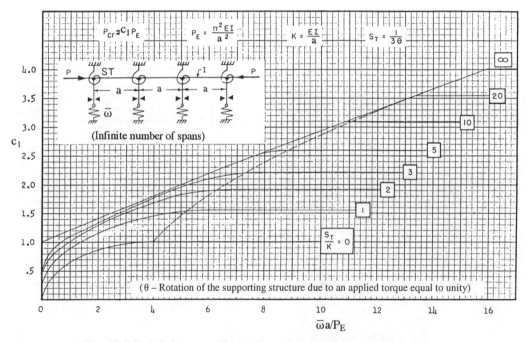

Fig. 10.2.7 A Column on Elastic Supports with Rotational Restraints

Some engineers design the continuous stringers in wings and fuselages for an allowable load:

$$P_{cr} = \frac{2\pi^2 EI}{a^2}$$ Eq. 10.2.5

According to Fig. 10.2.7, a critical load of this magnitude requires a minimum rotational stiffness of the supporting structure:

$S_T = 2.2K$ (S_T – Torsional constant, in.-lbs./rad)

10.3 STEPPED COLUMNS

Fig. 10.3.1 through 10.3.5 present column allowables for symmetrical and unsymmetrical stepped columns as a function of the computed Euler column load, P_E. If the stress in either portion of the column exceeds the proportional limit, the tangent modulus, E_t, should be used for E.

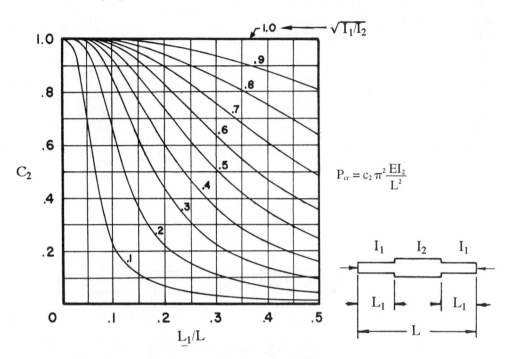

Fig. 10.3.1 Critical Loads for Symmetrical Stepped Columns ($E_1 = E_2$)

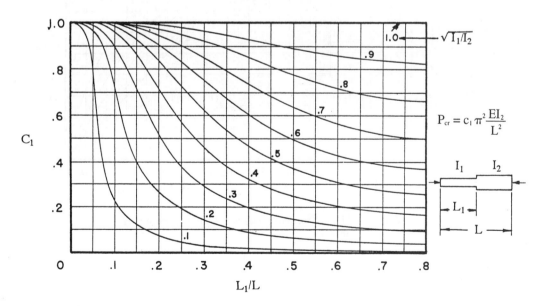

Fig. 10.3.2 Critical Loads for Unsymmetrical Stepped Columns ($E_1 = E_2$)

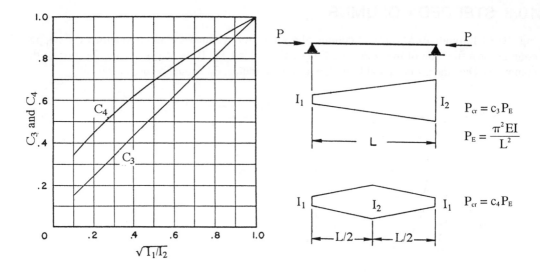

Fig. 10.3.3 Critical Loads for Tapered Columns ($E_1 = E_2$)

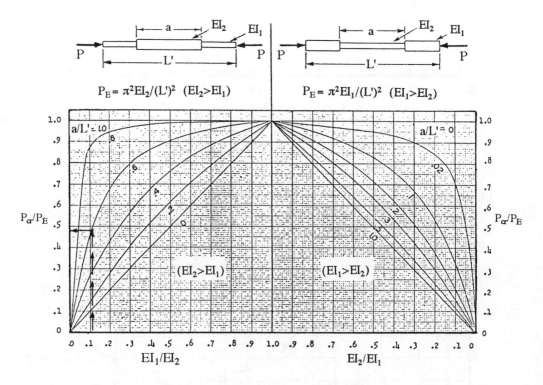

Fig. 10.3.4 Critical Loads for Symmetrical Stepped Columns ($E_1 \neq E_2$)

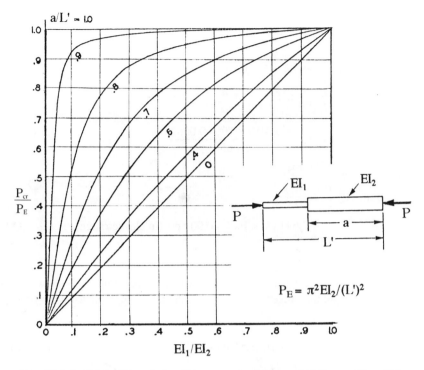

Fig. 10.3.5 Critical Loads for Unsymmetrical Stepped Columns ($E_1 \neq E_2$)

If both end parts of the stepped columns are not symmetrical, the curve shown in Fig. 10.3.4 may still be used to find the critical load. An estimate must be made for the point of maximum deflection of the column. Based on this assumption, a critical load can be calculated for each end of the column by assuming symmetry about the point of maximum deflection. An iteration process is necessary for estimating the point of maximum deflection until the two computed critical loads agree.

Example 1:

Determine the critical column load (P_{cr}) for a symmetrical stepped column of a control rod as shown. (This is a long column and the tangent modulus (E_t) should not be applied. An iteration process is not necessary).

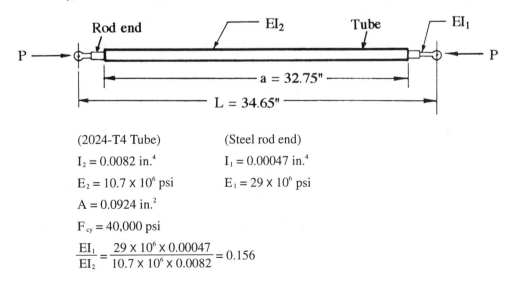

(2024-T4 Tube) (Steel rod end)

$I_2 = 0.0082$ in.^{4} $I_1 = 0.00047$ in.^{4}

$E_2 = 10.7 \times 10^6$ psi $E_1 = 29 \times 10^6$ psi

$A = 0.0924$ in.^{2}

$F_{cy} = 40,000$ psi

$$\frac{EI_1}{EI_2} = \frac{29 \times 10^6 \times 0.00047}{10.7 \times 10^6 \times 0.0082} = 0.156$$

$$\frac{a}{L} = \frac{32.75}{34.65} = 0.945 \text{ (use } c = 1.0 \text{ and } L' = L)$$

From Fig. 10.3.4, obtain $\frac{P_{cr}}{P_E} = 0.98$

Therefore, $P_{cr} = 0.98\left(\frac{\pi^2 EI_2}{L^2}\right)$

$$= 0.98\left(\frac{\pi^2 \times 10.7 \times 10^6 \times 0.0082}{34.65^2}\right)$$

$$= 706 \text{ lbs.}$$

Check stress: $\frac{P}{A} = \frac{706}{0.0924}$

$$= 7,641 \text{ psi} < F_{cy} = 40,000 \text{ psi (in elastic range)} \qquad \text{O.K.}$$

Example 2:

Determine the critical column load (P_{cr}) for a symmetrical stepped column as shown below:
(This is a short column and the tangent modulus (E_t) should be applied; An iteration process must be used).

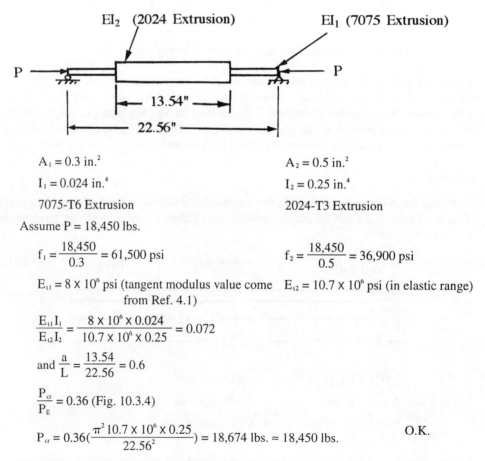

$A_1 = 0.3$ in.²	$A_2 = 0.5$ in.²
$I_1 = 0.024$ in.⁴	$I_2 = 0.25$ in.⁴
7075-T6 Extrusion	2024-T3 Extrusion

Assume P = 18,450 lbs.

$f_1 = \dfrac{18,450}{0.3} = 61,500$ psi $\qquad f_2 = \dfrac{18,450}{0.5} = 36,900$ psi

$E_{t1} = 8 \times 10^6$ psi (tangent modulus value come $E_{t2} = 10.7 \times 10^6$ psi (in elastic range)
from Ref. 4.1)

$$\frac{E_{t1} I_1}{E_{t2} I_2} = \frac{8 \times 10^6 \times 0.024}{10.7 \times 10^6 \times 0.25} = 0.072$$

and $\dfrac{a}{L} = \dfrac{13.54}{22.56} = 0.6$

$\dfrac{P_{cr}}{P_E} = 0.36$ (Fig. 10.3.4)

$P_{cr} = 0.36\left(\dfrac{\pi^2 \, 10.7 \times 10^6 \times 0.25}{22.56^2}\right) = 18,674$ lbs. ≈ 18,450 lbs. \qquad O.K.

The above calculated $P_{cr} = 18,674$ lbs. is very close to that of assumed value of P = 18,450 lbs. and therefore it is not necessry go to a second iteration.

Column Buckling

Example 3:

Determine the critical column load (P_{cr}) for an unsymmetrical stepped column as shown below and, for simplification of this calculation, use the following assumed $E_t I$ values.

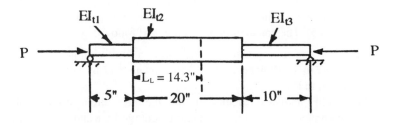

(Assume $E_{t1} I_1 = 70,000$ psi; $E_{t2} I_2 = 100,000$ psi; $E_{t3} I_3 = 40,000$ psi)

First trial, assuming $L_L = 14.3$ in. is the point of maximum deflection location and use end-fixity, c = 1 and, therefore, the effective column length L' = L:

Consider the left side:

$$\frac{a}{2} = L_L = 14.3 \text{ and } a = 2 \times L_L = 28.6$$

Fictitious entire column length: $L = 2 \times 5 + 28.6 = 38.6$

$$\frac{E_{t1} I_1}{E_{t2} I_2} = \frac{70,000}{100,000} = 0.7$$

$$\frac{a}{L} = \frac{28.6}{38.6} = 0.741 \text{ (use c = 1.0 and L' = L)}$$

$$\frac{P_{cr}}{P_E} = 0.977 \text{ (Fig. 10.3.4)}$$

$$P_{cr} = 0.977 (\frac{9.87 \times 10^5}{38.6^2}) = 647 \text{ lbs.}$$

Consider the right side:

$$\frac{a}{2} = L_R = (20 - L_L) = (20 - 14.3) \text{ and } a = 2(20 - 14.3) = 11.4$$

Fictitious entire column length: $L = 2 \times 5 + 11.4 = 31.4$

$$\frac{E_{t3} I_3}{E_{t2} I_2} = \frac{40,000}{100,000} = 0.4$$

$$\frac{a}{L} = \frac{11.4}{31.4} = 0.361$$

$$\frac{P_{cr}}{P_E} = 0.647 \text{ (From Fig. 10.3.4)}$$

$$P_{cr} = 0.647 (\frac{9.87 \times 10^5}{31.4^2}) = 648 \text{ lbs.} \quad \text{O.K.}$$

In general, the first trial will not work out and an iteration process must be continued until agreement is reached. Furthermore, the actual material tangent modulus (E_t) should be used following the same procedures as that of Example 2 instead of using assumed values.

Chapter 10.0

10.4 TAPERED COLUMN WITH VARIABLE CROSS-SECTION

Frequently compression columns of variable cross-sections occur in aircraft structures and may not be amenable to exact mathematical formulation, yet in the interest of efficient, economical design a comparatively accurate analysis is desired. The numerical procedure presented in this chapter is a calculation of the critical buckling load of a tapered column by a simplified method of successive approximation (Ref. 10.5). The calculation procedure is as follows:

- The first step is to generalize the solution by dividing the column into 4 (lower accuracy) to 10 (higher accuracy) increments and by assigning the load at the left end equal to P
- To simplify and save calculation time, the left column of the tabular form contains the "common terms" which are factored out until the end
- The moment of inertia (I) at each interval is entered as a function of I_o (a proportional ratio of the I_o) which is usually the smaller one of the column
- Assume a buckled shape, y_a deflection values along the column length
- Using the moment caused by the assumed y_a, calculate a new set of deflection y_{cal}
- Use the parabolic curve formula, shown in Fig. 10.4.1, to calculate the concentrated angle change

$$R_a = \frac{h}{12}(3.5a + 3b - 0.5c)$$

$$R_b = \frac{h}{12}(a + 10b + c)$$

$$R_c = \frac{h}{12}(3.5c + 3b - 0.5a)$$

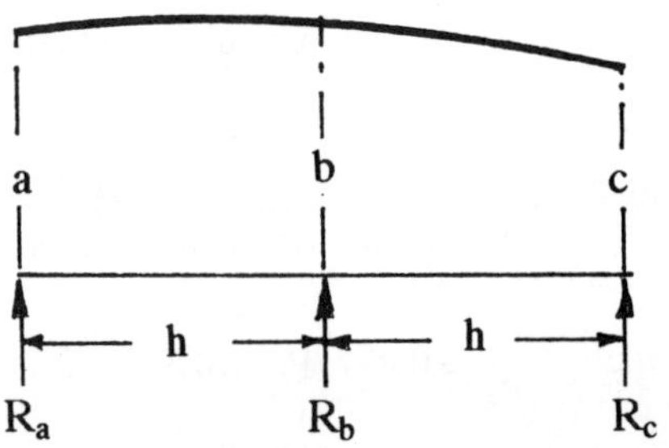

Fig. 10.4.1 Parabolic Variation Formulas

- If calculated y_{cal} are equal to assumed y_a, the initial assumed buckled shape was correct, then calculate the critical buckling load, P_{cr}, which is required to hold the deflected tapered column in this shape
- If the calculated deflections are not equal to the assumed one ($y_{cal} \neq y_a$), more exact value of P can be obtained by another iteration, using an assumed deflection at each point equal to the calculated deflection, y_{cal}, at that point, divided by the maximum calculated deflection
- Usually two or three iterations will give sufficient accuracy
- The sign conventions used are the same as those described in Chapter 2
- The calculation is set up in tabular form

Example:

Determine the critical column load for the following tapered beam. Assume the smallest moment of inertia (I_o) at the left end of the beam and divide it into four intervals, so that each interval is $h = \frac{L}{4}$.

Column Buckling

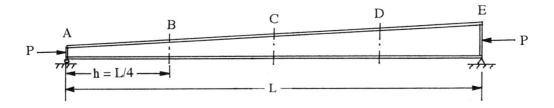

(a) First cycle of calculation:

	Common terms		A	B	C	D	E
(1) $I \to I_o$	I_o		1	3.6	6	8.4	11
(2) y_a	a		0	7	10	5	0
(3) M	Pa		0	7	10	5	0
(4) $\alpha = \dfrac{M}{EI}$	$\dfrac{Pa}{EI_o}$		0	-1.94	-1.67	-0.6	0
(5) α'	$\dfrac{Pah}{12EI_o}$		-5	-21.11	-19.21	-7.62	-0.95
(6) θ_a	$\dfrac{Pah}{12EI_o}$	50	45	23.89	4.68	-2.94	-3.89
(7) y_t	$\dfrac{Pah^2}{12EI_o}$		0	45	68.89	73.57	70.63
(8) y_c	$\dfrac{Pah^2}{12EI_o}$		0	-17.66	-35.32	-52.92	-70.63
(9) y_{cal}	$\dfrac{Pah^2}{12EI_o}$		0	27.34	33.57	20.65	0
(10) $\dfrac{y_a}{y_{cal}}$	$\dfrac{12EI_o}{Ph^2}$		0	0.256	0.298	0.242	0

$$\left(\dfrac{y_a}{y_{cal}}\right)_{av} = \dfrac{0.256 + 0.298 + 0.243}{3} = 0.266$$

Explanation of the steps for the first cycle:

Line (1) In this example, use the smallest moment of inertia of I_o at the left column as the common term and the rest of the moment of inertias are proportional to I_o.

The moment of inertia distribution are:

$I_A = I_o$; $I_B = 3.6 I_o$; $I_C = 6 I_o$; $I_D = 8.4 I_o$; $I_E = 11 I_o$;

Line (2) Use a half-sine wave curve for assuming deflections (in this example downward is positive) adjusted roughly to account for the variation in stiffness along the column.

Here an assumed deflection of $y_a = 0, 7, 10, 5, 0$ is used.

Line (3) Make computation easier by using column load of, P, as a common term.

Line (4) In the distributed angle change, $\alpha = (\dfrac{M}{EI})$, EI is carried as a common term.

The sign is always opposite to that of M because of the sign convention that is used. Here each interval is divided by its own moment of inertia, 'I'.

Chapter 10.0

$$\alpha_A = (\frac{M}{EI})_A = (\frac{M}{EI})_E = 0$$

$$\alpha_B = (\frac{M}{EI})_B = \frac{7}{3.6} = -1.94$$

$$\alpha_C = (\frac{M}{EI})_C = \frac{10}{6} = -1.17$$

$$\alpha_D = (\frac{M}{EI})_D = \frac{5}{8.4} = -0.6$$

Line (5)   Use the parabolic curve formulas shown in Fig. 10.4.1 to calculate the concentrated angle change α' and let $\frac{h}{12}$ be the common term:

$$\alpha'_A = \frac{h}{12}(3.5a + 3b - 0.5c)$$

$$= \frac{h}{12}[3.5(0) + 3(-1.94) - 0.5(-1.67)] = -5(\frac{h}{12})$$

$$\alpha'_B = \frac{h}{12}(a + 10b + c)$$

$$= \frac{h}{12}[0 + 10(-1.94) + (-1.67)] = -21.11(\frac{h}{12})$$

$$\alpha'_C = \frac{h}{12}(a + 10b + c)$$

$$= \frac{h}{12}[-1.94 + 10(-1.67) + (-0.6)] = -19.21(\frac{h}{12})$$

$$\alpha'_D = \frac{h}{12}(a + 10b + c)$$

$$= \frac{h}{12}[1.67 + 10(-0.6) + 0] = -7.62(\frac{h}{12})$$

$$\alpha'_E = \frac{h}{12}(3.5c + 3b - 0.5a)$$

$$= \frac{h}{12}[3.5(0) + 3(-0.6) - 0.5(-1.67)] = -0.95(\frac{h}{12})$$

Line (6)   The value of slope θ_a is completely arbitrary and any value can be chosen since the θ_a is not known at any intervals. Use $\theta_a = 50$ and $(\frac{h}{12})$ as the common term.

$$(\theta_a)_{AB} = \theta_a + \alpha'_A = 50 + (-5) = 45$$

$$(\theta_a)_{BC} = (\theta_a)_{AB} + \alpha'_B = 45 + (-21.11) = 23.89$$

$$(\theta_a)_{CD} = (\theta_a)_{BC} + \alpha'_C = 23.89 + (-19.21) = 4.68$$

$$(\theta_a)_{DE} = (\theta_a)_{CD} + \alpha'_D = 4.68 + (-7.62) = -2.94$$

Line (7)   Trial deflection, y_t, is calculated:

$$(y_t)_A = 0$$

$$(y_t)_B = (y_t)_A + (\theta_a)_{AB} = 0 + 45 = 45$$

$$(y_t)_C = (y_t)_B + (\theta_a)_{BC} = 45 + 23.89 = 68.89$$

$$(y_t)_D = (y_t)_C + (\theta_a)_{CD} = 68.89 + 4.68 = 73.57$$

$$(y_t)_E = (y_t)_D + (\theta_a)_{DE} = 73.57 - 2.94 = 70.63$$

Line (8) The trial value in Line (7) yield a deflection value at the right end (point 'E') of 70.63, and this cannot be correct since the deflection at that point must be zero. Use a linear correction, y_c, at every point in this line starting from point "E" as follows:

$$(y_c)_E = 70.63 - 70.63 = 0$$

$$(y_c)_D = -70.63 \left(\frac{3}{4}\right) = -52.97$$

$$(y_c)_C = -70.63 \left(\frac{2}{4}\right) = -35.32$$

$$(y_c)_B = -70.63 \left(\frac{1}{4}\right) = -17.66$$

$$(y_c)_A = 0$$

Line (9) The calculated deflection, $y_{cal} = y_t - y_c$
$(y_{cal})_A = 0$
$(y_{cal})_B = (y_t)_B - (y_c)_B = 45 - 17.66 = 27.34$
$(y_{cal})_C = (y_t)_C - (y_c)_C = 68.89 - 35.32 = 33.57$
$(y_{cal})_D = (y_t)_D - (y_c)_D = 73.57 - 52.97 = 20.6$
$(y_{cal})_E = 0$

Line (10) The calculated deflections, y_{cal}, are compared with the original assumed deflections, y_a:

- If the ratio of $\dfrac{y_a}{y_{cal}}$ is are the same at every point, the assumed values are correct

- The ratio of $\dfrac{y_a}{y_{cal}}$:

$$\left(\frac{y_a}{y_{cal}}\right)_B = \frac{7}{27.34} = 0.256$$

$$\left(\frac{y_a}{y_{cal}}\right)_C = \frac{10}{33.57} = 0.298$$

$$\left(\frac{y_a}{y_{cal}}\right)_D = \frac{5}{20.6} = 0.243$$

The average value $\left(\dfrac{y_a}{y_{cal}}\right)_{av} = \dfrac{0.256 + 0.298 + 0.243}{3} = 0.266$ and these values are reasonably close, so that P computed from them would be a good approximation to the buckling load:

$$\frac{y_a}{y_{cal}} = 1.0$$

From Line (10), the common term is:

$$0.266 \times \frac{12EI_o}{Ph^2} = 1.0$$

$$P = P_{cr} = 0.266 \times \frac{12EI_o}{h^2} = 51.07 \frac{EI_o}{L^2}$$

where: $h = \dfrac{L}{4}$

(b) Second cycle of calculation:

	Common terms		A	B	C	D	E
(1) $I \to I_o$	I_o		1	3.6	6	8.4	11
(2) y_a	a		0	8.1	10	6.2	0
(3) M	Pa		0	8.1	10	6.2	0
(4) $\alpha = \dfrac{M}{EI}$	$\dfrac{Pa}{EI_o}$		0	−2.25	−1.67	−0.74	0
(5) α'	$\dfrac{pah}{12EI_o}$		−7.59	−24.17	−19.69	−9.07	−1.38
(6) θ_a	$\dfrac{pah}{12EI_o}$	50	42.41	18.24	−1.45	−10.52	−11.9
(7) y_t	$\dfrac{pah^2}{12EI_o}$		0	42.41	60.65	59.2	48.68
(8) y_c	$\dfrac{pah^2}{12EI_o}$		0	−12.17	−24.34	−36.51	−48.68
(9) y_{cal}	$\dfrac{pah^2}{12EI_o}$		0	30.24	36.31	22.69	0
(10) $\dfrac{y_a}{y_{cal}}$	$\dfrac{12EI_o}{ph^2}$		0	0.268	0.275	0.273	0

$$\left(\dfrac{y_a}{y_{cal}}\right)_{av} = \dfrac{0.268 + 0.275 + 0.273}{3} = 0.272$$

Explanation of the steps for the second cycle:

Line (1) Same as the first cycle.

Line (2) Use values of $y_a = \dfrac{y_{cal}}{y_{(cal)max}}$ yields:

$(y_a)_A = 0$

$(y_a)_B = \dfrac{27.34}{33.57} = 0.81$ and use $(y_a)_B = 8.1$

$(y_a)_C = \dfrac{33.57}{33.57} = 1$ and use $(y_a)_C = 10$

$(y_a)_D = \dfrac{20.65}{33.57} = 0.62$ and use $(y_a)_D = 6.2$

$(y_a)_E = 0$

Line (3) Follow the same procedures as in the first cycle.

Line (4) Follow the same procedures as in the first cycle.

Line (5) Use the parabolic curve formulas shown in Fig. 10.4.1 to calculate the concentrated angle change α' and let $\dfrac{h}{12}$ be the common term:

$$\alpha'_A = \dfrac{h}{12}(3.5a + 3b − 0.5c)$$

$$= \dfrac{h}{12}[3.5(0) + 3(−2.25) − 0.5(−1.67)] = −7.59\left(\dfrac{h}{12}\right)$$

$$\alpha'_B = \frac{h}{12}(a + 10b + c)$$

$$= \frac{h}{12}[0 + 10(-2.25) + (-1.67)] = -24.17(\frac{h}{12})$$

$$\alpha'_C = \frac{h}{12}(a + 10b + c)$$

$$= \frac{h}{12}[-2.25 + 10(-1.67) + (-0.74)] = -19.69(\frac{h}{12})$$

$$\alpha'_D = \frac{h}{12}(a + 10b + c)$$

$$= \frac{h}{12}[1.67 + 10(-0.74) + 0] = -9.07(\frac{h}{12})$$

$$\alpha'_E = \frac{h}{12}(3.5c + 3b - 0.5a)$$

$$= \frac{h}{12}[3.5(0) + 3(-0.74) - 0.5(-1.67)] = -1.38(\frac{h}{12})$$

From Line (6) through Line (10):

Follow the same procedures as in the first cycle.

The average value $(\frac{y_a}{y_{cal}})_{av} = \frac{0.268 + 0.275 + 0.273}{3} = 0.272$ and from Line (10) the common term is:

$$0.272 \times \frac{12EI_o}{ph^2} = 1.0$$

$$P = P_{cr} = 0.272 \times \frac{12EI_o}{h^2} = 52.22\frac{EI_o}{L^2}$$

This is about 2% greater than the first cycle which indicates that the first cycle would have given a reasonable approximation of the critical buckling load, P_{cr}.

10.5 LATERAL BUCKLING OF A BEAM BENDING

Bending beams under certain conditions of loading and restraint can fail by lateral buckling in a manner similar to columns loaded in axial compression. However, it is conservative to obtain the buckling load by considering the compression side of the beam as a column since this approach neglects the torsional rigidity of the beam. A general solution to two types of beam instability are discussed in this chapter.

(A) EQUAL FLANGED I-BEAMS

The equal flanged I-beam, as shown in Fig. 10.5.1, and the applicable critical compression stress is defined by the following equation:

$$F_{cr} = K_1(\frac{L}{a})(\frac{h}{L})^2(\frac{I_y}{I_x}) \qquad \text{Eq. 10.5.1}$$

where: K_1 – Constant dependent on the types of loading and restraint as shown in Fig. 10.5.2
L – Length of the beam

$$a = \sqrt{\frac{2I_y h^2}{3J}}$$

$$J = \frac{2bt_f^3 + h_w t_w^3}{3}$$

b   – Flange width
t_f – Flange thickness
h_w – Depth of web
h   – Section depth between flange centroids
h_w – Depth of web
t_w – Web thickness
I_x – Moment of inertia about centroidal axis in the plane of the web
I_y – Moment of inertia about centroidal axis perpendicular to the plane of the web.

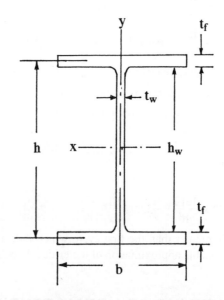

Fig. 10.5.1 Equal Flanged I-beam

Case	Side view	Top view	K_1	Case	Side view	Top view	K_1
1			$\frac{m_1}{4}(E)$	5			$\frac{m_5}{16}(E)$
2			$\frac{m_2}{4}(E)$	6			$\frac{m_6}{32}(E)$
3			$\frac{m_3}{16}(E)$	7			$\frac{m_7}{16}(E)$
4			$\frac{m_4}{32}(E)$	8			$\frac{m_8}{32}(E)$

(a) Lateral Stability Constant, K_1)

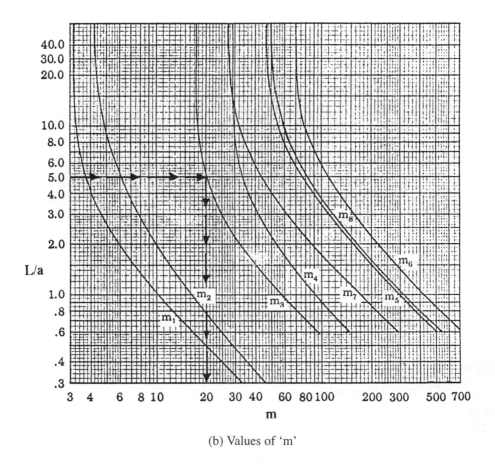

(b) Values of 'm'

Fig. 10.5.2 Lateral Stability Constant, 'K_1' and Value of 'm' for Equal Flange I-beam

Example:

Assume a beam having a span of L = 50 inches subjected to a concentrated load of P = 1,200 lbs. applied at the mid-span of a given aluminum I-beam (E = 10.7 × 106 psi) as shown below and determine the ciritcal compression stress in flange (assume the crippling stress is not critical).

(1) Determine the critical compression stress (F_{cr}) in the flange:

$$J = \frac{2bt_f^3 + h_w t_w^3}{3}$$

$$= \frac{2 \times 2 \times 0.125^3 + 1.75 \times 0.125^3}{3}$$

$$= 0.00374$$

h = 2 − 0.125 = 1.875 in.

h_w = 2 − 2 × 0.125 = 1.75 in.

$$a = \sqrt{\frac{2I_y h^2}{3J}}$$

$$= \sqrt{\frac{2 \times 0.154 \times 1.875^2}{3 \times 0.00358}} = 9.82$$

$$\frac{L}{a} = \frac{50}{9.82} = 5.09$$

(I_x = 0.495 in.^{4} and I_y = 0.154 in.^{4})

From Fig. 10.5.2(b) of Case 3, obtain $m_3 = 20$

From Fig. 10.5.2(a), $K_1 = m_3 \times \dfrac{E}{16} = 20 \times \dfrac{10.7 \times 10^6}{16} = 13.38 \times 10^6$

From Eq. 10.5.1:

$$F_{cr} = K_1 \left(\dfrac{L}{a}\right)\left(\dfrac{h}{L}\right)^2 \left(\dfrac{I_y}{I_x}\right)$$

$$= 13.38 \times 10^6 \left(\dfrac{50}{9.82}\right)\left(\dfrac{1.875}{50}\right)^2 \left(\dfrac{0.154}{0.495}\right)$$

$$= 13.38 \times 10^6 \times 5.09 \times 1,406 \times 10^{-3} \times 0.31$$

$$= 29,780 \text{ psi}$$

(2) Find the compression stress (f_c) in the flange due to bending:

$$f_b = \dfrac{(P \times \frac{L}{4})(\frac{h}{2})}{I_x} = \dfrac{(1,200 \times \frac{50}{4})(\frac{1.875}{2})}{0.495} = 28,409 \text{ psi}$$

(3) $MS = \dfrac{29,780}{28,409} - 1 = \underline{0.05}$ O.K.

(B) DEEP WEB BEAM WITHOUT FLANGES

The deep web beam, shown in Fig. 10.5.3, and the applicable critical stress may be expressed as:

The critical moment is:

$$M_{cr} = 0.0985 K_m E \left(\dfrac{ht^3}{L}\right) \qquad \text{Eq. 10.5.2}$$

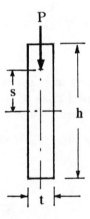

Fig. 10.5.3 Deep Web Beam

The critical stress is:

$$F_{cr} = K_f' E \left(\dfrac{t^2}{Lh}\right) \qquad \text{Eq. 10.5.3}$$

Where the value of K_f' is a corrected value for the applied load which is not loaded at the centroidal axis:

$$K_f' = K_f \left[1 - n\left(\dfrac{s}{L}\right)\right] \qquad \text{Eq. 10.5.4}$$

where: Constants of K_m, K_f and n value can be found in Fig. 10.5.4.

s – Distance from centroidal axis to the point of application of load:
- Negative if below the centroidal axis.
- Positive if above the centroidal axis.

L – Length of the beam.

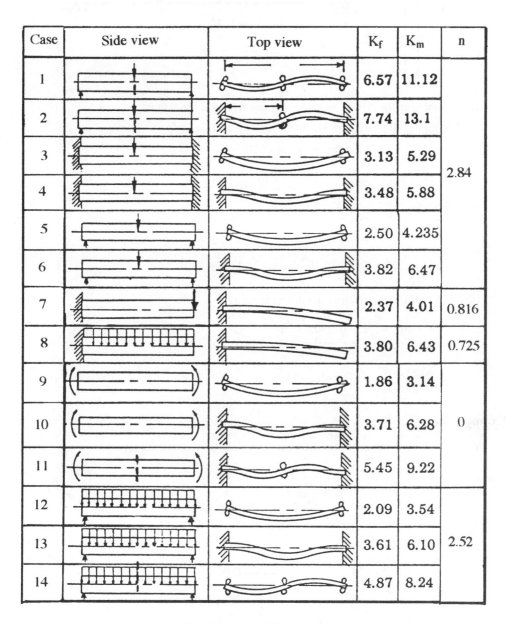

Case	Side view	Top view	K_f	K_m	n
1			6.57	11.12	
2			7.74	13.1	
3			3.13	5.29	2.84
4			3.48	5.88	
5			2.50	4.235	
6			3.82	6.47	
7			2.37	4.01	0.816
8			3.80	6.43	0.725
9			1.86	3.14	
10			3.71	6.28	0
11			5.45	9.22	
12			2.09	3.54	
13			3.61	6.10	2.52
14			4.87	8.24	

Fig. 10.5.4 Constants of K_m and K_f and n

Example 1:

Assume a beam having a span of L = 20 inches subjected a concentrated load of P = 600 lbs. applied at the mid-span of a given aluminum deep web beam (E = 10.7 x 10^6 psi) as shown below and determine the critical compression stress in the flange.

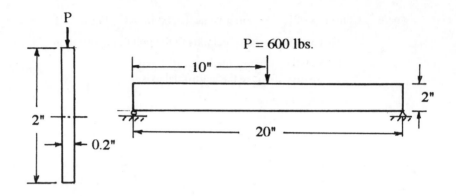

(1) Determine the critical compression stress (F_{cr}) in the flange:

From Fig. 10.5.4 (Case 5), obtain $K_f = 2.5$ and $n = 2.84$

From Eq. 10.5.4, the correct value of K_f':

$$K_f' = K_f[1 - n(\frac{s}{L})] = 2.5[1 - 2.84(\frac{1}{20})] = 2.15$$

From Eq. 10.5.3, use K_f':

$$F_{cr} = K_f'E(\frac{t^2}{Lh})$$

$$= 2.15 \times 10.7 \times 10^6 (\frac{0.2^2}{20 \times 2}) = 23,005 \text{ psi}$$

(2) Find compression stress (f_c) due to bending:

$$f_c = \frac{6(P \times \frac{L}{4})}{bh^2} = \frac{6(600 \times \frac{20}{4})}{0.2 \times 2^2} = 22,500 \text{ psi}$$

(3) $MS = \frac{23,005}{22,500} - 1 = 0.02$ O.K.

Example 2:

Given a shear clip as shown below, determine the allowable shear load (P_s).

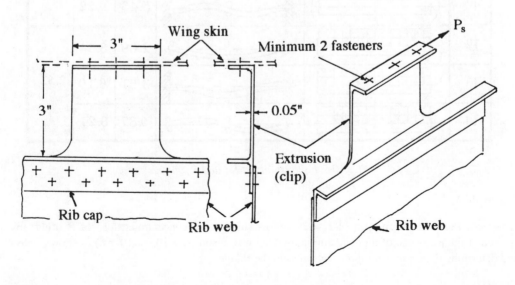

(1) Extruded clip used to connect wing rib cap to wing skin:

Assume a fictitious clip as a mirror image on other side as shown below, and assume both ends are fixed and L = 2b = 2 × 3 = 6.

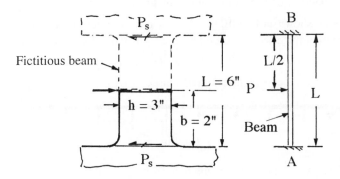

Use Case 2 from Fig. 10.5.4. to obtain $K_m = 13.1$. The critical moment (from Eq. 10.5.2) is:

$$M_{cr} = 0.0985 K_m E \left(\frac{ht^3}{L}\right)$$

$$= 0.0985 \times 13.1 \times 10.7 \times 10^6 \left(3 \times \frac{0.05^3}{6}\right)$$

$$= 863 \text{ in-lbs.}$$

The moment at Point A is $M = \frac{PL}{8}$ and,

$$P = \frac{8M}{L} = 8 \times \frac{863}{6} = 1,151 \text{ lbs.}$$

But here $P = 2P_s$ and, therefore, the allowable shear load is:

$$P_s = \frac{1,151}{2} = 576 \text{ lbs.}$$

(2) Assume a formed clip used to connect the fuselage frame to the outer skin:

$$K'_m = \frac{K_{m, \text{Case 1}} + K_{m, \text{Case 2}}}{2} = \frac{11.12 + 13.1}{2} = 12.11$$

$$M_{cr} = 0.0985 K_m' E \left(\frac{ht^3}{L}\right)$$

$$= 0.0985 \times 12.11 \times 10.7 \times 10^6 \left(3 \times \frac{0.05^3}{6}\right)$$

$$= 798 \text{ in-lbs.}$$

The moment at Point A, use $M = \frac{PL}{6}$ which is between fixed-end beam ($M = \frac{PL}{8}$) and a simply-supported beam ($M = \frac{PL}{4}$):

$$P = \frac{6M}{L} = \frac{6 \times 798}{6} = 798 \text{ lbs.}$$

$P = 2P_s$ and the allowable shear load is $P_s = \frac{798}{2} = 399 \text{ lbs.}$

10.6 BEAM-COLUMNS

A beam-column is a structural member that is subjected to both axial loads and bending moment. The bending moment may be caused by eccentric application of the axial load, initial curvature of the member, transverse loading, or any combination of these conditions.

Note that the beam-column shown in Fig. 10.6.1, is subjected to axial loads, P, and a transverse load, w (lbs/in.). Considering the effects of each of these loads individually forms the basis of the formulas to follow.

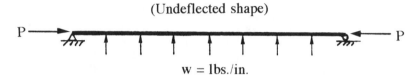

Fig. 10.6..1 Beam-Column under Transverse Loading

If the transverse load, w, is applied to the beam-column as a single condition, the member may be analyzed by conventional beam methods.

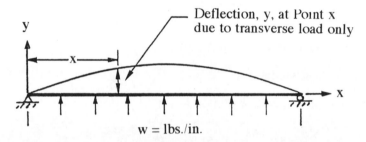

Fig. 10.6.2 Primary Deflection due to Transverse Loading

From Fig. 10.6.2, at a distance, x, from the left end there is a bending moment equal to $M = \frac{wLx}{2} - \frac{wx^2}{2}$ and at the same point there is a deflection equal to y. Both the moment and the deflection are found by conventional beam formulas and are known as the "primary" moment and deflection.

Using this deflected shape and applying the axial load, P, an additional moment equal to the load, P, multiplied by the eccentricity 'y' is incurred at point x. This additional moment increases the deflection 'y'. This in turn, leads to an even larger moment and hence a larger deflection. The additional moment and deflection caused by application of the axial load, as shown in Fig. 10.6.3, are known as the "secondary" moment and deflection.

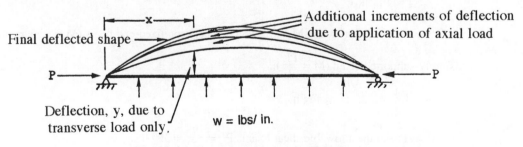

Fig. 10.6.3 Secondary Moment and Deflection

Column Buckling

There are two failure modes to be considered for beam-columns:
- If the axial load, P, is greater than the critical load, P_{cr}, then the member will fail by flexural instability
- If the axial load, P, combined with a transverse load such as a uniform loading, w, produces a stress level greater than the material allowable stress of the member, then the bending mode is critical

If neither of these modes of failure exceeds its critical value, the beam-column will reach a state of equilibrium. It is possible to represent increments of moment and deflection as some sort of a converging mathematical series.

(A) DESIGN EQUATIONS AND DATA

Design equations and data are given in Fig. 10.6.4 and additional equations can be found in Ref. 10.4 and Ref. 10.8.

Case	Loading	Moment
1	[simply supported beam with uniform load w over length L, axial load]	$(j = \sqrt{\frac{EI}{P}})$ $M_{center} = (\frac{wL^2}{8})\mu$
2	[simply supported beam with point load Q at L/2]	$M_{center} = (\frac{QL}{4})v$
3	[simply supported beam with end moments M]	$M_{center} = \frac{-M}{\cos(\frac{L}{2j})}$
4	[fixed-fixed beam with uniform load w]	$M_{center} = \frac{wL^2 \alpha''}{24}$ $M_A = M_B = -\frac{wL^2 \beta''}{12}$
5	[fixed-fixed beam with point load Q at L/2]	$M_{center} = (\frac{QL}{8})v''$ $M_A = M_B = -(\frac{QL}{8})v''$
6	[fixed-fixed beam with applied moment M at center]	$M_A = -(\frac{M}{4})(\frac{\alpha''}{\beta''})$ $M_B = (\frac{M}{4})(\frac{\alpha''}{\beta''})$
7	[propped cantilever: fixed at A, simply supported at other end, uniform load w]	$M_A = (\frac{wL^2}{8})(\frac{\gamma}{\beta})$

Case	Loading	Moment
8	 (cantilever with moment M at free end, pinned)	$M_A = -M\left(\dfrac{\alpha}{2\beta}\right)$
9	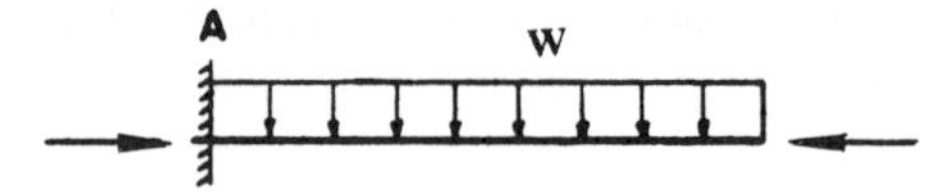 (cantilever with uniform load w)	$M_A = -\left(\dfrac{wL^2}{2}\right)\mu$
10	(cantilever with point load Q at free end)	$M_A = -(QL)\nu$
11	 (beam with moment M at end)	$M_{center} = \dfrac{-M}{\cos\left(\dfrac{L}{j}\right)}$
12	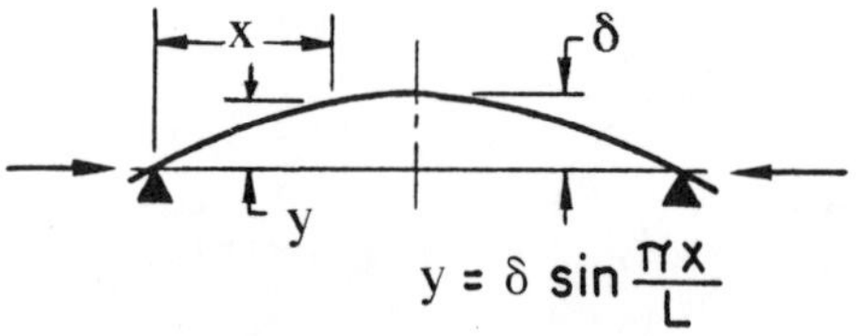 (simply supported, $y = \delta \sin\dfrac{\pi x}{L}$)	$M_{center} = \dfrac{P\delta}{1-\left(\dfrac{P}{P_{cr}}\right)}$ (Sine curve)
13	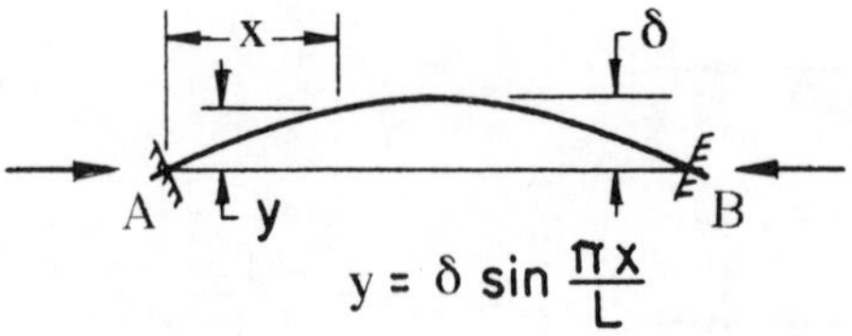 (fixed-fixed, $y = \delta \sin\dfrac{\pi x}{L}$)	$M_{center} = \dfrac{-P\delta}{1-\left(\dfrac{P}{P_E}\right)}\left[1 - \dfrac{2}{(\pi \cos\dfrac{L}{2j})\nu}\right]$ $M_A = M_B = \dfrac{P\delta}{1-\left(\dfrac{P}{P_E}\right)}\left[1 - \dfrac{2}{\pi \nu}\right]$ (Sine curve)
14	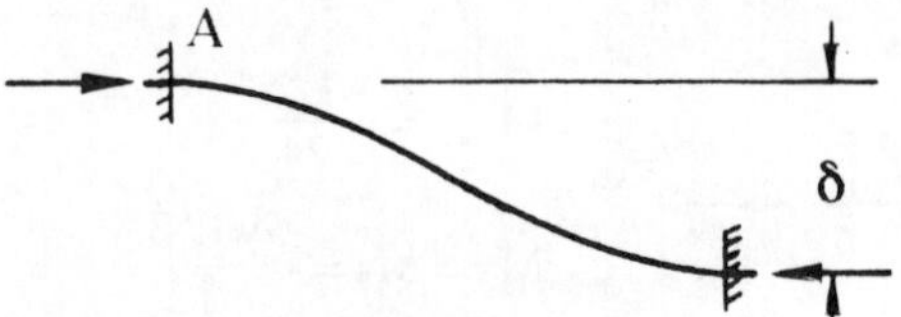 Forced deflection δ	$M_A = -\dfrac{6EI\delta}{L^2\beta''}$ $M_A = -\dfrac{6EI\delta}{L^2\beta''}$
15	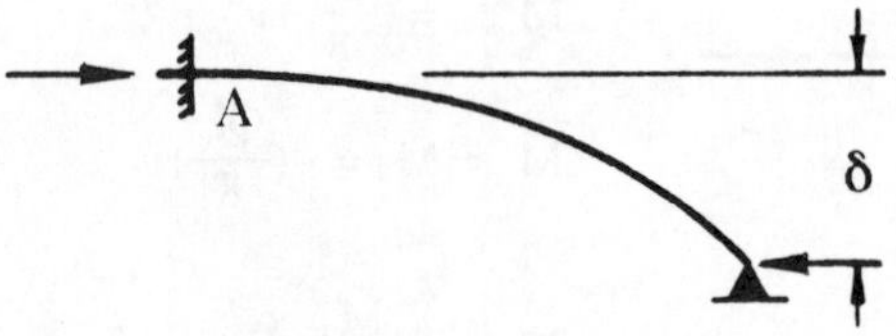 Forced deflection δ	$M_A = -\dfrac{3EI\delta}{L^2\beta}$

(The above beam-column coefficients can be found in Fig. 10.6.5 and Fig. 10.6.6)

Fig. 10.6.4 Beam-Column Equations

$\dfrac{L}{j}$	α	β	μ	ν	γ
0	∞	1.0	1.0	1.0	1.0
0.5	1.03	1.017	1.027	1.021	1.026
1.0	1.131	1.074	1.116	1.092	1.111
1.5	1.343	1.192	1.304	1.242	1.291
$\dfrac{\pi}{2}$	1.388	1.216	1.343	1.273	1.329
2.0	1.799	1.436	1.702	1.557	1.672
2.1	1.949	1.516	1.832	1.66	1.797
2.2	2.134	1.612	1.992	1.786	1.949
2.3	2.364	1.733	2.19	1.943	2.139
2.4	2.658	1.885	2.443	2.143	2.382
2.5	3.05	2.086	2.779	2.407	2.703
2.6	3.589	2.362	3.241	2.771	3.144
2.7	4.377	2.762	3.913	3.3	3.786
2.8	5.632	3.396	4.984	4.141	4.808
2.9	7.934	4.555	6.948	5.681	6.68
3.0	13.506	7.349	11.696	9.401	11.201
3.1	45.923	23.566	39.481	31.018	37.484
π	∞	∞	∞	∞	∞
3.2	−32.706	−15.74	−27.647	−22.395	−26.245
3.4	−7.425	−3.079	−6.065	−4.527	−5.738
3.6	−4.229	−1.457	−3.331	−2.381	−3.131
3.8	−2.996	−0.813	−2.267	−1.541	−2.111
4.0	−2.357	−0.46	−1.701	−1.093	−1.569
4.2	−1.979	−0.232	−1.352	−0.814	−1.234
4.4	−1.743	−0.065	−1.115	−0.624	−1.007
4.49	−1.668	0	−1.032	−0.557	−0.927
4.6	−1.596	0.068	−0.945	−0.487	−0.843
4.8	−1.515	0.185	−0.818	−0.382	−0.72
5.0	−1.491	0.298	−0.719	−0.299	−0.623
5.2	−1.528	0.417	−0.641	−0.231	−0.546
5.4	−1.644	0.559	−0.578	−0.175	−0.484
5.6	−1.889	0.754	−0.526	−0.127	−0.431
5.8	−2.405	1.075	−0.483	−0.085	−0.387
6.0	−3.746	1.802	−0.447	−0.048	−0.349
6.2	−11.803	5.881	−0.416	−0.013	−0.316
2π	∞	∞	−0.405	0	−0.304

Note: 1. $j = \sqrt{\dfrac{EI}{P}}$

2. $\bar{\mu}$ and $\bar{\nu}$ can be determined by taking μ and ν values for $\dfrac{2L}{j}$ instead of $\dfrac{L}{j}$; α" and β" can be found by taking α and β values for $\dfrac{L}{2j}$ instead of $\dfrac{L}{j}$

3. The above values are plotted in Fig. 10.6.6.

Fig. 10.6.5 Beam-Column Coefficients

Chapter 10.0

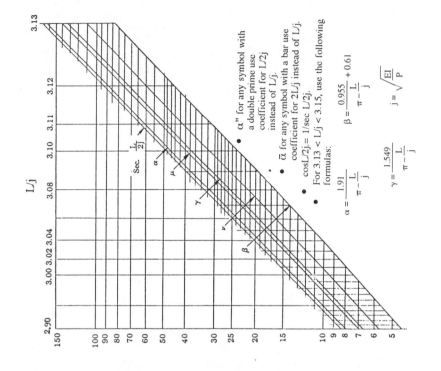

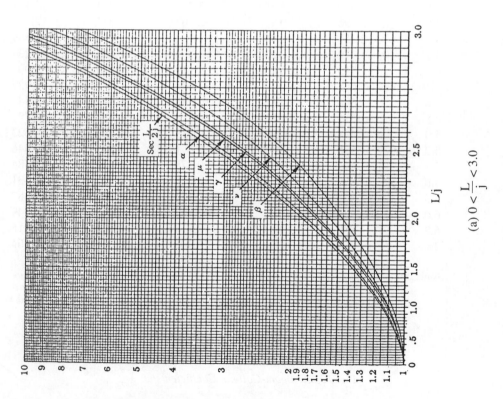

Fig. 10.6.6 Curves of Beam-Column Coefficients

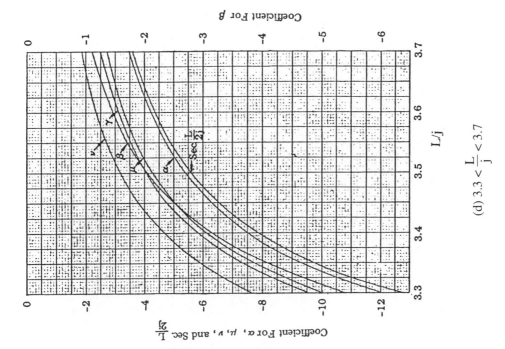

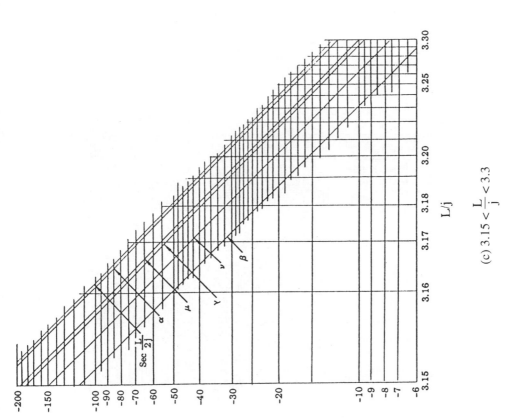

Fig. 10.6.6 Curves of Beam-Column Coefficients (cont'd)

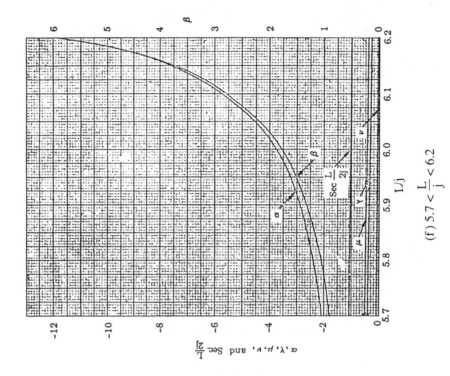
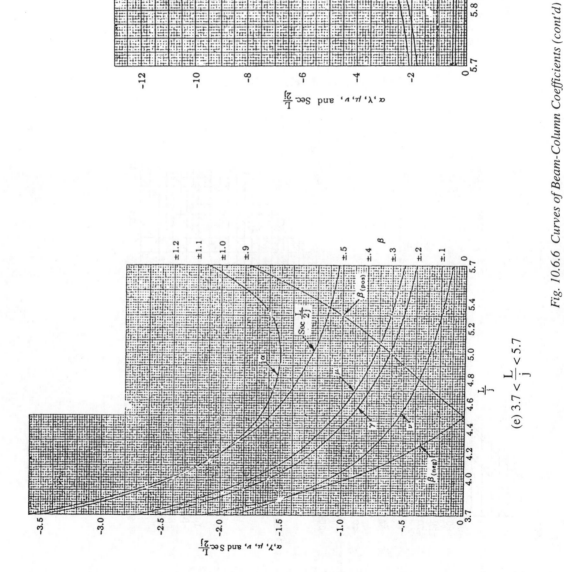

Fig. 10.6.6 Curves of Beam-Column Coefficients (cont'd)

(B) COEFFICIENTS [C_1, C_2, AND $f(w)$] METHOD

This method of analysis (Ref. 10.6) must consider the limitations of both beams and columns:

- The first step in a problem of this type is to calculate the critical column load, P_{cr} (or P_E)
- If the axial load is not greater than the critical column load, the combined loading effects must then be examined

The following equations should be used to calculate the effects of combined loads on a single span beam-column.

Total bending moment:

$$M = C_1 \sin\left(\frac{x}{j}\right) + C_2 \cos\left(\frac{x}{j}\right) + f(w) \qquad \text{Eq. 10.6.1}$$

Total shear:

$$V = \frac{dM}{dx} = \left(\frac{C_1}{j}\right)\cos\left(\frac{x}{j}\right) - \left(\frac{C_2}{j}\right)\sin\left(\frac{x}{j}\right) + f'(w) \qquad \text{Eq. 10.6.2}$$

Where: C_1, C_2 – Constants depend on the character of the transverse loading
$\quad\quad\quad\;\; f(w)$ – A term containing transverse load, w, and possibly j, x, and L, but neither the axial load, P nor the end moments, M_1 and M_2
$\quad\quad\quad\;\; f'(w)$ – Derivative of $f(w)$ with respect to x
$\quad\quad\quad\;\; x$ – Distance from left end of beam column
$\quad\quad\quad\;\; P$ – Applied axial load

$$j = \sqrt{\frac{EI}{P}}$$

The deflection and slope can be obtained by the following equations:

Total deflection:

$$\delta = \frac{M_o - M}{P} \qquad \text{Eq. 10.6.3}$$

Total slope:

$$\phi = \frac{V_o - V}{P} \qquad \text{Eq. 10.6.4}$$

The above primary bending moment (M_o) and shear (V_o) would be produced by the transverse loads without the axial load.

Fig. 10.6.7 gives values of C_1 C_2, and $f(w)$ for the most common cases of transverse loading. These values, used with the above equations, result in a high degree of accuracy provided the notes following the Fig. 10.6.7 are adhered to. Equations are given for the maximum moment whenever possible.

In general, the principle of superposition does not apply to a beam-column, since the bending moments due to a transverse load and an axial load acting simultaneously are not the same as the sum of the bending moments caused by those loads acting separately.

However, the method of superposition can be applied provided each of the superposed bending moments is the effect of the combined action of one or more of the transverse loads under the same axial load:

> On the other hand it will be found that the moments due to a transverse loading, w_1, acting simultaneously with an axial load, P, added to the bending moments due to another transverse loading, w_2, acting simultaneously with the same axial load, P, will be identical with the bending moments due to the transverse loadings w_1 and w_2 and the axial load, P, acting simultaneously

Chapter 10.0

Case	Loading	C_1	C_2	$f(w)$	Maximum Moment
1	Pin-end beam with uniform load 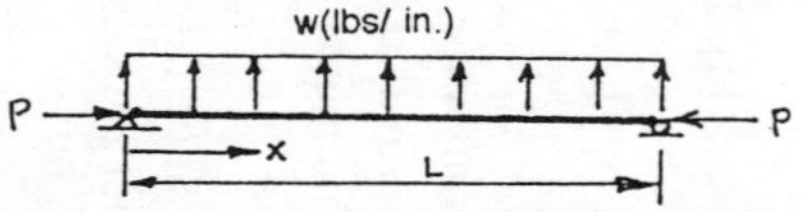	$\dfrac{wj^2[\cos(L/j) - 1]}{\sin(L/j)}$	$-wj^2$	wj^2	$M_{max} = wj^2[1 - \sec(L/2j)]$ (At midspan)
2	Uniform load plus un-equal end moments 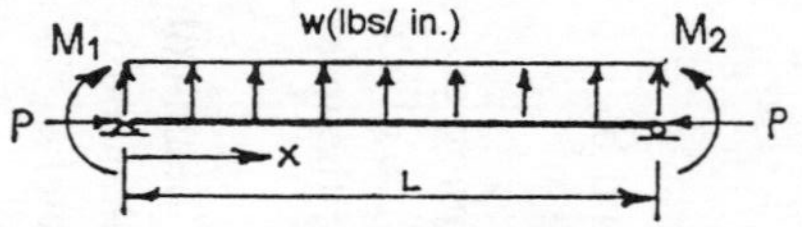	$\dfrac{(M_2 - wj^2) - (M_1 - wj^2)\cos(L/j)}{\sin(L/j)}$	$M_1 - wj^2$	wj^2	$M_{max} = \left[\dfrac{(M_1 - wj^2)}{\cos(x/j)}\right] + wj^2$ where: $\tan(x/j) = \dfrac{(M_2 - wj^2) - (M_1 - wj^2)\cos(L/j)}{(M_1 - wj^2)\sin(L/j)}$
3	Pin-end beam with partial uniformly distributed load 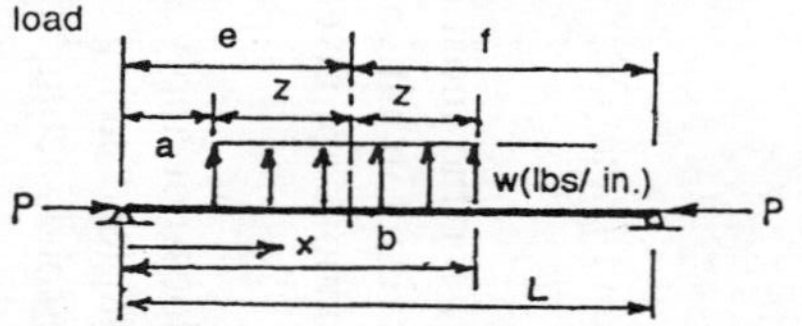	$x < a$: $-\dfrac{wj^2 \sin(z/j)\sin(f/j)}{\sin(L/j)}$ $a < x < b$: $\dfrac{2wj^2\sin(z/j)\sin(e/j)}{\tan(L/j)} - wj^2\sin(b/j)$ $b < x < L$: $\dfrac{2wj^2\sin(z/j)\sin(e/j)}{\tan(L/j)}$	0 $-wj^2\cos(a/j)$ $-2wj^2\sin(z/j)\sin(e/j)$	0 wj^2 0	(see Note 5)
4	Pin-end beam with symmetrical partial uniformly distributed load 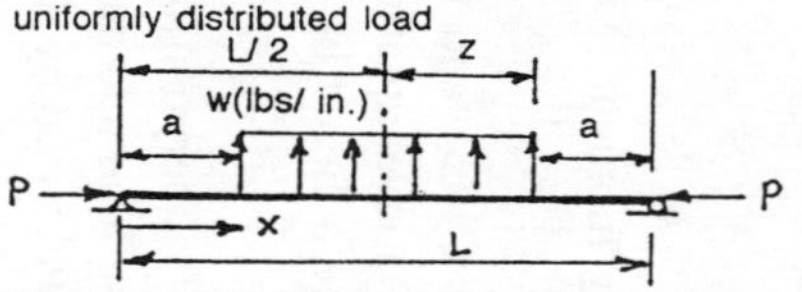	$x < a$: $-wj^2\sin(z/j)\sec(L/2j)$ $a < x < (L-a)$: $-wj^2\tan(L/2j)\cos(a/j)$ $(L-a) < x < L$: $wj^2\sin(z/j)\sec(L/2j)\cos(L/J)$	0 $-wj^2\cos(a/j)$ $-2wj^2\sin(z/j)\sin(L/2j)$	wj^2 0	$M_{max} = wj^2\left\{1 - \dfrac{[\cos(a/j)]}{[\cos(L/2j)]}\right\}$ (At midspan)
5	Pin-end beam with concentrated load at center 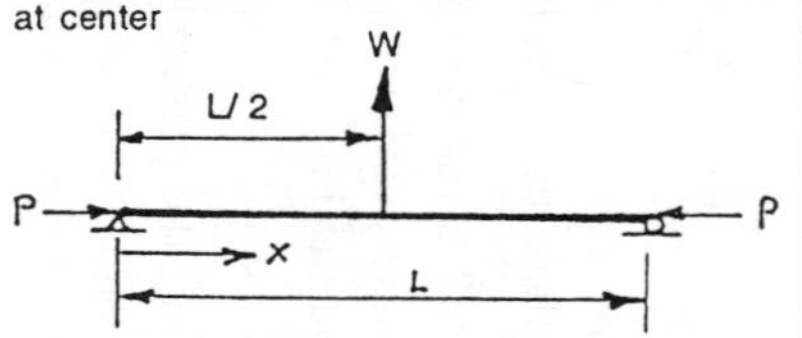	$x < L/2$: $\dfrac{wj\sin(z/j) - Wj\sec(L/2j)}{2}$ $x > L/2$: $\dfrac{Wj\cos(L/j)\sec(L/2j)}{2}$	0 $-Wj\sin(L/2j)$	0 0	$M_{max} = \left[\dfrac{C_1^2 + C_2^2}{C_2}\right]\cos(L/2j)$ (At midspan)

Fig. 10.6.7 Beam-Column with Axial Compression Load

Case	Loading	C_1	C_2	$f(w)$	Maximum Moment
6	Pin-end beam with two symmetrical concentrated loads 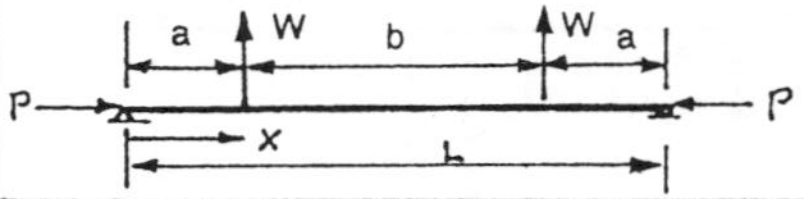	$x < a$: $-\dfrac{W j \cos(b/2j)}{\cos(L/2j)}$ $a < x < (L-a)$: $-W j \sin(a/j)\tan(L/2j)$ $(L-a) < x < L$: $\dfrac{W j \cos(L/j)\cos(b/2j)}{\cos(L/2j)}$	0	0	$-\dfrac{W j \sin(a/j)}{\cos(L/2j)}$ (At midspan)
7	Pin-end beam with concentrated load 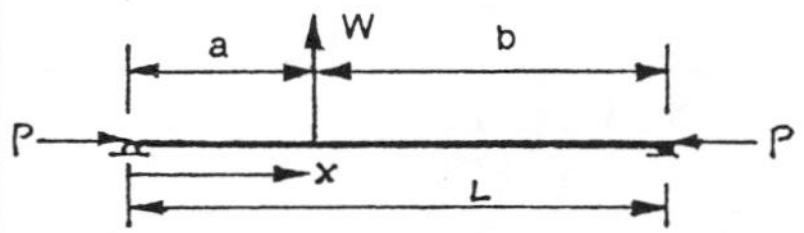	$x < a$: $-\dfrac{W j \sin(b/j)}{\sin(L/j)}$ $x > a$: $\dfrac{W j \sin(a/j)}{\tan(L/j)}$	$-W j \sin(a/j)$ $-\dfrac{W j \sin(L/j)\cos(b/2j)}{\cos(L/2j)}$ $-W j \sin(a/j)$	0 0 0	$M_{max} = \left[\dfrac{C_1^2 + C_2^2}{C_2}\right]\cos(x/j)$ where: $\tan(x/j) = C_1/C_2$
8	Pin-end beam with uniform increasing load 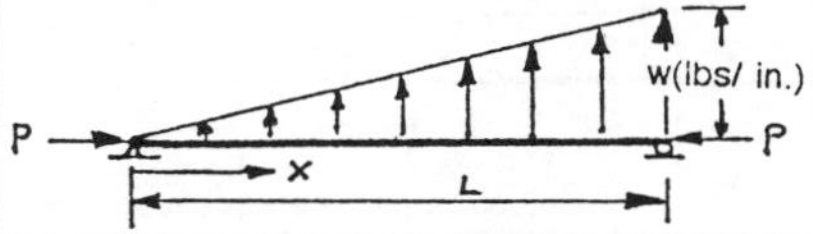	$-\dfrac{w j^2}{\sin(L/j)}$	0	$\dfrac{w j^2 x}{L}$	M_{max} occurs at $\cos(x/j) = (j/L)\sin(L/j)$ (Solve for x and x/j; used in general equation)
9	Pin-end beam with uniform decreasing load 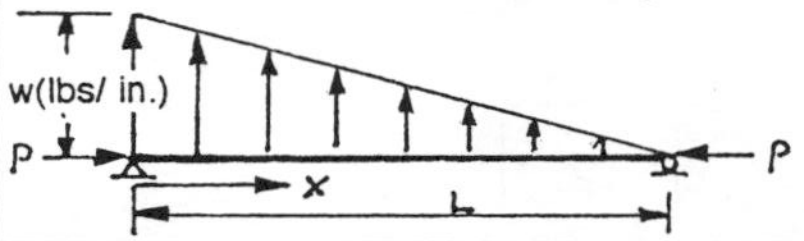	$\dfrac{w j^2}{\tan(L/j)}$	$-w j^2$	$w j^2 (1 - x/L)$	M_{max} occurs at $\cos(L-x)/j] = (j/L)\sin(L/j)$ (Solve for x and x/j; and use in general equation)
10	Un-equal end moments 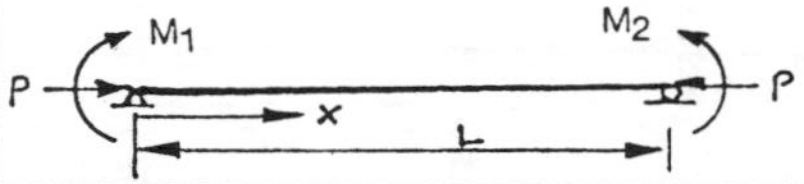	$\dfrac{M_2 - M_1 \cos(L/j)}{\sin(L/j)}$	M_1	0	$M_{max} = \dfrac{M_1}{\cos(x/j)}$ where: $\tan(x/j) = \dfrac{M_2 - M_1 \cos(L/j)}{M_1 \sin(L/j)}$
11	Equal end moments 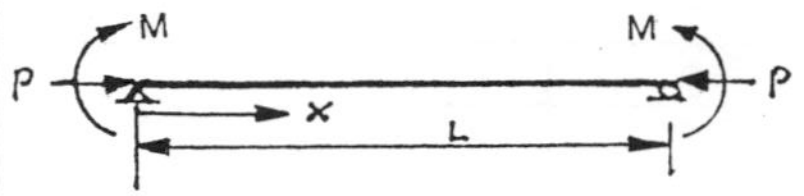	$M \tan(L/2j)$	M	0	$M_{max} = M/\cos(L/2j)$ (At midspan)

Fig. 10.6.7 Beam-Column with Axial Compression Load (cont'd)

Chapter 10.0

Case	Loading	C_1	C_2	$f(w)$	Maximum Moment
12	Fixed-end beam with uniform load 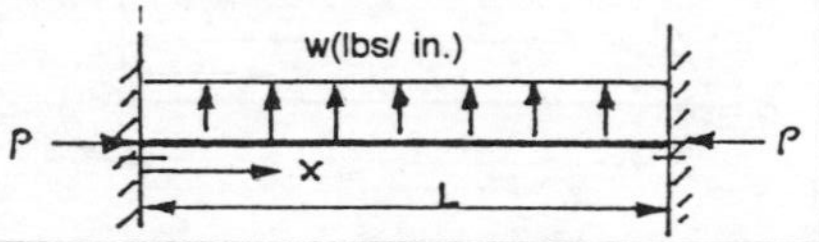	$\dfrac{-wjL}{2}$	$\dfrac{-wjL}{2\tan(L/2j)}$	wj^2	At $x = 0$: $M = wj^2\left[1 - \dfrac{L/2}{\tan(L/2j)}\right]$ At $x = L/2$: $M = -wj^2\left[\dfrac{(L/2)}{\sin(L/2j)} - 1\right]$
13	Fixed-end beam with concentrated load at center 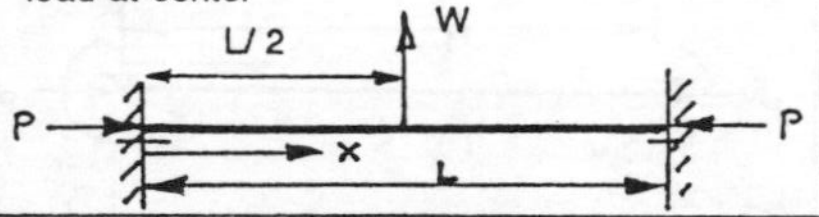	$x < L/2$: $\dfrac{-wj}{2}$ $x > L/2$: $\dfrac{wj}{2}[2\cos(L/2j) - 1]$	$\dfrac{wj}{2}\left\{\dfrac{1-\cos(L/2j)}{\sin(L/2j)}\right\}$ $\dfrac{wj}{2}\left\{\dfrac{\cos(L/j) - \cos(L/2j)}{\sin(L/2j)}\right\}$	0 0	At $x=0$: $M_{max} = \dfrac{wj}{2}\left\{\dfrac{1-\cos(L/2j)}{\sin(L/2j)}\right\}$ At $x = L/2$: $M_{max} = -M_{max}$ (at $x=0$)
14	Pin-end beam with concentrated moment 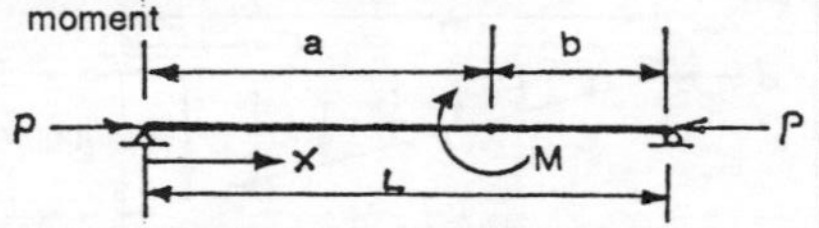	$x < a$: $\dfrac{-M\cos(b/j)}{\sin(L/j)}$ $x > a$: $\dfrac{-M\cos(a/j)}{\tan(L/j)}$	0 $M\cos(a/j)$	0 0	(see Note 5)
15	Fixed-end and pin-end beam with concentrated load at center 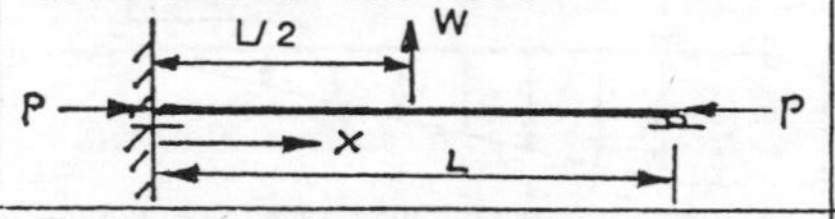	$x < L/2$: $-\dfrac{wj}{2}\left\{\dfrac{[\tan(L/j)]\sec(L/2j) - L}{j\tan(L/j) - L}\right\}$ $x > L/2$: $\dfrac{wj}{2}\left\{\dfrac{L + 2j\sin(L/2j) - 2L\cos(L/2j)}{j\tan(L/j) - L}\right\}$	$\dfrac{wL}{2}\left\{\dfrac{[\tan(L/j)]\sec(L/2j)-1]}{j\tan(L/j)-L}\right\}$ $\dfrac{wj}{2}\left\{\dfrac{L[\tan(L/j)]\sec(L/2j) - L}{\tan(L/j)} - 2\sin(L/2j)\right\}$	0 0	$M_{max} = \dfrac{wL}{2}\left\{\dfrac{[\tan(L/j)]\sec(L/2j)-1]}{j\tan(L/j)-L}\right\}$ (At $x = 0$)
16	Fixed-end and pin-end beam with uniform load 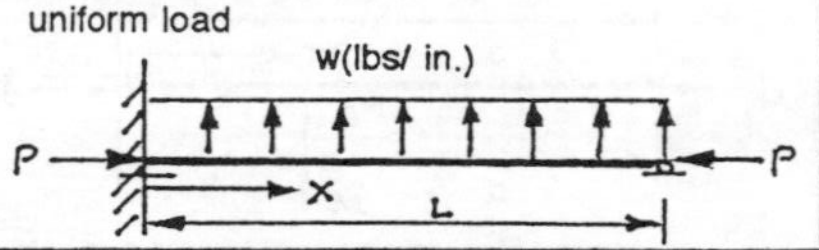	$x < L/2$: $-wLj\left\{\dfrac{\tan(L/2j) - L/2j}{\tan(L/j) - L/j}\right\}$ $x > L/2$: $wj^2\left[\dfrac{1-\cos(L/j)}{\sin(L/j)}\right]$	$\left[\dfrac{\tan(L/2j)-L/2j}{\tan(L/j)-L/j}\right] \times \tan(L/j)(wLj) - wj^2$	wj^2	$M_{max} = wLj\left\{\dfrac{\tan(L/2j) - L/2j}{\tan(L/j) - L/j}\right\}\tan(L/j)$ (At $x = 0$)
17	Pin-end beam with symmetrical triangle load 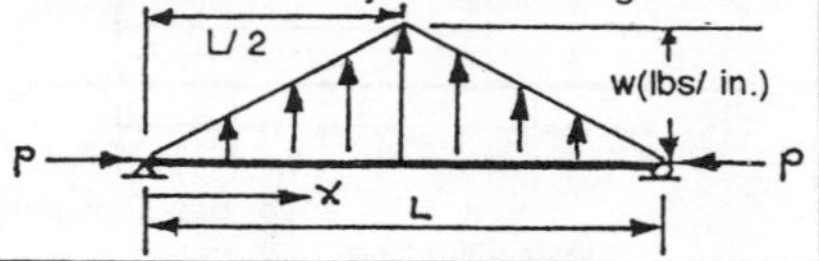	$x < L/2$: $\dfrac{-2wj^3}{L\cos(L/2j)}$ $x > L/2$: $\dfrac{2wj^3\cos(L/j)}{L\cos(L/2j)}$	$\dfrac{-4wj^3\sin(L/2j)}{L}$	$2wj^2 x/L$ $2wj^2(1 - x/L)$	$M_{max} = -2wj^3\tan(L/2j) + wj^2$ (At midspan)

Fig. 10.6.7 Beam-Column with Axial Compression Load

Case	Loading	C_1	C_2	$f(w)$	Maximum Moment
18	Cantilever beam with uniform load w(lbs/in.)				$M_{max} = wj\{[1-\sec(L/j)] + L\tan(L/j)\}$ (At $x=0$)
19	Cantilever beam with concentrated end load				$M_{max} = wj\tan(L/j)$ (At $x=0$)
20	Fixed-end beam with lateral displacement				At $x=0$: $M_{max} = \dfrac{a P \tan(L/2j)}{2[\tan(L/2j) - L/2j]}$ At $x=L$: $M_{max} = -M_{max}$ (at $x=0$)

Notes:
(1) Loading w (lbs/in.) or W (lbs) is positive when upward
(2) M is positive when producing compression in upper surface of the beam at the section being considered
(3) $j = \sqrt{EI/P}$ with a dimension of length
(4) All angles for trigonometric functions are in radians
(5) When the formula for the maximum moment is not provided in the table, methods of differential calculus may be employed, if applicable, to find the loaction of maximum moments or moments at several points in a span may be computed and a smooth curve then drawn through the plotted results. The same principle applies in the case of a complicated combination of loading.
(6) All points where concentrated loads or moments are acting should also be checked for maximum possible bending moments.
(7) Before the total stress can reach the material yield point a compression beam column may fail due to buckling. This instability failure is independent of lateral loads and the maximum, P, that the structure can sustain may be computed pertaining to the boundary condition without regard to lateral loads. A check using ultimate loads should always be made to insure that, P, is not beyond the critical value.
(8) It is recommended that all calculations should be carried to a least four significant figures.
(9) Ref. NACA TM 985 (1941) for many other loading cases.

Fig. 10.6.7 Beam-Column with Axial Compression Load

Chapter 10.0

- In order to write the proper formula for the bending moment on a beam-column subjected to two or more transverse loads, all that is necessary to add the values from Fig. 10.6.7 for each of the transverse loads to obtain the coefficients of C_1, C_2, and $f(w)$ to be used in Eq. 10.6.1 and Eq. 10.6.2

For example, using superposition of loading conditions of Case 1, Case 2, and Case 3 under the same axial load are as follows:

$$\text{Total } \Sigma C_1 = C_1 \text{ (from Case 1)} + C_1 \text{ (from Case 2)} + C_1 \text{ (from Case 3)}$$

$$\text{Total } \Sigma C_2 = C_2 \text{ (from Case 1)} + C_2 \text{ (from Case 2)} + C_2 \text{ (from Case 3)}$$

$$\text{Total } \Sigma f(w) = f(w) \text{ (from Case 1)} + f(w) \text{ (from Case 2)} + f(w) \text{ (from Case 3)}$$

Superposition can be applied in the same manner to determine the total shear (Eq. 10.6.2), deflection (Eq. 10.6.3) and slope (Eq. 10.6.4), due to an axial compressive load acting in combination with two or more transverse loads.

The following examples illustrate the use of Eq. 10.6.1 and the loading cases, as shown in Fig. 10.6.7.

Example 1:

Given: The I-beam is subjected to a transverse load of 30 lbs./in. and an axial load of 2,000 lbs. The following shows the various methods of analysis, pointing out the relative accuracy and work involved in each.

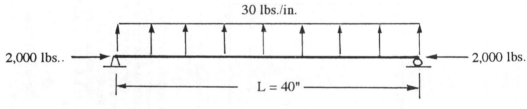

Use C_1, C_2 and $f(w)$	Use M_{max} equation	Use coefficient
[Fig. 10.6.7 (case 1)]	[Fig. 10.6.7 (case 1)]	[Fig. Fig. 10.6.4 (case 1)]
$j = \sqrt{\dfrac{EI}{P}}$	$j = \sqrt{\dfrac{EI}{P}} = 34.191$	$j = \sqrt{\dfrac{EI}{P}} = 34.191$
$= \sqrt{\dfrac{(10.3 \times 10^6)(0.227)}{2,000}}$		
$= \sqrt{1169}$		
$= 34.191$		
$\dfrac{L}{j} = \dfrac{40}{34.191} = 1.1699$		$\dfrac{L}{j} = 1.1699$
$\sin(\dfrac{L}{j}) = 0.9207$		
$\cos(\dfrac{L}{j}) = 0.3902$		
$\dfrac{L}{2j} = 0.5849$	$\dfrac{L}{2j} = 0.5849$	
$\sin(\dfrac{L}{2j}) = 0.5522$		

430

$\cos(\frac{L}{2j}) = .8337$ $\sec(\frac{L}{2j}) = \frac{1}{\cos(\frac{L}{2j})}$

$C_1 = \frac{wj^2[\cos(\frac{L}{j}) - 1]}{\sin(\frac{L}{j})}$

$= \frac{1}{0.8337}$

$= 1.1994$

$= \frac{(-30)(1169)(-0.6098)}{0.9207}$

$= 23,228$

$C_2 = -wj^2$

$= -(-30)(1169)$

$= 35,070$

$f(w) = -35,070$ From Fig. 10.6.4 (case 1):

$M = C_1\sin(\frac{x}{j}) + C_2\cos(\frac{x}{j}) + f(w)$ $M_{max} = wj^2[1 - \sec(\frac{L}{2j})]$ $M_{max} = (\frac{wL^2}{8})\mu$

M_{max} is at $x = \frac{L}{2}$ and then

$= (-30)(1169)(-0.1994)$ $= (\frac{30 \times 40^2}{8})1.18$

$= 6,993$ in-lbs.

$M_{max} = (23,228)(.5522) +$
        $(35,070)(.8337) + (-35,070)$

$= 7,080$ in-lbs

(The above value of $\mu = 1.18$ is obtained from Fig. 10.6.6)

$= 12,827 + 29,238 - 35,070$

$= 6,995$ in-lbs.

Example 2:

The following beam-column is subject to a transverse load of an uniformly varying load increasing to the right ($w_1 = 20$ lbs/in. at the left and $w_2 = 30$ lbs/in. at the right). This beam-column is a combination of loading Case 1 and 8 in Fig. 10.6.7.

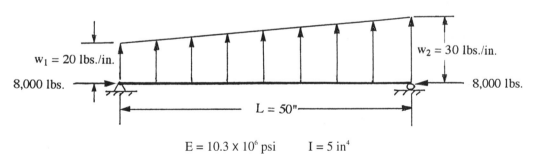

$E = 10.3 \times 10^6$ psi $I = 5$ in^4

Compute:

$j = \sqrt{\frac{EI}{P}} = \sqrt{\frac{(10.3 \times 10^6)(5)}{8,000}} = 80.23$

$\sin(\frac{L}{j}) = \sin(\frac{50}{80.23}) = \sin(0.623) = 0.5836$

$\cos(\frac{L}{j}) = 0.812$

Case 1 (see Fig. 10.6.7):

$$C_1 = \frac{wj^2[\cos(\frac{L}{j}) - 1]}{\sin(\frac{L}{j})}$$

$$= \frac{20 \times (80.23)^2 \times (0.812 - 1)}{0.5836} = -41,471$$

$$C_2 = -wj^2 = -20(80.23)^2 = -128,737$$

$$f(w) = wj^2 = 128,737$$

Case 8 (see Fig. 10.6.7):

$$C_1 = \frac{-wj^2}{\sin(\frac{L}{j})} = -\frac{10 \times 80.23^2}{0.5836} = -110,296$$

$$C_2 = 0$$

$$f(w) = \frac{wj^2 x}{L} = \frac{10(80.23)^2 x}{50} = 1,287x$$

Superposition each value of C_1, C_2 and $f(w)$:

Total ΣC_1 = C_1 (from Case 1) + C_1 (from Case 8)
= $-41,471 - 110,296 = -151,767$

Total ΣC_2 = C_2 (from Case 1) + C_2 (from Case 8)
= $-128,737 + 0 = -128,737$

Total $\Sigma f(w)$ = $f(w)$ (from Case 1) + $f(w)$ (from Case 8)
= $128,737 + 1,287x$

The resulting moment equation (from Eq. 10.6.1):

$$M = -151,767 \sin(\frac{x}{j}) - 128,737 \cos(\frac{x}{j}) + 128,737 + 1,287x \quad \text{Eq. A}$$

The max. moment occurs where $\frac{dM}{dx} = 0$:

$$\frac{dM}{dx} = \frac{-[151,767 \cos(\frac{x}{j})]}{j} + \frac{[128,737 \sin(\frac{x}{j})]}{j} + 1,287 = 0 \quad \text{Eq. B}$$

Solve for above Eq. B and obtain:

$$\sin(\frac{x}{j}) = 0.3161; \quad \cos(\frac{x}{j}) = 0.9487$$

$$\therefore \frac{x}{j} = 0.3217 \quad \text{and } x = (0.3217)(80.23) = 25.81 \text{ in.}$$

Substituting into Eq. A:

$$M_{max} = -151,767(0.3161) - 128,737(0.9487) + 128,737 + 1,287(25.81)$$
$$= -47,974 - 122,132 + 128,737 + 33,217$$

$$M_{max} = -8,152 \text{ in-lbs.}$$

Example 3:

The following beam column is subject to a transverse concentrated load W = 500 lbs. and two different end moments, namely M_1 = 4,000 in-lbs. and M_2 = 6,000 in-lbs. This beam-column is a combination of loading Case 7 and 10 in Fig. 10.6.7.

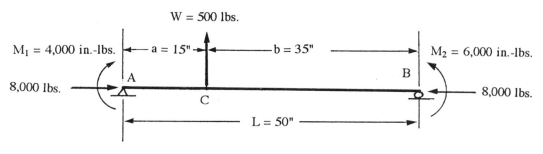

$$E = 10.3 \times 10^6 \text{ psi} \qquad I = 5 \text{ in}^4$$

Compute:

$$j = \sqrt{\frac{EI}{P}} = \sqrt{\frac{(10.3 \times 10^6)(5)}{8,000}} = 80.23; \quad \frac{L}{j} = \frac{50}{80.23} = 0.623$$

$$\sin(\frac{L}{j}) = \sin(\frac{50}{80.23}) = \sin(0.623) = 0.5836$$

$$\cos(\frac{L}{j}) = 0.812$$

$$\tan(\frac{L}{j}) = 0.7188$$

$$\sin(\frac{a}{j}) = \sin(\frac{15}{80.23}) = \sin(0.187) = 0.1859; \quad \frac{a}{j} = \frac{15}{80.23} = 0.187$$

$$\cos(\frac{a}{j}) = 0.9826$$

$$\sin(\frac{b}{j}) = \sin(\frac{35}{80.23}) = 0.4235$$

For x < a:

$$C_1 = \frac{-Wj\sin(\frac{b}{j})}{\sin(\frac{L}{j})} + \frac{M_2 - M_1\cos(\frac{L}{j})}{\sin(\frac{L}{j})}$$

$$= \frac{-500(80.23)(0.4235)}{0.5836} + \frac{6,000 - 4,000(0.812)}{0.5836}$$

$$= -24,395$$

$$C_2 = 0 + M_1 = 4,000$$

f(w) = 0 (for both cases)

$$M = -24,395\sin(\frac{x}{j}) + 4,000\cos(\frac{x}{j}) \qquad \text{Eq. A}$$

Chapter 10.0

For x > a:

$$C_1 = \frac{Wj\sin(\frac{a}{j})}{\tan(\frac{L}{j})} + \frac{M_2 - M_1\cos(\frac{L}{j})}{\sin(\frac{L}{j})}$$

$$= \frac{500(80.23)(0.1859)}{0.7188} + \frac{6,000 - 4,000(0.812)}{0.5836}$$

$$= 15,090$$

$$C_2 = -Wj\sin(\frac{a}{j}) + M_1$$

$$= -500(80.23)(0.1859) + 4,000 = -3,457$$

f(w) = 0 (for both cases)

$$M = 15,090\sin(\frac{x}{j}) - 3,457\cos(\frac{x}{j}) \qquad \text{Eq. B}$$

The maximum bending moment may occur:

(1) At either Point A or B, whichever is greater (which is 6,000 in-lbs at Point B)

(2) At Point C where the concentrated load, W, is acting and using Eq. A and x = a

$$M = -24,395\sin(\frac{x}{j}) + 4,000\cos(\frac{x}{j})$$

$$= -24,395(0.1859) + 4,000(0.9826) = -605 \text{ in-lbs}$$

(3) Check between Point A and C:

If this is the case, the maximum moment occurs where

$$\frac{dM}{dx} = 0 \text{ and differentiating Eq. A and equating it to zero,}$$

$$(\frac{-24,395}{j})\cos(\frac{x}{j}) - (\frac{4,000}{j})\sin(\frac{x}{j}) = 0$$

$$\tan(\frac{x}{j}) = \frac{-24,395}{4,000} = -6.099$$

$$\therefore \frac{x}{j} = 1.733 \text{ rad.}$$

Since Eq. A is valid in the interval $0 \leq \frac{x}{j} \leq \frac{a}{j} = 0.187$, therefore the maximum bending moment does not exist in the interval.

(4) Check between Point B and C:

Differentiating Eq. B and equating it to zero

$$(\frac{15,090}{j})\cos(\frac{x}{j}) + (\frac{3,457}{j})\sin(\frac{x}{j}) = 0$$

$$\tan(\frac{x}{j}) = \frac{-15,090}{3,457} = -4.365$$

$$\therefore \frac{x}{j} = 1.797 \text{ rad.}$$

The above $\frac{x}{j} = 1.79$ rad. is larger than $\frac{L}{j} = 0.623$. Since Eq. B is valid in the interval $\frac{a}{j} = 0.187 \leq \frac{x}{j} \leq \frac{L}{j} = 0.623$, the maximum bending moment does not occur in this interval.

The maximum moments:

$M_{max} = 6{,}000$ in-lbs at the Point B (right end)

(C) APPROXIMATION METHOD

In the design of a beam-column it is necessary to assume a value of EI before the values of $\frac{L}{j}$ can be calculated. If the bending moment shows the assumed beam cross section to be unsatisfactory, a new cross section must be assumed again, and the bending moment must be recalculated. It is obviously desirable to estimate the cross section required rather accurately before starting the bending moment calculations in order to avoid numerous trials.

The following approximation method per Eq. 10.6.5 of sizing enables the structural engineer to select a cross section before applying the equations of Fig. 10.6.1. If the bending moment in a beam-column with a compressive axial load, P, is compared to the bending moment in a simple beam with no axial load, the following approximate relationship is obtained:

$$M = \frac{M_o}{1 - \frac{P}{P_{cr}}}$$
Eq. 10.6.5

where: M – The bending moment in the beam-column
P – Applied compressive axial load
M_o – The primary bending moment (without the compressive axial load)
P_{cr} – The Euler column load, $P_E = \frac{\pi^2 EI}{L^2}$

The above equation is accurate enough for most practical design cases in which the end moments are zero.

Example 4:

Use the same values as given in Example 1

w = 30 lbs/in.   P = 2,000 lbs.
$E = 10.3 \times 10^6$ $I = 0.227$ in.^{4} L = 40 in.

The primary moment (max. moment at mid-span of x = 20 in.):

$$M_o = \frac{wL^2}{8} = \frac{(30)(40)^2}{8} = 6{,}000 \text{ in.-lbs}$$

The Euler column load:

$$P_{cr} = \frac{\pi^2 EI}{L^2} = \frac{\pi^2 (10.3 \times 10^6)(0.227)}{40^2} = 14{,}423 \text{ lbs.}$$

From Eq. 10.6.5, the approximate max. beam-column bending moment:

$$M = \frac{M_o}{1 - \frac{P}{P_{cr}}}$$

Chapter 10.0

$$M = \frac{6,000}{1 - \frac{2,000}{14.423}}$$

$$= 6,966 \text{ in.-lbs}$$

This value is very close to the exact value of 6,987 in-lbs as calculated in Example 1.

(D) BEAM-COLUMNS WITH ELASTIC SUPPORTS

Design equations and data are given in Fig. 10.6.8.

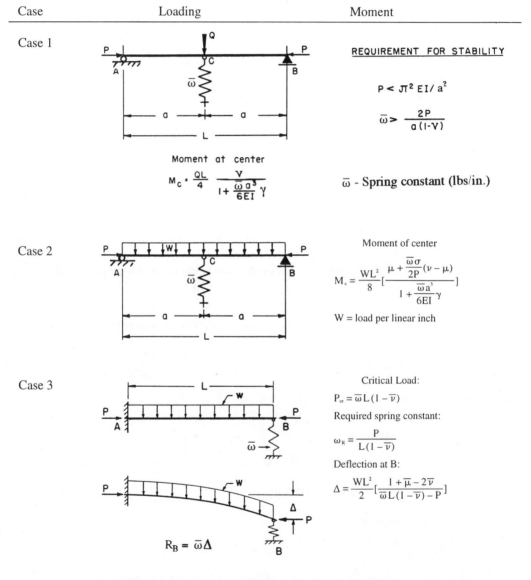

(The above beam-column coefficients can be found in Fig. 10.6.6)

Fig. 10.6.8 Beam-Column Equations with Elastic Supports

10.7 CRIPPLING STRESS

Compression in airframe members can be considered as instability problems and may be classified as:

- Column failure
- Local instability failure (usually referred to as crippling)

A perfect column is a member that is initially straight and has zero deflection up to a load P_{cr} at which point the member becomes unstable, as shown in Fig. 10.7.1. In an airframe the structural members are very seldom initially straight.

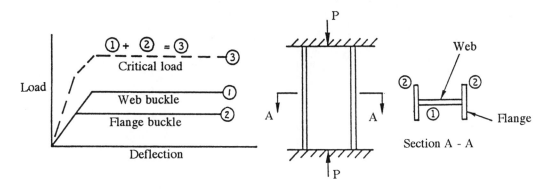

Fig. 10.7.1 Deflection vs. Load

When a crippling failure, as shown in Fig. 10.7.2, occurs on a formed section it appears as a local distortion. The more stable parts of the section continue to carry load and support the buckled parts until failure of the total section takes place. The initial buckling stress of the various elements of a section can be calculated, but the determination of the failing stress of the section is impossible to calculate mathematically.

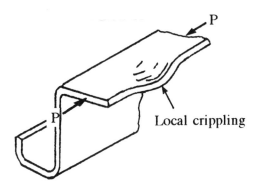

Fig. 10.7.2 Flange Crippling

Lacking a satisfactory theory for the prediction of crippling failure, it is necessary to

- Rely upon test results
- Use empirical methods

On the following pages empirical methods for predicting the crippling strength of extruded and formed sheet metal elements are presented. Allowable stresses computed by these methods agree reasonably well with test results.

Compressive crippling (for further information, refer to Ref. 10.9), also referred to as local buckling, is defined as an inelasticity of the cross section of a structural member in its own plane rather than along its longitudinal axis, as in column buckling. The maximum crippling stress of a member is a function of its cross section rather than its length.

The crippling stress for a given section is calculated as if the stress were uniform over the entire section. In reality, the stress is not uniform over the entire section; parts of the section buckle at a stress below the crippling stress with the result that the more stable areas, such as intersections and corners, reach a higher stress than the buckled members as shown in Fig. 10.7.3.

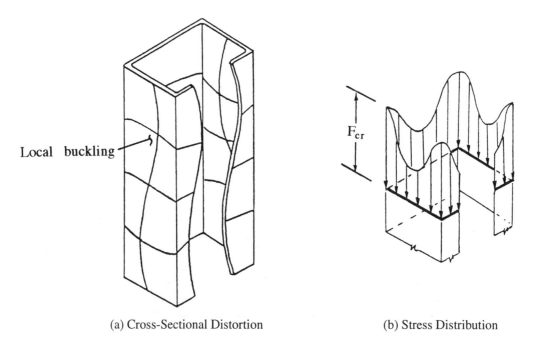

(a) Cross-Sectional Distortion   (b) Stress Distribution

Fig. 10.7.3 A Cross Section Subjected to Crippling Stress

This chapter presents data for buckling and crippling compression failures of aluminum alloys which are commonly used in airframe structures.

(A) TEST CURVES OF SPECIFIC MATERIALS

Formed and extruded sections are analyzed in the same manner, although different values are used for each.

(a) The section is broken down into individual segments, as shown in Fig. 10.7.4, and each segment has a width 'b' and a thickness 't' and will have either one or no edge free.

(b) The allowable crippling stress for each segment is found from the applicable material test curve of which can be selected from the typical curves shown in Fig. 10.7.6 and Fig. 10.7.7 for aluminum alloys that are commonly used in airframe structures.

(c) The allowable crippling stress for the entire section is computed by taking a weighted average of the allowables for each segment:

$$F_{cc} = \frac{(b_1 t_1 F_{cc1} + b_2 t_2 F_{cc2} + ...)}{(b_1 t_1 + b_2 t_2 + ...)}$$

$$F_{cc} = \frac{\Sigma b_n t_n F_{ccn}}{\Sigma b_n t_n}$$

Eq. 10.7.1

where: $b_1, b_2,$ — Lengths of the individual segments
  $t_1, t_2,$ — Individual segment thicknesses
  $F_{cc1}, F_{cc2},$ — Allowable crippling stresses of individual segments
    (see Fig. 10.7.6 and Fig. 10.7.7)

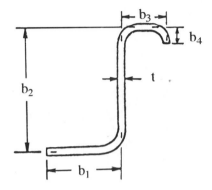

(a) Formed Section

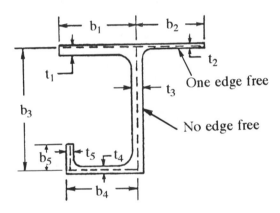

(b) Extruded Section

Fig. 10.7.4 Formed vs. Extruded Sections

The following Fig. 10.7.5 provides sufficient stability to adjacent formed flange segment of b_L.

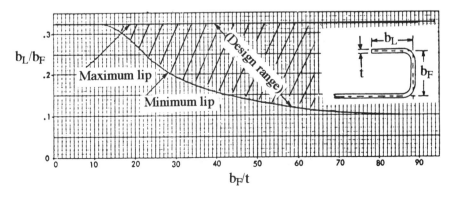

Fig. 10.7.5 Lip Criteria for Formed Sections

Chapter 10.0

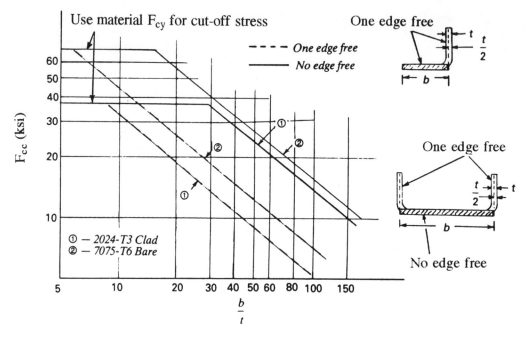

(2024-T3 Clad and 7075-T6 sheets)

Fig. 10.7.6 Crippling Stress of Formed Sections

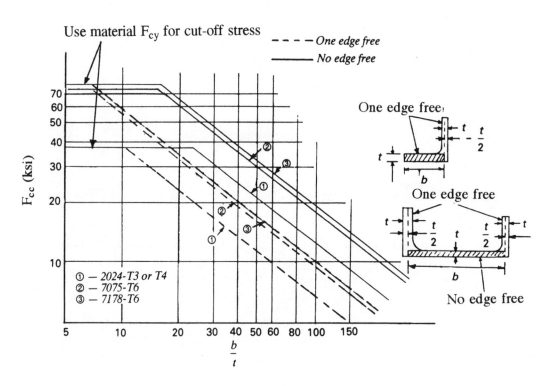

(2024-T3511, T4, T42, T81 and 7075-T6511 extrusions)

Fig. 10.7.7 Crippling Stress of Extruded Sections

The allowable crippling stress curves of Fig. 10.7.6 and Fig. 10.7.7 are semi-empirical curves based on test data. Actually there are more test design curves in terms of the formed and extruded sections geometrical properties and classified by alloy. These curves can be found in the handbook of any major aircraft manufacturing company. In the absence of the specific test data, the average values in good approximation as discussed later may be used.

Example 1:

Find crippling stress of the following formed section and material; 2024-T3 clad (F_{cy} = 38 ksi).

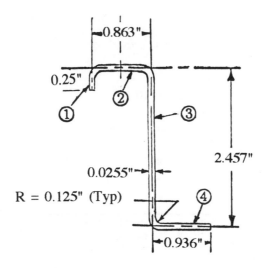

Segment	Free edges	b_n	t_n	$\dfrac{b_n}{t_n}$	$b_n t_n$	F_{ccn} (from Fig. 10.7.6)	$t_n b_n F_{ccn}$
①	1	0.25	.0255	9.8	0.0064	38	0.243
②	0	0.863	.0255	33.8	0.0221	33	0.726
③	0	2.457	.0255	96.4	0.0627	14	0.879
④	1	0.936	.0255	36.7	0.0239	11.7	0.28
					Σ 0.1151	22	2.128

From Eq. 10.7.1, the crippling stress is

$$F_{cc} = \frac{\Sigma b_n t_n F_{ccn}}{\Sigma b_n t_n} = \frac{2.128}{0.1151} = 18.5 \text{ ksi}$$

Determine whether the lip (segement ①) provides sufficient stability to adjacent flange segment

$$\frac{b_L}{b_F} = \frac{0.25}{0.863} = .29 \text{ and } \frac{b_F}{t} = \frac{0.863}{0.0255} = 33.8$$

From Fig. 10.7.5, it lies within the design range.   O.K.

(B) GENERAL SOLUTION OF FORMED SECTIONS

If the test curves for a specified material and shape (formed or extrusion) are not available, Fig. 10.7.9 gives the necessary parameters for a generalized method of computing crippling allowables of formed sections in terms of:

- Material modulus of elasticity (E)
- Compressive yield strength (F_{cy})
- The experimentally-derived crippling curve parameters of 'B' and 'm'

The following crippling equations and curves for predicting formed sections can be used:

(a) Zee and channel sections:

$$F_{cc} = B \left[\left(\frac{1}{\frac{A}{t^2}} \right)^m \right] \left[\sqrt[3]{E^m F_{cy}^{(3-m)}} \right] \qquad \text{Eq. 10.7.2}$$

(b) All other sections:

$$F_{cc} = B \left[\left(\frac{1}{\frac{A}{gt^2}} \right)^m \right] \left[\sqrt{E^m F_{cy}^{(2-m)}} \right] \qquad \text{Eq. 10.7.3}$$

where: B – Constant (see Fig. 10.7.9)
A – Cross section area
g – Number of flanges and cuts, see Fig. 10.7.8
t – Thickness
m – Slope of crippling curve (see Fig. 10.7.9)

Fig. 10.7.9 provides a general crippling design curves for formed sections to be used in sizing.

Cross-section	Number of flanges (ΣF)	Number of cuts (ΣC)	g = F + C
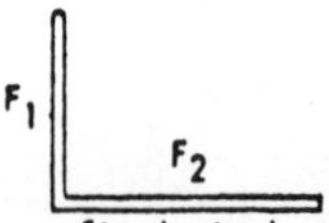 Simple Angle	2	0	2
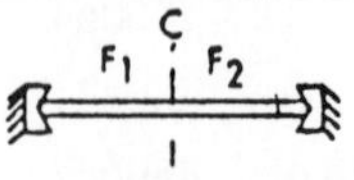 Plate	2	1	3
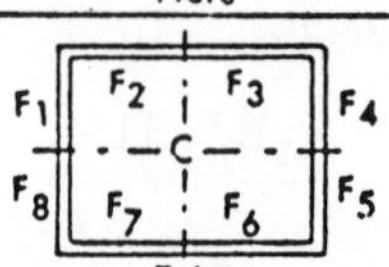 Tube	8	4	12
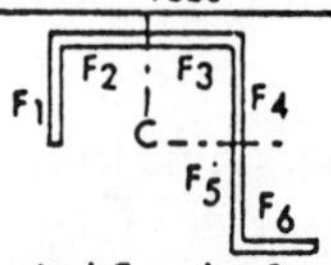 Typical Complex Section	6	2	8
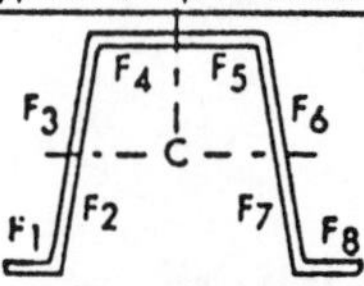 Hat (Complex Section)	8	3	11

Fig. 10.7.8 g Values for Various Cross Sections

Column Buckling

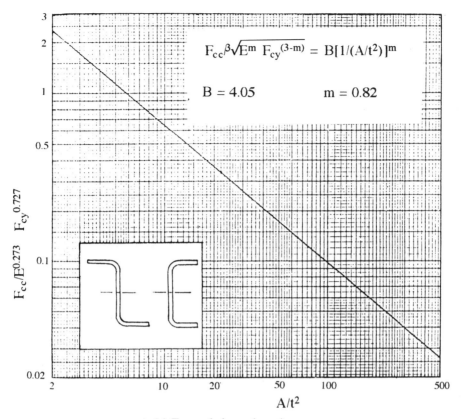

(a) Zee and channel sections

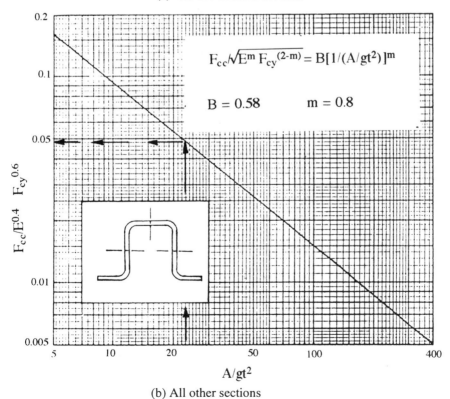

(b) All other sections

Fig. 10.7.9 Crippling Allowables for Formed Sections

Example 2:

Given a Zee section which is the same one used in Example 1, determine the crippling stress (F_{cc}).

Material: 2024-T3 clad; $E = 9.5 \times 10^6$ psi; $F_{cy} = 38,000$ psi

Cross-section area: $A = 0.1151$ in.^{2}; Thickness: $t = 0.0255$ in.

From Fig. 10.7.8: $g = 6$ flanges $+ 2$ cuts $= 8$

$$\frac{A}{gt^2} = \frac{0.1151}{8 \times 0.0255^2} = 23 \text{ and go to Fig. 10.7.9(b) to obtain the following value:}$$

$$\frac{F_{cc}}{E^{0.4} \times F_{cy}^{0.6}} = 0.048$$

$$F_{cc} = 0.048 (E^{0.4} \times F_{cy}^{0.6}) = 0.048 (9.5 \times 10^6)^{0.4} (38,000)^{0.6}$$
$$= 0.048 \times 618 \times 560 = 16,612 \text{ psi}$$

Compare this crippling stress value with $F_{cc} = 18,500$ psi from Example 1; this value is conservative by about 11%.

(C) GENERAL SOLUTION OF EXTRUDED SECTIONS

Since extrusions are available in a wide variety of cross-sectional shapes and usually have unequal flange thicknesses, crippling allowables cannot be as conveniently categorized as those for formed sections of uniform thickness. When aluminum alloys are plotted in the non-dimensional form as straight lines, as shown in Fig. 10.7.10, the plot shows good correlation with test data.

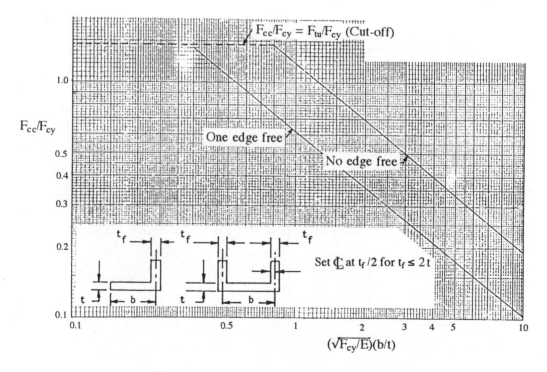

(This curves are applicable to all ductile aircraft materials)

Fig. 10.7.10 Non-dimensional Crippling Stress Curves for Extrusions

Example 3:

Given an extruded Zee-shape as shown below, determine the crippling stress (F_{cc}).

Material: 7075-T6 Extrusion

$E = 10.5 \times 10^6$ psi

$F_{cy} = 70$ ksi

$\sqrt{\dfrac{F_{cy}}{E}} = 0.082$

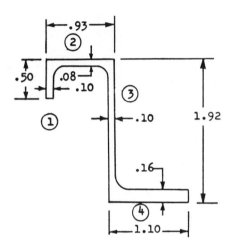

Seg't	Free edges	b_n	t_n	$b_n t_n$	$\dfrac{b_n}{t_n}$	$\sqrt{\dfrac{F_{cy}}{E}}$	$\sqrt{\dfrac{F_{cy}}{E}}(\dfrac{b_n}{t_n})$	$\dfrac{F_{ccn}}{F_{cy}}$ (from Fig. 10.7.10)	F_{ccn}	$t_n b_n F_{ccn}$
①	1	0.46	.1	0.046	4.6	0.082	0.38	1.3	(70)*	3.22
②	0	0.83	.08	0.066	10.4	0.082	0.85	1.35	(70)*	4.62
③	0	1.8	.10	0.18	18	0.082	1.48	0.87	60.9	10.96
④	1	1.05	.16	0.168	6.6	0.082	0.54	1.0	70	11.76
				$\Sigma b_n t_n = 0.46$						$\Sigma b_n t_n F_{ccn} = 30.56$

* Use cut-off stress by $F_{cy} = 70$ ksi.

From Eq. 10.7.1, the crippling stress is

$$F_{cc} = \dfrac{\Sigma b_n t_n F_{ccn}}{\Sigma b_n t_n} = \dfrac{30.56}{0.46} = 66.4 \text{ ksi}$$

(D) CROSS SECTION IN BENDING

Data for beam bending subjected to local crippling failure of the cross section in compression is given in Fig. 10.7.11 through Fig. 10.7.13.

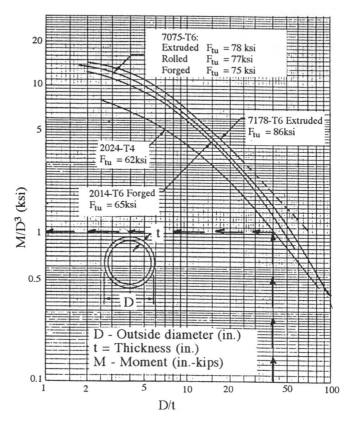

Fig. 10.7.11 Aluminum Alloy Tubing in Bending

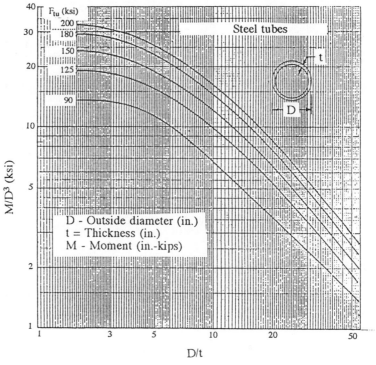

(Modulus of rupture bending stress (F_b) can be found in Fig. 9.8.16)

Fig. 10.7.12 Steel Tubing in Bending

Column Buckling

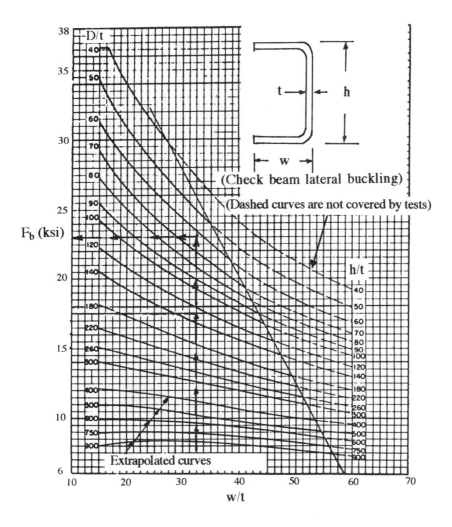

Fig. 10.7.13 Formed Channel Section (Aluminum Alloy 2024-T4 Clad) in Bending

Example 1:

Assume aluminum tube (2024-T4) as shown below and determine the allowable bending moment.

$$D = 2" \text{ and } t = 0.05"$$

$$\frac{D}{t} = \frac{2}{0.05} = 40$$

From Fig. 10.7.11, obtain $\frac{M}{D^3} = 1.02$ and:

$$M = 1.02 \times 2^3 = 8.16 \text{ in.-kips.}$$

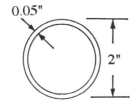

447

Chapter 10.0

Example 2:

Assume aluminum channel (2024-T4) as shown below and determine the allowable bending stress (F_b).

w = 2"; h = 4"; t = 0.063"

$\frac{w}{t} = \frac{2}{0.063} = 32$ and $\frac{h}{t} = \frac{4}{0.063} = 63$

Go to Fig. 10.7.13 and find

$F_b = 23$ ksi.

10.8 INTERACTION BETWEEN COLUMN AND CRIPPLING STRESS

There are two general types of primary column failure for stable columns and both of them are affected by the slenderness ratio as shown in Fig. 10.8.1. These types of failure are:

- Long columns, characterized by a high slenderness ratio, exhibit failure according to the classical Euler equation
- In the short to intermediate column range, characterized by low slenderness ratios, failure is due to either local crippling or block compression of the material at its compressive yield point

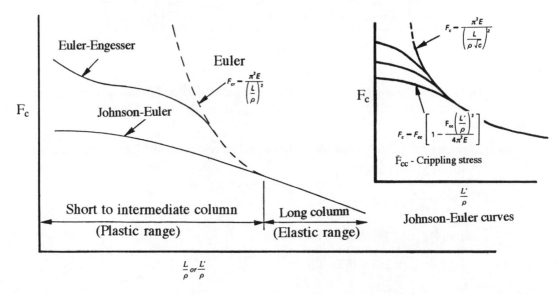

Fig. 10.8.1 Column Failures vs. Slenderness Ratio ($\frac{L'}{\rho}$)

Column Buckling

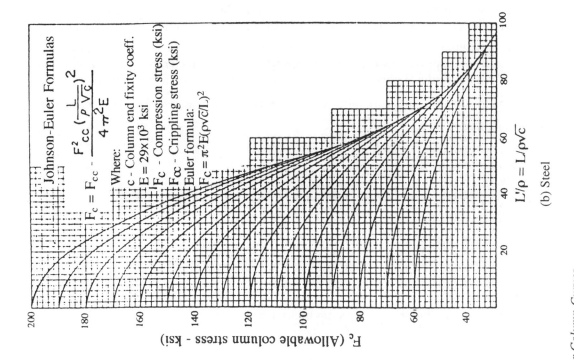

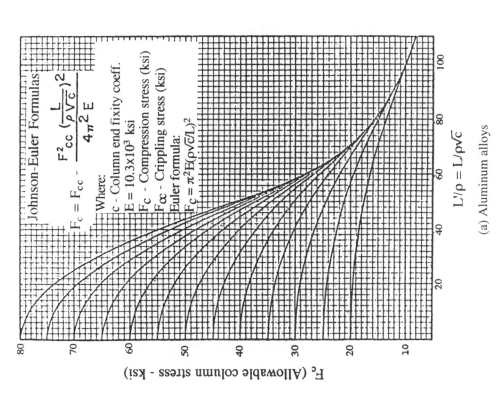

Fig. 10.8.2 Johnson-Euler Column Curves

To facilitate checking a column for failure in the flexural instability mode various empirical formulas have been developed from tests predicting short to intermediate column failure. A known Euler-Engesser equation does not cover the local crippling interaction, so for practical design the Johnson parabola (or called Johnson-Euler) curves should be used for primary column sizing.

For columns subject to local crippling, the critical stresses in a column are predicted to cause failure at a lower stress. This is due to the interaction between the primary flexural instability mode of failure and local crippling mode. To take account of this effect, an empirical column curve, the Johnson parabola, is recommended for the intermediate range of $\frac{L'}{\rho}$. The Johnson-Euler column formula is

$$F_c = F_{cc} - \frac{(F_{cc})^2 (\frac{L'}{\rho})^2}{4\pi^2 E}$$
Eq. 10.8.1

Fig. 10.8.2 presents Johnson-Euler column curves for aluminum alloys and steels.

This chart is used as follows:

(1) Determine the allowable crippling stress, F_{cc}

(2) Using the value of F_{cc}, enter the F_c scale and follow the Johnson-Euler curve as shown in Fig. 10.8.1 for this value to the $\frac{L'}{\rho}$ value

(3) Read the allowable column stress back on the F_c scale

Example:

Given an aluminum alloy, $\frac{L'}{\rho} = 42$, and $F_{cc} = 40,000$ psi

Entering the curve for aluminum alloys, Fig. 11.8.2(a) and following the curve for $F_{cc} = 40,000$ psi, read $F_c = 33,000$ psi from the vertical scale at left.

References

10.1 Timoshenko, S and Gere, J. M., "THEORY OF ELASTIC STABILITY", McGraw-Hill Book, Co., Inc., New York, NY, (1961).
10.2 Niu, C. Y., "AIRFRAME STRUCTURAL DESIGN", Hong Kong Conmilit Press Ltd., P.O. Box 23250, Wanchai Post Office, Hong Kong, (1988).
10.3 Bruhn, E. F., "ANALYSIS AND DESIGN OF FLIGHT VEHICLE STRUCTURES", Jacobs Publishers, 10585 N. Meridian St., Suite 220, Indianapolis, IN 46290, (1965).
10.4 Roark, R. J. and Young, W. C., "FORMULAS FOR STRESS AND STRAIN", McGraw-Hill Book Company, (1975).
10.5 Newmark, N. M., "Numerical Procedure for Computing Deflections, Moments and Buckling Loads", Proceedings, American Society of Civil Engineers, Vol. 68, No. 5, (May, 1942).
10.6 Niles, A. S. and Newell, J. S., "AIRPLANE STRUCTURES", John Wiley & Sons, Inc., (1938).
10.7 Ensrud, A. F., "The Elastic Pole", Lockheed Aircraft Corp., (1952).
10.8 Cassens, J., "Tables for Computing Various Cases of Beam Columns", NACA TM 985, (Aug., 1941).
10.9 Gerard, G., "The Crippling Strength of Compression Elements", Journal of The Aeronautical Science, (Jan. 1958). pp. 37-52.
10.10 Needham, R. A., "The Ultimate Strength of Aluminum-Alloy Formed Structural Shapes in Compression", Journal of The Aeronautical Sciences, (April, 1954). pp. 217-229.
10.11 Shanley, F. R., "Inelastic Column Theory", Journal of The Aeronautical Sciences, (May, 1947). pp. 261- 267.
10.12 Jones, W. R., "The Design of Beam-Columns", Aero Digest, (June, 1935). pp. 24-28.
10.13 Chajes, A. and Winter, G., "Torsional-Flexural Buckling of Thin-Walled Numbers", Journal of Structural Division, ASCE Proceedings, (August, 1965). p. 103.

Chapter 11.0

BUCKLING OF THIN SHEETS

11.1 INTRODUCTION

The stability of a plate supported on its edges and loaded by various types of in-plane loads has been solved for many types of boundary conditions. This chapter presents design data and curves for the determination of initial buckling stresses for flat and curved plates subjected to in-plane compression, shear, bending, and combinations of these stresses, for materials commonly used in airframe structures. It should be remembered that the practical skin-stringer panel constructions after the initial buckling of the plate will not take additional loads. Although the adjacent stringers will still withstand additional loads until the stringers reach their crippling stress (see Chapter 10).

The basic equation of plate buckling is derived from the Euler column equation (Eq. 10.2.3 in Chapter 10) which is

$$F_{cr} = \frac{\pi^2 E}{(\frac{L'}{\rho})^2}$$

$$\text{or } F_{cr} = \frac{c \pi^2 EI}{AL^2}$$

Eq. 11.1.1

The unit elongation (e) of a flat plate loaded in two directions (x, y) is shown in Fig. 11.1.1.

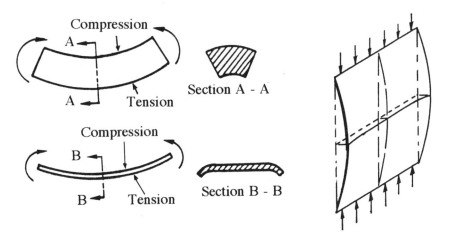

Fig. 11.1.1 Effect of Poisson's Ratio on a Flat Plate

The strain in the y-direction due to the effect of Poisson's ratio (μ) is:

$$e_y = (\frac{f_y}{E}) - \mu(\frac{f_x}{E})$$

Chapter 11.0

The stress in the x-direction is:
$$f_x = \mu f_y$$

So that the strain in y-direction becomes:
$$e_y = (\frac{f_y}{E}) - \mu^2(\frac{f_y}{E})$$

or $e_y = \frac{f_y(1-\mu^2)}{E}$

It follows that the stiffness of flat plates can be expressed as:
$$EI = (\frac{f_y}{e_y})I = \frac{EI}{(1-\mu^2)}$$

The Euler column equation of Eq. 11.1.1 for flat plate becomes:
$$F_{cr} = \frac{c\pi^2 EI}{(1-\mu^2)(AL^2)}$$

Since the moment of inertia of a plate is $I = \frac{bt^3}{12}$ and plate area is $A = bt$, when these two values are substituted into the equation above it becomes:

$$F_{cr} = \frac{k\pi^2 E}{12(1-\mu^2)}(\frac{t}{L})^2 \qquad \text{Eq. 11.1.2}$$

or $F_{cr} = KE(\frac{t}{L})^2$ 　　　　　　　　Eq. 11.1.3

where: L – Plate length (parallel to the load direction)
E – Modulus of elasticity
k – Buckling coefficient
c – End-fixity (see Chapter 10)
$K = \frac{k\pi^2}{12(1-\mu^2)} = 0.904k$ – Modified buckling coefficient for $\mu \approx 0.3$ (e.g., aluminum, steel alloys)

The two equations above are still the Euler equation, but they apply to a plate loaded as a column. For instance, a flat plate subjected to load on two ends which are hinged and the other two edges are free, as shown in Fig. 11.1.1:

　　K = 0.9 approximately

　　or K = 0.82 (without effect of Poisson's ratio)

When in-plane loads are applied to the edge of a flat plate, it will buckle at some critical load depending on the plate aspect ratio, plate thickness, and edge conditions, as shown in Fig, 11.1.2:

- Fig. 11.1.2(a) shows both that the unloaded side is free and that the plate acts as a column
- Fig. 11.1.2(b) shows that one unloaded side is free and the other side is restrained; this is referred to as a free flange
- Fig. 11.1.2(c) shows that both unloaded sides are restrained and this is referred to as a plate

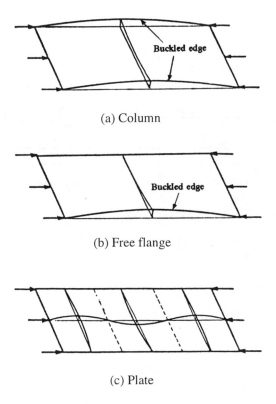

(a) Column

(b) Free flange

(c) Plate

Fig. 11.1.2 Plates with Various Edge Supports

11.2 GENERAL BUCKLING FORMULAS

Let the vertical edges of a flat plate be supported by vee groves so that they can rotate but must remain straight lines, as shown in Fig. 11.2.1. If the panel is to buckle, it must bend in two directions and the resistance to buckling is greatly increased; in fact the panel will now sustain four times the load previously carried when no edge support was provided. The buckling equation for a square plate with hinged supports on all four edges is:

$$F_{cr} = 3.62 \, E \left(\frac{t}{L}\right)^2 \qquad \text{Eq. 11.2.1}$$

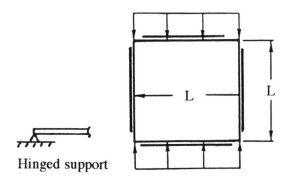

Hinged support

Fig. 11.2.1 Square Plate with Hinged Support on Four Edges

When the plate is lengthened in the direction of loading (L), the principal restraint against buckling is the bending of the plate across the minimum panel dimension 'b'. Fig. 11.2.2 shows the plate under in-plane compression loading buckles into three waves, each of them being square and acting in the same manner as the plate in Fig. 11.2.1.

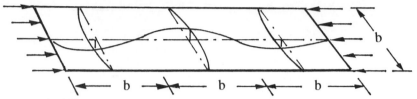

Fig. 11.2.2 Long Rectangular Plate with Four Hinged Support Edges

(A) LOADING AND EDGE CONDTIONS

It is seen that the minimum dimension or short side 'b' of the plate is the most important parameter in the buckling formula. The buckling equation for the rectangular plate shown in Fig. 11.2.2 is:

$$F_{cr} = 3.62 E \left(\frac{t}{b}\right)^2 \qquad \text{Eq. 11.2.2}$$

Therefore, the general buckling equation for both flat and curved plates is:

$$F_{cr} = \frac{k \eta_p \pi^2 E}{12(1-\mu^2)} \left(\frac{t}{b}\right)^2 \qquad \text{Eq. 11.1.3}$$

$$\text{or} \quad F_{cr} = K \eta_p E \left(\frac{t}{b}\right)^2 \qquad \text{Eq. 11.2.4}$$

The buckling coefficients (k and K) depend upon:

(a) Plate size (aspect ratio)

(b) Edge restraint (free, hinged, fixed, or rotational restraints) is shown in Fig. 11.2.3:
  - Free edge (F)   – Entirely free to deflect and rotate
  - Hinged edge (H) – Simply supported (SS) where the plate cannot deflect, but can freely rotate
  - Clamped edge (C) – Fixed support so that the plate cannot deflect or rotate
  - Edge rotational restraint (ε) having a degree of restraint between that of a hinged edge and a fixed edge

(c) Type of loading (in-plane compression, shear, or bending)

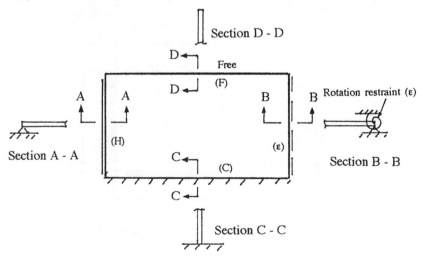

Fig. 11.2.3 Rectangular Plate with Various Edge Supports

For Eq. 11.2.3 and Eq. 11.2.4, where t (thickness) and b (width or short side) are dimensions of the plate it is established that these are general formulas and are valid with the selection of the proper buckling coefficients:

- k_c or K_c – Compression load
- k_s or K_s – Shear load
- k_b or K_b – bending load
- η_p – plasticity reduction factor

In this chapter select the proper design curve based on whether the equation being used contains a k or K value.

(A) EDGE ROTATIONAL RESTRAINTS

Rotational restraint along the the unloaded edges can vary from a simple support to a fixed (or clamped) support condition which is presented by the following equation:

$$\varepsilon = \frac{4 S_v b}{\eta_p D} \qquad \text{Eq. 11.2.5}$$

where: $D = \dfrac{E t^3}{12(1-\mu^2)}$ – Flexural rigidity of the plate (see Chapter 7.2)

$\quad$ b – Short side of the plate

$\quad$ S_v – Stiffness per unit length of the elastic restraining medium or the moment required to rotate a unit length through $\dfrac{1}{4}$ radian. This value is very difficult to determine and usually is based on the engineer's judgment

$\quad$ η_p – Plasticity reduction factor of material in plastic range.

Fig. 11.3.2 and Fig. 11.3.3 give ε coefficients for various degrees of rotational restraints under in-plane compression loading.

If there are two different coefficients on unloaded edges, i.e., k_1 and k_2, use the following approximate equation:

$$k_{av} = \sqrt{k_1 \times k_2} \qquad \text{Eq. 11.2.6}$$

The buckling coefficients k_1 and k_2 (or K_1 and K_2) are applicable to both compression and shear loading cases with different unloaded edge conditions.

Example:

Assume a plate with one unloaded edge which is hinged and all other edges are clamped; use Eq. 11.2.6 to find the shear buckling coefficient $k_{s,av}$ with an aspect ratio equal to $\dfrac{a}{b} = 3$.

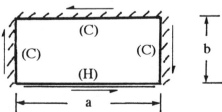

From Fig. 11.3.5, obtain $K_{s,①} = 8.7$ (Case ①) for the two unloaded edges which are clamped and $K_{s,③} = 6$ (Case ③) for the two unloaded edges which are hinged. Therefore, the average buckling coefficient in compression loading is

$$K_{s,av} = \sqrt{K_{s,①} \times K_{s,③}}$$
$$= \sqrt{8.7 \times 6} = 7.22$$

(B) PLASTIC REDUCTION FACTORS

Since the values of E and μ are not constant in the plastic range as they are in the elastic range, the plastic reduction factor (η_p) as defined below is used:

(a) Compression stress:

$$\eta_c = (\frac{E_t}{E})^{\frac{1}{2}} = [\frac{1}{1 + (\frac{0.002 E n}{F_{cy}})(\frac{F_{c,cr}}{F_{cy}})^{n-1}}]^{\frac{1}{2}} \qquad \text{Eq. 11.2.7}$$

where: n – Material shape parameter

(b) Shear stress:

$$\eta_s = (\frac{G_t}{G})^{\frac{1}{2}} = [\frac{1}{1 + (\frac{0.002 G n}{F_{sy}})(\frac{F_{s,cr}}{F_{sy}})^{n-1}}]^{\frac{1}{2}} \qquad \text{Eq. 11.2.8}$$

where: $G = \frac{E}{2(1 - \mu^2)}$ – Modulus of rigidity

$F_{sy} = 0.55 F_{cy}$

A tremendous amount of theoretical and empirical work has been done to obtain these values and Eq. 11.2.7 and Eq. 11.2.8 can be applied for any plate boundary condition. A simple method for obtaining the buckling stresses, $F_{c,cr}$, and $F_{s,cr}$ is shown in Fig. 11.2.4 and Fig. 11.2.5, respectively.

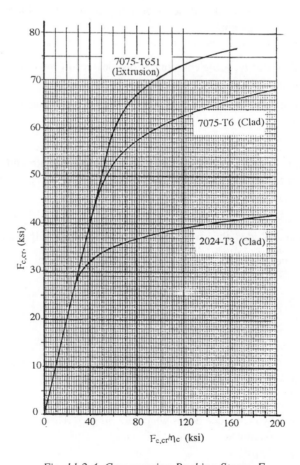

Fig. 11.2.4 Compression Bucking Stress $F_{c,cr}$

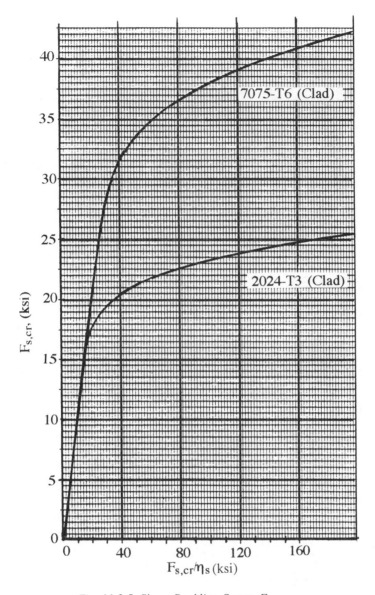

Fig. 11.2.5 Shear Buckling Stress $F_{s,cr}$

If design curves are not available they can be constructed for specific materials by the following procedure:

- Use Eq. 11.2.7 or Eq. 11.2.8 to construct the curves
- E, F_{cy} and n values can be obtained from Ref. 4.1
- If an n value is not available, use n = 20 which is a good approximation for most materials
- Assume a $F_{c,cr}$ value and calculate the η_p value several times and then plot them as shown in Fig. 11.2.4
- Shear curves, such as shown in Fig. 11.2.5, can be constructed in the same manner

Chapter 11.0

(C) CLADDING EFFECT

Cladding will reduce the plate buckling stress (F_{cr}) of an aluminum clad material. A simple method for accounting the presence of the cladding is to use bare material properties and reduce the material thickness used in Eq. 11.2.3 and Eq. 11.2.4. The actual thickness used to calculate buckling stress is found by multiplying the actual thickness by the cladding reduction factor (λ) as indicated in Fig. 11.2.6.

Material	Cladding	Plate thickness (in.)	Reduction facotr (λ)
2014	6053	t < 0.04	0.8
		t > 0.04	0.9
2024	1230	t < 0.064	0.9
		t > 0.064	0.95
7075	7072	All thickness	0.92

Fig. 11.2.6 Cladding Reduction Factor (λ) for Aluminum Clad Materials (Ref. 11.7)

11.3 FLAT PLATES

(A) COMPRESSION LOAD

The initial buckling stress for a flat plate under an in-plane compression load is:

$$F_{c,cr} = \frac{k_c \eta_c \pi^2 E}{12(1-\mu^2)} \left(\frac{t}{b}\right)^2 \qquad \text{Eq. 11.3.1}$$

$$\text{or} \quad F_{c,cr} = K_c \eta_c E \left(\frac{t}{b}\right)^2 \qquad \text{Eq. 11.3.2}$$

where: η_p – Plasticity reduction factor in compression load

Fig. 11.3.1 shows flat plate buckling coefficients (K_c) for in-plane compression loads for Eq. 11.3.2 and Fig. 11.3.2 through Fig. 11.3.4 shows flat plate buckling coefficients (k_c) for in-plane compression loads for Eq. 11.3.1.

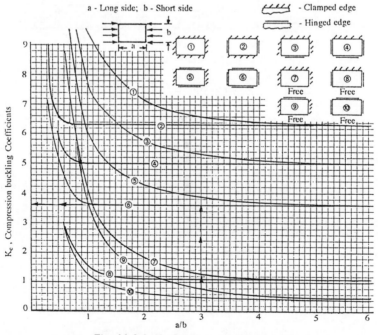

Fig. 11.3.1 K_c Coefficients (Compression)

Buckling of Thin Sheets

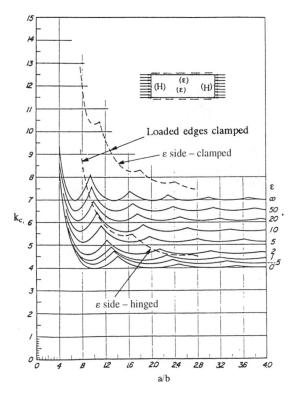

Fig. 11.3.2 k_c Coefficients with Various Edge Rotational Restraints (Compression)

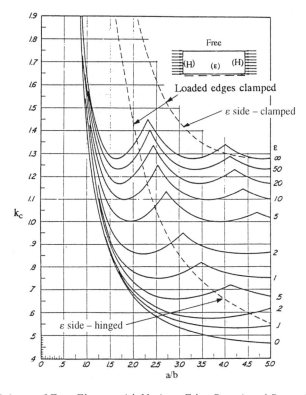

Fig. 11.3.3 k_c Coefficients of Free Flange with Various Edge Rotational Restraints (Compression)

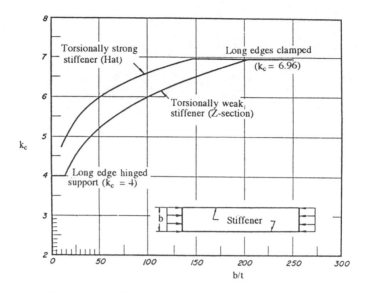

Fig. 11.3.4 k_c Coefficients for a Skin with Stiffeners on Two Sdies (Compression) (Ref. 11.7)

(B) SHEAR LOAD

The initial buckling stress for a flat plate under in-plane shear load is

$$F_{s,cr} = \frac{k_s \eta_s \pi^2 E}{12(1-\mu^2)} \left(\frac{t}{b}\right)^2 \qquad \text{Eq. 11.3.3}$$

$$\text{or} \quad F_{s,cr} = K_s \eta_s E \left(\frac{t}{b}\right)^2 \qquad \text{Eq. 11.3.4}$$

Fig. 11.3.5 shows flat plate buckling coefficients (K_s) for in-plane shear.

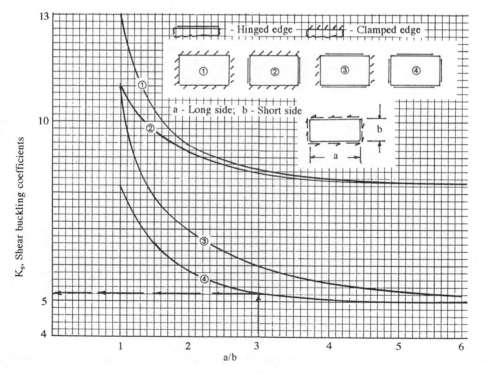

Fig. 11.3.5 K_s Coefficients (Shear)

(C) BENDING LOAD

The initial buckling stress for a flat plate under in-plane bending load is

$$F_{b,cr} = \frac{k_b \eta_c \pi^2 E}{12(1-\mu^2)} \left(\frac{t}{b}\right)^2 \qquad \text{Eq. 11.3.5}$$

$$\text{or} \quad F_{b,cr} = K_b \eta_c E \left(\frac{t}{b}\right)^2 \qquad \text{Eq. 11.3.6}$$

Fig. 11.3.6 shows flat plate buckling coefficients, K_b, for in-plane bending loads.

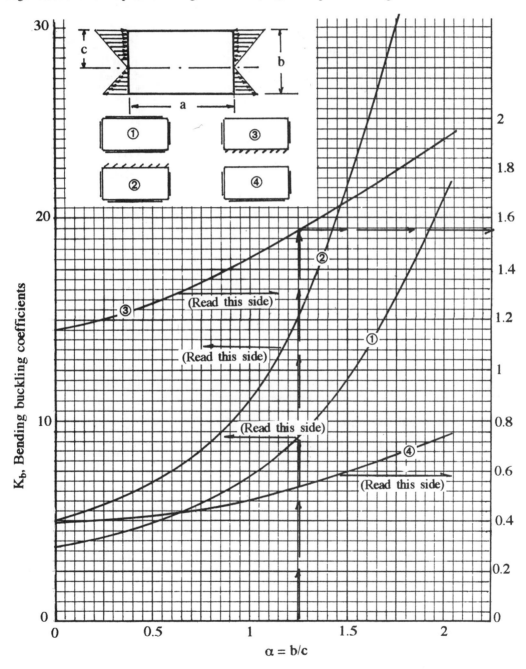

Fig. 11.3.6 K_b Coefficients (Bending) – $\frac{a}{b} > 1.0$

Example 1:

Given a plate as shown and determine buckling stress (F_{cr}) under various boundary conditions.
boundary conditions.

Material: 2024-T3 bare and $E = 10^7$ psi

a = 8"
b = 6" (short side)
t = 0.04"

$$\frac{a}{b} = \frac{8}{6} = 1.33$$

(1) Compression – all hinged edges:

$K_c = 3.6$ (see Fig. 11.3.1, Case ⑥)

From Eq. 11.3.2

$$\frac{F_{c,cr}}{\eta_c} = K_c E \left(\frac{t}{b}\right)^2$$

$$\frac{F_{c,cr}}{\eta_c} = 3.6 \times 10^7 \left(\frac{0.04}{6}\right)^2$$

$$= 1,600 \text{ psi}$$

From Fig. 11.2.4, the true buckling stress, $F_{c,cr} = 1,600$ psi

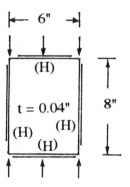

(2) Compression – one edge free:

$K_c = 0.9$ (see Fig. 11.3.1, Case ⑩)

From Eq. 11.3.2

$$\frac{F_{c,cr}}{\eta_c} = K_c E \left(\frac{t}{b}\right)^2$$

$$\frac{F_{c,cr}}{\eta_c} = 0.9 \times 10^7 \left(\frac{0.04}{6}\right)^2$$

$$= 400 \text{ psi}$$

From Fig. 11.2.4, the true buckling stress, $F_{c,cr} = 400$ psi

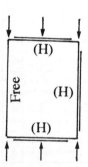

(3) Shear – all hinged edges:

$K_s = 6.7$ (see Fig. 11.3.5, Case ④)

From Eq. 11.3.4

$$\frac{F_{s,cr}}{\eta_s} = K_s E \left(\frac{t}{b}\right)^2$$

$$\frac{F_{s,cr}}{\eta_s} = 7 \times 10^7 \left(\frac{0.04}{6}\right)^2$$

$$= 3,111 \text{ psi}$$

From Fig. 11.2.5, the true buckling stress, $F_{c,cr} = 3,111$ psi

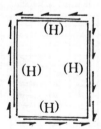

(4) Bending:

$$K_b = 21.6 \text{ (see Fig. 11.3.6, } \alpha = \frac{b}{c} = \frac{6}{3} = 2.0)$$

From Eq. 11.3.6

$$\frac{F_{b,cr}}{\eta_c} = K_b E \left(\frac{t}{b}\right)^2$$

$$\frac{F_{b,cr}}{\eta_c} = 21.6 \times 10^7 \left(\frac{0.04}{6}\right)^2$$

$$= 9{,}600 \text{ psi}$$

From Fig. 11.2.4, the true buckling stress, $F_{b,cr} = F_{c,cr} = 9{,}600$ psi

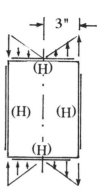

Example 2:

Given a plate with hinged edges as shown and determine buckling stress ($F_{c,cr}$) under an in-plane compression loading.

Material: 2024-T3 clad
$E = 10.5 \times 10^6$ psi
$F_{cy} = 39{,}000$ psi
$a = 9$ in.
$b = 3$ in. (short side)
$\frac{a}{b} = \frac{9}{3} = 3$

(1) $t = 0.125$ in.:

From Fig. 11.2.6, the cladding reduction factor, $\lambda = 0.95$, since $t > 0.064$ in.

The correct $t = 0.125 \times 0.95 = 0.119"$

$K_c = 3.6$ (see Fig. 11.3.1, Case ⑥)

From Eq. 11.3.2

$$F_{c,cr} = K_c \eta_c E \left(\frac{t}{b}\right)^2$$

$$\frac{F_{c,cr}}{\eta_c} = 3.6 \times 10.5 \times 10^6 \left(\frac{0.119}{3}\right)^2$$

$$= 59{,}476 \text{ psi}$$

From Fig. 11.2.4, obtain the true compression buckling stress, $F_{c,cr} = 35{,}000$ psi

(2) $t = 0.063$ in.:

From Fig. 11.2.6, the cladding reduction factor, $\lambda = 0.9$, since $t < 0.064$ in.

The correct $t = 0.063 \times 0.9 = 0.0567"$

$$\frac{F_{c,cr}}{\eta_c} = 3.6 \times 10.5 \times 10^6 \left(\frac{0.0567}{3}\right)^2$$

$$= 13{,}503 \text{ psi}$$

From Fig. 11.2.4, obtain $F_{c,cr} = 13{,}503$ psi (in elastic range and $\eta_c = 1.0$)

Example 3:

Use the same plate given in Example 2 with t = 0.125 in. and determine buckling stress ($F_{s,cr}$) under an in-plane shear loading.

K_s = 5.2 (see Fig. 11.3.5, Case 4) with $\frac{a}{b} = 3$

From Eq. 11.3.4

$$F_{s,cr} = K_s \eta_s E \left(\frac{t}{b}\right)^2$$

$$\frac{F_{s,cr}}{\eta_s} = 5.2 \times 10.5 \times 10^6 \left(\frac{0.119}{3}\right)^2$$

$$= 85,910 \text{ psi}$$

From Fig. 11.2.5, obtain the true shear buckling stress, $F_{s,cr}$ = 22,700 psi

Example 4:

Given a T-shaped extrusion as shown below, find the MS under an in-plane moment loading of M = 3,900 in.-lbs.

Material: 7075-T6 and E = 10.7 × 10⁶ psi

b = 1.625 − 0.094 = 1.53

c = 1.625 − 0.42 = 1.21

$\alpha = \dfrac{b}{c} = \dfrac{1.53}{1.21} = 1.26$

K_b = 1.55 (see Fig. 11.3.6, Case ③)

(7075-T6, Extrusion)

From Eq. 11.3.6:

$$F_{b,cr} = K_b \eta_c E \left(\frac{t}{b}\right)^2$$

$$\frac{F_{b,cr}}{\eta_c} = 1.55 \times 10.7 \times 10^6 \left(\frac{0.094}{1.53}\right)^2$$

$$= 62,602 \text{ psi}$$

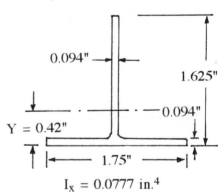

I_x = 0.0777 in.⁴

From Fig. 11.2.4, obtain the true compression buckling stress, $F_{b,cr} = F_{c,cr}$ = 61,000 psi

The bending stress,

$$f_b = \frac{Mc}{I_x} = \frac{3,900 \times 1.21}{0.0777} = 60,734 \text{ psi}$$

$$MS = \frac{61,000}{60,734} - 1 = \underline{0} \qquad \text{O.K.}$$

11.4 CURVED PLATES

The initial buckling stress for a curved plate is the same as that of a flat plate, the same equations (see Eq. 11.3.1 and Eq. 11.3.3) can be used for curved plates except that the buckling coefficients for a curved plate, k_c' and k_s' are used instead of k_c and k_s.

(A) COMPRESSION

$$F_{c,cr} = \frac{k_c' \eta_c \pi^2 E}{12(1-\mu^2)} \left(\frac{t}{b}\right)^2 \quad \text{(compression)} \qquad \text{Eq. 11.4.1}$$

$$F_{s,cr} = \frac{k_s' \eta_s \pi^2 E}{12(1-\mu^2)} \left(\frac{t}{b}\right)^2 \quad \text{(shear)} \qquad \text{Eq. 11.4.2}$$

The buckling coefficients of k_c' and k_s' depend on the curve radius-to-thickness ratio and on edge support conditions. The plasticity factors (η_c and η_s) and cladding reduction factor (λ) used in the flat plate calculations are also applicable to curved plates.

Fig. 11.4.1 gives buckling coefficient of k_c' for long curved plates under in-plane compression loading.

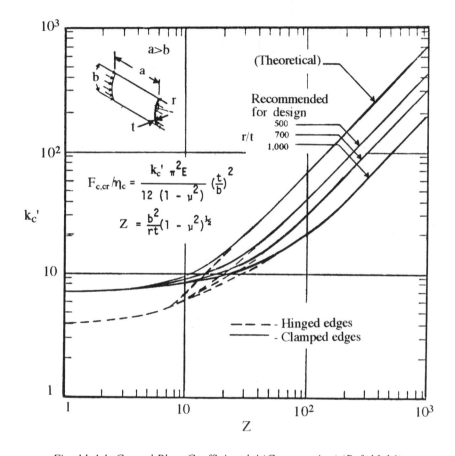

Fig. 11.4.1 Curved Plate Coefficient k_c' (Compression) (Ref. 11.14)

(B) SHEAR

Figures 11.4.2 through Fig. 11.4.5 give the buckling coefficient of k_s' for curved plates under in-plane shear loading.

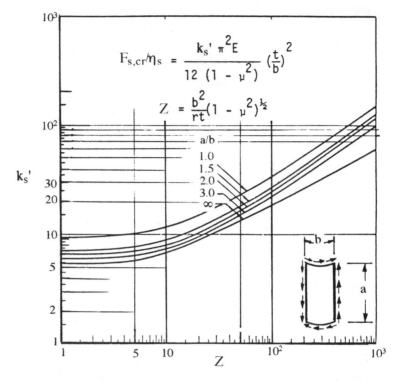

Fig. 11.4.2 Long Curved Plate Coefficient k_s' (Shear) – Four Edges are Hinged (Ref. 11.14)

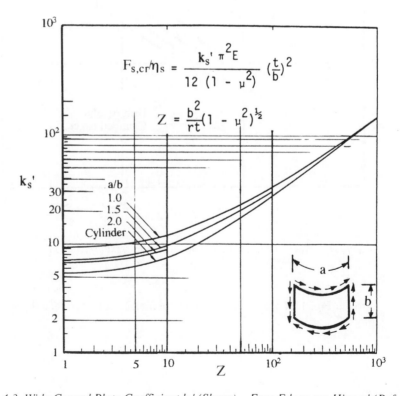

Fig. 11.4.3 Wide Curved Plate Coefficient k_s' (Shear) – Four Edges are Hinged (Ref. 11.14)

Buckling of Thin Sheets

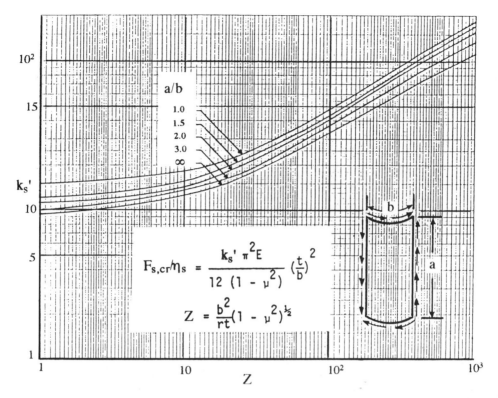

Fig. 11.4.4 Long Curved Plate Coefficient k_s' (Shear) – Four Edges are Clamped (Ref. 11.14)

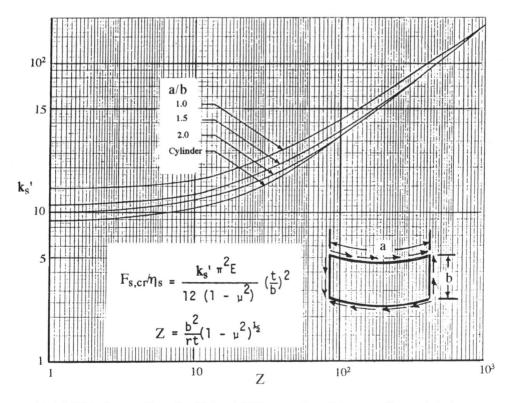

Fig. 11.4.5 Wide Curved Plate Coefficient k_s' (Shear) – Four Edges are Clamped (Ref. 11.14)

467

Chapter 11.0

Example 1:

Using the same plate data (t = 0.125") given in the previous Example 2 for a flat plate, consider a curved plate with a radius of R = 50 inches; determine buckling stress (F_{cr}) under the following in-plane loadings.

(1) Compression buckling stress ($F_{c,cr}$) under an in-plane compression loading:

From Fig. 11.2.6, the cladding reduction factor, = 0.95

Use the correct t = 0.125 × 0.95 = 0.119"

$$Z = \sqrt{(1-\mu^2)}\left(\frac{b^2}{Rt}\right)$$

$$Z = \sqrt{(1-0.33^2)}\left(\frac{3^2}{50 \times 0.119}\right)$$

$$= 0.944 \times 1.513 = 1.43$$

$k_c' = 4.1$ (see Fig. 11.4.1)

From Eq. 11.4.1

$$F_{c,cr} = \frac{k_c' \eta_c \pi^2 E}{12(1-\mu^2)}\left(\frac{t}{b}\right)^2$$

$$\frac{F_{c,cr}}{\eta_c} = \frac{4.1 \times \pi^2 \times 10.5 \times 10^6}{12(1-0.33^2)}\left(\frac{0.119}{3}\right)^2 = 62,520 \text{ psi}$$

From Fig. 11.2.4, obtain the true compression buckling stress, $F_{c,cr}$ = 35,500 psi

(2) Shear buckling stress ($F_{s,cr}$) under an in-plane shear loading:

$Z = 1.43$ and $\frac{a}{b} = \frac{9}{3} = 3$

$k_s' = 6$ (see Fig. 11.4.2)

From Eq. 11.4.2:

$$F_{s,cr} = \frac{k_s' \eta_s \pi^2 E}{12(1-\mu^2)}\left(\frac{t}{b}\right)^2$$

$$\frac{F_{s,cr}}{\eta_s} = \frac{6 \times \pi^2 \times 10.5 \times 10^6}{12(1-0.33^2)}\left(\frac{0.119}{3}\right)^2 = 91,492 \text{ psi}$$

From Fig. 11.2.5, obtain the true shear buckling stress, $F_{s,cr}$ = 23,000 psi

11.5 COMBINED LOADINGS

An example of the interaction curves of combinations of various types of loadings is shown in Fig. 11.5.1. The more frequently used interaction curves are given in Fig. 11.5.2 through Fig. 11.5.4. Interaction curves for buckling conditions can be constructed based on the stress ratio 'R', which is the ratio of the actual stress (f) to the allowable stress (F_{cr}):

$$R_c = \frac{f_c}{F_{c,cr}} \quad \text{(compression stress ratio)}$$

$$R_s = \frac{f_s}{F_{s,cr}} \quad \text{(shear stress ratio)}$$

$$R_b = \frac{f_b}{F_{b,cr}} \quad \text{(bending stress ratio)}$$

Buckling of Thin Sheets

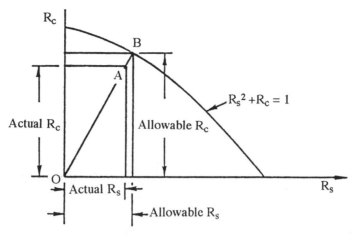

Fig. 11.5.1 Example of an Interaction Curve for Combined Compression (R_c) and Shear (R_s) Loadings

The margin of safety for a combination of various buckling stresses can be determined using an interaction curve:

$$MS = \frac{O\text{-}B}{O\text{-}A} - 1$$

Do not attempt to calculate a Margin of Safety (MS) without using these curves, because it is easy to obtain a false answer by using another method. Use the following interaction equations

Type of combined loadings	Equation	Fig. No.
Shear and compression	$R_s^2 + R_c = 1.0$	11.5.2
Shear and bending	$R_s^2 + R_b^2 = 1.0$	11.5.3
Bending and compression	$R_b^{1.75} + R_c = 1.0$	11.5.4

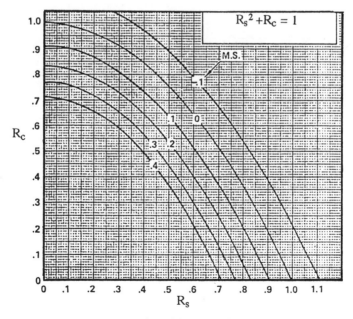

Fig. 11.5.2 Interaction Curves For Combined Shear and Compression Loading ($R_s^2 + R_c = 1$)

469

Chapter 11.0

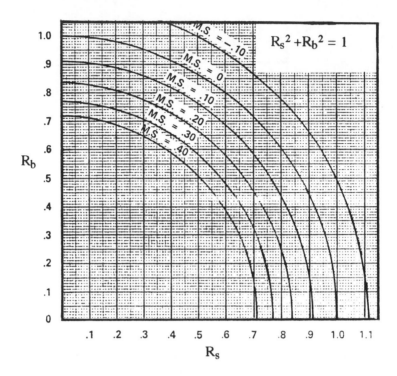

Fig. 11.5.3 Interaction Curves For Combined Shear and Bending Loading ($R_s^2 + R_b^2 = 1$)

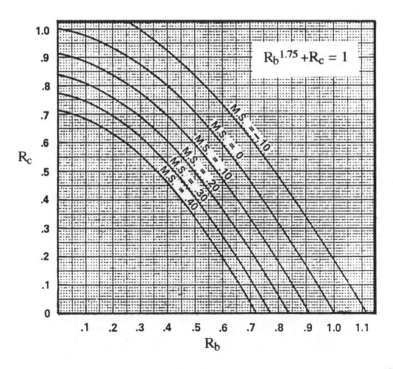

Fig. 11.5.4 Interaction Curves For Combined Bending and Compression Loading ($R_b^{1.75} + R_c = 1$)

Example:

Use the same plate (t = 0.125 in.) and conditions as given in Examples 2 and 3 in Chapter 11.3. Find the MS under the following compression (3,000 lbs./in.) and shear (1,000 lbs./in.) loadings.

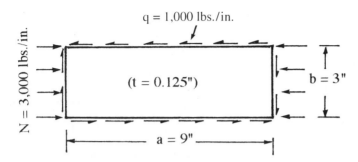

Use shear and compression (use $R_s^2 + R_c = 1$) interaction curves:

$F_{c,cr}$ = 35,000 psi and λ = 0.95 (see Example 2, in Chapter 11.3)

$F_{s,cr}$ = 22,700 psi and λ = 0.95 (see Example 3, in Chapter 11.3)

$$f_c = \frac{3,000}{0.125 \times 0.95} = 25,263 \text{ psi}$$

$$f_s = \frac{1,000}{0.125 \times 0.95} = 8,421 \text{ psi}$$

$$R_c = \frac{f_c}{F_{c,cr}} = \frac{25,263}{35,000} = 0.72$$

$$R_s = \frac{f_s}{F_{c,cr}} = \frac{8,421}{22,700} = 0.37$$

Use curves from Fig. 11.5.2

MS = <u>0.15</u> O.K.

References

11.1 Timoshenko, S and Gere, J. M., "THEORY OF ELASTIC STABILITY", McGraw-Hill Book, Co., Inc., New York, NY, (1961).
11.2 Shanley, F. R., "STRENGTH OF MATERIALS", McGraw-Hill Book, Co., Inc., New York, NY, (1957).
11.3 Lundquist, E. and Stowell, E., "Critical Compressive Stress for Flat Rectangular Plates Supported Along All Edges and Elastically Restrained Against Rotation Along the Unloaded Edges", NACA TN 733, (1942).
11.4 Stein, M. and Neff, J., "Buckling Stresses of Simply Supported Rectangular Flat Plates in Shear", NACA TN 1222, (1947).
11.5 Batdorf, S. B. and Stein, M., "Critical Combinations of Shear and Direct Stress For Simply Supported Rectangular Flat Plates", NACA TN 1223, (1947).
11.6 Budinasky, B. and Connor, R. W., "Buckling Stresses of Clamped Rectangular Flat Plates in Shear", NACA TN 1559, (1948).
11.7 Gerard, G. and Becker, H., "Handbook of Structural Stability (Part I – Buckling of Flat Plates)", NACA TN 3781, (July, 1957).
11.8 Becker, H., "Handbook of Structural Stability (Part II – Buckling of Composite Elements)", NACA TN 3782, (July, 1957).
11.9 Gerard, G. and Becker, H., "Handbook of Structural Stability (Part III – Buckling of Curved Plates and Shells)", NACA TN 3783, (August, 1957).
11.10 Gerard, G., "Handbook of Structural Stability (Part IV – Failure of Plates and Composite Elements)", NACA TN 3784, (August, 1957).
11.11 Gerard, G., "Handbook of Structural Stability (Part V – Compressive Strength of Flat Stiffened Panels)", NACA TN 3785, (August, 1957).

11.12 Gerard, G., "Handbook of Structural Stability (Part VI – Strength of Stiffened Curved Plates and Shells)", NACA TN 3786, (July, 1958).
11.13 Gerard, G., "Handbook of Structural Stability (Part VII – Strength of Thin Wing Construction)", NACA TN D-162, (September, 1959).
11.14 Schildsrout and Stein, "Critical Combinations of Shear and Direct Axial Stress for Curved Rectangular Panels", NACA TN 1928.
11.15 Bruhn, E. F., "ANALYSIS AND DESIGN OF FLIGHT VEHICLE STRUCTURES", Jacobs Publishers, 10585 N. Meridian St., Suite 220, Indianapolis, IN 46290, (1965).
11.16 Roark, R. J. and Young, W. C., "FORMULAS FOR STRESS AND STRAIN", McGraw-Hill Book Company, (1975).
11.17 Niu, C. Y., "AIRFRAME STRUCTURAL DESIGN", Hong Kong Conmilit Press Ltd., P.O. Box 23250, Wanchai Post Office, Hong Kong, (1988).
11.18 Shanley, F. R., "Lockheed Paper No. 28: Engineering Aspects of Buckling", Aircraft Engineering, (Jan., 1939). pp. 13-20.
11.19 Cozzone, F. P. and Melcon, M. A., "Nondimensional Buckling Curves – Their Development and Application", Journal of The Aeronautical Sciences, (Oct., 1946). pp. 511-517.
11.20 Fischel, J. R., "The Compressive Strength of Thin Aluminum Alloy Sheet in the Plasic Region", Jorunal of The Aeronautical Sciences, (August, 1941). pp. 373-375.
11.21 Pope, G. G., "The Buckling of Plates Tapered in Planform", Royal Aircraft Establishment, Report No. Structures 274, (Jan., 1969).

Chapter 12.0

SHEAR PANELS

12.1 INTRODUCTION

Diagonal tension webs are one of the most outstanding examples of methods used in airframe stress analysis and structural sizing. Standard structural practice has been to assume that the load-carrying capacity of a shear web terminates when the web buckles and stiffeners are used merely to raise the buckling stress of the web.

Airframe design assumes that a thin web with transverse stiffeners does not fail when it buckles. The web forms diagonal folds and functions as a series of tension straps while the stiffeners act as compression posts. The web-stiffener thus acts as a truss and may be capable of carrying loads far greater than those producing the buckling of the web.

(a) Shear resistant web:
A shear resistant web is one that does not buckle at its design load. In its most common usage, the shear resistant web becomes part of a built up beam where the design shear flow of the beam is less than the buckling shear flow for the individual web panels. Generally, this type of beam carries a weight penalty due to low stresses and is used only when an unbuckled web is a requirement for a special condition. This type of web design is comparatively heavy, which prevents its wide use in aerospace structures. Unless special design requirements must be met, e.g., no buckling at limit load.

(b) Pure diagonal tension web:
A pure diagonal tension web is one which buckles immediately under the application of load. It is the development of a structure in which buckling of the webs is permitted with the shear loads being carried by diagonal tension stresses in the web. However, this type of web is not applicable to airframe structures because it does not meet the requirements of either aerodynamic surface smoothness or structural durability for fatigue life.

(c) Diagonal tension (or incomplete diagonal tension) web:
A diagonal tension web does not buckle under small loads, but as the load increases the web will start to buckle without collapse until it reaches the design failure load.

The discussion in this chapter of the shear resistant web and the diagonal tension web (or skin), which is shown in Fig. 12.1.1, is confined to information required for structural sizing (and most of the material is taken directly from Ref. 12.1 and 12.2). The discussion is necessarily brief since the development of design equations is semi-empirical and the development of theoretical equations, beyond the scope of those presented, is not justified for the purposes of this chapter.

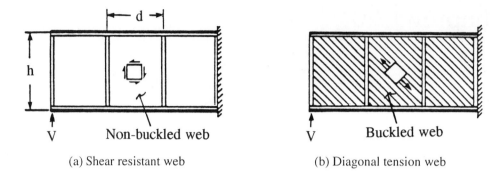

(a) Shear resistant web   (b) Diagonal tension web

Fig. 12.1.1 Shear beams

12.2 SHEAR RESISTANT WEBS

Shear resistant webs are those that do not buckle under ultimated shear load. Purely shear resistant webs are rare in airframe structural design but are typical in non-airframe structural design (for example, the girders used in bridge and building design). In shear resistant webs, the allowable web stresses can be increased only by

- Increasing web thickness
- Decreasing panel size by addition of stiffeners

Such a procedure leads to a relatively heavy design. The typical shear resistant beam has nearly all of its bending material concentrated at the caps. The web carries shear almost exclusively and adds little to the bending stiffness of the beam. From Eq. 6.5.2 of Chapter 6.0:

$$q = \frac{V}{h_e}$$

where: q – Shear flow (lbs/in.)
         V – Vertical shear load (lbs.)
         h_e – Beam depth between cap centroids

Eq. 6.5.2 is the engineering formula for determining shear flow in beams in which the cap area is large with respect to the web area, and the depth of web is large compared to the depth of beam. In shear resistant webs, stiffeners do not have direct influence on the stress distribution; their function is to prevent web buckling, and to introduce load into the web. The load acting on a unit element in the shear resistant web is shown in Fig. 12.2.1.

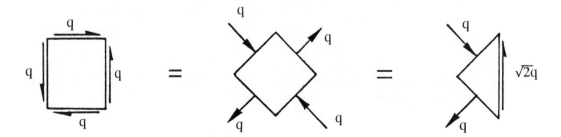

Fig. 12.2.1 Principal Stresses of a Shear Resistant Web

Note that the forces on the rotated element have resolved into tension and compression forces at right angles to each other. These are the "principal stresses" which can also be determined by the Mohr's circle method (see Fig. 5.2.2). Thus, the effect of the applied shear load is to induce a compressive stress in the web.

The sizing of the shear resistant web of a beam is a matter of trial and error; generally the applied shear flow (q) and full depth of beam are given and the structural engineer must determine or size web thickness (t), stiffener spacing (d), and minimum stiffener moment of inertia (I_u) required, consistent with practical spacing and thickness constraints

- Most ratios of $\frac{d}{h_e}$ fall between $0.2 \leftrightarrow 1.0$
- $\frac{d}{h_e}$ values should be as small as is practical for minimim weight
- Cost will increase because of the increased number of stiffeners

(A) WEB

When all of the beam parameters are known, two stress checks and a stability check are made as follows:

- Shear stress, f_s, computed by the formula $\frac{VQ}{It}$ at the beam neutral axis does not exceed $F_{s,all}$ at ultimate load
- The web net area shear stress does not exceed $F_{s,all}$ along the first line of web to cap (or chord) rivets or along a stiffener rivet pattern
- The shear stress present in the web does not exceed the critical buckling shear stress

For sizing, the hinged (or simply supported) edges should be used, unless sufficient test data is available to justify the use of higher edge restraint buckling coefficients (K_s or k_s, refer to Chapter 11).

Set the applied shear stress (f_s) equal to the buckling stress ($F_{s,cr}$) and then use Eq. 11.3.4 from Chapter 11:

$$F_{s,cr} = K_s \eta_s E (\frac{t}{b})^2$$

Assuming $\eta_s = 1.0$ for the first trial, then the web thickness is:

$$t = (\frac{q d_c^2}{K_s E})^{\frac{1}{3}} \qquad \text{Eq. 12.2.1}$$

where d_c – Clear stiffener spacing (see Fig. 12.4.3)

Fig. 12.2.2 and 12.2.3 show approximation curves to be used only as a method of close approximation. Knowing the shear flow, q, select the proper Figure and choose one of several web gage – stiffener spacing combinations. The straight line portion of these curves shows the elastic range. If a web thickness and stiffener spacing combination in the plastic range is chosen, an equivalent elastic thickness must also be determined. This is done by interpolating between the dashed lines for the point chosen. The equivalent thickness must be used in determining the required stiffener moment of inertia.

Chapter 12.0

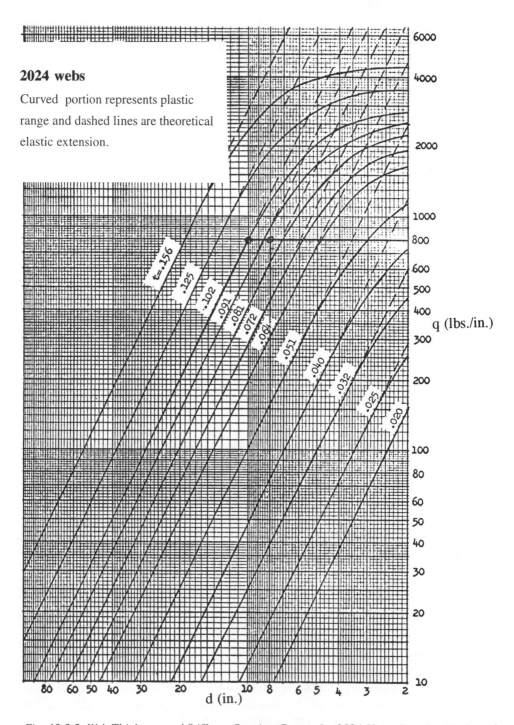

Fig. 12.2.2 Web Thickness and Stiffener Spacing Curves for 2024 Shear Resistant Webs

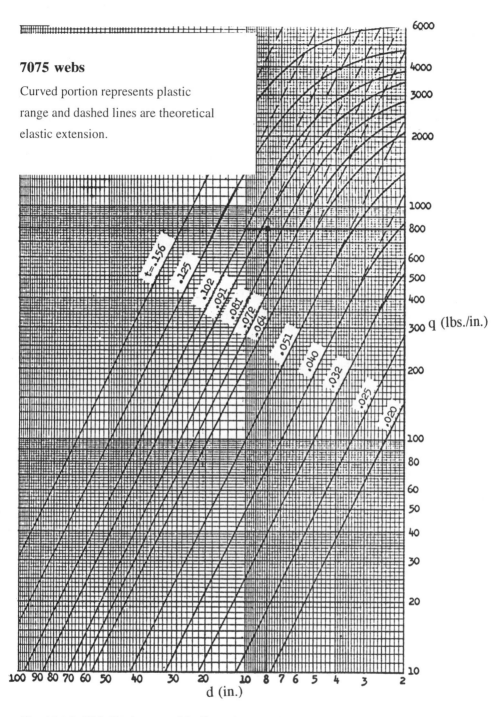

Fig. 12.2.3 Web Thickness and Stiffener Spacing Curves for 7075 Shear Resistant Webs

(B) BEAM CAP (FLANGE)

When the depth of the beam cap is small compared to the beam depth, and the bending stresses in the web are negligible, as would be the case for thin web beams, the beam cap stresses can be calculated from:

$$f_{cap} = \frac{M}{h_e A_{cap}}$$ Eq. 12.2.2

The beam caps must be checked for both column and crippling failure stress (see Chapter 10.0).

(C) STIFFENER

Stiffeners on shear resistant beams do not carry any structural load, but are used to prevent web buckling. However, if external loads are applied at the stiffener, they shall be checked for all modes of failure. After the stiffener spacing and web thickness are determined, refer to Fig. 12.2.4 for the minimum stiffener moment of inertia required for stability.

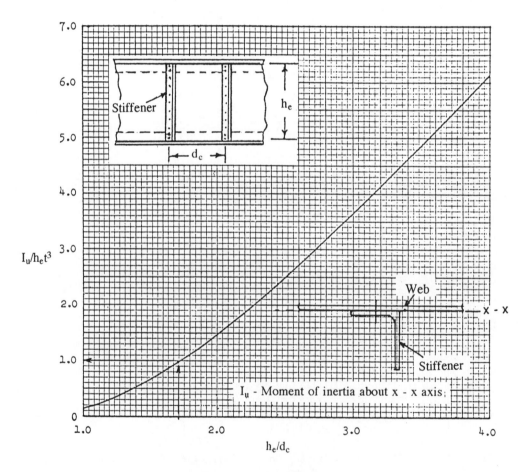

Fig. 12.2.4 Minimum Stiffener Moment of Inertia for Shear Resistant Webs

In addition to the moment of inertia requirement, certain $\frac{t_u}{t}$ relationships are recommended. t_u is the thickness of the stiffener leg which attaches to the web. Fig. 12.2.5 is to be used as a guide for shear resistant beam stiffeners.

Web Thickness (t)	Thickness of Attached Leg of Stiffener (t_u)
0.02	0.03
0.025	0.03
0.032	0.04
0.036	0.04
0.04	0.04
0.05	0.05
0.063	0.05
0.071	0.06
0.08	0.06
0.09	0.07
0.1	0.07
0.125	0.08
0.156	0.08
> 0.156	0.6t

Fig. 12.2.5 Thickness (t_u) of Attached Flange Stiffener (Shear Resistant Webs)

(D) RIVETS

(a) Web-to-cap:

- Initially, the rivet size and spacing (pitch) are determined by the web shear flow (q) but the rivet spacing must also be checked for inter-rivet buckling (and tank seal condition on wing fuel tanks)
- Generally, the pitch is 4-8D and refer to Chapter 9.0 for addtional design data

(b) Web-to-stiffener:

- Theoretically, there is no load transfer between the web and the stiffeners for a shear resistant web unless the shear flow changes between bays and then the rivets must resist this change in shear flow
- While web-to-cap attachments carry a running load per inch equal to $\frac{V}{h_e}$, the web-to-stiffener rivets are proportioned to stiffener size by the following empirical equation:

$$\frac{K A_u}{d_c} \text{ (lbs./in.)} \qquad \qquad \text{Eq. 12.2.3}$$

where: A_u – Area of stiffener
d_c – Stiffener spacing
K – 20,000 for 2024 webs
– 25,000 for 7075 and 7178 webs

(c) Stiffener-to-cap:

- Minimum of two rivets
- Use the next larger size than those used on web-to-cap attachment or use the same size lockbolts or Hi-Loks with higher shear strength (also check bearing strength)

Chapter 12.0

Example:

Size the shear resistant beam described below to carry an ultimate load of V = 10,000 lbs.

Material: ("B" values, see Fig. 4.3.4 and Fig. 4.3.5):

Web – 7075-T6 bare sheet
- F_{tu} = 80,000 psi
- F_{cy} = 71,000 psi
- F_{su} = 48,000 psi
- F_{bru} = 160,000 psi
- E = 10.5 × 10^6 psi

Cap and stiffeners –
7075-T6 Extrusion :
- F_{tu} = 82,000 psi
- F_{cy} = 74,000 psi
- F_{bru} = 148,000 psi
- E = 10.7 × 10^6 psi

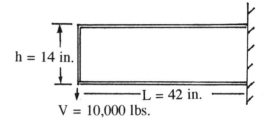

h = 14 in.
L = 42 in.
V = 10,000 lbs.

General Solution	Numerical calculation
1. For the first trial assume a value of h_e and determine shear flow, q	Assume h_e = 12 in. then $q = \dfrac{V}{h_e} = \dfrac{10,000}{12} = 833$ lbs./in.
2. From Fig. 12.2.2 or 12.2.3 pick a web thickness and stiffener spacing or use Eq. 12.2.1	Since the example calls for a 7075 alloy, use Fig. 12.2.3 at q = 833 lbs./in. Pick t = 0.091 in. (web thickness) and d_c = 8 in. (stiffener spacing) (Note: This is just one of several choices; thickness of t = 0.102 in. and stiffener spacing of 10 in. could have been used instead)
3. Size cap areas by the following formula: $A_{cap} = \dfrac{M_{max}}{h_e \times F_{cy}}$ Material: 7075-T6 Extrusion F_{cy} = 74,000 psi Assume $F_{cc} = F_{cy}$	$A_{cap} = \dfrac{10,000 \times 42}{12 \times 74,000}$ = 0.473 in.^{2}
4. Pick cap sections:	Use a standard T-section as shown below: 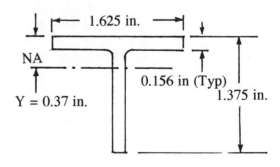 1.625 in. NA Y = 0.37 in. 0.156 in (Typ) 1.375 in.

480

Shear Panels

General Solution	Numerical calculation
	$A_{cap} = 0.437$ in.^{2}
	$Y = 0.37$ in.
	$I_x = 0.0725$ in.^{4}
	$r = 0.407$ in.
	($A_{cap} = 0.437$ in.^{2} meets the requirement, See step 13)
5. Determine an exact h_e and $\dfrac{h_e}{d_c}$	$h_e = h - 2Y = 14 - 2 \times 0.37 = 13.26$ in.
	$\dfrac{h_e}{d_c} = \dfrac{13.26}{8} = 1.66$
6. Knowing $\dfrac{h_e}{d_c}$ and t, use Fig. 12.2.4 to determine the required stiffener moment of inertia.	From Fig. 12.2.4 with $\dfrac{h_e}{d_c} = 1.66$, the required $\dfrac{I_u}{h_e t^3} = 1.0$. Therefore,
	$I_u = (1.0)(13.26)(0.091)^3 = 0.01$ in.^{4}
7. Pick a stiffener section and check that the minimum attached flange thickness (t_u) requirements from Fig. 12.2.5 are met.	Use single stiffeners (equal-leg angles) as follows:

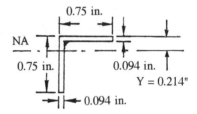

$A_u = 0.132$ in.^{2}
$Y = 0.214$ in.
$I_x = 0.0063$ in.^{4}
$\rho = 0.219$ in.

$I_u = I_x + AY^2 = 0.0123$ in.$^4 > 0.01$ in.^{4} O.K.

From Fig. 12.2.3, for a web thickness of $t = 0.091$ in. the recommended min. attached leg thickness (see Fig. 12.2.5) is 0.07 in. O.K.

| 8. Determine size and pattern of web-to-cap rivets | Actual shear flow, $q = \dfrac{V}{h_e} = \dfrac{10,000}{13.26}$ |

$= 754$ lbs./in.

Use 2017-T3(D), $\dfrac{3}{16}$ dia. rivets with an allowable of 1090 lbs/ rivet (Fig. 9.2.5). The bearing strength:

$P = 160,000 \ (80\%) \times 0.1875 \times 0.091$
$= 2,184$ lbs. (not critical).

Therefore, rivet pitch $= \dfrac{1,090}{754} = 1.45$ in.

Use 1.0 in. to avoid inter-rivet buckling.

Chapter 12.0

General Solution	Numerical calculation
9. Determine size and pattern of web-to-stiffener rivets from Eq. 12.2.3	$\dfrac{KA_u}{d_c} = \dfrac{(25{,}000)(0.132)}{8}$ $= 413$ lbs/in. Use 2017-T3(D), $\dfrac{1}{8}$ dia. rivets with shear allowable of 494 lbs/rivet (Fig. 9.2.5). Therefore, rivet pitch: $\dfrac{494}{413} = 1.2$ in. Use 0.75 in.
10. Make shear stress check of web-to-cap attachment	Rivet hole efficiency, $\eta = \dfrac{1.0 - (\tfrac{3}{16})}{1.0} = 0.81$ $F_{s,\,net} = \dfrac{q}{\eta t}$ $= \dfrac{754}{0.81 \times 0.091}$ $= 10{,}229$ psi $< F_{su} = 48{,}000$ psi O.K.
11. Check shear buckling stress of web	From Fig. 11.3.5 (Case ④) with $\dfrac{h_e}{d_c} = 1.66$, obtain shear buckling coefficient, $K_s = 6.3$ From Eq. 11.3.4 $F_{s,\,cr} = K_s \eta_s E \left(\dfrac{t}{b}\right)^2$ where: $b = d_c$ $\dfrac{F_{s,\,cr}}{\eta_s} = 6.3 \times 10.5 \times 10^6 \left(\dfrac{0.091}{8}\right)^2$ $\dfrac{F_{s,\,cr}}{\eta_s} = 8{,}559$ psi and from Fig. 11.2.5, $F_{s,\,cr} = 8{,}559$ psi and $\eta_s = 1.0$ $f_s = \dfrac{q}{t} = \dfrac{754}{0.091}$ $= 8{,}286$ psi $< F_{s,\,cr} = 8{,}559$ psi O.K.
12. Additional trials may be required until reasonable agreement of the values of f_s and $F_{s,\,cr}$ is reached.	
13. Check the cap stress	Since the actual $h_e = 13.26$ in., which is greater than the assumed $h_e = 12$ in., it is therefore not critical. The actual required cap area is: $A_{cap} = \dfrac{10{,}000 \times 42}{13.26 \times 74{,}000}$ $= 0.428$ in.$^2 < 0.437$ in.^{2}, see step 4 O.K.

12.3 PURE DIAGONAL TENSION WEBS

Pure diagonal tension webs are those that buckle immediately upon the application of shear load. The classic description of the action of the pure diagonal tension web compares its action to that of the redundant truss shown in Fig. 12.3.1.

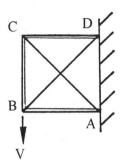
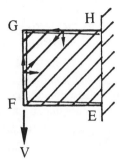

Fig. 12.3.1 Diagonal Members are Straps   *Fig. 12.3.2 Thin web*

If the diagonals CA and BD are thin straps or cables, CA will act in compression and BD in tension. Since, by definition, CA is incapable of carrying compression because it buckles, and the load is carried by DB in tension.

The pure diagonal tension beam of Fig. 12.3.2 is similar to that of Fig. 12.3.1, except that a thin web replaces the diagonals. Under the application of shear load, the web buckles in diagonal folds as shown. The web will carry tension in the direction of the buckles, but will not carry compression loads across the buckles. In fact, the compression load in the general direction GE causes the web to buckle.

The effect of the diagonal tension web on the stiffeners and caps can be pictured by imagining a number of wires acting in the direction of the buckles. The loads in the wires have vertical and horizontal components which tend to bring the four sides of the beam together. The vertical component loading on the top cap (or flange) tends to put that member in bending and shear. These loads are reacted by GF and HE in compression, where they are balanced out by the vertical loads on FE. Similarly, the horizontal component loading acts on GF and HE to produce bending and shear. The compression load on GH and FE is the result of the vertical component loading.

In order to make a qualitative evaluation of the tension field web, cut a one inch square of the web, as shown by the shear resistant web in Fig. 12.2.1. By definition, a pure tension field web can carry no compression. Therefore, the triangular element in Fig. 12.2.1. must be balanced with the compressive $q = 0$, and retain the vertical $q = \sqrt{2}\, q$ for a length of $\sqrt{2}$. The balanced triangular element is shown in Fig. 12.3.3.

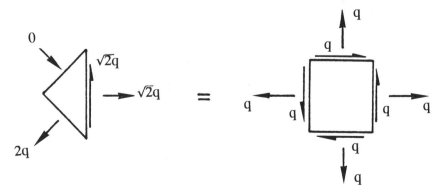

Fig. 12.3.3 Load Distribution of a Pure Diagonal Tension Web

A comparison of Fig. 12.2.1 and Fig. 12.3.3 shows that the difference between the diagonal tension web and the shear resistant web is that an additional tension force in two directions is added to the original web shear forces, to compensate for the lack of compressive stiffness.

12.4 DIAGONAL TENSION FLAT WEBS

Theories for shear resistant webs and pure diagonal tension webs, which are the two extreme cases of diagonal tension beam webs were previously presented. Diagonal tension webs fall between these two extremes.

In Chapter 12.3 it was assumed that the web of a pure tension field beam would not carry compression across the wrinkles, and that shear was resisted entirely by diagonal tension in the web. In acutal design this is not true, because the skin does not drop its compression load when it buckles.

When a yardstick buckles, as shown in Fig. 12.4.1(a), it maintains its buckling load even when subjected to large deflections. The load deflection curve for this column is shown in Fig. 12.4.1(b).

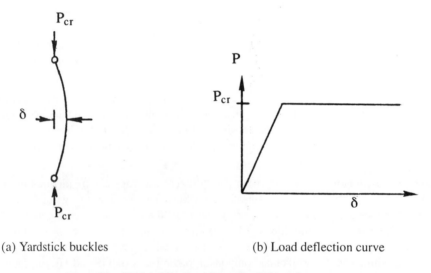

(a) Yardstick buckles (b) Load deflection curve

Fig. 12.4.1 The Buckling of a Yardstick

The unbuckled web of a shear beam is subject to equal tension and compression forces, as was developed and shown in Fig. 12.2.1. As the shear load is increased beyond the buckling point, the web buckles in compression and the additional shear load is carried as tension in the web and the final load picture for the web is as shown in Fig. 12.4.2.

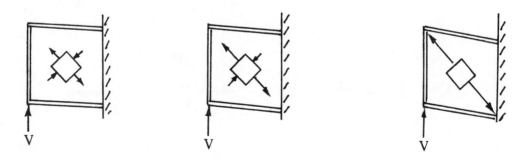

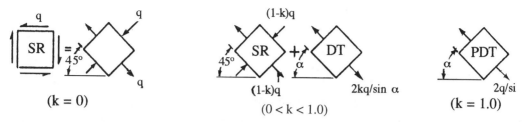

(Note: SR – Shear resistant web; DT – Diagonal tension web; PDT – Pure diagonal tension web)

Fig. 12.4.2 Web Stresses at Different Stages of the Diagonal Tension Buckled Web

Note that the element in the center of Fig. 12.4.2 is an intermediate case between shear resistant web and pure diagonal tension web. The element is in shear with a tension field load of $q - q_{cr}$ in two perpendicular directions. For the thin web used in airframe structures the diagonal tension can be divided into two parts:

- The part carried by shear (f_{SR})
- The remaining part carried by diagonal tension (f_{DT})

The total shear stress on a diagonal tension web can be written as:

$$f_s = f_{SR} + f_{DT}$$

where: $f_{SR} = (1 - k)f_s$ (Shear resistant stress)
$f_{DT} = kf_s$ (Diagonal tension stress)

'k' is the "diagonal-tension factor", and is defined as:

- If k = 0, Shear resistant web ($\frac{q}{q_{cr}} = 1.0$)
- If k = 1, Pure or near pure diagonal tension web
- If 1.0 > k > 0, Diagonal tension web (or semi-tension field)

The method most generally used at present for in-plane diagonal-tension web analysis is the one developed at the NACA and given in detail in NACA TN 2661 (Ref. 12.1). It is a combination of theory and empiricism which attempts to predict the stresses in the various components of the diagonal-tension web and also gives empirical design criteria for the various beam components. These design criteria are mainly concerned with guarding against web failure and stiffener failure in the various attachments.

(A) NACA TN 2661 METHOD

See Fig. 12.4.3 for an illustration of the various beam dimensions and nomenclature referred to below:

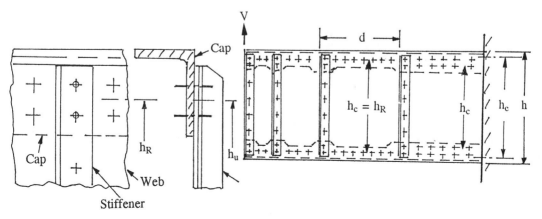

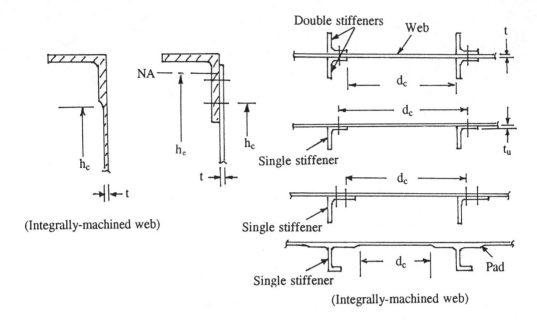

Fig. 12.4.3 Geometry of Typical Plane Diagonal-Tension Webs

A_{cap} – Cross-sectional area of cap (flange)
A_u – Cross-sectional area of stiffener
b – Width of outstanding leg of the striffener
d – Stiffener spacing
d_c – Clear stiffener spacing
e – Distance from median plane of web to centroid of stiffener (e = 0 for symmetrical double stiffener)
h – Full depth of the beam
h_c – Clear depth of web
h_e – Effective depth of beam measured between centroids of caps (or flanges)
h_R – Depth of beam between web-to-cap rivet pattern centroids.
h_u – Length of stiffener measured between the upper and lower centerlines of stiffener-to-cap (or flange) rivet patterns
I_C, I_T – Moments of inertia of compression and tension caps (or flanges) respectively
q – Applied web shear flow ($\frac{V}{h_e}$)
t – web thickness
t_F – Thickness of cap leg attached to the web
t_u – Thickness of stiffener leg directly attached to the web
V – Applied web shear load
ρ – Centroidal radius of gyration of stiffener cross section about an axis parallel to the web (no web area should be included)
ωd – Cap (flange) flexibility factor
c – Distance from web to outer fiber of stiffener
Q – Static moment of cross section of one stiffener about an axis in the median plane of the web

The following specific limitations should be observed when using this method to analyze diagonal tension webs:

- $115 < \dfrac{h_c}{t} < 1500$

- $0.2 < \dfrac{d_c}{h_c} < 1.0$
- $\dfrac{t_u}{t} > 0.6$

Test data is not available for beam dimensions falling outside the above limits and this method of diagonal-tension beam analysis should not be expected to apply in such cases. When the web is allowed to buckle under shear load, additional forces are generated within the beam which must be resisted by the stiffeners, caps, and rivets as shown in Fig. 12.4.4.

In practical shear beam design, the shear flows are different in the adjacent bays, an average values of shear flow (q) should be used in sizing the diagonal tension web. In general, there is no simple way to calculate the exact diagonal tension forces when shear flows vary from bay to bay in a structural beam. The method presented here is a good approximation that can be used for sizing.

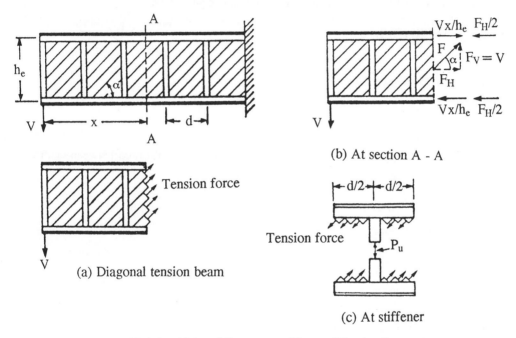

Fig. 12.4.4 Additional Forces on a Diagonal Tension Beam

The NACA TN 2661 method is recommended for actual design or sizing of diagonal tension web:

- This method is conservative and gives higher margins of safety
- In heavily loaded structures such as wing spars or bulkheads subjected to large external loads, the webs are usually much thicker and the diagonal tension factor (k) is much less
- It is recommended that the k factor at ultimate load design should be limited to a maximum value as shown in the following equations in order to avoid excessive wrinkling, permanent set at limit load and premature fatigue failure:

$$k = 0.78 - \sqrt{(t - 0.012)} \qquad \text{Eq. 12.4.1}$$

For fatigue critical webs such as wing spar webs and wing surface skins, use following equation:

$$\dfrac{f_s}{F_{s,cr}} \leq 5 \qquad \text{Eq. 12.4.2}$$

(**Note:** The criteria given are for reference **only**; engineers should follow the sizing policy of the company for which they are working)

- Take care not to exceed the maximum k value
- Care must be taken in the design of diagonal tension webs since the sizing of rivets is more complex compared to the design of shear resistant web
- The sizing procedure is by trial and error, with q and h as the given parameters
- Use of a thinner web thickness to carry shear load will result in lighter structural weight
- Use the smallest stiffener area possible to meet the diagonal tension requirement but also consider the lateral pressure loads from fuel, air, etc.
- Cap should be sized to carry the primary beam bending as well as the additional stresses (both axial and bending) caused by diagonal tension forces

It has been more than 40 years since NACA TN 2661 was published for use by airframe structural engineers. During the past decades, the major airframe companies have modified this method and incorporated their own version of the method into their design manuals.

(B) SIZING PROCEDURES

(a) Estimate the beam cap cross section areas:

From given moment (M)

$$A_{cap} \approx \frac{M}{h_e F_{cc}}$$

where: M – Given moment of the beam
where: h_e – Assume an effective depth of beam measured between centroids of caps (or flanges)
F_{cc} – Assume the crippling stress $F_{cc} = F_{cy}$

Select cap section and then determime the actual values of h_e, h_c, h_u.

Applied shear flow in the web:

$$q = \frac{V}{h_e} \qquad \text{Eq. 12.4.3}$$

(b) Diagonal tension factor (k)

Choose one of the criteria from Eq. 12.4.1 or Eq. 12.4.2 to determine the diagonal tension factor as follows:

$$k = 0.78 - \sqrt{(t - 0.012)} \qquad \text{(from Eq. 12.4.1)}$$

or $\dfrac{f_s}{F_{s,cr}} \leq 5$ (from Eq. 12.4.2)

(b1) Estimate the web thickness (t):

Use Eq. 11.3.4 from Chapter 11:

$$\frac{F_{s,cr}}{\eta_s} = K_s E \left(\frac{t}{d_c}\right)^2$$

Where d_c is clear stiffener spacing (see Fig. 12.4.3), d_c must be equal to or less than the clear depth of the web ($d_c \leq h_c$). The use of hinged edges is recommended when determining K_s.

(i) If choosing criteria from Eq. 12.4.2, assume $\eta_s = 1.0$:

Use $\dfrac{f_s}{F_{s,cr}} = 5$ in Fig. 12.4.6 and obtain k = 0.32. (for flat web).

$$F_{s,cr} = K_s E \left(\frac{t}{d_c}\right)^2 = 5 f_s = 5\left(\frac{q}{t}\right)$$

$$\therefore t \approx 1.15 \left(\frac{q d_c^2}{5 K_s E}\right)^{\frac{1}{3}} \qquad \text{Eq. 12.4.4}$$

(ii) If choosing criteria from Eq. 12.4.1, assume $\eta_s = 1.0$:

$$t \approx 1.07 \left(\frac{q}{F_{s,all}}\right) \qquad \text{Eq. 12.4.5}$$

Where $F_{s,all}$ from Fig. 12.4.15 with $k = 0.78$

(iii) Since $k > 0.2$ for most beam webs, the web thickness may be approximately determined from the following equation:

$$t \approx \frac{q}{F_s} \qquad \text{Eq. 12.4.6}$$

where: $F_s = 21,500$ psi (2024-T3)
$F_s = 27,500$ psi (7075-T6)

The above F_s values can be found from Fig. 12.4.5 with $k = 1.0$ (conservative).

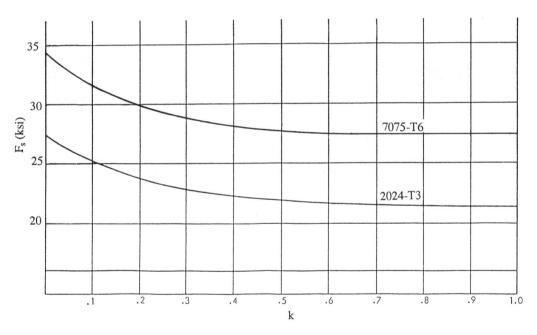

($F_{s,all}$ at $\alpha = 45°$ from Fig. 12.4.15)

Fig. 12.4.5 Allowable Maximum Web Shear Stress

(b2) Use the above t value to recalculate the shear buckling stress of the web ($F_{s,cr}$):

$$F_{s,cr} = K_s \eta_s E \left(\frac{t}{d_c}\right)^2$$

Use the $\frac{h_c}{d_c}$ ratio to find K_s in Fig. 11.3.5 Case ④; assume the four edges are hinged supports.

(b3) Determine the final diagonal tension factor (k):

The applied shear stress in the web:

$$f_s = \frac{V}{h_e t} = \frac{q}{t}$$ Eq. 12.4.7

(i) Calculate the value of $\frac{f_s}{F_{s,cr}}$

(ii) Use $\frac{f_s}{F_{s,cr}}$ value to determine the diagonal tension factor (k) from the curve in Fig.12.4.6 and check whether the k value meets the design criteria either from Eq. 12.4.1. or Eq. 12.4.2.

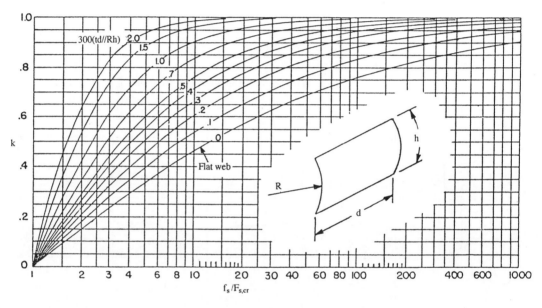

(If h > d, replace $\frac{td}{Rh}$ by $\frac{th}{Rd}$; if $\frac{d}{h}$ or $\frac{h}{d}$ > 2, use 2)

Fig. 12.4.6 Diagonal Tension Factor, k

(c) Stiffener (or upright)

(c1) Calculate the average compressive stress along the length of the stiffener in the median plane of the web:

Estimate cross-sectional area of stiffener (upright) by:

$$A_u \approx 30\% (dt)$$ Eq. 12.4.8

Effective cross-sectional area of a single stiffener (use A_u for double stiffeners):

$$A_{u,e} = \frac{A_u}{1 + (\frac{e}{\rho})^2}$$ Eq. 12.4.9

$$f_u = \frac{k f_s \tan \alpha}{\frac{A_u}{dt} + 0.5(1-k)}$$ Eq. 12.4.10

or use $f_u = (\frac{f_u}{f_s}) f_s$

where $\frac{f_u}{f_s}$ is given in Fig. 12.4.7

The above f_u value is based on the assumption of f_s, k and tan α to be the same in the bays on each side of the stiffener. If they are not, average values or the largest shear stress (f_s) value on the conservative side should be used.

(c2) Calculate the maximum stiffener stress in the median plane of the web:

$$f_{u,max} = (\frac{f_{u,max}}{f_u})f_u$$

where $\frac{f_{u,max}}{f_u}$ is determined from in Fig. 12.4.8, and f_u is from previous step

(c3) Determine F_{fu}, the stress causing a forced crippling failure in the stiffener, from Fig. 12.4.9. Compare the value of $f_{u,max}$ determined in Step (c2) with F_{fu}

$$MS = \frac{F_{fu}}{f_{u,max}} - 1 \quad \text{(forced crippling failure in stiffeners)}$$

(c4) Check against column failure in the stiffeners according to the following criteria:

(i) Double Stiffeners (use the equivalent pin-end column length):

$$L'' = (\frac{L''}{h_u})h_u$$

where $\frac{L''}{h_u}$ is determined from Fig. 12.4.10; from the curves in Fig. 10.8.2 to determine the allowable column stress F_c for the stiffener.

$$MS = \frac{F_c}{f_u} - 1 \quad [f_u \text{ determined in Step (c1)}]$$

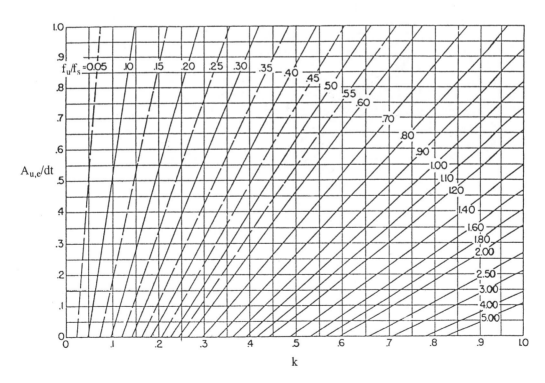

Fig. 12.4.7 Ratio of Stiffener Compressive Stress to Web Shear Stress

Chapter 12.0

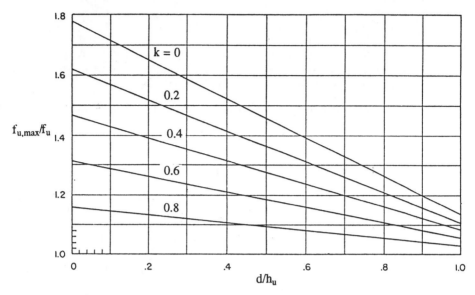

Note: For use on curved webs:

For ring, read abscissa as $\frac{d}{h}$; for stringer, read abscissa as $\frac{h}{d}$.

Fig. 12.4.8 Ratio of Maximum to Average Stiffener Stress

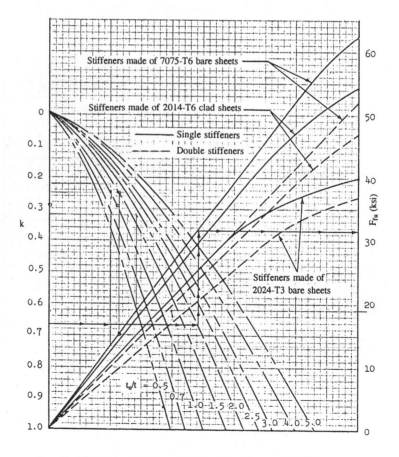

Fig. 12.4.9 Allowable Forced Crippling Stress of Stiffeners

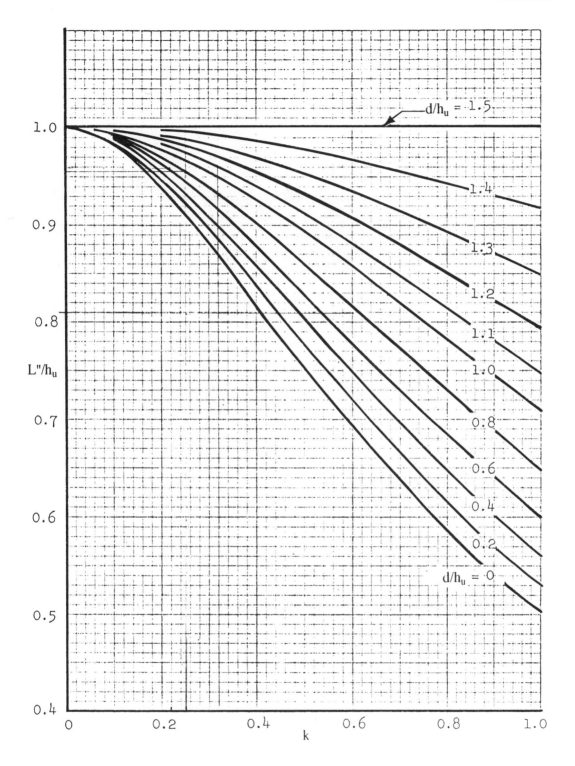

Fig. 12.4.10 Equivalent Pin-End Column Length of Double Stiffener

(ii) Single Stiffeners:

Use the following guidelines for determining the effective stiffener area:

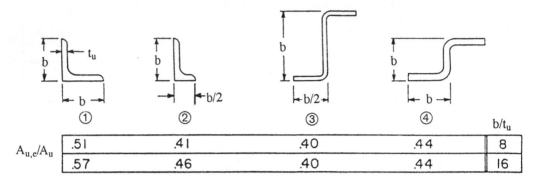

Fig. 12.4.11 Ratio of Effective and Actual Single Stiffener Area ($\frac{A_{u,e}}{A_u}$)

The equivalent pinned-end column length for a single stiffener is $L' = \frac{L''}{2}$; from the curves in Fig. 10.8.2, determine the allowable column stress F_c corresponding to the $\frac{L'}{\rho}$ value for the single stiffener (from Fig. 12.4.11). Compare this value of F_c with the average stiffener stress,

$$f_{u, av} = (\frac{A_{u,c}}{A_u})f_u \quad [f_u \text{ determined in Step (c1)}] \qquad \text{Eq. 12.4.11}$$

where: $\frac{A_{u,e}}{A_u}$ determined from Fig. 12.4.11

In addition, f_u must be less than the column allowable stress F_c of the stiffener.

$$MS = \frac{F_c}{f_{u, av}} - 1$$

where: F_c – Allowable compression stress from Fig. 10.8.2

(d) Maximum web shear stress ($f_{s, max}$)

(d1) Determine $\tan \alpha$ value from Fig. 12.4.12.

(d2) Use $\tan \alpha$ value to determine C_1 value from Fig. 12.4.13.

(d3) Use the following ωd value to determine the C_2 value from Fig. 12.4.14.

$$\omega d = 0.7d[\frac{t}{h_e(I_T + I_c)}]^{\frac{1}{4}} \qquad \text{Eq. 12.4.12}$$

(d4) Calculate the maximum shear stress in the web:

$$f_{s, max} = f_s(1 + k^2 C_1)(1 + kC_2) \qquad \text{Eq. 12.4.13}$$

where C_1 and C_2 represent the effects of tension wrinkling and beam cap sagging, respectively.

(d5) Obtain the allowable maximum web shear stress $F_{s, all}$ from Fig. 12.4.15.

(d6) Margin of safety:

$$MS = \frac{F_{s, all}}{f_{s, max}} - 1$$

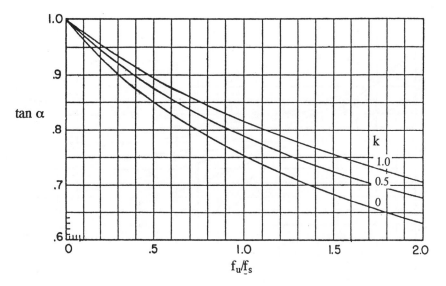

Fig. 12.4.12 Angle of Diagonal Tension Web, tan α

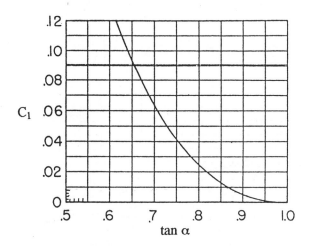

Fig. 12.4.13 Angle Factor, C_1

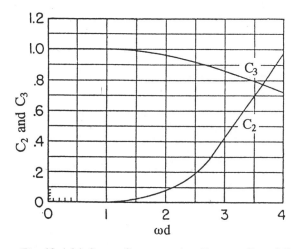

Fig. 12.4.14 Stress Concentration Factors, C_2 and C_3

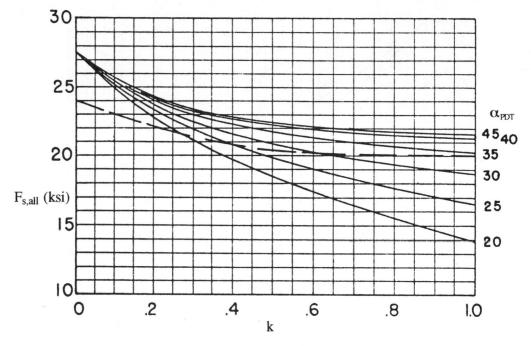

(a) 2024-T3 (F_{tu} = 62 ksi), dashed line is yield stress.

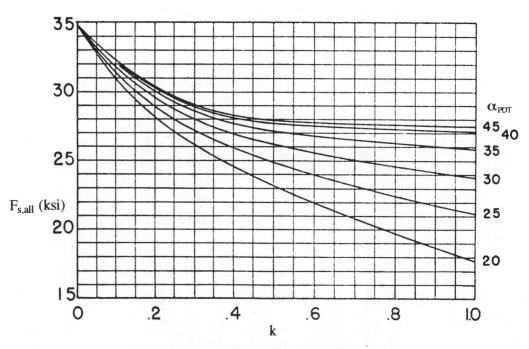

(b) 7075-T6 (F_{tu} = 72 ksi), dashed line is yield stress

Fig. 12.4.15 Allowable Maximum Web Shear Stress, $F_{s,all}$

(e) Rivets

    (e1) Web-to-cap rivets transmit the following running shear load:

$$q_R = f_s t(1 + 0.414k) \qquad \text{Eq. 12.4.14}$$

    where k is determined in Step (b3)

    (e2) Stiffener-to-cap rivets transmit the following column load:

        (i) Single stiffener:

$$\text{or } P_u = f_u A_{u,c} \qquad \text{Eq. 12.4.15}$$

        (ii) Double stiffeners:

$$\text{or } P_u = f_u A_u \qquad \text{Eq. 12.4.16}$$

    where f_u is determined in Step (c1)

    (e3) Stiffener-to-web rivets must satisfy the following criteria:

        (i) For single stiffeners:

        The rivet pitch must be small enough to prevent inter-rivet buckling of the web at stress equal to $f_{u,\text{max}}$ [see Step (c2)]. The inter-rivet buckling curves are given in Chapter 13.

        (ii) For double stiffeners:

        Size the rivets to resist the following running shear load to avoid shear failure:

$$q_u = \frac{2.5 F_{cy} Q}{b L'} \qquad \text{Eq. 12.4.17}$$

        Where L' value is determined in Step (c4), and F_{cy} is the compression yield stress of the stiffener. If the actual running shear strength q_A of the rivets is less than the required value q_u, then the allowable column stress F_c determined in Step (c4) should be multiplied by the reduction factor given in Fig. 12.4.16.

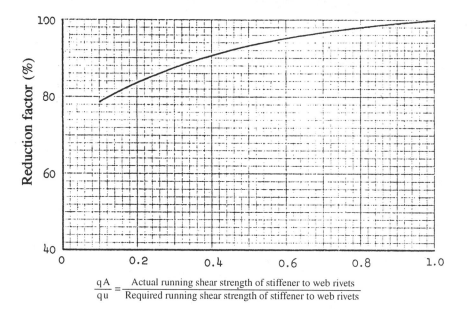

$$\frac{qA}{qu} = \frac{\text{Actual running shear strength of stiffener to web rivets}}{\text{Required running shear strength of stiffener to web rivets}}$$

Fig. 12.4.16 Column Strength Reduction Factor for Double Stiffeners

Chapter 12.0

(iii) The required tension running load of the rivets per inch is:

Single stiffener:

$$0.22\, t\, F_{tu} \qquad \text{Eq. 12.4.18}$$

Double stiffeners:

$$0.15\, t\, F_{tu} \qquad \text{Eq. 12.4.19}$$

where F_{tu} is the ultimate tensile stress of the web material

(f) Cap

(f1) The compressive stress (f_b) in a cap caused by beam bending moment is

$$f_b = \frac{M}{h_e A_{cap}} \qquad \text{Eq. 12.4.20}$$

(f2) The compressive stress in a cap caused by diagonal tension (f_F) is

$$f_F = \frac{k\,V}{2\,A_F \tan\alpha} \qquad \text{Eq. 12.4.21}$$

where $\tan\alpha$ is obtained in Step (d1) or given in Fig. 12.4.12

(f3) The maximum secondary bending moment (M_{max}) in the cap (over a stiffener) is

$$M_{max} = \frac{k\, C_3\, f_s\, t\, d^2 \tan\alpha}{12} \qquad \text{Eq. 12.4.22}$$

where: C_3 is given in Fig. 12.4.14

The secondary bending stress (f_{sb}):

$$f_{sb} = \frac{M_{max} \times c}{I_{x,\,cap}} \qquad \text{Eq. 12.4.23}$$

where: c is the distance of extreme fiber from NA of the beam cap cross section

$I_{x,\,cap}$ is the moment of inertia of the beam cap cross section

Note: The secondary bending moment halfway between stiffeners is $\frac{M_{max}}{2}$.

(f4) The margin of safety is

$$MS = \frac{1}{\dfrac{f_b + f_F}{F_{cc}} + \dfrac{f_{sb}}{F_{tu}}} - 1$$

where: F_{cc} – Crippling stress of the cap section (see Chapter 10.7)
F_{tu} – Ultimate tensile stress of the material

Example:

Use diagonal tension web analysis per the NACA TN 2661 method to size the beam panel D-E-K-J (bay ④), as given below, for an ultimate shear load V = 30,000 lbs.

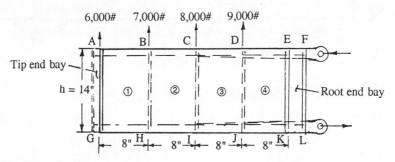

Basic dimensions:

Beam depth, h = 14 in. and stiffener spacing, d = 8 in.
Use single stiffener design and protruding head fasteners.

Materials: Web – 7075-T6 bare sheet (from Fig. 4.3.4, 'B' values):

$E = 10.5 \times 10^6$ psi $F_{tu} = 80,000$ psi
$F_{cy} = 71,000$ psi $F_{bru} = 160,000$ psi (dry pin value)
$F_{su} = 48,000$ psi

Caps and stiffeners – 7075-T6 Extrusion (from Fig. 4.3.5, 'B' values):

$E = 10.7 \times 10^6$ psi $F_{tu} = 82,000$ psi
$F_{cy} = 74,000$ psi $F_{bru} = 148,000$ psi (dry pin value)
$F_{su} = 44,000$ psi

Sizing procedures for bay ④ (use the same sizing procedures as previously described):

(a) Estimate the beam cap cross section area at E:

From given moment at section E-K:

M = 6,000 × 32 + 7,000 × 24 + 8,000 × 16 + 9,000 × 8 = 560,000 in.-lbs.

Assume $h_e = 13$ in. and use $F_{cc} = F_{cy} = 74,000$ psi to estimate cap section:

$$A_{cap} \approx \frac{M}{h_e F_{cc}} \approx \frac{560,000}{13 \times 74,000} \approx 0.582 \text{ in.}^2$$

Select cap section as shown below:

$A_{cap} = 0.918$ in.^{2}
$I_x = 0.6$ in.^{4}
$Y = 0.959$ in.
$\rho = 0.809$ in.

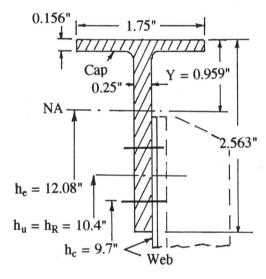

Determine the following values:

$h_e = 14 - 2 \times 0.959 = 12.08$ in.
$h_c = 9.7$ in.
$h_u = h_R = 10.4$ in.

Recalculate the beam cap area:

$$A_{cap} \approx \frac{M}{h_e F_{cc}} \approx \frac{560,000}{12.08 \times 74,000} \approx 0.626 \text{ in.}^2 < 0.918 \text{ in.}^2 \text{ (selected)} \quad \text{O.K.}$$

Note: This is just one of several choices and the reasonable choice is to select a cross section with shorter vertical flange. This choice is to obtain deeper h_e value to achieve lighter weight and also give the stiffener support closer to horizontal flange in case of lateral pressure on stiffeners such as fuel pressure.

Therefore, the applied shear flow in the web:

$$q = \frac{30,000}{12.08} = 2,483 \text{ lbs./in.}$$

Chapter 12.0

(b) Diagonal tension factor (k)

This a wing spar and is a fatigue critical area and use the criteria from Eq. 12.4.2 to determine the diagonal tension factor are as follows:

(b1) Estimate the web thickness (t):

Aspect ratio of $\dfrac{h_c}{d_c} = \dfrac{9.7}{8} = 1.21$ and entering Fig. 11.3.5 (Case ④), obtain $K_s = 7.4$ and from Eq. 12.4.4

$$t \approx 1.15 \left(\dfrac{q d_c^2}{5 K_s E}\right)^{\frac{1}{3}} \approx 1.15 \left(\dfrac{2{,}483 \times 8^2}{5 \times 7.4 \times 10.5 \times 10^6}\right)^{\frac{1}{3}} \approx 0.085''$$

(b2) Use the t value to recalculate the shear buckling stress of the web ($F_{s,cr}$):

$$\dfrac{F_{s,cr}}{\eta_s} = K_s E \left(\dfrac{t}{d_c}\right)^2 = 7.4 \times 10.5 \times 10^6 \left(\dfrac{0.085}{8}\right)^2 = 8{,}772 \text{ psi}$$

From Fig. 11.2.5, obtain $F_{s,cr} = 8{,}772$ psi ($\eta_s = 1.0$).

(b3) Determine the diagonal tension factor (k):

The applied shear stress in the web from Eq. 12.4.7:

$$f_s = \dfrac{q}{t} = \dfrac{2{,}483}{0.085} = 29{,}212 \text{ psi}$$

(i) Calculate $\dfrac{f_s}{F_{s,cr}} = \dfrac{29{,}212}{8{,}772} = 3.33$

(ii) Enter $\dfrac{f_s}{F_{s,cr}} = 3.33$ value into Fig. 12.4.6 and obtain

k = 0.25.

(c) Stiffener

(c1) Calculate the average, along the length of the stiffener, of the stiffener compressive stress in the median plane of the web:

Estimate the cross-sectional area of stiffener (upright) from Eq. 12.4.8:

$A_u \approx 30\% (dt) \approx 30\% \times 8 \times 0.085 \approx 0.204 \text{ in.}^2$

Select a cross-sectional area of single stiffener (skin not included) as shown below:

$A_u = 0.234$ in.^{2}
$I_x = 0.021$ in.^{4}
$Y = 0.29$ in.
$\rho = 0.298$ in.

e = 0.333" 1.0" Y = 0.29"
NA
1.0"
Stiffener 0.125"
0.125"
0.085"

Calculate the effective cross-sectional area of stiffener from Eq. 12.4.9:

$$A_{u,e} = \dfrac{A_u}{1 + \left(\dfrac{e}{\rho}\right)^2} = \dfrac{0.234}{1 + \left(\dfrac{0.333}{0.298}\right)^2} = 0.104$$

$$\dfrac{A_{u,e}}{dt} = \dfrac{0.104}{8 \times 0.085} = 0.153$$

Use $\frac{A_{u,e}}{dt} = 0.153$ and $k = 0.25$ and from Fig. 12.4.7 obtain $\frac{f_u}{f_s} = 0.42$

$\therefore f_u = 0.42 \times 29{,}212 = 12{,}269$ psi

(c2) Calculate the maximum stiffener stress in the median plane of the web:

Use $\frac{d}{h_u} = \frac{8}{10.4} = 0.77$ and $k = 0.25$ and from Fig. 12.4.8 obtain $\frac{f_{u,max}}{f_u} = 1.22$

$\therefore f_{u,max} = 1.22 \times 12{,}269 = 14{,}968$ psi

(c3) Determine F_{fu}, the stress causing a forced crippling failure in the stiffener:

Use $\frac{t_u}{t} = \frac{0.125}{0.085} = 1.47$ and $k = 0.25$ and from Fig. 12.4.9 obtain

$f_{fu} = 15{,}000$ psi.

$MS = \frac{F_{fu}}{f_{u,max}} - 1 = \frac{15{,}000}{14{,}968} - 1 = 0$ O.K.

(c4) Check against column failure in the stiffeners according to the following criteria:

(ii) Single Stiffeners:

Calculate the stiffener crippling stress (F_{cc}):

Segment	Free edges	b_n	t_n	$\frac{b_n}{t_n}$	$b_n t_n$	F_{ccn} (from Fig. 10.7.7)	$t_n b_n F_{ccn}$
①	1	0.938	0.125	7.5	0.117	74,000	8,658
②	1	0.938	0.125	7.5	0.117	74,000	8,658
					Σ 0.234		17,316

$F_{cc} = \frac{\Sigma b_n t_n F_{ccn}}{\Sigma b_n t_n} = \frac{17{,}316}{0.234} = 74{,}000$ psi

Note: It is not necessary to calculate the crippling stress for this symmetric angle section since the two legs have the same F_{ccn} values. Their crippling stress is $F_{cc} = F_{ccn}$.

Use $\frac{d}{h_u} = \frac{8}{10.4} = 0.77$ and $k = 0.25$ and from Fig. 12.4.10 obtain $\frac{L''}{h_u} = 0.955$.

$\therefore L'' = 0.955 \times 10.4 = 9.93$ in.

Use half the length of the L" to find the stiffener column stress:

$\frac{L'}{\rho} = \frac{\frac{L''}{2}}{\rho} = \frac{\frac{9.93}{2}}{0.298} = 16.7$

Use $\frac{L'}{\rho} = 16.7$ and $F_{cc} = 74{,}000$ psi from Fig. 10.8.2(a) to obtain

$F_c = 71{,}000$ psi.

Use $\frac{b}{t_u} = \frac{1.0}{0.125} = 8$ in Fig. 12.4.11 (Case ①) to obtain

$\frac{A_{u,e}}{A_u} = 0.51$

$f_{u,av} = \left(\frac{A_{u,e}}{A_u}\right) f_u = 0.51 \times 12{,}269 = 6{,}257$ psi

$MS = \frac{F_c}{f_{u,av}} - 1 = \frac{71{,}000}{6{,}257} - 1 = $ high O.K.

Chapter 12.0

(d) Maximum web shear stress

(d1) Determine tan α value:

Use $\dfrac{f_u}{f_s} = \dfrac{12{,}269}{29{,}212} = 0.42$ and k = 0.25 and from Fig. 12.4.12 obtain

tan α = 0.88 or α = 41.2°

(d2) Use tan α = 0.88 value in Fig. 12.4.13 to obtain:

$C_1 = 0.008$

(d3) Determine C_2 value from Eq. 12.4.12:

$$\omega d = 0.7 d \left[\dfrac{t}{h_e(I_T + I_C)}\right]^{\frac{1}{4}}$$

$$= 0.7 \times 8 \left[\dfrac{0.085}{12.08(0.6 + 0.6)}\right]^{\frac{1}{4}} = 1.55$$

From Fig. 12.4.14, obtain:

$C_2 = 0.02$

(d4) Calculate the maximum shear stress in the web from Eq. 12.4.13:

$f_{s,\,max} = f_s(1 + k^2 C_1)(1 + k C_2)$

$= 29{,}212(1 + 0.25^2 \times 0.008)(1 + 0.25 \times 0.02)$

$= 29{,}212 \times 1.006 = 29{,}387$ psi

(d5) Obtain the allowable maximum web shear stress $F_{s,\,all}$ from Fig. 12.4.15. (use k = 0.25 and α = 41.2°)

$f_{s,\,all} = 29{,}700$ psi

(d6) Margin of safety:

$$MS = \dfrac{F_{s,\,all}}{f_{s,\,max}} - 1$$

$$= \dfrac{29{,}700}{29{,}387} - 1 = \underline{0.01} \qquad\qquad\text{O.K.}$$

(e) Rivets

Use protruding head rivets for this spar beam design.

(e1) Web-to-cap rivets transmit the following running shear load from Eq. 12.4.14:

$q_R = f_s t (1 + 0.414 k)$

$q_R = 29{,}212 \times 0.085 (1 + 0.414 \times 0.25)$

$= 2{,}740$ lbs./in.

Use a double row of rivets, $D = \dfrac{3}{16}"$ (E aluminum rivets) with rivet spacing (pitch) of s = 0.9 in. with a shear allowable, $P_{s,\,all} = 1{,}230$ lbs./rivet (see Fig. 9.2.5)

Check web bearing strength (cap vertical flange of t = 0.25" is not critical):

$P_{b,\,all} = 80\% \times 160{,}000 \times 0.085 \times 0.1875 = 2{,}040$ lbs $> P_{b,\,all} = 1{,}230$

where 80% is the reduction factor from the dry pin value to the wet pin; see Chapter 9.2.

Therefore, rivet shear is critical and the margin of safety for this double row of rivets is:

$$MS = \dfrac{2 P_{s,\,all}}{q_R \times s} - 1 = \dfrac{2 \times 1{,}230}{2{,}740 \times 0.9} - 1 = \underline{0.0} \qquad\qquad\text{O.K.}$$

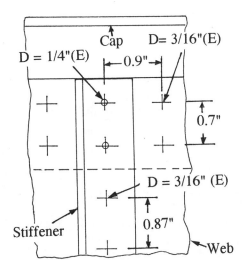

(e2) Stiffener-to-cap rivets transmit the following column load from Eq. 12.4.15:

$P_u = f_u A_{u,e}$
$= 12,269 \times 0.104 = 1,276$ lbs./in.

Use two Hi-Loks with $D = \frac{3}{16}"$ ($P_{s,all} = 2,623$ lbs., see Fig. 9.2.5) or use two larger rivets with $D = \frac{1}{4}"$ (E) and $P_{s,all} = 2,230$ lbs. (see Fig. 9.2.5).

$$MS = \frac{2 P_{s,all}}{P_u} - 1 = \frac{2 \times 2,230}{1,276} - 1 = \underline{high} \qquad O.K.$$

(It may appear that only one fastener is necessary, but it is customary to always have at least two fasteners at this location)

(e3) Stiffener-to-web rivets must satisfy the following criteria:

(i) For single stiffeners (used in this example):

$f_{u,max} = 14,968$ psi (from step (c2))

Use rivet $D = \frac{3}{16}"$ (E) with spacing s = 0.87" (tension running load is critical, see following calculation); inter-rivet buckling should be checked under the compression stress of $f_{u,max} = 14,968$ psi.

(ii) The required tension running load of the rivets per inch is:

Single stiffener from Eq. 12.4.18:

$P_t = 0.22 \, t \, F_{tu}$
$= 0.22 \times 0.085 \times 80,000 = 1,496$ lbs./in.

Use rivet $D = \frac{3}{16}"$ (E); the tension allowable load is $P_{t,all} = 1,304$ lbs. (see Fig. 9.2.11) for the rivet spacing

$$s = \frac{1,304}{1,496} = 0.87"$$

(f) Caps

Calculate the cap [see sketch in Step (a)] crippling stress (F_{cc}):

Segment	Free edges	b_n	t_n	$\dfrac{b_n}{t_n}$	$b_n t_n$	F_{ccn} (from Fig. 10.7.7)	$t_n b_n F_{ccn}$
①	1	0.875	0.156	5.6	0.137	74,000	10,138
②	1	0.875	0.156	5.6	0.137	74,000	10,138
③	1	2.485	0.25	9.94	0.621	58,000	36,018
					Σ 0.895		56,294

$$F_{cc} = \frac{\Sigma b_n t_n F_{ccn}}{\Sigma b_n t_n} = \frac{56,294}{0.895} = 62,898 \text{ psi}$$

(f1) The compressive stress in a cap caused by beam bending moment from Eq. 12.4.20:

$$f_b = \frac{M}{h_e A_{cap}} = \frac{560,000}{12.08 \times 0.918} = 50,498 \text{ psi}$$

(f2) The compressive stress in a cap caused by diagonal tension from Eq. 12.4.21:

$$f_F = \frac{kV}{2 A_{cap} \tan \alpha}$$

$$= \frac{0.25 \times 30,000}{2 \times 0.918 \times 0.88} = 4,642 \text{ psi}$$

where $\tan \alpha = 0.88$ from Step (d1)

(f3) The maximum secondary bending moment in the cap (over a stiffener) from Eq. 12.4.22:

$$M_{max} = \frac{k C_3 f_s t d^2 \tan \alpha}{12}$$

$$M_{max} = \frac{0.25 \times 0.98 \times 29,212 \times 0.085 \times 8^2 \times 0.88}{12} = 2,855 \text{ in.-lbs.}$$

$C_3 = 0.98$ obtained from Fig. 12.4.14

The secondary bending stress from Eq. 12.4.23:

$$f_{sb} = \frac{M_{max} \times c}{I_{x, cap}} = \frac{2,855 \times (2.563 - 0.959)}{0.6} = 7,632 \text{ psi}$$

(f4) The margin of safety is

$$MS = \frac{1}{\dfrac{f_b + f_F}{F_{cc}} + \dfrac{f_{sb}}{F_{tu}}} - 1$$

$$= \frac{1}{\dfrac{50,498 + 4,642}{62,898} + \dfrac{7,632}{82,000}} - 1 = \underline{0.03} \qquad \text{O.K.}$$

12.5 DIAGONAL TENSION CURVED WEBS

The fuselage of an aircraft is a good example of diagonal tension in a curved web since it carries the design loads in spite of a considerable degree of skin buckling. Comparing this structure to a flat web beam, the stringers correspond to the beam cap (flange or chord), the fuselage frames (or rings) correspond to the uprights and the skin corresponds to the web, as shown in Fig. 12.5.1 through Fig. 12.5.3. Thus,

- No pressurization
- Stringers carry axial loads
- Stringers have normal loading that tends to bend them inward between supporting frames
- Frames support the stringers as inward loading which puts them in "hoop compression" and the frames also divide the skin panels into shorter lengths
- The skin carries shear loads

The development of diagonal tension theory in curved skins is complicated by the following:

(a) If the fuselage was built as a polygonal cylinder and subjected to torque loads [see Fig. 12.5.1(a)], the theory of diagonal tension would evidently be applicable and require only minor modifications from the flat web design version.

(b) If the fuselage was built with a circular section skin, but polygonal frames were used [see Fig. 12.5.1(b), which is the floating frame design shown in Fig. 12.5.3(b)], the skin would begin to "flatten" after buckling and would approach the shape of the polygonal cylinder more and more as the load increased.

(c) The theory of diagonal tension for flat webs is not directly applicable to curved webs, but is usable when some modifications are made.

In an actual fuselage, the frames are circular, not polygonal [see Fig. 12.5.2(c)]; consequently, all the tension diagonals of one sheet bay cannot lie in one flat web, even when the diagonal tension is fully developed; an additional complication therefore exists.

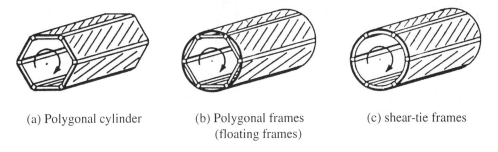

(a) Polygonal cylinder (b) Polygonal frames (c) shear-tie frames
                         (floating frames)

Fig. 12.5.1 Diagonal-Tension Effects in Curved skin

Fuselage pressurization requirements set the skin thickness and often dictate the frame sizes. The diagonal tension method for curved webs takes into consideration not only the diagonal tension effect, but also the additional forces of fuselage bending, torsion, and pressurization.

(A) TYPES OF PANELS

Generally, the curved skin in diagonal tension is one of two types, as shown in Fig. 12.5.2:

(a) Type A ($\frac{d}{h} > 1$):

This type has an arrangement which results in the skin panels being longer in the axial direction, d, than in the circumferential direction, h. This is typical of the skin-stringer panel used for transport fuselages, and wing and empennage surfaces.

(b) Type B ($\frac{d}{h} < 1$):

This type of curved skin structure may be referred to as the longeron construction. Its main characteristic is that the skin panels are long in the circumferential direction. Typically, this type of structure consists of a few axial members (a minimum of three but more usually 4 to 8 for a fail-safe design) and a large number of closely spaced formers (usually a light frame (former) structure compared to the fuselage frame).

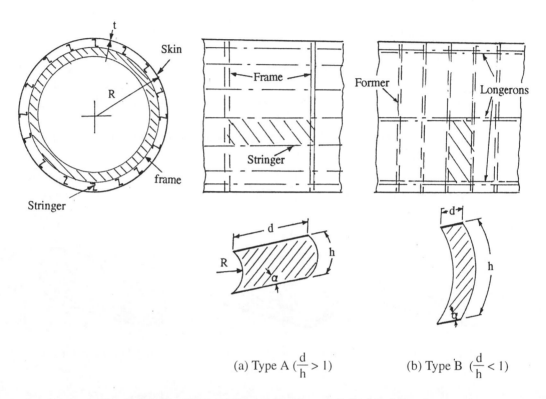

(a) Type A ($\frac{d}{h} > 1$)   (b) Type B ($\frac{d}{h} < 1$)

Fig. 12.5.2 Two Types of Curved Panels on a Cylindrical Fuselage

The skin-stringer panel is usually found, for example, in fuselage structures where there are relatively few large "cutouts" to disrupt the stringer continuity, as is typical of transport aircraft. The longeron type structure is more efficient and suitable where large numbers of "quick-access" doors or other "cutouts" are necessary to allow rapid serviceing. Therefore, longeron type fuselage are more usually found in fighter aircraft.

Transport fuselage construction utilizes the following:

- Shear-tie frames, as shown in Fig. 12.5.3(a), which can transfer shear load from frame to fuselage skin
- Floating frames, as shown in Fig. 12.5.3(b), which do not have the capability of shear transfer

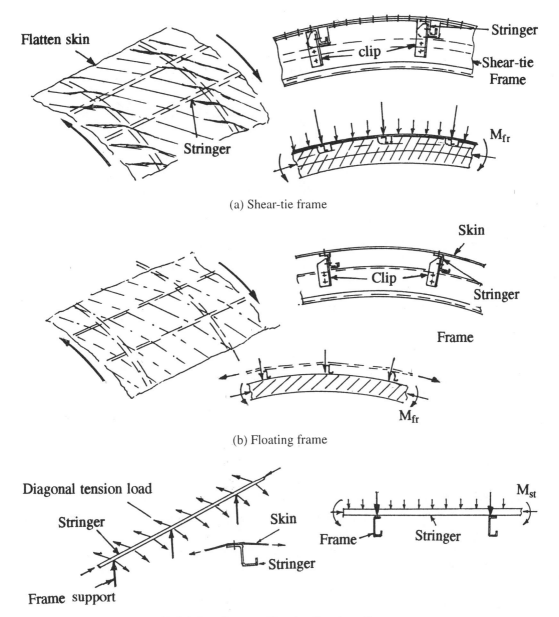

Fig. 12.5.3 Loadings on Circular Fuselage Components

(B) LOADING CASES

In reality, in aircraft fuselage design it is the case that the loads consist of combined fuselage bending moment, vertical shear load, and torsion, as shown in Fig. 12.5.4, which will occur as follows:

- Below the neutral axis – A greater amount of skin buckling occurs due to combined compression and shear loading (due to fuselage down bending moments) on the skin; thus there is a larger diagonal tension load on the stringer and frames
- Above the neutral axis – Less skin buckling occurs due to the tension strain produced by fuselage down bending moments

- Near the neutral axis – There is little or no strain on skins due to fuselage down or up bending moments but buckling will occur because of the combination of maximum fuselage vertical shear and torsional loads.

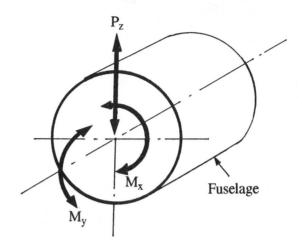

Fig. 12.5.4 Cylindrical Fuselage Under Various Loads

In an actual fuselage design, the applied loading is more complex. Instead of simply an applied torsion, there is also a vertical and perhaps a sideward set of loads producing shears that vary from panel to panel.

(C) INITIAL BUCKLING OF CURVED SKINS

Curved skins carry fuselage bending moment which produces axial compression and tension loads in the stringers as well as in the skin. In fuselage design, it is generally assumed that shear load is being applied at the same time as the compressive or tension axial load under zero cabin pressure.

- The skin will buckle at a lower amount of applied shear load since they are now also strained axially in compression
- This can be obtained from an interaction buckling equation of which combine shear and compression or tension stress

The skin buckles according to the following interaction equation with shear stress (f_s) and compression stress (f_c):

(a) Shear buckling stress under shear stress (f_s) and compression stress (f_c):

$$\frac{f_c}{F_{c,cr}} + (\frac{f_s}{F_{s,cr}})^2 = 1.0 \qquad \text{Eq. 12.5.1}$$

where: f_c = given axial compression or tension stress

f_s = given shear stress ($\frac{q}{t}$)

Where the initial buckling stress (Type A panel, see Fig. 12.5.2) for a curved skin is:

$$F_{c,cr} = \frac{k_c' \eta_c \pi^2 E}{12(1-\mu^2)} (\frac{t}{h})^2 \qquad \text{(from Eq. 11.4.1)}$$

$$F_{s,cr} = \frac{k_s' \eta_s \pi^2 E}{12(1-\mu^2)} (\frac{t}{h})^2 \qquad \text{(from Eq. 11.4.2)}$$

(Replace above h by d for Type B panel)

Use four hinged supports as edge conditions for the values of k_c' (from Fig. 11.4.1) and k_s' (from Fig. 11.4.2 through 11.4.5). The following is the reduced shear buckling stress of a curved skin under axial compression stress:

$$F'_{s,cr} = F_{s,cr} R_c \qquad \text{Eq. 12.5.2}$$

where:

$$R_c = \frac{-\frac{B}{A} + \sqrt{(\frac{B}{A})^2 + 4}}{2} \qquad \text{Eq. 12.5.3}$$

$$A = \frac{F_{c,cr}}{F_{s,cr}}$$

$$B = \frac{f_c}{f_s}$$

(**Note:** R_c is always less than 1.0)

(b) Shear buckling stress under shear stress (f_s) and tension stress (f_t):

$$F''_{s,cr} = F_{s,cr} R_t \qquad \text{Eq. 12.5.4}$$

$$R_t = \frac{f_t}{2 F_{s,cr}} \qquad \text{Eq. 12.5.5}$$

(**Note:** R_t is always greater than 1.0 and the value of $F''_{s,cr}$ will become several times greater than $F_{s,cr}$, thus, if it is large enough, skin buckling will not occur)

(D) SIZING PROCEDURES

(a) The values used are defined as follows:

 p – Cabin pressure (see determining cabin pressure in Chapter 3.8)
 f_c – Fuselage axial compression stress on stringer
 f_t – Fuselage axial tension stress on stringer
 q – Fuselage skin shear flow (or $f_s = \frac{q}{t}$)
 $f_{h,fr}$ – Fuselage frame circumferential axial stress
 F_{fat} – Fatigue allowable tension stress which is from 10,000 to 18,000 psi depending on number of flights per design life of the airframe
 R – Fuselage radius
 d – Stringer length (fuselage frame spacing)
 h – Stringer spacing

(b) Determine skin thickness.

The minimum fuselage skin under fatigue requirement for cabin pressure (p):

$$t_{min} = \frac{pR}{F_{fat}} \qquad \text{Eq. 12.5.6}$$

(c) Determine the skin buckling stress of $F_{c,cr}$ (from Eq. 11.4.1) or $F_{s,cr}$ (from Eq. 11.4.2).

(d) Determine reduced skin shear buckling stress of $F'_{s,cr}$ (from Eq. 12.5.2) or increased skin shear buckling stress of $F''_{s,cr}$ (from Eq. 12.5.4).

(e) Diagonal tension factor (k) for a curved skin can be found in Fig. 12.4.6 with $\frac{f_s}{F'_{s,cr}}$ or $\frac{f_s}{F''_{s,cr}}$.

(f) Allowable shear stress ($F_{s,all}$) of the curved skin is given from the empirical equation:

$$F_{s,all} = F_s(0.65 + \Delta) \qquad \text{Eq. 12.5.7}$$

where: F_s – Allowable skin shear stress from Fig. 12.4.5

$$\Delta = 0.3 \tanh \frac{A_{fr}}{dt} + 0.1 \tanh \frac{A_{st}}{ht} \qquad \text{Eq. 12.5.8}$$

The "Δ" value can be obtained from Fig. 12.5.5.

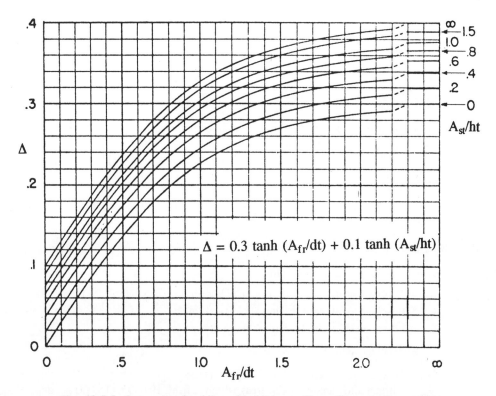

Fig. 12.5.5 *Correction for Allowable Ultimate Shear Stress in Curved Skin*

$$MS = \frac{F_{s,all}}{f_s} - 1$$

(g) Stringer and frame stress:

Functionally, the stringers (st) as well as the frame (fr) of a cylinder fuselage under shear or torsion load, act like the stiffeners of a flat web shear beam.

(g1) Average stringer stress (f_{st}) due to the diagonal tension effect:

$$f_{st} = \frac{-kf_s \cot \alpha}{\frac{A_{st}}{ht} + 0.5(1-k)(R_c \text{ or } R_t)} \qquad \text{Eq. 12.5.9}$$

(g2) Average frame stress (f_{fr}) due to the diagonal tension effect for shear-tie and floating frames (see Fig. 12.5.3):

Shear-tie frame:

$$f_{fr} = \frac{-kf_s \tan \alpha}{\frac{A_{fr}}{dt} + 0.5(1-k)} \qquad \text{Eq. 12.5.10}$$

Floating frame:

$$f_{fr} = \frac{-k f_s \tan \alpha}{\dfrac{A_{fr}}{dt}}$$ Eq. 12.5.11

(g3) Diagonal tension angle (α):

The $\tan \alpha$ values are defined as below:

$$\tan^2 \alpha = \frac{\varepsilon_{sk} - \varepsilon_{st}}{\varepsilon_{sk} - \varepsilon_{fr} + \dfrac{1}{24}\left(\dfrac{h}{R}\right)^2} \quad \text{(for } d > h\text{)}$$ Eq. 12.5.12

or $$\tan^2 \alpha = \frac{\varepsilon_{sk} - \varepsilon_{st}}{\varepsilon_{sk} - \varepsilon_{fr} + \dfrac{1}{8}\left(\dfrac{d}{R}\right)^2} \quad \text{(for } h > d\text{)}$$ Eq. 12.5.13

where the skin strain (ε_{sk}) is given below or can be obtained from Fig. 12.5.6:

$$\varepsilon_{sk} = \left(\frac{f_s}{E_{sk}}\right)\left[\frac{2k}{\sin 2\alpha} + \sin 2\alpha (1-k)(1+\mu)\right]$$ Eq. 12.5.14

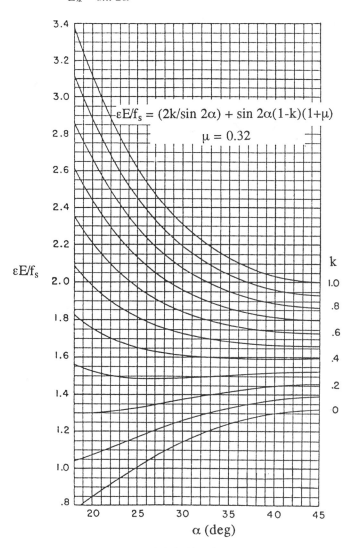

Fig. 12.5.6 Skin Strain, ε_{sk}

Chapter 12.0

$$\varepsilon_{sk} = \frac{f_{st}}{E_{st}} \quad \text{(stringer strain)} \qquad \text{Eq. 12.5.15}$$

$$\varepsilon_{fr} = \frac{f_{fr}}{E_{fr}} \quad \text{(frame strain)} \qquad \text{Eq. 12.5.16}$$

Solve Eq. 12.5.9 through 12.5.11 by assuming a value of α; also solve for f_{st} or f_{fr} and then from Eq. 12.5.12 or 12.5.13 obtain the calculated α values, until the assumed value of α reasonably agrees with the calculated one.

The evaluation of the α value is as follows:

- The angle of α is dependent on the ratio of $\frac{h}{R} \cdot \frac{d}{R}$ or $\frac{d}{R}$
- For a large R, which has little effect on the angle of α, assuming α is 45° is a good approximation (applications: transport fuselage, wing and empennage surface panels)
- For a smaller R, assume α to be between 20° and 30° (applications: commuter and general aviation aircraft fuselage panels)

(h) Stringer stress :

(h1) Calculate the maximum stringer stress ($f_{st, max}$) in the same manner as the flat web of sizing procedures in Step (C2) (replace f_u with f_{st}):

$$f_{st, max} = (\frac{f_{st, max}}{f_{st}}) f_{st} \qquad \text{Eq. 12.5.17}$$

where $\frac{f_{st, max}}{f_{st}} = \frac{f_{u, max}}{f_u}$ which is given in Fig. 12.4.8.

(h2) Secondary bending moment (M_{st}) on stringers:

Secondary bending moment tends to bow the stringer inward between frames:

$$M_{st} = \frac{k f_s h t d^2 \tan \alpha}{24R} \qquad \text{Eq. 12.5.18}$$

[This is an empirical equation which occurs at the frames (skin in tension) and also halfway between the **frames (skin in compression)**]

The stringer bending stresses are given below:

$$f_{sb} = \frac{M_{st} c}{I_x} \quad \text{(between frames)}$$

$$f'_{sb} = \frac{M_{st} c'}{I_x} \quad \text{(at the frame)}$$

where: c' – the distance between stringer NA to the most remote fiber of the outstanding flange away from skin
c – the distance between stringer NA to the most remote fiber of the skin
I_x – moment of inertia of the stringer section with 30t width of the skin

(h3) Stringer forced crippling stress (effective width of skin equal to 30t should be included with the stringer when calcualating their section properties):

For forced crippling stress, the following should be considered:

- Axial stress due to diagonal tension effect
- Bending stress due to bending moment because of the diagonal tension effect

Use the following interaction equation:

$$\text{MS} = \frac{1}{\dfrac{f_c + f_{sb}}{F_{cc}} + \dfrac{f_{st,\,max}}{F_{f,\,st}}} - 1 \qquad \text{Eq. 12.5.19}$$

where: f_c – Given applied axial stress on the stringer section
f_{sb} – Compression stress casued by the secondary bending moment (M_{st}) between frames (see Eq. 12.5.18)
F_{cc} – Crippling stress of the stringer section (see Chapter 10.7)
$f_{st,\,max}$ – Maximum stringer stress defined by Eq. 12.5.17
$F_{f,\,st}$ – Allowable forced crippling stress from Fig. 12.4.9 (replace t_u with t_{st} and F_{fu} with $F_{f,\,st}$); use $t_u = 3t_u$ for hat section stringer

(h4) Stringer column failure:

At the junction of the stringer and the frame:

$$\text{MS} = \frac{1}{\dfrac{f_c + f_{st} + f'_{sb}}{F'_{cc}}} - 1 \qquad \text{Eq. 12.5.20}$$

Between (at mid-span of the stringer) frames:

$$\text{MS} = \frac{1}{\dfrac{f_c + f_{st} + f_{sb}}{F_c}} - 1 \qquad \text{Eq. 12.5.20}$$

where: F_c – Compression allowable stress [from Eq. 10.8.2(a)]; it is assumed that both ends of the stringer are considered to be fixed end (c = 4) to calculate the effective column length (L')
F'_{cc} – Crippling stress of stringer cap (away from the skin)
f_{sb} and f'_{sb} – Bending stress from Step (h2)
f_{st} – From Eq. 12.5.9 [Step (g2)]

(i) Frame stresses (the stress due to the diagonal tension effect causes hoop compression in the frame):

(i1) Shear-tie frame:

Maximum frame stress ($f_{fr,\,max}$):

Obtained by the same manner as that for the upright in flat web analysis from Fig. 12.4.8 (replace f_u with f_{fr})

$$f_{fr,\,max} = \left(\frac{f_{fr,\,max}}{f_{fr}}\right) f_{fr} \qquad \text{Eq. 12.5.22}$$

where $\dfrac{f_{fr,\,max}}{f_{fr}} = \dfrac{f_{u,\,max}}{f_u}$ which is given in Fig. 12.4.8.

Use following MS equation:

$$\text{MS} = \frac{1}{\dfrac{f_{h,\,fr}}{F_{cc}} + \dfrac{f_{fr,\,max}}{F_{fr}}} - 1 \qquad \text{Eq. 12.5.23}$$

where: $f_{h,\,fr}$ – Applied frame compression stress (if it is in tension, $f_{h,\,fr} = 0$) from other than diagonal tension effect
F_{cc} – Frame crippling stress (see Chapter 10.7)
F_{fr} – Allowable frame stress (Fig. 12.4.9, replace t_u with t_{fr} and f_u with f_{fr})

(i2) Floating frame (not subject to forced crippling by the skin):

Bending moment at the junction with the stringer (the secondary bending moment at mid-bay is $\frac{M_{fr}}{2}$):

$$M_{fr} = \frac{k f_s h^2 t d \tan \alpha}{12R}$$ Eq. 12.5.24

Use the following MS equation:

$$MS = \frac{1}{\frac{f_{fr} + f_b}{F_{cc}}} - 1$$ Eq. 12.5.25

where: f_{fr} – Frame compression stress due to the diagonal tension effect from Eq. 12.5.10 or 12.5.11
f_b – Bending stress from bending moment per Eq. 12.5.24
F_{cc} – Frame crippling stress (see Chapter 10.7)

(j) Rivet loads

(j1) Shear loads – At the edge of a curved skin attached to a stringer:

$$q_R = q[1 + k(\frac{1}{\cos \alpha} - 1)]$$ Eq. 12.5.26

(At the edge of a curved skin attached to a frame, replace $\cos \alpha$ with $\sin \alpha$)

If the skin is continuous across a stringer, then the fasteners must carry the difference in shear flow between bays (e.g., splices on fuselage and wing surface panels).

(j2) Tension loads – Prying forces on the rivets which pop off the rivet heads or cause pull-through of the skin (particularly with countersunk rivets). Use of Eq. 12.4.18 is recommended (conservative).

$$0.22 \, t \, F_{tu}$$

Observations:
- Check for rivet shear-off at areas of skin discontinuity (e.g., web splices; cutouts and frame attachments, etc.)
- Around cutouts the skins should be made either shear resistant or to have a smaller diagonal tension factor (k) which will relieve the diagonal tension effect along the edge members and the rivets

(k) General instability check:

Beware of the collapse of a cylindrical fuselage from general instability. Use the following parameter (from Ref. 12.5) in Fig. 12.5.7 to obtain instability shear stress ($F_{s, inst}$):

$$F_{s, inst} = \frac{(\rho_{st} \rho_{fr})^{\frac{7}{8}}}{(dh)^{\frac{1}{2}} R^{\frac{3}{4}}}$$ Eq. 12.5.27

Radii of gyration of ρ_{st} (stringer) and ρ_{fr} (frame) should include the full width of the skin.

$$MS = \frac{F_{s, inst}}{f_s} - 1$$

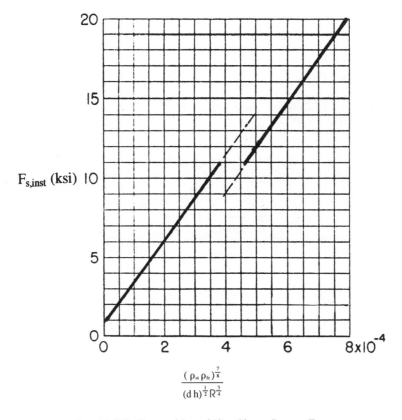

Fig. 12.5.7 *General Instability Shear Stress*, $F_{s,inst}$

Example:

Use diagonal tension web analysis per the NACA TN 2661 method to size the curved skin of a fuselage skin-stringer panel as given below:

(a) Given the following data:

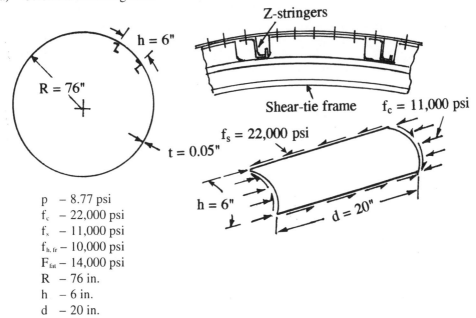

p – 8.77 psi
f_c – 22,000 psi
f_s – 11,000 psi
$f_{h.fr}$ – 10,000 psi
F_{fat} – 14,000 psi
R – 76 in.
h – 6 in.
d – 20 in.

Material: ('B' values, see Fig. 4.3.4 and Fig. 4.3.5):

Web – 2024-T3 sheet:
- F_{tu} = 65,000 psi
- F_{cy} = 40,000 psi
- F_{su} = 40,000 psi
- F_{bru} = 131,000 psi
- $E = 10.7 \times 10^6$ psi
- $\mu = 0.33$

Stringer and frame (7075-T6 sheet):
- F_{tu} = 80,000 psi
- F_{cy} = 71,000 psi
- F_{su} = 48,000 psi
- F_{bru} = 160,000 psi
- $E = 10.5 \times 10^6$ psi
- $\mu = 0.33$

Section properties of stringer:

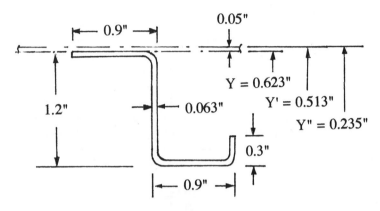

A = 0.196 in.²	A' = 0.271 in.²	A" = 0.496 in.²
Y = 0.632 in.	Y' = 0.513 in.	Y" = 0.235 in.
I_x = 0.0435 in.⁴	I'_x = 0.067 in.⁴	$I"_x$ = 0.0946 in.⁴
ρ = 0.471 in.	ρ' = 0.497 in.	$\rho"$ = 0.437 in.
(Stringer excluding skin)	(stringer with 30t width skin)	(stringer with full width skin)

Section properties of frame :

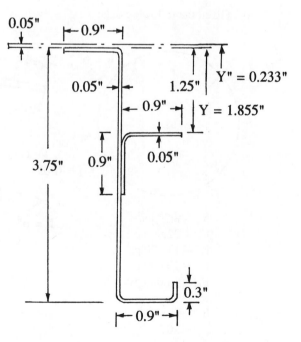

A_{st} = 0.372 in.²
Y = 1.855 in.
I_x = 0.564 in.⁴
ρ = 1.23 in.
(excluding skin)

A" = 1.844 in.²
Y" = 0.233 in.
$I"_x$ = 1.689 in.⁴
$\rho"$ = 0.769 in.
(with full width of skin)

Shear Panels

(b) Determine skin thickness:

The minimum fuselage skin under fatigue requirements from Eq. 12.5.6:
$$t_{min} = \frac{8.77 \times 76}{14,000} = 0.048 \text{ in.}$$

Use skin thickness, $t = 0.05$ in.

(c) Determine the skin buckling stress $F_{c,cr}$ (from Eq. 11.4.1) or $F_{s,cr}$ (from Eq. 11.4.2).

Panel aspect ratio: $\frac{a}{b} = \frac{d}{h} = \frac{20}{6} = 3.33$ and

(1) Compression buckling stress ($F_{c,cr}$):
$$Z = \sqrt{(1-\mu^2)}\left(\frac{b^2}{Rt}\right)$$
$$Z = \sqrt{(1-0.33^2)}\left(\frac{6^2}{76 \times 0.05}\right) = 8.94$$

$k_c' = 5.6$ (see Fig. 11.4.1)

From Eq. 11.4.1
$$F_{c,cr} = \frac{k_c' \eta_c \pi^2 E}{12(1-\mu^2)}\left(\frac{t}{h}\right)^2$$
$$\frac{F_{c,cr}}{\eta_c} = \frac{5.6 \times \pi^2 \times 10.7 \times 10^6}{12(1-0.33^2)}\left(\frac{0.05}{6}\right)^2 = 3,841 \text{ psi}$$

From Fig. 11.2.4, obtain the true compression buckling stress ($\eta_c = 1$):
$$F_{c,cr} = 3,841 \text{ psi}$$

(2) Shear buckling stress ($F_{s,cr}$):

$Z = 8.94$ and $\frac{a}{b} = \frac{d}{h} = \frac{20}{6} = 3.33$

$k_s' = 7.0$ (see Fig. 11.4.2)

From Eq. 11.4.2:
$$F_{s,cr} = \frac{k_s' \eta_p \pi^2 E}{12(1-\mu^2)}\left(\frac{t}{b}\right)^2$$
$$\frac{F_{s,cr}}{\eta_s} = \frac{7 \times \pi^2 \times 10.7 \times 10^6}{12(1-0.33^2)}\left(\frac{0.05}{6}\right)^2 = 4,801 \text{ psi}$$

From Fig. 11.2.5, obtain the true shear buckling stress ($\eta_s = 1$):
$$F_{s,cr} = 4,115 \text{ psi}$$

(d) Determine the reduced skin shear buckling stress $F'_{s,cr}$ (from Eq. 12.5.2) due to axial compression stress (f_c):

$$A = \frac{F_{c,cr}}{F_{s,cr}} = \frac{3,841}{4,801} = 0.8$$

$$B = \frac{f_c}{f_s} = \frac{22,000}{11,000} = 2$$

From Eq. 12.5.3:
$$R_c = \frac{-\frac{B}{A} + \sqrt{\left(\frac{B}{A}\right)^2 + 4}}{2}$$

$$R_c = \frac{-\frac{2}{0.8} + \sqrt{(\frac{2}{0.8})^2 + 4}}{2} = 0.35$$

From Eq. 12.5.2:

$$F'_{s,cr} = F_{s,cr} R_c = 4,801 \times 0.35 = 1,680 \text{ psi}$$

(e) The diagonal tension factor (k) for a curved skin can be found from Fig. 12.4.6 with

$$\frac{f_s}{F'_{s,cr}} = \frac{11,000}{1,680} = 6.55$$

(since $\frac{d}{h} = \frac{20}{6} = 3.33 > 2.0$, use $\frac{d}{h} = 2.0$, see Fig. 12.4.6)

$$300(\frac{t}{R})(\frac{d}{h}) = 300(\frac{0.05}{76})(2.0) = 0.395$$

From Fig. 12.4.6, obtain k = 0.63

(f) Allowable shear stress ($F_{s,all}$):

Assume the angle of diagonal tension α = 45° (to be checked later) with k = 0.63 and from Fig. 12.4.5 obtain:

F = 22,000 psi

$$\frac{A_{st}}{ht} = \frac{0.196}{6 \times 0.05} = 0.65$$

$$\frac{A_{fr}}{dt} = \frac{0.372}{20 \times 0.05} = 0.37$$

From Fig. 12.5.5, Δ = 0.162

From Eq. 12.5.7:

$$F_{s,all} = F_s(0.65 + \Delta) = 22,000(0.65 + 0.162) = 17,864 \text{ psi}$$

$$MS = \frac{F_{s,all}}{f_s} - 1 = \frac{17,864}{11,000} - 1 = \underline{0.62} \qquad \text{O.K.}$$

(g) Stringer and frame stress:

Assume an angle of diagonal tension, α = 45°

(g1) Average stringer stress (f_{st} – skin under both shear and compression) from Eq. 12.5.9:

$$f_{st} = \frac{-k f_s \cot \alpha}{\frac{A_{st}}{ht} + 0.5(1-k)(R_c)}$$

$$= \frac{-0.63 \times 11,000 \times \cot 45°}{\frac{0.196}{6 \times 0.05} + 0.5(1-0.63)(0.35)} = -9,652 \text{ psi (in compression)}$$

(g2) Average frame stress (f_{fr}) from Eq. 12.5.10 (this is a shear-tie frame):

$$f_{fr} = \frac{-k f_s \tan \alpha}{\frac{A_{fr}}{ht} + 0.5(1-k)}$$

$$= \frac{-0.63 \times 11,000 \times \tan 45°}{\frac{0.372}{20 \times 0.05} + 0.5(1-0.63)} = -12,442 \text{ psi (in compression)}$$

(g3) Diagonal tension angle (α):

From Eq. 12.5.14:

$$\varepsilon_{sk} = (\frac{f_s}{E_{sk}})[\frac{2k}{\sin 2\alpha} + \sin 2\alpha(1-k)(1+\mu)]$$

$$= (\frac{11,000}{10.5 \times 10^6})[\frac{2 \times 0.63}{\sin 2 \times 45°} + \sin 2 \times 45°(1-0.63)(1+0.33)] = 1.84 \times 10^{-3}$$

From Eq. 12.5.15:

$$\varepsilon_{st} = \frac{f_{st}}{E_{st}} = \frac{-9,652}{10.5 \times 10^6} = -0.919 \times 10^{-3}$$

From Eq. 12.5.16:

$$\varepsilon_{fr} = \frac{f_{fr}}{E_{fr}} = \frac{-12,442}{10.5 \times 10^6} = -1.185 \times 10^{-3}$$

The α values are defined in Eq. 12.5.12 (for d > h) as below:

$$\tan^2\alpha = \frac{\varepsilon_{sk} - \varepsilon_{st}}{\varepsilon_{sk} - \varepsilon_{fr} + \frac{1}{24}(\frac{h}{R})^2}$$

$$= \frac{1.84 \times 10^{-3} - (-0.919 \times 10^{-3})}{1.84 \times 10^{-3} - (-1.185 \times 10^{-3}) + \frac{1}{24}(\frac{6}{76})^2} = 0.839$$

$\tan\alpha = 0.91$ or $\alpha = 42.3°$

The angle of diagonal tension reasonably agrees with the assumed $\alpha = 45°$; therefore, use 45° for this example calculation.

(h) Stringer stress:

(h1) By entering k = 0.63 and $\frac{h}{d} = \frac{6}{20} = 0.3$ into the curves in Fig. 12.4.8 obtain $(\frac{f_{u,max}}{f_u})f_u = 1.22$. Let $(\frac{f_{u,max}}{f_u})f_u = (\frac{f_{st,max}}{f_{st}})f_{st}$ and use Eq. 12.5.17 to obtain the maximum stringer stress ($f_{st,max}$):

$$(\frac{f_{st,max}}{f_{st}})f_{st} = 1.22$$

$f_{st,max} = 1.22 \times (-9,652) = -11,775$ psi (compression)

(h2) Secondary bending moment (M_{st}) on stringers between frames from Eq. 12.5.18:

$$M_{st} = \frac{k f_s h t d^2 \tan\alpha}{24 R}$$

$$M_{st} = \frac{0.63 \times 11,000 \times 6 \times 0.05 \times 20^2 \tan 45°}{24 \times 76} = 456 \text{ in.-lbs.}$$

The stringer bending stress (compression on the stringer outstanding flange which does not attach to the skin) at frames:

$$f'_{sb} = \frac{M_{st}(1.25 - Y')}{I'_x} = \frac{456 \times (1.25 - 0.513)}{0.067} = 5,016 \text{ psi}$$

Chapter 12.0

The stringer bending stress (compression on skin) between frames (at center span):

$$f_{sh} = \frac{M_{st}Y'}{I'_x} = \frac{456 \times 0.513}{0.067} = 3,491 \text{ psi}$$

where: $c = Y'$ and $c' = 1.25 - Y'$

(The stringer section properties include the effective skin width of 30t)

(h3) Stringer forced crippling stress (Effective skin width of 30t is included with the Z-stringer):

(1) Calculate the stiffener crippling stress (F_{cc} of 7075-T6 sheet) (skin not included):

Segment	Free edges	b_n	t_n	$\frac{b_n}{t_n}$	$b_n t_n$	F_{ccn} (from Fig. 10.7.6)	$t_n b_n F_{ccn}$
①	1	0.869	0.063	13.8	0.0547	37,000	2,024
②	0	1.137	0.063	18	0.0716	64,000	4,582
③	0	0.837	0.063	13.3	0.0527	71,000	3,742
④	1	0.269	0.063	4.3	0.0169	71,000	1.2
					Σ 0.1959		11,548

$$F_{cc} = \frac{\Sigma b_n t_n F_{ccn}}{\Sigma b_n t_n}$$

$$= \frac{11,548}{0.1959}$$

$$= 58,948 \text{ psi}$$

Check lip section of ④:

$$\frac{b_L}{b_F} = \frac{0.269}{0.837} = 0.32$$

$$\frac{b_F}{t_{st}} = \frac{0.837}{0.063} = 13.3$$

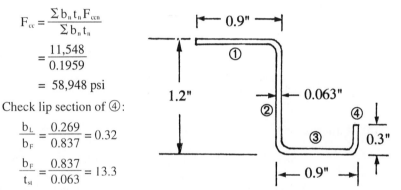

From Fig. 10.7.5, the values are within the design range.   O.K.

(2) Forced crippling stress of the stringer (7075-T6 sheet):

Enter $\frac{t_u}{t} = \frac{t_{st}}{t} = \frac{0.063}{0.05} = 1.26$ and $k = 0.63$ into curves in Fig. 12.4.9, obtain,

$F_{fu} = F_{f,st} = 26,000$ psi

(3) Allowable compression stress (F_c) of the stringer (7075-T6 sheet):

$$\frac{L'}{\rho} = \frac{\frac{20}{\sqrt{4}}}{0.497} = 20.1 \text{ and } F_{cc} = 58,948 \text{ psi}$$

From Fig. 10.8.2(a), obtain $F_c = 55,500$ psi

(4) Use the interaction equation from Eq. 12.5.19:

$$MS = \frac{1}{\frac{f_c + f_{sh}}{F_{cc}} + \frac{f_{st,max}}{F_{f,st}}} - 1$$

$$= \frac{1}{\frac{22,000 + 5,016}{58,948} + \frac{11,775}{26,000}} - 1 = \underline{0.1} \qquad \text{O.K.}$$

(h4) Stringer column failure:

At the junction of the stringer and the frame from Eq. 12.5.20:

$$MS = \frac{1}{\frac{f_c + f_{st} + f'_{sb}}{F'_{cc}}} - 1$$

$$= \frac{1}{\frac{22{,}000 + 9{,}652 + 5{,}016}{71{,}000}} - 1 = \underline{0.94} \qquad \text{O.K.}$$

$F'_{cc} = 71{,}000$ psi which is the value of F_{ccn} of segment ③ is obtained from Step (h3).

Between frames (mid-span of the stringer) from Eq. 12.5.21:

$$MS = \frac{1}{\frac{f_c + f_{st} + f_{sb}}{F_c}} - 1$$

$$= \frac{1}{\frac{22{,}000 + 9{,}652 + 3{,}491}{55{,}500}} - 1 = \underline{0.58} \qquad \text{O.K.}$$

(i) Frame stresses:

(i1) Shear-tie frame –

By entering $k = 0.63$ and $\frac{h}{d} = \frac{6}{20} = 0.3$ into the curves in Fig. 12.4.8 obtain $(\frac{f_{u,\,max}}{f_u})f_u = 1.22$. Let $(\frac{f_{u,\,max}}{f_u})f_u = (\frac{f_{fr,\,max}}{f_{fr}})f_{fr}$ and use Eq. 12.5.22 to obtain the maximum stringer stress ($f_{fr,\,max}$):

$$(\frac{f_{fr,\,max}}{f_{fr}})f_{fr} = 1.22$$

$f_{fr,\,max} = 1.22 \times 12{,}442 = 15{,}179$ psi

where $\frac{f_{st,\,max}}{f_{fr}} = \frac{f_{u,\,max}}{f_u}$ in Fig. 12.4.8

(i2) Calculate the shear-tie frame crippling stress (F_{cc}) (skin not included):

Segment edges	Free	b_n	t_n	$\frac{b_n}{t_n}$	$b_n t_n$	F_{ccn} (from Fig. 10.7.6)	$t_n b_n F_{ccn}$
①	1	0.875	0.05	17.5	0.04375	30,000	1,312
②	0	3.7	0.05	74	0.185	20,000	3,700
③	1	0.875	0.05	17.5	0.04375	30,000	1,312
④	0	0.875	0.05	17.5	0.04375	64,800	2,835
⑤	0	0.85	0.05	17	0.04375	65,500	2,866
⑥	1	0.275	0.05	5.5	0.01375	71,000	976
					Σ 0.374		13,001

$$F_{cc} = \frac{\Sigma b_n t_n F_{ccn}}{\Sigma b_n t_n}$$

$$= \frac{13{,}001}{0.374} = \underline{34{,}762} \text{ psi}$$

Chapter 12.0

Check lip section of ⑥:

$\frac{b_L}{b_F} = \frac{0.275}{0.875} = 0.31$

$\frac{b_F}{t_{fr}} = \frac{0.875}{0.05} = 17.5$

From Fig. 10.7.5, the values are within the design range.   O.K.

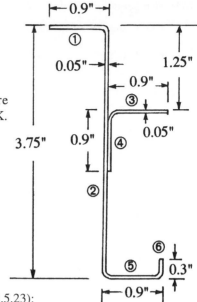

(i3) Use following MS equation (Eq. 12.5.23):

$$MS = \frac{1}{\frac{f_{b,fr}}{F_{cc}} + \frac{f_{fr,max}}{F_{fr}}} - 1$$

$$= \frac{1}{\frac{10,000}{34,762} + \frac{15,179}{24,100}} - 1 = \frac{1}{0.288 + 0.63} - 1 = \underline{0.09} \quad \text{O.K.}$$

where: $F_{fu} = F_{fr} = 24,100$ psi from Fig. 12.4.9 by using $\frac{t_u}{t} = \frac{t_{fr}}{t} = \frac{0.05}{0.05} = 1$

and k = 0.63

(j) Rivet loads:

(j1) Assume skin splices occur at stringer; obtain shear loads from Eq. 12.5.26:

$$q_R = q[1 + k(\frac{1}{\cos \alpha} - 1)]$$

$$= 11,000 \times 0.05[1 + 0.63(\frac{1}{\cos 45°} - 1)] = 694 \text{ lbs./in.}$$

Double or staggered rows (some use triple rows) of rivets are typically used to design the transport longitudinal and circumferential splices for fatigue hoop tension stress. Therefore, there is plenty of extra strength to take care of the $q_R = 694$ lbs./in. shear flow.

(j2) Tension loads (In this example a single row of rivets is used between skin and stringer as well as skin and frame) from Eq. 12.4.18:

$0.22 \, t \, F_{tu} = 0.22 \times 0.05 \times 65,000 = 715$ lbs./in.

Note: Local pad is required to increase the pull-through strength of the countersunk rivets.

(k) General instability check:

Calculate the following parameter from Eq. 12.5.27:

$\rho_{st} = 0.437$ in. (use stringer with full width of skin, $\rho'' = 0.437$ in.)

$\rho_{fr} = 0.769$ in. (use frame with full width of skin, $\rho'' = 0.769$ in.)

$$F_{s,\,inst} = \frac{(\rho_{st}\rho_{fr})^{\frac{7}{8}}}{(d\,h)^{\frac{1}{2}}R^{\frac{3}{4}}} = \frac{(0.437 \times 0.769)^{\frac{7}{8}}}{(6 \times 20)^{\frac{1}{2}}(76)^{\frac{3}{4}}} = 13.66 \times 10^{-4}$$

From Fig. 12.5.7, obtain $F_{s,\,inst} \approx 30{,}000$ psi

$$MS = \frac{F_{s,\,inst}}{f_s} - 1 = \frac{30{,}000}{22{,}000} - 1 = \underline{0.36} \qquad \text{O.K.}$$

12.6 DIAGONAL TENSION EFFECT AT END BAYS AND SPLICES

Diagonal tension forces produce vertical forces which cause sagging of the beam caps (or horizontal members in a cut-out) and are reacted by the vertical stiffeners; these forces also produce horizontal forces which must be reacted by the end stiffener (or vertical members in a cut-out), as shown in Fig. 12.6.1, to avoid excess end stiffener bending and sagging. Since the diagonal tension effect results in an inward pull load (w_h) on the end stiffener or members, it produces bending as well as the axial compression load.

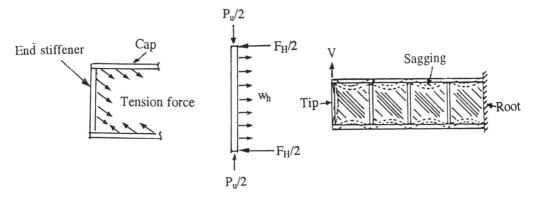

Fig. 12.6.1 Sagging of a Diagonal Tension Beam

(A) END BAY

The end bay of a diagonal tension web beam must be specially handled since the web is discontinuous at the end and the tension component of web stress must be transferred to the end stiffener as well as the caps. The end stiffener must be considerably heavier than the others, or at least supported by additional stub member(s) or doublers to form a built-up beam, as shown in Fig. 12.6.2. This approach is frequently used in the design of end bays for wing rib webs.

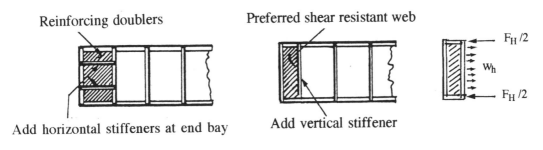

(a) Add Horizontal Stub Members   (b) Add Closely Spaced Vertical Stiffeners

Fig. 12.6.2 End Bay Reinforcement of a Diagonal Tension Beam

Chapter 12.0

The distributed loads on an end stiffener are:

(a) Horizontal uniform load on a single end stiffener (see Fig. 12.6.1):

$$w_h = kq \cot \alpha \quad \text{(lbs./in.)} \qquad \text{Eq. 12.6.1}$$

The maximum bending moment (refer to Fig. 10.6.4 case 1):

$$M_E = (\frac{w_h h_e^2}{8})\mu$$

End stiffeners act as a simply supported beam-column, as shown in Fig. 12.6.1.

(b) If a built-up beam design is used, as shown in Fig. 12.6.2(b), only consider the bending moment (M_E); the column load of P_u can be ignored because it does not have a significant effect on this type of construction.

Generally, the end bay web structures of the beam are configured as shown in Fig. 12.6.2(b) so that the web is more shear resistant or the diagonal tension effect is lessened:

- Add closely spaced vertical stiffeners without adding doublers
- Increase web thickness or add doublers
- Add horizontal stub member(s) in the end bay
- A combination of the above

The location of the stiffener will affect end bay reinforcement, as shown in Fig. 12.6.3; there are basically two types of configurations:

- A non-joggled stiffener is attached on the reverse side of the attached flange of the cap, as shown in Fig. 12.6.3(a); most engineers prefer this design
- A joggled stiffener is attached on the same side as the attached flange of the cap, as shown in Fig. 12.6.3(b)

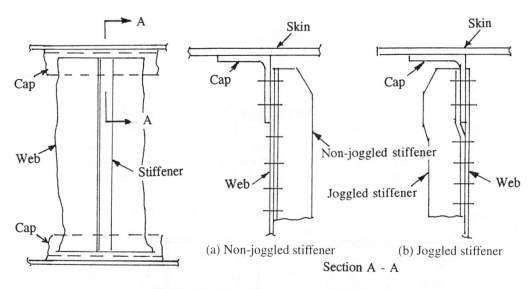

Fig. 12.6.3 Affected by Location of Stiffener

Example:

Assume there is a fitting at the spar cap at F and L to transfer beam bending and shear loads (q = 2,483 lbs./in.) as given in the example calculation in Chapter 12.4. Design an end bay to carry loads resulting from the diagonal tension effect.

Shear Panels

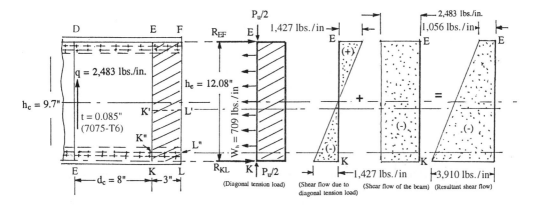

(a) Shear flow in panel D-E-K-J is q = 2,483 lbs./in. from Step (a) of previous example in Chapter 12.4 and assume the shear flow in panel E-F-L-K is q = −2,483 lbs./in. (observe the load sign conventions from Fig. 2.5.1).

(b) Shear flow at this panel due to diagonal tension effect:

Horizontal uniform load on a single end stiffener from Eq. 12.6.1:

w_h = kq cot α
    = 0.25 × 2,483 × cot 41.2° = 709 lbs./in.

$$R_{EF} = R_{KL} = \frac{w_h \times h_e}{2} = \frac{709 \times 12.08}{2} = 4,282 \text{ lbs.}$$

Shear flow due to diagonal tension effect (w_h):

Shear flow along edge of EF:

$$q_{h, EF} = \frac{4,282}{3} = 1,427 \text{ lbs./in.}$$

Shear flow along edge of KL:

$q_{h, KL}$ = −1,427 lbs./in.

(c) Final shear flow at EF:

q_{EF} = −q + $q_{h, EF}$ = −2,483 + 1,427 = −1,056 lbs./in.   Eq. (A)

q_{KL} = −q − $q_{h, KL}$ = −2,483 − 1,427 = −3,910 lbs./in.   Eq. (B)

Observations:

- The root end bay E-F-L-K web will have higher shear loads in the lower web than in the upper; this phenomenon occurs in diagonal tension web analysis. The tip end bay A-B-H-G web will be the opposite.
- Check for reverse loading (downward loading in this example calculation) which may produce a higher shear load at the upper web
- In this example case the bending stress due to diagonal tension loading (w_h) in stiffener E-K is small and is, therefore, ignored

(d) Check the shear buckling stress ($F_{s, cr}$) of the web (7075-T6 sheet, F_{su} = 48,000 psi) at the panel E-F-L-K:

Aspect ratio $\frac{h_c}{d_{c, KL}} = \frac{9.7}{3} = 3.2$ and from Fig. 11.3.5 (Case ④), obtain K_s = 5.2.

525

Chapter 12.0

From Eq. 11.3.4:

$$\frac{F_{s,cr}}{\eta_s} = K_s E \left(\frac{t}{d_{c,KL}}\right)^2 = 5.2 \times 10.7 \times 10^6 \left(\frac{0.085}{3}\right)^2 = 44,666 \text{ psi} \qquad \text{Eq. (C)}$$

From Fig. 11.2.5, obtain $F_{s,cr(EFLK)} = 32,500$ psi (in plastic range)

$$q_{cr(EFLK)} = 0.085 \times 32,500 = 2,763 \text{ lbs./in.} \qquad \text{Eq. (D)}$$

The shear buckling load of $q_{cr(EFLK)} = 2,763$ lbs./in. is not enough to meet the requirement of shear resistant web, with $q_{KL} = 3,910$ lbs./in.

(e) Design the end bay panel E-F-L-K:

The following is one of many conservative options for designing the end bay as a shear resistant web:

- Locate K' point (horizontal stiffener is installed here) in the following diagram: the shear flow is equal to $q_{cr(EFLK)} = 2,763$ lbs./in. [from Eq. (D)] and the distance K'-K" = 3.66" ($h_{K'K"} = 3.66"$)
- Shear flow and dimensions along E-K are shown in the following diagram:

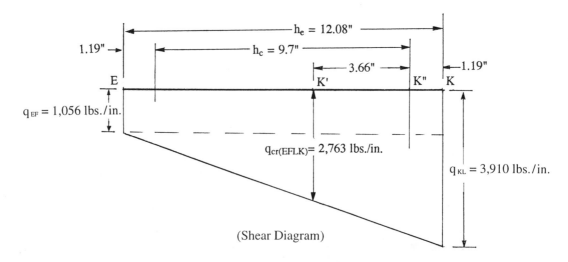

(Shear Diagram)

- Compute the web thickness in bay K'-L'-L-K:

  Aspect ratio $\dfrac{h_{K'K"}}{d_{c,KL}} = \dfrac{3.66}{3} = 1.22$ and from Fig. 11.3.5 (Case ④), obtain:

  $K_s = 7.35$

  Assume the required thickness t' and and use the same shear buckling stress $\dfrac{F_{s,cr}}{\eta_s} = 44,666$ psi [from Eq. (C)]. The entire web of the panel E-F-L-K will have the same shear deformation and thus will meet the structural requirements of equilibrium and compatibility (refer to Chapter 5.3)

  $$\frac{F_{s,cr}}{\eta_s} = K_s E \left(\frac{t'}{d_{c,KL}}\right)^2 = 7.35 \times 10.7 \times 10^6 \left(\frac{t'}{3}\right)^2 = 44,666 \text{ psi} \qquad \text{Eq. (E)}$$

  $\therefore$ t' = 0.071"

- The calculated thickness t' = 0.071" < t = 0.085" (at panel of K'-L'-L-K) which means that the existing web is not buckled and it is O.K. to use this same thickness (t = 0.085")
- The location of edge K'L' can be moved slightly toward edge of EF as long as the buckling stress remains $\dfrac{F_{s,cr}}{\eta_s} \approx 44,666$ psi [from Eq. (C)]

Try $h_{K'K''} = 4$ and aspect ratio $\frac{h_{K'K''}}{d_{c,KL}} = \frac{4}{3} = 1.33$ and from Fig. 11.3.5 (Case ④), obtain $K_s = 7$.

From Eq. (E):

$$\frac{F_{s,cr}}{\eta_s} = K_s E \left(\frac{t''}{d_{c,KL}}\right)^2 = 7 \times 10.7 \times 10^6 \left(\frac{t'}{3}\right)^2 = 44,666 \text{ psi} \qquad \text{Eq. (F)}$$

$\therefore t'' = 0.073''$

The value of $t'' = 0.073''$ is O.K. ($< 0.085''$) and, if desired, it could be increased to enlarge the aspect ratio ($\frac{h_{K'K''}}{d_{c,KL}}$) by using a value of $h_{K'K''}$ greater than 4".

The resulting end bay design is sketched below :

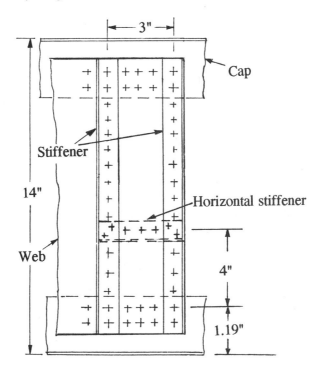

(B) SPLICES

Splices are commonly used in airframe design, e.g., wing and empennage spanwise splices, fuselage longitudinal and circumferential splices, etc., as shown in Fig. 12.6.4. Attention must be given in splice design to diagonal tension effects at areas such as wing tip areas and empennage, fuselage tail, and control surfaces in which thinner skins are under higher shear loads due to torsion forces.

The common design considerations at splices are as follows:

- For wing fuel tanks use double or staggered rows of fasteners (some designs use triple rows for fuselage splices in pressurized fuselage cabin)
- Provide local pads to prevent knife edge with countersunk fastener (see Fig. 9.1.5)
- Design bearing to be critical unless fastener shear and bearing are both high
- Use closer fastener spacing to prevent fuel leakage (Hi-Lok fasteners may be required) and inter-rivet buckling

Splice design recommendations:

- Since splices are heavy, expensive and cause fatigue problems, they should not be used unless required by production or special design requirements
- Since the curvature of the wing and empennage surface panels is relatively large, they may be analyzed using the flat web method for preliminary sizing
- Fuselage circumferential splices (see Section A – A of Fig. 12.6.4) and wing (or empennage) chordwise splices (which should be avoided except near the tip area where the tension or compression stress is not the primary design load case) are analyzed in the same manner as the intermediate frame (or rib cap) of the curved skin method except at discontinuities of the skin where the load is transferred through fasteners, stringers or bulkheads
- Fuselage longitudinal splices (see Section B – B of Fig. 12.6.4) and wing or empennage spanwise splices (see Section C – C of Fig. 12.6.4) are analyzed in the same manner as the intermediate stringer of the curved skin method except at discontinuities of the skin in which the load is transferred through fasteners and stringers
- Spanwise splices at a wing spar (see Section D – D of Fig. 12.6.4) should be analyzed in the same manner as the beam cap in the flat web method
- Spar web splices (see Section E – E of Fig. 12.6.4) are analyzed in the same manner in the same manner as the intermediate stiffener of the flat web method except at discontinuities of the web in which the load is transferred through fasteners and splice stiffeners. Attention should be given to the lower portion or the spar because the tension stress in the cap can induce fatigue problems

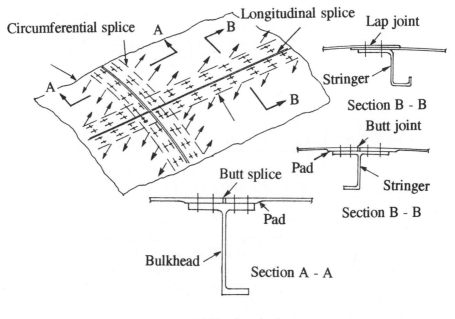

(a) Fuselage body

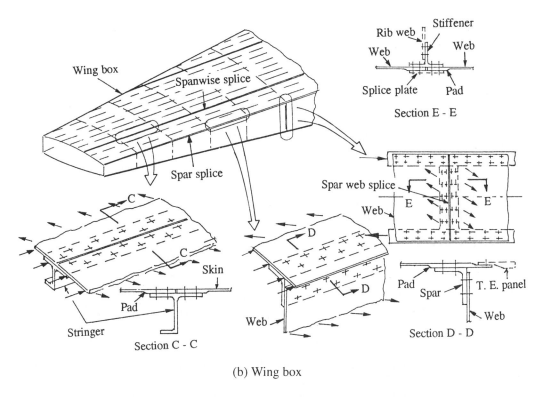

(b) Wing box

Fig. 12.6.4 Splice Configurations on Wing Box and Fuselage Body

References

12.1 P. Kuhn and J. Peterson, "A Summary of Diagonal Tension, Part I, Methods of Analysis", NACA TN 2661, May, 1952.
12.2 P. Kuhn and J. Peterson, "A Summary of Diagonal Tension, Part II, Experimental Evidence", NACA TN 2662, May, 1952.
12.3 P. Kuhn, "STRESS IN AIRCRAFT AND SHELL STRUCTURES", McGraw-Hill Book Company, Inc., New York, NY, 1956.
12.4 Bruhn, E. F., "ANALYSIS AND DESIGN OF FLIGHT VEHICLE STRUCTURES", Jacobs Publishers, 10585 N. Meridian St., Suite 220, Indianapolis, IN 46290, (1965).
12.5 Dunn, L. G., "Some investigations of the General Instability of Stiffened Metal Cylinders. VIII – Stiffened Metal Cylinders Subjected to Pure Torsion", NACA TN 1197, 1947.

Chapter 13.0

CUTOUTS

13.1 INTRODUCTION

Airframe engineers view any cutouts in airframe structures with disfavor because the necessary reinforcement of the cutout increases costs and adds weight to the overall design. In addition, the design and sizing of cutouts is a difficult process since it is an area of stress concentration, a problem area for both static and fatigue strength and there is insufficient design data.

Cutouts are essential in airframe structures to provide the following:

- Lightening holes in webs, as shown in Fig. 13.1.1(a), are frequently used to save structural weight in cases of minimum gage thickness requirements
- Passages for wire bundles, hydraulic lines, control linkages, fluids, etc. (small holes), as shown in Fig. 13.1.1(a) and (b)
- Accessibility for final assembly and maintenance (e.g., manholes in wing lower surfaces, crawl holes in wing ribs, etc.), as shown in Fig. 13.1.1(c) and Fig. 13.1.2
- Inspection for maintenance (medium-sized cutouts called hand holes), as shown in Fig. 13.1.1(b)

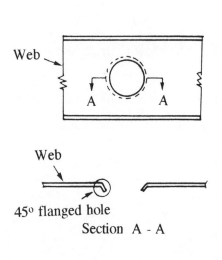

(a) Standard round 45° flanged holes

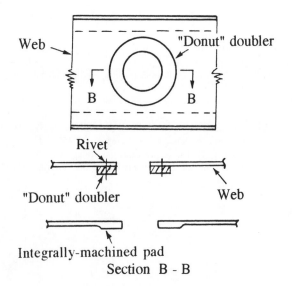

(b) Doubler around holes

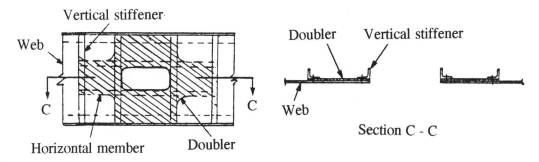

(c) Large framing cutout

Fig. 13.1.1 Cutouts in Shear Beams

While access holes are a necessity, they should be kept as small as possible to meet the minimum standard requirements, whether commercial or military, as shown in Fig. 13.1.2.

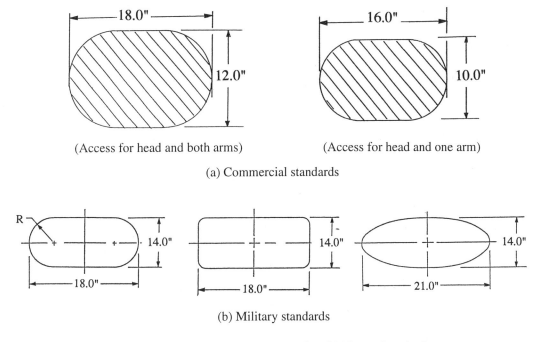

Fig. 13.1.2 Cutouts for Commercial and Military Standards

Flanged holes (either round or elliptical), as shown in Fig. 13.1.2(a) and Fig. 13.1.3, are a simple and lower cost method for fabricating access holes. Elliptical flanged holes are typically used for crawl holes in lightly or moderately loaded rib webs in wing and empennage applications. The strength of panels containing these holes is determined by using empirical design curves and data rather than stress analysis.

Chapter 13.0

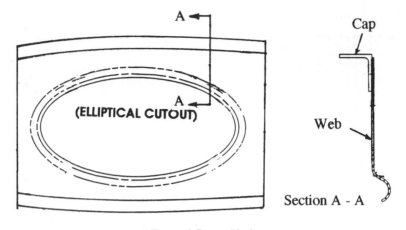

(a) Formed flanged holes

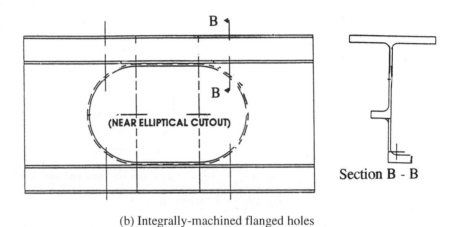

(b) Integrally-machined flanged holes

Fig. 13.1.3 Elliptical Crawl Holes

The following very large cutout applications are considered to be special and complicated structural design cases which require much more thorough analysis (e.g., finite element modeling analysis) and detail design:

- Passenger doors
- Cargo doors
- Fuselage nose and rear loading and unloading doors

(A) REINFORCING CUTOUTS FOR PRIMARY LOAD CASES

(a) Cutout reinforcements in shear web beams (see Fig. 13.1.1):

- Use standard round or elliptical holes that have a 45° flange (or equivalent flange) in lightly loaded shear cases
- Use a "donut" ring doubler to reinforce cutouts in shear webs [see Fig. 13.1.1(b)]
- Integrally-machined ring doublers have the same or greater stiffness than the "donut" ring shown in Fig. 13.1.1(b)
- Built-up or machined framed hole for large cutout
- Wing surface access door (e.g., 10" x 18")

(b) Cutout reinforcements in axial tension or compression loaded panels (see Fig. 13.1.4, the example of a wing surface access door):

- Use a machined pad to increase thickness locally on wing surface skins to mount access doors or fuel probes as well as to meet strength requirements
- Machine a thicker pad around the cutout with a gentle taper in spanwise direction to reduce the stress concentration and eccentricity and if necessary terminate one or two stringers for both skin-stringer and integrally-stiffened panels
- Machine a thicker pad around the cutout and extend it to the next cutout in the spanwise direction if the next cutout is in the adjacent bay of the transport wing surface (stringers are not used in this region)

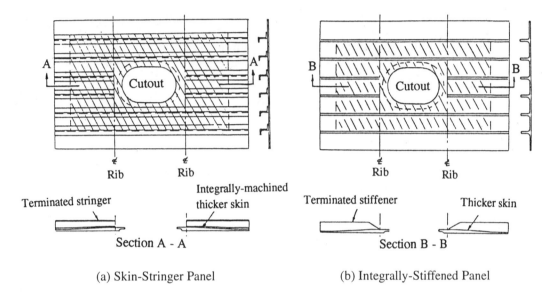

Fig. 13.1.4 Cutouts in Axially Loaded Panels

(B) CATEGORIES OF CUTOUT USAGE

(a) Cutouts in shear beams (shear load in web, see Fig. 13.1.1):

- Lightly loaded beams (e.g., intercostals, ribs, formers, etc.)
- Moderately loaded beams (e.g., floor beams, control surface box spars, auxiliary spars, etc.)
- Heavily loaded beams (e.g., wing and empennage spars, bulkhead ribs, side openings in fuselages, etc.)

(b) Cutouts in stiffened panels (axial or compression load, see Fig. 13.1.4):

- Wing or empennage surface panels
- Upper and lower fuselage openings (e.g., service doors, emergency exit, etc.)

(c) Other examples of cutouts on a transport fuselage, as shown in Fig. 13.1.5:

- Fuselage window (e.g., 9" x 12.5")
- Fuselage passenger entry door (e.g., 34" x 72, 42" x 76", etc.)
- Upper deck freight door (e.g., 120" x 73")
- Emergency exit (refer to FAR 25.807 for requirements)
- Fuselage nose and/or rear cargo loading cutouts (cargo and military transports)

533

Chapter 13.0

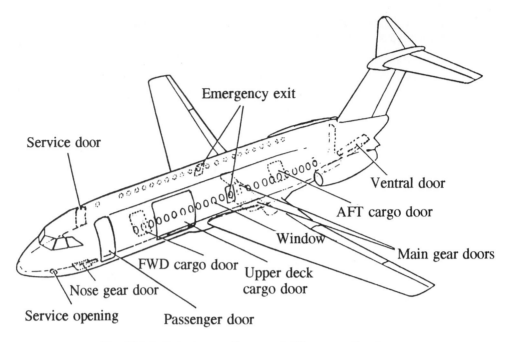

Fig. 13.1.5 Cutouts on a Commercial Transport Fuselage

(C) **CUTOUT ANALYSIS METHODS**

- No clear-cut analytical procedure for sizing cutouts is available
- With the help of existing test data and empirical reports an experienced engineer can select acceptable proportions
- Often the proportions of panels in question are sufficiently close to previously tested panels so that a direct comparison can be made
- Doubtful cases should be tested
- There is still not sufficient test data available for the sizing of various cutouts in airframe structures and extensive test programs are required to substantiate cutout analysis methods for new aircraft programs

13.2 UNSTIFFENED-WEB SHEAR BEAMS

Small cutouts or lightening holes (see Fig. 13.2.1) are frequently used in the lightly loaded and shallow beams (use formers) used in fuselage frames, intercostal beams, etc. These holes also serve for the passage of wire bundles, control cables, and hydraulic lines. The unstiffened-web shear beam carries light loads and it is a low cost construction which uses flanged holes (plain holes are not recommended because of the potential for crack growth around the edge of hole in service). This chapter provides empirical design data on standard round holes for designer use. Since a small amount of normal loading (i.e., uniformly distributed air loads) is imposed on these webs an extra margin of safety above the shear load is required. The allowable shear design data is based on the assumption of pure shear in the web only and a stiffener must be provided whenever a normal concentrated load is introduced into the beam, as shown in Fig. 13.3.1 and Fig. 13.3.4.

Cutouts

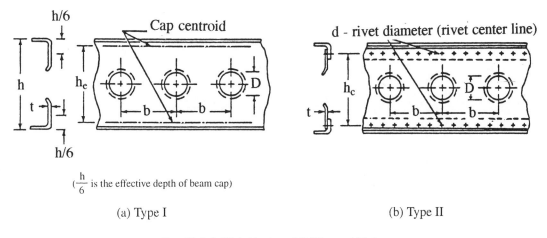

($\frac{h}{6}$ is the effective depth of beam cap)

(a) Type I (b) Type II

Fig. 13.2.1 Web Having 45° Flanged Holes

It may be advantageous from a weight standpoint to omit web stiffeners and, instead, introduce a series of standard flanged lightening holes. NACA (Ref. 13.3) developed an empirical formula for a web with round 45° flanged lightening holes (see Fig. 13.2.2), and the formula gives the allowable shear flow for webs having this type of hole.

D (inch)	2.0	2.5	3.0	3.5	4.0	4.5	5.0	6.0
f (inch)	.25	.3	.4	.45	.5	.5	.5	.55

Fig. 13.2.2 Dimensions of 45° Flanged Holes

Sizing procedures:

(a) The limits for this design are as follows:
- Use the standard 45° flanged holes shown in Fig. 13.2.2 (unstiffened-web)
- $0.016'' \leq t \leq 0.125''$
- $0.25 \leq \frac{D}{h_e} \leq 0.75$
- $0.3 \leq \frac{D}{b} \leq 0.7$
- $40 \leq \frac{h_e}{t} \leq 250$

(b) Ultimate allowable shear stress:

$$F_s = K_1 F_o \quad \text{Eq. 13.2.1}$$

where: F_o – Allowable gross shear stress given in Fig. 13.2.3
K_1 – Reduction factor given in Fig. 13.2.3

or $q_{all} = K_1 F_o t$ Eq. 13.2.2

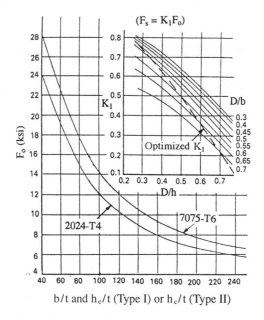

Fig. 13.2.3 Allowable Gross Shear Stress, F_o and Correction Factor, K_l

(c) The net shear stresses:

(i) At the web between holes:

$$f_s = \frac{q}{t}\left(\frac{b}{b-D}\right) \qquad \text{Eq. 13.2.3}$$

(ii) The vertical net web at holes:

$$f_s = \frac{q}{t}\left(\frac{h_e}{h_e-D}\right) \qquad \text{Eq. 13.2.4}$$

(d) Web-to-cap rivets (for Type II beam):

Because of the non-uniformity of shear stress caused by the holds, the web to flange rivets should be designed by the greater of:

$$q_R = 1.25\, q \qquad \text{Eq. 13.2.5}$$

$$\text{or}\quad q_R = 0.67\left(\frac{b}{b-D}\right) q \qquad \text{Eq. 13.2.6}$$

(e) For optimum design:

- When the hole diameter, D, and hole spacing, b, are chosen, generally $\frac{D}{h_e} \approx 0.25$ and $\frac{D}{b} \approx 0.45$.
- When web height, h_e, hole diameter, D, and shear loading, q, are given, generally the lightest weight of the beam is for $\frac{D}{b} \approx 0.45$.

Example 1:

Given a formed channel section with thickness t = 0.04" and material 7075-T6 sheet, as shown below, to carry q = 350 lbs./in. shear flow. Determine the hole diameter D and spacing b.

h = 3.0" and h_e = 2.7"

Cutouts

7075-T6 sheet (from Fig. 4.3.4):

$F_{su} = 48,000$ psi

$F_{tu} = 80,000$ psi

$F_{bru} = 160,000$ psi

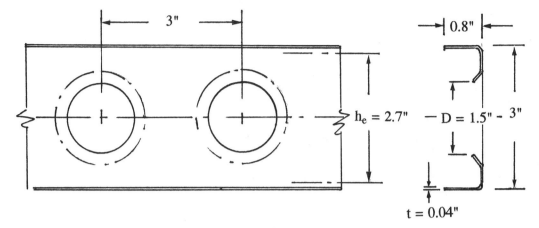

(a) The limits for this design are as follows:

$0.016" \leq t = 0.04" \leq 0.125"$ O.K.

$0.25 \leq \dfrac{D}{b} \leq 0.75$ (to be determined later)

$0.3 \leq \dfrac{D}{d_c} \leq 0.7$ (to be determined later)

$40 \leq \dfrac{h_e}{t} = \dfrac{2.7}{0.04} = 68 \leq 250$ O.K.

(b) Ultimate allowable shear stress:

$f_s = \dfrac{q}{t} = \dfrac{350}{0.04} = 8,750$ psi

Use $\dfrac{h_e}{t} = 68$ and from Fig. 13.2.3, obtain:

$F_o = 20,000$ psi and assume $F_s = f_s = 8,750$ psi

From Eq. 13.2.1:

$F_s = K_1 F_o$ or $K_1 = \dfrac{F_s}{F_o} = \dfrac{8,750}{20,000} = 0.44$

Optimize the lightening holes (from Fig. 13.2.3 using optimized K_1 curve):

$\dfrac{D}{h_e} = 0.54 \quad \therefore D = 0.54 \times 2.7 = 1.46"$

$\dfrac{D}{b} = 0.62 \quad \therefore b = \dfrac{1.46}{0.62} = 2.4"$

Use hole diameter, $D = 1.5"$ and $b = 3"$ (the reason for using a wider b value is to provide adequate space between holes for installing clips; e.g., for use in mounting interior fuselage equipment).

Use: $\dfrac{h_e}{t} = \dfrac{2.7}{0.04} = 68$ or $\dfrac{b}{t} = \dfrac{3}{0.04} = 75$

$$\frac{D}{h_e} = \frac{1.5}{2.7} = 0.56 \text{ and } \frac{D}{b} = \frac{1.5}{3} = 0.5$$

From Fig. 13.2.3, obtain:

$F_o = 18,500$ psi and $K_1 = 0.51$

From Eq. 13.2.2: $q_{all} = K_1 F_o t = 0.51 \times 18,500 \times 0.04 = 377$ lbs./in.

$$MS = \frac{q_{all}}{q} - 1 = \frac{377}{350} - 1 = \underline{0.08} \qquad \text{O.K.}$$

Check the following limits:

$$0.25 \leq \frac{D}{h_e} = \frac{1.5}{2.7} = 0.56 \leq 0.75 \qquad \text{O.K.}$$

$$0.3 \leq \frac{D}{b} = \frac{1.5}{3} = 0.5 \leq 0.7 \qquad \text{O.K.}$$

(c) The net shear stresses:

(i) At the web between holes from Eq. 13.2.3:

$$f_s = \frac{q}{t}(\frac{b}{b-D}) = \frac{350}{0.04}(\frac{3}{3-1.5}) = 17,500 \text{ psi} < F_{su} = 48,000 \text{ psi} \quad \text{O.K.}$$

(ii) The vertical net web at holes from Eq. 13.2.4:

$$f_s = \frac{q}{t}(\frac{h_e}{h_e-D}) = \frac{350}{0.04}(\frac{2.7}{2.7-1.5}) = 19,688 \text{ psi} < F_{su} = 48,000 \text{ psi} \quad \text{O.K.}$$

13.3 STIFFENED-WEB SHEAR BEAMS

Moderate cutouts, as shown in Fig. 13.3.1 and Fig. 13.3.4, are generally used on a deeper web which can carry higher shear loads than that of web having small cutouts. The bead-flanged holes, shown in Fig. 13.3.2, are recommended for use on deeper webs because they are less likely to have problems with edge cracking than the 45° flanged holes, shown in Fig. 13.2.2. However, the bead-flanged holes are more expensive to fabricate and difficult to repair. Shear web beams which are stiffened by vertical beads are lower in cost than beams reinforced with vertical stiffeners (see Fig. 13.3.4).

(A) BEAD-STIFFENED WEB HAVING BEAD-FLANGED HOLES

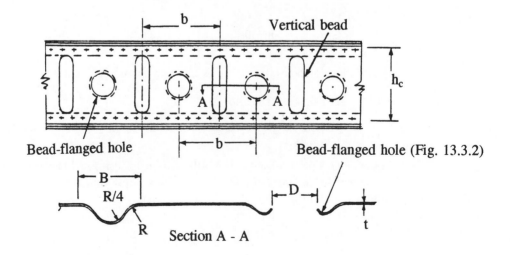

t (in)	B (in)	R (in)
0.02	0.95	0.32
0.032	1.16	0.52
0.04	1.27	0.64
0.05	1.42	0.81
0.063	1.55	1.02

t (in)	B (in)	R (in)
0.073	1.65	1.15
0.08	1.73	1.3
0.09	1.8	1.45
0.1	1.9	1.6
0.125	2.12	2.0

Fig. 13.3.1 Beam has Vertical Bead-Stiffened web with round bead-flanged holes

D_o (Inch)	D (Inch)	a (Inch)
1.7	0.8	0.2
1.95	1.05	0.2
2.65	1.7	0.25
3.0	2.05	0.25
3.65	2.7	0.25
3.9	2.95	0.25
4.95	3.8	0.4
5.95	4.8	0.4
6.95	5.8	0.4
7.44	6.3	0.4
7.95	6.8	0.4
8.95	7.8	0.4
9.45	8.3	0.4

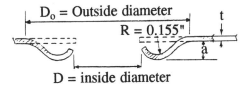

Fig. 13.3.2 Dimensions of Bead-Flanged Holes

The allowable shear flows shown in Fig. 13.3.3 are based on the design of the vertical bead stiffened-web with round bead-flanged holes shown in Fig. 13.3.1. The vertical beads are designed to extend as close to the beam caps as assembly will allow. Fig. 13.3.3 gives the allowable shear flows based on the following limiting conditions:

- Bead-stiffened web per Fig. 13.3.1
- Bead-flanged holes per Fig. 13.3.2
- $t \leq 0.072"$
- $\dfrac{D_o}{h_c} \approx 0.6$
- $b \approx h_c$

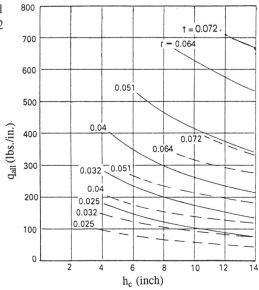

——— Allowable gross shear stress: $q_{s,all}$ is shown by solid lines
--------- Buckling shear stress: $q_{s,cr}$ is shown by dashed lines

Fig. 13.3.3 Allowable Shear Flow ($q_{s,all}$) for Clad 2024-T4 and 7075-T6 Web

Example 2:

Determine the beam web thickness (t), stiffener spacing (b), and hole diameter (D) as shown below to carry a shear flow of q = 645 lbs./in.

Web material: 7075-T6 sheet (data from Fig. 4.3.4)

F_{su} = 48,000 psi F_{tu} = 80,000 psi F_{bru} = 160,000 psi

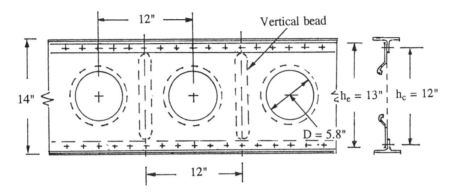

(a) Check limiting conditions:

Use bead-stiffened web per Fig. 13.3.1

Use bead-flanged holes per Fig. 13.3.2

Assume vertical bead-stiffener spacing, b = h_c = 12" and web thickness t = 0.072"

Go to Fig. 13.3.2 and choose hole dimensions of D = 5.8" and D_o = 6.95".

$$\frac{D_o}{h_c} = \frac{6.95}{12} = 0.58 \approx 0.6$$ O.K.

(b) Margin of safety (ultimate load):

With h_c = 12 in., from Fig. 13.3.3, obtain q_{all} = 710 lbs./in.

$$MS = \frac{q_{all}}{q} - 1 = \frac{710}{645} - 1 = \underline{0.1}$$ O.K.

(B) STIFFENER-STIFFENED WEB HAVING 45° FLANGED HOLES

The lightening holes are centered between single angle stiffeners as shown in Fig. 13.3.4. The web thickness is light and function of the stiffeners is to provide stiffness and accessibility when loads are fairly low, such as in the case of ribs, spars, and formers for control surfaces.

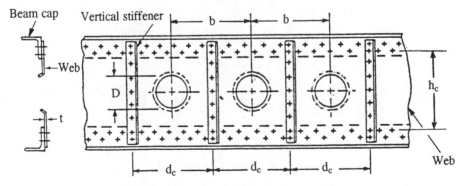

Fig. 13.3.4 Beam Web with Vertical Stiffeners and Round 45° Flanged Holes

Sizing procedures:

(a) Limiting conditions:
- Use the standard 45° flanged holes shown in Fig. 13.2.2 (stiffener – stiffened web)
- $0.025" \leq t \leq 0.125"$
- $0.235 \leq \dfrac{d_c}{h_c} \leq 1.0$
- $115 \leq \dfrac{h_c}{t} \leq 1500$
- $0.3 \leq \dfrac{D}{h_c} \leq 0.7$

(b) Shear failure between holes:

$$\dfrac{D}{h_c} \leq 0.85 - 0.1\left(\dfrac{h_c}{d_c}\right) \qquad \text{Eq. 13.3.1}$$

(c) Single angle stiffener must satisfy the following conditions:

$$t_o \geq t$$

$$\dfrac{A_o}{d_c t} \geq 0.385 - 0.08\left(\dfrac{d_c}{h_c}\right)^3 \qquad \text{Eq. 13.3.2}$$

$$I_o \geq \dfrac{f_s t \, d_c \, h_c^3}{10^8 (h_c - D)} \qquad \text{Eq. 13.3.3}$$

where I_o is the moment of inertia of the stiffener (formed or extruded section as shown below) about its natural axis (NA) and parallel to the skin line. The approximate equation is:

$$I \approx \dfrac{t_o b_{o2}^3 (4 b_{o1} + b_{o2})}{12 (b_{o1} + b_{o2})} \qquad \text{Eq. 13.3.4}$$

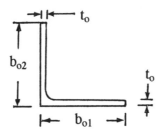

(d) The maximum stiffener width-to-thickness ($\dfrac{b_{o1}}{t_o}$ and $\dfrac{b_{o2}}{t_o}$) ratio can be found in Fig. 13.3.5

t_o	0.032	0.040	0.050	0.063	0.071	0.080	0.090	0.100	0.125
$\dfrac{b_{o1}}{t_o}$	12.5	12.5	11.5	10.0	9.5	9.0	9.0	9.0	9.0
$\dfrac{b_{o2}}{t_o}$	16.0	16.0	15.0	13.0	12.0	12.0	12.0	12.0	12.0

Fig. 13.3.5 Stiffener Width-to-Thickness Ratio

(e) The ultimate allowable gross shear stress:

$$F_s = K_2 F_o \quad \text{Eq. 13.3.5}$$

where F_o – Allowable gross shear stress without hole from Fig. 13.2.3
K_2 – Reduction factor from Fig. 13.3.6

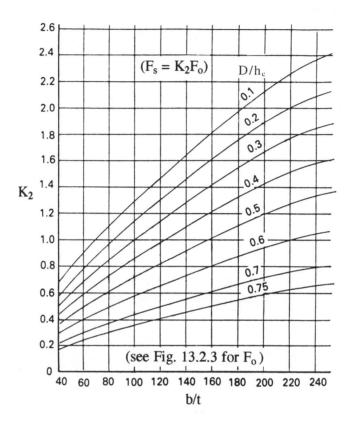

(Based on 45° flanged hole per Fig. 13.2.2)

Fig. 13.3.6 Correction Factor, K_2

(f) Rivets:
  (i) Web-to-cap rivets:

  $$q_R = 1.25 f_s t \left(\frac{h_c}{h_c - D}\right) \quad \text{Eq. 13.3.6}$$

  (ii) Cap-to-stiffener rivets:

  $$P_{st} = \frac{0.0024 A_o f_s d_c}{t} \left(\frac{h_c}{h_c - D}\right) \quad \text{Eq. 13.3.7}$$

  (iii) Web-to-stiffener rivets should conform to Fig. 13.3.7:

  For single angle stiffener, the required tensile strength of the rivets per inch:

  $$P_t = 0.20 F_{tu} t \quad \text{Eq. 13.3.8}$$

  where: F_{tu} – Ultimate tensile stress of the web material

  This criterion is satisfied by using MS 20470 aluminum rivets as specified in Fig. 13.3.7.

Cutouts

Web thickness (t)	0.025	0.032	0.040	0.051	0.064	0.072	0.081	0.091	0.102	0.125
Rivet	AD-4	AD-4	AD-4	AD-5	AD-5	AD-6	AD-6	DD-6	DD-6	DD-8
Rivet diameter (d)	$\frac{1}{8}$	$\frac{1}{8}$	$\frac{1}{8}$	$\frac{5}{32}$	$\frac{5}{32}$	$\frac{3}{16}$	$\frac{3}{16}$	$\frac{3}{16}$	$\frac{3}{16}$	$\frac{1}{4}$
Rivet spacing (s)	$\frac{1}{2}$	$\frac{5}{8}$	$\frac{5}{8}$	$\frac{7}{8}$	$\frac{3}{4}$	1.0	$\frac{7}{8}$	1.0	1.0	$1\frac{3}{8}$

Fig. 13.3.7 Web-to-Stiffener Rivet Spacing (2024-T4 and MS 20470 rivets)

Example 3:

Determine the beam web thickness with hole diameter $D = 5"$ in every bay of this beam as shown below to carry a shear flow of $q = 1,008$ lbs./in.

Web material: 7075-T6 sheet

(data from Fig. 4.3.4)

$F_{su} = 48,000$ psi

$F_{tu} = 80,000$ psi

$F_{bru} = 160,000$ psi

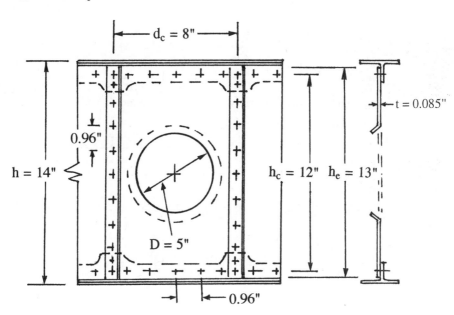

(a) Check limiting conditions:
- Use the standard 45° flanged lightening holes shown in Fig. 13.2.2
- $0.025" \leq t \leq 0.125"$ (to be checked later)
- $0.235 \leq \dfrac{d_c}{h_c} = \dfrac{8}{12} = 0.67 \leq 1.0$ O.K.
- $115 \leq \dfrac{h_c}{t} \leq 1500$ (to be checked later)

543

Optimize the web thickness (t):

t	$b = d_c$	$\dfrac{d_c}{t}$	F_o (Fig. 13.2.3)	K_2 (Fig. 13.3.6)	F_s ③ × ④	f_s $q = \dfrac{1{,}008}{t}$
①	②	②/①	③	④	⑤	⑥
0.063	8	127	11,500	1.0	11,500	16,000
0.08	8	100	14,200	0.84	11,930	12,600
0.085	8	94	15,000	0.81	12,150	11,860
0.09	8	89	15,700	0.78	12,250	11,200
0.1	8	80	17,500	0.72	12,600	10,080

The lightest web sizing is such that column ⑤ equals that of column ⑥. Therefore, the proper value of web thickness (t) is between 0.08" and 0.085"; in this calculation use t = 0.085"

(b) Shear failure between holes from Eq. 13.3.1:

$$\dfrac{D}{h_c} \leq 0.85 - 0.1\left(\dfrac{h_c}{d_c}\right)$$

$$0.85 - 0.1\left(\dfrac{h_c}{d_c}\right) = 0.85 - 0.1\left(\dfrac{12}{8}\right) = 0.7$$

$$\dfrac{D}{h_c} = \dfrac{5}{12} = 0.42 \leq 0.7 \qquad \text{O.K.}$$

(c) Single angle stiffener must satisfy the following conditions:

$$t_o = 0.125 > t = 0.085" \qquad \text{O.K.}$$

The minimum requirement for stiffener area from Eq. 13.3.2:

$$A_o = d_c t \left[0.385 - 0.08\left(\dfrac{d_c}{h_c}\right)^3\right]$$

$$= 8 \times 0.085\left[0.385 - 0.08\left(\dfrac{8}{12}\right)^3\right] = 0.246 \text{ in.}^2$$

The minimum requirement for moment of inertia from Eq. 13.3.3:

$$I_o \geq \dfrac{f_s t d_c h_c^3}{10^8 (h_c - D)} = \dfrac{11{,}860 \times 0.085 \times 8 \times 12^3}{10^8 (12 - 5)} = 0.02 \text{ in.}^4$$

The stiffener properties are:

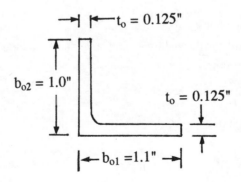

The stiffener area:

$$A_o = 1.1 \times 0.125 + (1.0 - 0.125) \times 0.125$$
$$= 0.247 > 0.246 \text{ in.}^2 \quad \text{O.K.}$$

Here $b_{o1} = 1.1''$ and $b_{o2} = 1.0''$
Moment of inertia calculation from Eq. 13.3.4:

$$I = \frac{t_o b_{o2}^3 (4b_{o1} + b_{o2})}{12(b_{o1} + b_{o2})}$$

$$= \frac{0.125 \times 1^3 (4 \times 1.1 + 1.0)}{12(1.1 + 1.0)} = 0.0268 \text{ in.}^4 > 0.02 \text{ in.}^4 \quad \text{O.K.}$$

(d) The maximum stiffener width-to-thickness ($\frac{b_{o1}}{t_o}$ and $\frac{b_{o2}}{t_o}$) ratio from Fig. 13.3.5

$$t_o = 0.125''$$

$$\frac{b_{o1}}{t_o} = \frac{1.1}{0.125} = 8.8 < 9.0 \quad \text{O.K.}$$

$$\frac{b_{o2}}{t_o} = \frac{1.0}{0.125} = 8 < 12.0 \quad \text{O.K.}$$

(e) The ultimate allowable gross shear stress with $t = 0.085''$:

$$F_s = 12,150 \text{ psi [from Step (a) or use Eq. 13.3.5]}$$

$$MS = \frac{F_s}{f_s} - 1 = \frac{12,150}{11,860} - 1 = \underline{0.02} \quad \text{O.K.}$$

(f) Rivets:

(i) Web-to-cap rivets from Eq. 13.3.6:

$$q_R = 1.25 f_s t \left(\frac{h_c}{h_c - D}\right)$$

$$= 1.25 \times 11,860 \times 0.085 \left(\frac{12}{12 - 5}\right) = 2,160 \text{ lbs./in.}$$

Use double rows of rivet of diameter, $d = \frac{3}{16}''$ (E – rivet) with spacing $s = 0.96''$
($P_{s, all} = 1,230$ lbs./rivet per Fig. 9.2.5; bearing is not critical)

$$MS = \frac{2 \times P_{s, all}}{q_R} - 1 = \frac{2 \times 1,230}{2,160} - 1 = \underline{0.14} \quad \text{O.K.}$$

(ii) Cap-to-stiffener rivets from Eq. 13.3.7:

$$P_u = \frac{0.0024 A_o f_s d_c}{t} \left(\frac{h_c}{h_c - D}\right)$$

$$= \frac{0.0024 \times 0.247 \times 11,860 \times 8}{0.085} \left(\frac{12}{12 - 5}\right) = 1,134 \text{ lbs.}$$

Use 2 of rivet $d = \frac{1}{4}''$ (E) ($P_{s, all} = 2,230$ lbs./rivet per Fig. 9.2.5; bearing is not critical) with spacing $s = 1.0''$

$$MS = \frac{2 \times P_{s, all}}{q_R} - 1 = \frac{2 \times 2,230}{1,134} - 1 = \text{high} \quad \text{O.K.}$$

(iii) Web-to-stiffener rivets should conform to Fig. 13.3.7:

For single angle stiffener, the required tensile strength of the rivets per inch:

$$P_t = 0.20 F_{tu} t = 0.20 \times 80,000 \times 0.085 = 1,360 \text{ lbs./in.}$$

Use rivet $d = \frac{3}{16}"$ (E); tension allowable load is $P_{t, all} = 1,304$ lbs. (see Fig. 9.2.11)

with rivet spacing, $s = \frac{1,304}{1,360} = 0.96"$

13.4 STIFFENED WEB HAVING DOUBLER-REINFORCED HOLES

A stiffener-stiffened web having doubler-reinforced round holes is shown in Fig. 13.4.1:

- The diameter of the hole (D) ≤ 50% of the beam depth
- Ring doubler width (W) and hole diameter (D) ratios of $\frac{W}{D} = 0.35$ to 0.5

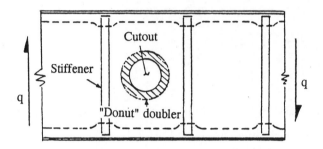

Fig. 13.4.1 Typical Stiffened Web with Doubler-Reinforced Holes

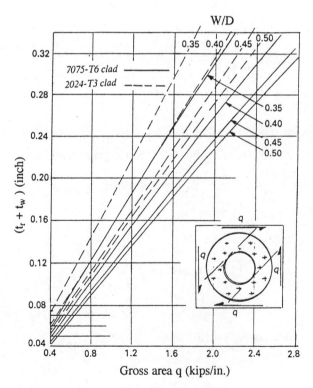

(2024-T3 and 7075-T6 of clad aluminum alloys)

Fig. 13.4.2 Shear Web Hole Reinforcement Doublers

Sizing procedures:

(a) Assume rivet size and pattern to determine doubler width (W)

(b) Determine doubler thickness from Fig. 13.4.2 by giving $\frac{W}{D}$ and shear flow, q.

(c) Rivet pattern, as shown in Fig. 13.4.3:

$$q_R = 2q \left(\frac{t_r}{t_r + 0.8 t_w} \right)$$ 	Eq. 13.4.1

where: q_R – Rivet pattern between tangent lines to develop a running load strength/inch
W – Width of the "donut" ring doubler
q – Web shear flow (lbs./in.)
t_r – "Donut" ring thickness
t_w – web thickness

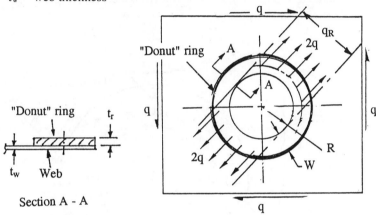

Fig. 13.4.3 Calculation of Rivet Pattern for "Donut" Ring

Example:

Given the web thickness of t = 0.085" (7075-T6) determine the "donut" ring thickness with hole diameter D = 5" to carry a shear flow of q = 2,483 lbs./in.

(a) Determine doubler width (W):

Assume double rows of rivets, diameter $d = \frac{1}{4}$" (E) ($P_{s,\,all}$ = 2,230 lbs./rivet per Fig. 9.2.5; bearing is not critical) with rivet spacing s = 1.0"

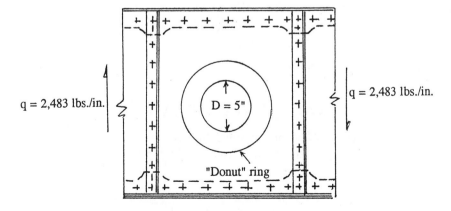

(b) Determine doubler thickness:

Use $\dfrac{W}{D} = \dfrac{2.06}{5} = 0.41$ and from Fig. 13.4.2 obtain $t_w + t_r = 0.31''$

∴ $t_r = 0.31 - 0.085 = 0.225''$ (doubler thickness)

(c) Rivet pattern from Eq. 13.4.1:

$$q_R = 2q\left(\dfrac{t_r}{t_r + 0.8 t_w}\right)$$

$$= 2 \times 2{,}483\left(\dfrac{0.225}{0.225 + 0.8 \times 0.085}\right) = 3{,}814 \text{ lbs./in.}$$

The total load on rivets with hole $D = 5''$:

$P_R = 5 \times 3{,}814 = 19{,}070$ lbs.

There are a total number of n = 8 rivets between tangent lines (see sketch below) to develop rivet load strength:

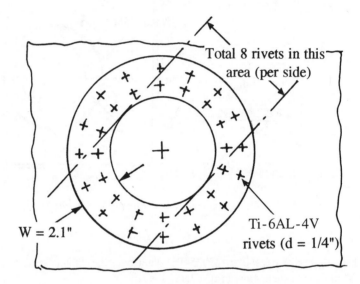

Use fastener $d = \dfrac{1}{4}''$ (0.25") rivets and the shear allowable is:

$P_{s,all} = 4{,}660$ lbs. [Rivet material – Ti-6AL-4V, data from Fig. 9.2.5]

The bearing allowable on web (7075-T6 sheet, $t_w = 0.085''$):

$P_{b,all} = 160{,}000\,(80\%) \times 0.25 \times 0.085 = 2{,}720$ lbs. (not critical)

(80% is bearing reduction, see Chapter 9.2)

$$\text{MS} = \dfrac{n P_{s,all}}{P_R} = \dfrac{8 \times 2{,}720}{19{,}070} - 1 = \underline{0.14} \qquad\qquad \text{O.K.}$$

13.5 WEB CUTOUT WITH BENT DOUBLER

A rectangular doubler or a "bent" doubler must be provided when a rectangular cutout is required in shallow beam webs to carry the shear load, as shown in Fig. 13.5.1.

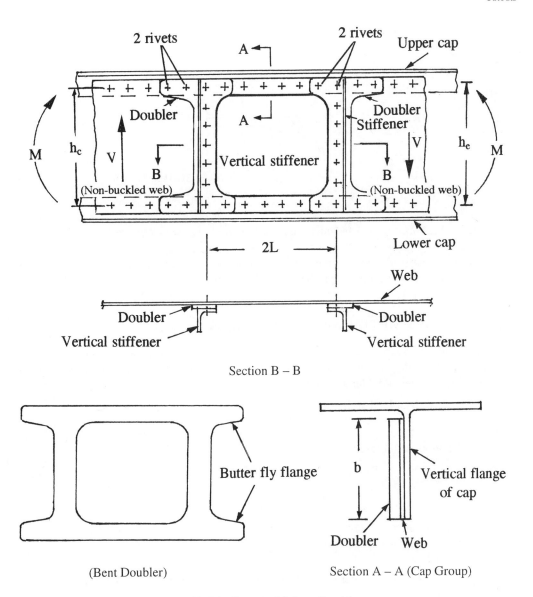

Fig. 13.5.1 Cutout with Bent Doubler

The loading imposed in the cutout region on the doubler, web and beam cap (cap-group) should be analyzed as a frame in bending. The bending moment, axial loads, and shears at any section of the frame follow as a matter of statics and should be checked as beam-column. Some designs may use a deeper outstanding flange (including caps, outstanding flange, web, and doubler, see Fig. 13.5.1) at the upper cap (assumed to be in compression) to provide greater stiffness to take care of the compression load. The deeper flange will also carry a greater portion of the total shear load.

Design requirements:

- Design shear resistant web on both sides of the cutout
- Use bent-doubler shape as shown in Fig. 13.5.1
- Vertical stiffeners are required on both sides of the cutout
- The minimum moment of inertia (I_u – from Fig. 12.2.4) required for the vertical stiffener must include the doubler as shown in Fig. 13.5.2
- Use minimum fastener spacing per Fig. 9.12.4 in the region of the cutout

- Use at least two fasteners to attach vertical stiffener through vertical flange of cap

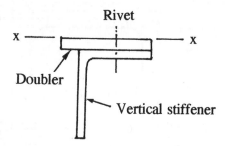

Fig. 13.5.2 Minimum Moment of Inertia of Vertical Stiffener (Refer to Fig. 12.2.4)

The following sizing procedures assume that the shear force (V) and beam bending moment (M) are positive and that the moment produces compression in the upper cap group, as shown in Section A – A of Fig. 13.5.1.

(A) CASE I – SYMMETRICAL BENT DOUBLER

On symmetrical bent doublers, the moment of inertia of the upper cap group (I_U) is equal to that of the lower cap group (I_L), as shown in Fig. 13.5.1 at Section A – A. The cap group includes the beam cap, web, and bent doubler.

Vertical shear load at the cutout:

$$V = qh_e \quad \text{Eq. 13.5.1}$$

Cap group axial load:

$$P_H = \frac{M}{h_e} \quad \text{Eq. 13.5.2}$$

where: M – Beam bending moment

Axial compression stress in upper cap group:

$$f_{c,U} = \frac{P_H}{A_U} \quad \text{Eq. 13.5.3}$$

where: A_U – Upper cap group area (includes cap, web, and doubler, see Fig. 13.5.1)

Axial tension stress in lower cap group:

$$f_{t,L} = \frac{P_H}{A_L} \quad \text{Eq. 13.5.4}$$

where: A_L – Lower cap group area (includes cap, web, and doubler)

(a) Upper cap group is analyzed as beam-column, as shown in Fig. 13.5.3:

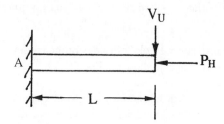

Fig. 13.5.3 Cap Group Acts as Beam-Column

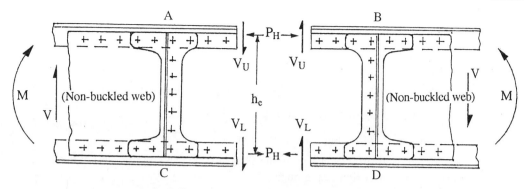

Fig. 13.5.3 Cap Group Acts as Beam-Column (cont'd)

End moment of the cap group due to beam shear load (V):

$$M_{s,A} = -M_{s,B} = V L (v)$$ Eq. 13.5.5

The above moment equation is derived from Case 10 of Fig. 10.6.4 and the coefficient "v" is obtained from Fig. 10.6.5 or Fig. 10.6.6.

The compression stress on the vertical flange (see Section A-A of Fig. 13.5.1) due to above bending moment:

$$f_{sb,A} = \frac{M_{s,A}(c)}{I_U}$$ Eq. 13.5.6

where: $M_{s,A}$ – Bending moment from Eq. 13.5.5 for whichever side of the cutout produces compression in the vertical flange of the cap group
 c – Extreme fiber of the cap group vertical flange in compression
 I_U – Moment of inertia of the upper cap group

(b) One-edge-free plate having buckling stress in compression:

The effective laminated thickness (includes cap vertical flange, web, and doubler, as shown in Section A-A of Fig. 13.5.1):

$$t_e = 0.8 (t_f + t_w + t_d)$$

where: t_f – Vertical flange thickness of upper cap
 t_w – Web thickness
 t_d – Doubler thickness

Average buckling coefficients in compression (K_c):

$$K_c = \frac{K_{c,\circled{7}} + K_{c,\circled{10}}}{2} = \frac{1.1 + 0.4}{2} = 0.75$$

where: $K_{c,\circled{7}}$ – Buckling coefficient equal to 1.1 from Case ⑦ of Fig. 11.3.1
 $K_{c,\circled{10}}$ – Buckling coefficient equal to 0.4 from Case ⑩ of Fig. 11.3.1

Compression buckling stress from Eq. 11.3.2:

$$\frac{F_{c,cr}}{\eta_c} = 0.75 E \left(\frac{t_e}{b}\right)^2$$ Eq. 13.5.7

(c) One-edge-free plate having buckling stress in bending:

Average buckling coefficients in bending (K_b):

$$K_b = \frac{K_{b,\circled{3}} + K_{b,\circled{4}}}{2}$$

where: $K_{b,\circled{3}}$ – Buckling coefficient from Case ③ of Fig. 11.3.6
 $K_{b,\circled{4}}$ – Buckling coefficient from Case ④ of Fig. 11.3.6

Bending buckling stress from Eq. 11.3.6:

$$\frac{F_{b,cr}}{\eta_c} = K_b E \left(\frac{t_e}{b}\right)^2 \qquad \text{Eq. 13.5.8}$$

(d) Crippling stress (F_{cc}) of upper cap (excluding web and doubler):

Determine crippling stress calculation per Eq. 10.7.1.

(e) Margin of safety (MS) of upper cap group at 'A':

The compression stress is critical in the vertical flange and, in this case, it occurs at the upper left and lower right cap groups.

Use the following interaction equation to find MS:

$R_b^{1.75} + R_c = 1.0$ (Determine MS from interaction curves in Fig. 11.5.4)

where: $R_b = \dfrac{f_{sb,A}}{F_{b,cr}}$

$R_c = \dfrac{f_{c,U}}{(F_{cc} \text{ or } F_{c,cr} \text{ whichever is smaller})}$

$f_{sb,A}$ – Bending stress from the bending moment, $M_{s,A}$, which produces the compression stress on the vertical flange of the cap group, from Eq. 13.5.6

$f_{c,U}$ – Axial compression stress on the upper cap group due to beam bending moment (M) from Eq. 13.5.3

$F_{c,cr}$ – From Eq. 13.5.7

$F_{b,cr}$ – From Eq. 13.5.8

F_{cc} – Crippling stress of the upper cap only (per Eq. 10.7.1)

Also check MS of the upper cap group at 'B' (generally not critical).

(f) Check MS of the lower cap group at 'C':

This is the maximum tension stress area:

$$MS = \frac{F_{tu}}{f_{sb,C} + f_{t,L}} - 1$$

where: $f_{sb,C}$ – Bending stress from the bending moment, $M_{s,C}$, which produces the tension stress on the outstanding flange of the cap group

$f_{t,L}$ – Axial tension stress on the lower cap section due to beam bending moment (M) from Eq. 13.5.4

(g) Check MS of the lower cap group at 'D':

$$MS = \frac{F_{b,cr}}{f_{sb,D} - f_{t,L}} - 1$$

where: $f_{sb,D}$ – Bending stress from the bending moment, $M_{s,D}$, which produces the compression stress on the vertical flange of the cap group

$F_{b,cr}$ – From Eq. 13.5.8

(h) Check the end fasteners of the butterfly flanges on the bent doubler:

Axial load on doubler butterfly flange is:

$$P_d = \left[\frac{\text{Doubler area } (A_d)}{\text{Area of upper cap group } (A_U)}\right] P_H \qquad \text{Eq. 13.5.9}$$

$$MS = \frac{P_{all}}{\dfrac{P_d}{n}} - 1 \geq 0.2$$

where: P_{all} – Use the smallest fastener value of $P_{s,all}$ (shear) or $P_{b,all}$ (bearing on web)
n – Number of fasteners

The extra 20% margin is to take care of the local moment (M_d) produced in the butterfly flange from the moment of the cap group ($M_{sb,A}$), see Fig. 13.5.4.

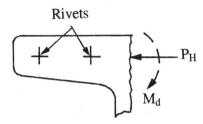

Fig. 13.5.4 Loads on Butterfly Flange of Bent Doubler

(i) Check the combined section of vertical stiffener and attached doubler:

Minimum stiffener cross-sectional moment of inertia (doubler and stiffener):

From Fig. 12.2.4 obtain the minimum moment of inertia (I_u).

Determine crippling stress ($F_{cc,st}$) of stiffener.

Compression allowable of the combined section of doubler and stiffener (use stiffener crippling stress $F_{cc,st}$):

$F_{d,st} = F_{cc,st}$

Maximum compression stress:

$$f_{c(d,st)} = \frac{V_U}{A_{d,st}}$$

$$MS = \frac{F_{d,st}}{f_{c(d,st)}} - 1$$

where: $A_{d,st}$ is the total cross-sectional area of both doubler and stiffener

Stability check for combined section:

The stability compression stress:

$$F_{cr(d,st)} = \frac{5.3\,E}{(\frac{h_e}{\rho_{d,st}})^2}$$ Eq. 13.5.10

$$MS = \frac{F_{cr(d,st)}}{f_{c(d,st)}} - 1$$

(j) Fastener pattern in cutout region (use min. rivet spacing, per Fig. 9.12.4 to avoid inter-rivet buckling).

(B) CASE II – UNSYMMETRICAL BENT DOUBLER

Unsymmetrical bent doubler (generally having a deeper vertical flange on the upper cap group, $I_U > I_L$, see sketch shown in Example calculation):

Vertical shear load distributed to upper cap group:

$$V_U = (\frac{I_U}{I_U + I_L})V$$ Eq. 13.5.11

where: I_U – moment of inertia of upper cap group
I_L – moment of inertia of lower cap group

Chapter 13.0

Vertical shear load distributed to lower cap group:

$$V_L = \left(\frac{I_L}{I_U + I_L}\right) V \qquad \text{Eq. 13.5.12}$$

Horizontal axial load, P_H, (upper and lower cap are equal) due to beam bending moment (M); use the same equation from Eq. 13.5.2:

$$P_H = \frac{M}{h_e}$$

Upper cap group axial compression stress:

$$f_{c,U} = \frac{P_H}{A_U}$$

(a) Cap group analyzed as beam-column:

End moment on upper cap:

$$M_{s,A} = -M_{s,B} = V_U L(\nu) \qquad \text{Eq. 13.5.13}$$

The above moment equation is derived from Case 10 of Fig. 10.6.4, and coefficient ν is obtained from Fig. 10.6.5 or Fig. 10.6.6.

End moment on lower cap:

$$M_{s,C} = -M_{s,D} = V_L L \qquad \text{Eq. 13.5.14}$$

The compression stress on the vertical flange due to bending moment from Eq. 13.5.6:

$$f_{sb,A} = \frac{M_{s,A}(c)}{I_U}$$

(b) One-edge-free plate having buckling stress in compression:

The effective laminated thickness (including vertical flange of cap, web, and doubler):

$$t_e = 0.8 (t_f + t_w + t_d)$$

where: t_f – Vertical flange thickness of cap
t_w – Web thickness
t_d – Doubler thickness

From Eq. 13.5.7:

$$\frac{F_{c,cr}}{\eta_c} = 0.75 E \left(\frac{t_e}{b}\right)^2$$

(c) One-edge-free plate having buckling stress in bending:

Average buckling coefficients in bending (K_b):

$$K_b = \frac{K_{b,③} + K_{b,④}}{2}$$

where: $K_{b,③}$ – Buckling coefficient from Case ③ of Fig. 11.3.6
$K_{b,④}$ – Buckling coefficient from Case ④ of Fig. 11.3.6

From Eq. 13.5.8:

$$\frac{F_{b,cr}}{\eta_c} = K_b E \left(\frac{t_e}{b}\right)^2$$

(d) Crippling stress (F_{cc}) of upper cap (excluding web and doubler):

Crippling stress calculation per Eq. 10.7.1.

(e) Margin of safety (MS) of upper cap group at 'A':

The compression stress is critical in the vertical flange and, in this case, it is at the lefthand end of the upper cap group and righthand end of the lower cap group.

Use the following interaction equation to find MS:

$R_b^{1.75} + R_c = 1.0$ (Determine MS from Fig. 11.5.4)

where: $R_b = \dfrac{f_{sb,A}}{F_{b,cr}}$

$R_c = \dfrac{f_{c,U}}{(F_{cc} \text{ or } F_{c,cr} \text{ whichever is smaller})}$

$f_{sb,A}$ – Bending stress from the bending moment Ms,A (from Eq. 13.5.13) which produces the compression stress on the vertical flange of the cap group

$f_{c,U}$ – Axial compression stress on the upper cap group due to beam bending moment (M) from Eq. 13.5.3

$F_{c,cr}$ – From Eq. 13.5.7

$F_{b,cr}$ – From Eq. 13.5.8

F_{cc} – Crippling stress of the upper cap only

Also check MS of the upper cap group at 'B' (generally not critical).

(f) Check MS of the lower cap group at 'C':

This is the maximum tension stress:

$MS = \dfrac{F_{tu}}{f_{sb,C} + f_{t,L}} - 1$

where: $f_{sb,C}$ – Bending stress from the bending moment, $M_{s,C}$ (from Eq. 13.5.14) which produces the tension stress on the vertical flange of the cap

(g) Check MS of the lower cap group at 'D':

$MS = \dfrac{F_{b,cr}}{f_{sb,D} - f_{t,L}} - 1$

where: $f_{t,L}$ – Axial tension stress on the lower cap section due to beam bending moment (M) from Eq. 13.5.4

$f_{sb,D}$ – Bending stress from the bending moment $M_{s,D}$ (similar manner as that of Eq. 13.5.5) which produces the compression stress on the vertical flange of the cap

$F_{b,cr}$ – From Eq. 13.5.8

(h) Check end fasteners on the butterfly flanges of the bent doubler:

[See Step (h) of Case I of symmetrical bent doubler]

(i) Check the combined section of vertical stiffener with attached doubler:

[See Step (i) of Case I of symmetrical bent-doubler]

(j) Fastener pattern in cutout region (use minimum rivet spacing, per Fig. 9.12.4 to avoid inter-rivet buckling).

Example:

Size an unsymmetrical bent cutout in a floor beam web, as shown below, to carry a vertical shear V = 1,323 lbs. and bending moment, M = 45,938 in.-lbs.

7075-T6 sheet (web and doubler):	7075-T6 Extrusion (caps and stiffeners):
(from Fig. 4.3.4):	(from Fig. 4.3.5):
$F_{tu} = 80,000$ psi	$F_{tu} = 82,000$ psi
$F_{su} = 48,000$ psi	$F_{su} = 44,000$ psi
$F_{bru} = 160,000$ psi	$F_{bru} = 148,000$ psi
$E = 10.5 \times 10^6$ psi	$E = 10.7 \times 10^6$ psi

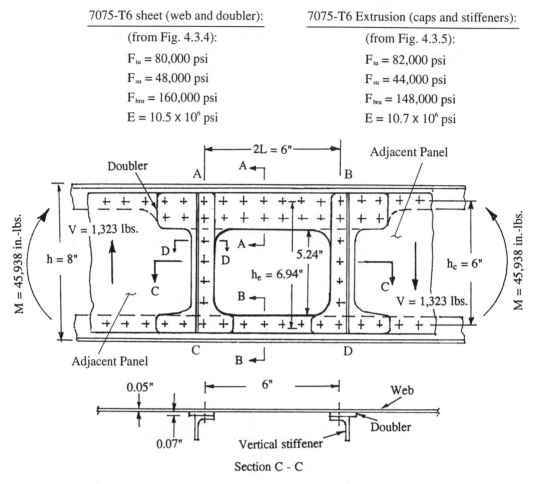

Assume the thickness of bent doubler: $t_d = 0.07$ in. (7075-T6 plate, same as web).

Properties of upper cap group are given below:

$I_U = 0.078$ in.^{4}

$A_U = 0.433$ in.^{2}

$Y_U = 0.645$ in.

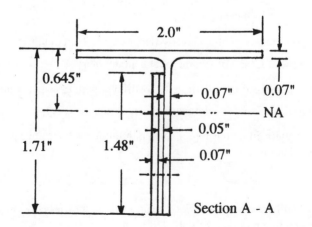

Properties of lower cap group are given below:

$I_L = 0.015$ in.^{4}
$A_L = 0.235$ in.^{2}
$Y_L = 0.417$ in.
$h_e = 8 - 0.645 - 0.417$
$\quad = 6.94$ in.

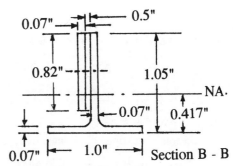

Section B - B

Vertical shear load distributed to upper cap group from Eq. 13.5.11:

$$V_U = \left(\frac{I_U}{I_U + I_L}\right) V = \left(\frac{0.078}{0.078 + 0.015}\right) \times 1,323 = 1,110 \text{ lbs.}$$

Vertical shear load distributed to lower cap group from Eq. 13.5.12:

$$V_L = \left(\frac{I_L}{I_U + I_L}\right) V = \left(\frac{0.015}{0.078 + 0.015}\right) \times 1,323 = 213 \text{ lbs.}$$

Horizontal axial load, P_H, from Eq. 13.5.3:

$$P_H = \frac{M}{h_e} = \frac{45,938}{6.94} = 6,619 \text{ lbs.}$$

Upper cap group axial compression stress:

$$f_{c,U} = \frac{P_H}{A_U} = \frac{6,619}{0.433} = 15,286 \text{ psi}$$

(a) Cap group analyzed as beam-column:

$$j = \sqrt{\frac{EI}{P_H}} = \sqrt{\frac{10.5 \times 10^6 \times 0.078}{6,619}} = 11.12$$

$$\frac{L}{j} = \frac{3}{11.12} = 0.27$$

Go to Fig. 10.6.5 or Fig. 10.6.6, obtain $\nu = 1.01$

End moment on upper cap at 'A' from Eq. 13.5.13:

$\quad M_{s,A} = V_U L (\nu) = 1,110 \times 3 (1.01) = 3,363$ in.-lbs.

End moment on lower cap group at 'C' from Eq. 13.5.14:

$\quad M_{s,C} = V_L L = 213 \times 3 = 639$ in.-lbs.

The compression stress on vertical flange due to bending moment from Eq. 13.5.6:

$$f_{sb,A} = \frac{M_{s,A} c}{I_U} = \frac{3,363 \times (1.71 - 0.645)}{0.078} = 45,918 \text{ psi (compression)}$$

(b) One edge-free plate having buckling stress in compression:

$\quad t_e = 0.8 (t_f + t_w + t_d) = 0.8 (0.07 + 0.05 + 0.07) = 0.152$ in.

From Eq. 13.5.7:

$$\frac{F_{c,cr}}{\eta_c} = 0.75 E \left(\frac{t_e}{b}\right)^2$$

$$\frac{F_{c,cr}}{\eta_c} = 0.75 \times 10.5 \times 10^6 \left(\frac{0.152}{1.48}\right)^2 = 83,064 \text{ psi}$$

Chapter 13.0

From Fig. 11.2.4 (use 7075-T6 extrusion curves equivalent to 7075-T6 sheet):

$F_{c,cr} = 68,000$ psi

(c) One-edge-free plate having buckling stress in bending:

$\alpha = \dfrac{1.065}{1.48} = 0.72$ and go to Fig. 11.3.6 to obtain:

$K_{b,③} = 1.35$ (case ③)

$K_{b,④} = 0.44$ (case ④)

Average buckling coefficients in bending (K_b):

$K_b = \dfrac{K_{b,③} + K_{b,④}}{2} = \dfrac{1.35 + 0.44}{2} = 0.9$

From Eq. 13.5.8:

$\dfrac{F_{b,cr}}{\eta_c} = 0.9 \times 10.5 \times 10^6 (\dfrac{0.152}{1.48})^2 = 99,677$ psi

From Fig. 11.2.4 (assume 7075-T6 extrusion curves equivalent to 7075-T6 sheet):

$F_{b,cr} = 70,500$ psi

(d) Crippling stress (F_{cc}) of upper cap (excluding web and doubler):

Segment	Free edges	b_n	t_n	$\dfrac{b_n}{t_n}$	$b_n t_n$	F_{ccn} (from Fig. 10.7.7)	$t_n b_n F_{ccn}$
①	1	1.0	0.07	14.3	0.07	43,500	3,045
②	1	1.0	0.07	14.3	0.07	43,500	3,045
③	1	1.675	0.07	23.9	0.117	28,500	3,335
				Σ	0.257		9,425

$F_{cc} = \dfrac{\Sigma b_n t_n F_{ccn}}{\Sigma b_n t_n} = \dfrac{9,425}{0.257} = 36,673$ psi

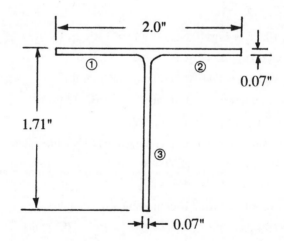

Margin of safety (MS) of upper cap group at 'A':

$R_b = \dfrac{f_{sb,A}}{F_{b,cr}} = \dfrac{45,918}{70,500} = 0.65$

$$R_c = \frac{f_{c,U}}{F_{cc}} = \frac{15,286}{36,673} = 0.42$$

Find MS from Fig. 11.5.4 ($R_b^{1.75} + R_c = 1.0$), obtain:

MS = 0.09   O.K.

Check upper cap group at 'B':

$f_{sb,B} = 45,918$ psi (tension)

$f_{c,U} = 15,286$ psi (compression)

The maximum tension stress at vertical flange of the cap group:

$f_{sb,B} - f_{c,U} = 45,918 - 15,286$
$= 30,632$ psi $< F_{tu} = 80,000$ psi   NOT CRITICAL

(e) Check MS of the lower cap group at 'C':

End moment:

$M_{s,C} = V_L L = 213 \times 3 = 639$ in.-lbs.

The bending compression stress on vertical flange from Eq. 13.5.6:

$$f_{sb,C} = \frac{M_{s,C}(c)}{I_U} = \frac{639 \times (1.05 - 0.417)}{0.015} = 26,966 \text{ psi (tension)}$$

$$f_{t,L} = \frac{6,619}{0.235} = 28,166 \text{ psi (tension)}$$

$f_{sb,C} + f_{t,L} = 26,966 + 28,166 = 55,132$ psi

$$MS = \frac{F_{tu}}{f_{sb,C} + f_{t,L}} - 1 = \frac{80,000}{55,132} - 1 = \underline{0.45} \quad \text{O.K.}$$

(f) Check MS of the lower cap group at 'D':

$f_{sb,D} = 26,966$ psi (compression)

$f_{t,L} = 28,166$ psi (tension)

$-f_{sb,D} + f_{t,L} = -26,966 + 28,166 = 1,200$ psi $< F_{tu} = 80,000$ psi   NOT CRITICAL

(g) Check end fasteners of butterfly flanges of bent doubler:

Axial compression load in upper butterfly flange ($t_d = 0.07"$) is:

$$P_d = (\frac{1.48 \times 0.07}{0.433}) \times 6,619 = 1,584 \text{ lbs.}$$

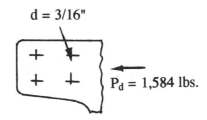

There are four $d = \frac{3}{16}"$ (0.1875") rivets and the rivet shear allowable is:

$P_{s,all} = 1,230$ lbs. (Rivet material – 7075-T731(E), data from Fig. 9.2.5)

Since the doubler thickness $t_d = 0.07" > t_w = 0.05"$, the critical bearing allowable on web (7075-T6 sheet, $t_w = 0.05"$):

$P_{b, all} = 160,000\,(80\%) \times 0.1875 \times 0.05 = 1,200$ lbs. (critical)

(80% is bearing reduction, see Chapter 9.2)

$$MS = \frac{P_{b, all}}{\frac{P_d}{n}} - 1 = \frac{1,200}{\frac{1,584}{4}} - 1 = \underline{high} \qquad \text{O.K.}$$

Axial tension load in lower butterfly flange ($t_d = 0.07"$) is:

$$P_d = \left(\frac{0.82 \times 0.07}{0.235}\right) \times 6,619 = 1,617 \text{ lbs.}$$

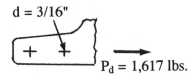

There are two $d = \frac{3}{16}"$ rivets:

$$MS = \frac{P_{b, all}}{\frac{P_d}{n}} - 1 = \frac{1,200}{\frac{1,617}{2}} - 1 = \underline{0.48} > 0.2 \qquad \text{O.K.}$$

(h) Check combined section of vertical stiffener with doubler:

Sectional properties of combined section of doubler and stiffener as shown below:

$I_{(d, st)\,x-x} = 0.0115$ in.^{4}

$A_{d, st} = 0.115$ in.^{2}

$\rho_{d, st} = 0.102$ in.

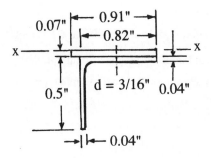

Section D - D

Minimum moment of inertia (combined section of doubler and stiffener):

Assume for the adjacent panel, $\frac{h_e}{d_c} = 1.0$ and from Fig. 12.2.4 obtain:

$$\frac{I_u}{h_e t^3} = 0.15$$

$I_u = 0.15\, h_e t^3 = 0.15 \times 6.94 \times 0.05^3$

$\quad = 0.0001$ in.$^4 < I_{(d, st)\,x-x} = 0.0115$ in.^{4} \qquad O.K.

Crippling stress ($F_{cc, st}$) of stiffener:

Segment	Free edges	b_n	t_n	$\dfrac{b_n}{t_n}$	$b_n t_n$	F_{ccn} (from Fig. 10.7.7)	$b_n t_n F_{ccn}$
①	1	0.8	0.04	20	0.032	32,000	1,024
②	1	0.48	0.04	12	0.0192	50,000	960
					Σ 0.0512 = A_{st}		1,984

$$F_{cc, st} = \frac{\Sigma b_n t_n F_{ccn}}{\Sigma b_n t_n} = \frac{1,984}{0.0512} = 38,750 \text{ psi}$$

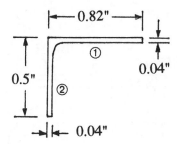

Compression stress in combined section:

$F_{d, st}$ = 38,750 psi (use $F_{cc, st}$ value as allowable compression stress)

Maximum compression stress on stiffener including doubler:

$$f_{c(d, st)} = \frac{V_U}{A_{d, st}} = \frac{1,110}{0.115} = 9,652 \text{ psi}$$

$$MS = \frac{F_{cc, st}}{f_{c(d, st)}} - 1 = \frac{38,750}{9,652} - 1 = \underline{\text{High}} \qquad \text{O.K.}$$

Combined section stability check:

The stability compression stress from Eq. 13.5.10:

$$F_{cr(d, st)} = \frac{5.3 \, E}{(\dfrac{h_e}{\rho_{d, st}})^2} = \frac{5.3 \times 10.5 \times 10^6}{(\dfrac{6.94}{0.102})^2} = 12,021 \text{ psi}$$

$$MS = \frac{F_{cr(d, st)}}{f_{c(d, st)}} - 1 = \frac{12,021}{9,652} - 1 = \underline{0.24} \qquad \text{O.K.}$$

(i) Fastener pattern in cutout region (use min. spacing, s = 0.75" for rivet d = $\dfrac{3}{16}$", per Fig. 9.12.4):

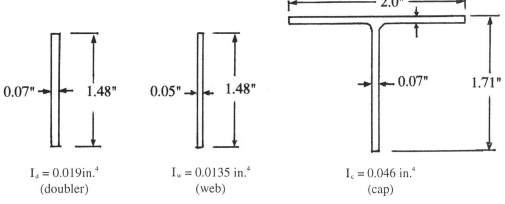

$I_d = 0.019$ in.^{4} (doubler) $I_w = 0.0135$ in.^{4} (web) $I_c = 0.046$ in.^{4} (cap)

Fastener loads between vertical flange of upper cap and web:

Use the following approximate calculation to obtain the upper cap load distribution from $V_U = 1,110$ lbs. is

Load on cap:

$$V_{U,c} = V_U(\frac{I_c}{I_d + I_w + I_c}) = 1,110(\frac{0.046}{0.019 + 0.0135 + 0.046}) = 650 \text{ lbs.}$$

Assume there are two effective rivets of $d = \frac{3}{16}"$ which pass through vertical flange of the cap and that the vertical stiffener which takes the load of $V_{U,c} = 650$ lbs.
Rivet material – 7075-T731(E), see Fig. 9.2.5, $P_{s,all} = 1,230$ lbs.

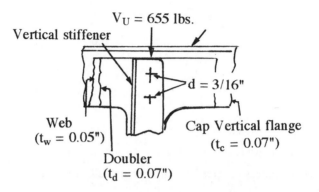

The bearing allowable on the web ($t_w = 0.05"$, assume web react this load):

$P_{b,all} = 148,000\,(80\%) \times 0.1875 \times 0.05$

$= 1,110$ lbs. (critical)

$$MS = \frac{2 \times P_{b,all}}{V_{U,c}} - 1 = \frac{2 \times 1,110}{650} - 1 = \underline{High} \qquad\qquad O.K.$$

Fastener loads (V = 1,323 lbs.) between web and doubler:

There are 7 effective rivets of $d = \frac{3}{16}"$ [Rivet material - 7075-T731(E)] and bearing allowable on the attached web ($t_w = 0.05"$) is 1,110 lbs. (see above)

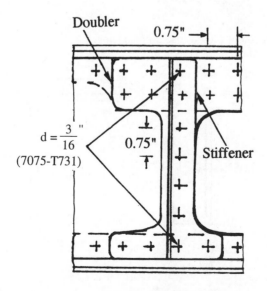

$$MS = \frac{2 \times P_{b,all}}{V_{U,c}} - 1 = \frac{2 \times 1,110}{650} - 1 = \underline{High} \qquad O.K.$$

(Note: Rivet diameter $d = \frac{5}{32}"$ could have been used instead of $d = \frac{3}{16}"$ in the second trial sizing).

13.6 FRAMED CUTOUTS IN DEEP SHEAR BEAMS

Large cutouts provide crawl holes needed for purposes of final assembly and/or inspection and repair. Such large holes occur in shear webs on wing and empennage ribs as follows:

- Moderately loaded bulkhead webs are generally a built-up beams with a cutout, as shown in Fig. 13.6.1(a)
- Highly loaded bulkhead webs are generally a machined panel with a cutout, as shown in Fig. 13.6.1(b)

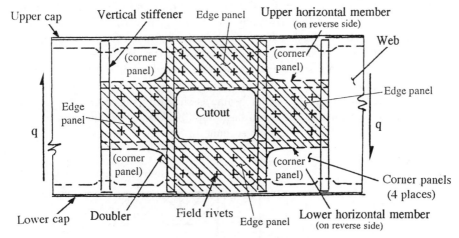

(Shading indicates area where doublers are required)

(a) Built-Up Rectangular Cutouts

(b) Machined Bulkhead with Rectangular Cutout

Fig. 13.6.1 Large Cutouts

Design recommendations for doubler installation are as follows:

- A doubler [see Fig. 13.6.1(a)] or additional machined thickness [see Fig. 13.6.1(b)] is required to make up the material in the cutout
- Keep the basic beam web thickness in corner panels (don't reduce material thickness)
- Use a one-piece doubler with the same sheet stock thickness at edge panels (around all four edges of the cutout) and install horizontal members on the reverse side of the beam to avoid interference with the existing vertical stiffeners [see Fig. 13.6.1(a) and Fig. 13.6.2]
- Add a doubler around the edge panels of the cutout to make the web non-buckling
- Additional shear strength can be obtained by adding stub member(s) to the edge panels to make the web a non-buckling
- Use at least two fasteners at the intersection between the vertical stiffeners and horizontal members and where the vertical stiffeners attach to both upper and lower beam caps
- Use of moderate radius sizes on doublers is required to avoid potential fatigue cracks
- In cases where a large doubler is required around the edges of a cutout, it is recommended that field rivets, as shown in Fig. 13.6.1(a), be installed between the existing web and the doubler to combine them as one effective web rather than two separate webs in order to improve shear flow resistance
- Be sure there are sufficient rivets along the edges of the cutout and where the doubler and web attach to the upper and lower caps to take the high shear loads
- An integrally-machined shear beam, as shown in Fig. 13.6.1(b), will provide a lightweight structure because it has a high buckling shear but extra margin of web thickness must be considered to account for future web repairs

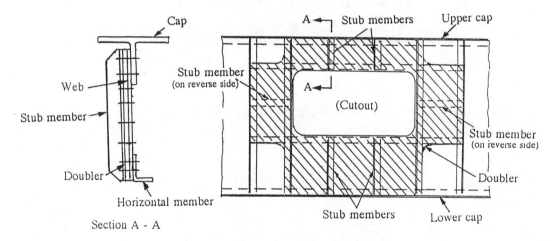

(Use one-piece doubler and install stub membersb)

Fig. 13.6.2 Doubler Installation

(A) FULLY FRAMED CUTOUTS

To eliminate redundancy in a framed cutout, as shown in Fig. 13.6.3, a good approximation method is to assume that the shear flow is the same:

- In the panels above and below the framed cutout
- In the panels to the left and right of the framed cutout
- In the corner panels

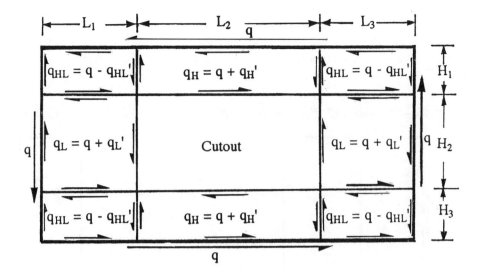

Fig. 13.6.3 Shear Flows Distribution at Full-Framing Cutout

The redistribution of shear flows due to cutout, q, is calculated from following equations:

(a) The shear flow in the webs above and below the cutout web:

$$q_H' = q\left(\frac{H_2}{H_1 + H_3}\right)$$ Eq. 13.6.1

(b) The shear flow in the webs to the left and right of the cutout web:

$$q_L' = q\left(\frac{L_2}{L_1 + L_3}\right)$$ Eq. 13.6.2

(c) The shear flow in the corners of the cutout web:

$$q_{HL}' = q_H'\left(\frac{L_2}{L_1 + L_3}\right)$$ Eq. 13.6.3

The final shear flows in the panels are as follows:

(a) The shear flow in the webs above and below the cutout web

$$q_H = q + q_H'$$ Eq. 13.6.4

(b) The shear flow in the webs to the left and right of the cutout web:

$$q_L = q + q_L'$$ Eq. 13.6.5

(c) The shear flow in the corners of the cutout web:

$$q_{HL} = q - q_{HL}'$$ Eq. 13.6.6

Chapter 13.0

Example 1:

For the single cantilever beam shown below, determine the shear flows and the variation in loads in the vertical stiffeners and horizontal members under a shear load, V = 12,000 lbs.

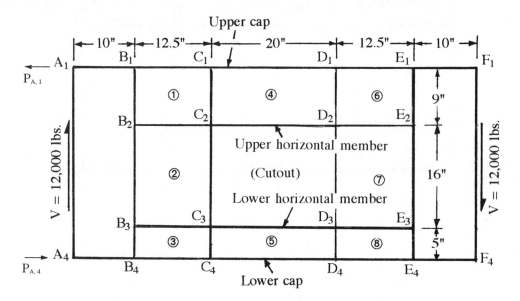

(a) The shear flow, q, in the web:

$$q = \frac{12,000}{9 + 16 + 5} = 400 \text{ lbs./in.}$$

(b) The redistribution of shear flows in the adjacent webs of the cutout:

$$q_H' = q\left(\frac{H_2}{H_1 + H_3}\right) = 400\left(\frac{16}{9 + 5}\right) = 457.1 \text{ lbs./in.}$$

$$q_L' = q\left(\frac{L_2}{L_1 + L_3}\right) = 400\left(\frac{20}{12.5 + 12.5}\right) = 320 \text{ lbs./in.}$$

$$q_{HL}' = q_H'\left(\frac{L_2}{L_1 + L_3}\right) = 457.1\left(\frac{20}{12.5 + 12.5}\right) = 365.7 \text{ lbs./in.}$$

(c) The final shear flow in the adjacent webs of the cutout:

The shear flow in panel ④ and ⑤ from Eq. 13.6.4:

$$q_④ = q_⑤ = q + q_H' = 400 + 457.1 = 857.1 \text{ lbs./in.}$$

The shear flow in panel ② and ⑦ from Eq. 13.6.5:

$$q_② = q_⑦ = q + q_L' = 400 + 320 = 720 \text{ lbs./in.}$$

The shear flow in panels ①, ③, ⑥, and ⑧ from Eq. 13.6.6:

$$q_① = q_③ = q_⑥ = q_⑧ = q - q_{HL}' = 400 - 365.7 = 34.3 \text{ lbs./in.}$$

(d) The final shear flows are sketched below:

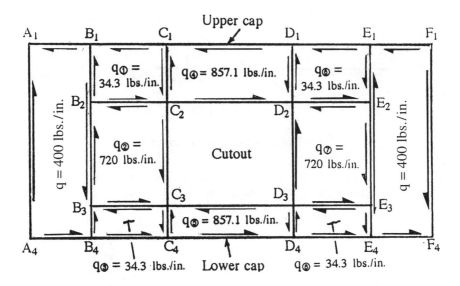

(e) The axial tension load distribution along the beam cap (only the upper cap is shown, the lower cap is opposite):

$P_{E,1} = q \times 10 = 400 \times 10 = 4,000$ lbs.

$P_{D,1} = P_{E,1} + q_{⑥} \times L_3 = 4,000 + 34.3 \times 12.5 = 4,429$ lbs.

$P_{C,1} = P_{D,1} + q_{④} \times L_2 = 4,429 + 857.1 \times 20 = 21,571$ lbs.

$P_{B,1} = P_{C,1} + q_{①} \times L_1 = 21,571 + 34.3 \times 12.5 = 22,000$ lbs.

$P_{A,1} = P_{B,1} + q \times 10 = 22,000 + 400 \times 10 = 26,000$ lbs.

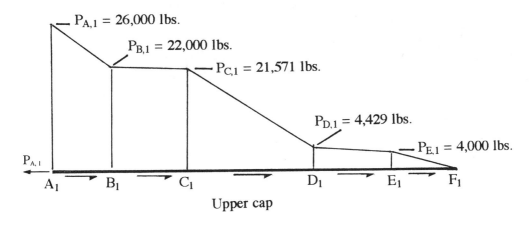

(f) The axial load distribution along the horizontal members (only the upper horizontal member is shown, the lower member is opposite):

$P_{C,2} = (q_{①} + q_{②}) \times L_1 = (34.3 - 720) \times 12.5 = -8,571$ lbs.

$P_{D,2} = P_{C,2} + q_{④} \times L_2 = -8,571 + 857.1 \times 20 = 8,571$ lbs.

567

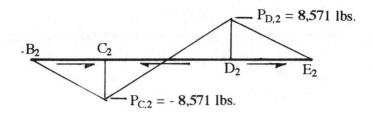

Upper horizontal member

(g) The axial load distribution along the vertical stiffeners (only those at locations of 'B' and 'C' are shown, those at location 'D' and 'E' are opposite):

Stiffener at location 'B':

$P_{B.2} = (q - q_①) \times H_1$
$= (400 - 34.3) \times 9$
$= 3,291$ lbs.

$P_{B.3} = P_{B.2} + (q - q_②) \times H_2$
$= 3,291 + (400 - 720) \times 16$
$= -1,829$ lbs.

Vertical stiffener at 'B'

Stiffener at location 'C':

$P_{C.2} = (q_① - q_④) \times H_1$
$= (34.3 - 857.1) \times 9$
$= -7,405.2$ lbs.

$P_{C.3} = P_{C.2} + q_② \times H_2$
$= -7,405.2 + 720 \times 16$
$= 4,114.8$ lbs.

Vertical stiffener at 'C'

Example 2:

Assume a section, as shown below, taken from the side of a fuselage with a window cutout. Treating the forward end as being fixed and assuming that the stringer loads at the forward end will be distributed in proportion to the $\frac{My}{I}$ beam theory.

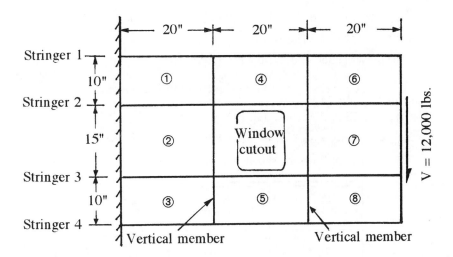

I - is the moment of inertia (assuming unit stringer areas):

$$2 \times (1.0 \times 17.5^2 + 1.0 \times 7.5^2) = 725 \text{ in.}^4$$

(a) Without the cutout the shear flow in bays ①, ④, ⑥, ③, ⑤, and ⑧ are:

The first moment of area, $Q_{1-2} = 1.0 \times 17.5 = 17.5$ in.³ and from Eq. 6.5.2 the shear flow equation (eliminate the 'b' value from the equation):

$$q_{1-2} = q_① = q_④ = q_⑥ = \frac{VQ_{1-2}}{I} = \frac{12,000 \times 17.5}{725} = 289.66 \text{ lbs./in.}$$

Similarly:

$$q_{3-4} = q_③ = q_⑤ = q_⑧ = 289.66 \text{ lbs./in.}$$

(b) In bays between stringers 2 and 3:

The first moment of area, $Q = 1.0 \times 17.5 + 1.0 \times 7.5 = 25$ in.³

$$q_{2-3} = \frac{VQ_{2-3}}{I} = \frac{12,000 \times 25}{725} = 413.79 \text{ lbs./in.}$$

(c) For the bay with the cutout the shear flow of $q_{2-3} = 413.79$ lbs./in. must be distributed to the adjacent bays:

For bays ④ and ⑤:

The redistribution of q_{2-3}, from Eq. 13.6.1:

$$q_H' = q_{2-3} \left(\frac{H_2}{H_1 + H_3} \right) = 413.79 \left(\frac{15}{10 + 10} \right) = 310.34 \text{ lbs./in.}$$

The final shear flow (from Eq. 13.6.4):

$$q_④ = q_⑤ = q_{1-2} + q_H' = 289.66 + 310.34 = 600 \text{ lbs./in.}$$

For bays ② and ⑦:

The redistribution of q_{2-3}, from Eq. 13.6.2:

$$q_L' = q_{2-3}\left(\frac{L_2}{L_1+L_3}\right) = 413.79\left(\frac{20}{20+20}\right) = 206.9 \text{ lbs./in.}$$

The final shear flow (from Eq. 13.6.5):

$$q_② = q_⑦ = q_{2-3} + q_L' = 413.79 + 206.9 = 620.69 \text{ lbs./in.}$$

For the corner bays ①, ③, ⑥, and ⑧:

The redistribution of q_{2-3}, from Eq. 13.6.3:

$$q_{HL}' = q_H'\left(\frac{L_2}{L_1+L_3}\right) = 310.34\left(\frac{20}{20+20}\right) = 155.17 \text{ lbs./in.}$$

The final shear flow (from Eq. 13.6.6):

$$q_① = q_③ = q_⑥ = q_⑧ = q_{1-2} - q_{HL}' = 289.66 - 155.17 = 134.49 \text{ lbs./in.}$$

The final shear flows are shown below:

$q_① =$ 134.49 lbs./in.	$q_④ =$ 600 lbs./in.	$q_⑤ =$ 134.49 lbs./in.
$q_② =$ 620.69 lbs./in.	Cutout	$q_⑦ =$ 620.69 lbs./in.
$q_③ =$ 134.49 lbs./in.	$q_⑥ =$ 600 lbs./in.	$q_⑧ =$ 134.49 lbs./in.

(d) These axial loads on stringers and vertical members can be calculated in the same manner as that shown in Example 1.

(B) HALF-FRAMED CUTOUTS

The design of the half-framed cutout, shown in Fig. 13.6.5, (in which the horizontal members do not extend into the bay on the right side of the cutout) is not recommended for use except for special cases under moderate or light shear loads.

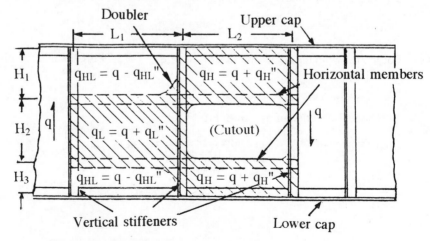

Fig. 13.6.5 Shear Flow Distribution for Half-framed Cutout

The redistribution of shear flow due to cutout, q, is calculated from following equations:

(a) The shear flow in the webs above and below the cutout web:

$$q_H'' = q\left(\frac{H_2}{H_1 + H_3}\right)$$ Eq. 13.6.7

(b) The shear flow in the webs to the left and right of the cutout web:

$$q_L'' = q\left(\frac{L_2}{L_1}\right)$$ Eq. 13.6.8

(c) The shear flow in the corners of cutout web:

$$q_{HL}'' = q_H''\left(\frac{L_2}{L_1}\right)$$ Eq. 13.6.9

The final shear flow in each panel is as follows:

(a) The shear flow in the webs above and below the cutout web:

$$q_H = q + q_H''$$ Eq. 13.6.10

(b) The shear flow in the webs to the left and right of the cutout web:

$$q_L = q + q_L''$$ Eq. 13.6.11

(c) The shear flow in the corners of cutout web:

$$q_{HL} = q - q_{HL}''$$ Eq. 13.6.12

(C) QUARTER-FRAMED CUTOUTS

The quarter-framed cutout design shown in Fig. 13.6.6 is not recommended for use in airframe applications under primary loading, such as wing and empennage spars, due to lack of fail-safe features in the lower cap over the cutout region. This design may be used only in non-vital or secondary structures.

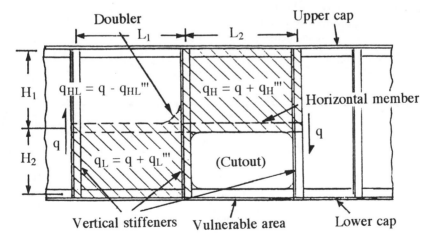

Fig. 13.6.6 Shear Flow Distribution for Quarter-framed Cutout

The redistribution of shear flow due to cutout, q, is calculated from following equations:

(a) The shear flow in the webs above and below the cutout web:

$$q_H''' = q\left(\frac{H_2}{H_1}\right)$$ Eq. 13.6.13

(b) The shear flow in the webs to the left and right of the cutout web:

$$q_L''' = q\left(\frac{L_2}{L_1}\right)$$

Eq. 13.6.14

(c) The shear flow in the corners of cutout web:

$$q_{HL}''' = q_H'''\left(\frac{L_2}{L_1}\right)$$

Eq. 13.6.15

The final shear flow in each panel is as follows:

(a) The shear flow in the webs above and below the cutout web

$$q_H = q + q_H'''$$

Eq. 13.6.16

(b) The shear flow in the webs to the left and right of the cutout web:

$$q_L = q + q_L'''$$

Eq. 13.6.17

(c) The shear flow in the corners of cutout web:

$$q_{HL} = q - q_{HL}'''$$

Eq. 13.6.18

(D) BUCKLED-WEB EFFECT

If any web in the region of the framed cutout has buckled into diagonal tension, the design or sizing procedures discussed in Chapter 12.4 and 12.5 should be followed. The edge members bordering a cutout (in the same manner as an end bay of a beam) must carry the bending due to the running loads per inch (w_h or w_v, see Fig. 13.6.7) from diagonal tension buckling loads. In a buckled web the running load produces bending in edge members (vertical stiffeners and horizontal members) and is given by the formulas shown in Fig. 10.6.4, Case 1.

$$w_v = k\,q\,\tan\alpha \text{ along horizontal members} \qquad \text{Eq. 13.6.19}$$

$$w_h = k\,q\,\cot\alpha \text{ along vertical members} \qquad \text{Eq. 13.6.20}$$

where: k – Diagonal tension factor
q – Shear flow of the web
α – Diagonal tension angle

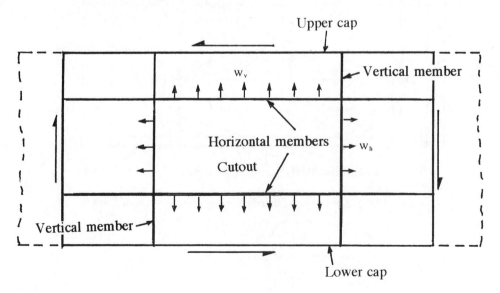

Fig. 13.6.7 Cutout with Buckled-webs

There are several ways of dealing with the edge member subjected to bending from the running loads of w_h or w_v:

- It is preferable to have a non-buckling web ($w_h = w_v = 0$)
- Simply beef up or strengthen the edge member so that it can take the running loads (this is inefficient for the edge panels with long span)
- Reduce the diagonal tension factor, k, and thereby reduce the running load
- Provide additional stud member(s) to support the edge members as well as the beam cap in large edge panels, as shown in Fig. 13.6.2, and thereby reduce the bending moment on edge members due to running loads (this requires additional parts)

The reinforcements which surround a cutout are designed in a similar manner to that of the end bay in diagonal tension web described in Chapter 12.6. However, in the case of cutouts, the addition of doublers or an increase in web thickness around the four edge panels of the cutout (not the corner webs) is necessary to make up for the material lost in the cutout.

13.7 CUTOUTS IN SKIN-STRINGER PANELS UNDER AXIAL LOAD

Cutouts in aircraft primary load-carrying (compression or tension loads) structures constitute one of the most troublesome problems in stress analysis. In order to reduce the labor of analyzing such panels, the three-stringer method (a simplified of cutout analysis described in Reference 13.6 and shown in Fig. 13.7.1) may be utilized. The number of stringers is reduced for purposes of combining a number of stringers into a single substitute stringer. If this cutout is designed for axial load, the local buckling has to be investigated carefully. For simplified solution of cutout analysis in plane shear loads, see Reference 13.7 and 13.8.

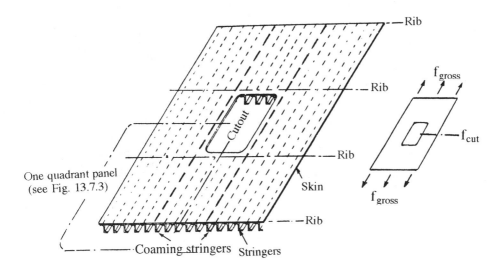

Fig. 13.7.1 Skin-stringer Panel with Cutout

In practice, the coaming stringer is designed to have a higher moment of inertia to increase its buckling strength and a smaller cross-sectional area to reduce and absorb local axial loads caused by the cutout. However, stringer next to the coaming stringer in the cutout area should have larger cross-sectional area to absorb a portion of the coaming stringer load. In addition, the local skin thickness around the cutout area should also be beefed up such that it can redistribute both axial and skin shear loads to the adjacent continuous stringers (A_1) in addition to the coaming stringer (A_2).

Chapter 13.0

The following design considerations apply (see Fig. 13.7.2 and Fig. 13.7.3):

- Increase moment of inertia (I) at coaming stringers with the minimum cross-sectional area (A_2) needed for axial loads
- Minimize eccentricity along the stringer length including the skin
- Beef up skin thickness around cutout
- Cut stringers at cutout (requires detail design to reduce the stress concentration)

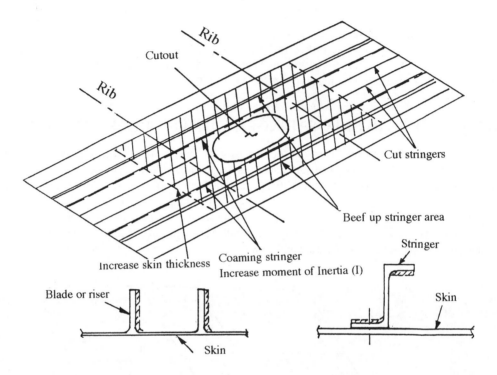

Fig. 13.7.2 Design Considerations Around Cutout

Ground rules for the three-stringer method (simplified analysis of cutout in area of axial load):

- If it is assumed that the panel is symmetrical about both axes, the analysis can then be confined to one quadrant
- The cross-sectional areas of the stringers and the skin do not vary in the longitudinal direction
- When the panel is very long, the stringer stresses are uniform at large longitudinal distances from the cutout
- The coaming stringer (designated A_2) remains an individual stringer in the substitute structure, as shown in Fig. 13.7.3

The significant stresses in sizing:

- Axial stresses (f_2) in coaming stringers (stringer No. 2)
- Shear flows or shear stresses ($f_{s,23}$) between stringers No. 2 and 3

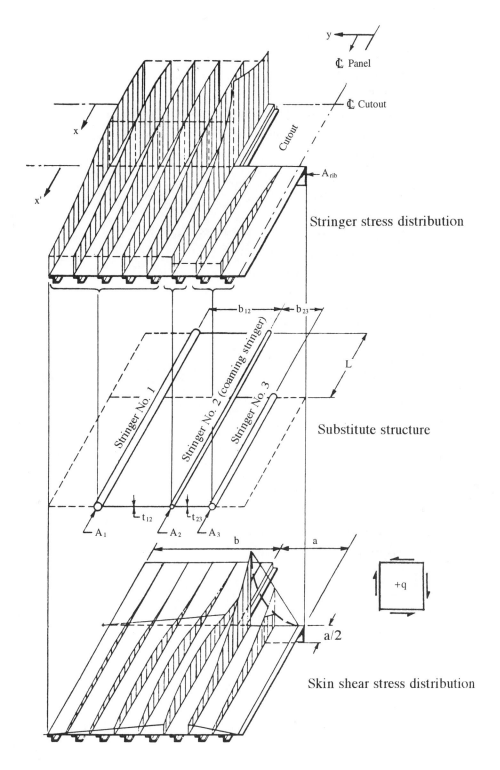

Fig. 13.7.3 Schematic Representation of a Substitute Structure for One Quadrant Panel

(A) SYMBOLS

(Refer to Fig. 13.7.3)

A_1 Effective cross-sectional area of all continuous stringers (stringer No. 1), exclusive of coaming stringer

A_2 Effective cross-sectional area of coaming stringers (stringer No. 2)

A_3 Effective cross-sectional area of all cut stringers (stringer No. 3)

A_{rib} Cross-sectional area of coaming rib at the edge of cutout

$$B = \sqrt{\frac{K_1^2 + K_2^2 + 2K}{K_1^2 + K_2^2 + 2K - \frac{K_3 K_4}{K_1^2}}}$$

C_o Stress-excess factor for cutout of zero length, see Fig. 13.7.4

D $\sqrt{K_1^2 + K_2^2 + 2K}$ (see Fig. 13.7.6)

E Young's Modulus of Elasticity

G Modulus of rigidity (shear modulus)

$$K_1^2 = \frac{G t_1}{E b_1}\left(\frac{1}{A_1} + \frac{1}{A_2}\right)$$

$$K_2^2 = \frac{G t_2}{E b_2}\left(\frac{1}{A_2} + \frac{1}{A_3}\right)$$

$$K_3 = \frac{G t_2}{E b_1 A_2}$$

$$K_4 = \frac{G t_1}{E b_2 A_2}$$

K $\sqrt{K_1^2 K_2^2 - K_3 K_4}$

L Half length (spanwise) of the cutout

R Stress reduction factor to take care of change in length of cutout, see Fig. 13.7.5

b_{12} Distance between centroid of A_1 and A_2

b_{23} Distance between centroid of A_2 and A_3

$r_1 = \dfrac{f_{s,23R} t_{23}}{A_3 f_{gross}}$ Rate-of-decay factor

$r_2 = \dfrac{f_{s,23R} t_{23} - f_{s,12R} t_{12}}{A_2 (f_{s,23R} - f_{gross})}$ Rate-of-decay factor

$r_3 = \dfrac{G f_{2R}}{E b_{23} f_{s,23R}}$ Rate-of-decay factor

t_{12} Thickness of continuous skin between stringers No. 1 and 2

t_{23} Thickness of discontinuous skin between stringers No. 2 and 3

x Spanwise (or longitudinal) distance, origin is at the middle of the cutout, see Fig. 13.7.3

x' Spanwise (or longitudinal) distance, origin is at the coaming rib station of the cutout, see Fig. 13.7.3

y Chordwise distance, origin is at the middle of the cutout

f_{gross} Average axial stress in the gross section

f_1 Axial stress in substitute continuous stringer (No. 1)

f_2 Axial stress in substitute coaming stringer (No. 2)

f_3 Axial stress in substitute cut stringer (No. 3)

f_{rib} Axial stress in coaming rib (chordwise)

f_{cut} Average axial stress in net section (in cut section)

$f_{s,12}$ Shear stress in substitute continuous skin in t_{12}

$f_{s,23}$ Shear stress in substitute discontinuous skin in t_{23}

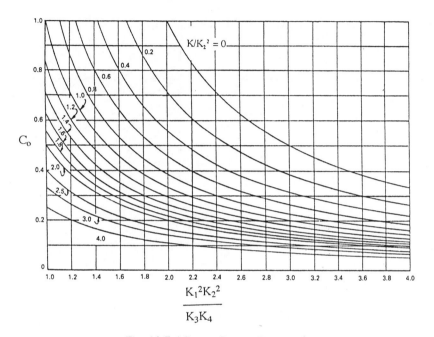

Fig. 13.7.4 Stress-Excess Factor, C_o

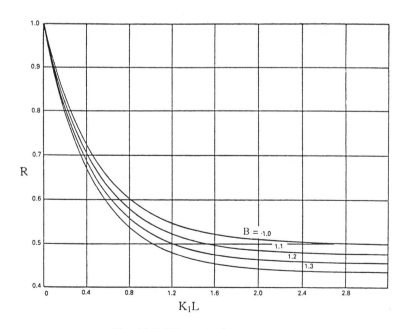

Fig. 13.7.5 Stress Reduction Factor, R

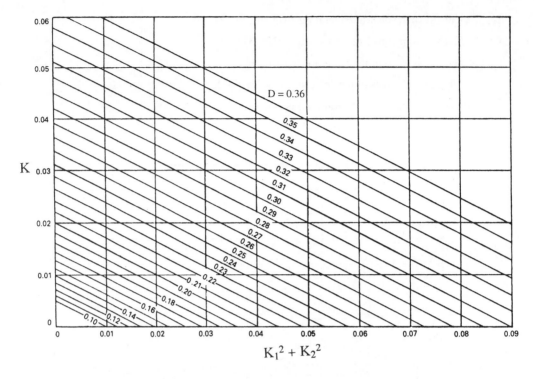

Fig. 13.7.6 Factor, D

(B) ANALYSIS PROCEDURES

(a) Compute the following parameters:

$$K_1^2; \quad K_1 = \sqrt{K_1^2}; \quad K_2^2; \quad K_3; \quad K_4; \quad K; \quad r_1; \quad r_2; \quad r_3$$

Find values of C_o from Fig. 13.7.4 and R from Fig. 13.7.5.

(b) Stress at the coaming rib station (x = L).

Stress in stringer No. 1:

$$f_{1R} = f_{cut}(1 - \frac{RC_o A_2}{A_1}) \quad \text{Eq. 13.7.1}$$

Stress in stringer No. 2:

$$f_{2R} = f_{cut}(1 + RC_o) \quad \text{Eq. 13.7.2}$$

Shear flow in the panel between stringers No. 1 and 2:

$$q_{12R} = f_{s,12R} t_{12} = f_{cut} RC_o A_2 K_1 \tanh K_1 L \quad \text{Eq. 13.7.3}$$

Shear flow in the panel between stringers No. 2 and 3:

$$q_{23R} = f_{s,23R} t_{23} = f_{cut} A_2 (\frac{K_4}{D})(1 + RC_o + \frac{K_1^2}{K}) \quad \text{Eq. 13.7.4}$$

where: D from Fig. 13.7.6

- f_{1R} and f_{2R} are the max. values of f_1 and f_2, respectively
- f_1 reaches its max. at the centerline of the cutout
- $f_{s,12}$ reaches its max. in the gross section at the section where $f_1 = f_2 \neq f_{gross}$ as shown in Fig. 13.7.3

(c) Stress at the net section (cutout region) of the substitute structure (the x origin is at the middle of the cutout, see Fig. 13.7.3).

The stresses in stringer No. 1:

$$f_1 = f_{cut}(1 - \frac{RC_o A_2 \cosh K_1 x}{A_1 \cosh K_1 L})$$
Eq. 13.7.5

The stresses in stringer No. 2:

$$f_2 = f_{cut}(1 + RC_o \frac{\cosh K_1 x}{\cosh K_1 L})$$
Eq. 13.7.6

The shear flow in panel t_{12}:

$$q_{12} = f_{s,12} t_{12} = f_{cut} RC_o A_1 K_1 (\frac{\sinh K_1 x}{\cosh K_1 L})$$
Eq. 13.7.7

(The shear flow decreases rapidly to zero at the centerline of the cutout, when $x = 0$)

(d) Stress in the gross section (x' origin is at the coaming rib station of the cutout), see Fig. 13.7.3.

The stress in cut stringer (No. 3):

$$f_3 = f_{gross}(1 - e^{-r_1 x'})$$
Eq. 13.7.8

The stress in the coaming stringer (No. 2):

$$f_2 = f_{gross} + (f_{2R} - f_{gross}) e^{-r_2 x'}$$
Eq. 13.7.9

The stress in the continuous stringer (No. 1):

$$f_1 = f_{gross} + \frac{A_2}{A_1}(f_{gross} - f_2) + \frac{A_3}{A_1}(f_{gross} - f_3)$$
Eq. 13.7.10

The shear flow in the skin:

Between stringers No. 1 and 2:

$$q_{12} = f_{s,12} t_{12} = f_{s,23R} t_{23} e^{-r_1 x'} - (f_{s,23R} t_{23} - f_{s,12R} t_{12}) e^{-r_2 x'}$$
Eq. 13.7.11

Between stringers No. 2 and 3:

$$q_{23} = f_{s,23R} t_{23} e^{-r_3 x'}$$
Eq. 13.7.12

(e) Stress in the actual structure:

(1) The force acting on a substitute stringer is distributed over the corresponding actual stringers chordwise (y – direction) according to the hyperbolic cosine law.
(2) The shear stresses ($f_{s,12}$) in the substitute structure equal the shear stresses, in the first continuous skin panel adjacent to the coaming stringer (No. 2).
(3) The chordwise distribution of stringer stresses is uniform.
(4) The chordwise distribution of shear stresses tapers linearly from τ_1 to zero at the edge of the panel.
(5) The chordwise distribution of shear stresses in the cut skin panel varies linearly from ($f_{s,23}$), adjacent to the coaming stringer, to zero at the centerline of the panel.
(6) Chordwise distribution of shear stresses at the coaming rib station varies as a cubic parabola.

Chapter 13.0

Example:

For the rectangular cutout shown below, under a gross tension stress, $f_{gross} = 40$ ksi, determine the axial stresses in the stringers and shear stresses in the skin.

Materials: $F_{tu} = 80$ ksi $F_{su} = 48$ ksi $E = 10.5 \times 10^6$ ksi $G = 3.9 \times 10^6$ ksi

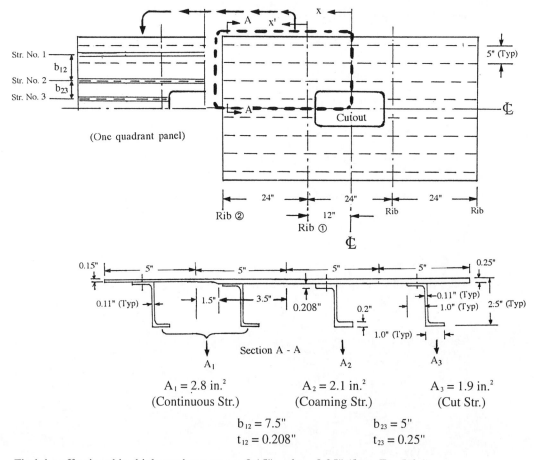

$A_1 = 2.8$ in.^{2}
(Continuous Str.)

$A_2 = 2.1$ in.^{2}
(Coaming Str.)

$A_3 = 1.9$ in.^{2}
(Cut Str.)

$b_{12} = 7.5"$
$t_{12} = 0.208"$

$b_{23} = 5"$
$t_{23} = 0.25"$

Find the effective skin thickness between $t = 0.15"$ and $t = 0.25"$ (from Eq. 5.6.4):

$$t_{12} = \frac{5}{\frac{1.5}{0.15} + \frac{3.5}{0.25}} = 0.208"$$

(a) Compute the following parameters:

The given gross tension stress: $f_{gross} = 40$ ksi

The net stress at cutout:

$$f_{cut} = f_{gross}(\frac{A_1 + A_2 + A_3}{A_1 + A_2}) = 40(\frac{2.8 + 2.1 + 1.9}{2.8 + 2.1}) = 55.51 \text{ ksi}$$

$$K_1^2 = \frac{Gt_{12}}{Eb_{12}}(\frac{1}{A_1} + \frac{1}{A_2})$$

$$= \frac{3.9 \times 10^6 \times 0.208}{10.5 \times 10^6 \times 7.5}(\frac{1}{2.8} + \frac{1}{2.1}) = 0.0086$$

$$K_1 = \sqrt{K_1^2} = \sqrt{0.086} = 0.0927$$

$$K_2^2 = \frac{Gt_{23}}{Eb_{23}}(\frac{1}{A_2} + \frac{1}{A_3})$$

$$= \frac{3.9 \times 10^6 \times 0.25}{10.5 \times 10^6 \times 5}(\frac{1}{2.1} + \frac{1}{1.9}) = 0.0186$$

$$K_3 = \frac{Gt_{23}}{Eb_{12}A_2} = \frac{3.9 \times 10^6 \times 0.25}{10.5 \times 10^6 \times 7.5 \times 2.1} = 0.0059$$

$$K_4 = \frac{Gt_{12}}{Eb_{23}A_2} = \frac{3.9 \times 10^6 \times 0.208}{10.5 \times 10^6 \times 5 \times 2.1} = 0.00736$$

$$K = \sqrt{K_1^2 K_2^2 - K_3 K_4} = \sqrt{0.0086 \times 0.0186 - 0.0059 \times 0.00736} = 0.0108$$

$$K_1^2 + K_2^2 = 0.0086 + 0.0186 = 0.0272$$

$$\frac{K_1^2 K_2^2}{K_3 K_4} = \frac{0.0086 \times 0.0186}{0.0059 \times 0.00736} = 3.684$$

$$\frac{K}{K_2^2} = \frac{0.0108}{0.0186} = 0.581$$

From Fig. 13.7.4 obtain $C_o = 0.21$

$$K_1 L = 0.0927 \times 12 = 1.112$$

$$B = \sqrt{\frac{K_1^2 + K_2^2 + 2K}{K_1^2 + K_2^2 + 2K - \frac{K_3 K_4}{K_1^2}}}$$

$$= \sqrt{\frac{0.0086 + 0.0186 + 2 \times 0.0108}{0.0086 + 0.0186 + 2 \times 0.0108 - 0.0059 \times \frac{0.00736}{0.0086}}} = 1.056$$

From Fig. 13.7.5 obtain $R = 0.55$

From Fig. 13.7.6 obtain $D = 0.22$

(b) Stress at rib station ① ($x = L = 12''$).

Axial stress at stringer No. 1 (from Eq. 13.7.1):

$$f_{1R} = f_{cut}(1 - \frac{RC_o A_2}{A_1})$$

$$= 55.51(1 - \frac{0.55 \times 0.21 \times 2.1}{2.8}) = 51.7 \text{ ksi}$$

Axial stress at stringer No. 2 (from Eq. 13.7.2):

$$f_{2R} = f_{cut}(1 + RC_o)$$

$$= 55.51(1 + 0.55 \times 0.21) = 61.9 \text{ ksi}$$

Shear flow in the panel between stringers No. 1 and 2 (from Eq. 13.7.3):

$$q_{12R} = f_{cut} RC_o A_2 K_1 \tanh K_1 L$$

$$= 55.51 \times 0.55 \times 0.21 \times 2.1 \times 0.0927 \tanh 1.112 = 1.005 \text{ kips/in.}$$

$$f_{s, 12R} = \frac{q_{12R}}{t_{12}} = \frac{1.005}{0.208} = 4.83 \text{ ksi}$$

Shear flow in the panel between stringers No. 2 and 3 (from Eq. 13.7.4):

$$q_{23R} = f_{cut} A_2 (\frac{K_4}{D})(1 + RC_o + \frac{K_1^2}{K})$$

$$= 55.51 \times 2.1(\frac{0.00736}{0.22})(1 + 0.55 \times 0.21 + \frac{0.0086}{0.0108}) = 7.42 \text{ kips/in.}$$

Chapter 13.0

$$f_{s,23R} = \frac{q_{23R}}{t_{23}} = \frac{7.42}{0.25} = 29.68 \text{ ksi}$$

(c) Stress at the net section (cutout region) of the substitute structure (the x origin is at the middle of the cutout, see Fig. 13.7.3).

The stresses in stringer No. 1 (from Eq. 13.7.5):

$$f_1 = f_{cut}(1 - \frac{RC_o A_2 \cosh K_1 x}{A_1 \cosh K_1 L})$$

$$= 55.51(1 - \frac{0.55 \times 0.21 \times 2.1 \cosh K_1 x}{2.8 \times \cosh 1.112})$$

$$= 55.51(1 - 0.0514 \cosh 0.0927x)$$

At the middle of the cutout (x = 0):

$f_1 = 52.66$ ksi

At the coaming rib station ① (x = 12):

$f_1 = 50.7$ ksi (same as f_{1R})

The stresses in stringer No. 2 (from Eq. 13.7.6):

$$f_2 = f_{cut}(1 + RC_o \frac{\cosh K_1 x}{\cosh K_1 L})$$

$$= 55.51(1 + 0.55 \times 0.21 \frac{\cosh K_1 x}{\cosh 1.112})$$

$$= 55.51(1 + 0.0685 \cosh 0.0927x)$$

At the middle of the cutout (x = 0):

$f_2 = 59.31$ ksi

At the coaming rib station ① (x = 12):

$f_2 = 61.9$ ksi (same as f_{2R})

The shear flow at the panel t_{12} (from Eq. 13.7.7):

$$q_{12} = f_{cut} RC_o A_2 K_1 (\frac{\sinh K_1 x}{\cosh K_1 L})$$

$$= 55.51 \times 0.55 \times 0.21 \times 2.1 \times 0.0927 (\frac{\sinh 0.0927x}{\cosh 1.112})$$

$$= 0.741 \sinh 0.0927x$$

At the middle of the cutout (x = 0), the shear flow:

$$q_{12} = 0 \qquad f_{s,12} = \frac{q_{12}}{t_{12}} = \frac{0}{0.208} = 0$$

At the coaming rib station ① (x = 12), the shear flow:

$$q_{12} = 1.005 \text{ kips/in.} \qquad f_{s,12} = \frac{q_{12}}{t_{12}} = \frac{1.005}{0.208} = 4.83 \text{ ksi (same as } f_{s,12R})$$

(d) Stress in the gross section (x' origin is at the coaming rib station of the cutout):

Compute the following rate-of-decay factors:

$$r_1 = \frac{f_{s,23R} t_{23}}{A_3 f_{gross}} = \frac{7.42}{1.9 \times 40} = 0.0976$$

$$r_2 = \frac{f_{s,23R} t_{23} - f_{s,12R} t_{12}}{A_2 (f_{s,23R} - f_{gross})} = \frac{7.42 - 1.005}{2.1(61.9 - 40)} = 0.139$$

$$r_3 = \frac{Gf_{2R}}{Eb_{23}f_{s,23R}} = \frac{3.9 \times 10^6 \times 61.9}{10.5 \times 10^6 \times 5 \times 29.68} = 0.155$$

The stress in the cut stringer (No. 3) (from Eq. 13.7.8):

$$f_3 = f_{gross}(1 - e^{-r_1 x'}) = 40(1 - e^{-0.0976x'})$$

The stress in the coaming stringer (from Eq. 13.7.9):

$$f_2 = f_{gross} + (f_{2R} - f_{gross})e^{-r_2 x'} = 40 + (61.9 - 40)e^{-0.139x'} = 40 + 21.9e^{-0.139x'}$$

At the coaming rib station ① ($x' = 0$), the stringer stresses:

$$f_3 = 0 \qquad f_2 = 61.9 \text{ ksi}$$

The stress in the continuous stringer (No. 1) (from Eq. 13.7.10):

$$f_1 = f_{gross} + \frac{A_2}{A_1}(f_{gross} - f_2) + \frac{A_3}{A_1}(f_{gross} - f_3)$$

$$= 40 + \frac{2.1}{2.8}(40 - 61.9) + \frac{1.9}{2.8}(40 - 0)$$

$$= 50.72 \text{ ksi (the value is close to } f_{1R} = 50.7 \text{ ksi)} \qquad\qquad \text{O.K.}$$

At the rib station ② $x' = 24$, the stringer stresses:

$$f_3 = 36.16 \qquad f_2 = 40.78 \text{ ksi}$$

$$f_1 = f_{gross} + \frac{A_2}{A_1}(f_{gross} - f_2) + \frac{A_3}{A_1}(f_{gross} - f_3)$$

$$= 40 + \frac{2.1}{2.8}(40 - 40.78) + \frac{1.9}{2.8}(40 - 36.16) = 42 \text{ ksi}$$

The shear flow in the skin:

Shear flow between stringers No. 1 and 2 (from Eq. 13.7.11):

$$q_{12} = f_{s,23R} t_{23} e^{-r_1 x'} - (f_{s,23R} t_{23} - f_{s,12R} t_{12}) e^{-r_2 x'}$$
$$= 7.42 e^{-0.0976 x'} - (7.42 - 1.005) e^{-0.139 x'}$$

Shear flow between stringers No. 2 and 3 (from Eq. 13.7.12):

$$q_{23} = 7.42 e^{-0.155 x'}$$

At the coaming rib station $x' = 0$, the shear flows are:

$$q_{12} = 7.42 e^{-0.0976 x'} - (7.42 - 1.005) e^{-0.139 x'}$$
$$= 7.42 - (7.42 - 1.005) = 1.005 \text{ lbs./in. (same as } q_{12R})$$

$$q_{23} = 7.42 e^{-0.155 x'} = 7.42 \text{ lbs./in. (same as } q_{23R})$$

At the rib station ② $x' = 24$, the shear flows are:

$$q_{12} = 7.42 e^{-0.0976 \times 24} - (7.42 - 1.005) e^{-0.139 \times 24}$$
$$= 7.42 \times 0.0961 - (7.42 - 1.005) \times 0.0356 = 0.485 \text{ lbs./in.}$$

$$f_{s,12} = \frac{q_{12}}{t_{12}} = \frac{0.485}{0.208} = 2.33 \text{ ksi}$$

$$q_{23} = 7.42 e^{-0.155 x'} = 7.42 e^{-0.155 \times 24} = 0.18 \text{ lbs./in.}$$

$$f_{s,23} = \frac{q_{23}}{t_{23}} = \frac{0.18}{0.25} = 0.72 \text{ ksi}$$

(d) Stress in the actual structure.

The involved stresses in the sizing of the given structure are the axial stress distribution in coaming stringer and the shear stress distribution between stringers No. 2 and 3. In this case, substitute stringers No. 2 and 3 are individual stringers in the actual structure. These stresses are plotted below:

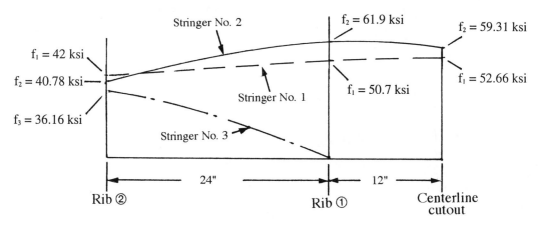

Stringer stress (ksi)

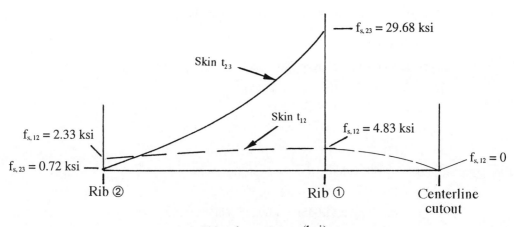

Skin shear stress (ksi)

(e) Discussions:

The maximum tension stress of $f_2 = 61.9$ ksi for the coaming stringer is too high, it should be approximately 45 to 55 ksi to meet fatigue life requirements (see Fig. 15.1.2). If the cutout is at the wing upper surface and the gross stress will be in compression, $f_{gross} = -40$ ksi, this will produce a compression stress, $f_2 = -59.31$ ksi at the coaming stringer at the centerline of the cutout. This stress is important to consider when sizing the coaming stringer for compression buckling.

13.8 LARGE CUTOUTS IN CURVED SKIN-STRINGER PANEL (FUSELAGE)

The following simplified method (from Reference 13.7) is applicable for developing the preliminary sizing of a plug-type door, as shown in Fig. 13.8.1, with an aspect ratio of approximately $\frac{L}{h} \approx 0.6$. The design approach for a very large cargo door cutout on the upper deck of a transport fuselage is discussed in Reference 13.7.

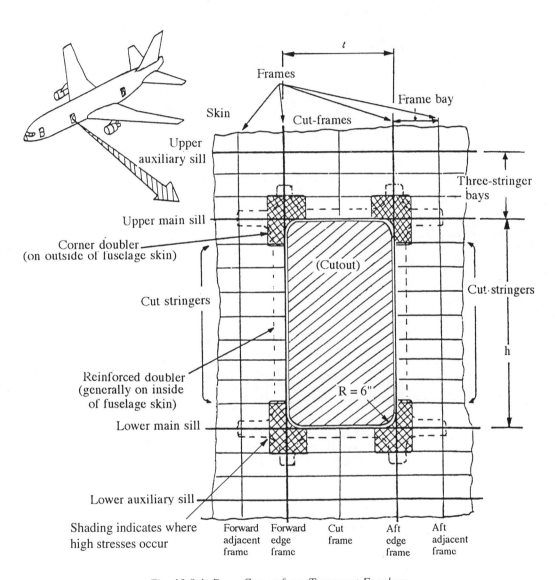

Fig. 13.8.1 Door Cutout for a Transport Fuselage

The following design guidelines apply:

(a) Structure definitions:
- Fuselage of nearly constant radius
- One frame bay forward and aft of the cutout edge frame members
- Two or three stringer bays above and below the cutout

585

(b) Internal load redistribution:
   (i) To clearly define the load cases involving redistribution of internal loads, each case should be considered independently and then combined to formulate the most critical design condition on each of the structural elements.
   (ii) Design load cases:
   - Case I Fuselage skin shears – flight conditions
   - Case II Cut stringer loads – flight conditions
   - Case III Longitudinal and circumferential tension loads – cabin pressurization condition
   - Case IV Plug-type door with doorstops – cabin pressurization condition

(c) Skin panels:
   - Skin panel thickness should be based on the average panel shear flow
   - The panel is assumed to be made up of skin or skin and doublers without inclusion of strap locally over frames and sills
   - The panel thickness should be such that the gross area shear stress should be lower than 25 ksi, when aluminum skins are used (fatigue consideration)

(d) Frame or sill outer chord (cap) members (see Fig. 13.8.2):

   The outer chord consists of the frame or sill cap, strap and effective skin. For preliminary sizing, consider the following:
   - Frame and sill axial loads act on the outer chord only
   - Bending loads ($\frac{M}{h'}$) may combine with the axial loads to produce the same type of loading (tension or compression) in the outer chord

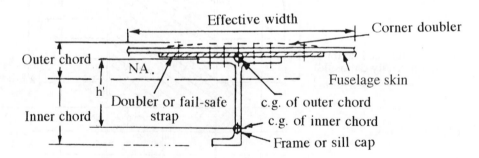

Fig. 13.8.2 Frame or Sill Chord Member

(e) Frame or sill inner chord (cap) members (see Fig. 13.8.2):
   - The inner chord is assumed to carry bending load only
   - If the inner chord ultimate tension stress is very high, the chord must be a fail-safe design
   - The $\frac{b}{t}$ ratio for the outstanding flange should be maintained at less than 12 to prevent compression crippling buckling

(f) Straps:
- Straps are included over the main sill and edge frame members around the cutout to provide required outer chord for redistribution of ultimate loads in addition to providing a fail-safe load path
- The determination of strap width and thickness should be based on the critical combination of axial and bending loads in the outer chord
- The area of the strap, skin, and doubler, in the immediate vicinity of the main sill and edge frame junctions shall keep a gross area shear stress low to meet fatigue life

(g) Skin attachments:

Attachments shall be good for 1.25 times q_{max} (max. ultimate shear flow in the skin).

(A) LOAD DISTRIBUTION DUE TO FUSELAGE SKIN SHEAR

(a) The redistribution of constant shear flows (q_o) which occurs in the vicinity of a cutout is shown in Fig. 13.8.3 (refer to Eq. 13.6.1 to 13.6.6). Such redistribution would occur at the entrance door on the upper deck of a narrow or wide body transport fuselage.

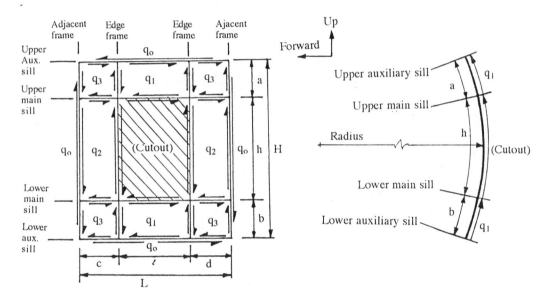

Fig. 13.8.3 Shear Flow Redistribution due to Constant Shear Flow (q_o)

$$q_1 = (1 + \frac{h}{a + b})q_o \quad \text{Eq. 13.8.1}$$

$$q_2 = (1 + \frac{\ell}{c + d})q_o \quad \text{Eq. 13.8.2}$$

$$q_3 = [1 - (\frac{\ell}{c + d})(\frac{h}{a + b}) - 1] q_o \quad \text{Eq. 13.8.3}$$

(b) The redistribution of variable shear flows (q_a, q_b, q_c, and q_d) which occurs in the vicinity of a cutout is shown in Fig. 13.8.4. This variable shear flow redistribution is different from that of previous constant shear flow (q_o), and would occur, for example, at the cargo door below the floor beam of a narrow or wide body transport fuselage.

Chapter 13.0

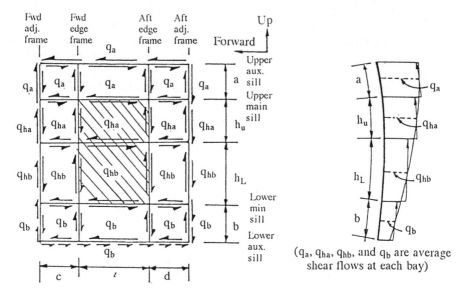

(a) Variable Shear Flows

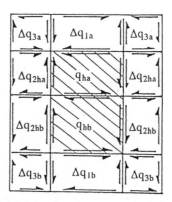

(b) Δq Shear Flow Distribution

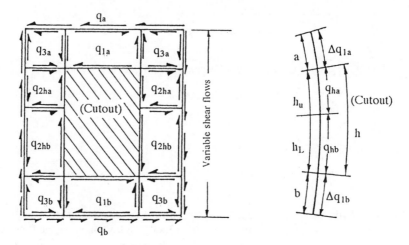

(c) Final Shear Flow Distribution

(The direction of the shear flows shown would be reversed for a cutout above the N.A. of the fuselage)

Fig. 13.8.4 Shear Flow Redistribution due to Variable Shear Flows

$$q_{1a} = q_a + \Delta q_{1a} \qquad \text{Eq. 13.8.4}$$

$$q_{1b} = q_b + \Delta q_{1b} \qquad \text{Eq. 13.8.5}$$

$$q_{2ha} = q_{ha} + \Delta q_{2ha} \qquad \text{Eq. 13.8.6}$$

$$q_{2hb} = q_{hb} + \Delta q_{2hb} \qquad \text{Eq. 13.8.7}$$

$$q_{3a} = \Delta q_{3a} - q_a \qquad \text{Eq. 13.8.8}$$

$$q_{3b} = \Delta q_{3b} - q_b \qquad \text{Eq. 13.8.9}$$

where: $\Delta q_{1a} = \dfrac{q_b h_b}{a}$; $\quad \Delta q_{2ha} = \dfrac{q_{ha} \ell}{c+d}$; $\quad \Delta q_{3a} = \dfrac{q_{2ha} h_b}{a}$

$\Delta q_{1b} = \dfrac{q_{hb} h_c}{b}$; $\quad \Delta q_{2hb} = \dfrac{q_{hb} \ell}{c+d}$; $\quad \Delta q_{3b} = \dfrac{q_{2hb} h_c}{b}$

$h_b = \dfrac{a\,h}{a+b}$; $\quad h_c = h - h_b$

(c) For shear flow distribution and sill reactions on the forward edge frame, as shown in Fig. 13.8.5, (the aft edge frame can be determined by the same method):

  (i) Consider a 3:1 distribution of frame shear flows when two stringer bays exist between the main and auxiliary sills.

  (ii) Consider a 5:3:1 distribution of frame shear flows when three stringer bays exist between the main and auxiliary sills. The 5:3:1 implies that the applied frame shear (average) in the upper, center, and lower portions of the frame are proportioned $\dfrac{5q}{3}$, q, and $\dfrac{q}{3}$, as shown in Fig. 13.8.5.

  (iii) Assume $R_2 = 2R_1$ at upper sills and $R_3 = 2R_4$ at lower sills.

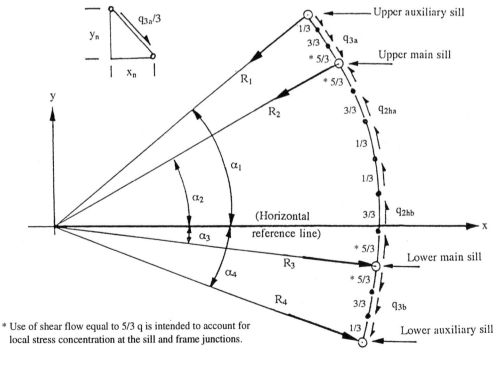

* Use of shear flow equal to $5/3\,q$ is intended to account for local stress concentration at the sill and frame junctions.

(The values of q_{3a}, q_{2ha}, q_{2hb}, and q_{3b} are obtained from Fig. 13.8.4)

Fig. 13.8.5 Load Distribution at Forward Edge Frame for a Cutout of Three-Stringer Bays

(ix) Summation of vertical forces = 0:

$$R_1 \sin \alpha_1 + 2R_1 \sin \alpha_2 + 2R_4 \sin \alpha_3 + R_4 \sin \alpha_4 + \frac{q_{3a} y_1}{3} + q_{3a} y_2 + \frac{5 q_{3a} y_3}{3} - \frac{5 q_{2ha} y_4}{3}$$

$$- q_{2ha} y_5 - \frac{q_{2ha} y_6}{3} - \frac{q_{2hb} y_7}{3} - q_{2hb} y_8 - \frac{5 q_{2hb} y_9}{3} + \frac{5 q_{3b} y_{10}}{3} + q_{3b} y_{11} + \frac{q_{3b} y_{12}}{3} = 0$$

Eq. 13.8.10

(x) Summation of horizontal forces = 0:

$$R_1 \cos \alpha_1 + 2R_1 \cos \alpha_2 - 2R_4 \cos \alpha_3 - R_4 \cos \alpha_4 - \frac{q_{3a} x_1}{3} - q_{3a} x_2 - \frac{5 q_{3a} x_3}{3} + \frac{5 q_{2ha} x_4}{3}$$

$$+ q_{2ha} x_5 + \frac{q_{2ha} x_6}{3} + \frac{q_{2hb} x_7}{3} + q_{2hb} x_8 - \frac{5 q_{2hb} x_9}{3} + \frac{5 q_{3b} x_{10}}{3} + q_{3b} x_{11} + \frac{q_{3b} x_{12}}{3} = 0$$

Eq. 13.8.11

(xi) With the sill reactions at the forward edge frame known, as determined from the above equations, the bending moments at any point on the frame can be determined.

(xii) The aft edge frame reactions can be obtained by the same method as for the forward edge frame.

(xiii) With sill reactions known on both side of the edge frames, sill balancing loads in adjacent frames are obtained as shown in Fig. 13.8.6(a). Bending moment (BM) in the adjacent frames are obtained by direct proportion of radial loads as follows:

$$BM_{(Fwd\ adjacent\ frame)} = \left[\frac{R_{(Fwd\ adjacent\ frame)}}{R_{(Fwd\ edge\ frame)}}\right] \times BM_{(Fwd\ edge\ frame)}$$

Eq. 13.8.12

For practical structural considerations, it is assumed that the sill member extends two bays on each side of the cutout, as shown in Fig. 13.8.6(b):

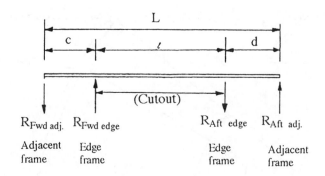

(a) One bay on each side

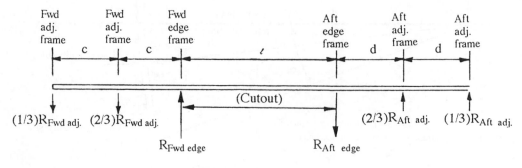

(b) Two bays on each side

Fig. 13.8.6 Sill Extending over Each Side of the Cutout

(B) LOAD DISTRIBUTION DUE TO FUSELAGE BENDING

The cut-stringer loads are assumed to be generally diffused into the main sills over one frame bay on either side of the cutout, as shown in Fig. 13.8.7. This approach results in somewhat higher panel shear loads and frame axial loads than actually occur in practice. However, this conservative approach is necessary to meet the requirements of fatigue design. If the cut-stringer loads are not the same on either side of the cutout, use average stringer loads at centerline of cutout for this method.

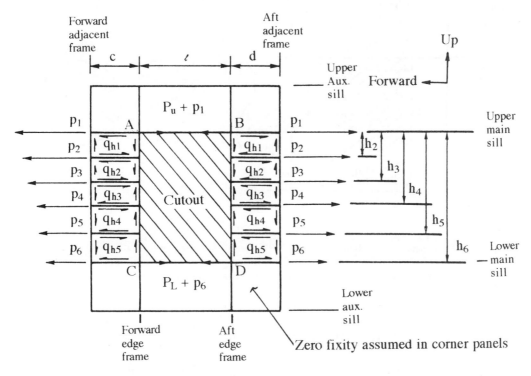

Fig. 13.8.7 Additional Shear Flows from Cut-Stringer Loads

Obtain P_L and P_u by moment and force balance:

$$P_L = \frac{p_2 \times h_2 + p_3 \times h_3 + p_4 \times h_4 + p_5 \times h_5}{h_6}$$ Eq. 13.8.13

$$P_u = (p_2 + p_3 + p_4 + p_5) - P_L$$ Eq. 13.8.14

Skin shear flows at forward panel between forward edge frame and forward adjacent frame:

$$q_{h1} = \frac{P_u}{c}$$ Eq. 13.8.15

$$q_{h2} = q_{h1} - \frac{p_2}{c}$$ Eq. 13.8.16

$$q_{h3} = q_{h2} - \frac{p_3}{c}$$ Eq. 13.8.17

•
•
•

Etc.

(Note: Shear flows for aft panel are obtained in the same manner but care should be taken in establishing shear flow direction)

(C) LOAD DISTRIBUTION DUE TO FUSELAGE CABIN PRESSURIZATION

(a) Panels above and below the cutout:

Hoop tension loads above and below the cutout are redistributed into the edge frames via shear flows in the panels between the main and auxiliary sills, as shown in Fig. 13.8.8.

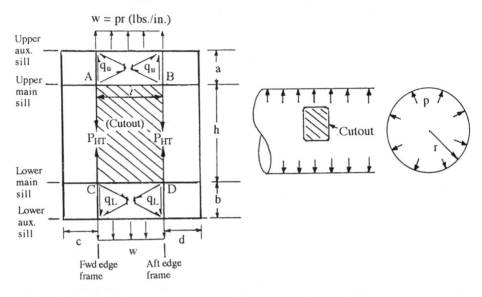

Fig. 13.8.8 Hoop Tension Loads in the Edge Frames due to Cabin Pressurization

w = pr (hoop tension running load – lbs./in.)

q_u = 0 at the upper panel center varies to $q_u = \dfrac{w\ell}{2a} = \dfrac{pr\ell}{2a}$ at the edge frame

q_L = 0 at the lower panel center varies to $q_L = \dfrac{w\ell}{2b} = \dfrac{pr\ell}{2b}$ at the edge frame

$P_{HT} = \dfrac{w\ell}{2} = \dfrac{pr\ell}{2}$ hoop tension load in edge frames

(b) Panels at the side (left and right) of the cutout:

Longitudinal tension load (P_{LT}) in the upper and lower main sill, as shown in Fig. 13.8.9.

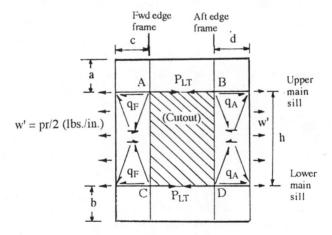

Fig. 13.8.9 Tension Loads in the Upper and Lower Main Sill due to Cabin Pressurization

Cutouts

$w' = \dfrac{pr}{2}$ (longitudinal tension running load – lbs./in.)

$q_F = 0$ at the left side panel center varies to $q_F = \dfrac{w'h}{2c} = \dfrac{prh}{4c}$ at the main sill

$q_A = 0$ at the right side panel center varies to $q_A = \dfrac{w'h}{2d} = \dfrac{prh}{2d}$ at the main sill

$P_{LT} = \dfrac{w'h}{2} = \dfrac{prh}{4}$ longitudinal tension load in main sills

(c) Corner panels of the cutout:

Due to the shear flows in the corner panels, as shown in Fig. 13.8.10, the incremental axial loads in the frame and sill members are based on corner panel fixity. Distribution factors (see Eq. 13.8.18, Eq. 13.8.19 and Fig. 13.8.11) are used which are based on the constant moment of inertia of the panels, i.e., panels above, below, and to the left and right of the cutout (usually the values of a, b, c, and d are approximately equal).

$K_1 = \dfrac{h}{h + \ell}$ (at panels above and below the cutout)   Eq. 13.8.18

$K_2 = \dfrac{\ell}{h + \ell}$ (at panels to the left and right of the cutout)   Eq. 13.8.19

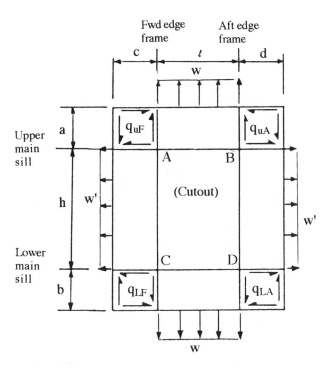

q_{uF} – Shear flow at upper and forward corner panel
q_{LF} – Shear flow at lower and forward corner panel
q_{uA} – Shear flow at upper and aft corner panel
q_{LA} – Shear flow at lower and aft corner panel

Fig. 13.8.10 Corner Shear Flows due to Cabin Pressurization

Distribution factors	K_1	K_2
FEM (+ moment)	$\dfrac{w\ell^2}{12}$	$-\dfrac{w'h^2}{12}$
Correction values	$-K_1(M_{uL} + M_{FA})$	$-K_2(M_{uL} + M_{FA})$
Adjusted FEM	$M'_{uL} = \dfrac{w\ell^2}{12} - K_1(M_{uL} + M_{FA})$	$M'_{FA} = -\dfrac{w'h^2}{12} - K_2(M_{uL} + M_{FA})$

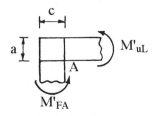

Fig. 13.8.11 Adjusted Fixed-End Moments (FEM)

Fixed-end moment at upper or lower panels:

$$M_{uL} = \frac{w\ell^2}{12}$$ Eq. 13.8.20

Fixed-end moment at forward or aft side panels:

$$M_{FA} = \frac{w'h^2}{12}$$ Eq. 13.8.21

Shear flow in upper forward corner panel:

$$q_{uF} = \frac{50\% \text{ (Adjusted FEM)}}{a \times c}$$

$$= \frac{50\%[\frac{w\ell^2}{12} - K_1(M_{uL} + M_{FA})]}{a \times c}$$ Eq. 13.8.22

Shear flows in the other corner (q_{LF}, q_{uA}, and q_{LA}) are generated in the same manner based on their respective geometry. The axial loads in the sills and frames are obtained by summation of q's.

(d) Balancing loads on upper and lower panel of the cutout:

Fig. 13.8.12 shows that the hoop tension forces (pr) produce radial component load (force direction of R_{u1} and R_{u2} toward the center of the circular fuselage) which creates the bending moment of the sills and reactions on the edge frame from the sill reaction. Fig. 13.8.12 shows a unit force (1.0 psi) of hoop tension load and Fig. 13.8.13 shows the integration of this unit force over half of the door width ($\dfrac{L}{2}$).

Cutouts

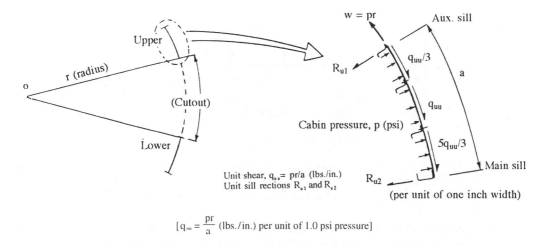

$$\left[q_{uu} = \frac{pr}{a} \text{ (lbs./in.) per unit of 1.0 psi pressure}\right]$$

Fig. 13.8.12 Balancing Loads on Upper Panels of the Cutout (Unit Cabin pressure = 1.0 psi)

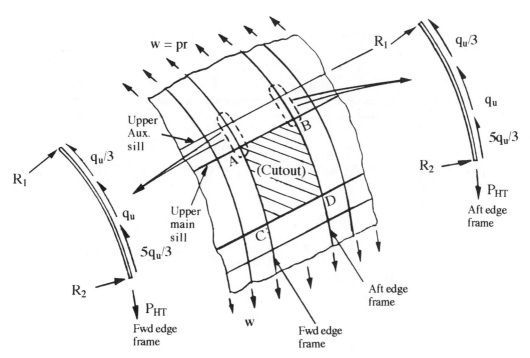

(The shear flow q_u (lbs./in.) is calculated in the same manner as that shown in Fig. 13.8.8 and the shear flow, q_L, on edge frames between the lower main sill and the lower auxiliary sill are obtained in the same way)

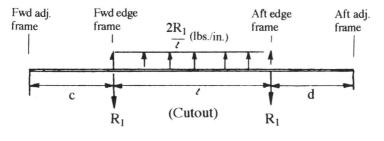

(a) Sill loads – upper auxiliary sill

595

Chapter 13.0

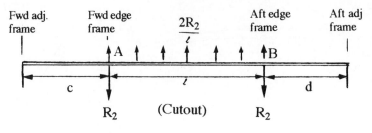

(b) Sill loads – upper main sill

(Load on the lower sills is obtained in the same way)

Fig. 13.8.13 Radial Loads on Upper Sills of the Cutout

(e) Corner panel fixity induces frame and sill loads in a ΔP_{HT} load in the frame (tension in edge frames and compression in adjacent frames) plus local frame and sill bending due to frame curvature, as shown in Fig. 13.8.14.

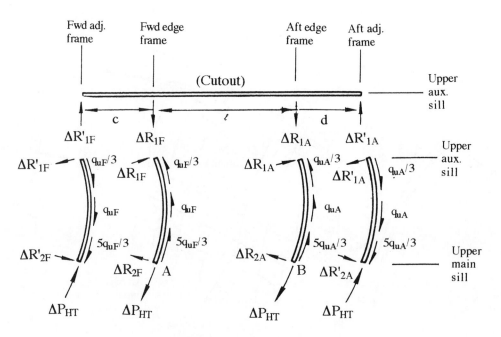

ΔR_{1A}, $\Delta R'_{1A}$, ΔR_{1F} and $\Delta R'_{1F}$ are radial loads

Note: The loading effects on the frames and sills, when combined with the loads developed in Fig. 13.8.13, constitute the total loading due to pressurization and fixity effects.

Fig. 13.8.14 Induced Frame and Sill Radial Loads Resulting from Corner Shear Flows (q_{uF} and q_{uA})

(f) Doorstop (plug-type door) redistribution effects:

The cabin pressure load on the door is redistributed to the forward and aft edge frames by means of doorstop fittings, as shown in Fig. 13.8.15.

Cutouts

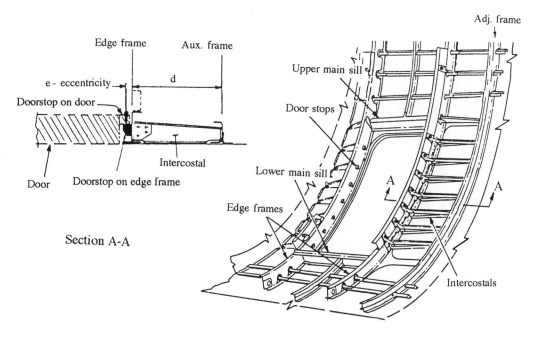

Fig. 13.8.15 Doorstops on either Side of Edge Frame (Fuselage Lower Cargo Entrance Cutout)

Since the doorstops are somewhat eccentric from the edge frame, a Δ moment is produced at the edge frame. The following design considerations apply:

- These Δ moments are reacted by providing intercostals between the edge frames and the reversed loads on the adjacent frames
- Assume the point loads introduced at the doorstops are uniformly distributed on the frames, thus introducing no localized bending on the frames
- In general, frame bending moment from doorstop load effects will be small compared to bending moment from flight load conditions

Example:

For the upper deck of a transport fuselage made from aluminum alloys, assume a 49" x 45" cutout with a radius r = 74" under the three load conditions given below, and determine the internal load redistribution in the structures surrounding the cutout.

Given load conditions at the centerline of the cutout:

(1) Constant shear flow, q_o = 450 lbs./in.

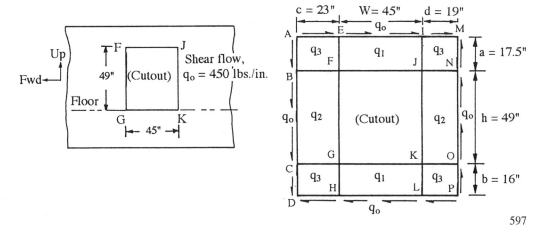

597

(2) Stringer axial loads, due to the given fuselage bending moment shown below:

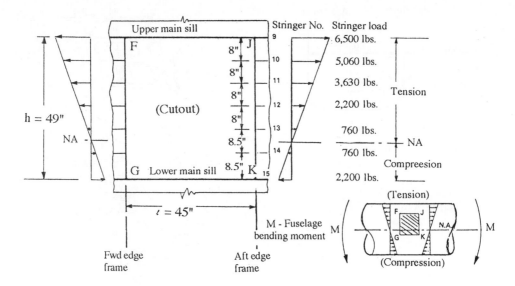

(3) Cabin pressurization, p = 8.6 psi

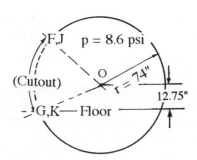

Design approaches:

    (1) Assume two stringer bays between the main and auxiliary sills.

    (2) Assume one frame bay on either side of the cutout (between edge frame and adjacent frame).

(A) Under shear flow $q_o = 450$ lbs./in. (assume a constant shear flow at the cutout):
(from Eq. 13.8.1 to 13.8.3)

$$q_{o,1} = (1 + \frac{h}{a+b})q_o = (1 + \frac{49}{17.5 + 16}) \times 450 = 1,108.2 \text{ lbs./in.}$$

$$q_{o,2} = (1 + \frac{\ell}{c+d})q_o = (1 + \frac{45}{23 + 19}) \times 450 = 932.1 \text{ lbs./in.}$$

$$q_{o,3} = [(\frac{\ell}{c+d})(\frac{h}{a+b}) - 1]q_o = [(\frac{45}{23 + 19})(\frac{49}{17.5 + 16}) - 1] \times 450 = 255.2 \text{ lbs./in.}$$

The shear flows are sketched below:

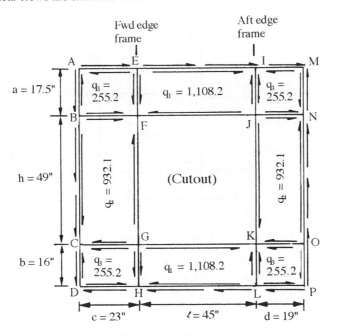

(Above shear flows are in lbs./in.)

Note: Frame and sill axial loads are determined by summation of shear flows (by the same method as that shown in the example in Section 13.6) and loads are carried in the frame or sill outer chord as defined in Fig. 13.8.2

(B) Under the given stringer loads (from fuselage bending moment):

Take the moment center about point K to obtain the upper main sill reaction load, S_9:

$$49 \times S_9 = -760 \times 8.5 + 760 \times 17 + 2{,}200 \times 25 + 3{,}630 \times 33 + 5{,}060 \times 41 + 6{,}500 \times 49$$

$S_9 = 14{,}433$ lbs. (tension)

Take the moment center about point J to obtain the **lower main sill reaction load, S_{15}**:

$S_{15} = 757$ lbs. (compression)

Shear flow below stringer No. 9 due to diffusion of axial loads:

$$q_{9,10} = \frac{14{,}433 - 6{,}500}{19} = 418 \text{ lbs./in. (at aft edge frame)}$$

$$q_{9,10} = \frac{14{,}433 - 6{,}500}{23} = 345 \text{ lbs./in. (at forward edge frame)}$$

Shear flow above stringer No. 15:

$$q_{14,15} = \frac{757 - (-2{,}200)}{19} = 156 \text{ lbs./in. (at aft edge frame)}$$

$$q_{14,15} = \frac{757 - (-2{,}200)}{23} = 129 \text{ lbs./in. (at forward edge frame)}$$

The shear flows are sketched below:

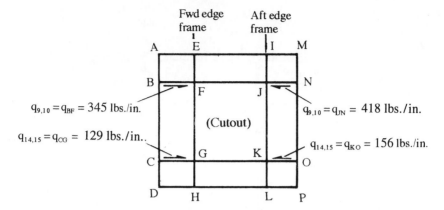

Forward edge frame shear flows (between upper main and auxiliary sills):

$0.5a \times q_{EF} + 0.5a \times 3q_{EF} = a \times (q_1 + q_3)$

$q_{EF} = \dfrac{q_1 + q_3}{2} = \dfrac{1,108.2 + 255.2}{2} = 681.7$ lbs./in.

$3q_{EF} = 3 \times 681.7 = 2,045.1$ lbs./in.

Similarly for forward edge frame shear flows (between lower main and auxiliary sills):

$\dfrac{h}{4}(3q_{FG} + q_{FG} + q_{FG} + 3q_{FG}) = q_2 \times h$

$q_{FG} = \dfrac{932.1}{2} = 466.05$ lbs./in.

$3q_{FG} = 3 \times 466.05 = 1,398.15$ lbs./in.

The shear flows for the forward edge frame are sketched below:

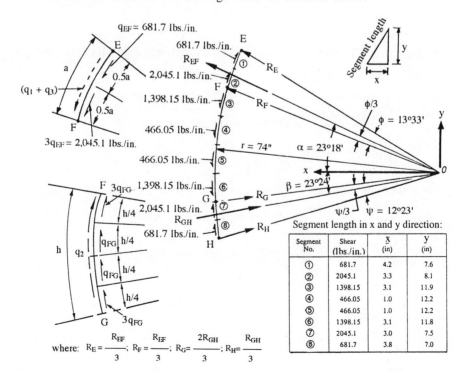

Segment length in x and y direction:

Segment No.	Shear (lbs./in.)	x (in)	y (in)
①	681.7	4.2	7.6
②	2045.1	3.3	8.1
③	1398.15	3.1	11.9
④	466.05	1.0	12.2
⑤	466.05	1.0	12.2
⑥	1398.15	3.1	11.8
⑦	2045.1	3.0	7.5
⑧	681.7	3.8	7.0

where: $R_E = \dfrac{R_{EF}}{3}$; $R_F = \dfrac{R_{EF}}{3}$; $R_G = \dfrac{2R_{GH}}{3}$; $R_H = \dfrac{R_{GH}}{3}$

The sill reactions R_{EF} and R_{GH} are located at one-third the distance between the main and auxiliary sills and are measured from the main sill.

Vertical equilibrium of the frame (from Eq. 13.8.10):

$$R_{EF} \sin \alpha + R_{GH} \sin \beta + 681.7 y_1 + 2,045.1 y_2 - 1,398.15 y_3 - 466.05 y_4 - 466.05 y_5 - 1,398.15 y_6 + 2,045.1 y_7 + 681.7 y_8 = 0$$

Rearrange the above equation:

$$R_{EF} \sin \alpha + R_{GH} \sin \beta + 681.7(y_1 + y_8) + 2,045.1(y_2 + y_7) - 1,398.15(y_3 + y_6) - 466.05(y_4 + y_5) = 0$$

or $R_{EF} \sin 23°18' + R_{GH} \sin 23°24' + 681.7(7.6 + 7.0) + 2,045.1(8.1 + 7.5) - 1,398.15(11.9 + 11.8) - 466.05(12.2 + 12.2) = 0$

$$0.396 R_{EF} + 0.397 R_{GH} = 2,651.4 \qquad \text{Eq. A}$$

Horizontal equilibrium of the frame (from Eq. 13.8.11):

$$-R_{EF} \cos \alpha + R_{GH} \cos \beta + 681.7 x_1 + 2,045.1 x_2 - 1,398.15 x_3 - 466.05 x_4 - 466.05 x_5 + 1,398.15 x_6 - 2,045.1 x_7 - 681.7 x_8 = 0$$

Rearrange the above equation:

$$-R_{EF} \cos \alpha + R_{GH} \cos \beta + 681.7(x_1 - x_8) + 2,045.1(x_2 - x_7) - 1,398.15(x_3 - x_6) - 466.05(x_4 + x_5) = 0$$

or $-R_{EF} \cos 23°18' + R_{GH} \cos 23°24' + 681.7(4.2 - 3.8) + 2,045.1(3.3 - 3.0) - 1,398.15(3.1 - 3.1) - 466.05(1.0 + 1.0) = 0$

$$-0.918 R_{EF} + 0.918 R_{GH} = 45.9 \qquad \text{Eq. B}$$

Solution of Eq. A and Eq. B yields

$$R_{EF} = 3,322.73 \text{ lbs. and } R_{GH} = 3,372.73 \text{ lbs.}$$

and therefore:

$$R_E = \frac{R_{EF}}{3} = \frac{3,322.73}{3} = 1,107.58 \text{ lbs./in.}$$

$$R_F = \frac{2 R_{EF}}{3} = 2,215.15 \text{ lbs./in.}$$

$$R_G = \frac{2 R_{GH}}{3} = \frac{2 \times 3,372.73}{3} = 2,248.49 \text{ lbs./in.}$$

$$R_H = \frac{R_{GH}}{3} = 1,124.24 \text{ lbs./in.}$$

In a similar manner, establish the loads on the remaining frames (forward adjacent, aft edge and aft adjacent) and then draw shear flows and bending moment diagrams based on the reactions for the main and auxiliary sills. The bending moment and end loads in the frames can be calculated in the same way. These load diagrams are based on flight conditions only. In addition, the effects arising from cabin pressurization must also be considered.

(C) Cabin pressurization:

Use running load on doorstop and intercostals based on a unit hoop tension pressure of 1.0 psi (the actual doorstop spacing is about the same as that of cut stringer spacing):

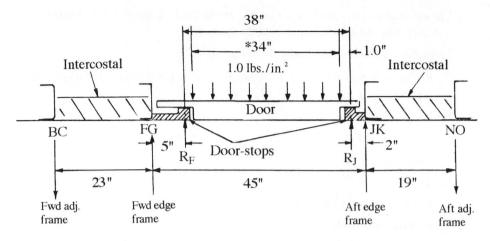

* This is the pressure width of the door

$$R_F = \frac{35 \times 18.5}{38} = 17.05 \text{ lbs./in.} \qquad R_A = \frac{35 \times 19.5}{38} = 17.95 \text{ lbs./in.}$$

Reaction loads on forward edge and adjacent frames:

$$R_{BC} = \frac{17.05 \times 5}{23} = 3.71 \text{ lbs./in.} \qquad R_{FG} = \frac{17.05 \times (23 + 5)}{23} = 20.76 \text{ lbs./in.}$$

Reaction loads on forward edge and adjacent frames:

$$R_{JK} = \frac{17.95 \times 21}{19} = 19.85 \text{ lbs./in.} \qquad R_{NO} = \frac{17.95 \times 2}{19} = 1.9 \text{ lbs./in.}$$

Forward edge frame at cabin pressure of 2.0 × 8.6 = 17.2 psi (where 2.0 is the pressure factor and 8.6 psi is the cabin pressure differential):

$w = R_{FG} \times 17.2 = 357$ lbs./in.

$P_{HT} = 357 \times 74 = 26,418$ lbs.

$$A_1 = \frac{74^2}{2}[(6°46')\frac{\pi}{180} - \sin 6°46'] = 0.739 \text{ in.}^2$$

$$A_2 = \frac{74^2}{2}[(13°33')\frac{\pi}{180} - \sin 13°33'] - 0.752 = 5.285 \text{ in.}^2$$

$$q_u(\frac{a}{2}) + (\frac{q_u}{3})(\frac{a}{2}) = P_{HT}$$

$$q_u(\frac{17.5}{2}) + (\frac{q_u}{3})(\frac{17.5}{2}) = 26,418 \text{ lbs.}$$

$\therefore q_u = 2,264.4$ lbs./in. and $\frac{q_u}{3} = 754.8$ lbs./in.

Geometry of forward edge frame shown below:

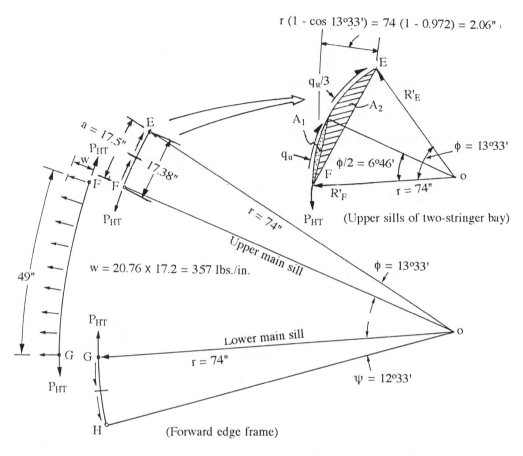

(Upper sills of two-stringer bay)

(Forward edge frame)

Taking moments about F:

$$2 \times A_1 \times q_u + 2 \times A_2\left(\frac{q_u}{3}\right) - 17.38\left(\cos\frac{\phi}{2}\right)R_E' = 0; \text{ where } \frac{\phi}{2} = 6°46'$$

$$2 \times 0.739 \times 2{,}264.4 + 2 \times 5.285 \times 754.8 - 17.26\, R_E' = 0$$

$$R_E' = 656.14 \text{ lbs.}$$

Taking moments about E:

$$2 \times A_1 \times \left(\frac{q_u}{3}\right) + 2 \times A_2 \times q_u - 2.06 \times P_{HT} + 17.38\,(\cos 6°46')\,R_F' = 0$$

$$2 \times 0.739 \times 754.8 + 2 \times 5.285 \times 2{,}264.4 - 26{,}418 \times 2.06 + 17.26\, R_F' = 0$$

$$R_F' = 1{,}701.67 \text{ lbs.}$$

Similarly the shear flows and reaction loads may be established. The loads at the other frames, i.e., forward adjacent, aft edge and aft adjacent frames may be computed by the same method.

The bending moments of sill panels and frame panels due to cabin pressure of $2.0 \times 8.6 = 17.2$ psi can be generated by:

$$w = pr = 17.2 \times 74 = 1{,}272.8 \text{ lbs./in.}$$

$$w' = \frac{pr}{2} = \frac{17.2 \times 74}{2} = 636.4 \text{ lbs./in.}$$

Distribution factor (based on constant moment of inertia of sill panels and frame panels around cutout):

From Eq. 13.8.18:

$$K_1 = \frac{h}{h+\ell} = \frac{49}{49+45} = 0.521$$

Forward and aft side panels (from Eq. 13.8.19):

$$K_2 = \frac{\ell}{h+\ell} = \frac{45}{49+45} = 0.479$$

Determine fixed-end moment (FEM):

For long side (edge frame panel B-F-G-C) (from Eq. 13.8.21):

$$M_{FA} = \frac{w'h^2}{12} = \frac{636.4 \times 49^2}{12} = 127,333 \text{ in.-lbs.}$$

For short side (sill panel E-I-J-F) (from Eq. 13.8.20):

$$M_{UL} = \frac{w\ell^2}{12} = \frac{1,272.8 \times 45^2}{12} = 214,785 \text{ in.-lbs.}$$

Consider a corner panel A-E-F-B (by the moment distribution method described in Chapter 5.5):

Distribution factors	$K_1 = 0.521$	$K_2 = 0.479$
FEM	214,785	−127,333
Correction values	−45,562	−41,890
Adjusted FEM	169,223	−169,223

FEM at corner panel is sketched below:

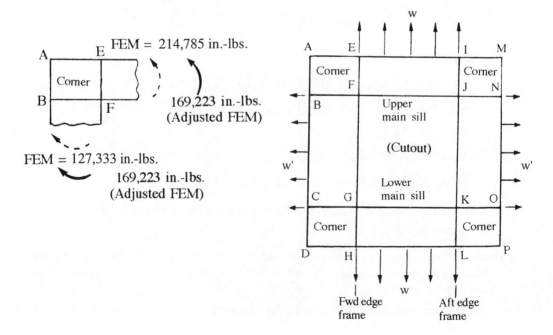

Assume that the deformation of the corner panel will permit only 50% fixity. Thus, for the forward top corner (A-B-F-E) the shear flow is given in Eq. 13.8.22 by:

$$\frac{169{,}223 \times 50\%}{17.5 \times 23} = 210.2 \text{ lbs./in.}$$

and the shear flows for all other corners may be found in a similar manner.

Consider the shear flows in the structure bounded by forward and aft adjacent frames and main and auxiliary sills (A-B-F-J-N-M-I-E-A):

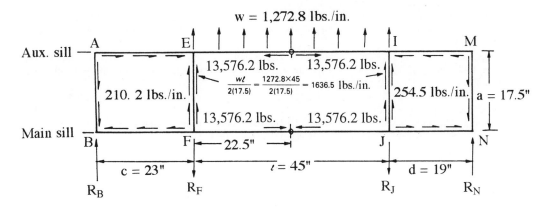

Reactions at edge frames:

$R_F = (1{,}636.5 + 210.2)(17.5) = 32{,}317$ lbs.

$R_J = (1{,}636.5 + 254.5)(17.5) = 33{,}093$ lbs.

Reactions at adjacent frames:

$R_B = (210.2)(17.5) = 3{,}678.5$ lbs.

$R_N = (254.5)(17.5) = 4{,}453.8$ lbs.

Axial loads in main and auxiliary sills at center of cutout:

$$\frac{32{,}317(\frac{45}{2}) - 3{,}678.5(23 + \frac{45}{2}) - 1{,}272.8(\frac{22.5^2}{2})}{17.5} = 13{,}576.2 \text{ lbs.}$$

[**Note:** The similar calculation of reaction load (R_F) is required for the running load of w' = 636.4 lbs./in. due to longitudinal pressure will be sheared to the main sills (upper and lower) to give additional loads]

A similar procedure can be adopted for the lower panels, i.e., between the main and auxiliary sills and forward and aft frames (C-D-H-L-P-O-K-G-C).

Hence, panel shear flows due to a pressure factor of 2.0 could be drawn. If a pressure factor of 3.0 is the critical case, these shears may be obtained by simple ratio (3:2) for shear panels. However, if a pressure factor of 2.0 plus flight load conditions is the design case, all loading may be obtained by simple superposition of the flight and pressure cases.

(d) It is now possible to draw shear flows, bending moment, and axial load diagrams for all sills and frames and finally the cutout structure can be sized.

References

13.1 Kuhn, Paul, "STRESSES IN AIRCRAFT AND SHELL STRUCTURES", McGraw-Hill Book Company, Inc., New York, NY, 1956.

13.2 Kuhn, Paul, "The Strength and Stiffness of Shear Webs with and Without Lightening Holes", NACA WR L-402, (June, 1942).

13.3 Kuhn, Paul, "The Strength and Stiffness of Shear Webs with Round Lightening Holes Having 45° Flanges", NACA WR L-323, (Dec., 1942).

13.4 Anevi, G., "Experimental Investigation of Shear Strength and Shear Deformation of Unstiffened Beams of 24ST Alclad with and without Flanged Lightening Holes", Sweden, SAAB TN-29, SAAB Aircraft Company, Sweden, (Oct., 1954).

13.5 Rosecrans, R., "A Method for Calculating Stresses in Torsion-Box Covers with Cutouts", NACA TN 2290,

13.6 Kuhn, Paul, Duberg, J. E. and Diskin, J. H., "Stresses Around Rectangular Cut-outs in Skin-stringer Panels Under Axial Load - II", NACA WR L368 (ARR 3J02), (Oct., 1943).

13.7 Niu, C. Y., "AIRFRAME STRUCTURAL DESIGN", Hong Kong Conmilit Press Ltd., P.O. Box 23250, Wanchai Post Office, Hong Kong, (1988).

13.8 Moggio, E, M., and Brilmyer, H. G., "A Method for Estimation of Maximum Stresses Around a Small Rectangular Cutout in a Sheet-Stringer Panel in Shear", NACA L4D27, (April, 1944).

Chapter 14.0

COMPRESSION PANELS

14.1 INTRODUCTION

Failure of a panel under compression axial load involves a vast array of problems ranging from properties of material to initial instability and post-buckling phenomena. Since the 1950's, NACA has published a series entitled "Handbook of Structural Stability" (see Reference 14.8 – 14.14) as well as enormous reports regarding these issues. This chapter will not cover all aspects of this subject but will present an analysis procedure which is as simple as possible for use in preliminary sizing.

Fig. 14.1.1 gives a structural efficiency (optimum values) comparison for stringer configurations but, in practical design, this is not the only important aspect in final selection of a stringer configuration for a compression panel. The final selection should based on:

- Structural efficiency
- Contour compatibility
- Ease of manufacturing, installation and final assembly
- Ease of repair and maintenance
- Load transfer between skin and stringer (skin-stringer panels)
- Damage tolerance
- Ease to splice
- Compatibility of attachment and/or shear ties (clips) to fuselage frame or wing rib

In view of the current tendency to employ skin-stringer panels and integrally-stiffened panels for major airframe components, it is desirable to have a rapid method of achieving an optimum design. Such a method must be based on the compression axial load because it is the primary design load (The interaction between axial and shear loads is beyond the scope of this discussion and a detailed analysis of this interaction is so complicated it would require computer analysis). When panel sizing is dependent on the domination of axial compression loads, e.g., inboard span of wing and empennage boxes. it is sufficient to use the primary axial loads to size a panel (ignore the small magnitude of shear load in skin and normal pressure load on the panel surface) for preliminary design purposes.

By the same manner, when the domination of shear loads is in the panel, e.g., outboard span of wing and empennage boxes, refer to Chapter 12.0.

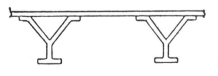

$\varepsilon_{max.} = 1.23$

(a) Y-section

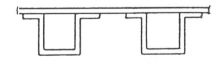

$\varepsilon_{max.} = 0.928$

(b) Hat Section

Chapter 14.0

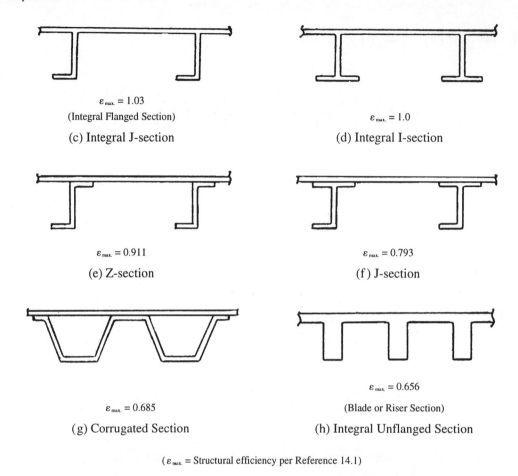

($\varepsilon_{max.}$ = Structural efficiency per Reference 14.1)

Fig. 14.1.1 Structural Efficiency (Common Extruded Stringer Configurations)

Fig. 14.1.1 shows some typical skin-stringer panel designs which are used on existing airframes to take axial compression loads, especially for machined stringers which can carry very high axial compression load, e.g., wing and empennage box structures.

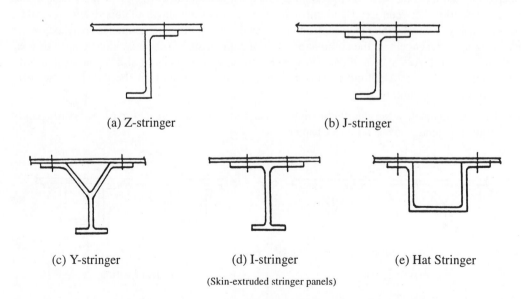

(Skin-extruded stringer panels)

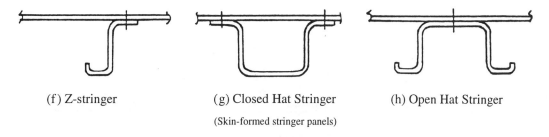

(f) Z-stringer   (g) Closed Hat Stringer   (h) Open Hat Stringer

(Skin-formed stringer panels)

Fig. 14.1.2 Typical Skin-Stringer Panels

The considerations for adopting skin-stringer panels are:

- The Z-stringer [Fig. 14.1.2(a)] and J-stringer [Fig. 14.1.2(b)] are the most popular configurations utilized in recent transport structural design, especially the Z-stringer due to its high structural efficiency
- The J-stringer is not as efficient as Z-stringer but it has good fail-safe characteristics due to the double row of fasteners attaching the stringer to the skin
- The J-stringer is also used to splice spanwise wing skins or fuselage longitudinal skin joints
- The Y-stringer [Fig. 14.1.2(c)] and I-stringer [Fig. 14.1.2(d)] are used on some transports but there is some difficulty in attaching these stringers to rib structures in a wing box
- The Y-stringer has the highest structural efficiency but the problem of corrosion in areas where inspection is very difficult has limited its use
- The hat-stringer [Fig. 14.1.2(e)] is generally not acceptable because it has the same corrosion inspection problem as the Y-stringer but it could be used on upper wing surfaces as a fuel tank vent passage as well as a load-carrying member
- The Z-stringer [Fig. 14.1.2(f)] and open hat stringer [Fig. 14.1.2(g)] are frequently seen on existing aircraft fuselage construction
- The closed hat stringer [Fig. 14.1.2(g)] has high structural efficiency but is not acceptable due to the corrosion inspection problem

In airframe applications, the most significant advantages of using integrally-stiffened panels, as shown in Fig. 14.1.3, over comparable riveted skin-stringer panel are:

- Reduction in the amount of sealing material for pressurized shell structures, e.g., wing fuel tank, cabin pressurized fuselage, etc.
- Increase in allowable compression loads by elimination of attached stringer flanges
- Increased structural efficiencies through the use of integral doublers
- Improved performance through smooth exterior surfaces by reduction in number of attachments and more-buckling resistant characteristics of skin

Fig. 14.1.3, shows the most typically used integrally-stiffened panels on airframe structure:

- The unflanged section [Fig. 14.1.3(a)] shows a simple panel and low cost fabrication
- The flanged section [Fig. 14.1.3(b) and (c)] provides higher structural efficiency than (a) but requires complicated machining

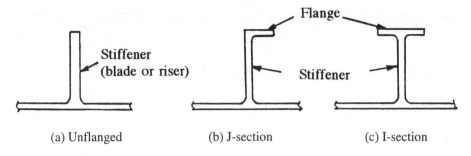

(a) Unflanged (b) J-section (c) I-section

Fig. 14.1.3 Typical Integrally-Stiffened Panels

ORTHOGRID AND ISOGRID PANELS

Orthogrid and isogrid panels, as shown in Fig. 14.1.4, are used on large space boosters and space vehicles to provide lightweight, economical, and efficient structures for aerospace use. These panels will not be discussed in depth in this chapter but Reference 14.29 provides detailed analysis information and design data. Recent study shows that it has great potential in future composite structure applications.

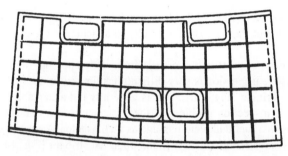

(a) Orthogrid (Waffle) Panels

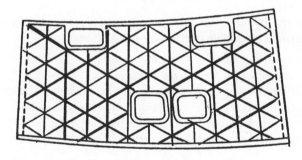

(b) Isogrid Panel

Fig. 14.1.4 Typical Geodesic Panels

14.2 EFFECTIVE WIDTHS

The effective width of skin (see Reference 14.3) is that portion of skin supported by a stringer in a skin-stringer construction that does not buckle when subjected to axial compression load (refer Eq. 11.2.2)

$$F_{cr, skin} = 3.62 E \left(\frac{t}{b}\right)^2$$

Buckling of the skin alone does not constitute a panel failure; in fact, the panel will carry additional load up to the stress at which the stringer (or stiffener) starts to fail. As the stringer stress is increased beyond the skin buckling stress ($F_{cr, skin}$), the skin adjacent to the stringers will carry additional stress because of the support given by the stringers. It is noted that the stress at the center of the panel does not exceed the initial buckling stress no matter how high the stress becomes at the stringer.

It is seen that the skin is most effective at the stringers, where support against buckling exists. At a given stress, the effective width (b_e), as shown in Fig. 14.2.1, is equal to the panel width at which buckling will just begin.

$$b_e = \left(\sqrt{\frac{K_c E}{F_{st}}}\right) t \qquad \text{Eq. 14.2.1}$$

where: F_{st} – Stringer or stiffener allowable stress (generally use the stringer crippling stress, F_{cc})

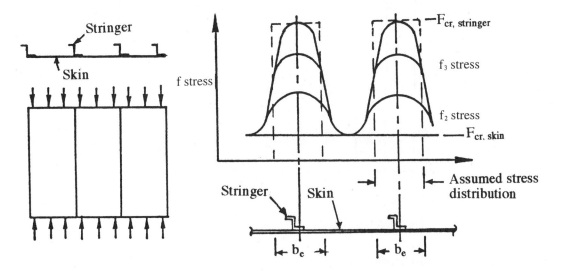

Fig. 14.2.1 Effective Width of a Skin-Stringer Panel

A buckled panel exerts a twisting force on the stringers. The direction of this twisting moment reverses with each change in buckle direction. These reverses occur at frequent intervals along the length of the stringer and the net twisting moment on the stringer is zero. For a large panel with a thin skin, as shown in Fig. 14.2.2(a), the torsional stiffness of a stringer is large in comparison to the force tending to twist it (e.g., wing panel near wing tip arear). This effect produces a fixed edge condition for the panel and the compression buckling constant, $K_c = 6.32$

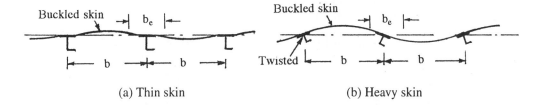

(a) Thin skin (b) Heavy skin

Fig. 14.2.3 Twisting Forces on the Stringers

A narrow panel with heavy skin (e.g., wing panels near wing root areas) produces buckling forces so great that the stringer will twist locally, as shown in Fig. 14.2.2(b). This panel will act as if it had hinged edges and the buckling constant, $K_c = 3.62$

Therefore, it is seen that there are two limits for K_c values. It has been found from tests:

$$K_c = 3.62 \text{ for } \frac{b}{t} < 40$$

$$K_c = 6.32 \text{ for } \frac{b}{t} > 110$$

Between above two values there is a gradual transition from $K_c = 3.62$ to $K_c = 6.32$, as plotted in Fig. 14.2.3.

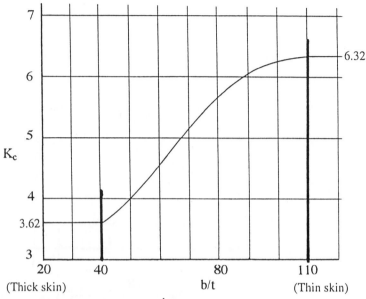

Fig. 14.2.3 K_c Value vs. $\frac{b}{t}$ Ratio for Skin-Stringer Panel

Reduction factors which should be considered are:

- For values of $\frac{b}{t} < 40$, a reduction factor should be considered for the effect in the plasticity range which will reduce the initial buckling stress of relatively narrow panels or, in terms of effective width, reducing the amount of skin acting with the stringer (refer to Chapter 11.2(B) for this effect)
- Material cladding effect (refer to Fig. 11.2.6 for cladding reduction factor, λ)

Example:

Assume allowable crippling stress of the stringers, $F_{cc} = 25,000$ psi and the material $E = 10.5 \times 10^6$ psi. Determine the skin effective width (b_e) of the No. 2 stringer.

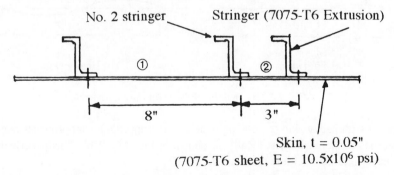

From Fig. 14.2.3, read:

$(\frac{b}{t})_① = \frac{8}{0.05} = 160 \rightarrow K_c = 6.32$

$(\frac{b}{t})_② = \frac{3}{0.05} = 60 \rightarrow K_c = 4.6$

The effective width is found using Eq. 14.2.1 (use $F_{st} = F_{cc} = 25,000$ psi):

$b_{e①} = (\sqrt{\frac{K_c E}{F_{st}}}) t = (\sqrt{\frac{6.32 \times 10.5 \times 10^6}{25,000}}) \times 0.05 = 2.58''$

$b_{e②} = (\sqrt{\frac{K_c E}{F_{st}}}) t = (\sqrt{\frac{4.6 \times 10.5 \times 10^6}{25,000}}) \times 0.05 = 2.2''$

The total effective width of the No. 2 stringer is:

$\frac{(b_{e①} + b_{e②})}{2} = \frac{2.58 + 2.2}{2} = 2.39''$

14.3 INTER-RIVET BUCKLING

Inter-rivet buckling (see Reference 14.4) is of special importance on compression members with skin attached by rivets, as shown in Fig. 14.3.1. Skins on a compression panel which have stringers (or stiffeners) attached by means of rivets are subject to local failure by outward buckling of the skin between rivets. The skin stress at which this phenomenon occurs is called the inter-rivet buckling stress (F_{ir}). Examples of a structure where the spacing of these attachments is important are:

- Upper surface of a wing box
- Fuselage stringers
- Fuselage skin-support frames

If the skin buckles between rivets, the calculated effective width is erroneous and the structure is much less efficient.

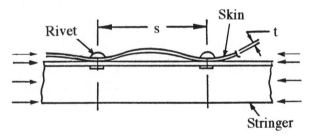

Fig. 14.3.1 Inter-rivet Buckling

The inter-rivet buckling theory is the same as that for buckling of a flat plate or sheet as a column (refer to Chapters 10 and 11) with two fixed edges at each end (at rivet).

The inter-rivet buckling equation (refer to Eq. 11.1.3) for a flat skin with unloaded free edges are is:

$$F_{ir} = 0.9c\, E\, (\frac{t}{s})^2 \qquad \text{Eq. 14.3.1}$$

where: t – Skin thickness
s – Rivet spacing (equivalent to column length)
c – End-fixity coefficient

Eq. 14.3.1 has been plotted in Fig. 14.3.2 and Fig. 14.3.3 for universal head (c = 4.0) aluminum rivets which provides for inter-rivet buckling in common aluminum alloy skins. Suitable correction factor must be made if other than universal head rivets are used:

Chapter 14.0

$$\sqrt{\frac{c}{4}}$$
Eq. 14.3.2

where: c = 3.0 for brazier heads

c = 1.0 for countersunk rivets

Normally the skin-stringer panel is designed so that the rivet spacing is derived from the crippling stress of the stringer ($F_{ir} = F_{cc}$).

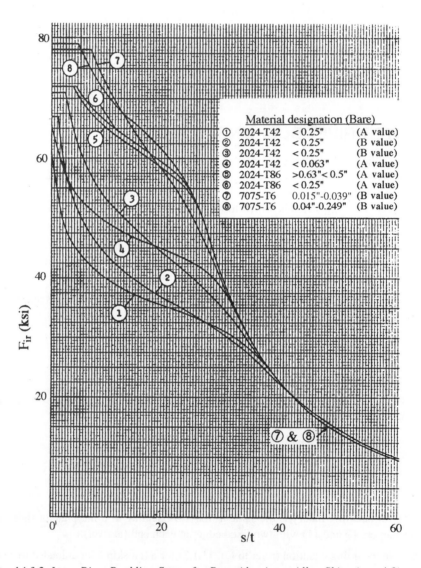

Fig. 14.3.2 Inter-Rivet Buckling Stress for Bare Aluminum Alloy Skins (c = 4.0)

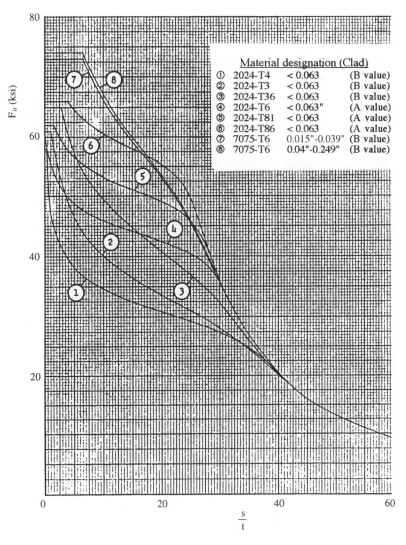

Fig. 14.3.3 Inter-Rivet Buckling Stress for Clad Aluminum Alloy Skins (c = 4.0)

Example:

Obtain the rivet spacing for countersunk head rivets from the following given data:

Stringer crippling stress, F_{cc} = 32 ksi

Skin thickness, t = 0.05"; material is 7075-T6 bare (non-clad material).

Using Fig. 14.3.2 with $F_{ir} = F_{cc}$ = 32 ksi, go across horizontally to curve ⑧ for 7075-T6 material.

Go down vertically to read the rivet spacing ratio $\frac{s}{t}$ = 33.5 (for universal head rivets, c = 4.0).

s = 33.5 × 0.05 = 1.68"

For countersunk head rivets, c = 1.0.

The rivet spacing of countersunk head rivet (see Eq. 14.3.2) is:

$s = 1.68 \sqrt{\frac{1.0}{4.0}} = 0.84"$

14.4 SKIN-STRINGER PANELS

For thin skins supported by sturdy stiffeners, the initial buckling stress can usually be calculated assuming that the skin buckles between stringers with some rotational restraint by the stringers. When the skin is thick the initial buckling stress of the skin may be comparable to the failure stress of the stringer and both may approach the yield stress of the material. It may no longer be possible to regard the stringer as "sturdy" and it becomes necessary to take into account the flexural, torsional, and local deformation of the stringer.

The method of analysis or sizing presented in this section is applicable to the following panel configurations:

- J-stringer-skin panels (stringer and skin may be of the same or different materials)
- Z-stringer-skin panels (stringer and skin may be of the same or different materials)
- Z-integrally-stiffened panels (one material)

(A) FAILURE STRESSES

The failure modes of a skin-stringer panel are illustrated in Fig. 14.4.1.

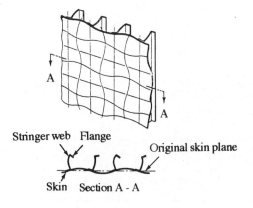

(a) Initial Buckling (Skin Buckling)

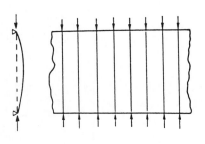

(b) Flexural Instability (Euler Mode)

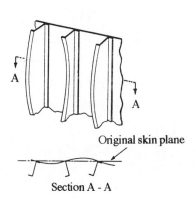

(c) Torsional Instability

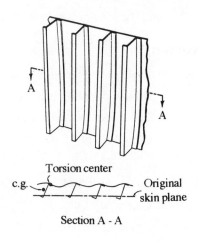

(d) Flexural and Torsional Instability

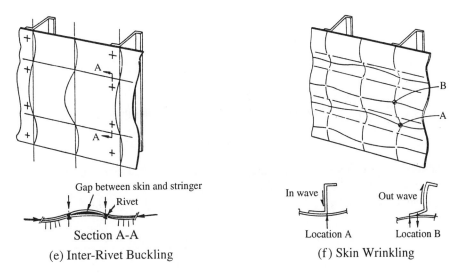

Gap between skin and stringer
Rivet
Section A-A
(e) Inter-Rivet Buckling

In wave — Location A
Out wave — Location B
(f) Skin Wrinkling

Fig. 14.4.1 Failure Modes of a Skin-Stringer Panel

The local buckling mode is a mixture of skin buckling, local instability and torsional instability, but the predominant type of buckling is dictated by the geometry of the skin-stringer construction used. Typical results are shown in Fig. 14.4.2 in which the non-dimensional buckling stress ratio of $\frac{f_i}{f_o}$ is plotted against $\frac{A_{st}}{bt}$ for various values of $\frac{t_w}{t}$.

- f_o is the buckling stress of the skin assuming hinged-edges along the stringers
- f_i is the actual initial buckling stress
- The upper portions of the curves correspond to a localized instability in skin
- The lower portions of the curves correspond to a torsional-and-lateral type of instability in the stringer
- The change of slope in the curves takes place when these two types of initial buckling occur at the same stress

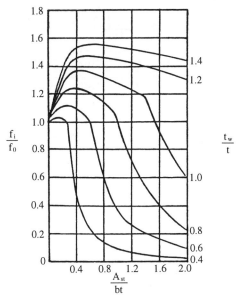

Fig. 14.4.2 Initial Buckling Stress of a Skin-Stringer Panel ($\frac{b_f}{b_w} = 0.3$, see Fig. 14.4.3)

The Farrar's efficiency factor (F) accounts for a pure flexural instability (assume flexural-torsional coupling is small):

$$F = f(\frac{L}{NE_t})^{\frac{1}{2}}$$ Eq. 14.4.1

where: f – Failure stress of skin-stringer panel
N – End-load per inch width of skin-stringer panel
E_t – Tangent modulus
L – Length of the panel (rib or frame spacing)

F is a function of $\frac{A_{st}}{bt}$ and $\frac{t_w}{t}$ and it is a measure of the structural efficiency of the skin-stringer panel. The quantity $F = f(\sqrt{\frac{L}{NE_t}})$ is plotted against $\frac{A_{st}}{bt}$ and $\frac{t_w}{t}$ in Fig. 14.4.3.

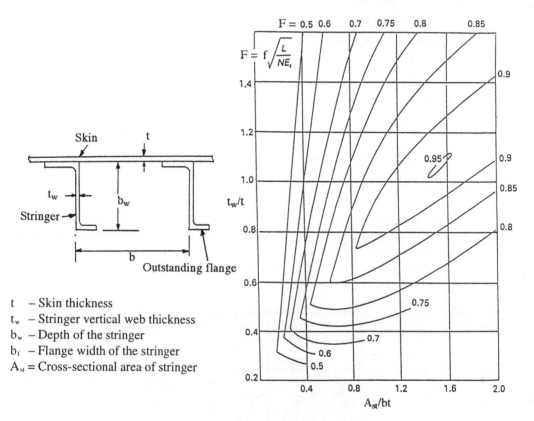

Fig. 14.4.3 *Farrar's Efficiency Factor (F) for Z-Stringer Panels (Reference 14.2)*

It is seen that an optimum value of $\frac{A_s}{bt}$ and $\frac{t_w}{t}$ exists, at which for a given N, E and L the stress realized will be a maximum. For this optimum design with $\frac{A_s}{bt} = 1.5$ and $\frac{t_w}{t} = 1.05$ then:

$$f = 0.95(\frac{NE_t}{L})^{\frac{1}{2}}$$ Eq. 14.4.2

which means that the maximum value of F = 0.95 is achieved for Z-stringer panels. In practice the full theoretical value of "F" is not achieved and experimental results indicate that about 90% is an average realized value. The results of some similar tests for stringer sections other than Z-stringers are given in Fig. 14.4.4.

Type of panel	Theoretical best value of "F"	Realized value of "F"
Z-stringer section	0.95	0.88
Hat stringer section	0.96	0.89
Y-stringer section	1.25	1.15

Fig. 14.4.4 Realized Values of Farrar's Efficiency Factor "F" vs. Various Stringer Sections

(B) GEOMETRY OF SKIN-STRINGER PANEL

Fig. 14.4.5 shows the definitive geometry for a skin-stringer panel.

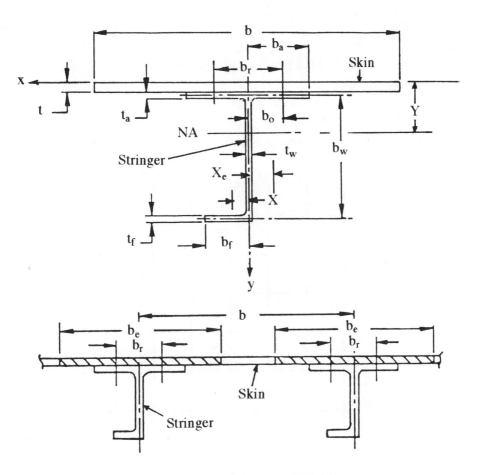

Fig. 14.4.5 Panel Geometry Definition

Notations: P – Total axial load in panel (lbs. or kips)   N – Panel axial load intensity (lbs./in.)

A_{sk} – Area of the skin (in.^{2}) A_{st} – Area of the stringer (in.^{2})

A – Panel area with full effective skin width (b) (in.^{2})

A_e – Panel area with effective skin width (b_e) (in.^{2})

F_c – Allowable compression stress (with effective skin width) in panel (psi or ksi)

$\dfrac{A_{st}}{A_{sk}}$ – Panel stiffening ratio

(C) PLASTICITY CORRECTION FACTORS FOR PANELS

Plasticity correction (or reduction) factors:

(a) For the same materials (generally used for integrally-stiffened panels):

- Compression stress: $\eta_c = (\frac{E_t}{E})^{\frac{1}{2}}$ Eq. 14.4.3(a)

- Bending stress: $\eta_b = (\frac{E_t}{E})$ Eq. 14.4.3(b)

(b) For skin-stringer panels, use the £ ratio:

$$£ = \frac{F_c}{F_t} = \frac{\pi^2}{(\frac{L'}{\rho_{xx}})^2}$$ Eq. 14.4.4

£ is plotted in Figs. 14.4.6 to 14.4.8 and which considers the following parameters:
- Section geometry
- Different material on skin and stringer
- Effective skin width
- Stiffening ratio, $\frac{A_{st}}{A_{sk}}$ (where: A_{st} – stringer area and A_{sk} – skin area)

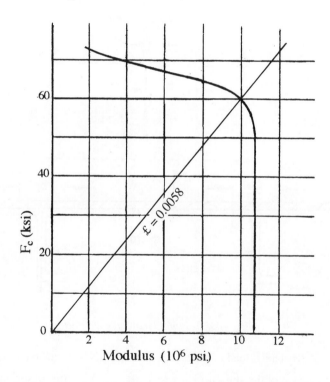

Fig. 14.4.6 £ Used on The Same Material (7075-T6 Extrusion)

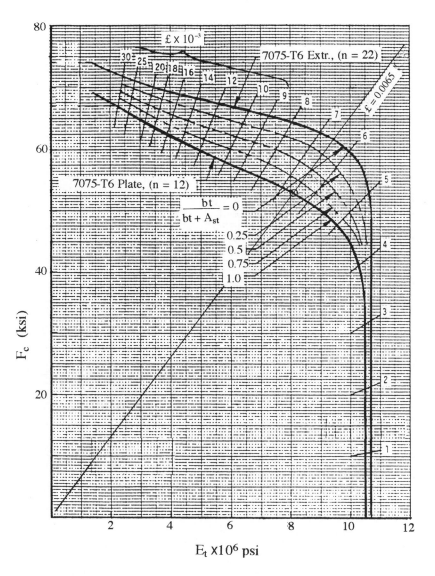

Fig. 14.4.7 £ Used for Different Materials (7075-T651 Plate and 7075-T6 Extrusion)

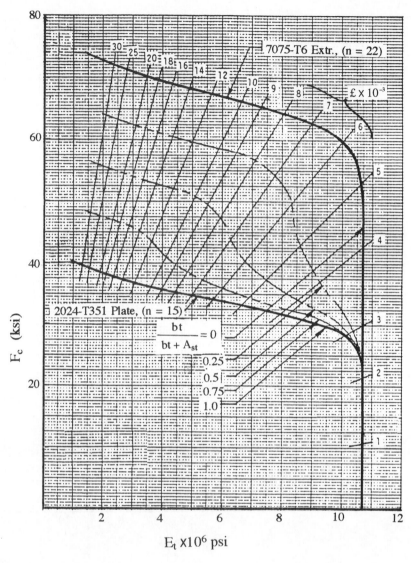

Fig. 14.4.8 £ Used for Different Materials (2024-T351 Plate and 7075-T6 Extrusion)

(D) APPROXIMATE EFFECTIVE WIDTH

For preliminary sizing purposes, effective width can be calculated from the information in Section 14.2 or approximate values can be derived from the curves shown in Fig. 14.4.9.

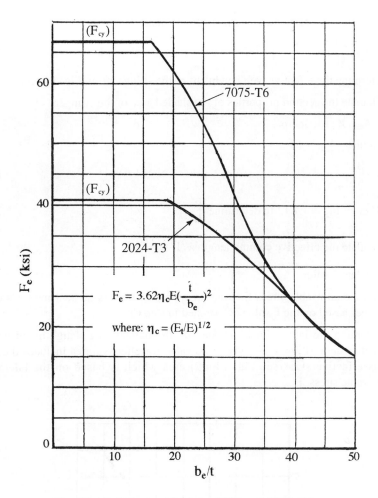

Fig. 14.4.9 Effective Width for 2024 and 7075 Skins

(E) LATERAL INSTABILITY OF OUTSTANDING STRINGER FLANGES

Lateral instability in stringer outstanding flanges, see Fig. 14.4.10 (shaded area), is treated as a beam on an elastic support. The effective web height, h, is obtained from Fig. 14.4.11.

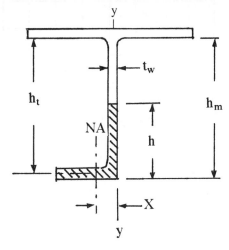

Fig. 14.4.10 Effective Stringer Web

(1) Determine 'β' based on the stringer dimensions:

$$\beta = \frac{Et_w^3}{4h_t^3}$$ Eq. 14.4.5

(2) Determine the effective web height 'h' of the stringer from Fig. 14.4.11

(3) Calculate the section properties of the shaded area of the stringer:

$A_{c,f}$, X, I_{yy}, and ρ_{yy}

(4) Calculate

$$\frac{\beta L^4}{16EI_{yy}}$$ Eq. 14.4.6

and enter Fig. 14.4.12 to determine $\frac{L'}{L}$

(5) Calculate the effective column length as $L' = (\frac{L'}{L})L$.

(6) Calculate £ value.

(7) Read the allowable stress F_c from an appropriate stress-tangent modulus curve for the stringer based on the £ value determined in step (6).

(8) Determine the lateral buckling load on the outstanding flange based on the section effective area as $P = F_c A_e$ where the value for effective area includes the stringer area plus effective skin (see Fig. 14.4.9) area which is based on the lateral instability allowable stress, F_c.

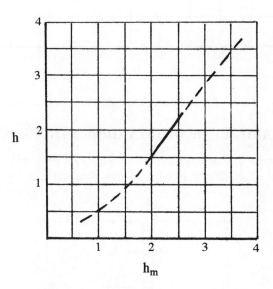

Fig. 14.4.11 Effective Stringer Web Height (h) for Lateral Instability

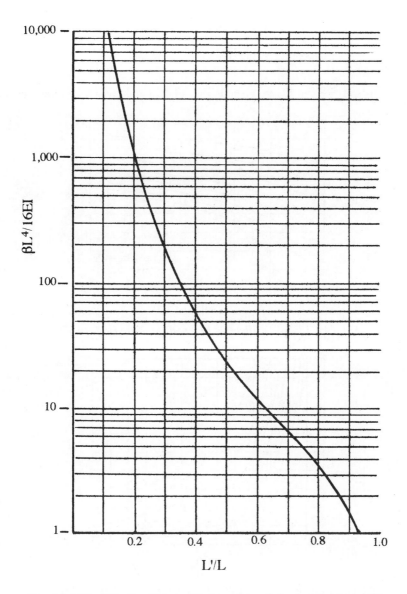

Fig. 14.4.12 Effective Stringer Column Length for Lateral Instability

(F) ATTACHMENT CRITERIA

Attachment criteria – The tension force on the attachment is given below:

$$P_r = \left(\frac{E}{1-\mu^2}\right)\left[\frac{1}{\left(\frac{b_{oe}}{t_a}\right)^3}\right]\left[\frac{3\left(\frac{b_{oe}}{t_a}\right)+\left(\frac{b_w}{t_w}\right)}{3\left(\frac{b_{oe}}{t_w}\right)+4\left(\frac{b_w}{t_w}\right)}\right]\left(\frac{t}{5}\right)(s) \qquad \text{Eq. 14.4.7}$$

- P_r – Rivet tension force (lbs.) for Z-stringer [e.g., see Fig. 14.1.2(a) and (f)]; $\frac{P_r}{2}$ for J or I-stringer [e.g., see Fig. 14.1.2(b) and (d)]
- s – Rivet spacing (pitch); b_{oe} – Effective rivet offset
- $\frac{b_{oe}}{t_a}$ – Effective rivet offset ratio

- Attachment spacing must satisfy the criteria given:

  $\frac{s}{b} < \frac{1.27}{\sqrt{K_m}}$ where K_m is skin wrinkling coefficient

- Values of $\frac{b_{oe}}{t_a}$ and K_m are given in Fig. 14.4.13 and Fig. 14.4.14, respectively

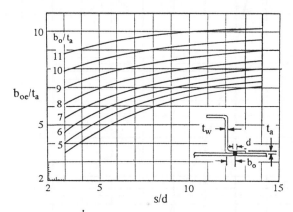

Fig. 14.4.13 $\frac{b_{oe}}{t_a}$ – Effective Rivet Offset Ratio (Ref. 14.12)

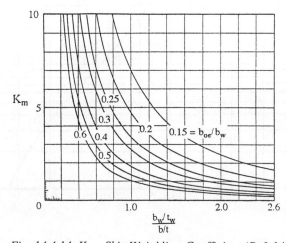

Fig. 14.4.14 K_m – Skin Wrinkling Coefficient (Ref. 14.12)

(G) BEAM-COLUMN EFFECT

Beam-column panel design considerations:

(a) Use of the following effective fixity to determine panel bending moments is recommended (exact fixity is difficulty to obtain):

  - Use c = 2 for moment at the center span of the panel, see Fig. 14.4.15
  - Use c = 3.5 for moment over supports, see Fig. 14.4.16

(b) Use c = 1.5 for typical wing panel under axial compression load because:

  - Some fixity is supplied by the rib tie to the panel
  - Wing panel has contour, but allowables are based on flat panel

Procedure for calculating the beam-column panel stress:

(1) Assume an applied axial stress, f_c, for this stress obtained from a stress-tangent modulus curve (use appropriate curve from Fig. 14.4.6 through Fig. 14.4.8)

(2) Find b_e from Fig. 14.4.10 at the assumed stress, f_c.

(3) Calculate section properties for the combined stringer and effective skin: A_e, Y, I_{xx}, and ρ_{xx}.

(4) Calculate the applied load $P = f_c A_e$

(5) Calculate torsional spring constant:
$$\mu_T = \frac{PL^2}{E_t I_{xx}} \qquad \text{Eq. 14.4.8}$$

(6) From Fig. 14.4.16 or Fig. 14.4.17 read off $\dfrac{M}{w_o L^2}$.

(7) Calculate:
$$M = (\frac{M}{w_o L^2}) w_o L^2$$

(8) Calculate $f_b = \dfrac{My}{I_{xx}}$ where y is the appropriate distance from the neutral axis (NA) to the outer fiber that is in compression.

(9) Calculate the combined stress $f = f_c + f_b$ and compare this stress to the appropriate allowable stress in step (1).

(10) Repeat steps (1) through (9) until agreement is obtained. The resulting stress is defined as the beam-column allowable stress, $F_{b,c}$.

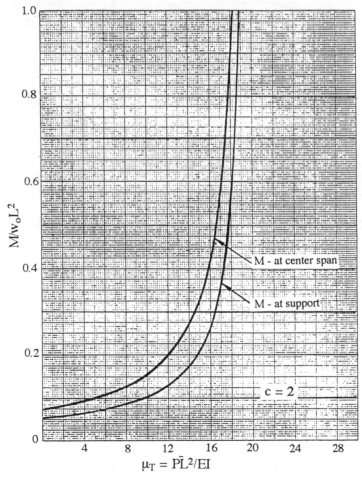

Fig. 14.4.15 Beam-Column Bending Moment, c = 2

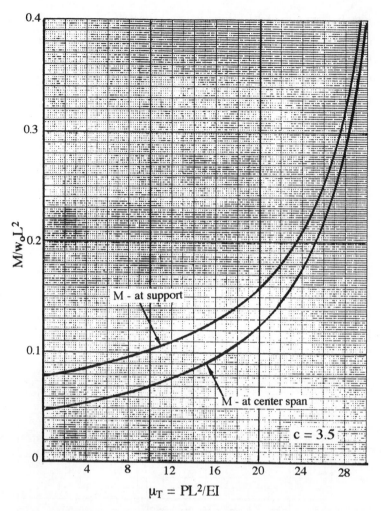

Fig. 14.4.16 Beam-Column Bending Moment, c = 3.5

(H) PRACTICAL DESIGN GUIDELINES IN PRELIMINARY SIZING

(a) Comparison efficiencies for various stringer shapes is given in Fig. 14.4.17.

PANEL TYPE	IDEAL*		PRACTICAL DESIGN	
	$\dfrac{A_{st}}{A_{sk}}$	EFFICIENCY	$\dfrac{A_{st}}{A_{sk}}$	EFFICIENCY
 ⊥ (Y)	2.16	1.23	0.5	—
 ⊥ (T)	1.3	1.03	0.5	0.82
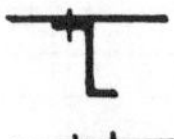 ⊥ (J)	1.47	0.911	0.5	0.68
 ⊥ (hat)	1.28	0.793	0.5	0.58

(* See Fig. 14.1.1)

Fig. 14.4.17 Stiffened Panel Efficiencies

Compression Panels

(b) Maximum efficiencies shown in Fig. 14.4.17 correspond to stiffening ratios ($\frac{A_{st}}{A_{sk}}$) which are much higher than can normally be used for high aspect ratio transport wings.

(c) Wing flutter and fuselage pressurization requirements dictate a thicker skin than that dictated by compression load requirements.

(d) A compromise stiffening ratio that meets flutter and pressurization requirements, as well as fail-safe requirements, is about 0.5. This are on the order of 20-30% less than maximum efficiencies.

(e) Use Fig. 14.4.18 through Fig. 14.4.21 are for preliminary sizing of 7075-T6 stringer-skin panels and integrally-stiffened panels:

- At low values of structural index ($\frac{N}{L'}$), it is more efficient to let the skin buckle ($b_e < b$)

- At high values of structural index ($\frac{N}{L'}$), the most efficiency is obtained when the skin is not buckled ($b_e = b$)

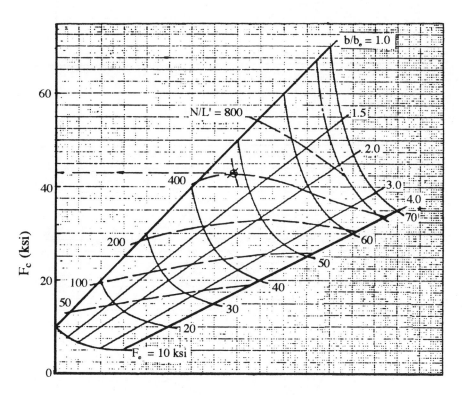

(7075-T6 plate: $E = 10.5 \times 10^6$ psi and $n = 12$)

Fig. 14.4.18 Design Curves for Stringer-Skin Panels ($\frac{A_{st}}{A_{sk}} = 0.5$)

Chapter 14.0

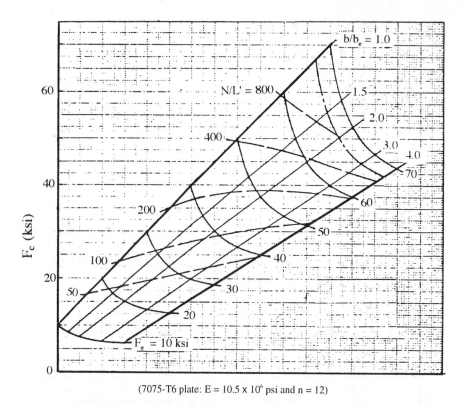

(7075-T6 plate: E = 10.5 x 10⁶ psi and n = 12)

Fig. 14.4.19 Design Curves for Stringer-Skin Panels ($\frac{A_{st}}{A_{sk}} = 1.0$)

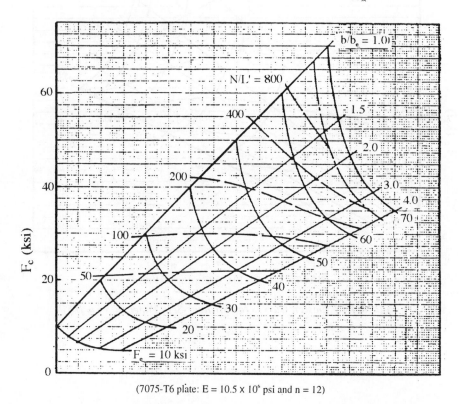

(7075-T6 plate: E = 10.5 x 10⁶ psi and n = 12)

Fig. 14.4.20 Design Curves for Integrally-Stiffened Panels ($\frac{A_{st}}{A_{sk}} = 0.5$)

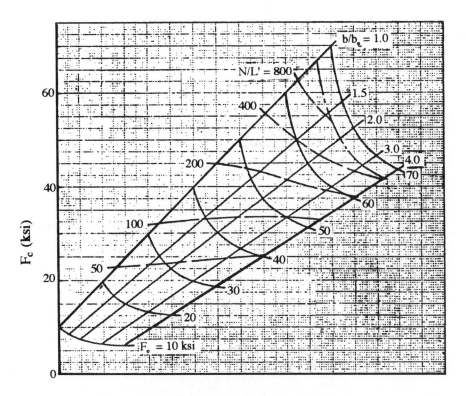

(7075-T6 plate: E = 10.5 × 10⁶ psi and n = 12)

Fig. 14.4.21 Design Curves for Integrally-Stiffened Panels ($\frac{A_{st}}{A_{sk}} = 1.0$)

(f) Twisting instability and its interaction with bending instability is neglected due to the difficulty of determining spring restraints and the effect of plasticity factors

(g) Wing and empennage structures:

- Use stiffening ratio, $\frac{A_{st}}{A_{sk}} \approx 0.5$
- A stiffening ratio less than 0.5 should not be used because of panel damage tolerance considerations

(h) Pressurized fuselage structures:

- Use stiffening ratio, $\frac{A_{st}}{A_{sk}} \approx 0.3$
- A high stiffening ratio will not, generally, produce lightweight structures because cabin pressure load requirements usually set the skin thickness higher than that required for axial compression loads

(i) Set $t_w \geq t_f$ (let t_w slightly greater than t_f)

(j) Set stringer $\frac{b}{t}$ ratio ($\frac{b}{t}$ is defined for crippling stress in Chapter 10.7) such that the sum of element crippling stress is greater than the compression panel allowable stress.

(k) Use the following values in design:

- $\frac{b_w}{t_a} < 10$

Chapter 14.0

- $\dfrac{b_w}{t_w} < 18$ to 22
- $\dfrac{b_f}{t_f} < 6$ to 8

(l) Set $t_a \geq 0.7 t_s$ (to prevent forced crippling)

(m) Set $\dfrac{b_f}{b_w} = 0.4$ to 0.5 (to prevent rolling of the stringer)

(n) Set b_o as small as possible (to prevent skin wrinkling but leave enough space for possible future repairs, see Fig. 16.1.2).

(o) Attachment requirements:
- Rivet spacing (s) = 4 to 6 D (preferable for wing and empennage panels)
- Rivet spacing (s) = 6 to 7 D (preferable for fuselage panels)

(p) Optimum element dimensions are shown in Fig. 14.4.22 and practical dimensions are shown in Fig. 14.4.23.

Elements	Applied equations	Remarks
t	$t = \dfrac{N(\dfrac{b}{b_e})}{F_e[1 + (\dfrac{A_{st}}{A_{sk}})(\dfrac{b}{b_e})]}$	
b_e	$(\dfrac{b_e}{t}) t$	$\dfrac{b_e}{t}$ from Fig. 14.4.9
b	$(\dfrac{b}{b_e}) b_e$	
b_a	Use $2.08 t + 0.688$ if $t \leq 0.3$ Use 1.312 if $t > 0.3$	$b_a = 0$ for integral panels
t_a	$0.7 t$	$t_a = 0$ for integral panels
A_{st}	$(\dfrac{A_{st}}{A_{sk}}) bt$	$\dfrac{A_{st}}{A_{sk}}$ from Fig. 14.4.17
b_w	$b_w = [(\dfrac{b_w}{t_w})(\dfrac{A_{st} - 2b_a t_a}{1.327})]^{\frac{1}{2}}$	Use $\dfrac{b_w}{t_w} \approx \dfrac{b_e}{t}$
t_w	$\dfrac{b_w}{\dfrac{b_w}{t_w}}$	
b_f	$0.327 b_w$	
t_f	$\approx t_w$	

(**Note:** All above dimensions should be checked by practical dimensions (see Fig. 14.4.23) and repairability requirements (see Chapter 16.1) before finalizing the panel cross section)

Fig. 14.4.22 Optimum Element Dimensions

Compression Panels

Elements	Design practice
$\dfrac{b_a}{t_a}$	10 or less
$\dfrac{b_w}{t_w}$	18 – 22
$\dfrac{b_f}{t_f}$	6 – 8
$\dfrac{A_{st}}{A_{sk}}$	0.5
t_a	0.7t
$\dfrac{b_f}{b_w}$	0.4

Fig. 14.4.23 Practical Dimensions

Example:

Size a J-stringer stiffened panel for an upper wing surface using the following given information:

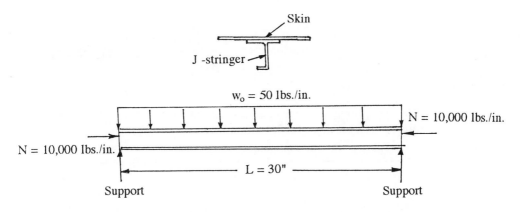

Materials:

    Machined-skin (7075-T651 plate material properties of t = 0.25", see Fig. 4.3.4):

    F_{tu} = 80,000 psi; F_{cy} = 72,000 psi;   E = 10.5 x 10^6 psi; μ = 0.33

    Machined-stringer (7075-T6 extrusion material properties of t = 0.5", see Fig. 4.3.5):

    F_{tu} = 85,000 psi; F_{cy} = 77,000 psi;   E = 10.7 x 10^6 psi; μ = 0.33

Load intensity: N = 10,000 lbs./in.

Panel length (between rib supports):   L = 30"

Surface pressure load (use running load, w_o = 50 lbs./in.)

Step 1   Compute panel dimensions:

    Design requirements: $\dfrac{A_{st}}{A_{sk}} = 0.5$ (see Fig. 14.4.17)

    Use c = 1.5 [see discussion in Section 14.4(G)] and $L' = \dfrac{L}{\sqrt{c}} = \dfrac{30}{\sqrt{1.5}} = 24.5$.

    $\dfrac{N}{L'} = \dfrac{10,000}{24.5} = 408$

Use $\frac{N}{L'} = 408$, from Fig. 14.4.18 obtain optimum $\frac{b}{b_e} = 1.2$ and effective stress of $F_e = 48,500$ psi. (the optimum value is at the apex of the curve of $\frac{N}{L'} = 408$).

From Fig. 14.4.22 obtain the following optimum elements:

$$t = \frac{N(\frac{b}{b_e})}{F_e[1 + (\frac{A_{st}}{A_{sk}})(\frac{b}{b_e})]} = \frac{10,000 \times 1.2}{48,500(1 + 0.5 \times 1.2)} = 0.155"$$

$b_e = (\frac{b_e}{t})t = 27.3(0.155) = 4.23"$ (See Fig. 14.4.9)

$t_a = 0.7t = 0.7 \times 0.155 = 0.108"$ (See Fig. 14.4.23)

$A_{st} = (\frac{A_{st}}{A_{sk}})(bt) = 0.5 \times 5.08 \times 0.155 = 0.394"$ (See Fig. 14.4.22)

$b_w = [(\frac{b_w}{t_w})(\frac{A_{st} - 2b_a t_a}{1.327})]^{\frac{1}{2}}$ (See Fig. 14.4.22)

$\quad = [(27.3)(\frac{0.394 - 2 \times 1.01 \times 0.108}{1.327})]^{\frac{1}{2}} = 1.9"$

$t_w = \frac{b_w}{\frac{b_w}{t_w}} = \frac{1.9}{27.3} = 0.07"$

$b_f = 0.327 \times b_w = 0.327 \times 1.9 = 0.621"$

$t_f = t_w = 0.07"$

(See Fig. 14.4.2)

The above optimum element dimensions are then checked against the practical dimensions (see Fig. 14.4.23) shown below:

Elements	Optimum	Practical	Comments
$\frac{b_a}{t_a} = \frac{1.01}{0.108} =$	9.35	10 or less	O.K.
$\frac{b_w}{t_w} = \frac{1.9}{0.07} =$	27.14	18 – 22	Too high
$\frac{b_f}{t_f} = \frac{0.621}{0.07} =$	8.87	6 – 8	High
$\frac{A_{st}}{A_{sk}}$	0.5	0.5	O.K.
t_a	0.108	$0.7t = 0.7 \times 155 = 0.109$	O.K.
$\frac{b_f}{b_w} = \frac{0.621}{1.9} =$	0.327	0.4	Low

The resulting ratios which meet the design guidelines are:

$\frac{b_w}{t_w} = 19.7$

$\frac{b_f}{t_f} = 7.81$

$\frac{b_f}{b_w} = 0.4$

The dimensions of this panel are shown below:

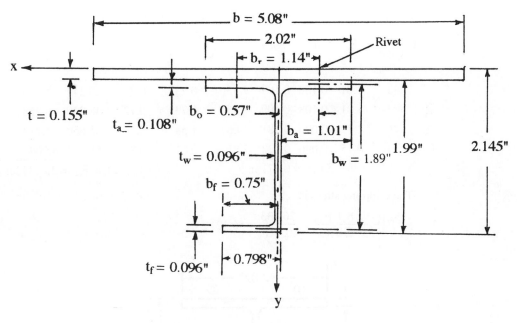

Step 2 Effective width of the skin (b_e) and panel section properties:

With $F_c = 48,500$ psi (see Step 1), from Fig. 14.4.9 obtain $\frac{b_e}{t} = 27.3$

$b_e = 27.3 \times 0.155 = 4.23"$

Assume [use MS20426(DD) fastener $d = \frac{3}{16}"$] rivet spacing at 5d [$s = 5 \times (\frac{3}{16}")$ = 0.94"] and edge distance, $e = 2d + \frac{1}{16}" = 0.44"$. Therefore, $b_o = b_a - 0.44 = 1.01 - 0.44 = 0.57"$

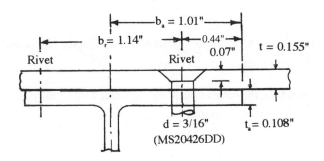

$b_r = 2 \times b_o = 2 \times 0.57 = 1.14"$ (distance between double rows of rivets)

The total effective width $b_e' = b_e + b_r = 4.23 + 1.14 = 5.37"$ and this value is greater than the stringer spacing ($b = 5.08"$), so that:

$b_e' = b = 5.08"$ (full skin width is effective)

Panel section properties (use full skin width, $b = 5.08"$):

$A = 1.254$ In.^{2} $Y = 0.371$ in.

$I_{xx} = 0.455$ in.^{4} $\rho_{xx} = 0.602$ in.

$L' = \frac{L}{\sqrt{c}} = \frac{30}{\sqrt{1.5}} = 24.5$ in. $\frac{L'}{\rho_{xx}} = \frac{24.5}{0.602} = 40.7$

Compute the stiffener (7075-T6 extrusion) crippling stress (F_{cc}):

Seg't	Free edges	b_n	t_n	$b_n t_n$	$\dfrac{b_n}{t_n}$	$\sqrt{\dfrac{F_{cy}}{E}}(\dfrac{b_n}{t_n})$	$\dfrac{F_{ccn}}{F_{cy}}$ (from Fig. 10.7.10)	F_{ccn}	$t_n b_n F_{ccn}$
①	1	1.01	0.108	0.109	9.35	0.73	0.72	48,240	5,258
②	1	1.01	0.108	0.109	9.35	0.73	0.72	48,240	5,258
③	0	1.89	0.096	0.181	19.68	1.54	0.98	65,660	11,913
④	1	0.75	0.096	0.072	7.81	0.61	0.83	55,610	4,004
				$\Sigma b_n t_n = 0.471$ in.^{2}				$\Sigma t_n b_n F_{ccn} =$	26,433 lbs.

The crippling stress is:

$$F_{cc} = \frac{\Sigma b_n t_n F_{ccn}}{\Sigma b_n t_n} = \frac{26{,}433}{0.471} = 56{,}121 \text{ psi}$$

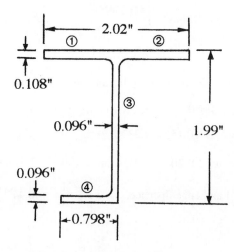

Use Eq. 10.8.1 or from Fig. 10.8.2(a) in Chapter 10.8 with $F_{cc} = 56{,}121$ psi and $\dfrac{L'}{\rho_{xx}} = 40.7$, obtain $F_c = 43{,}500$ psi

Therefore, the column buckling allowable load of the panel is:

$$N_{all} = \frac{F_c A}{b} = \frac{43{,}500 \times 1.254}{5.08} = 10{,}738 \text{ lbs./in.}$$

Step 3 Check panel plasticity correction factor:

From Eq. 14.4.4

$$\pounds = \frac{\pi^2}{(\dfrac{L'}{\rho_{xx}})^2} = \frac{\pi^2}{40.7^2} = 0.006$$

$$\frac{A_{sk}}{A_{sk} + A_{st}} = \frac{5.08 \times 0.155}{5.08 \times 0.155 + 0.471} = 0.63$$

From Fig. 14.4.7 find $F_c = 54{,}000$ psi which is greater than $F_c = 48{,}500$ psi. Further iteration is required by assuming $F_c = 54{,}000$ psi as the next approximate stress and from Fig. 14.4.9 obtain $\dfrac{b_e}{t} = 24$ and

$$b_e = 24 \times 0.155 = 3.72"$$

The total effective width $b_e' = b_e + b_r = 3.72 + 1.14 = 4.86"$ (double rows of rivets) and this value is smaller than the stringer spacing ($b = 5.08"$), so use:

$b_e' = 4.86"$ (not full skin width is effective)

Panel section properties (use effective skin width, $b_e' = 4.86"$):

$A_e = 1.224$ in.^{2} $Y_e = 0.406$ in.

$I_{e,xx} = 0.442$ in.^{4} $\rho_{e,xx} = 0.628$ in.

$L' = \dfrac{L}{\sqrt{c}} = \dfrac{30}{\sqrt{1.5}} = 24.5$ in. $\dfrac{L'}{\rho_{e,xx}} = \dfrac{24.5}{0.628} = 39$

From Eq. 14.4.4 again:

$$\pounds = \dfrac{\pi^2}{(\dfrac{L'}{\rho_{xx}})^2} = \dfrac{\pi^2}{39^2} = 0.0065$$

$$\dfrac{A_{sk}}{A_{sk} + A_{st}} = \dfrac{4.86 \times 0.155}{4.86 \times 0.155 + 0.471} = 0.62$$

From Fig. 14.4.7 find $F_c = 56,000$ psi which is close enough to the assumed value of $F_c = 54,000$ psi. Therefore, the allowable load intensity (conservative) is

$$N_{all} = \dfrac{F_c A_e}{b} = \dfrac{54,000 \times 1.224}{5.08} = 13,011 \text{ lbs./in.}$$

Step 4   Stringer (7075-T6 extrusion) lateral instability and attachment tension load:

Determine the effective stringer web height from Fig. 14.4.11 with given $h_m = 1.882"$, obtain $h = 1.33$.

Calculate the section properties for the shaded section:

$A_{e,f} = 0.159$ in.^{2} $X = 0.186$ in.

$I_{yy} = 0.01$ in.^{4} $\rho_{yy} = 0.225$ in.

From Eq. 14.4.5:

$$\beta = \dfrac{Et_w^3}{4h_t^3} = \dfrac{10.5 \times 10^6 \times 0.096^3}{4 \times 1.834^3} = 376$$

and from Eq. 14.4.6

$$\dfrac{\beta L^4}{16 E I_{yy}} = \dfrac{376 \times 30^4}{16 \times 10.5 \times 10^6 \times 0.01} = 181$$

Use the above value and from Fig. 14.4.12 obtain $\dfrac{L'}{L} = 0.31$.

$$L' = (\dfrac{L'}{L})L = 0.31 \times 30 = 9.3$$

and $\dfrac{L'}{\rho_{yy}} = \dfrac{9.3}{0.225} = 41.3$

From Eq. 14.4.4, Calculate:

$$\pounds \text{ ratio} = \dfrac{\pi^2}{(\dfrac{L'}{\rho_{xx}})^2} = \dfrac{\pi^2}{41.3^2} = 0.0058$$

From Fig. 14.4.6 obtain $F_c = 59,000$ psi and using this value, from Fig. 14.4.9, obtain $\frac{b_e}{t} = 22$. Therefore,

$$b = 22 \times 0.155 = 3.41"$$

The total skin effective width:

$$b_e' = b_e + b_r = 3.41 + 1.14 = 4.55" < b = 5.08" \text{ (not full effective skin width)}$$

The panel area, $A = 1.149$ in.^{2}

Therefore, the column buckling allowable load of the panel is:

$$N_{all} = \frac{F_c A}{b} = \frac{59,000 \times 1.149}{5.08} = 13,345 \text{ lbs./in.}$$

Use $\frac{b_o}{t_a} = \frac{0.57}{0.108} = 5.28$ and $\frac{s}{d} = 5$ (see Step 2), from Fig. 14.4.13 obtain $\frac{b_{oe}}{t_a} = 5.3$. And the rivet tension load from Eq. 14.4.7:

$$P_r = \left(\frac{E}{1-\mu^2}\right)\left[\frac{1}{\left(\frac{b_{oe}}{t_a}\right)^3}\right]\left[\frac{3\left(\frac{b_{oe}}{t_a}\right) + \left(\frac{b_w}{t_w}\right)}{3\left(\frac{b_{oe}}{t_a}\right) + 4\left(\frac{b_w}{t_w}\right)}\right]\left(\frac{t}{5}\right)(s)$$

$$= \left(\frac{10.5 \times 10^6}{1 - 0.33^2}\right)\left[\frac{1}{5.3^3}\right]\left[\frac{3 \times 5.3 + 19.7}{3 \times 5.3 + 4 \times 19.7}\right]\left(\frac{0.155}{5}\right)(0.94)$$

$$= 867 \text{ lbs.}$$

There are two rows of MS20426DD ($d = \frac{3}{16}"$) rivets and the allowable tension load is 1,312 lbs./rivet (see Fig. 9.2.13). Therefore, it has an ample margin of safety.

Calculate:

$$\frac{b_{oe}}{t_a} = 5.3 \text{ and therefore } b_{oe} = 5.3 \times 0.108 = 0.572$$

$$\frac{b_{oe}}{b_w} = \frac{0.572}{1.9} = 0.3; \quad \frac{\frac{b_w}{t_w}}{\frac{b}{t}} = \frac{\frac{1.9}{0.096}}{\frac{5.08}{0.155}} = 0.6$$

From Fig. 14.4.14 obtain $K_m = 5.8$ and check the criteria:

$$\frac{s}{b} = \frac{0.94}{5.08} = 0.185 < \frac{1.27}{\sqrt{5.8}} = 0.527 \text{ (satisfactory)}$$

Therefore, the rivet strength is not critical (check shear buckling and damage tolerant design conditions).

Step 5 Panel beam-column (up-bending at panel center span – compression in skin):

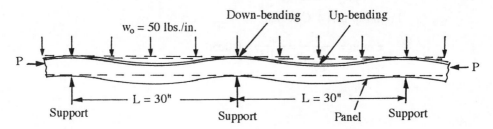

5(A) Assume an arbitrary value of $f_c = 40,000$ psi ($E_t = 10.3 \times 10^6$ plate (skin) from stress-strain curves of Fig. 14.4.7) and from Step 2 obtain panel section properties (full skin width, b = 5.08"):

$A = 1.254$ In.² $\qquad Y = 0.371$ in.

$I_{xx} = 0.455$ in.⁴ $\qquad \rho_{xx} = 0.602$ in.

The total axial load is:

$P = f_c \times A = 40,000 \times 1.254 = 50,160$ lbs.

5(B) From Eq. 14.4.8:

$$\mu_T = \frac{PL^2}{E_t I_{xx}} = \frac{50,160 \times 30^2}{10.3 \times 10^6 \times 0.455} = 9.63$$

Enter Fig. 14.4.15 with fixity c = 2 [see discussion in Section 14.4(G)], obtain:

$$\frac{M}{w_o L^2} = 0.149$$

$M = 0.149 \times w_o L^2 = 0.149 \times 50 \times 30^2 = 6,705$ in.-lbs.

The bending compression stress (in skin) is:

$$f_b = \frac{MY}{I_{xx}} = \frac{6,705 \times 0.371}{0.455} = 5,467 \text{ psi}$$

5(C) The maximum compression stress (in skin) is:

$f_c + f_b = 40,000 + 5,467 = 45,467$ psi which is greater than the allowable stress, $F_c = 43,500$ psi (from Step 2) by 1,967 psi.

5(A') Repeat Steps 5(A) through 5(C), and for this trial use the following assumed stress:

$$f_c' = 40,000 - \frac{1,967}{2} = 39,017 \text{ psi } (E_t = 10.35 \times 10^6 \text{ plate (skin)}$$

from Fig. 14.4.7)

The total axial load is:

$P = f_c' \times A = 39,017 \times 1.254 = 48,927$ lbs.

5(B') From Eq. 14.4.8:

$$\mu_T = \frac{PL^2}{E_t I_{xx}} = \frac{48,927 \times 30^2}{10.35 \times 10^6 \times 0.455} = 9.4$$

From Fig. 14.4.15, obtain $\frac{M}{w_o L^2} = 0.144$

$M = 0.144 \times w_o L^2 = 0.144 \times 50 \times 30^2 = 6,480$ in.-lbs.

The bending compression stress (in skin) is:

$$f_b' = \frac{MY}{I_{xx}} = \frac{6,480 \times 0.371}{0.455} = 5,284 \text{ psi}$$

5(C') The allowable compression stress and load intensity (critical in skin):

$f_c' + f_b' = 39,017 + 5,284 = 44,301$ psi which is greater than the allowable stress, $F_c = 43,500$ psi (from Step 2) by 801 psi. This is reasonably close so that the axial stress with lateral running load of $w_o = 50$ lbs./in. is:

Chapter 14.0

$$F_{b,c}' = 39,017 - 801 = 38,216 \text{ psi}$$

or the compression load intensity:

$$N_{all} = \frac{F_{b,c}'A}{b} = \frac{38,216 \times 1.254}{5.08} = 9,434 \text{ lbs./in.}$$

Step 6 Panel beam-column (down-bending at supports, compression in stringer outstanding flange, see sketch in Step 5):

6(A) Assume $f_c = F_{b,c}' = 38,216$ psi [calculated in Step 5(C') this is equal to the allowable panel beam-column between supports] for which the tangent modulus, $E_t = 10.37 \times 10^6$ of plate (skin) from Fig. 14.4.7 when set $f_c = F_c = 38,216$ psi. Panel section properties (full skin width, b = 5.08") from Step 2:

$A = 1.254$ in.² $\qquad\qquad Y = 0.371$ in.

$I_{xx} = 0.455$ in.⁴ $\qquad\qquad \rho_{xx} = 0.602$ in.

The total axial load is:

$$P = f_c \times A = 38,216 \times 1.254 = 47,923 \text{ lbs.}$$

6(B) From Eq. 14.4.8:

$$\mu_T = \frac{PL^2}{E_t I_{xx}} = \frac{47,923 \times 30^2}{10.37 \times 10^6 \times 0.455} = 9.1$$

From Fig. 14.4.16 with fixity c = 3.5 [see discussion in Section 14.4(G)], obtain: $\dfrac{M}{w_o L^2} = 0.1$

$$M = 0.1 \times w_o L^2 = 0.1 \times 50 \times 30^2 = 4,500 \text{ in.-lbs.}$$

The bending compression stress (in stringer outstanding flange) is:

$$f_b = \frac{M(2.145 - Y)}{I_{xx}} = \frac{4,500 \times (2.145 - 0.371)}{0.455} = 17,545 \text{ psi}$$

6(C) The compression stress (critical in stringer outstanding flange):

$f_c + f_b = 38,216 + 17,545 = 55,761$ psi which is slightly greater than the allowable stress of the stringer outstanding flange (t_f) where the crippling stress, $F_{cc} = 55,610$ psi (from Step 2 see crippling stress calculation of segment ④ only), but close enough. No further iteration is required.

Step 7 Summary of the final panel allowable loads:

The allowable load taken is the lowest load as determined from Step 2 through Step 6. The maximum compression stress [from Step 5(C')] is the most critical case which is in the skin (t) of the panel center span with lateral running load of $w_o = 50$ lbs./in.:

$F_{b,c} = 38,216$ psi

or the load intensity: $N_{all} = 9,434$ lbs./in.

The above allowable load of $N_{all} = 9,434$ lbs./in. is slightly less (5.7%) than that of applied load of N = 10,000 lbs./in. Therefore, it may be considered as good enough for preliminary sizing; otherwise, further iteration of all steps is required.

In actual design, these values would be considered together with local shear stress in the skin and then the allowable stress would have to be less than $F_{b,c} = 38,216$ psi [or 9,434 lbs./in., see Step 5(C')].

Compression Panels

(I) DESIGN CURVES FOR QUICK RESULTS

A set of design curves which yield quick results are presented in Fig. 14.4.24 and 14.4.25 (the skin shear loads are omitted). Before using these curves, carefully read the instructions for each figure. These curves are used for comparison panel.

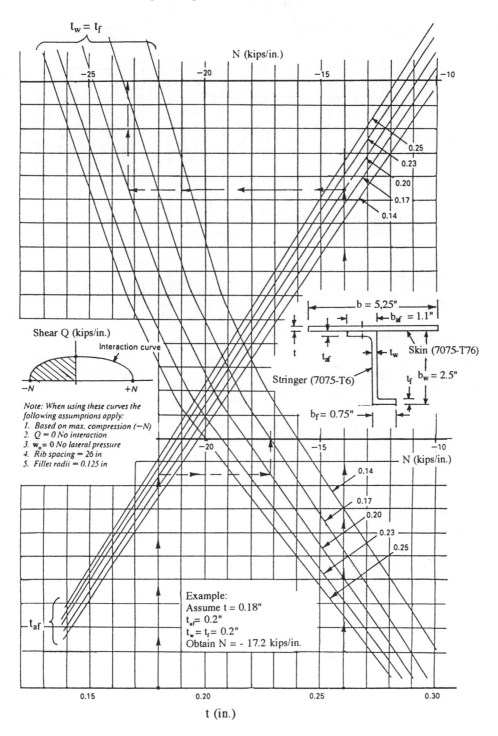

Fig. 14.4.24 Design Curves for Skin-Extruded Stringer Panels (Wing Boxes)

Chapter 14.0

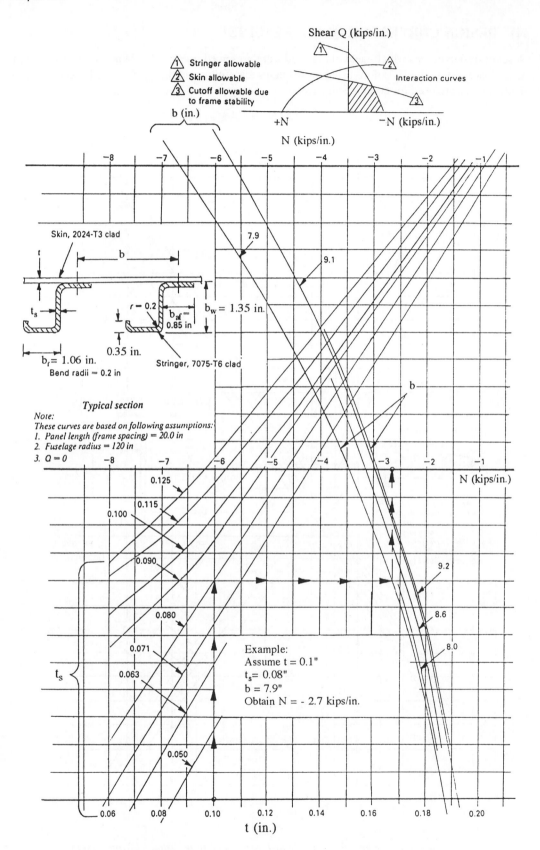

Fig. 14.4.25 Design Curves for Skin-Formed Stringer Panels (Fuselage)

14.5 STURDY INTEGRALLY-STIFFENED PANELS

A compact stiffener is described as "sturdy" when it is not subject to local buckling and therefore only the axial rigidities of the stiffener influence the behavior of the integrally-stiffened panel. For a general method of analysis refer to Chapter 14.4 (for flanged integrally-stiffened panels) or Chapter 14.6 (for unflanged integrally-stiffened panels).

The initial local buckling stress for plates or stiffeners under compression load (refer Eq. 11.3.1) is given by:

$$\frac{F_{c,cr}}{\eta_c} = \frac{k_c \pi^2 E}{12(1-\mu^2)} \left(\frac{t}{b}\right)^2 \quad \text{Eq. 14.5.1}$$

where: t – Skin thickness
        b – Stiffener spacing
        E – Modulus of elasticity
        μ – Poisson's ratio (use 0.3 for aluminum and steel)
        η_c – Plasticity reduction factor (see Fig. 11.2.4)
        k_c – Compression buckling coefficient (see Fig. 14.5.1 to 14.5.5)

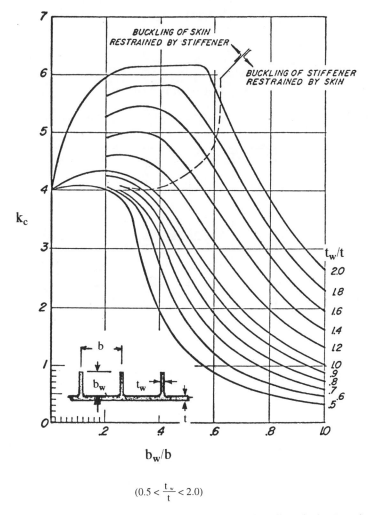

$(0.5 < \frac{t_w}{t} < 2.0)$

Fig. 14.5.1 k_c Coefficients for Infinitely Wide Idealized Unflanged-Stiffened Flat Panels (Ref. 14.9)

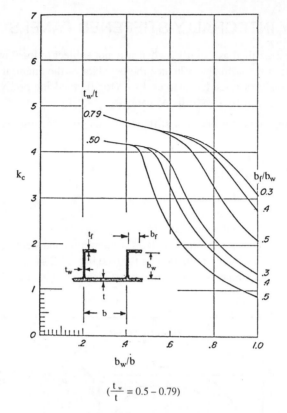

($\frac{t_w}{t} = 0.5 - 0.79$)

Fig. 14.5.2 k_c Coefficients for Infinitely Wide Idealized Z-Stiffened Flat Panels (Ref. 14.9)

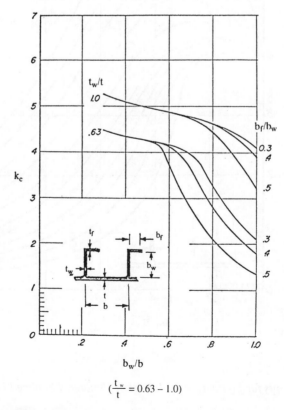

($\frac{t_w}{t} = 0.63 - 1.0$)

Fig. 14.5.3 k_c Coefficients for Infinitely Wide Idealized Z-Stiffened Flat Panels (Ref. 14.9)

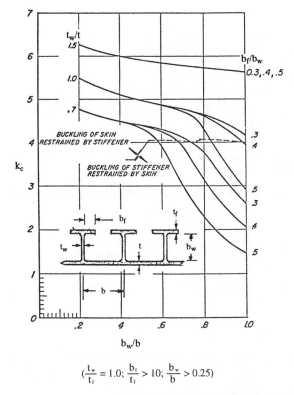

$(\frac{t_w}{t_f} = 1.0; \frac{b_f}{t_f} > 10; \frac{b_w}{b} > 0.25)$

Fig. 14.5.4 k_c Coefficients for Infinitely Wide Idealized I-Stiffened Flat Panels (Ref. 14.9)

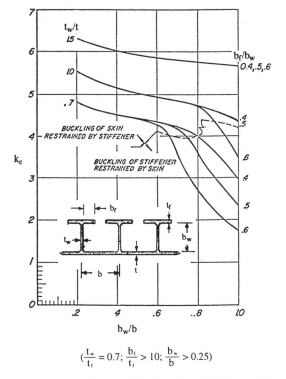

$(\frac{t_w}{t_f} = 0.7; \frac{b_f}{t_f} > 10; \frac{b_w}{b} > 0.25)$

Fig. 14.5.5 k_c Coefficients for Infinitely Wide Idealized I-Stiffened Flat Panels (Ref. 14.9)

Example:

Determine the initial buckling stress of the Z-stiffened panel with the following given information:

Material: 7075-T6 Extrusion

$$\frac{b_f}{b_w} = \frac{0.5}{1.5} = 0.333$$

$$\frac{b_w}{b} = \frac{1.5}{3.0} = 0.5$$

$$\frac{t_w}{t} = \frac{0.0625}{0.125} = 0.5$$

From Fig. 14.5.2, obtain $k_c = 3.5$ and substitute this value in Eq. 14.5.1:

$$\frac{F_{c,cr}}{\eta_c} = \frac{k_c \pi^2 E}{12(1-\mu^2)} \left(\frac{t}{b}\right)^2 \quad \text{Eq. 14.5.1}$$

$$= \frac{3.5 \times \pi^2 \times 10.7 \times 10^6}{12(1-0.3^2)} \left(\frac{0.125}{3}\right)^2 = 58{,}763 \text{ psi}$$

From Fig. 11.2.4 obtain $F_{c,cr} = 58{,}000$ psi (The reduction due to stress in plasticity range).

14.6 UNFLANGED (BLADE) INTEGRALLY-STIFFENED PANELS

An unflanged integrally-stiffened panel is shown in Fig. 14.6.1; the assumptions for analyzing such panel (per Reference 14.32) are as follows (see Chapter 14.4 for the method of analysis for flanged integrally-stiffened panel):

- The panels are assumed to be sufficiently wide to allow their treatment as a simple column, e.g., no restraint is imposed on the longitudinal edges of the panels
- Each panel is assumed in the analysis to be pin-ended over its bay length (L); but account may be taken of end fixity by regarding L as the effective pin-ended length rather than as the actual bay length
- The ribs or frames are assumed to impose no restraint on the buckling

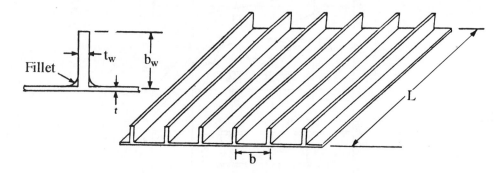

Fig. 14.6.1 Typical Unflanged Integrally-Stiffened Panel (Blade Panel)

Notations:
- t – Skin thickness
- b_w – Stiffener depth
- f – Applied stress
- f_E – Euler instability stress
- b – Stiffener spacing
- L – Pin-ended bay length
- f_i – Panel section initial buckling stress

f_o – Initial buckling stress of a long plate of stiffener spacing, b, and skin thickness, t, simply supported along its edges = $3.62 E_t (\frac{t}{b})^2$

N – Load per unit panel width $\quad\quad$ E_t – Tangent modulus

ρ – Radius of gyration of skin-stiffener panel about its own neutral axis

$$F = f(\frac{L}{NE_t})^{\frac{1}{2}} \quad \text{(Farrar's efficiency factor, see Eq. 14.4.1)}$$

$$J_1 = b(\frac{E_t}{NL^3})^{\frac{1}{4}}; \quad J_2 = t_w(\frac{E_t}{NL})^{\frac{1}{2}}; \quad J_3 = b_w(\frac{E_t}{NL^3})^{\frac{1}{4}}; \quad J_4 = t(\frac{E_t}{NL})^{\frac{1}{2}}$$

$$R_b = \frac{b_w}{b}; \quad\quad R_t = \frac{t_w}{t}$$

The initial buckling stresses of a variety of skin-stiffener combinations are plotted in Fig. 14.6.2 as ratios of $\frac{f_i}{f_o}$ and these buckling stresses take full account of the interaction between skin and stiffener buckling stress.

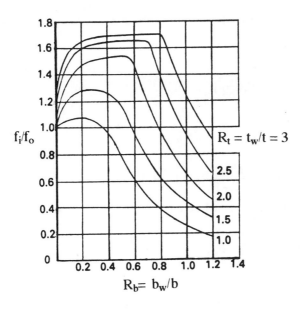

Fig. 14.6.2 Initial Buckling Stress

For a given material, for which the relationship between f and E_t is known, Farrar's equation of

$$F = f(\frac{L}{NE_t})^{\frac{1}{2}} \quad \text{or} \quad f = F(\frac{NE_t}{L})^{\frac{1}{2}} \quad \text{(refer to Eq. 14.4.1)}$$

may be used to plot curves of 'f' against the structural index $\frac{N}{L}$ for various values of F; this has been done in Fig. 14.6.3 for 2024-T3 extrusion material.

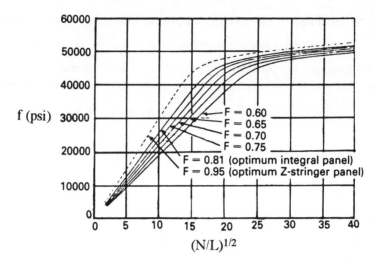

(Similar curves for 7075-T6 extrusion or other materials can be generated by using Eq. 14.4.1)

Fig. 14.6.3 Panel Stress vs. Structural Index $(\frac{N}{L})$ for 2024-T3 Extrusion

Initial buckling stress:

$$f_i = (\frac{f_i}{f_o})[3.62 E_t (\frac{t}{b})^2] \qquad \text{Eq. 14.6.1}$$

Euler instability stress:

$$f_E = \pi^2 E_t (\frac{\rho}{L})^2 \qquad \text{Eq. 14.6.2}$$

where: $\rho^2 = \frac{b^2 R_b^3 R_t}{12(1 + R_b R_t)}(4 + R_b R_t)^2$

Relating the stress in the panel to the load intensity:

$$f = \frac{N}{t(1 + R_b R_t)} \qquad \text{Eq. 14.6.3}$$

Imposing the condition:

$$f = f_i = f_E$$

Take (Eq. 14.6.1) × (Eq. 14.6.2) × (Eq. 14.6.3)² obtain:

$$f^4 = \pi^2 E_t^2 (\frac{3.62 \rho^2 f_i}{f_o b^2 L^2})[\frac{N^2}{(1 + R_b R_t)^2}]$$

Hence, taking the fourth root of both sides:

$$f = F(\frac{NE_t}{L})^{\frac{1}{2}} \qquad \text{Eq. 14.6.4}$$

where Farrar's efficiency factor:

$$F = 1.314 \frac{R_b^3 R_t (4 + R_b R_t)^{\frac{1}{4}}}{(1 + R_b R_t)} (\frac{f_i}{f_o})^{\frac{1}{4}} \qquad \text{Eq. 14.6.5}$$

(A) OPTIMUM PANEL DESIGN

From Eq. 14.6.3 and Eq. 14.6.4:

$$t = \frac{1}{F(1 + R_b R_t)}(\frac{NL}{E_t})^{\frac{1}{2}} = J_4(\frac{NL}{E_t})^{\frac{1}{2}} \qquad \text{Eq. 14.6.6}$$

For the optimum design $R_b = 0.65$, $R_t = 2.25$, and $F = 0.81$, therefore:

$$t = 0.501 \left(\frac{NL}{E_t}\right)^{\frac{1}{2}} \qquad \text{Eq. 14.6.7}$$

From Eq. 14.6.2, Eq. 14.6.4, Eq. 14.6.5, and Eq. 14.6.6:

$$b = 1.103 \frac{\sqrt{F}(1 + R_b R_t)}{[R_b^3 R_t (4 + R_b R_t)]^{\frac{1}{2}}} \left(\frac{NL^3}{E_t}\right)^{\frac{1}{4}} = J_1 \left(\frac{NL^3}{E_t}\right)^{\frac{1}{4}} \qquad \text{Eq. 14.6.8}$$

For the optimum design:

$$b = 1.33 \left(\frac{NL^3}{E_t}\right)^{\frac{1}{4}} \qquad \text{Eq. 14.6.9}$$

t_w may be calculated from:

$$b_w = 1.33 R_b b \qquad \text{Eq. 14.6.10}$$

$$t_w = R_t t \qquad \text{Eq. 14.6.11}$$

The optimum design should be:

$$b_w = 0.65 b \qquad \text{Eq. 14.6.12}$$

$$t_w = 2.25 t \qquad \text{Eq. 14.6.13}$$

(B) DESIGN LIMITATION – STIFFENER THICKNESS (t_w) AND SPACING (b)

To keep value of t_w and b constant for ease of machining, even though N and L will vary:

$$J_1 = b \left(\frac{E_t}{NL^3}\right)^{\frac{1}{4}} \qquad \text{Eq. 14.6.14}$$

$$J_2 = t_w \left(\frac{E_t}{NL}\right)^{\frac{1}{2}} \qquad \text{Eq. 14.6.15}$$

The two equations shown above have been plotted in Fig. 14.6.4

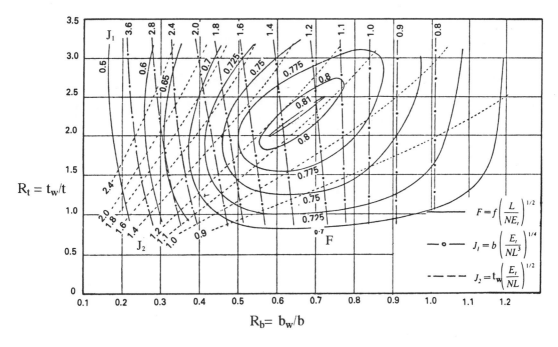

Fig. 14.6.4 *Design Curves for Limitations of Stiffener Thickness (t_w) and Spacing (b)*

To design an optimized panel under a given load intensity, N (lbs./in.):

Use optimized values: F = 0.81; R_t = 2.25; R_b = 0.65

(1) From Fig. 14.6.3 find the 'f' value corresponding to the $\frac{N}{L}$ using F = 0.81 for 2024-T3 extrusion. For other materials, use Eq. 14.4.1 to generate a set of curves similar to those shown in Fig. 14.6.3.

(2) Find the E_t value corresponding to 'f' from tangent modulus curve of the material.

(3) Determine t, b, b_w and t_w from Eq. 14.6.7 and Eq. 14.6.13.

At next section (could be a continuous section) of this panel under a different load intensity of N' (Use prime to represent the next section):

(4) Assume a value of E'_t (e.g., for a first approximation make E'_t = E).

(5) Let b' = b and $t'_w = t_w$, and then calculate J'_1 and J'_2.

(6) Determine F' from Fig. 14.6.4.

(7) Determine f' from Fig. 14.6.3.

(8) Find tangent modulus of E'_t corresponding to f'.

(9) Repeat steps (4) to (8) until the values of E'_t in steps (4) and (8) coincide. Repetition will be unnecessary if the material is in elastic range.

(10) From Fig. 14.6.4 find the intersection of R'_b and R'_t:

$$b'_w = R'_b b \text{ and } t' = \frac{t_w}{R'_t}$$

where b and t_w are the same values as those of the optimum values calculated before.

(C) DESIGN LIMITATION – SKIN THICKNESS (t) AND STIFFENER DEPTH (b_w)

Fig. 14.6.5 gives:

$$J_3 = b_w \left(\frac{E_t}{NL^3}\right)^{\frac{1}{4}}$$

$$J_4 = t \left(\frac{E_t}{NL}\right)^{\frac{1}{2}}$$

These equations have been prepared to facilitate the optimal design of panels.

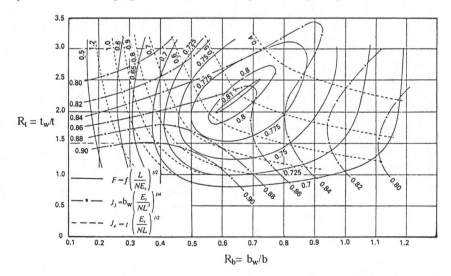

Fig. 14.6.5 Design Curves for Limitations of Skin Thickness (t) and Stiffener Depth (b_w)

(D) DESIGN CURVE FOR QUICK RESULTS

To reach the ideal panel material distribution, it will be necessary to be able to vary the skin thickness, stiffener spacing (usually kept constant), and the stiffener dimensions.

Fig. 14.6.6 presents a series of curves for designing unflanged integrally-stiffened panels from 7075-T76 extrusions. The N allowables (with skin shear load, Q = 0) are generated from the following:

- Fixed values of length (L), stiffener spacing (b) and fillet radius (r_f)
- Variable dimensions for stiffener depth (b_w), stiffener thickness (t_w), and skin thickness (t)

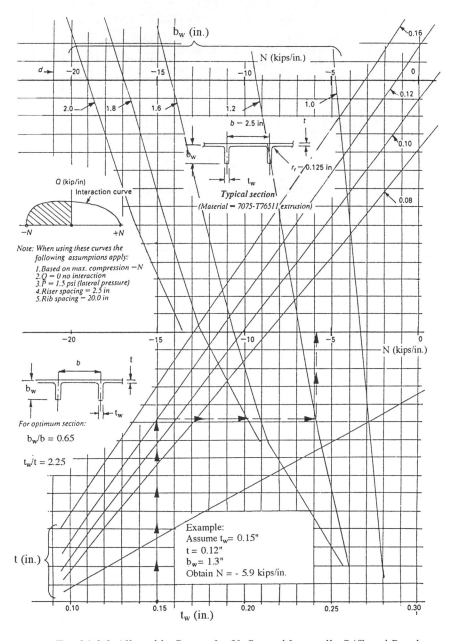

Fig. 14.6.6 Allowable Curves for Unflanged Integrally-Stiffened Panel

Example:

Design an optimized panel consisting of an unflanged integrally-stiffened panel under compression load intensity, N = 5,000 lbs./in.

    Panel length: L = 20"

    Panel material: 2024-T3 extrusion; the tangent modulus curve is provided below:

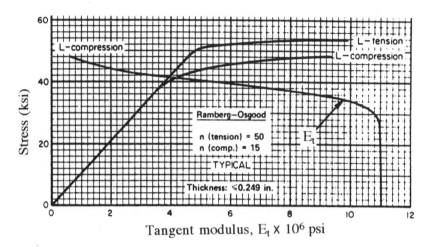

(Tangent modulus curve for 2024-T3 extrusion)

(a) Use optimized values: F = 0.81; b_w = 0.65b (see Eq. 14.6.12); t_w = 2.25t (see Eq. 14.6.13)

$(\frac{N}{L})^{\frac{1}{2}} = (\frac{5,000}{20})^{\frac{1}{2}} = 15.8$, and from Fig. 14.6.3, obtain compression stress, f = 40,000 psi and the corresponding tangent modulus $E_t = 5.4 \times 10^6$ psi (see tangent modulus curve above)

From Eq. 14.6 7:

$$t = 0.501(\frac{NL}{E_t})^{\frac{1}{2}} = 0.501(\frac{5,000 \times 20}{5.4 \times 10^6})^{\frac{1}{2}} = 0.068"$$

From Eq. 14.6.13: t_w = 2.25t = 2.25 × 0.068 = 0.153"

From Eq. 14.6.9:

$$b = 1.33(\frac{NL^3}{E_t})^{\frac{1}{4}} = 1.33(\frac{5,000 \times 20^3}{5.4 \times 10^6})^{\frac{1}{4}} = 2.194"$$

From Eq. 14.6.12: b_w = 0.65b = 0.65 × 2.194 = 1.426"

The optimum design panel is sketched below:

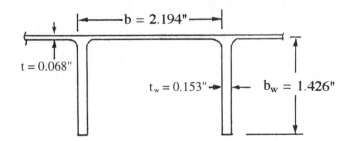

(b) At next section under load intensity, N' = 4,000 lbs./in.:

With limitations on stiffener thickness (t_w) and spacing (b)

Namely b' = b = 2.194" and t'_w = t_w = 0.153"

Assume E'_t = E = 10 × 10^6 psi (in elastic range)

$$J'_1 = b\left(\frac{E_t}{NL^3}\right)^{\frac{1}{4}} = 2.194\left(\frac{10 \times 10^6}{4,000 \times 20^3}\right)^{\frac{1}{4}} = 1.64$$

$$J'_2 = t_w\left(\frac{E_t}{NL}\right)^{\frac{1}{2}} = 0.153\left(\frac{10 \times 10^6}{4,000 \times 20}\right)^{\frac{1}{2}} = 1.71$$

Go to Fig. 14.6.4, obtain:

F' = 0.72 (use N' = 4,000 lbs./in.)

$R_b' = \frac{b_w}{b'} = 0.49$ ∴ b_w' = 0.49 × b' = 0.49 × 2.194 = 1.075"

$R_t' = \frac{t_w}{t'} = 3.1$ ∴ t' = $\frac{0.153}{3.1}$ = 0.049"

Calculate $\left(\frac{N'}{L}\right)^{\frac{1}{2}} = \left(\frac{4,000}{20}\right)^{\frac{1}{2}}$ = 14.1 and from Fig. 14.6.3, obtain compression stress, f = 34,000 psi and the corresponding tangent modulus E_t' ≈ 10 × 10^6 psi (see tangent modulus curve). Therefore, no iteration is required.

The panel is sketched below:

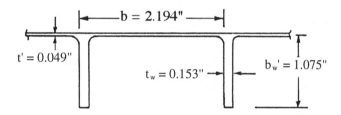

References

14.1 Emero, D. H. and Spunt, L., "Wing Box Optimization under Combined Shear and Bending", Journal of The Aeronautical Sciences, (Feb., 1966). pp. 130-141.

14.2 Farrar, D. J., "The Design of Compression Structures for Minimum Weight", Journal of The Royal Aeronautical Society, (Nov., 1949). pp. 1041-1052.

14.3 Fischel, R., "Effective Width in Stiffened Panels under Compression", Journal of The Aeronautical Sciences, (Mar., 1940). pp. 213-216.

14.4 Prescott, J., "Buckling Between Rivets", Aircraft Engineering, (April, 1941). p. 104.

14.5 Peery, David, "AIRCRAFT STRUCTURES", McGraw-Hill Book Company, Inc., (1950).

14.6 Kuhn, Paul, "STRESSES IN AIRCRAFT AND SHELL STRUCTURES", McGraw-Hill Book Company, Inc., New York, NY, 1956.

14.7 Bruhn, E. F., "ANALYSIS AND DESIGN OF FLIGHT VEHICLE STRUCTURES", Jacobs Publishers, 10585N. Meridian St., Suite 220, Indianapolis, IN 46290, (1965).

14.8 Gerard, G. and Becker, H., "Handbook of Structural Stability (Part I – Buckling of Flat Plates)", NACA TN 3781, (July, 1957).

14.9 Becker, H., "Handbook of Structural Stability (Part II – Buckling of Compression Elements)", NACA TN 3782, (July, 1957).

14.10 Gerard, G. and Becker, H., "Handbook of Structural Stability (Part III – Buckling of Curved Plates and Shells)", NACA TN 3783, (August, 1957).

14.11 Gerard, G., "Handbook of Structural Stability (Part IV – Failure of Plates and Composite Elements)", NACA 14.12
14.12 Gerard, G., "Handbook of Structural Stability (Part V – Compressive Strength of Flat Stiffened Panels)", NACA TN 3785, (August, 1957).
14.13 Gerard, G., "Handbook of Structural Stability (Part VI – Strength of Stiffened Curved Plates and Shells)", NACA TN 3786, (August, 1957).
14.14 Gerard, G., "Handbook of Structural Stability (Part VII – Strength of Thin Wing Construction)", NACA TN D-162, (Sept., 1959).
14.15 Hickman, W. A. and Dow, N. F., "Data on the Compressive Strength of 75S-T6 Aluminum-Alloy Flat Panels with Longitudinal Extruded Z-Section Stiffeners", NACA TN 1829, (Mar., 1949).
14.16 Hickman, W. A. and Dow, N. F., "Data on the Compressive Strength of 75S-T6 Aluminum-Alloy Flat Panels Having Small, Thin, Widely Spaced, Longitudinal Extruded Z-Section Stiffeners", NACA TN 1978, (Nov., 1949).
14.17 Rossman, C. A., Bartone, L. M., and Dobrowski, C. V., "Compressive Strength of Flat Panels with Z-Section Stiffeners", NACA RR 373, (Feb., 1944).
14.18 Niles, A. S., "Tests of Flat Panels with Four Types of Stiffeners", NACA TN 882, (Jan., 1943).
14.19 Schuette, E. H., Barab, S. and McCracken, H. L., "Compressive Strength of 24-T Aluminum-Alloy Flat Panels with Longitudinal Formed Hat-Section Stiffeners", NACA TN 1157, (Dec., 1946).
14.20 Holt, M. and Feil, G. W., "Comparative Tests on Extruded 14S-T and Extruded 24S-T Hat-Shape Stiffener Sections", NACA TN 1172, (Mar., 1947).
14.21 Dow, N. F. and Hickman, W. A., "Design Charts for Flat Compression Panels Having Longitudinal Extruded Y-Section Stiffeners and Comparison with Panels Having Formed Z-Section Stiffeners", NACA TN 1389, (Aug., 1947).
14.22 Gallaher, G. L. and Boughan, R. B., "A Method of Calculating the Compressive Strength of Z-Stiffened Panels that Develop Local Instability", NACA TN 1482, (Nov., 1947).
14.23 Hickman, W. A. and Dow, N. F., "Compressive Strength of 24S-T Aluminum-Alloy Flat Panels with Longitudinal Formed Hat-Section Stiffeners Having Four Ratios of Stiffener Thickness to Skin Thickness", NACA TN 1553, (Mar., 1948).
14.24 Schildsrout and Stein, "Critical Combinations of Shear and Direct Axial Stress for Curved Rectangular Panels", NACA TN 1928.
14.25 Argyris, J. H., "Flexure-Torsion Failure of Panels", Aircraft Engineering, Volume XXVI, (1954). pp. 174-184 and 213-219.
14.26 Semonian, J. W. and Peterson, J. P., "An Analysis of the Stability and Ultimate Compressive Strength of Short-Sheet-Panels with Special Reference to the Influence of Riveted Connection Between Sheet and Stringer", NACA TN 1255, (1956).
14.27 Rothwell, A., "Coupled Modes in the Buckling of Panels with Z-Section Stringer in Compression", Journal of The Royal Aeronautical Society, (Feb., 1968). pp. 159-163.
14.28 Cox, H. L., "The Application of the Theory of Stability in Structural Design", Journal of The Royal Aeronautical Society, Vol. 62, (July, 195-). pp. 498-512.
14.29 NASA CR-124075, "Isogrid Design Handbook", NASA-GEORGE C. MARSHALL SPACE FLIGHT CENTER, (Feb., 1973).
14.30 Gallaher and Boughan, "A Method for Calculating the Compressive Strength of Z-Stiffened Panels that Develop Local Instability", NACA TN 1482, (1947).
14.31 Boughan and Baab, "Charts for Calculating the Critical Compressive Stress for Local Instability of Idealized T Stiffened Panels", NACA WRL-204, (1944).
14.32 Catchpole, E. J., "The Optimum Design of Compression Surfaces Having Unflanged Integral Stiffeners", Journal of The Royal Aeronautical Society, Vol. 58, (Nov., 1954). pp. 765-768.
14.33 Saelman, B., "Basic Design and Producibility Considerations for Integrally Stiffened Structures", Machine Design, (Mar., 1955). pp. 197-203.

Chapter 15.0

DAMAGE TOLERANT PANELS (TENSION)

15.1 INTRODUCTION

The purpose of this chapter is to give the structural engineer a better understanding of fatigue, crack propagation and fail-safe design. The structural engineer's primary responsibility is to see that structures are designed to have adequate strength and rigidity at minimum weight. The airframer must also give customer an aircraft which will meet the customer's operational life requirements. This means that certain precautions must be taken against premature fatigue failure. Fatigue criteria are generally specified in customer contracts.

GLOSSARY:

  Fail-safe – Structure will support designated loads in spite of the failure of any single member or partial damage to a large part of the structure.

  Fatigue life – The number of repetitions (cycles or flights) of load or stress of a specified character that a specimen sustains before a specific failure occurs.

  Fatigue quality index (K) – A measure of the "fatigue quality" (generally the minimum value of 4.0 is the best but for practical applications 4.5 is used) of local conditions on a structure or component affected by the following factors:

  – Stress concentration
  – Load concentration
  – Eccentricity
  – Fretting fatigue
  – Surface finish
  – Corrosion fatigue
  – Damage

  Safe-life – The component must remain crack free during its service life.

  Scatter factor – A factor used to reduce the calculated fatigue life, time interval of crack growth, or verification testing of a safe-life structure (it is generally 3.0-5.0). The scatter factor is affected by the following factors:

  – A confidence level factor due to the size of the test sample establishing the fatigue performance
  – Number of test samples
  – An environmental factor that gives some allowable for environmental load history
  – A risk factor that depends on whether the structure is for safe-life or fail-safe capability

  S-N curves – (or F-N curves) Alternating or maximum stress (S or F) is plotted versus the number of cycles (N) to failure, as shown in Figures A and B in the example calculation given in Chapter 15.3. In this book use F instead of S which is used by fracture mechanics engineer.

Chapter 15.0

Since it is impossible to eliminate stress concentrations, it is, therefore, necessary to evaluate their effects in order to establish the fatigue life of a structure. The three items of major importance in determining the fatigue life of a given structure are material selection, the stress level, and the number of cycles to be tolerated at various stress levels (loading spectrum).

To ensure that an aircraft will perform satisfactorily in service, the structure is designed for the following four main failure modes:

- Static ultimate strength (including the yield strength requirement)
- Fatigue life of the structure (crack initiation)
- Fatigue life of the damaged structure (inspection interval)
- Static residual strength of the damaged structure

The damage tolerance of a structure is based on the progress of degradation/damage accumulation until a finite crack occurs, and the crack propagates until the failure process culminates in a complete failure of the structure, as shown in Fig. 15.1.1.

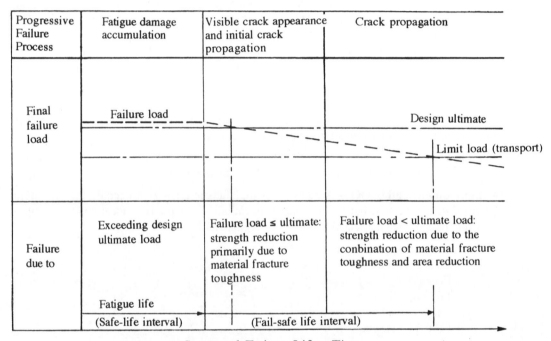

Fig. 15.1.1 Progressive Failure of a Structural Element

The fail-safe design principle, that is, adequate safety after some degree of damage, has reduced the fatigue problem from the safety level to the economic level. The following considerations are essential to the development of a damage tolerant design:

(a) Material selection:

The fatigue strength, crack growth tendency, and residual strength of the material receive primary consideration because these properties actually determine the size of the primary structure. The 2000 series of aluminum alloy has good fatigue properties and has exhibited satisfactory service experience in the past.

(b) The fastener load distribution:
- Accuracy of fastener load distribution, especially for the end (or first) fastener
- Fastener flexibility characteristics and hole filling characteristics, along with the joint configuration, have a primary influence upon local stress levels
- An improvement in fatigue performance can be obtained by using interference fit of fastener installation
- Fastener load distribution in area of stringer run-out (e.g., wing upper and lower skin-stringer panels)

(c) Ultimate design cut-off stress:
- Is usually selected for each airframe based on past experience, coupon and component testing, and fatigue analysis of representative points in the structure
- Depends on the severity of the loading spectrum
- Is represented by using the ultimate cut-off stress (in the neighborhood of 45,000 to 55,000 psi in commercial transport lower wing surface), as shown in Fig. 15.1.2

Aircraft	Lower wing surface	
	Material	Max. ultimate gross stress (psi)
B707-321	2024	55,000
B727-100	2024-T351	50,000
B737-200	2024-T351	45,000
B747	2024-T3 Skin-stringer panels	55,000
C-135A		55,000
B52A-D	7075-T6 Skin-stringer panels	55,000
B52(Orig.)	7178 Skin-stringer panels	65,000
B52G-H	2024-T3 Skin-stringer panels	49,000
DC-8	7075-T6 Skin-stringer panels	49,000
DC-8-55		45,000
DC-9	2024 Skin-stringer panels	44,000
DC-10	2024 Skin-stringer panels	52,000
CV880	7075 Skin-stringer panels	50,000
CV990	2024 Skin-stringer panels	49,900
L-1011	7075-T76 Skin-stringer panels	45,000
C-141A	7075-T6 Integrally-stiffened panels	55,000
S-3A		46,000
Jetstar		55,000
C-130E		55,000
P-3C		55,000
188A	7075-T6 Integrally-stiffened panels	50,000

Fig. 15.1.2 Ultimate Cut-off Stress of Lower Wing Surfaces

Based on past experience, fatigue design (considered during in detail design) alone is not enough to safeguard airframe structural integrity and additional design concepts must be considered:

- Crack growth analysis – Assume crack could occur and use crack growth analysis to setup inspection time intervals (e.g., a major overhaul maintenance "D" check is required every 4 to 5 years for commercial transports) such that inspection will occur before crack could reach critical length
- Fail-safe design – Residual strength at the critical crack length (under limit load)

15.2 THE STRESS CYCLE AND THE LOAD SPECTRUM

(A) THE STRESS CYCLE

The stress cycle is the stress-time function which is repeated periodically and identically, as shown in Fig. 15.2.1.

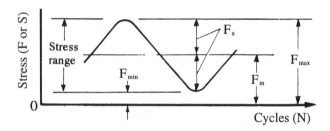

Fig. 15.2.1 Stress Cycle (N)

- F_{max} (maximum stress) – The highest algebraic value of stress in the stress cycle, for either tensile stress (+) or compressive stress (–)
- F_{min} (minimum stress) – The lowest algebraic value of stress in the stress cycle, for either tensile stress (+) or compressive stress (–)
- F_a (alternating stress or variable stress):

$$F_a = \frac{F_{max} - F_{min}}{2}$$ Eq. 15.2.1

- F_m (mean stress):

$$F_m = \frac{F_{max} + F_{min}}{2}$$ Eq. 15.2.2

- R (stress ratio):

$$R = \frac{F_{min}}{F_{max}}$$ Eq. 15.2.3

(B) THE LOAD SPECTRUM

The loading environment is a collection of data based on a statistical assessment of the load exposure of a structure. Gust, maneuvers, landing impact, ground roll or taxi, climb cruise, descent, etc., all have a part in defining the load history input. Fig. 15.2.2 illustrates the flight profiles of transport and fighter aircraft.

Damage Tolerant Panels (Tension)

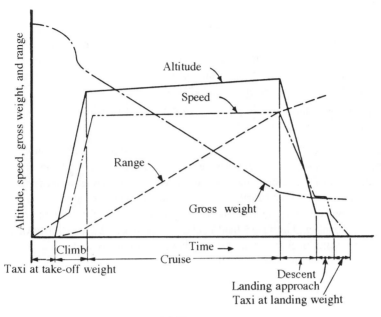

(a) Transports

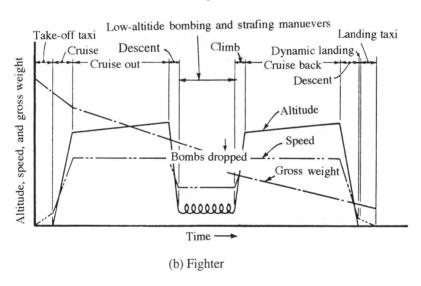

(b) Fighter

Fig. 15.2.2 Typical Aircraft Flight Profiles

Many discrete loading distributions are used to give the actual number of cycles to be applied at particular load levels for the purpose of representing a cumulative loading or stress spectrum in fatigue analysis and tests. The procedure for converting a cumulative stress spectrum to a discrete stress spectrum is shown in the example calculation in Chapter 15.3.

In power-spectral gust load analysis (see Reference 3.4), which is used to describe turbulence exposure, the atmosphere is defined in terms of its own characteristics:

- Generalizes description of the atmosphere in terms of frequencies and the coincident power in each frequency level
- Requires definition of the elastic response characteristics of the aircraft in terms of a transfer function which will translate the defined gust spectrum into a specific aircraft response

The stress range of the ground-air-ground cycle (GAG), as shown in Fig. 15.2.3, for the wing or the pressurized fuselage is probably the largest and most frequently applied single load cycle for most transport aircraft. Every flight regardless of extent of maneuvering and turbulence exposure adds load variability and fatigue damage to the structure. Maneuvers and turbulence are merely superimposed upon the 1.0g levels on the ground or in the air so that the maximum stress range of the combined loads is taken as a measure of the GAG cycle:

- Short flights have a preponderance of GAG load cycles
- Long flights have fewer GAG load cycles

During the early phases of design, the design ultimate cut-off stresses (see Fig. 15.1.2) must be established for the various structural components. The determination of these stresses must be based on consideration of the anticipated loading history to provide assurance that the design life will be achieved. The use of GAG load cycles provides a relatively rapid means for establishing the permissible design ultimate tension stress for preliminary sizing of structures subjected to complex spectra of loading. The GAG load cycle is a measure of the severity of the anticipated loading spectra and it produces the same fatigue damage to the structure as would be produced if the complete spectra of loading were considered.

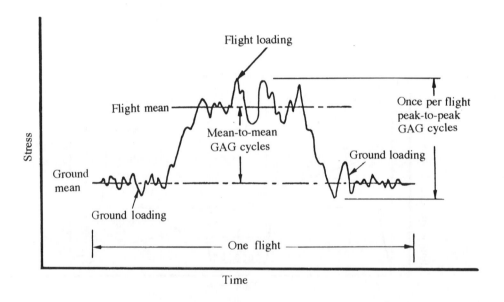

Fig. 15.2.3 Ground-Air-Ground (GAG) Loading

15.3 STRUCTURAL LIFE PREDICTION (SAFE-LIFE)

The simplest and most practical technique for predicting fatigue performance is the Palmgren-Miner hypothesis. The hypothesis contends that fatigue damage incurred at a given stress level is proportional to the number of cycles applied at that stress level divided by the total number of cycles required to cause failure at the same level. If the repeated loads are continued at the same level until failure occurs, the cycle ratio will be equal to one. When fatigue loading involves many levels of stress amplitude, the total damage is a sum of the different cycle ratios and failure should still occur when the cycle ratio sum equals one, as shown in Eq. 15.3.1. and Fig. 15.3.1.

$$D = \sum_{i=1}^{k} \left(\frac{n_i}{N_i}\right) \qquad \text{Eq. 15.3.1}$$

where: D – Fatigue life utilization ratio
  n_i – Number of loading cycles at the ith stress level
  N_i – Number of loading cycles to failure for the ith stress level based on S-N curve data for the applicable material and stress concentration factor
  k – Number of stress levels considered in the analysis

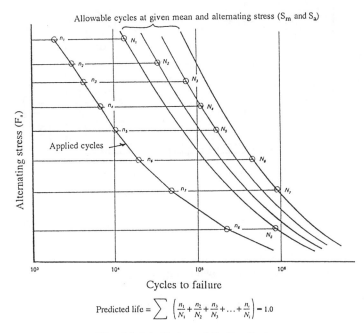

Predicted life = $\sum \left(\frac{n_1}{N_1} + \frac{n_2}{N_2} + \frac{n_3}{N_3} + ... + \frac{n_i}{N_i} \right) = 1.0$

Fig. 15.3.1 Fatigue Life Prediction

Example:

Use the Palmgren-Miner cumulative damage equation (Eq. 15.3.1) to predict the fatigue life of a part which is aluminum 2024-T3 material with the following given data:

  Use scatter factor = 3.0 for this calculation.

  Use fatifue quality index, K = 4.0 and the matreial S-N (where F = S stress) curves as shown below:

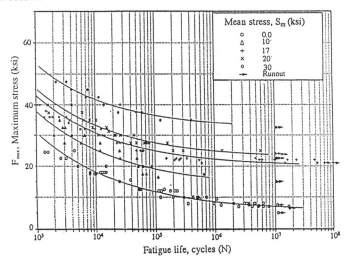

Fig. A S-N Curves [F_{max} vs. cycles (N)] for 2024-T3 with K = 4.0 (from Fig. 9.12.16)

Chapter 15.0

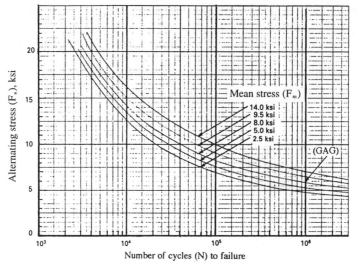

[**Note:** These curves are generated from S-N curves with F_a vs. cycles (N) shown in previous Fig. A]

Fig. B S-N Curves [F_a vs. cycles (N)] for 2024-T3 with K = 4.0

The loading spectrum applied to the structural component, consisting of 5 loading sequences arranged in flight-by-flight load cycles is shown in Fig. C.

(a) Determine the GAG stresses for every 1,000 flights

F_m ①	F_a ②	n ③	F_{max} ④ = ① + ②	F_{min} ⑤ = ① − ②
8.0	3.7	70	11.7	4.3
	7.0	10	15	1.0
	8.3	150	16.3	− 0.3
	9.1	30	17.1	− 1.1
9.5	2.1	50	11.6	7.4
	7.9	140	17.4	1.6
	13.0	150	22.5	− 3.5
	13.8	40	23.3	− 4.3
2.5	12.5	70	15.0	−10.0
	13.5	60	16.0	−11.0
	15.1	40	17.6	−12.6
	15.8	40	18.3	−13.3
14.0	7.0	150	21.0	7.0
	7.8	70	21.8	6.2
5.0	9.6	180	14.6	− 4.6
	10.3	120	15.3	− 5.3
	10.9	50	15.9	− 5.9
	11.7	30	16.7	− 6.7

(Total 1,450 cycles)

Fig. C Stress Spectrum (ksi) Occurring Every 1,000 Flights

Damage Tolerant Panels (Tension)

S_{max}	n	Σn	F_{min}	n	Σn
23.3	40	40	−13.3	40	40
22.5	150	190	−12.6	40	80
21.8	70	260	−11.0	60	140
21.0	150	410	−10.0	70	210
18.3	40	450	− 6.7	30	240
17.6	40	490	− 5.9	50	290
17.4	140	630	− 5.3	120	410
17.1	30	660	− 4.6	180	590
16.7	30	690	− 4.3	40	630
16.3	150	840	− 3.5	150	780
16.0	60	900	− 1.1	30	810
15.9	50	950	− 0.3	150	960
15.3	120	1,070	1.0	10	970
15.0	70	1,140	1.6	140	1,110
15.0	10	1,150	4.3	70	1,180
14.6	180	1,330	6.2	70	1,250
11.7	70	1,400	7.0	150	1,400
11.6	50	1,450	7.4	50	1,450

Fig. D Translation of the Max. Stress (F_{max}) and Min. Stress (F_{min}) vs. Σn (Cycles)

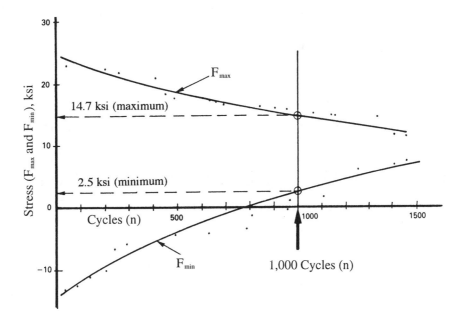

Fig. E Plots of the Max. and Min. Stresses (F_{max} & F_{min}) vs. Σ Cycles (Data from Fig. D)

The alternating stress (from Eq. 15.2.1):

$$F_a = \frac{F_{max} - F_{min}}{2} = \frac{14.7 - 2.5}{2} = \pm 6.1 \text{ ksi}$$

where: $F_{max} = 14.7$ ksi and $F_{min} = 2.5$ ksi from Fig. E

(b) Calculate the fatigue life utilization ratio (D):

F_m	F_a (From Fig. C)	n	N (From Fig. B)	$\frac{n}{N}$
8.0	3.7	70	–	–
	7.0	10	270,000	0
	8.3	150	100,000	0.0015
	9.1	30	64,000	0.00047
9.5	2.1	50	–	–
	7.9	140	240,000	0.00058
	13.0	150	18,000	0.00833
	13.8	40	14,000	0.00286
2.5	12.5	70	10,500	0.0067
	13.5	60	8,200	0.00732
	15.1	40	6,000	0.00667
	15.8	40	5,000	0.008
14.0	7.0	150	1,100,000	0.00014
	7.8	70	430,000	0.000163
5.0	9.6	180	36,000	0.005
	10.3	120	26,000	0.00462
	10.9	50	21,000	0.00238
	11.7	30	16,000	0.00188
(Add GAG cyclic loads from Fig. E):				
8.6	6.1	1,000	1,000,000	0.001
				$D = \Sigma\, 0.05761$

(c) Predict fatigue life:

$$\text{The predicted life} = \frac{1,000}{D \times \text{(scatter factor)}} = \frac{1,000}{0.05761 \times 3.0} = 5,786 \text{ flights}$$

15.4 STRUCTURAL CRACK GROWTH (INSPECTION INTERVAL)

To determine the structural crack growth behavior start with the given:

- Initial crack size
- The stress spectra at the location being analyzed
- The crack propagation and fracture toughness properties of the material

Crack growth rate depends on:

- Crack growth characteristics of the material, as illustrated in Fig. 15.4.1
- Environmental effects (rate of growth increases in humidity)
- Loading frequency
- Crack geometry (e.g., edge crack, surface flaw, through crack, etc.)
- Structural arrangement (e.g., skin-stringer panels, integrally-stiffened panels, etc.)
- Periodic high load tends to slow or retard the crack growth rate

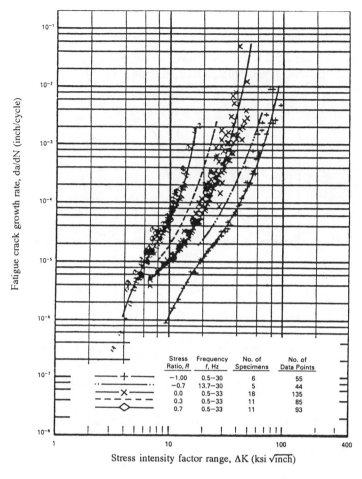

Fig. 15.4.1 Crack Growth Data for 0.09" thickness of 7075-T6 (Ref. 4.1)

The following is an approximate method for a given spectrum loading. The crack growth is calculated in crack length increments per flight basis, as shown in Fig. 15.4.2. See Chapter 15.5 of Reference 15.10 for constant amplitude load case.

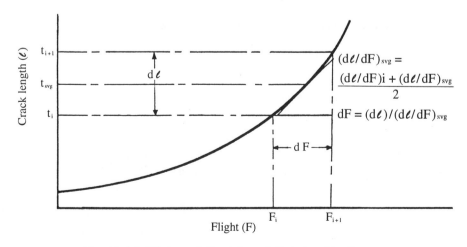

Fig. 15.4.2 Flight and Crack Length Relationship Curve

Chapter 15.0

The calculations are performed in the following steps:

Step 1: Assume crack length, ℓ_i (in.)

Step 2: Calculate ΔK (crack growth stress intensity factor) for each cycle in the spectrum (the example spectrum is shown in Fig. 15.4.3)

$$\Delta K = F_{max}(\sqrt{1-R})\sqrt{\ell} \text{ (Aluminum material)} \qquad \text{Eq. 15.4.1}$$

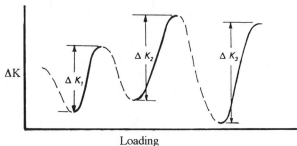

Fig. 15.4.3 ΔK vs. Spectrum Loading

Step 3: Calculate number of cycles for each ΔK_i

$$\text{Each flight} = \frac{\text{Total number of cycles}}{\text{Number of flights}} = \text{(cycles/flight)}$$

Step 4: Determine $\frac{d\ell}{dN}$ (e.g., from Fig. 15.4.1 where $\frac{d\ell}{dN} = \frac{da}{dN}$) for each ΔK (from Step 2)

Step 5: Determine $\Delta \ell_i = \frac{d\ell}{dN} \times \text{(cycles/flight)}$ for each stress cycle in the spectrum

Step 6: Sum $\Delta \ell_i$'s for all cycles in the flight:

$$\frac{\Sigma \Delta \ell_i}{\text{Flight}}$$

Step 7: Calculate the average crack growth rate per flight (F):

$$\frac{\frac{d\ell_1}{dF} + \frac{d\ell_2}{dF}}{2} = (\frac{d\ell}{dF})_{avg}$$

Step 8: Calculate number of flights versus increment of crack growth:

$$\text{Number of flights} = \frac{\Delta \ell}{(\frac{d\ell}{dF})_{avg}}$$

Step 9: Repeat calculations for all increments of crack growth.

Example:

Calculate the crack growth for one increment of growth in 7075-T6 aluminum alloy (see Fig. 15.4.1); the stress spectra are shown in columns ①, ② and ③ in Fig. A. Calculate the number of flights for the crack to increase from 1.0" to 1.5" (neglect the effect of compression stress in crack growth).

(a) Use the following tabular calculation for convenience:

Flight conditions	F_{max} (ksi) ①	F_{min} (ksi) ②	No. of cycles per 8,000 flights ③	Cycles per flight ④ = ③/8,000	R Stress ratio ⑤ = ②/①	⑥ = ①$\sqrt{1-R}$
Maneuver	12.5	7.2	102 × 10³	12.750	0.576	8.14
	16.1	7.2	57 × 10³	7.125	0.447	11.97
	19.6	7.2	38 × 10³	4.75	0.367	15.59
	23.0	7.2	28 × 10³	3.5	0.313	19.06
	26.6	7.2	14 × 10³	1.75	0.27	22.71
Gust	13.5	−0.6	16.2 × 10³	2.025	0	13.5
	14.8	−2.1	1.73 × 10³	0.216	0 (zero for negative values)	14.8
	16.0	−3.5	0.5 × 10³	0.063	0	16.0
	17.8	−4.8	0.13 × 10³	0.016	0	17.8
GAG	25.0	−1.0	8.0 × 10³	1.0	0	25

Fig. A Tabular Calculation

Flight conditions	ΔK (ksi $\sqrt{in.}$) ⑦* = ⑥ × $\sqrt{\ell}$	$(\frac{d\ell}{dN}) \times 10^{-5}$ (inch/cycle) From Fig. 15.4.1 ⑧	$\Delta \ell \times 10^{-5}$ (inch/flight) ⑨ = ④ × ⑧
Maneuver	8.14	0.6	7.65
	(9.97)	(1.0)	(12.75)
	11.97	1.5	10.69
	(14.66)	(2.2)	(15.68)
	15.59	3.0	14.25
	(19.09)	(6.0)	(28.5)
	19.06	6.0	21.0
	(23.34)	(11.0)	(38.5)
	22.71	11.0	19.25
	(27.81)	(13.0)	(22.75)
Gust	13.5	1.0	2.03
	(16.53)	(2.2)	(4.46)
	14.8	2.5	0.54
	(18.13)	(4.5)	(0.97)
	16.0	3.0	0.19
	(19.6)	(6.0)	(0.38)
	17.8	4.0	0.06
	(21.8)	(10.0)	(0.16)
GAG	25	20.0	20.0
	(30.62)	(40.0)	(40.0)

$$\Sigma \Delta \ell = \frac{95.7 \times 10^{-5}}{(164.2 \times 10^{-5})}$$

Notes: The above calculated values: xxx for crack length $\ell_1 = 1.0"$
(xxx) for crack length $\ell_2 = 1.5"$

Fig. B Tabular Calculation

(b) Average crack growth rate:

$$\frac{95.7 \times 10^{-5} + 164.2 \times 10^{-5}}{2} = 129.95 \times 10^{-5} \text{ (in./flight)}$$

(c) Number of flights for crack to grow from 1.0" to 1.5":

$$\frac{0.5}{129.95 \times 10^{-5}} = 385 \text{ flights (F)}$$

15.5 RESIDUAL STRENGTH (FAIL-SAFE DESIGN)

A large percentage of airframe structures exhibit fail-safe capability without having been specifically designed for such a characteristic. Because loading conditions in service seldom exceed limit loads it is not necessary to size the structure to be fail-safe for ultimate design loads. Therefore, the limit loads are considered to be ultimate loads for fail-safe design as specified by U.S. FAR 25.571 and 25.573 for commercial transports. For military airframe requirements refer to Reference 15.2.

The sizing or design procedures are as follows:

- Establish fail-safe design criteria and **required stress level**
- Specify fail-safe damage limits
- Select materials with high fracture toughness characteristics
- Specify fail-safe design stress allowables
- Specify methods of fail-safe analysis
- Establish fail-safe design load levels
- Size structure to meet damage limit requirements
- Require proof tests

The sizing methods to retain residual strength are as follows:

(a) The gross skin area stress without reinforcements (e.g., stringers or stiffeners):

$$F_g = \frac{K}{(\sqrt{\frac{\pi}{2}})(\sqrt{\ell})\alpha} \qquad \text{Eq. 15.5.1}$$

where: $K = F_g(\sqrt{\pi a})\alpha$ – Critical stress intensity factor in ksi $\sqrt{\text{in.}}$ (e.g., Fig. 4.5.1 in Chapter 4.0)
F_g – Skin gross area stress (ksi)
ℓ – Through crack length ($\ell = 2a$) (in.)
α – Correction factor for skin width (generally use $\alpha = 1.0$)

(b) The basic calculation for the ultimate residual strength of a damaged structure with reinforcement is given below:

$$F_g = \frac{\beta K}{(\sqrt{\frac{\pi}{2}})(\sqrt{\ell})\alpha} \qquad \text{Eq. 15.5.2}$$

where: β – Reinforcement factor (from Fig. 15.5.1)

Damage Tolerant Panels (Tension)

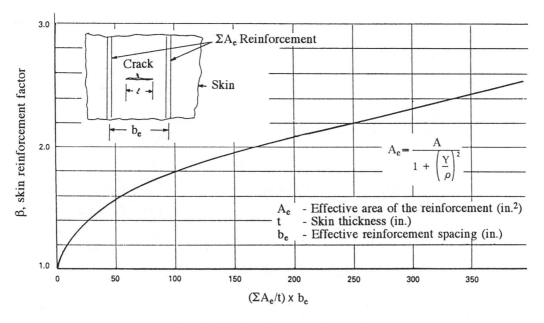

Fig. 15.5.1 Reinforcement Factor (β)

Example:

Determine the fail-safe allowable stress for the damaged skin-stringer panel (7075-T6) shown below; stringer spacing equals 10 inches (crack in skin with crack length, $\ell = 10$ inches).

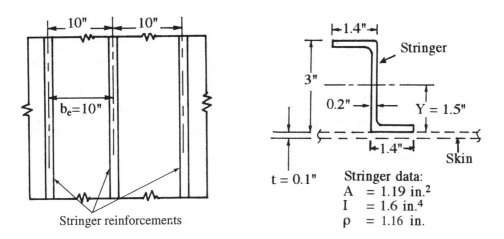

(a) Compute the effective area of the Z-stringer:

$$A_e = \frac{A}{1 + (\frac{Y}{\rho})^2} = \frac{1.19}{1 + (\frac{1.5}{1.16})^2} = 0.45 \text{ in.}^2$$

$$(\frac{\Sigma A_e}{t})(b_e) = (\frac{2 \times 0.45}{0.1})(10) = 90$$

Use the computed value in Fig. 15.5.1 to obtain the reinforcement factor $\beta = 1.75$

K = 60 ksi (from Fig. 4.5.1)

669

(b) The damaged allowable stress with $l = 10"$ crack length between stringers (from Eq. 15.5.2):

$$F_g = \frac{\beta K}{(\sqrt{\frac{\pi}{2}})(\sqrt{l})\alpha} = \frac{1.75 \times 60}{(\sqrt{\frac{\pi}{2}})(\sqrt{10})1.0} = 26.5 \text{ ksi}$$

(c) If the crack length increase to $l = 20"$ and the stringer over the cracked skin is broken:

$$(\frac{\Sigma A_e}{t})(b_e) = (\frac{2 \times 0.45}{0.1})(20) = 180 \text{ and from Fig. 15.5.1, obtain } \beta = 2.05$$

$$F_g = \frac{\beta K}{(\sqrt{\frac{\pi}{2}})(\sqrt{l})\alpha} = \frac{2.05 \times 60}{(\sqrt{\frac{\pi}{2}})(\sqrt{20})1.0} = 21.95 \text{ ksi}$$

15.6 RESIDUAL STRENGTH OF BEAM ASSEMBLY

In general the shear web is not considered to be as critical as a tension panel for damage tolerant structures. However, for example, the wing spar web is considered to be a primary structure and a crack in the web is a major concern in wing design. Therefore the structural engineer must understand the characteristics and phenomenon of crack growth in webs for both fatigue life and fail-safe residual strength as well as static strength. Fig. 15.6.1 illustrates a web crack which starts at the lower beam cap (assume lower cap is not broken). Assuming there is an up-bending moment acting on the wing box, the crack propagates upward for a small distance and then has the tendency to turn diagonally inboard. As crack goes upward the bending tension stress (spanwise) becomes smaller and is replaced by diagonal tension stress from the shear load in the web.

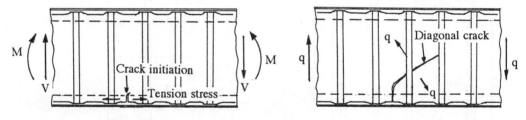

(a) Crack Initiation due to Tension Stress   (b) Diagonal Crack Propagation due to Shear

Fig. 15.6.1 Cracked Web on a Beam Assembly (Built-up Construction)

The following recommendations should be considered when designing a beam assembly:
- First, size the assembly based on the static strength
- Use ductile material for stiffeners and closer spacing
- Use increased thickness for the flange which attaches the stiffener to the web
- For integrally-machined spar design, as shown in Fig. 15.6.2, provide a spanwise auxiliary flange located about $\frac{1}{3}$ to $\frac{1}{4}$ of the beam depth from the lower flange to stop and/or retard crack growth and to provide residual strength when the lower flange and web are cracked
- Use Finite Element Analysis to model a cracked web to determine load distribution in this highly redundant structure
- Check stiffener attachment strength for those stiffeners attached to spar caps
- Verify the design by use of damage tolerance test for both fatigue and fail-safe residual strength

Damage Tolerant Panels (Tension)

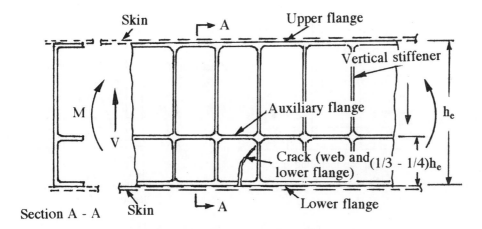

Fig. 15.6.2 Integrally-Machined Beam

The sizing procedure which is presented is derived from the test results of a beam assembly. The design limitations of this method are as follows:

- Valid for built-up beam assemblies (stiffeners are fastened on web) only
- The crack, which is vertical initially, will become diagonal
- The max. uncracked depth (h_{un}) ≤ h_e (beam depth between cap centroids)
- There is an affected stiffener in the path of the crack
- Aluminum alloys

(a) Calculate the following fictitious loads:

The residual strength of a shear web assembly is a combination of the strengths of the web, vertical stiffeners, and lower spar cap, shown in Fig. 15.6.3; the total fictitious residual strength is:

$$V' = V'_{cap} + V'_{st} + V'_{web} \qquad \text{Eq. 15.6.1}$$

where: V' – Total shear load (lbs.) capability
V'_{cap} – Shear load carried by lower spar cap
V'_{st} – Shear load carried by stiffener
V'_{web} – Shear load carried by uncracked portion of the web

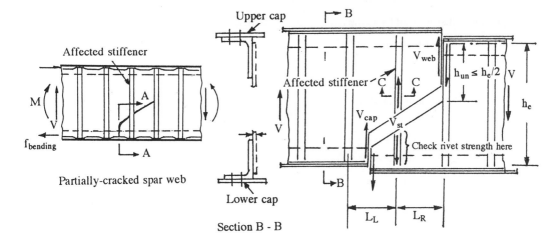

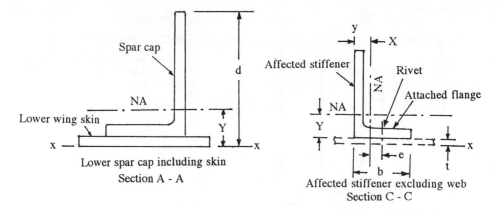

Fig. 15.6.3 Beam Assembly with Partially-Cracked Web

(a1) Fictitious load fraction on cracked web (V'_{web}):

$$\frac{V'_{web}}{V'} = 26.1 \left(\frac{h_e t}{L_R + L_L}\right) \qquad \text{Eq. 15.6.2}$$

where: h_e – Beam depth between cap centroids
 t – Web thickness
 L_R – Stiffener spacing on right-hand side
 L_L – Stiffener spacing on left-hand side

(The unit of the constant 26.1 is 1/in.)

(a2) Fictitious load fraction on affected stiffener (V'_{st}):

$$\frac{V'_{st}}{V'} = 0.067 \, h \, t \, L_R \qquad \text{Eq. 15.6.3}$$

(The unit of the constant 0.067 is $1/in.^3$)

(a3) Fictitious load fraction on lower spar cap (V'_{cap}):

$$\frac{V'_{cap}}{V'} = \frac{1,158.6 \, I_{cap}}{(L_R + L_L)^3} \qquad \text{Eq. 15.6.4}$$

where: I_{cap} – The moment of inertia of the cap including wing surface skin

(The unit of the constant of 1,158.6 is 1/in.)

(b) Load distribution (V is the applied shear load):

(b1) Load on cracked web (V_{cap}):

$$V_{web} = V \left(\frac{V'_{web}}{V'}\right) \qquad \text{Eq. 15.6.5}$$

(b2) Load on affected stiffener (V_{st}):

$$V_{st} = V \left(\frac{V'_{st}}{V'}\right) \qquad \text{Eq. 15.6.6}$$

(b3) Load on lower spar cap (V_{cap}):

$$V_{cap} = V \left(\frac{V'_{cap}}{V'}\right) \qquad \text{Eq. 15.6.7}$$

(c) MS of uncracked portion of the beam web (V_{web}):

The maximum shear stress in the uncracked portion of the web is:

$$f_{web} = \frac{V_{web}}{t\, h_{un}}$$

Eq. 15.6.8

where: h_{un} – Height of uncracked portion of web

$$MS_{web} = \frac{F_{su}\,\eta}{f_{web}} - 1$$

where: η – Rivet hole efficiency factor in the web

(d) MS of stiffener (V_{st}):

The maximum tension stress on stiffener is:

$$f_{st} = \frac{V_{st}}{\eta\, A_{st}} + \frac{V_{st}\, e\,(b - X)}{I_{yy}} + \frac{V_{st}\, Y^2}{I_{xx}}$$

Eq. 15.6.9

where: η – Rivet hole efficiency factor in the stiffener

$$MS_{st} = \frac{F_{tu}}{f_{st}} - 1$$

Check rivet strength between affected stiffener and web in the area below the crack line, as shown in Fig. 15.6.3.

(e) MS of lower spar cap (V_{cap}):

The maximum stress on lower cap is:

$$f_{cap} = f_{bending} + \frac{M_e\, Y}{I_{xx}}$$

Eq. 15.6.10

where: $f_{bending}$ – The bending stress at the centroid of the beam cap
$L_e = 0.453\, L_L$ – The effective moment arm of the cap
$M_e = V_{cap} \times L_e$ – The equivalent moment

$$MS_{cap} = \frac{F_{tu}}{f_{cap}} - 1$$

Example:

A rear beam web assembly is designed to carry a limit load, V = 50,000 lbs., and has a cracked web as shown below.

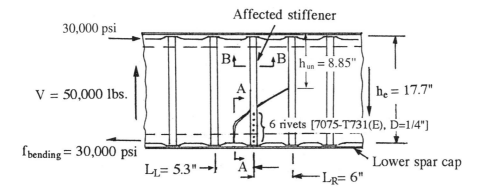

673

Chapter 15.0

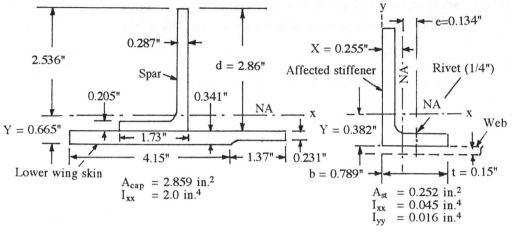

Section A - A
Excluding web

Section B - B
Excluding web

$L_R = 6"$ $L_L = 5.3"$

$h_e = 17.7"$ $h_{un} = \dfrac{17.7}{2} = 8.85"$ $t = 0.15"$

$f_{bending} = 30,000$ psi (limit load bending stress at the lower cap)

Web material (2024-T3 bare plate, see Fig. 4.3.3):

   $F_{tu} = 66,000$ psi $F_{su} = 41,000$ psi $F_{bru} = 133,000$ psi

Spar cap material (2024-T3 extrusion, see Ref. 4.1):

   $F_{tu} = 62,000$ psi $F_{su} = 32,000$ psi $F_{bru} = 101,000$ psi

Stiffener material (7075-T6 extrusion, see Fig. 4.3.5):

   $F_{tu} = 82,000$ psi $F_{su} = 44,000$ psi $F_{bru} = 148,000$ psi

(a) Calculate the following fictitious loads:

From Eq. 15.6.2:

$$V'_{web} = 26.1V'(\dfrac{h_e t}{L_R + L_L}) = 26.1V'(\dfrac{17.7 \times 0.15}{6 + 5.3}) = 6.13V'$$

From Eq. 15.6.3:

$$V'_{st} = 0.067V' h_e t L_R = 0.067V' \times 17.7 \times 0.15 \times 6 = 1.07V'$$

From Eq. 15.6.4:

$$V'_{cap} = \dfrac{1,158.6V' I_{cap}}{(L_R + L_L)^3} = \dfrac{1,158.6V' \times 2.0}{(6 + 5.3)^3} = 1.61V'$$

(b) Load distribution:

$$V' = V'_{web} + V'_{st} + V'_{cap} = 6.13V' + 1.07V' + 1.61V' = 8.81V'$$

From Eq. 15.6.5, obtain load on cracked web (V_{web}):

$$V_{web} = V(\dfrac{V'_{web}}{V'}) = 50,000(\dfrac{6.13V'}{8.81V'}) = 34,790 \text{ lbs.}$$

(V is the total applied shear load at the beam assembly)

From Eq. 15.6.6, obtain load on affected stiffener (V_{st}):

$$V_{st} = V\left(\frac{V'_{st}}{V'}\right) = 50{,}000\left(\frac{1.07V'}{8.81V'}\right) = 6{,}073 \text{ lbs.}$$

From Eq. 15.6.7, obtain load on lower spar cap (V_{cap}):

$$V_{cap} = V\left(\frac{V'_{cap}}{V'}\right) = 50{,}000\left(\frac{1.61V'}{8.81V'}\right) = 9{,}137 \text{ lbs.}$$

(c) MS of uncracked portion of the beam web (V_{web}):

From Eq. 15.6.8, obtain the maximum shear stress on the uncracked web (2024-T3, plate):

$$f_{web} = \frac{V_{web}}{t\,h_{un}} = \frac{34{,}790}{0.15 \times 8.85} = 26{,}207 \text{ psi}$$

$$MS_{web} = \frac{41{,}000 \times 0.75}{26{,}207} - 1 = \underline{0.17} \qquad \text{O.K.}$$

use rivet hole efficiency; $\eta = 75\%$

(d) MS of affected stiffener (V_{st}):

From Eq. 15.6.9, obtain the maximum stress on affected stiffener (7075-T6 Extrusion):

Use rivet hole efficiency; $\eta = 85\%$ ($A_{st} - \frac{1}{4}''$ dia. rivet hole)

$$f_{st} = \frac{V_{st}}{\eta A_{st}} + \frac{V_{st}\,e\,(b-X)}{I_{yy}} + \frac{V_{st}\,Y^2}{I_{xx}}$$

$$= \frac{6{,}073}{0.85 \times 0.252} + \frac{6{,}073 \times 0.134 \times (0.789 - 0.255)}{0.016} + \frac{6{,}073 \times 0.382^2}{0.045}$$

$$= 28{,}352 + 27{,}160 + 19{,}693 = 75{,}205 \text{ psi}$$

$$MS_{st} = \frac{82{,}000}{75{,}205} - 1 = \underline{0.09} \qquad \text{O.K.}$$

Check rivet strength at the lower portion of the affected stiffener; there are six 7075-T731(E) rivets with dia. = $\frac{1}{4}''$ and the shear allowable is $P_{s,\,all} = 2{,}230$ lbs./rivet (see Fig. 9.2.5).

The rivet bearing strength for the web material (2024-T3):

$P_{b,\,all} = 0.25 \times 0.15 \times 133{,}000 \times 80\% = 3{,}990$ lbs./rivet $> P_{s,\,all}$ NOT CRITICAL

[80% is the bearing reduction factor (see Chapter 9.2) when rivets are installed wet]

$$MS_{rivet} = \frac{6 \times P_{s,\,all}}{V_{st}} - 1 = \frac{6 \times 2{,}230}{6{,}073} - 1 = \underline{\text{High}} \qquad \text{O.K.}$$

(e) MS of lower spar cap (V_{cap}):

From Eq. 15.6.10, obtain the maximum stress on the lower cap (2024-T3 Extrusion):

$L_e = 0.453 L_L = 0.453 \times 5.3 = 2.4''$

$M_e = V_{cap} \times L_e = 9{,}137 \times 2.4 = 21{,}929$ in.-lbs.

$$f_{cap} = f_{bending} + \frac{M_e Y}{I_{xx}} = 30{,}000 + \frac{21{,}929 \times 2.536}{2.0} = 57{,}806 \text{ psi}$$

$$MS_{cap} = \frac{62{,}000}{57{,}806} - 1 = \underline{0.07} \qquad \text{O.K.}$$

References

15.1 Federal Aviation Regulations (FAR), Vol. III, Part 25.571 – Airworthiness Standards: Transport Category. "14 Code of Federal Regulations (14 CFR)", U.S. Government Printing Office, Washington, D. C., (1992).

15.2 MIL-A-83444 (USAF), "Airplane Damage Tolerance Requirements", (July, 1974).

15.3 Osgood, C. C., "FATIGUE DESIGN", John Wiley & Sons, Inc., New York, NY, (1970).

15.4 Broek, David, "The Practical Use of Fracture Mechanics", Kluwer Academic Publishers, 101 Philip Drive, Norwell, MA 02061, (1988).

15.5 Anon., "Fatigue and Stress Corrosion – Manual for Designers", Lockheed-California Company, Burbank, CA, (1968).

15.6 Ekvall, J. C., Brussat, T. R., Liu, A. F., and Creager, M., "Preliminary Design of Aircraft Structures to Meet Structural Integrity Requirements", Journal of Aircraft, (March, 1974). pp. 136-143.

15.7 Mackey, D. J. and Simons, H., "Structural Development of The L-1011 Tri-star", AIAA Paper No. 72-776, (1972).

15.8 Anon., "Specialists Meeting on Design Against Fatigue", AGARD-CP-141, (Oct., 1971).

15.9 Anon., "Problems with Fatigue in Aircraft", Proceedings of the Eighth ICAF Symposium (ICAF Doc. 801) Held at Lausanne, 2-5 June, (1975).

15.10 Niu, C. Y., "AIRFRAME STRUCTURAL DESIGN", Hong Kong Conmilit Press Ltd., P.O. Box 23250, Wanchai Post Office, Hong Kong, (1988).

15.11 Chang, J.B., Engle, R.M., and Hiyama, R.M., "Application of an Improved Crack Growth Prediction Methodology on Structural Preliminary Design", ASTM STP761, American Society for Testing and Materials, PP. 278-295, (1982).

Chapter 16.0

STRUCTURAL SALVAGE AND REPAIRS

16.1 INTRODUCTION

Airframe damage does not always result from human error but may occur because of a material fatigue crack, corrosion, etc., on an in-service or aging aircraft. The existing structural repair manuals for commercial aircraft [or T.O. (Technical order) for military aircraft] give repair engineers very simple and elementary repair methods which are appropriate for small areas of damage only. Any major repair rework must be approved by the original aircraft manufacturer and, in some cases, the repair must be performed by the original manufacturer.

To be an airframe liaison and repair engineer one must not only be familiar with detail design and have years of practical experience in airframe work but one must also understand stress analysis. For repair work, the engineer must understand how to repair different types of structures, in different areas, and under different environments. The accessibility in the repair area is a major concern not only to accomplish the repair but also to reinforce members in the location of the repair or retrofit. For example, repair work is frequently performed inside a wing or empennage box where the working space is restricted. No universal repair method exists; each repair case must be treated as an individual design case. In other words, most repair cases are much more complicated than starting a new design from scratch.

(A) CLASSIFICATION OF AIRFRAME VITALITY

It is important that the liaison engineer be able to recognize the classification of airframe vitality, as shown in Fig. 16.1.1, to which the salvage or repair parts belong before evaluating the problem.

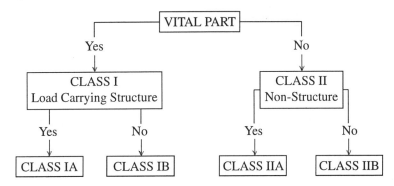

Fig. 16.1.1 Classification of Airframe Vitality

Classification of airframe structure:

(a) Class I – Vital structures, parts, and equipment are those whose failure:
- By any means, could cause loss of the aircraft
- Could cause loss of control of the aircraft

Class IA – Single load path
Class IB – Multi-load path

(b) Class II – Non-vital structures, parts, and equipment are those whose failure:

Class IIA – Could cause injury to personnel on the ground or in the air

Class IIB:
- Essentially non-structural components whose failure would not effect the efficiency or serviceability of the system
- No formal stress analysis is required for airworthiness

Also structural salvage and repair dispositions must be approved by the cognizant structural engineer or designated representative. In no case shall a negative margin of safety be tolerated.

(B) REPAIRABILITY

Repairability means the structure can be repaired while the aircraft is in service; therefore, during the design stage the final size of the structural dimensions should allow sufficient space for future repair using fasteners, see Fig. 16.1.2.

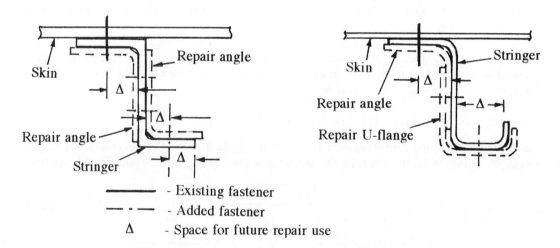

Fig. 16.1.2 Sufficient Space for Repair

Various arrangements of repair elements on typical structural sections are shown in Fig. 16.1.3:
- When two or more repair elements are used on a member, they should overlap and be joined by fasteners throughout the length of the overlap
- The free-edged flange of a repair element shall have at least the same flange width as the corresponding flange of the existing member

Structural Salvage and Repairs

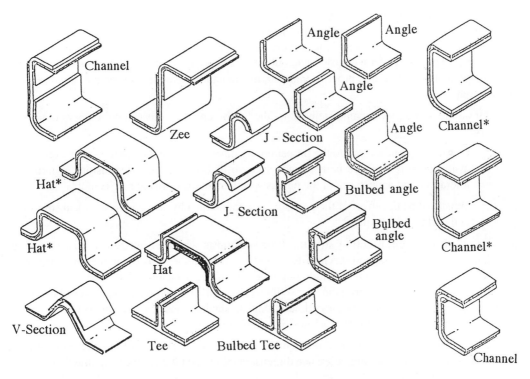

* These repair configurations are not recommended because they may require shims.

Fig. 16.1.3 Arrangement of Repair Members and Elements

16.2 STRUCTURAL SALVAGE

Most structural salvage (e.g., disposition, acceptance, or repair) occurs in in-production aircraft, especially in the early period of prototype airframe (in learning curve stage) but should be kept in lower level scrap rate (e.g., 5% or less) in the downstream of production.

Structures are salvaged based on the ability of the repair to provide strength equal to an original part, so that the original margin of safety (MS) is maintained. This is done to provide for possible retroactive changes such as future increases in gross weight, etc. Also, it may require scrapping parts which have a positive margin of safety.

Problems with the following are typical causes for structural salvage:

- Rivet or bolt hole distance
- Fit of fasteners in hole
- Misdrilling of fastener hole
- Introduction of stress risers
- Introduction of eccentricities
- Fillet radii
- Abrupt changes in section
- Heat treatment of the part
- Mis-machining or under cut
- Cracks caused by forming
- Plating (e.g., anodized aluminum and cadmium plated steel)

Process specifications, manufacturing process standards, and operation documents spell out how airframe parts are to be manufactured and the sequence of operations to be followed. Unfortunately, all these things can go wrong! Here is where the knowledge and judgment of the liaison engineers becomes important. To render a decision and outline a rework procedure, the liaison engineer must combine the talents of designer, process engineer, manufacturing engineer, detective, mind reader, and diplomat.

When for any reason a part becomes rejectable, the liaison engineer must find the answers to two questions:

(a) Can the part be restored to the proper configuration, i.e., are the facilities and techniques available to rework the part, if necessary, so that the physical dimensions, symmetry, tolerances and appearance of the part as specified on the drawing can be met?

(b) What will be the effect of the proposed rework on the physical and mechanical strength and stiffness of the part?

Any salvage disposition affecting structural airworthiness which is not reworked to provide equivalent strength, stiffness, and service life should be made on the basis of engineering judgment and analysis. This judgment must include potentially significant factors which can affect service life such as stiffness, fatigue, corrosion susceptibility, etc. Adverse combinations of the above may dictate rejection of the salvage, or special additional salvage actions (e.g., repair or rework) even though the strength margins of safety are positive or acceptable.

Salvage should be based on matching the equivalent strength and service life of the original part fabricated to at least the minimum tolerance dimensions shown on the engineering drawing.

Salvage and/or repaired parts must maintain the following as a minimum:

- Ultimate strength capability
- Equivalent structural stiffness
- Adequate service life, taking into account corrosion, stress corrosion, temperature, etc.
- Specified control surface unbalance limits, if applicable
- Required fail-safe damage tolerance
- In no case shall negative margins of safety be tolerated

It is better from a business standpoint, and also from the standpoint of the reputation of the aircraft manufacturer to discover the source of a recurring manufacturing error and correct it rather than to continually make repairs. It is the responsibility of the engineer approving the rework to make certain that appropriate action is being taken by design and manufacturing to rectify any situation which results in recurring rework.

(A) SALVAGE POLICY

A salvage policy is required because:

(a) Structures must be salvaged on the basis of providing equivalent strength and life to that of a normal part without a reduction of the margin of safety (MS).

(b) A reduction in strength is governed by a 10% margin of safety policy

(c) The reasons for the 10% margin of safety policy are that:

- It provides for retroactive changes such as future aircraft gross weight increases below or equal to 10%
- It should not be necessary to go through salvage dispositions to determine if a specific airframe has areas which are locally weaker (those of margin of safety less than 10%)
- It can save substantial rework cost and downtime

(B) REWORKING OF MATERIAL PROCESSES

The material process by which material properties are most greatly affected is heat treatment. Any operation involving changes in temperature for the purpose of altering the properties of an alloy is a "heat treat" operation (the term "heat treat" refers to a hardening operation as opposed to annealing). Material processes are not within the scope of this chapter but a general explanation of two types of aluminum heat treatment are given to aid understanding of the problems in salvage:

(a) Softening operations, called annealing, make the alloy more formable or are used to remove the effects of previous heat treatment.

(b) Hardening operations, which increase the strength of the material, involve:
- Solution heat treatment (heating to a relatively high temperature)
- Quenching
- Aging (holding at relatively low temperature)

(c) There are some conditions under which full properties cannot be restored to the material:
- Too many heat treats on clad alloys
- The solution temperature has gone too high
- A necessary reheat treat which cannot bring the material back to the proper temper
- In many cases, involving 2014 and 7075, a reheat treat is not possible because of the severe distortion resulting from quenching parts (such as integrally-stiffened wing panels)

(d) High heat treat steels:
- Most high heat treat parts and assemblies are manufactured by subcontractors and sent to the aircraft manufacturer in a finished condition.
- Occasions arise where reworking is necessary on these assemblies; any rework or salvage operation should conform to specification requirements for the original processing
- When it is impossible to conform 100% to the specification, it is advisable to contact the company Material and Processes Department or the subcontractor

Therefore, if an alloy in a heat treated temper (i.e., -T4, -T6, -T76, -T73 etc.) is for any reason annealed, the part must be reheat treated to restore it to a hardened condition. However, in the few cases, where the alloy is in a cold worked condition such as 2024-T3, annealing will lower the strength below the design allowables. The reduced allowables must be approved by stress engineering.

16.3 SALVAGE EXAMPLES

(A) SHORT EDGE DISTANCE FOR END FASTENER

When there is a short edge distance (E.D.) for an end (or first) fastener, as shown in Fig. 16.3.1, the following procedures are recommended:

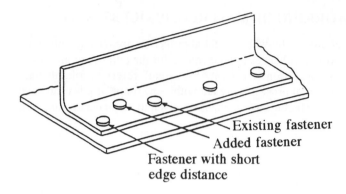

(a) Add Fastener

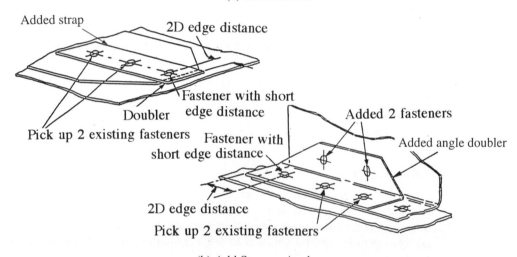

(b) Add Strap or Angle

Fig. 16.3.1 Short Edge Distance (E.D.) for End Fastener

Method A – Add fastener, see Fig. 16.3.1(a):

- Install additional fastener between existing fasteners
- Stagger the added fastener as far as possible from centerline of existing fasteners and maintain minimum edge distance and spacing
- Do not remove discrepant fastener
- Use fastener of same type and size as existing fasteners

Method B – Add strap or angle, see Fig. 16.3.1(b):

- A strap of the same material, one gage thicker with the same contour as the discrepant part
- An angle of the same material, one gage thicker
- Length of strap or angle should be sufficient to pick up at least two fasteners adjacent to the discrepant short E.D. fastener
- Width of angle should provide space for at least two similar fasteners through the web or vertical flange of the discrepant part

(B) SHORT EDGE DISTANCE FOR END FASTENER AT JOGGLE

When there is a short edge distance (E.D.) for an end fastener at a joggle, as shown in Fig. 16.3.2, the following procedures are recommended:

Structural Salvage and Repairs

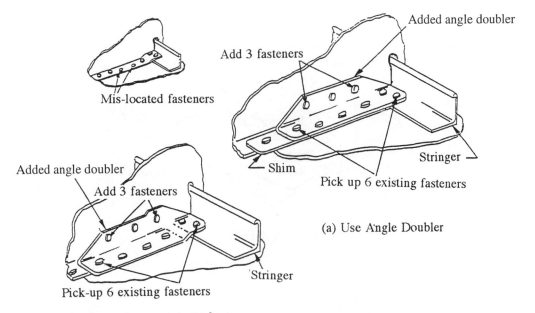

(a) Use Angle Doubler

(b) Use Joggled Angle Doubler

Fig. 16.3.2 Short Edge Distance (E.D.) for End Fastener

Method A – Add angle doubler, see Fig. 16.3.2(a):

- Use same type and one gage thicker than damaged part
- Use sufficient length for installation of two rivets at each end past mislocated rivets
- Use width sufficient to provide adequate edge distance to mislocated rivets
- Angle doubler formed to standard bend radius and contoured to flange, if necessary
- Use shim which is same thickness as stiffener flange and long enough to pick up additional rivets

Method B – Add joggled angle doubler, see Fig. 16.3.2(b):

- Drill out mislocated rivets and adjacent rivets
- Install joggled angle doubler by nesting inside flange
- Attach doubler by riveting to flange and web, using identical rivets and spacing through both flanges of doubler
- Maintain minimum edge distance for all rivets

(C) TRIMMING TO PROVIDE CLEARANCES

When it is appropriate to trim to provide clearance because of the interference of a local misdrilled fastener with a structural flange, as shown in Fig. 16.3.3, the following procedures are recommended:

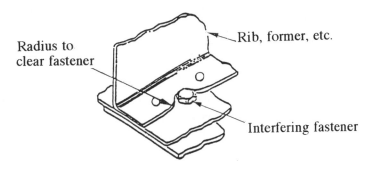

(a) Trimming Clearance

683

Chapter 16.0

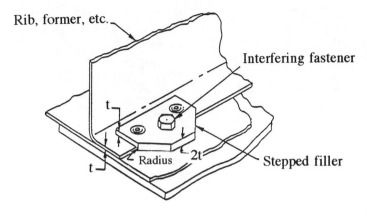

(b) With Stepped Filler

Fig. 16.3.3 Trimming Edge to Provide Fastener Clearance

Method A – Trimming clearance, see Fig. 16.3.3(a):

- Remove the fastener which interferes with structural flange
- Substitute non-interfering fastener as specified by engineering
- Radius of trimmed flange must clear the interfering fastener a min. of 0.03"
- Min. radius should be 0.25"; re-finish raw edges
- Maintain minimum edge distance for all rivets

Method B – With stepped filler, see Fig. 16.3.3(b):

- Follows all the steps given above in Method A
- Make stepped filler of the same material same as flange and twice as thick as flange
- Attach the stepped filler to pick up interfering fastener and one existing fastener on each side or add one fastener equally spaced on each side
- Maintain fastener spacing and edge distance requirements

(D) INTERFERENCE OF FASTENERS WITH BEND RADII

When fasteners interfere with bend radii, as shown in Fig. 16.3.4, the following procedures are recommended:

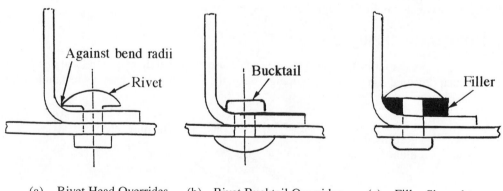

(a) Rivet Head Overrides Bend Radius (Unacceptable)

(b) Rivet Bucktail Overrides Bend Radius (Acceptable)

(c) Filler Shaped to Fit Radius (Preferred)

Fig. 16.3.4 Short Edge Distance for End Fastener

Method A – Reverse fastener installation, see Fig. 16.3.4(b):
- Rivet bucktail may override radius, provided bucking bar does not cut flange
- The hole must not be located in the radius

Method B – use filler, see Fig. 16.3.4(c):
- Filler plate material and finish must be same as that of the flange
- Length and width of filler must be sufficient to pick up discrepant fastener
- Thickness of filler must be equal to bend radius plus 0.02" minimum
- Maintain fastener edge distance requirement

(E) PROPER USE OF SHIMS/FILLERS

Use of shims is generally not recommended for salvage and/or repair except in a few special situations such as mounting fittings and mismatch in final assembly (see Fig. 9.11.5)

(a) Extend shim beyond the jointed members in direction of load path, as shown in Fig. 16.3.5(a).

(b) When possible, attach shim to the continuous member beyond the joint with the same rivet size and pattern as exists in jointed member.

(c) Shim both sides of the member or fitting, as shown in Fig. 16.3.5(b), to give equal load distribution. The use of a shim only on one side of a member will cause eccentric loading of the member and overload the non-shimmed side and cause fatigue problems.

(d) To avoid excessive eccentricities and stress concentrations, taper thickness of shim when shim thickness exceeds 0.064" (refer to Fig. 9.11.1)

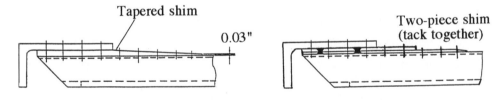

(a) Shim Used in Assembly

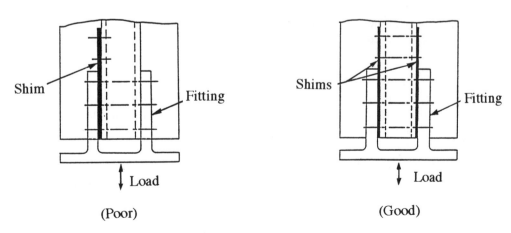

(b) Shim Used on Heavy Fitting

Fig. 16.3.5 Shim Applications

16.4 REPAIR CONSIDERATIONS

A repair must restore the capability of a part to withstand design ultimate loads (without limitations unless otherwise specified, e.g., fail-safe residual strength) and must restore the full service life of the part. The airframe safety system interaction of the repair is shown in Fig. 16.4.1.

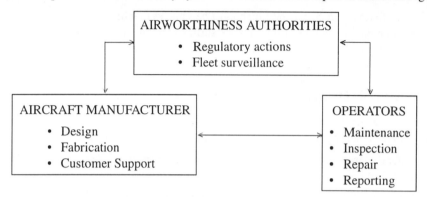

Fig. 16.4.1 Structural Repair Safety Interaction

Knowledge of structural design and analysis will assist the repair engineer in devising standard repairs and procedures (e.g., structural repair manuals, military T. O. manuals, etc.) which will permit the repair of parts or assemblies where airworthiness can be restored. To arrive at the proper solution to a repair problem, the following information should be obtained:

(a) The location and function (e.g., aerodynamic smoothness) of the repair part

(b) Determine classification (see Fig. 16.1.1) of the part to obtain the quality required to sell the disposition or repair to the Material Review Board (MRB):

The following are important rules for repairs:

- The most important requirement is make repair work as simple as possible and do not consider using shims specially where space may be restrictive (e.g., in the wing box)
- Saving weight is not the most important design requirement as it is in new design
- Use the simplest riveting system which will achieve reasonable quality, e.g., Hi-Lok fasteners
- Fatigue and damage tolerance must be considered along with static strength
- Taper both ends of a patch to reduce the end fastener load
- Use over-sized fasteners when reinstalling a removed fastener in the same fastener hole

It is the responsibility of the repair engineer to:

- Avoid overdependence on the stress engineer for assistance on repair problems
- Have a repair in mind and sufficient reason for it before approaching the stress engineer
- Remember that the primary concern in all repair work is the fit, function, and performance of the structure
- Realize that in many cases a repair may require aircraft downtime (cost)

Damage is defined as any deformation or reduction in cross section of a structural member or skin of the airframe. In general, damage to the airframe is categorized as:

- Negligible (no repair required)
- Repairable
- Necessitating replacement

(A) CONSIDERATIONS OF DAMAGE TOLERANCE IN THE EVALUATION OF REPAIRS

In general, any airframe repair can degrade fatigue life if proper care in repair design details is not taken. In the past, the evaluation of repairs was based on equal or better static strength only without any consideration of fatigue life. This approach can easily lead to a design with static over-strength but considerable loss in fatigue quality compared to the original structure.

(a) Repair structural life improvement:

From the Example of Chapter 9.12, it is can be seen that a doubler with tapered ends yields a 40% improvement in fatigue life over a doubler without tapering. It is pointed out that the repaired structure is sensitive to the doubler thickness and the magnitude of the end (first) fastener load. In order to provide some guidance to structural repairs, a number of typical configurations should be analyzed and tested. Finite element modeling or displacement compatibility analysis should be performed to obtain the critical fastener load for a variety of skins and repair doublers in a lap splice configuration (a common configuration in structural repair as shown in Fig. 16.4.2). After obtaining the end fastener load and then use the Severity Factor (the fatigue analysis approach described in Chapter 9.12) to compute the repair structural life.

(b) Review of method of analyses:

The major aircraft manufacturers have developed sophisticated fracture mechanics analysis methods to solve for a variety of materials, geometry and load spectra variations. These methods are used to design new airframes and to meet structural damage tolerance requirements. These methods can be also used to assess standard repairs and to update structural repair manuals. However, a simple and conservative analysis of repairs (refer to Ref. 16.1) is more than sufficient and the additional sophistication is not necessary. The simple method for joint fatigue analysis which was discussed in Chapter 9.12 can be used to assess structural repairs.

(c) Improvement in inspectability:

Any repair doubler applied externally to a repaired structure, such as those shown in Fig. 16.4.2(a) and (c), will decrease external inspectability for the fuselage skin at the critical first or end fastener row. Skin cracking at the first fastener will be hidden externally (could be inspected internally with removal of the lining or by sophisticated NDI – non-destructive inspection) by the doubler. The inspectability of this repair can be improved by placing the secondary doubler inside and the primary doubler outside, as shown in Fig. 16.4.2(b) and (d).

(d) Damage tolerant repair structures:

- All major repair structures certified to FAR 25.571 Amendment 45 or AC 91-56 must be evaluated for damage tolerance
- New thresholds must be established for the repaired structure
- The repaired structure should be inspected at the determined frequencies after the threshold is established

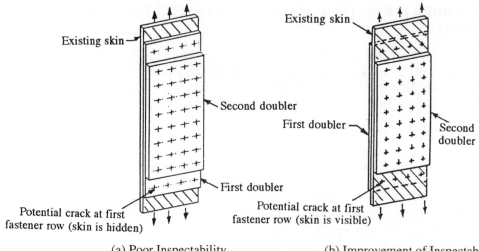

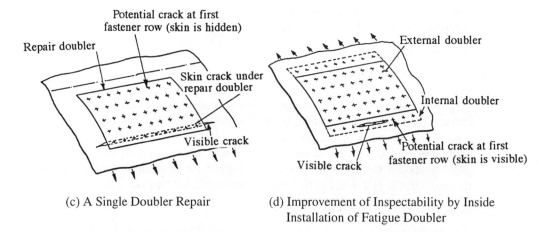

Fig. 16.4.2 Improvement of Fatigue Life and Inspectability

(B) EFFECT OF MIXING DIFFERENT MATERIALS

The first choice is always to use the same material as that of the member being repaired. When dissimilar materials are combined to form an axial load-carrying member, the load on the member must be distributed to the elements based on equal strains (see Chapter 5.1):

(a) The ultimate tension load-carrying capacity of the combination depends on the ultimate elongation (e_u) of the less ductile material (see material elongation effect in Chapter 4.2).

(b) Even aluminum alloys have differing elongations (the material which breaks first has the smallest elongation as shown in the Example calculation in Chapter 4.2).

Repairs with using steel materials:

(a) Steel can be used as a permanent repair for specific applications.

(b) Thickness of steel ($E_s = 29 \times 10^6$ psi) must be reduced proportionally from the aluminum ($E_a = 10.5 \times 10^6$ psi) thickness by ratio of their modulus of elasticity ($\frac{E_a}{E_s}$); attention also should be paid to the material characteristics of yield strength (F_{cy}) and elongation (ductility)

(c) Where the use of aluminum requires thickness would make the installation of fasteners impractical:

- To restore strength in areas where the use of aluminum would result in excessive material stack up
- In nested angles due to short edge margins
- Fasteners in the radius
- Fasteners interference problems

(d) The most popular steel materials for repair are:

- 4130, 4340, 15-5PH
- 301 Cress $\frac{1}{2}$ hard (F_{su} = 82 ski) used on shear webs on bulkheads, beams, and frames made from 2024-T3
- 301 Cress should not be used for repair of fitting flanges or formed and extruded angles made from 7000 series aluminum, unless the grain direction of the 301 is known [when using $\frac{1}{2}$ hard condition for example: F_{tu} = 151 ski (L) and F_{ty} = 69 ksi (L); F_{tu} = 152 ski (LT) and F_{ty} = 116 ksi (LT); data from Ref. 4.1 indicates that the LT grain direction is stronger than the L grain direction]

(C) STIFFNESS EFFECT

Avoid over-strength repair (stiffness greater the original structure) which has a detrimental effect on the fatigue life of the repaired structure. The stiffness comparison is shown in Fig. 16.4.3 and they are:

(a) Fig. 16.4.3(a) shows a part with a hole or damaged area in it (without patch), where loads are forced to "go-around" the weak spot and the maximum stress occurs.

(b) Fig. 16.4.3(b) shows the use of a patch that is more flexible than the original structure and which may not carry its share of the load. This can cause a overload in the surrounding material leading to premature fatigue failure.

(c) Fig. 16.4.3(c) shows the use of a patch that is more stiff than the original structure and which will carry more than its share of the load. This can cause a overload leading to the patch material (at the first fastener row) in premature fatigue failure.

(d) Fig. 16.4.3(d) shows a well-made repair patch which carries the load across the hole.

(e) Fig. 16.4.3(e) and (f) show patches with fasteners that fail to transfer the full load because of loose fits and/or fastener deformation.

(f) The computation of stiffness is not a simple matter and it is function of:

- Material of the doubler(s)
- Numbers of fasteners and fastener flexibility (fastener spring constant, see Appendix B)
- Type of fastener, fastener fit in the hole, and the quality of the fastener installation, etc.
- Double or single shear joint, see Fig. 9.12.13 in Chapter 9.12
- Fastener clamp-up force (preloading)

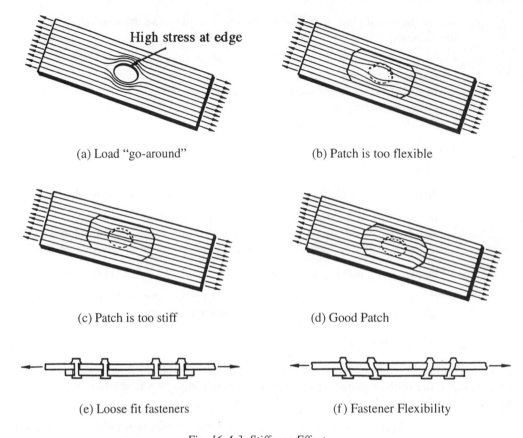

Fig. 16.4.3 Stiffness Effect

(D) NON-FLUSH AND FLUSH SURFACE REPAIRS

(a) Flush repairs are more complex and costly than other repair operations:
- Generally, the leading edge of the wing and empennage, designated as Zone I and II in Fig. 16.4.4(a), cannot tolerate non-flush repairs since they would affect the aerodynamic lift force significantly
- Practically, whether a repair is flush or non-flush and how flush the repair must be dependent on aerodynamic characteristics
- Some external surfaces, such as Zone III of Fig. 16.4.4(a) and Zone II and III of Fig. 16.4.4(b), can be considered for non-flush repair based on the repair engineer's judgment
- In the forward area of the fuselage [most critical is forward of the wing as shown in Zone I of Fig. 16.4.4(b)] is generally required for flush repair,

(b) Non-flush repair is a simple and low cost repair operation:
- Thin repair doublers can be used since countersunk head fasteners are not required
- Use in the areas where it does not significantly affect aerodynamic characteristics, e.g. in the tail area of the aft fuselage
- Use on the interiors of wing and empennage boxes, and on areas of the fuselage where aerodynamic smoothness is not required, e.g., spars, ribs, stringers, frames, floors, etc.
- In the trailing edge portion of the wing and empennage (refer to the Structural Repair Manual for the aircraft)

- It may be used in Zone II, Zone III, and the aft portion of Zone I, as shown in Fig. 16.4.4(b), for the high-time aircraft (refer to the Structural Repair Manual for the aircraft)
- If the doubler is fastened to the existing skin with countersunk fasteners, countersunk washers are required in the countersunk hole between doubler and skin, as shown in Fig. 16.4.5

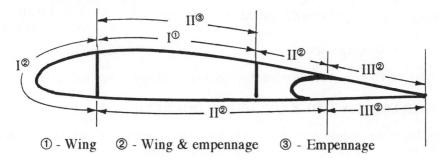

① - Wing   ② - Wing & empennage   ③ - Empennage

(a) Wing and Empennage Airfoil Surfaces

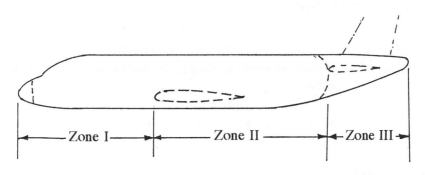

(b) Fuselage Body Surfaces

(Most critical area is Zone I which requires flush repair)

Fig. 16.4.4 Critical Areas Require Flush Repairs

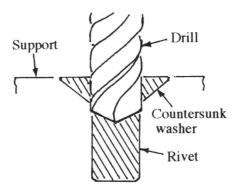

(a) Countersunk Washer Can be made from Countersunk Rivet

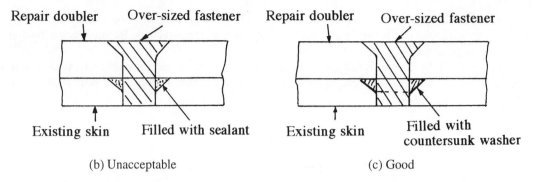

(b) Unacceptable (c) Good

Fig. 16.4.5 Use of Countersunk Washers between Doubler and Existing Skin

(E) REPAIR USING A THICK ONE-PIECE DOUBLER

The fatigue life of the reinforced structure can be improved by tapering or steps on the ends of the doubler. Usually this method is used by aircraft manufacturers since they have the facilities to perform this repair, see Fig. 16.4.6.

The repair procedure is as follows:

(a) Tapering of slope must be gentle, or steps used, to accommodate fastener heads.

(b) Total cross-sectional area is approximately 30% or greater than that of the replaced area.

(c) Enough rows of fasteners should be used in the doubler to fully develop the ultimate strength of the doubler.

(d) It is generally desirable to use two additional rows of fasteners on each end of the doubler to allow for fastener flexibility (stiffness), poor installation, bad holes, etc. in the repair.

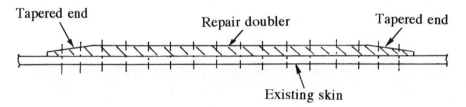

Fig. 16.4.6 One-Piece Doubler

(F) REPAIR USING MULTIPLE THIN DOUBLERS

(a) Use same material as damaged member.

(b) The fatigue life of the reinforced structure can be also improved by using multiple thin doublers. This method is the most popular for repair because it is much easier to accomplish, see Fig. 16.4.7, Fig. 16.4.8, and Fig. 16.4.9.

(c) In general, the thickness of the first doubler (a fatigue doubler is directly attached to the existing structure which requires repair) is based on the following (see example in Fig. 16.4.8):

- Standard gage 0.025" if damaged member thickness is less than 0.05"
- Standard gage 0.032" if damaged member thickness is greater than 0.05"
- The thickness of a fatigue doubler is a function of the thickness of the existing structure

(d) Total cross-sectional area is about 30% greater than the replaced area

(e) Sufficient rows of fasteners should be used in the doublers to fully develop the ultimate strength of the doublers.

(f) It is generally desirable to use 3 – 4 additional rows of fasteners (depending on how many multiple doublers are used in the patch; sometimes more than 4 rows of fasteners are required) on each end to allow for fastener flexibility (stiffness), poor installation, bad holes, etc. in the repair.

(g) Add two rows of fasteners on the first doubler beyond the second doubler on each end to reduce the end fastener load.

(h) Maintain a rivet spacing of 4D (minimum) to avoid inter-rivet compression buckling.

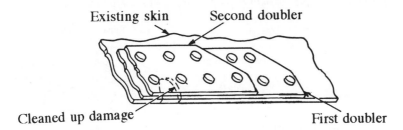

(a) Skin Damage Repair

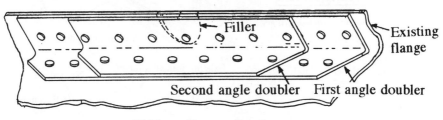

(b) Flange Damage Repair

Fig. 16.4.7 Repair Using Multiple Thin Doublers

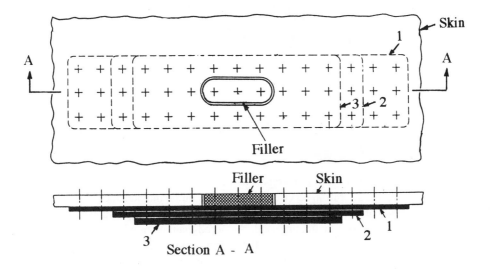

Existing Structural Skin Thickness (in.)	Standard Thickness (in.)		
	First Doubler (No. 1)	No. 2	No. 3
0.032 – 0.05	0.025	0.032	–
0.05 – 0.08	0.032	0.071	–
0.08 – 0.1	0.032	0.032	0.05
0.1 – 0.125	0.04	0.04	0.071
0.125 – 0.16	0.05	0.05	0.09
0.16 – 0.19	0.05	0.063	0.125
0.19 – 0.225	0.063	0.125	0.125
0.225 – 0.25	0.071	0.125	0.16
0.25 – 0.3	0.08	0.16	0.16

Fig. 16.4.8 Examples of Skin Repair Using Multiple Thin Doublers

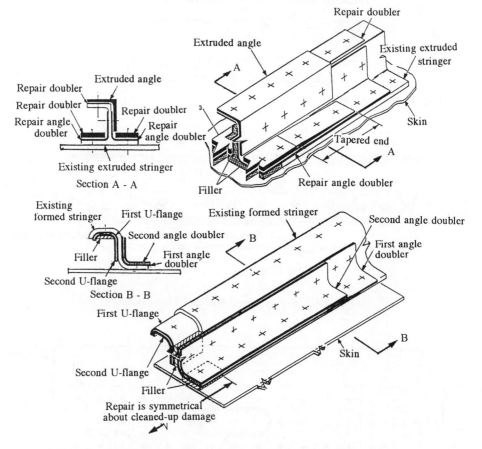

Fig. 16.4.9 Stringer Repair Using Multiple Thin Doublers

(G) TEMPORARY REPAIRS

- To permit a ferry flight to a repair facility
- To keep the aircraft in service until next scheduled maintenance
- An external patch may be used until a permanent flush repair can be accomplished
- When non-flush repairs are used, the aerodynamic smoothness will be improved if thick plates are chamfered and fillet sealant is applied
- Chamfer edges of repair plates 30 to 40 degrees, down to 0.03 inch minimum all-around thickness

- Extend the fillet material approximately 5 times the thickness of the repair plate, and smooth to a taper

(H) NEGLIGIBLE DAMAGE

The permissible limits of negligible damage for skin panels and bulkhead webs are given in Fig. 16.4.10:

- Adjacent damages shall be separated by at least four times the maximum dimension of the largest damage as measured between adjacent edges or rims
- Stop-drilling cracks is permissible
- Circle-cutting to remove the entire crack is preferable

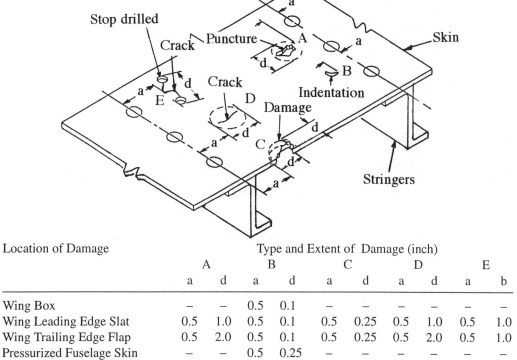

Location of Damage	Type and Extent of Damage (inch)									
	A		B		C		D		E	
	a	d	a	d	a	d	a	d	a	b
Wing Box	–	–	0.5	0.1	–	–	–	–	–	–
Wing Leading Edge Slat	0.5	1.0	0.5	0.1	0.5	0.25	0.5	1.0	0.5	1.0
Wing Trailing Edge Flap	0.5	2.0	0.5	0.1	0.5	0.25	0.5	2.0	0.5	1.0
Pressurized Fuselage Skin	–	–	0.5	0.25	–	–	–	–	–	–
Non-Pressurized Fuselage Skin	0.5	2.0	0.5	0.25	0.5	0.25	0.5	2.0	0.5	2.0
Stabilizer and Fin	0.5	1.0	0.5	0.1	0.5	0.25	0.5	1.0	0.5	1.0
Ailerons	0.5	0.5	0.5	0.1	0.5	0.1	0.5	0.5	0.5	0.5
Rudders	0.5	2.0	0.5	0.25	0.5	0.25	0.5	2.0	0.5	2.0
Fairing	0.5	2.0	0.5	0.25	0.5	0.5	0.5	2.0	0.5	2.0

(1) a – Min. distance from row of fasteners.
(2) d – Max. extent of trimmed damage (diameter).
(3) b – Max. distance of stop drilled holes from a crack.
(4) A – Skin puncture trimmed to smooth round hole.
(5) B – Small indentation free from cracks and abrasions.
(6) C – Damage at edge of sheet trimmed and faired.
(7) D – Crack trimmed to smooth round hole.
(8) E – Crack stop-drilled at extremities.
  Drill diameter is approximately 3 times thickness of the material,
  but not less than 0.098" (No. 40) nor greater than 0.25".
(9) Min. distance between damages is 4 times diameter (d) of greater damage.
(10) Plug holes with non-structural patches.

Fig. 16.4.10 Typical Examples of Negligible Damage

(I) **TYPICAL REPAIR PROCEDURES**

(1) Evaluate the extent of the damage.
(2) Prepare a repair as shown in Aircraft Structural Manual or military T. O., or supplement with approved repair drawing.
(3) Fabricate repair parts.
(4) Provide structural support to maintain alignment of structure during the repair.
(5) Position repair parts and mark structure for removal or rework of damaged material.
(6) Remove paint, primer and sealant material from the affected area.
(7) Position parts, clamp, and drill fastener holes.
(8) Remove repair parts and deburr all holes, and break all sheet metal edges. All edges should be smooth.
(9) Clean parts, apply corrosion preventive treatment, prime and paint.
(10) Install parts with required sealants.
(11) Apply protective coatings.
(12) Apply aerodynamic smoother on external gaps, or to fair external irregularities.
(13) Apply exterior touch-up paint
(14) Perform pressure or leak checks, if applicable
(15) Re-balance control surfaces.

16.5 REPAIR EXAMPLES

(A) TAPERING OF REINFORCEMENTS

(a) When possible, reinforcement angles, stiffeners and flat doublers over longitudinal members must be tapered so that the transfer of load will be gradual.

(b) An abrupt discontinuity increases the tendency of the skin to wrinkle or crack.

(c) Reinforced material should be tapered symmetrically wherever possible.

(d) Check the tapered area for inter-rivet compression buckling.

(B) STANDARD OVERSIZE FASTENERS

(a) The standard oversize fastener diameters are $\frac{1}{32}$" and $\frac{1}{64}$", edge distance, spacing of the fasteners and net section of material permitting.

(b) When flush fasteners are involved and material thickness permits re-countersink for the larger size fastener.

(c) When thickness does not permit re-countersinking, and the fastener is an aluminum rivet or Jo-bolt:

- Drill out oversized hole for the next size diameter
- Do not re-countersink the sheet to take the full head of the rivet
- Leave countersink as is or just increase slightly, and mill the head flush, as shown in Fig. 16.5.1.

Structural Salvage and Repairs

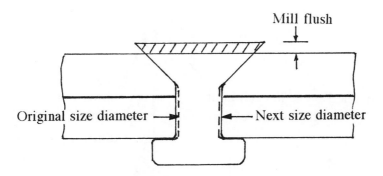

Fig. 16.5.1 Mill Countersunk Head Flush

(d) When the over-sized hole have dimpled sheets:
- Re-dimpling is not permitted because it causes work-hardening of the material and possible cracks
- If it is permissible, clean out the rejected hole for the next size diameter
- Leave remaining dimpled hole as is and install aluminum rivet and mill flush

(C) OVERSIZE FASTENER WITH COUNTERSUNK-DIMPLED WASHER

Countersunk dimpled washers are used for oversize countersunk holes where a flush-type fastener needed to be removed and a regular countersunk fastener re-installed, as shown in Fig. 16.5.2,. The following procedures are recommended:

- Use dimpled washer size to accommodate oversize countersunk fastener
- Dimple shape with standard tool and trim off excess as required to make a flush fit

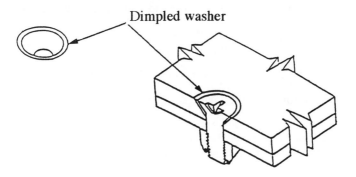

Fig. 16.5.2 Use of Countersunk Dimpled Washer

(D) USE PLUGS

Plugs are used in the repair or salvage of misdrilled holes. The stress concentration varies, depending on the configuration of the misdrilled hole, as shown in Fig. 16.5.3.

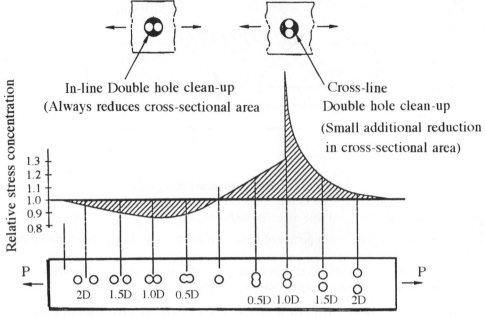

Fig. 16.5.3 Misdrilled Holes Versus Stress Concentration

When using plugs (freeze plugs) to repair damaged fastener holes, as shown in Fig. 16.5.4, keep in mind the following design considerations:

(a) The plug does not restore material to the basic section; it simply acts as a medium to reduce stress concentrations.

(b) Elongated holes should be hand-filled round and reamed for an oversize fastener if possible

(c) Do not use eyebrow-shaped plugs.

(d) Larger holes can be plugged and fasteners can be installed through the plug.

(e) Plug should be positively restrained against falling out or being blown out by internal pressure:

- Fastener should be light countersunk on both heads when possible
- Washers and adjacent structure may sometimes be used to provide a satisfactory means of restraint
- Design to a controlled interference fit (0.35 – 0.4% interference) to achieve durability
- Use 7075-T6 plug with 0.0035-0.0040 inch interference
- Install by immersing in liquid nitrogen approximate –320°F for about 10 minutes
- Hole finish is 63 √
- Minimum wall thickness equal to 0.05 inch

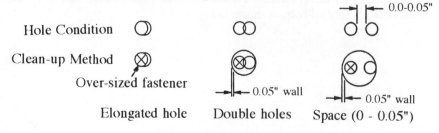

(a) Hole Condition and Clean-up Method

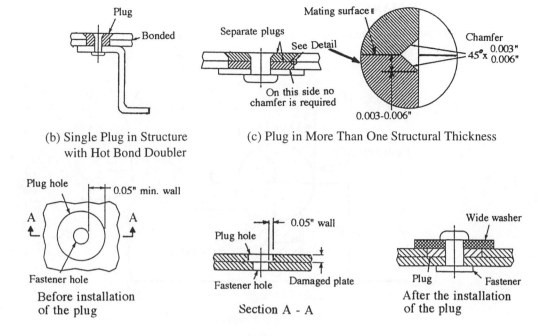

(b) Single Plug in Structure with Hot Bond Doubler

(c) Plug in More Than One Structural Thickness

(d) Plug in one Damaged Plate as Shown

Fig. 16.5.4 Using Plug to Repair Damaged Fastener Holes

(E) REPAIR FASTENER HOLE USING BUSHING

Repair of a fastener hole using a standard bushing is shown in Fig. 16.5.5.

- Use a standard flanged bushing
- Use a standard bushing wall thickness, e.g., the minimum bushing wall thickness of a steel bushing is 0.032 inch for $\frac{3}{16}$ and $\frac{1}{4}$ inch diameter fasteners
- The outside diameter of bushing should not exceed the outside diameter of the fastener head. When this is a necessary condition and the fastener is in shear, use over-sized washer under the fastener head and nut
- Chamfer the ends of the bushing and housing bores to prevent damage to main structure when bushing is pushed in place

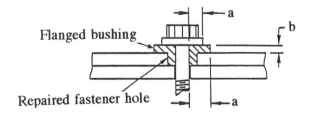

a – Sufficient bearing area above the below the flange
b – Thickness should be adequate to resist bending and shear where flange extends beyond edge of fastener head or where the bushing wall thickness is large

Fig. 16.5.5 Repair of Fastener Hole Using Flanged Bushing

(F) SHORT EDGE MARGINS

(a) Allowable shear-out loads between fasteners and edge, as shown in Fig. 16.5.6, and the edge margin (EM) are given below:

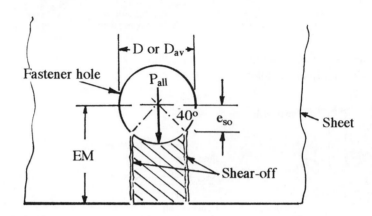

Fig. 16.5.6 Edge Margin (EM)

$$EM = \frac{P_{all}}{2t F_{su}} + e_{so} \qquad \text{Eq. 16.5.1}$$

Effective fillet shear-out material:

$$e_{so} = \frac{D_{av} \cos 40°}{2} \qquad \text{Eq. 16.5.2}$$

where: P_{all} – Fastener allowable load (smaller of $P_{s,all}$ and $P_{b,all}$, see Chapter 9.2)
t – Skin or plate thickness
F_{su} – Skin or plate ultimate shear allowable
D_{av} – Average fastener hole of countersunk fastener only

- When plate thickness $t \leq 0.5 D_{av}$, recommended using aluminum rivets
- When plate thickness $t \geq 0.5 D_{av}$, recommended using steel or titanium fasteners
- EM should be $\geq 1.5 D_{av}$
- To determinate the e_{so} values for protruding head ($D_{av} = D$) fastener:

D	$\frac{5}{32}$	$\frac{3}{16}$	$\frac{1}{4}$	$\frac{5}{16}$	$\frac{3}{8}$	$\frac{7}{16}$	$\frac{1}{2}$
e_{so}	0.06	0.07	0.096	0.12	0.144	0.168	0.192

(b) Average diameters for countersunk holes (D_{av}) are shown in Fig. 16.5.7 through Fig. 16.5.9.

Sheet Thickness (inch)	Fastener Diameter (D, inch)							
	0.19 ($\frac{3}{16}$)	0.25 ($\frac{1}{4}$)	0.3125 ($\frac{5}{16}$)	0.375 ($\frac{3}{8}$)	0.4375 ($\frac{7}{16}$)	0.5 ($\frac{1}{2}$)	0.5625 ($\frac{9}{16}$)	0.625 ($\frac{5}{8}$)
0.04	0.351	0.474	0.602	0.73	0.858	0.986	1.109	1.237
0.045	0.345	0.468	0.596	0.724	0.852	0.98	1.103	1.231
0.05	0.339	0.462	0.59	0.718	0.846	0.974	1.097	1.225
0.056	0.332	0.455	0.583	0.711	0.839	0.967	1.09	1.218
0.063	0.324	0.447	0.575	0.703	0.831	0.959	1.802	1.209
0.071	0.314	0.437	0.565	0.693	0.821	0.949	1.072	1.2
0.08	0.304	0.427	0.555	0.683	0.811	0.939	1.062	1.189
0.09	0.291	0.415	0.543	0.671	0.799	0.927	1.05	1.177
0.1	0.281	0.403	0.531	0.659	0.787	0.915	1.038	1.165
0.112	<u>0.272</u>	0.388	0.516	0.644	0.772	0.9	1.023	1.151
0.125	0.264	0.375	0.501	0.629	0.757	0.885	1.008	1.135
0.14	0.256	0.362	0.483	0.611	0.739	0.867	0.99	1.117
0.16	0.248	<u>0.349</u>	0.462	0.587	0.715	0.843	0.966	1.094
0.18	0.242	0.338	0.446	0.564	0.691	0.819	0.942	1.07
0.19	0.24	0.334	0.439	0.554	0.68	0.808	0.931	1.058
0.2	0.238	0.33	<u>0.433</u>	0.546	0.668	0.796	0.919	1.046
0.224	0.233	0.322	0.42	<u>0.528</u>	0.643	0.766	0.89	1.017
0.25	0.229	0.315	0.409	0.512	<u>0.622</u>	0.739	0.86	0.974
0.313	0.222	0.302	0.391	0.485	0.586	<u>0.692</u>	<u>0.801</u>	0.919
0.375	0.217	0.294	0.378	0.468	0.562	0.661	0.762	<u>0.872</u>
0.5	0.211	0.284	0.363	0.446	0.532	0.622	0.713	0.811
0.625	0.208	0.278	0.353	0.432	0.514	0.599	0.684	0.775

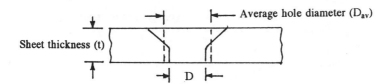

(1) Based on max. head and hole diameter; head below flush is 0.005"
(2) Sheet thickness/fastener diameter ratio above the underline results in countersunk height > $\frac{2}{3}$ – do not use these values for fatigue critical areas.

(100° countersunk tension head Hi-Lok, screws, lockbolts)

Fig. 16.5.7 Average Hole Diameter (D_{av}) for Countersunk Tension Head Fasteners

Sheet Thickness (inch)	Fastener Diameter (D, inch)					
	0.1875 ($\frac{3}{16}$)	0.25 ($\frac{1}{4}$)	0.3125 ($\frac{5}{16}$)	0.375 ($\frac{3}{8}$)	0.4375 ($\frac{7}{16}$)	0.5 ($\frac{1}{2}$)
0.04	0.262	0.356	0.436	0.523	0.631	0.718
0.045	0.256	0.35	0.43	0.517	0.625	0.712
0.05	0.25	0.344	0.424	0.511	0.619	0.706
0.056	0.241	0.337	0.417	0.504	0.612	0.699
0.063	0.235	0.329	0.409	0.496	0.604	0.691
0.071	0.23	0.32	0.399	0.486	0.594	0.681
0.08	0.226	0.312	0.389	0.475	0.583	0.67
0.09	0.222	0.305	0.381	0.465	0.572	0.659
0.1	0.218	0.3	0.375	0.456	0.56	0.647
0.112	0.215	0.294	0.367	0.447	0.547	0.632
0.125	0.213	0.29	0.361	0.44	0.536	0.619
0.14	0.21	0.285	0.356	0.433	0.525	0.606
0.16	0.208	0.281	0.351	0.425	0.514	0.593
0.18	0.206	0.277	0.346	0.42	0.506	0.582
0.19	0.205	0.276	0.345	0.417	0.502	0.578
0.2	0.204	0.275	0.343	0.415	0.499	0.574
0.224	0.202	0.272	0.34	0.411	0.492	0.566
0.25	0.201	0.27	0.337	0.407	0.486	0.559
0.313	0.199	0.266	0.332	0.401	0.476	0.547
0.375	0.197	0.263	0.328	0.356	0.47	0.539
0.5	0.195	0.26	0.324	0.391	0.462	0.529
0.625	0.194	0.258	0.322	0.388	0.457	0.523

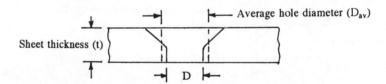

(1) Based on max. head and hole diameter; head below flush is 0.005"
(2) Sheet thickness/fastener diameter ratio above the underline results in countersunk height > $\frac{2}{3}$ – do not use these values for fatigue critical areas.

(100° countersunk shear head Hi-Lok, screws, lockbolts)

Fig. 16.5.8 Average Hole Diameter (D_{av}) for Countersunk Shear Head Fasteners

Sheet Thickness (inch)	Fastener Diameter (D, inch)									
	MS 20426					MS 20427				
	0.1562 ($\frac{5}{32}$)	0.1875 ($\frac{3}{16}$)	0.25 ($\frac{1}{4}$)	0.3125 ($\frac{5}{16}$)	0.375 ($\frac{3}{8}$)	0.1562 ($\frac{5}{32}$)	0.1875 ($\frac{3}{16}$)	0.25 ($\frac{1}{4}$)	0.3125 ($\frac{5}{16}$)	0.375 ($\frac{3}{8}$)
0.04	0.242	0.309	0.432	0.52	0.65	0.25	0.318	0.441	0.529	0.66
0.045	0.236	0.303	0.426	0.514	0.644	0.244	0.312	0.435	0.524	0.654
0.05	0.23	0.297	0.42	0.508	0.638	0.238	0.306	0.429	0.518	0.648
0.056	0.223	0.29	0.413	0.501	0.631	0.231	0.299	0.422	0.511	0.641
0.063	0.216	0.282	0.405	0.493	0.623	0.227	0.29	0.414	0.502	0.632
0.071	0.21	0.272	0.395	0.483	0.613	0.219	0.281	0.404	0.493	0.623
0.08	<u>0.204</u>	0.263	0.385	0.473	0.603	0.212	0.274	0.394	0.482	0.612
0.09	0.199	0.254	0.373	0.461	0.591	<u>0.206</u>	0.265	0.382	0.47	0.6
0.1	0.195	<u>0.248</u>	0.361	0.449	0.579	0.202	0.257	0.37	0.458	0.588
0.112	0.191	0.242	0.349	0.435	0.565	0.197	<u>0.25</u>	0.362	0.444	0.574
0.125	0.188	0.236	0.339	0.422	0.549	0.193	0.244	0.351	0.435	0.558
0.14	0.185	0.231	<u>0.33</u>	0.411	0.531	0.189	0.238	<u>0.34</u>	0.422	0.541
0.16	0.182	0.226	0.32	0.399	0.512	0.186	0.232	0.329	<u>0.409</u>	0.526
0.18	0.179	0.222	0.313	0.389	0.497	0.183	0.228	0.321	0.399	0.509
0.19	0.178	0.221	0.31	0.386	0.491	0.181	0.226	0.317	0.394	0.503
0.2	0.177	0.219	0.307	0.382	<u>0.485</u>	0.18	0.224	0.314	0.39	<u>0.496</u>
0.224	0.175	0.216	0.301	0.375	0.474	0.178	0.22	0.308	0.382	0.484
0.25	0.173	0.213	0.256	0.369	0.464	0.176	0.217	0.302	0.375	0.473

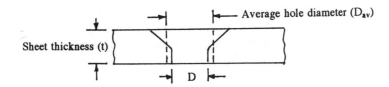

(1) Based on max. head and hole diameter; head below flush is 0.005"
(2) Sheet thickness/fastener diameter ratio above the underline results in countersunk height > $\frac{2}{3}$ – do not use these values for fatigue critical areas.

(100° countersunk MS 20426 and MS 20427 rivets)

Fig. 16.5.9 Average Hole Diameter (D_{av}) for Countersunk Aluminum Rivets

(G) MINIMUM FASTENER SPACING

Joint splices commonly have single-row, multi-row, or staggered-row fastener patterns as shown in Fig. 16.5.10 and Fig. 16.5.11. A splice joint consists of a row of fasteners, and new rows of fasteners are occasionally added as part of a repair. The new row is added directly behind the existing row, either staggered or forming a double-row; the row spacing must be such that the splice material will not shear-off between fasteners (L_{dou} or L_{sta}) and rows (S_{dou} or S_{sta}).

(a) Double-row pattern:

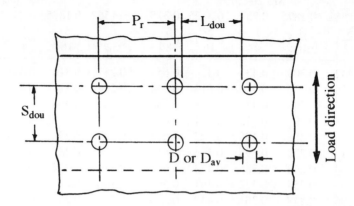

Fig. 16.5.10 Tension and Shear-Off Failure within a Double-Row Pattern

(a1) Tension failure occurs at the net section (L_{dou}) of a double-row or multi-row joint:

$$L_{dou} = p_r - D_{av},\qquad \text{Eq. 16.5.3}$$

(a2) Shear-out failure at the row spacing (S_{dou}) of double-row joint:

$$S_{dou} = \frac{P}{2tF_{su}} + 2e_{so} \qquad \text{Eq. 16.5.4}$$

where: P – Load on fastener (use the smallest load value from $P_{s,all}$ – fastener shear-off, $P_{b,all}$ – material bearing load, or P_{ult} design load)
e_{so} – Effective fillet shear-out material, see Eq. 16.5.2
t – Sheet thickness
F_{su} – Ultimate shear allowable of sheet material

(b) Staggered-row pattern:

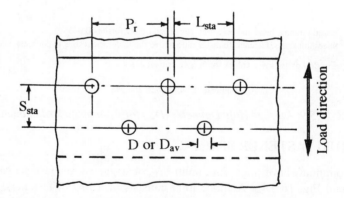

Fig. 16.5.11 Tension and Shear-Off Failure within a Staggered-Row Pattern

(b1) Tension failure in the net fastener spacing (L_{sta}) of a staggered-row joint occurs as follows:

- A decrease in the net section will occur unless the new row of fasteners is set back sufficiently (greater S_{sta})

- The net spacing (L_{sta}):

$$L_{sta} = p_r - 2D_{av} + \frac{S_{sta}^2}{p_r}$$ Eq. 16.5.5

where: D_{av} – Average hole dia. for countersunk head fastener; use D for protruding head fastener; Use $D_{av} = \frac{D_{av,1} + D_{av,2}}{2}$ if fasteners are different in size

p_r – Rivet pitch
S_{sta} – Given row spacing

- If the new L_{sta} is equal to or greater than the original $L_{dou} = P_r - D_{av}$, in the double rows of fasteners, the original $L_{dou} = L_{sta}$ should still be used and the added row will not adversely affect the net spacing stresses in the joint
- To calculate the net shear-off, use $L_{sta} = p_r - D_{av}$, (also used for double rows)

(b2) Shear-out failure between row spacing (S_{sta}) of staggered-row joint:

The Eq. 16.5.3 is also applicable to determine the row spacing (S_{sta}) with the given net spacing (L_{sta}) value of the joint.

The minimum row spacing from Eq. 16.5.5 can be solved by the graphical method, as shown in Fig. 16.5.12.

(1) Compute the hole-out percentage:

$$A_{hole} = \frac{D_{av}}{p_r}$$ Eq. 16.5.6

(2) Using a given effective hole-out percentage, from Fig. 16.5.12 obtain an "x" factor

(3) The minimum row spacing:

$$S_{sta} = \text{"x"} A_{hole}$$ Eq. 16.5.7

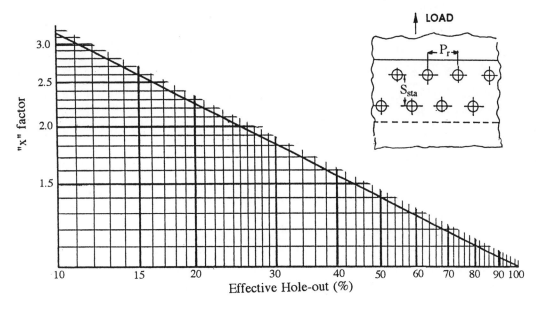

Fig. 16.5.12 *"x" Factor for Staggered-Row Spacing (S_{sta})*

Example 1:

Determine the min. fastener row spacing (S_{dou}) for a double-row splice with $\frac{3}{16}$" dia. countersunk shear head Hi-Loks.

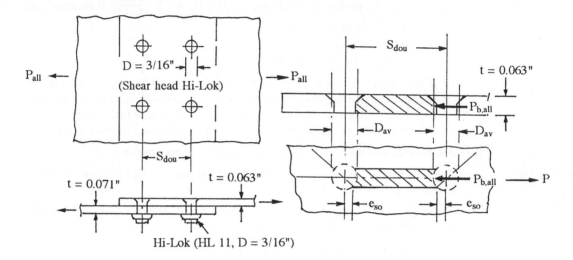

Hi-Lok (HL 11, D = 3/16")

(1) Fastener shear allowable, $P_{s,all}$ = 2,623 lbs./Hi-lok (see Fig. 9.2.5)

(2) Sheet material (2024-T3 sheet): F_{su} = 40,000 psi $\quad F_{bru}$ = 131,000 psi

(3) Countersunk head in t = 0.063" sheet of 2024-T3 with fastener bearing load:

$P_{b,all}$ = 1,393 × 80% = 1,114 lbs./fastener for 7075-T6 sheets (see Fig. 9.2.9)

Assume 15% reduction for 2024-T3 sheets from 7075-T6 sheets:

∴ $P_{b,all}$ = 1,114 × 85% = 947 lbs./fastener

(4) Solid shank is in t = 0.071" sheet of 2024-T3 with fastener bearing load:

$P_{b,all}$ = 131,000 × 0.1875 × 0.071 × 80% = 1,395 lbs./fastener

(The 80% value is a bearing reduction factor, see Chapter 9.2)

The smallest load is $P_{b,all}$ = 947 lbs. (critical on sheet, t = 0.063") and it is used in the following example calculation

From Eq. 16.5.2, the effective fillet shear-out material:

$$e_{so} = \frac{D_{av} \cos 40°}{2} = \frac{0.235 \cos 40°}{2} = 0.09"$$

where D_{av} = 0.235" (see Fig. 16.5.8)

From Eq. 16.5.1, the min. row spacing is:

$$S_{dou} = \frac{P_{b,all}}{2t F_{su}} + 2e_{so} = \frac{947}{2 \times 0.063 \times 40,000} + 2 \times 0.09 = 0.37"$$

(**Note:** In general if 3D min. spacing is maintained no shear-out problem will exist)

Example 2:

Use the same splice given in Example 1. Determine the min. fastener row spacing (S_{sta}) for a staggered-row splice with $\frac{3}{16}$" dia. countersunk shear head Hi-Loks and fastener spacing, p_r = 1.0".

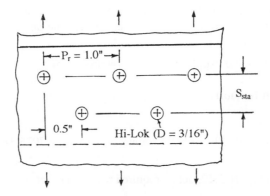

(1) Fastener shear allowable, $P_{s,all}$ = 2,623 lbs./Hi-Lok (see Fig. 9.2.5)
(2) Compute the hole-out percentage (from Eq. 16.5.6) at t = 0.063":

$$A_{hole} = \frac{0.235}{1.0} = 23.5\%$$

where D_{av} = 0.235" (see Fig. 16.5.8)

(3) From Fig. 16.5.12 obtain "x" = 2.05
(4) The minimum row spacing (from Eq. 16.5.7):

$$S_{sta} = \text{"x"} \, A_{hole} = 2.05 \times 0.235 = 0.48"$$

(H) PANEL SKIN AND BEAM WEB

The repair method shown in Fig. 16.5.13 is applicable for damage which is approximately 3 inches in diameter or less. The distribution of fasteners around the hole depends upon the quantity of fasteners required for the number of rows and the spacing of the fasteners.

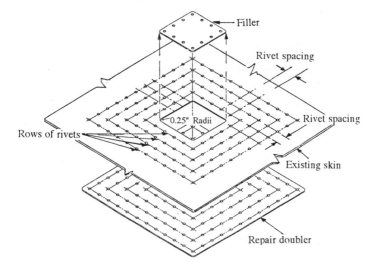

(1) Clean up damaged area
(2) Use the same material as the existing skin for the repair doubler but the thickness of the doubler should be one gage thicker than that of the skin.
(3) A filler is used for aerodynamic smoothness; use one row of rivets of the same type, size, and spacing as on the repair doubler.

Fig. 16.5.13 Repair of a Damaged Hole

Chapter 16.0

The following is a quick method for skins in tension and shear applications and it is used mainly for static load conditions. It may be used for tension fatigue applications but sample tests for airworthiness must be conducted.

The following example will illustrate the calculation procedures:

Example 1 (Skin in tension):

(a) Skin in tension:

(a1) Assume MS 20426(AD5) countersunk rivets in 7075-T6 clad sheet with thickness, t = 0.05"

(a2) From Fig. 16.5.19, obtain required rows of fasteners = 5.5

(a3) Determine desired rivet spacing:
- For protruding head rivets in plain holes, spacing shall be no less than 4 times rivet diameter
- For flush-head rivets in countersunk holes, spacing shall be no less than given in Fig. 16.5.22
- Since this example uses MS 20426(AD5) flush head rivet in 0.05" thick sheet, from Fig. 16.5.22 the minimum spacing = 0.96"

(a4) 5.5 × 0.96 = 5.28 rows of rivets; use 6 rows

(a5) The rivet spacing shall be no more than $\frac{6}{5.5}$ = 1.09" within each row and between rows

Note: When considering fatigue (assume this is a pressurized fuselage skin) the repair doubler should be divided into two doublers (shown in (b) in the sketch on next page):

- Check the rivet shear and bearing strength allowables (7075-T6 clad, F_{bru} = 148,000 psi) for the existing skin, t_{sk} = 0.05"

The rivet shear allowable:

$P_{s, all}$ = 596 lbs. [from Fig. 9.2.5, MS 20426(AD5)]

Since the countersunk head is in the existing skin and the bearing strength allowable [MS 20426(AD5) rivet in 7075-T6 clad sheet] is not available in this book, the closest one from Fig. 9.2.13 [MS 20426(AD5) rivet in 7075-T73 sheet, F_{bru} = 134,000 psi] which is:

$P_{b, all}$ = 413 lbs

- Assume use of the same material as the existing skin (7075-T6 clad, F_{bru} = 148,000 psi) for the first doubler with the thickness, t_1 = 0.025"

$P_{b, all}$ = 148,000 (80%) × 0.1563 × 0.025
= 463 lbs. > 413 lbs   O.K.

- The second doubler thickness = 0.04"
- Fastener distribution should be:
  - 2 rows on 0.025" fatigue doubler
  - 4 rows on both doublers

(a) Skin Repair for Static Strength

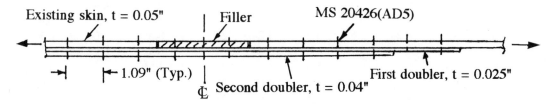

(b) Skin Repair for Fatigue Strength

Example 2 (Skin in shear):

(b) Skin in shear

Use the given data from Example 1 for the skin in shear:

(b1) through (b4) are same as (a1) through (a4).

(b5) Multiply the required number of rows obtained from Step (a4) by a factor of 0.6:

5.28 × 0.6 = 3.17 rows of MS 20426(AD5) rivets for a shear application. Use 4 rows.

(b6) Rivet spacing shall be:

$$\frac{4}{5.5} \times 1.67 = 1.2''$$

where the 1.67 is a multiplying factor

Sheet Thickness	Number of Rows of Protruding Head Rivets Required (Rivet spacing = 1.0 inch)								
	MS 20470						NAS1398		
↓ (t, inch)	AD3 ($\frac{3}{32}''$)	AD4 ($\frac{1}{8}''$)	AD5 ($\frac{5}{32}''$)	AD6 ($\frac{3}{16}''$)	DD6 ($\frac{3}{16}''$)	DD8 ($\frac{1}{4}''$)	D4 ($\frac{1}{8}''$)	D5 ($\frac{5}{32}''$)	D6 ($\frac{3}{16}''$)
0.025	5.0	3.3	–	–	–	–	4.5	–	–
0.032	6.3	3.6	2.7	2.9	2.2	–	4.9	3.7	3
0.04	7.8	4.4	2.9	2.9	2.2	–	5.3	4.0	3.1
0.05	10.0	5.6	3.6	3.1	2.3	1.7	5.6	4.3	3.3
0.063	12.6	7.0	4.5	3.1	2.3	1.6	7.0	4.5	3.6
0.071	–	7.9	5.1	3.5	2.6	1.6	7.9	5.1	3.7
0.08	–	8.9	5.8	3.9	2.9	1.6	–	–	4.0
0.09	–	–	6.5	4.5	3.3	1.8	–	–	4.5
0.1	–	–	–7.3	5.0	3.7	2.0	–	–	–
0.125	–	–	–	6.3	4.6	2.5	–	–	–
0.16	–	–	–	–	–	3.1	–	–	–

(**Note:** Rivet codes, see Fig. 9.2.5)

Fig. 16.5.14 Rivet (MS 20470, NAS 1398) Requirements for 2024-T4 (clad)

Sheet Thickness (t, inch)	Number of Rows of Flush Head Rivets Required (Rivet spacing = 1.0 inch)									
	MS 20426						NAS1399		
	AD3 ($\frac{3}{32}"$)	AD4 ($\frac{1}{8}"$)	AD5 ($\frac{5}{32}"$)	AD6 ($\frac{3}{16}"$)	DD6 ($\frac{3}{16}"$)	DD8 ($\frac{1}{4}"$)	D4 ($\frac{1}{8}"$)	D5 ($\frac{5}{32}"$)	D6 ($\frac{3}{16}"$)
0.032	7.6	5.0	–	–	–	–	–	–	–
0.04	8.8	5.5	4.1	–	–	–	10.7	–	–
0.05	10.5	6.4	4.5	3.8	2.8	–	9.2	8.4	–
0.063	12.6	7.5	5.2	4.1	3.0	2.1	8.3	7.4	6.8
0.071	–	8.2	5.6	4.3	3.2	2.1	8.5	7.0	6.3
0.08	–	–	6.1	4.6	3.4	2.2	8.9	6.7	6.0
0.09	–	–	6.7	5.0	3.7	2.3	10.0	6.9	5.5
0.1	–	–	–	5.4	4.0	2.5	–	7.3	5.6
0.125	–	–	–	6.5	4.8	2.9	–	9.1	6.3
0.16	–	–	–	7.6	5.6	3.3	–	–	–

(**Note:** Rivet codes, see Fig. 9.2.5)

Fig. 16.5.15 Rivet (MS 20426, NAS 1399) Requirements for 2024-T4 (clad)

Sheet Thickness (t, inch)	Number of Rows of Protruding Head Rivets Required (Rivet spacing = 1.0 inch)									
	MS 20470						NAS1398		
	AD3 ($\frac{3}{32}"$)	AD4 ($\frac{1}{8}"$)	AD5 ($\frac{5}{32}"$)	AD6 ($\frac{3}{16}"$)	DD6 ($\frac{3}{16}"$)	DD8 ($\frac{1}{4}"$)	D4 ($\frac{1}{8}"$)	D5 ($\frac{5}{32}"$)	D6 ($\frac{3}{16}"$)
0.025	5.3	3.5	–	–	–	–	4.8	3.9	–
0.032	6.1	3.8	2.8	3.1	2.3	–	5.2	3.9	3.2
0.04	8.2	4.6	3.1	3.1	2.3	–	5.6	4.2	3.3
0.05	10.5	5.9	3.8	3.2	2.4	2.3	5.9	4.5	3.5
0.063	13.9	7.7	5.0	3.5	2.5	1.8	7.7	5.0	3.9
0.071	–	8.7	5.7	3.9	2.9	1.8	8.7	5.7	4.1
0.08	–	10.1	6.5	4.4	3.3	1.9	–	–	4.5
0.09	–	–	7.2	4.8	3.6	2.0	–	–	4.9
0.1	–	–	8.2	5.4	4.0	2.2	–	–	–
0.125	–	–	–	6.7	5.0	2.7	–	–	–
0.16	–	–	–	–	–	3.4	–	–	–

(**Note:** Rivet codes, see Fig. 9.2.5)

Fig. 16.5.16 Rivet (MS 20470, NAS 1398) Requirements for Clad 2024-T3, -T6 and -T81

Structural Salvage and Repairs

Sheet Thickness (t, inch)	Number of Rows of Flush Head Rivets Required (Rivet spacing = 1.0 inch)								
	MS 20426						NAS1399		
	AD3 ($\frac{3}{32}$")	AD4 ($\frac{1}{8}$")	AD5 ($\frac{5}{32}$")	AD6 ($\frac{3}{16}$")	DD6 ($\frac{3}{16}$")	DD8 ($\frac{1}{4}$")	D4 ($\frac{1}{8}$")	D5 ($\frac{5}{32}$")	D6 ($\frac{3}{16}$")
0.032	8.0	5.2	–	–	–	–	–	–	–
0.04	9.3	5.8	4.3	–	–	–	11.3	–	–
0.05	11.1	6.7	4.7	4.0	3.0	–	9.7	8.9	–
0.063	14.1	8.3	5.7	4.5	3.4	2.3	9.2	8.2	7.6
0.071	–	9.1	6.2	4.8	3.6	3.2	9.4	7.7	7.0
0.08	–	–	7.0	5.2	3.9	2.5	10.1	7.6	6.8
0.09	–	–	7.4	5.5	4.1	2.6	11.1	7.6	6.2
0.1	–	–	–	5.9	4.4	2.7	–	8.0	6.2
0.125	–	–	–	7.0	5.2	3.1	–	9.9	6.8
0.16	–	–	–	8.3	6.2	3.6	–	–	–

(**Note:** Rivet codes, see Fig. 9.2.5)

Fig. 16.5.17 Rivet (MS 20426, NAS 1399) Requirements for Clad 2024-T3, -T6 and -T81

Sheet Thickness (t, inch)	Number of Rows of Protruding Head Rivets Required (Rivet spacing = 1.0 inch)								
	MS 20470						NAS1398		
	AD3 ($\frac{3}{32}$")	AD4 ($\frac{1}{8}$")	AD5 ($\frac{5}{32}$")	AD6 ($\frac{3}{16}$")	DD6 ($\frac{3}{16}$")	DD8 ($\frac{1}{4}$")	D4 ($\frac{1}{8}$")	D5 ($\frac{5}{32}$")	D6 ($\frac{3}{16}$")
0.025	6.2	4.1	–	–	–	–	5.6	4.5	–
0.032	7.7	4.4	3.3	3.6	2.7	–	6.0	4.5	3.7
0.04	9.6	5.4	3.6	3.7	2.8	–	6.5	4.9	3.8
0.05	12.3	6.9	4.5	3.9	2.9	2.1	6.9	5.2	4.1
0.063	15.9	8.9	5.7	4.0	2.9	2.1	8.9	5.7	4.5
0.071	–	10.0	6.5	4.4	3.2	2.1	10.0	6.5	4.7
0.08	–	11.2	7.3	5.1	3.7	2.1	–	–	5.0
0.09	–	–	8.2	5.6	4.1	2.3	–	–	5.7
0.1	–	–	9.2	6.3	4.6	2.6	–	–	–
0.125	–	–	–	7.8	5.7	3.1	–	–	–
0.16	–	–	–	–	–	3.9	–	–	–

(**Note:** Rivet codes, see Fig. 9.2.5)

Fig. 16.5.18 Rivet (MS 20470, NAS 1398) Requirements for Clad 2024-T86 and 7075-T6

Sheet Thickness ↓ (t, inch)	Number of Rows of Flush Head Rivets Required (Rivet spacing = 1.0 inch)								
	MS 20426						NAS1399		
	AD3 ($\frac{3}{32}$")	AD4 ($\frac{1}{8}$")	AD5 ($\frac{5}{32}$")	AD6 ($\frac{3}{16}$")	DD6 ($\frac{3}{16}$")	DD8 ($\frac{1}{4}$")	D4 ($\frac{1}{8}$")	D5 ($\frac{5}{32}$")	D6 ($\frac{3}{16}$")
0.032	9.4	6.1	–	–	–	–	–	–	–
0.04	10.8	6.8	5.0	–	–	–	13.2	–	–
0.05	13.0	7.8	5.5	4.8	3.5	–	11.3	10.3	–
0.063	15.9	9.5	6.6	5.4	3.9	2.6	10.5	9.3	8.6
0.071	–	10.4	7.1	5.6	4.1	2.7	10.8	8.8	8.0
0.08	–	–	7.8	6.0	4.4	2.8	11.2	8.5	7.5
0.09	–	–	8.5	6.5	4.7	2.9	12.6	8.7	7.1
0.1	–	–	–	7.0	5.1	3.1	–	9.2	7.1
0.125	–	–	–	8.1	5.9	3.6	–	11.3	7.8
0.16	–	–	–	9.8	7.1	4.2	–	–	–

(**Note:** Rivet codes, see Fig. 9.2.5)

Fig. 16.5.19 Rivet (MS 20426, NAS 1399) Requirements for Clad 2024-T86 and 7075-T6

Sheet Thickness ↓ (t, inch)	Number of Rows of Protruding Head Rivets Required (Rivet spacing = 1.0 inch)								
	MS 20470						NAS1398		
	AD3 ($\frac{3}{32}$")	AD4 ($\frac{1}{8}$")	AD5 ($\frac{5}{32}$")	AD6 ($\frac{3}{16}$")	DD6 ($\frac{3}{16}$")	DD8 ($\frac{1}{4}$")	D4 ($\frac{1}{8}$")	D5 ($\frac{5}{32}$")	D6 ($\frac{3}{16}$")
0.025	6.3	4.5	–	–	–	–	6.3	5.0	–
0.032	8.6	5.0	3.6	4.0	3.0	–	6.7	5.1	4.2
0.04	10.7	6.0	4.0	4.0	3.0	–	7.3	5.5	4.3
0.05	13.7	7.6	5.0	4.2	3.1	2.2	7.7	5.8	4.6
0.063	17.2	9.6	6.2	4.3	3.1	2.2	9.6	6.2	4.9
0.071	–	10.8	7.0	4.7	3.5	2.2	10.8	7.0	5.1
0.08	–	12.2	7.9	5.4	4.0	2.2	–	–	5.5
0.09	–	–	8.9	6.0	4.5	2.5	–	–	6.1
0.1	–	–	10.0	6.7	5.0	2.8	–	–	–
0.125	–	–	–	8.2	6.1	3.4	–	–	–
0.16	–	–	–	–	–	4.3	–	–	–

(**Note:** Rivet codes, see Fig. 9.2.5)

Fig. 16.5.20 Rivet (MS 20470, NAS 1398) Requirements for 7075-T6 (Bare)

Sheet Thickness	Number of Rows of Flush Head Rivets Required (Rivet spacing = 1.0 inch)								
	MS 20426						NAS1399		
(t, inch)	AD3 ($\frac{3}{32}"$)	AD4 ($\frac{1}{8}"$)	AD5 ($\frac{5}{32}"$)	AD6 ($\frac{3}{16}"$)	DD6 ($\frac{3}{16}"$)	DD8 ($\frac{1}{4}"$)	D4 ($\frac{1}{8}"$)	D5 ($\frac{5}{32}"$)	D6 ($\frac{3}{16}"$)
0.032	10.5	6.8	–	–	–	–	–	–	–
0.04	12.1	7.5	5.6	–	–	–	14.7	–	–
0.05	14.5	8.7	6.2	5.2	3.9	–	12.6	11.5	–
0.063	17.3	10.3	7.1	5.6	4.2	2.9	11.4	10.1	9.4
0.071	–	11.3	7.7	5.9	4.4	2.9	11.7	9.5	8.7
0.08	–	–	8.4	6.3	4.7	3.0	12.2	9.2	8.2
0.09	–	–	9.2	6.9	5.1	3.2	13.7	9.4	7.7
0.1	–	–	–	7.4	5.5	3.4	–	10.0	7.7
0.125	–	–	–	8.6	6.4	3.9	–	12.3	8.5
0.16	–	–	–	10.4	7.7	4.5	–	–	–

(**Note:** Rivet codes, see Fig. 9.2.5)

Fig. 16.5.21 Rivet (MS 20426, NAS 1399) Requirements for 7075-T6 (Bare)

Rivet Dia. (inch)	Skin Thickness (inch)											
	0.025	0.032	0.036	0.04	0.05	0.063	0.071	0.08	0.09	0.1	0.125	0.16
	(Spacing per inch given below)											
$\frac{3}{32}$	0.67	0.61	0.59	0.57	0.53	0.5	0.49	–	–	–	–	–
$\frac{1}{8}$	0.9	0.81	0.79	0.75	0.7	0.66	0.64	0.63	0.62	–	–	–
$\frac{5}{32}$	1.28	1.14	1.04	1.04	0.96	0.89	0.86	0.83	0.81	–	–	–
$\frac{3}{16}$	–	1.56	1.46	1.39	1.27	1.16	1.12	1.08	1.04	1.01	0.96	0.92
$\frac{1}{4}$	–	–	–	–	–	1.84	1.75	1.67	1.6	1.54	1.44	1.35

Fig. 16.5.22 Minimum Permissible Spacing for 100° Flush (Countersunk) Rivets

16.6 REPAIRS FOR CORROSION DAMAGE

Improper trimming of corrosion or damaged edges can result in stress concentrations of 3.5 to 4 times that of the basic stress of the original structure, as shown in Fig. 16.6.1. The following procedures are recommended for this repair:

- A 1.0 radius together with a 20:1 slope is recommended for use on all pressurized fuselages and other fatigue critical applications
- A 10:1 slope gives a significantly higher stress rise and should only be considered for lightly loaded or non-structural applications
- The same principles can be applied to blending out gouges on the surface of the parts
- Note that stress concentration factors decrease as θ and R increase

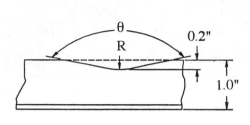
(a) Stringer Flange Edge Trim

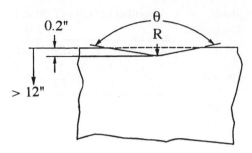
(b) Skin Panel Edge Trim

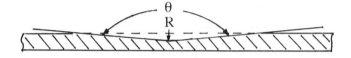
(c) Part surface blendout

	1.0" Wide Stringer Flange			For The Wide Panel		
Radius (R) →	0.25	0.5	1.0	0.25	0.5	1.0
5:1 Slope (θ = 157°)	1.42	1.28	1.18	1.6	1.47	1.35
10:1 Slope (θ = 168°)	1.32	1.22	1.15	1.4	1.35	1.3
20:1 Slope (θ = 174°)	1.15	1.1	1.06	1.22	1.2	1.1
30:1 Slope (θ = 176°)	1.15	1.1	1.04	1.18	1.17	1.08

Fig. 16.6.1 Stress Concentration Comparisons for Corrosion or Notch Removals

(A) REPAIR FOR REDUCTION IN THICKNESS OF MATERIAL

- The cross-sectional area of repair material (doubler) plus the cross-sectional area of the remaining thickness minus the area removed by corrosion clean-out should be at least 1.5 times the original cross-sectional area
- A standard thickness (0.025 or 0.032 inch) "fatigue doubler" should always be used against the skin with a repair doubler on top of the fatigue doubler
- A general rule of thumb for the total thickness of external repair doublers required is to take one-half the skin thickness and add the maximum depth of corrosion clean-out

Example:

A 0.076" thick wing skin has 0.036" deep corrosion that must be repaired. Assume the doubler width is 1.0" and the material is 7075-T6 bare sheet.

F_{tu} = 78,000 psi and F_{bru} = 156,000 psi for sheet t = 0.032"

F_{tu} = 80,000 psi and F_{bru} = 160,000 psi for sheet t = 0.050"

(a) Total doubler thickness:

$$\frac{0.076}{2} + 0.036 = 0.074"$$

(b) Fatigue doubler: use 0.032"

(c) Repair doubler thickness:

$$0.074 - 0.032 = 0.042"$$

(d) The nearest standard sheet thickness is 0.05": use 0.05"

(e) Final repair sheets are:

0.032 + 0.05 = 0.082"

(f) Total doubler strength:

0.032 × 78,000 + 0.05 × 80,000 = 6,496 lbs.

(g) Sufficient rows of fasteners must be used in the doublers to fully develop the ultimate strength of the doublers.

- Assume use of $\frac{3}{16}$" titanium Hi-Loks with a shear strength of $P_{s,all}$ = 2,632 lbs.
- Bearing strength on 0.032" doubler is:

  $P_{b,all}$ = 156,000 × 80% × 0.1875 × 0.032 = 749 lbs.
- Bearing strength on 0.05" doubler is:

  $P_{b,all}$ = 160,000 × 80% × 0.1875 × 0.05 = 1,200 lbs.

  (**Note:** The 80% is the material bearing reduction factor used when a fastener is installed wet, see Chapter 9.2)

(h) On the fatigue doubler extend two rows of fasteners beyond the repair doubler.

- There will be two rows of fasteners through the 0.032" fatigue doubler beyond the edge of the 0.05" doubler. These will be good for:

  2 × 749 = 1,498 lbs.
- The fasteners which go through both doublers (0.032" and 0.05") must be good for:

  6,496 − 1,498 = 4,998 lbs.
- Number of rows required for doubler of 0.05" is:

  $$\frac{4,998}{1,200} = 4.17$$

  The next higher number is 5 rows

(i) Use 3 additional rows of fasteners to allow for poor installation, bad holes, etc. in the repair.

Thus this installation has 8 rows of fasteners which go through both doublers and an additional 2 rows at each ends which only go through the 0.032" fatigue doubler, as sketched below:

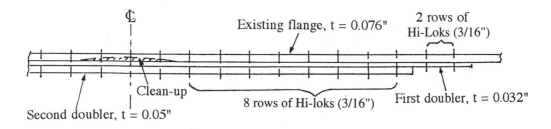

(B) REPAIR FOR REDUCTION IN FLANGE WIDTH

A partial repair, as shown in Fig. 16.6.2, should have at least one row of rivets which are spaced uniformly across the length of the repair angle doubler to provide continuous fasteners to the existing member.

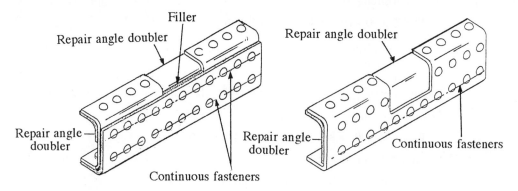

Fig. 16.6.2 Repair for Flange Reduction

16.7 REPAIR OF SKIN-STRINGER PANELS

(A) INSERTION

Generally insertion is used when:

- Reinforced portion is relatively long
- Reinforced member has a complex shape
- Interference between the reinforced member and adjacent structure members is to be avoided

Damage is repaired by (refer to Fig. 16.7.1):

- Cut away a portion of the damaged stringer
- Insert a length of member identical in shape and material to the damaged part in its place
- Splice doublers at each end of the inserted member provide load-transfer continuity between the existing part and the inserted member

For example, for a damaged fuselage stringer, a stringer insert is installed to splice a fuselage stringer where repair would interfere with a frame, thereby necessitating rework of the frame cutout area.

- An alternate method is to cut off a length of the existing stringer which passes through the frame cutout
- Install a new piece of stringer for the portion of the stringer removed
- A splice is then used at each end of the inserted stringer

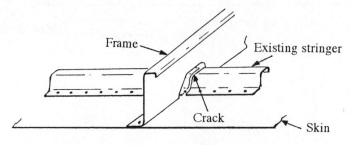

(a) Cracked Stringer before Repair

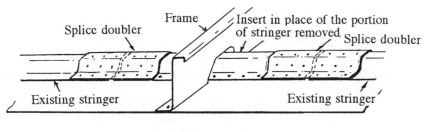

(b) After Repair

Fig. 16.7.1 Repair of a Fuselage Stringer Which Interferes with Frame

The repair of a fuselage skin and stringer shown in Fig. 16.7.2 is a non-flush repair. This repair requires the fatigue doubler inside of the skin and, therefore, the stringer has to be cut to a required length to fit this doubler. Use a new stringer of the same material and size (or save the one that was cut), insert in its original place, and splice or use the method as shown in Fig. 16.7.2.

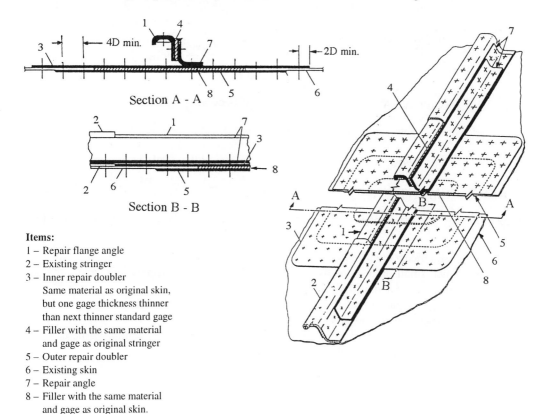

Items:
1 – Repair flange angle
2 – Existing stringer
3 – Inner repair doubler
   Same material as original skin, but one gage thickness thinner than next thinner standard gage
4 – Filler with the same material and gage as original stringer
5 – Outer repair doubler
6 – Existing skin
7 – Repair angle
8 – Filler with the same material and gage as original skin.

Fig. 16.7.2 Repair of Damage to Fuselage Skin and Stringer

For damage on a wing skin which is located at an undamaged stringer, as shown in Fig. 16.7.3, the flush repair of the skin for aerodynamic smoothness is required:

- Cut off the damaged skin area and at the same time cut a length off the existing stringer which is slightly longer than the cut skin area
- Prepare a reinforcement skin doubler (slightly thicker or one gage thicker than the damaged skin) which is the same length as that of the cut portion of the existing stringer

- Prepare an identical reinforcement stringer from the same material or use the cut portion (if not damaged) of the existing stringer
- A filler plate of the same size as the cutaway skin area is fastened on the doubler to obtain an aerodynamically smooth surface
- A splice joint is used at each end of the reinforcement stringer and around the splice doubler

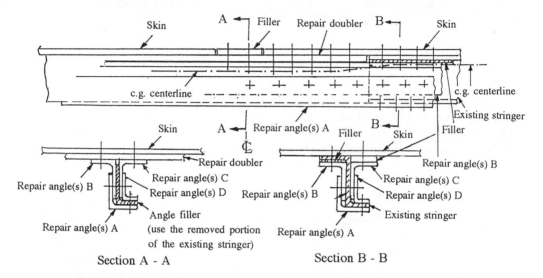

Note: Repair angle and doubler may consist of multiple members.

Fig. 16.7.3 Flush Repair for a Damaged Wing Skin

(B) REPLACEMENT

For the following types of damage or situations it is necessary to replace affected parts rather than repair:

- A significant damage cannot be repaired
- When small parts, such as doublers, splices, tension fittings, lugs, and clips, are damaged beyond negligible limits
- Damage to parts of relatively short length necessitates replacement because the repair of these parts is generally impractical
- Some highly stressed parts cannot be reinforced because the reinforced parts would not have an adequate margin of safety
- Parts with configurations not adaptable to practicable repair (e.g., machined or forged parts with complex shapes) must be replaced when damaged

16.8 STRUCTURAL RETROFIT

Structural retrofit occurs in the airframe when that particular type of aircraft is still in production. Retrofit involves adding reinforcement material to weak areas to increase the load capability to meet an increased aircraft growth weight. Generally, the reinforcement members are long members as opposed to the short doublers used in structural repairs.

General approaches for structural retrofit include:

- Generally, stringers or stiffeners (e.g., box structures of wing, empennage, fuselage, etc.) are reinforced rather than skin when bending (EI) is critical

- Increasing skin thickness is rare except for the need to increase the torsional stiffness (GJ) of box structures (add additional skin externally)
- The most common reinforcements to increase bending areas (EI) are straps and angle shapes, as shown in Fig. 16.8.1

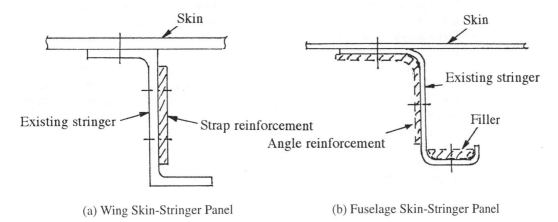

(a) Wing Skin-Stringer Panel   (b) Fuselage Skin-Stringer Panel

Fig. 16.8.1 Retrofit Skin-Stringer Panel Reinforced Using Straps or Angles

(A) WING AND EMPENNAGE SURFACE PANELS

The following retrofit guidelines apply to a wing box:

- Reinforce surface panels by adding straps on the interior stringer flanges, as shown in Fig. 16.8.1(a).
- Use straps rather than other shapes, e.g., angles, because the angle does not have the flexibility which is an essential requirement for a long member which must be placed in the wing box through lower access holes
- It is generally impractical to use thin doublers on the exterior surface of wings because it affects the airfoil characteristics; if use of doubler is allowed, protruding head fasteners are recommended
- Under compression load strap thickness is dictated by inter-rivet buckling (see Chapter 14.3)
- Aluminum should be the first choice of material (for aluminum structures)
- Select steel straps only when space and fastener edge distance are a problem
- Straps are the ideal reinforcing members, owing to their very high bending flexibility (EI) and torsional flexibility (GJ), for providing sufficient axial strength of cross-sectional reinforcement
- Use of Hi-Lok fasteners to fasten straps to stringers is preferred because of ease of installation

(B) FUSELAGE PANELS

The following retrofit guidelines apply to fuselage body:

- Generally, stringers are reinforced rather than skin
- Add straps and/or angle reinforcements [see Fig. 16.8.1(b)] using the same material as the stringer
- Since the working space is not so restricted, use angle reinforcements whenever possible
- Add filler bar on the outstanding flange [see Fig. 16.8.1(b)] of stringer or frame to increase cross-sectional moment of inertia (I) as well as area

Chapter 16.0

(C) SPAR AND RIB WEBS

The following retrofit guidelines apply to adding a skin doubler on existing spar and rib webs, as shown in Fig. 16.8.2:

- Local vertical stiffeners have to be removed prior to installation of doubler
- Use over-sized fasteners when re-installing the original stiffeners
- Field-fasteners are required to attach the doubler to the existing web to prevent premature buckling of the thin doubler
- The use of field-fasteners is complicated [assessment is usually based on the judgment of an experienced engineer or use the method per Example 2 of Chapter 16.5(H)] since conventional structural analysis is not applicable because it involves many parameters which are function of:
  - Fastener material, type, size, and spacing
  - The material properties and thickness of the doubler and the existing web
  - Sealant or bonding between doubler and the existing web
- In fatigue critical areas, e.g., caps of the wing spar, butterfly flanges with a tapered thickness are used on a thicker doubler

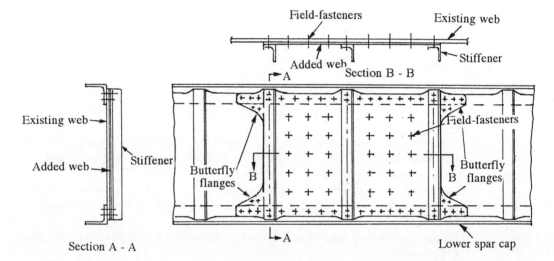

Fig. 16.8.2 Spar Web Reinforcement Using Skin Doubler

References

16.1　Swift, T., "Repairs to Damage Tolerant Aircraft", Presented to International Symposium on Structural Integrity of Aging Airplanes. Atlanta, Georgia. March 20-22, 1990.
16.2　Anon., "Engineering Liaison Reference Manual", Boeing Company, 1994.
16.3　Anon., "Structural Repair Course for Airline Engineers", The Boeing Company, 1994.
16.4　Anon., "Notes for Structural Repair Considerations", Training course notes No. TO501 (LX3118), Lockheed Aeronautical Systems Company (Burbank, California), Lockheed Technical Institute.
16.5　Anon., "Salvage Repair – Standard Factory Operation", Lockheed Corp.
16.6　Anon., "AGARD AG 278 Volume 1, Aircraft Corrosion: Causes and Case Histories", ARARD-AG-278, July, (1985).
16.7　Anon., "Corrosion Control for Aircraft", NAVWEPS 01-1A-509, Published under direction of the chief of the Bureau of Naval Weapons. May 1, (1965).
16.8　Freeman, M. G., "Cost-Effective Designs Ease Maintenance Problems", Commuter Air, Dec., (1982). pp. 44-45.
16.9　Ransden, J. M. and Marsden John, "Caring for the High-Time Jet", Flight International, Sept. 3, (1988). pp. 153-156.
16.10　Ransden, J. M., "The Long-Life Structure", Flight International, Sept. 17, (1988). pp. 56-59.
16.11　Ransden, J. M., "Ageing Jet Care", Flight International, July, 15, (1989). pp. 39-42..
16.12　Anon., "Extending an Airliner's Life", Aerospace Engineering, December, (1989). pp. 13-15.
16.13　Anon., "Higher Maintenance Standards Sought for World's Aging Fleet", Aviation Week & Space Technology, July 2, (1990). pp. 60-63.
16.14　DeMeis, Richard, "Aging Aircraft", Aerospace America, July, (1989). pp. 38-42.

Chapter 17.0

STRUCTURAL TEST SETUP

17.1 INTRODUCTION

As mentioned previously, the airframer must give the customer an aircraft which will meet the customer's operational life requirements as well as the requirements of static ultimate strength. To achieve this goal and to satisfy the certifying agency, use a group test matrix, as shown in Fig. 17.1.1.

Type of Test	Static	Fatigue	Fail-safe	Functional	Sonic
Coupon	•	•			
Element	•				
Component	•	•	•	•	•
Full scale	•	•	•	Iron bird	

Fig. 17.1.1 Group Test Matrix

In terms of total impact, the aircraft fatigue test is probably the most important in the test program (see Reference 17.7). Also, the fatigue test is extremely valuable for developing inspection and maintenance procedures. The fatigue test objectives are:

- To provide basic data on the methods of analysis used
- To determine the effects of various factors affecting the fatigue life of the structure such as materials and processes, fabrication procedures, environmental conditions, etc.
- To establish design allowable stresses for the materials to be used in the structure including processing effects
- To evaluate and compare the fatigue quality of structural design details
- To substantiate that the structure will meet or exceed the design life requirement of the airframe
- To locate fatigue-critical areas of the aircraft at the earliest possible time so any improvements can be incorporated early in production
- To develop inspection and maintenance procedures for the customer
- To provide test data for analytical studies to assist in determining times interval for inspection and maintenance

17.2 TESTING THE LOAD SPECTRUM

In aircraft fatigue tests, it is, of course, necessary to apply a load spectrum representative of the intended usage. These loads are applied in truncated load cycles to reduce the time of testing, such that the tests conducted over a period of a few weeks/months represent as near as possible the complete lifetime load spectrum. A reasonable representation of the various levels of stress, such as ground stresses and maneuver stresses, and the corresponding GAG cycles must be included, as shown in Fig. 17.2.1.

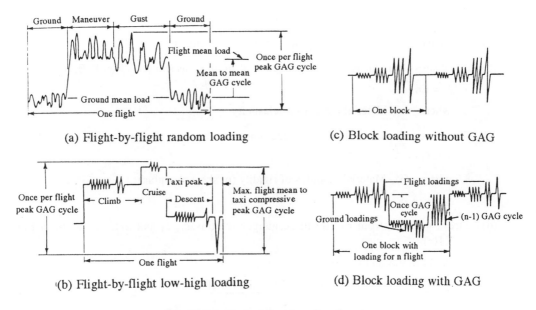

Fig. 17.2.1 Typical Test Loading Spectra

17.3 STRAIN GAUGES

Strain gauges are used as follows:

- To measure the strain (stress) at critical areas
- Expected readings of these gauges are predicted by calculation for each of the tests which are to be performed
- During the testing the gauges are read continuously by computerized equipment so that a running check is kept of the increase in stress at each station as the load is increased
- If any of the gauges indicates a stress significantly in excess of the predicted value, the test is halted until the discrepancy is explained

Three types of strain gauges are used on airframe fatigue and static tests for measuring strain (ε). These are axial, shear, and rosette gauges, as shown in Fig. 17.3.1.

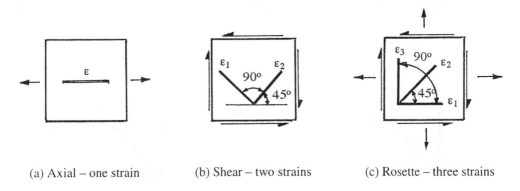

Fig. 17.3.1 Three Types of Strain Gauges (Positive Sign Shown)

Strain gauge measurements (minus sign represents compression strain or stress):

(a) All gauges measure in micro-inches per inch or 10^{-6} in./in.

(b) Axial stress from measured one gauge (axial strain ε), as shown in Fig. 17.3.1(a):

$$f_a = E\varepsilon$$

where: E – Material modulus of elasticity

(c) Shear stress from measured two gauges (strains ε_1 and ε_2), as shown in Fig. 17.3.1(b):

$$f_s = G\gamma$$

where: G – Material modulus of rigidity (or shear modulus)

$$\gamma = \varepsilon_1 + \varepsilon_2$$

(d) Stress measure from rosette three gauges (strains ε_1, ε_2 and ε_3), as shown in Fig. 17.3.1(c):

- Axial strains and stresses:

$$\varepsilon_{max} = \frac{\varepsilon_1 + \varepsilon_3}{2} + \frac{\sqrt{(\varepsilon_1 - \varepsilon_3)^2 + [2\varepsilon_2 - (\varepsilon_1 + \varepsilon_3)]^2}}{2} \qquad \text{Eq. 17.3.1(a)}$$

$$\varepsilon_{max} = \frac{\varepsilon_1 + \varepsilon_3}{2} + \frac{\gamma_{max}}{2} \qquad \text{Eq. 17.3.1(b)}$$

$$\varepsilon_{min} = \frac{\varepsilon_1 + \varepsilon_3}{2} - \frac{\sqrt{(\varepsilon_1 - \varepsilon_3)^2 + [2\varepsilon_2 - (\varepsilon_1 + \varepsilon_3)]^2}}{2} \qquad \text{Eq. 17.3.2(a)}$$

$$\varepsilon_{min} = \frac{\varepsilon_1 + \varepsilon_3}{2} - \frac{\gamma_{max}}{2} \qquad \text{Eq. 17.3.2(b)}$$

$$f_{max} = \frac{E}{(1 - \mu^2)}(\varepsilon_{max} + \mu\varepsilon_{min}) \qquad \text{Eq. 17.3.3}$$

$$f_{min} = \frac{E}{(1 - \mu^2)}(\varepsilon_{min} + \mu\varepsilon_{max}) \qquad \text{Eq. 17.3.4}$$

- Shear strain and stress:

$$\gamma_{max} = \sqrt{(\varepsilon_1 - \varepsilon_3)^2 + [2\varepsilon_2 - (\varepsilon_1 + \varepsilon_3)]^2} \qquad \text{Eq. 17.3.5}$$

$$f_{s, max} = \frac{E}{2(1 + \mu)}(\gamma_{max}) = G\gamma_{max} \qquad \text{Eq. 17.3.6}$$

- Inclined angle of ε_{max} or f_{max}:

$$\phi = \left(\frac{1}{2}\right)\tan^{-1}\left[\frac{2\varepsilon_2 - (\varepsilon_1 + \varepsilon_3)}{\varepsilon_1 - \varepsilon_3}\right] \text{ (in degrees)} \qquad \text{Eq. 17.3.7}$$

Sign convention for inclined angle of ϕ is shown in Fig. 17.3.2.

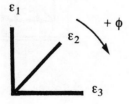
(a) Clockwise Numbering

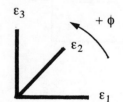
(b) Counterclockwise Numbering

(**Note:** ϕ is always less than 45° and is measured algebraically from the greater of ε_1 or ε_3)

Fig. 17.3.2 Sign Convention of Inclined Angle ϕ

Example:

The strain gauge readings for a set of loads applied on a component of 7075-T6 plate are shown below:

Material properties (from Fig. 4.3.4):

$E = 10.3 \times 10^6$ psi (tension)

$\mu = 0.33$

Strain gauge readings:

$\varepsilon_1 = 2,850 \times 10^{-6}$ in./in.

$\varepsilon_2 = 650 \times 10^{-6}$ in./in.

$\varepsilon_3 = 1,020 \times 10^{-6}$ in./in.

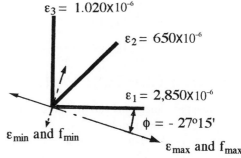

(Counterclockwise numbering)

(a) The maximum shear strain, from Eq. 17.3.5:

$$\gamma_{max} = \sqrt{(\varepsilon_1 - \varepsilon_3)^2 + [2\varepsilon_2 - (\varepsilon_1 + \varepsilon_3)]^2}$$
$$= \sqrt{(2,850 \times 10^{-6} - 1,020 \times 10^{-6})^2 + [2 \times 650 \times 10^{-6} - (2,850 \times 10^{-6} + 1,020 \times 10^{-6})]^2}$$
$$= 3,155 \times 10^{-6} \text{ rad.}$$

(b) The maximum axial strain, from Eq. 17.3.1(b):

$$\varepsilon_{max} = \frac{\varepsilon_1 + \varepsilon_3}{2} + \frac{\gamma_{max}}{2}$$
$$= \frac{2,850 \times 10^{-6} + 1,020 \times 10^{-6}}{2} + \frac{3,155 \times 10^{-6}}{2} = 3,513 \times 10^{-6} \text{ in./in.}$$

(c) The minimum axial strain, from Eq. 17.3.2(b):

$$\varepsilon_{min} = \frac{\varepsilon_1 + \varepsilon_3}{2} - \frac{\gamma_{max}}{2}$$
$$= \frac{2,850 \times 10^{-6} + 1,020 \times 10^{-6}}{2} - \frac{3,155 \times 10^{-6}}{2} = 358 \times 10^{-6} \text{ in./in.}$$

(d) The inclined angle of f, from Eq. 17.3.7:

$$\phi = (\frac{1}{2}) \tan^{-1}[\frac{2\varepsilon_2 - (\varepsilon_1 + \varepsilon_3)}{\varepsilon_1 - \varepsilon_3}]$$
$$= (\frac{1}{2}) \tan^{-1}[\frac{2 \times 650 \times 10^{-6} - (2,850 \times 10^{-6} + 1,020 \times 10^{-6})}{2,850 \times 10^{-6} - 1,020 \times 10^{-6}}]$$
$$= (\frac{1}{2}) \tan^{-1}[-1.4] = -27°15' \text{ (clockwise from } \varepsilon_1)$$

(e) The maximum axial stress from Eq. 17.3.3:

$$f_{max} = \frac{E}{(1-\mu^2)}(\varepsilon_{max} + \mu\varepsilon_{min})$$
$$= \frac{10.3 \times 10^6}{(1-0.33^2)}(3,513 \times 10^{-6} + 0.33 \times 358 \times 10^{-6}) = 41,976 \text{ psi}$$

(f) The minimum axial stress, from Eq. 17.3.4:

$$f_{min} = \frac{E}{(1-\mu^2)}(\varepsilon_{min} + \mu\varepsilon_{max})$$
$$= \frac{10.3 \times 10^6}{(1-0.33^2)}(358 \times 10^{-6} + 0.33 \times 3,513 \times 10^{-6}) = 17,540 \text{ psi}$$

Chapter 17.0

(g) The maximum shear stress, from Eq. 17.3.6:

$$f_{s,max} = \frac{E}{2(1+\mu)}(\gamma_{max})$$

$$= \frac{10.3 \times 10^6}{2(1+0.33)}(3,155 \times 10^{-6}) = 12,454 \text{ psi}$$

(**Note:** The above values can be obtained by using the Mohr's Circle graphic method)

17.4 FATIGUE TEST PANEL SPECIMEN

Before a particular configuration of a tension panel is finally selected for a project, various fatigue test panels, as shown in Fig. 17.4.1, are required to verify the fatigue life. Compared to compression tests, testing of fatigue panels is a costly time-consuming process. The following recommendations should be followed:

- Test section should have the same configuration as the proposed panel
- Keep panel cross-sectional centerline (or c.g.) along the panel length as straight as possible to avoid the occurrence of secondary bending stress
- Total panel length should provide enough buckling strength to react the maximum compression limit load
- Increased panel thickness – machined or bonded doublers are required where panel is bolted to fixture
- Use high strength tension bolts (usually use 12-point) to attach fixture to both ends of the panel; applied loads are transferred to the panel by friction force (use a friction coefficient of 0.08) rather than conventional bolt bearing strength
- Take careful in the design of the fastener pattern at the transition section to avoid premature failure
- Install axial strain gauges (similar to that of compression panel test as shown in Fig. 17.4.2) on every outstanding flanges of the stringer at test section to monitor the secondary bending stress due to improperly by applied loads from fixtures

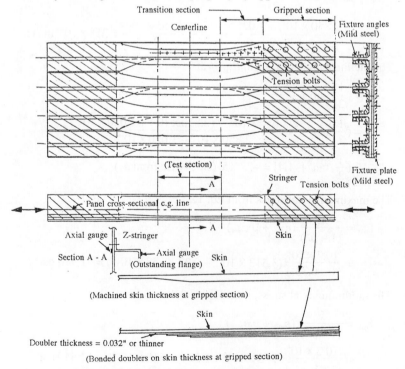

Fig. 17.4.1 Fatigue Test Panel Specimen

Structural Test Setup

Example:

Design of a fatigue test specimen of a skin-stringer panel of 2024-T3 material is shown below:

Material properties:

$F_{cy} = 40,000$ psi; $E_c = 10.7 \times 10^6$ psi

Maximum limit loads:

$P_{Limit, T} = 74,000$ lbs.

(Maximum tension load)

$P_{Limit, C} = 30,000$ lbs.

(Maximum compression load)

Test section (basic configuration):

Area: $A = 1.731$ in.^{2}

Moment of inertial: $I = 1.093$ in.^{4}

N.A. (y-axis): $Y = 0.531$ in.

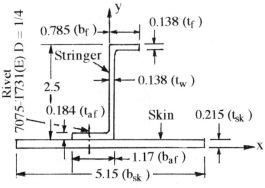

Section A-A (Cross Section of Test Section)

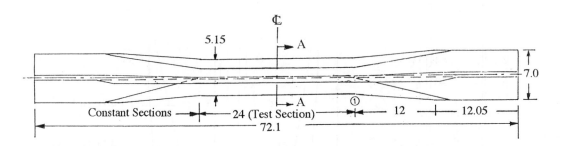

(Detail dimensions are shown in Fig. A and B, all dimensions are in inches)

(a) 12-point tension bolt (H.T. = 160 ksi) and their friction forces (use friction coefficient, $\mu = 0.08$ and standard torque):

12-point Bolt Diameter (in.)	Friction Force (lbs.) $2\mu \times P_{ten}{}^* \times 75\% = P_{Friction}{}^{**}$
7/16	0.16 x 18,120 x 75% = 2,174
1/2	0.16 x 24,570 x 75% = 2,948
5/8	0.16 x 39,520 x 75% = 4,742
3/4	0.16 x 57,700 x 75% = 6,924

*   Bolt tension load from Fig. 9.2.6.
**   Friction forces are based on the friction coefficient of 2 x 0.08 = 0.16 (double faces) as shown:
The value of $P_{friction}$ can be obtained by a given standard torque (provided by bolt supplier) on a bolt and apply a pull force of "P" until slippage occur. Use $P_{Friction} = 85\%P$

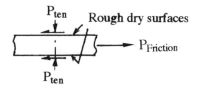

Chapter 17.0

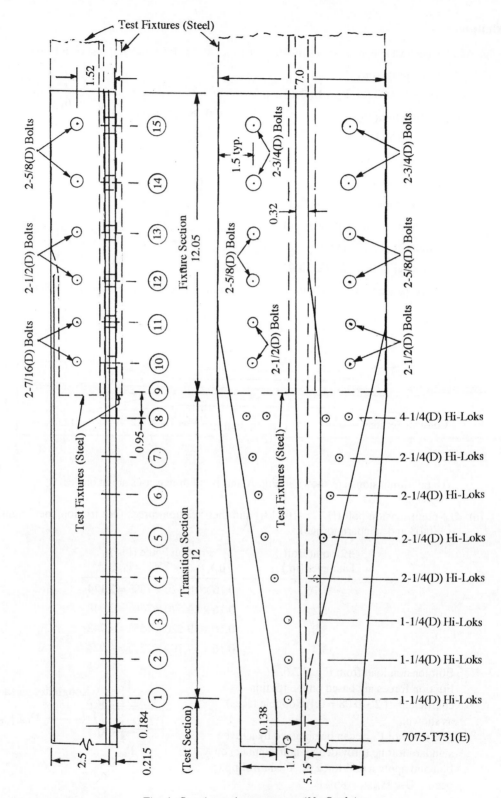

Fig. A Specimen Arrangement (No Scale)

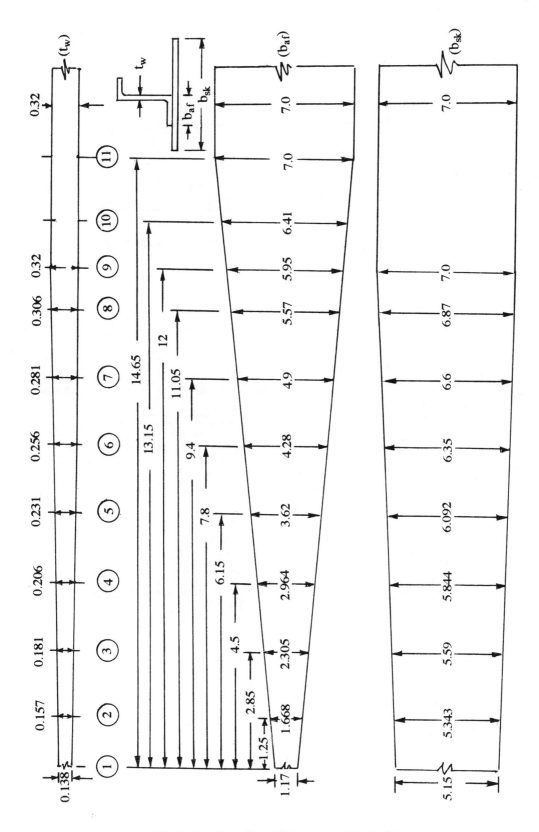

Fig. B Specimen Detail Dimensions (No Scale)

(b) Section properties (see Fig. A and B for detail dimensions):
(b – segment width; h – segment height; y – c.g. of the segment along y-axis)

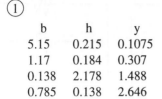

b	h	y
5.15	0.215	0.1075
1.17	0.184	0.307
0.138	2.178	1.488
0.785	0.138	2.646

Moment of inertia: $I_1 = 1.093$ in.^{4}

N.A. (y-axis): $Y_1 = 0.531$ in.

Area: $A_1 = 1.73$ in.^{2}

$k_1 = \dfrac{\text{skin area}}{\text{stringer area}} = \dfrac{1.105}{0.626} = 1.765$

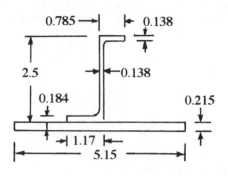

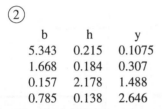

b	h	y
5.343	0.215	0.1075
1.668	0.184	0.307
0.157	2.178	1.488
0.785	0.138	2.646

Moment of inertia: $I_2 = 1.16$ in.^{4}

N.A. (y-axis): $Y_2 = 0.532$ in.

Area : $A_2 = 1.906$ in.^{2}

$k_2 = \dfrac{\text{skin area}}{\text{stringer area}} = \dfrac{1.148}{0.758} = 1.515$

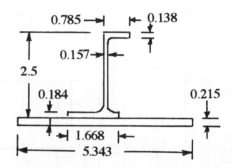

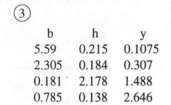

b	h	y
5.59	0.215	0.1075
2.305	0.184	0.307
0.181	2.178	1.488
0.785	0.138	2.646

Moment of inertia: $I_3 = 1.244$ in.^{4}

N.A. (y-axis): $Y_3 = 0.532$ in.

Area: $A_3 = 2.129$ in.^{2}

$k_3 = \dfrac{\text{skin area}}{\text{stringer area}} = \dfrac{1.201}{0.928} = 1.295$

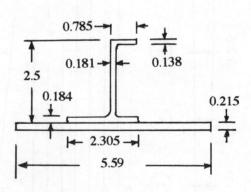

④

b	h	y
5.844	0.215	0.1075
2.964	0.184	0.307
0.206	2.178	1.488
0.785	0.138	2.646

Moment of inertia: $I_4 = 1.322$ in.^{4}

N.A. (y-axis): $Y_4 = 0.533$ in.

Area: $A_4 = 2.359$ in.^{2}

$k_4 = \dfrac{\text{skin area}}{\text{stringer area}} = \dfrac{1.255}{1.104} = 1.136$

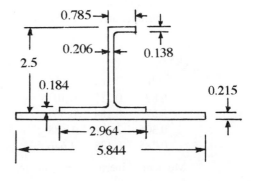

⑤

b	h	y
6.089	0.215	0.1075
3.62	0.184	0.307
0.231	2.178	1.488
0.785	0.138	2.646

Moment of inertia: $I_5 = 1.42$ in.^{4}

N.A. (y-axis): $Y_5 = 0.533$ in.

Area: $A_5 = 2.589$ in.^{2}

$k_5 = \dfrac{\text{skin area}}{\text{stringer area}} = \dfrac{1.31}{1.279} = 1.025$

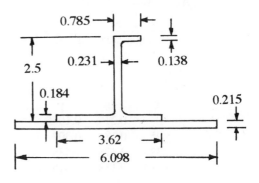

⑥

b	h	y
6.35	0.215	0.1075
4.28	0.184	0.307
0.256	2.178	1.488
0.785	0.138	2.646

Moment of inertia: $I_6 = 1.508$ in.^{4}

N.A. (y-axis): $Y_6 = 0.534$ in.

Area : $A_6 = 2.819$ in.^{2}

$k_6 = \dfrac{\text{skin area}}{\text{stringer area}} = \dfrac{1.365}{1.454} = 0.938$

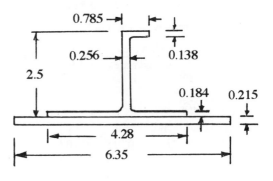

⑦

b	h	y
6.6	0.215	0.1075
4.9	0.184	0.307
0.281	2.178	1.488
0.785	0.138	2.646

Moment of inertia: $I_7 = 1.595$ in.^{4}

N.A. (y-axis): $Y_7 = 0.535$ in.

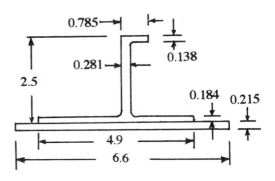

Area: $A_7 = 3.041$ in.^{2}

$k_7 = \dfrac{\text{skin area}}{\text{stringer area}} = \dfrac{1.42}{1.621} = 0..875$

⑧

b	h	y
6.87	0.215	0.1075
5.57	0.184	0.307
0.306	2.178	1.488
0.785	0.138	2.646

Moment of inertia: $I_8 = 1.683$ in.^{4}

N.A. (y-axis): $Y_8 = 0.535$ in.

Area: $A_8 = 3.277$ in.^{2}

$k_8 = \dfrac{\text{skin area}}{\text{stringer area}} = \dfrac{1.475}{1.802} = 0.818$

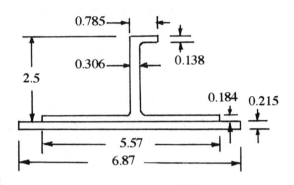

⑨

b	h	y
7.0	0.215	0.1075
5.95	0.184	0.307
0.32	2.178	1.488
0.785	0.138	2.646

Moment of inertia: $I_9 = 1.732$ in.^{4}

N.A. (y-axis): $Y_9 = 0.535$ in.

Area: $A_9 = 3.405$ in.^{2}

$k_9 = \dfrac{\text{skin area}}{\text{stringer area}} = \dfrac{1.505}{1.9} = 0.793$

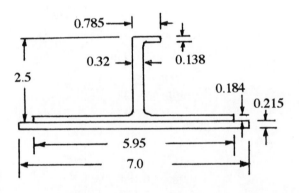

⑩

b	h	y
7.0	0.215	0.1075
6.41	0.184	0.307
0.32	2.178	1.488
0.785	0.138	2.646

Moment of initial: $I_{10} = 1.737$ in.^{4}

N.A. (y-axis): $Y_{10} = 0.529$ in.

Area: $A_{10} = 3.49$ in.^{2}

$k_{10} = \dfrac{\text{skin area}}{\text{stringer area}} = \dfrac{1.505}{1.985} = 0.758$

(See plot in Fig. C)

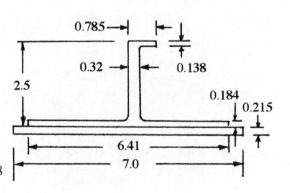

(c) N.A. (y-axis) of bolt friction forces for each location:
The following calculations are for checking the N.A. for each location of friction force which should be approximately in line with test specimen.

⑩ = ⑪

$$Y_{10} = \frac{\Sigma(P_{Friction}) \times (y)}{\Sigma(P_{Friction})}$$

$$= \frac{2{,}174 \times 1.52 + 2 \times 2{,}948 \times 0.2}{2{,}174 + 2 \times 2{,}948}$$

$= 0.556$ in.   Say O.K.

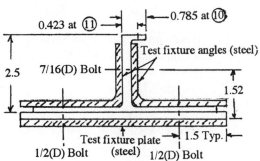

⑫ = ⑬

$$Y_{12} = \frac{\Sigma(P_{Friction}) \times (y)}{\Sigma(P_{Friction})}$$

$$= \frac{2{,}948 \times 1.52 + 2 \times 2{,}742 \times 0.2}{2{,}948 + 2 \times 4{,}742}$$

$= 0.513$ in.   O.K.

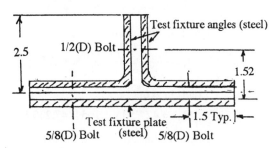

⑭ = ⑮

$$Y_{14} = \frac{\Sigma(P_{Friction}) \times (y)}{\Sigma(P_{Friction})}$$

$$= \frac{4{,}742 \times 1.52 + 2 \times 6{,}924 \times 0.2}{4{,}742 + 2 \times 6{,}924}$$

$= 0.537$ in.   O.K.

(See plot in Fig. C)

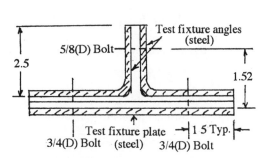

(d) The resultant N.A. (y-axis) of total bolt friction forces:

⑩ ⑪ ⑫ ⑬ ⑭ and ⑮

$$Y_{Bolts} = \frac{\Sigma(P_{Friction}) \times (Y)}{\Sigma(P_{Friction})}$$

$$= \frac{2[(2{,}174 + 2{,}948 + 4{,}742) \times 1.52 + 2(2{,}948 + 4{,}742 + 6{,}924) \times 0.2]}{2[(2{,}174 + 2{,}948 + 4{,}742) + 2(2{,}948 + 4{,}742 + 6{,}924)]} = 0.533 \text{ in.}$$

(See plot in Fig. C)

Chapter 17.0

(e) Total bolt friction forces:

$$\Sigma P_{Friction} = 2[(2,174 + 2,948 + 4,742) + 2(2,948 + 4,742 + 6,924)] = 78,184 \text{ lbs.}$$

$$MS_{Friction} = \frac{\Sigma P_{Friction}}{P_{Limit,T}} - 1 = \frac{78,184}{74,000} - 1 = \underline{0.06} \qquad \text{O.K.}$$

(f) Transition section (from sections ① to ⑨) – Check fastener shear loads (assume its bearing load is not ctritical):
[Fatigue analysis, which neglects in this example, should be conducted per method from Chapter 9.12(D)]

Maximum tension load: $P_{Limit,T} = 74,000$ lbs.

Between sections ① and ② (1 Hi-Loks fastener between these two sections):

$A_1 = 1.731$ in.^{2} $\qquad\qquad\qquad\qquad$ $A_2 = 1.906$ in.^{2}

Skin area: $A_{sk} = 1.105$ in.^{2} $\qquad\qquad$ Skin area: $A_{sk} = 1.148$ in.^{2}

Stringer area: $A_{st} = 0.626$ in.^{2} $\qquad$ Stringer area: $A_{st} = 0.758$ in.^{2}

Cross-sectional stresses:

$$f_1 = \frac{P_{Limit,T}}{A_1} = \frac{74,000}{1.731} = 42,750 \text{ psi} \qquad f_2 = \frac{P_{Limit,T}}{A_1} = \frac{74,000}{1,906} = 38,825 \text{ psi}$$

Loads in skin:

$P_{1,sk} = 42,750 \times 1.105 = 47,239$ lbs. $\qquad P_{2,sk} = 38,825 \times 1.148 = 44,571$ lbs.

Shear load at fastener (1 Hi-Lok fastener, $P_{all} = 4,660$ lbs., see Fig. 9.2.5):

$$P_{\Delta,sk} = P_{1,sk} - P_{2,sk} = 47,239 - 44,571 = 2,668 \text{ lbs.}$$

$$MS_{fastener} = \frac{P_{all, fastener}}{P_{\Delta,sk}} - 1 = \frac{4,660}{2,668} - 1 = \underline{0.75} \qquad \text{O.K.}$$

Between sections ② and ③ (1 Hi-Loks fastener between these two sections):

$A_2 = 1.906$ in.^{2} $\qquad\qquad\qquad\qquad$ $A_3 = 2.129$ in.^{2}

Skin area: $A_{sk} = 1.148$ in.^{2} $\qquad\qquad$ Skin area: $A_{sk} = 1.201$ in.^{2}

Stringer area: $A_{st} = 0.758$ in.^{2} $\qquad$ Stringer area: $A_{st} = 0.928$ in.^{2}

Cross-sectional stresses:

$$f_2 = \frac{P_{Limit,T}}{A_2} = \frac{74,000}{1.906} = 38,825 \text{ psi} \qquad f_3 = \frac{P_{Limit,T}}{A_3} = \frac{74,000}{2.129} = 34,758 \text{ psi}$$

Cross-sectional loads at skin:

$P_{2,sk} = 38,825 \times 1.148 = 44,571$ lbs. $\qquad P_{3,sk} = 34,758 \times 1.201 = 41,744$ lbs.

Shear load at fastener (1 Hi-Lok fastener, $P_{all} = 4,660$ lbs., see Fig. 9.2.5):

$$P_{\Delta,sk} = P_{2,sk} - P_{3,sk} = 44,571 - 41,744 = 2,827 \text{ lbs.}$$

$$MS_{fastener} = \frac{P_{all, fastener}}{P_{\Delta,sk}} - 1 = \frac{4,660}{2,827} - 1 = \underline{0.65} \qquad \text{O.K.}$$

(Sections from ③ to ⑨ are not critical)

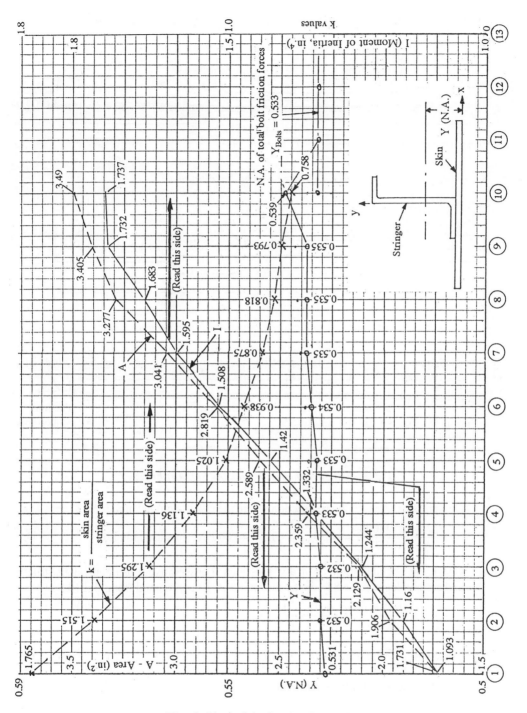

Fig. C Plots of the Section Properties

(h) Check over-all buckling under maximum compression load, $P_{Limit,C}$ = 30,000 lbs.:

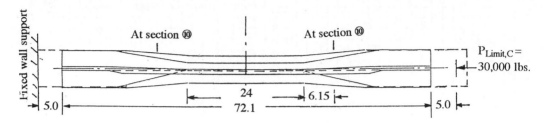

Fig. D Test Specimen Installation

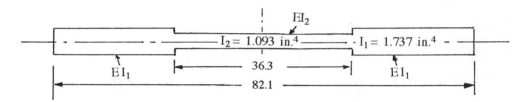

Fig. E Equivalent Structure for Column Buckling Strength Check

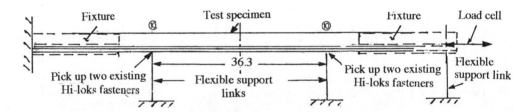

Fig. F Test Specimen with Flexible Support Links

Compression stress at the test section where the cross-sectional area is 1.732 in.²

$$\frac{P_{Limit,C}}{A_1} = \frac{30,000}{1.731} = 17,331 \text{ psi} < F_{cy} = 40,000 \text{ psi (in elastic range)}$$

Check the bucking strength of this specimen (conservatively use equivalent structure and column length, L = 82.1 in. as shown in Fig. E) per method from Fig. 10.3.4:

$$\frac{EI_2}{EI_1} = \frac{(10.7 \times 10^6)(1.093)}{(10.7 \times 10^6)(1.737)} = \frac{11.7}{18.59} = 0.63$$

Considering one end fixed and one pinned end to obtain $\frac{1}{\sqrt{c}} = 0.7$ (see Fig. 10.2.1):

$$L' = \frac{L}{\sqrt{c}} = 82.1 \times 0.7 = 57.47 \text{ in.}$$

$$\frac{a}{L'} = \frac{36.3}{57.47} = 0.63$$

From Fig. 10.3.4 obtain $\frac{P_{cr}}{P_E} = 0.65$:

where: $P_E = \frac{\pi^2 EI}{(L')^2} = \frac{\pi^2 \times 10.7 \times 10^6 \times 1.093}{57.47^2} = 34,948 \text{ lbs.}$

$\therefore P_{cr} = 0.65 \times 34,948 = 22,716 \text{ lbs.} < P_{Limit,C} = 30,000 \text{ lbs.}$

- Install a flexible support link (similar link shown in Fig. 17.6.2) at the load cell location
- Since the calculated critical load of $P_{cr} = 22,716$ lbs. which is less than $P_{Limit, c} = 30,000$ lbs., this specimen will buckle
- It requires the installation of two flexible support links as shown in Fig. F. to prevent lateral buckling
- These links are also used to support the structural weight of the specimen

(g) Test fixtures discussion:

- Test fixtures use low H. T. steel (i.e., 160 ksi or lower) for better fatigue characteristic as well as low cost
- Fatigue life should be longer than that of the basic test section
- Reasonable thickness to provide better contacted surfaces
- One end is fixed and other is horizontally free to move, by using flexible support link, for applying load to this specimen

17.5 COMPRESSION TEST PANEL SPECIMEN

Compared to the fatigue test, the compression test is a less expensive test, as shown in Fig. 17.5.1. This test is generally used to verify the method of analysis for the compression panel allowable.

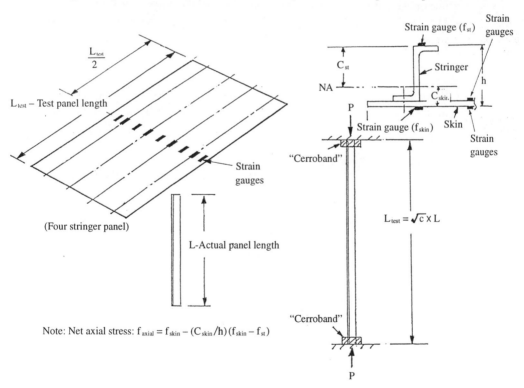

Fig. 17.5.1 Compression Test Panel Specimen

The following recommendations should be considered when designing a panel for use in compression testing:

(a) Generally use three to four stringers (or integrally-machined stiffeners) with the same cross-sections and a constant skin thickness throughout the panel.

(b) Use the end fixity of a hinge support (c = 1.0, see Fig. 10.2.1) for the actual panel (L).

(c) Use a reinforcement method, such as "Cerroband", at each end to avoid premature failure:

- This reinforcement will increase the panel column end-fixity, therefore, the reinforcement details must be specified to determine the end-fixity value by testing. For example, this value is approximately, c = 3.75 (Refer to NACA report TN3064)
- The test panel length:

$$L_{test} = \sqrt{c} \times L \qquad \text{Eq. 17.5.1}$$

where L is the actual panel length

For example, for a wing panel length (i.e., wing rib spacing distance) of 25 inches. The test panel length should be $L_{test} = \sqrt{3.75} \times 25 = 48.41$ inches

(d) The placement of strain gauges can be seen in Fig. 17.5.1:

- There are two axial gauges on each stringer at the center ($\frac{L_{test}}{2}$) of the tested panel; one on the outstanding flange of the stringer and the other on the skin
- The axial stress can be calculated from readings of these two gauges (strain or stress):

$$f_{axial} = f_{skin} - (\frac{C_{skin}}{h})(f_{skin} - f_{st}) \qquad \text{Eq. 17.5.2}$$

The stress of f_{skin} and f_{st} should be equal when no bending occurs; this is an important means of calibrating the test set-up to insure the correct alignment of the test jig.

- There are two axial gauges on the skin between stringers at the center ($\frac{L_{test}}{2}$) of the tested panel to check skin buckling
- The strain gauges are also used to check the alignment of the test panel at the beginning of the test procedure

17.6 SHEAR TEST PANEL SPECIMEN

This shear test is an in-plane static shear test of the panel skin or beam web assembly, as shown in Fig. 17.6.1, which is loaded in tension only. Based on the strength of material theory, the shear load can be divided into two diagonal components, one in tension and other in compression; the compression load will cause the skin to buckle, as shown in Fig. 12.2.1 and 12.4.2. This is the simplest set-up used by airframe engineers to accomplish this type of test.

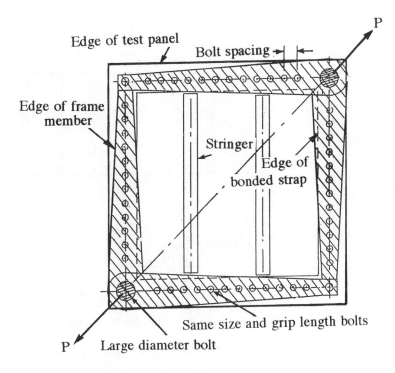

Fig. 17.6.1 Shear Panel Test Specimen

The shear test panel specimen should be constructed as follows (see Fig. 17.6.1):

- Shear test is accomplished by applying the load in tension in one diagonal direction
- Frame members of the jigs are made of mild steel
- The skin thickness of the panel edges which are mounted to the frame members should be beefed up by the bonding of additional straps to avoid skin net section shear failure
- Taper the width of the frame members rather than the thickness to avoid the use of various bolt grip lengths
- Taper the frame members to obtain uniform shear distribution along the edge of the test panel

The wing spar web test specimen should be constructed as follows (see Fig. 17.6.2):

- This is a complicated and costly test
- Panel consists of a built-up beam with beam caps and web stiffeners (fastened or integrally-machined shear panel)
- Use of a full scale section (generally used) for the test section is recommended, but a portion of full scale could be used
- End support and transition section must be strong enough to avoid premature failure during the fatigue (or static) test operation
- Use sufficient lateral support links to the keep test specimen in alignment

Chapter 17.0

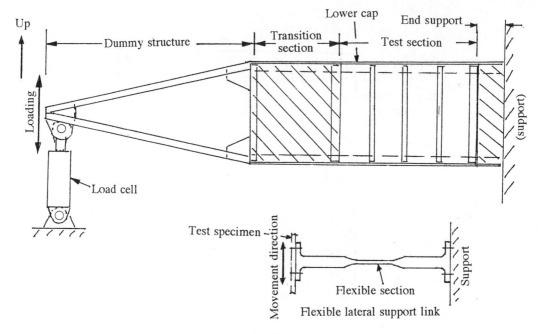

(Installed upside down for loading convenience purposes)

Fig. 17.6.2 A Test Set-Up of a Beam Specimen Subjected to a Fatigue Shear Test

17.7 TEST SPECIMENS FOR PRIMARY JOINTS

Typically, testing of wing or horizontal tail root joints is accomplished using a three-dimensional component test specimen which is subjected to either fatigue or static loading. This is not a simple matter, as shown in Fig. 17.7.1 (a). It is recommended to simplify the actual joint configuration by use of a test configuration [see Fig. 17.7.1(b)] which avoids the wing dihedral angle and swept angle. This simplified configuration still can yield reasonable results for preliminary design selection. Structural engineers can use these test results and extrapolate to get results which are very close to those of the more complicated tests. Obviously, the best results would always be obtained from a three-dimensional full-scale wing box rather than a single component test specemen as shown in Fig. 17.7.1(a).

The reasons of using simplified test configuration are as follows:

- The use of the vertical and horizontal support members which are used on the 3-D test specimen to react the component loads due to the effect of dihedral and swept angle should be avoided since they are difficult to control
- The deflection of these support members under the applied loads on the component test specimen will cause the redistribution or shift of stresses in the test panel when the applied loads vary during the test operations (**wasted effort!**)
- The conclusion is that it is impossible to obtain reasonable test results from a three-dimensional component test specimen

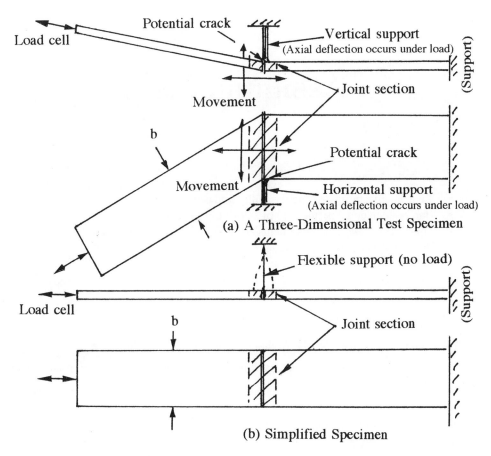

Fig. 17.7.1 Test Specimens for a Primary Wing Root Joint

References

17.1 Hetenyi, "Handbook of Experimental Stress Analysis", John Wiley & Sons, Inc., New York, NY, (1950).
17.2 "Performance Tests of Wire Strain Gages" refer to NACA reports TN's 954, 978, 997, 1042, 1318, and 1456.
17.3 Anon., "Notes on Strain Gages and Rosettes", Structures Bulletin (V.2, N.4), Lockheed Corp., 1956.
17.4 Vann, F. W., "A300B Static and Fatigue Tests", Aircraft Engineering, (Dec., 1973).
17.5 Stone, M., "Airworthiness Philosophy Developed From Full-scale Testing", Douglas Paper No. 6099. Biannual Meeting of The International Committee on Aeronautical Fatigue, London, England, (July, 1973).
17.6 Stone, M., "DC-10 Full-scale Fatigue Test Program", Douglas Aircraft Company, McDonnell-Douglas Corp.
17.7 Niu, C. Y., "AIRFRAME STRUCTURAL DESIGN", Hong Kong Conmilit Press Ltd., P.O. Box 23250, Wanchai Post Office, Hong Kong, (1988).

Assignments

Notes:
- Take manufacturing ($) and durability into consideration
- Hand calculation with the help of a desk-top calculator only
- Input from engineering judgments and assumptions may be required
- Provide sketches (if it is necessary)
- Correct answers may vary (there may be more than one)
- Asterisks (*) indicate essential assignments

ASSIGNMENT 1.0 (CHAPTER 1.0 GENERAL OVERVIEW)

1.1 Comment on "A message from the Author" on the front page of this book.

1.2 Why are engineering judgments necessary in airframe stress analysis or preliminary sizing?

1.3* State your positive and negative thoughts about computer usage in airframe stress analysis, e.g., finite element analysis, CAD/CAM, black-box analysis, etc.

1.4 How can aircraft manufacturing costs be reduced? Give at least three design cases (simple sketches are encouraged):

1.5 What are the most important design considerations in practical airframe structural design?

1.6 Do you always have a restrictive space problem to size a structure? Give examples and explain why.

ASSIGNMENT 2.0 (CHAPTER 2.0 SIZING PROCEDURES)

2.1 What is the difference between preliminary sizing and formal stress analysis?

2.2 Why are formal stress reports needed? Does this report cover structural fatigue and damage tolerance analysis?

2.3* Set up a temporary table of contents for a wing rib assembly (see Assignment 8.3) for a formal stress report which will be used in Assignments 3.3, 8.3, 9.3, 10.2, 12.2, 13.1, and 14.2.

ASSIGNMENT 3.0 (CHAPTER 3.0 EXTERNAL LOADS)

3.1 Why does a stress engineer need basic knowledge about aircraft loads?

3.2 Briefly explain the following aircraft loads:

External load –	Internal load –	Limit load –	Ultimate load –
Static load –	Dynamic load –	Fatigue load –	Fail-safe load –

3.3* Given a commercial transport wing (cantilever beam with two engines mounted on the wing as shown below) with the maximum take-off gross weight, W_G = 160,000 lbs. and a wing span of 1,350 inches (wing taper ratio, λ = 0.3). Assume
- Total fuel weight $W_F = 0.2 W_G$, (assume fuel tank is between root and tip)
- Total wing (including L. E and T. E.) structural weight $W_w = 0.14 W_G$
- Total engine (including pylons) weight W_E (2 engines) = $0.06 W_G$
- W_F and W_w weight distribution between wing root and tip using wing taper ratio (λ)
- Use uniform air load distribution (lbs./in.) across the entire wing span (spanwise)

Assume a single-cell wing box with the "flexural center" line at the center of the cross section of the wing box. Generate limit or ultimate load diagrams along the half wing span (between wing root and tip, see similar curves shown in Fig. 3.2.1) under a vertical gust limit load factor n = 3.8 g and W_F = 0 condition (critical for wing up-bending of wing inboard portion; the inertia force of engines, wing structures are downward which relieve wing up-bending):

(a) Bending moment (M_x)
(b) Shear load (S_z)
(c) Torsion load (M_y) = 0

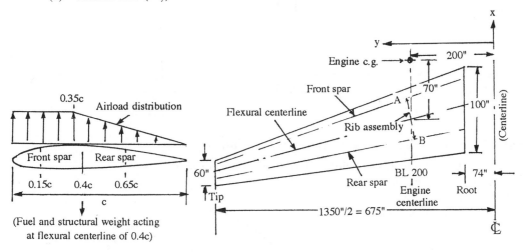

ASSIGNMENT 4.0 (CHAPTER 4.0 MATERIAL PROPERTIES)

4.1 What are the most important considerations when selecting airframe materials?

4.2 Have you considered a material stress-strain curve in your stress analysis? If the answer is yes or no, explain.

4.3* Explain the choice of 2000 or 7000 series aluminum material for use in airframe design. Give examples.

4.4* Why would you choose the following major material forms for use in airframe design?

(a) Sheet stock –   (b) Extrusion –   (c) Forging –

4.5 List materials for which the yield strength is less than $1.5 F_{tu}$.

4.6 Give examples of material grain directions and allowables of "A", "B", and "S" values.

ASSIGNMENT 5.0 (CHAPTER 5.0 STRUCTURAL ANALYSIS)

5.1 Explain the selection of statically determinate or indeterminate structures. Use the following examples to explain advantages, disadvantages, and design considerations:

Airframe Stress Analysis and Sizing

(a) Determinate structure on landing gear supports.
(b) Indeterminate structure on engine pylon supports.
(c) Indeterminate structure on skin-stringer panels of wings and fuselages.

5.2 Find the reaction loads of the following statically indeterminate beam.

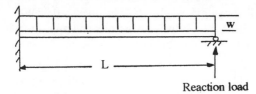

Reaction load

5.3 How many degrees of redundancy are in the following frame structure? Find M_A and M_D, when

(a) $E_2I_2 = 0$
(b) $E_2I_2 = E_1I_1$
(c) $E_2I_2 = \infty$

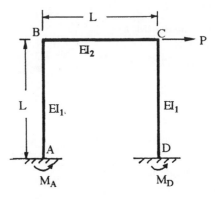

5.4* Given the fuselage cross section and data shown below:

(a) Double lobe cross section (radius: $r_{CU} = 74"$ and $r_{CL} = 70.75"$)
(b) Frame spacing = 20"
(c) Cabin pressure = 9 psi
(d) Floor load = 45 lbs./ft.^{2}

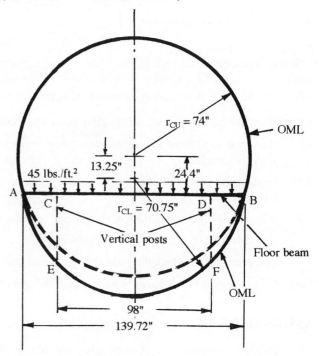

744

Determine the bending moments of the floor beam for the following load conditions (use of reasonable assumptions is necessary):

(a) Without vertical posts
(b) With vertical posts

[**Note:** For simplicity, use outer mold line (OML) as the structural centerline]

ASSIGNMENT 6.0 (CHAPTER 6.0 BEAM STRESS)

6.1* In beam theory:

Bending stress: $f_b = \dfrac{My}{I}$

Shear stress: $f_s = \dfrac{VQ}{It}$

Torsional stress: $f_{ts} = \dfrac{nT}{J}$

Find the max. bending stress, shear stress, and torsional stress of the following three cross sections:

(a) Rectangular section (b) I – section (c) Cylindrical section

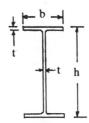
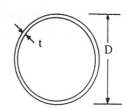

6.2 Use the Mohr's Circle graphic method to determine the maximum ($f_{x,\,max}$), minimum ($f_{y,\,max}$) and maximum shear ($f_{xy,\,max}$) stresses from the following given stresses:

(a) $f_x = 50$ ksi
$f_y = -30$ ksi
$f_{xy} = 30$ ksi

(b) $f_x = 50$ ksi
$f_y = -30$ ksi
$f_{xy} = 0$ ksi

6.3 Determine the shear flow in each panel shown in the sketch below:

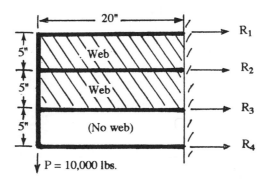

6.4 Determine which shape is stronger in torsion, if both have the same cross sectional thickness and area. Assume both are "thin sections" (thickness).

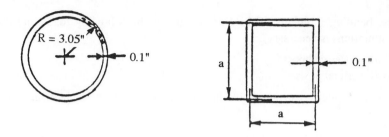

ASSIGNMENT 7.0 (CHAPTER 7.0 PLATE AND SHELL STRESS)

7.1 Compare the use of stiffened panels vs. honeycomb panels for a flat bulkhead under fuselage cabin pressure load.

7.2 Should stiffeners be placed on the pressure side or the other side of a flat bulkhead?

7.3 Why use a hoop-tension (vs. hoop-compression) design for a hemispherical bulkhead?

7.4* Resize the Example calculation of a honeycomb panel in Chapter 7.2 using the following parameters:

(a) Maximum deflection = 0.25"
(b) Use quarter-turn fasteners with a spacing of 3" (assume no pry-load on the fasteners)

ASSIGNMENT 8.0 (CHAPTER 8.0 BOX BEAMS)

8.1 For the conditions shown below, determine the shear flow of the web and cap axial loads at Section A-A.

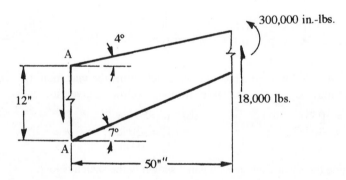

8.2 For the section given below, loaded as shown, determine the shear flow in the webs.

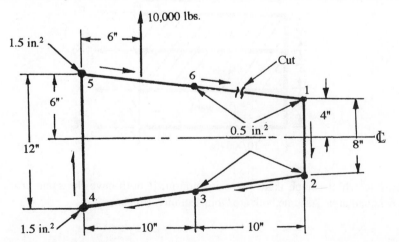

8.3* For the wing section given below use the loads (M_x, M_y, and P_z) as determinated in the BL 200 calculation in Assignment 3.3. Determine the shear flow in the skins and webs. Use a minimum of 6 lumped stringers for this assignment (8 is preferable). See Chapter 5.6 for lumping stringers.

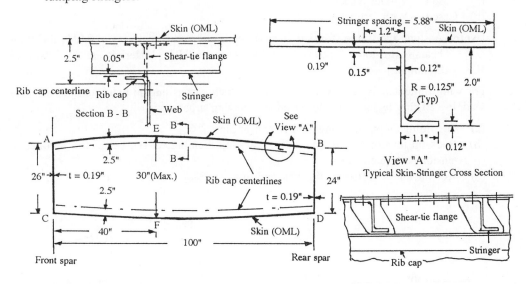

ASSIGNMENT 9.0 (CHAPTER 9.0 JOINTS AND FITTINGS)

9.1 Design a rivet joint to use for the installation of a floor beam to the fuselage frame.

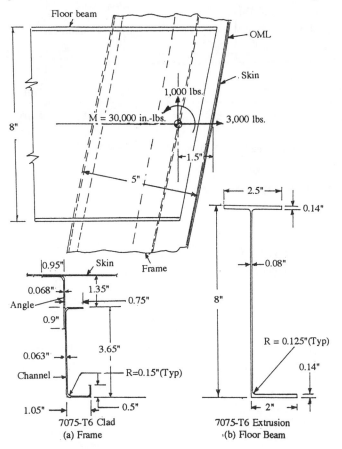

9.2* Use the loads obtained from Assignment 5.4 to size a rivet joint used for the installation of a floor beam to the fuselage frame (at points A and B, as was done in Assignment 9.1) for each of the following conditions:

(a) Case I – Without vertical posts.
(b) Case II – With vertical posts.

[**Note:** Assume the frame (formed sheets) depth is 3" and thickness is 0.05"; the floor beam (extrusion) depth is 5" and thickness is 0.05". Materials are 7075-T6 bare aluminum]

9.3* For wing rib assembly (see Assignment 8.3) which carries engine pylon tension ultimate load, P = 20 kips, by tension bolts, size the following structures:

(a) Lug fitting (at pt. H), using a tubular shear pin.
(b) Back-up tension fitting (including back-up member, G-H), select the number of tension bolts, e.g., 2 or 4.
(c) Web thickness, stiffeners, rib caps, and shear-tie flanges in the area of G-B-D-H

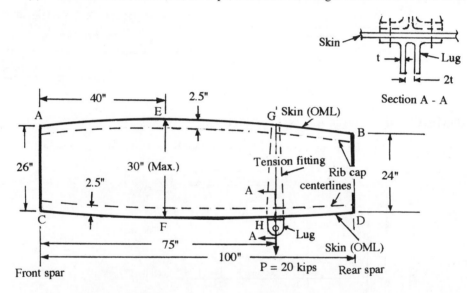

(Refer to sketch for Assignment 8.3 for additional data)

ASSIGNMENT 10.0 (CHAPTER 10.0 COLUMN BUCKLING)

10.1 Size the following members of the beam shown below.

(a) Upper cap
(b) Vertical stiffener A-D
(c) Vertical stiffener B-E

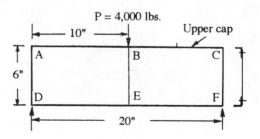

10.2* Using rib assembly shown in Assignment 8.3, size the connection between stringer and rib cap and web vertical stiffeners (beam-column action) under a crushing load condition (ignore the wing surface air and fuel pressure loading). See Chapter 6.9 for method of calculating crushing load.

- Assume rib spacing = 25"
- Use the M_x from Assignment 3.3 at BL 200
- Assume the cross-sectional geometry (consist of skin and stringer areas), shown in View "A" of Assignment 8.3, is identical for both upper and lower surfaces
- Assume spar cap area is equal to stringer area

ASSIGNMENT 11.0 (CHAPTER 11.0 BUCKLING OF THIN SHEETS)

11.1 Find the buckling MS for the panel shown below. Panel is flat 2024-T3 clad with a 0.081" thickness.

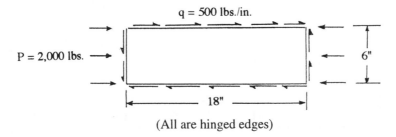

(All are hinged edges)

11.2* Fuselage is made up of thin skin panels fastened by longitudinal stringers and transverse frames. Frame spacing is 20", stringer spacing is 7", skin thickness is 0.051", and fuselage radius is 74". All materials are 2024-T4. Assuming pinned edges, determine the stress at which the panel will buckle in (a) shear and (b) compression.

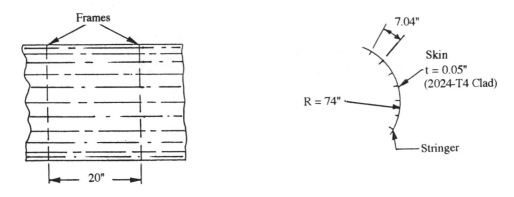

ASSIGNMENT 12.0 (CHAPTER 12.0 SHEAR PANELS)

12.1 Determine the web thickness and beam cap and single stiffener sizes required for the landing gear beam shown below. The stiffeners and beam caps are 7075-T6 extrusions and the web is 7075-T6 aluminum sheet.

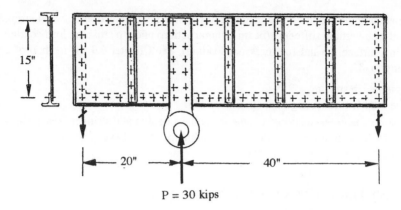

12.2* Determine the web thickness and beam cap and single stiffener sizes (see Assignment 8.3) required for a rib assembly with the air loading shown below. The sizing from Assignment 8.3 will remained unchanged except if they are critical under the following air loading. The stiffeners and rib caps are 7075-T6 extrusions and the web is 7075-T6 aluminum sheet.

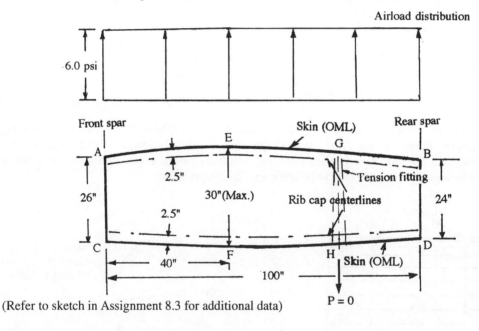

(Refer to sketch in Assignment 8.3 for additional data)

ASSIGNMENT 13.0 (CHAPTER 13.0 CUTOUTS)

13.1* Size a 12" × 18" rectangular cutout at the center of the rib web under the following loading conditions. The sizes from Assignment 12.2 will remain unchanged except if they are critical under the following combined loadings

(a) Crushing load (see Assignment 10.2).
(b) Air pressure loads (see Assignment 12.2).

Assignments

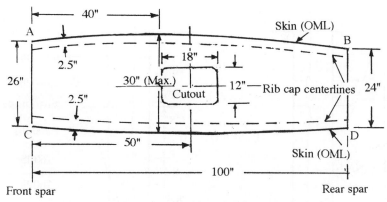

(Refer to sketch for Assignment 8.3 for additional data)

13.2 Determine the rectangular cutout load distribution as given in the Example calculation in Chapter 13.7 except use three cut-stringers (Str. No. 3) instead of two.

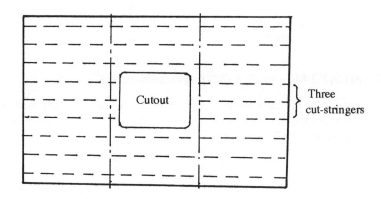

13.3 Determine the internal load distribution around the fuselage cutout structures as given in the Example calculation in Chapter 13.8 except use radius r = 122" (is) instead of 74" (was).

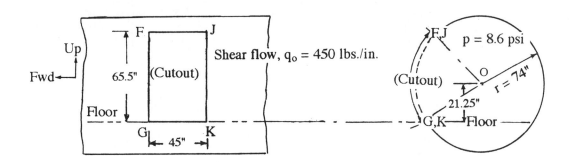

751

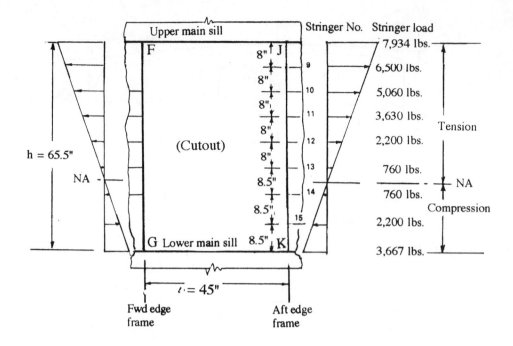

ASSIGNMENT 14.0 (CHAPTER 14.0 COMPRESSION PANELS)

14.1 Determine the effective width of skin (between A and B) of a skin-stringer panel as shown below when the panel is subjected to compression stresses (skin is 7075-T6 sheet and stringers are 7075-T6 extrusions).

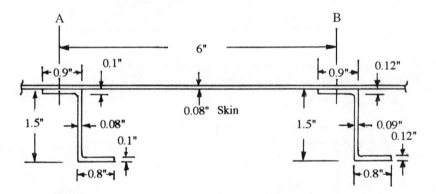

14.2* Determine the MS of the wing skin-stringer panel shown below (see View "A" of Assignment 8.3), under axial load (derived from M_x of Assignment 3.3). If the MS is negative, size an appropriate cross section. The skin is 7075-T6 plate and the stringer is a 7075-T6 extrusion.

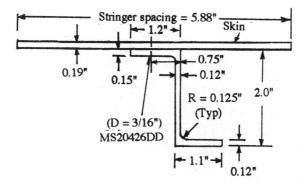

14.3* Generate similar curves as shown in Fig. 14.6.3 for 7075-T6 extrusion.

ASSIGNMENT 15.0 [CHAPTER 15.0 DAMAGE TOLERANT PANELS (IN TENSION)]

15.1* Determine the number of flights for the panel and the stress spectra as given in the Example calculation in Chapter 15.4 for crack growth length from 1.0" to 10" (assume the increment crack length is 0.5").

15.2* Use the same Example calculation from Chapter 15.5 except use a skin material of 2024-T3 and the two stringer configurations shown below:

  (a) Thickness of the outstanding flange is 0.4" (see Configuration A)
  (b) Thickness of the attaching flange is 0.4" (see Configuration B)

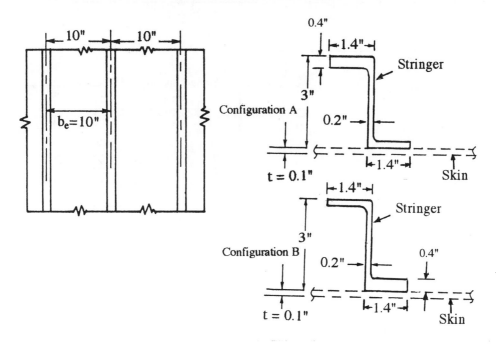

ASSIGNMENT 16.0 (CHAPTER 16.0 STRUCTURAL SALVAGE AND REPAIRS)

16.1 Determine the minimum fastener row spacing (S_{sta}) for a staggered-row splice with $\frac{3}{16}$" dia. countersunk shear head titanium Hi-Loks for the repair of the one fastener misdrilled as shown. Plates are 2024-T3 aluminum.

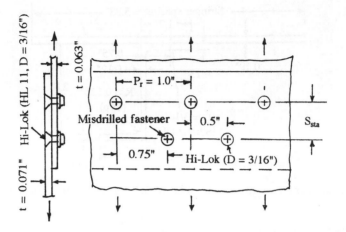

16.2* The wing lower surface has a 3-inch-diameter damaged area as shown below; repair (refer to Fig. 16.7.3) to meet fatigue requirements. Use a first doubler (0.032" fatigue doubler) to splice the skin and stringer for this repair.

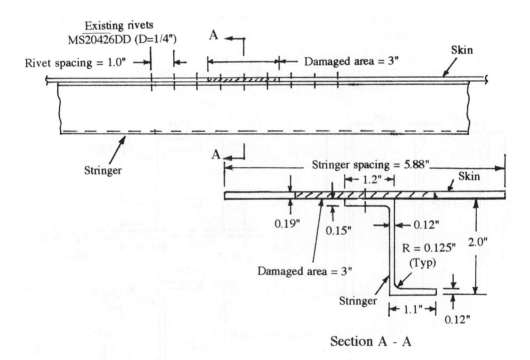

Section A - A

ASSIGNMENT 17.0 (CHAPTER 17.0 STRUCTURAL TEST SETUP)

17.1* Use the Mohr's Circle graphic method to determine the maximum, minimum and shear stresses from the strain gauge readings given in the Example calculation in Chapter 17.3.

17.2* Why is a dummy structure used in the specimen set-up (see example shown in Fig. 17.6.2). Give one example.

APPENDIX A

Conversion Factors

LENGTH :
    mil (0.001 inch) = 25.4 micron
    in. = 2.54 cm (centimeter)
    ft. = 0.3048 m (meter)
    ft. = 0.333 yd (yard)

AREA :
    in.^{2} = 6.452 cm^2
    ft.^{2} = 0.0929 m^2

VOLUME :
    in.^{3} = 16.387 cm^3 U.S. gallon = 3785 cm^3 = 3.785 liters
    ft.^{3} = 0.02832 m^3 = 231 in.^{3}

FORCE AND PRESSURE :
    psi = 6.8948 kpa
    ksi = 6.8948 Mpa
    Msi = 6.8948 Gpa
    Hg (32°F) = 3.3864 kpa

MASS :
    lb. = 0.4536 kg

DENSITY :
    lb/in.^{3} = 27.68 g/cc
    lb/ft.^{3} = 0.6243 kg/m^3

THERMAL EXPANSION :
    in./in./°F x 10^{-6} = 1.8 K x 10^{-6}

IMPACT ENERGY :
    ft.-lb. = 1.3558 J

COMMONLY USED SI PREFIXES :
    G – 10^9 micron – 10^{-6}
    M – 10^6
    k – 10^3

APPENDIX B

Fastener Data

B.1 FLUSH RIVET HEAD HEIGHT

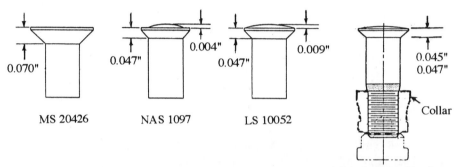

Comparison of D = 3/16" Shear Head Rivets   Hi-Lok (Shear Head)

Type of Rivet	Diameter (in.)						
	$\frac{4}{32}$	$\frac{5}{32}$	$\frac{3}{16}$	$\frac{1}{4}$	$\frac{5}{16}$	$\frac{3}{8}$	$\frac{7}{16}$
MS 20426	0.042	0.055	0.070	0.095	–	–	–
NAS 1097 & LS 10052	0.028	0.036	0.047	0.063	–	–	–
Hi-Lok (Tension Heads)	–	0.0685	0.0785	0.1060	0.1330	0.1600	0.1865
	–	0.0700	0.0805	0.1080	0.1350	0.1620	0.1895
Hi-Lok (Shear Heads)	–	0.039	0.045	0.059	0.066	0.076	0.094
	–	0.041	0.047	0.061	0.068	0.078	0.097

B.2 HI-LOK TENSION ALLOWABLES

Type	Diameter (in.)					
	$\frac{5}{32}$	$\frac{3}{16}$	$\frac{1}{4}$	$\frac{5}{16}$	$\frac{3}{8}$	$\frac{7}{16}$
(Hi-Lok Protruding Heads)						
Material: Ti-6Al-4V (F_{su} = 95 ksi)						
Tension Heads	2,180	3,180	5,800	9,200	14,000	18,900
Shear Heads	1,940	2,500	4,300	6,300	8,700	12,100
(Hi-Lok Flush Heads)						
Material: Ti-6Al-4V (F_{su} = 95 ksi)						
Tension Heads	2,180	3,180	5,800	9,200	14,000	18,900
Shear Heads	1,290	2,000	3,700	5,000	7,200	10,000

Type of Collar	Diameter (in.)					
	$\frac{5}{32}$	$\frac{3}{16}$	$\frac{1}{4}$	$\frac{5}{16}$	$\frac{3}{8}$	$\frac{7}{16}$
(Aluminum: 2024-T6)	1,400	1,600	3,000	5,000	7,000	9,500
(Steel: 303 Se)	2,300	2,750	5,000	8,300	12,700	19,000
(Steel: 17-4PH)	–	4,000	7,500	11,750	18,000	23,000

(**Note:** Check the current data and information from Hi-Shear Corp.)

B.3 EQUATION FOR FASTENER SPRING CONSTANTS

(Reference source: Swift, T., "Repairs to Damage Tolerant Aircraft", Presented to International Symposium on Structural Integrity of Aging Airplanes, Atlanta, GA. March 20-22, 1990)

$$k = \frac{ED}{[A + B(\frac{D}{t_d} + \frac{D}{t_s})]}$$ Eq. B.3.1

where: k – Fastener spring constant (lbs./in.)
E – Modulus of skin and doubler
D – Fastener diameter
t_s – Skin thickness
t_d – Doubler thickness
A – 5 for aluminum fasteners
  1.667 for steel fasteners
B – 0.8 for aluminum fasteners
  0.86 for steel fasteners

B.4 HYPOTHETICAL EQUATIONS FOR FASTENER SPRING CONSTANTS

(Reference source: "Preliminary Investigation of the Loads Carried by Individual Bolts in Bolted Joints", NACA TN 1051, 1946).

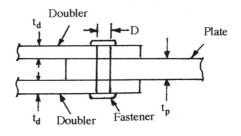

$$t_{av} = \frac{2t_d + t_p}{2}$$

C – Fastener constant (in./lbs.)
$k = \frac{1}{C}$ – Fastener spring constant (lbs./in.)
E – Modulus of elasticity of fastener

Appendix B

Case I – For steel doublers and plates with steel fasteners:

$$C = \frac{8}{t_{av} E} \{0.13 (\frac{t_{av}}{D})^2 [2.12 + (\frac{t_{av}}{D})^2] + 1.0\}$$ Eq. B.4.1

Case II – For aluminum doublers and plates with aluminum fasteners:
(Same as Case I)

Case III – For aluminum doublers and plates with steel fasteners:

$$C = \frac{8}{t_{av} E} \{0.13 (\frac{t_{av}}{D})^2 [2.12 + (\frac{t_{av}}{D})^2] + 1.87\}$$ Eq. B.4.2

Case IV – For aluminum plates and steel doublers with steel fasteners:

$$C = \frac{8}{t_{av} E} \{0.13 (\frac{t_{av}}{D})^2 [2.12 + (\frac{t_{av}}{D})^2] + 1.43\}$$ Eq. B.4.3

Case V – For aluminum plates and steel doublers with aluminum fasteners:

$$C = \frac{8}{t_{av} E} \{0.13 (\frac{t_{av}}{D})^2 [2.12 + (\frac{t_{av}}{D})^2] + 0.84\}$$ Eq. B.4.4

Case VI – For aluminum plates and doublers with titanium fasteners:

$$C = \frac{8}{t_{av} E} \{0.133 (\frac{t_{av}}{D})^2 [2.06 + (\frac{t_{av}}{D})^2] + 1.242\}$$ Eq. B.4.5

Case VII – For aluminum plates and titanium doublers with titanium fasteners:

$$C = \frac{8}{t_{av} E} \{0.1325 (\frac{t_{av}}{D})^2 [2.06 + (\frac{t_{av}}{D})^2] + 1.1125\}$$ Eq. B.4.6

B.5 FASTENER SPRING CONSTANTS – ALUMINUM RIVETS

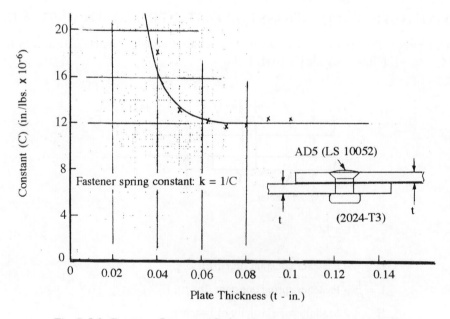

Fig. B.5.1 Fastener Constant (C) for Aluminum Plates and AD5 Rivet

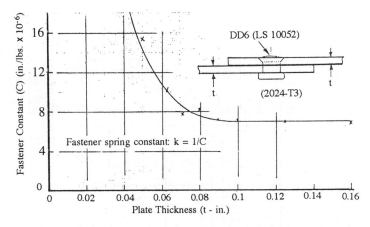

Fig. B.5.2 Fastener Constant (C) for Aluminum Plates and DD6 Rivet

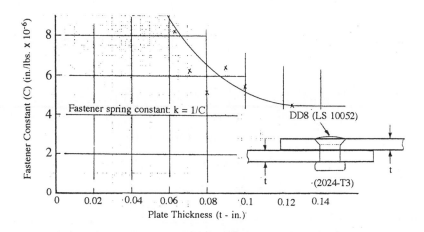

Fig. B.5.3 Fastener Constant (C) for Aluminum Plates and DD8 Rivet

B.6 FASTENER SPRING CONSTANTS – HI-LOKS

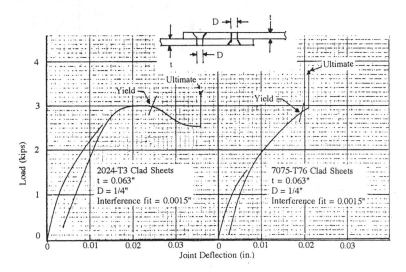

Fig. B.6.1 Load-Deflection Curve (Steel Hi-Lok HL319, $D = \frac{1}{4}"$; Aluminum, $t = 0.063"$)

Appendix B

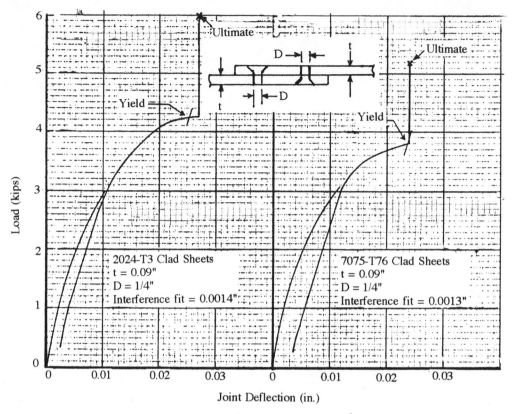

Fig. B.6.2 Load-Deflection Curve (Steel Hi-Lok HL319, $D = \frac{1}{4}"$; Aluminum, $t = 0.09"$)

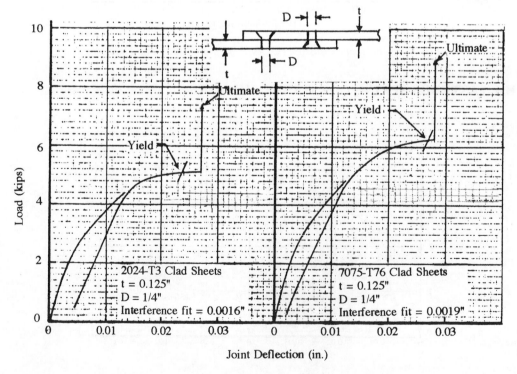

Fig. B.6.3 Load-Deflection Curve (Steel Hi-Lok HL319, $D = \frac{1}{4}"$; Aluminum, $t = 0.125"$)

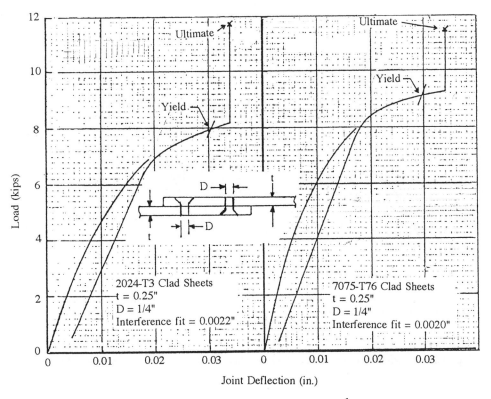

Fig. B.6.4 Load-Deflection Curve (Steel Hi-Lok HL319, $D = \frac{1}{4}''$; Aluminum, t = 0.25")

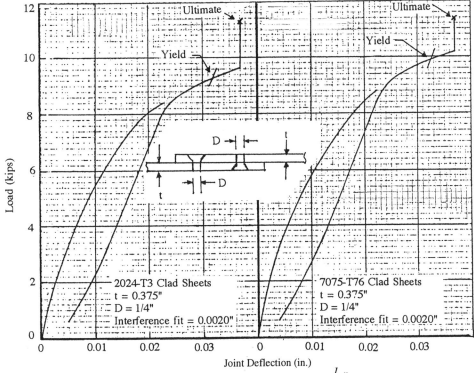

Fig. B.6.5 Load-Deflection Curve (Steel Hi-Lok HL319, $D = \frac{1}{4}''$; Aluminum, t = 0.375")

Appendix B

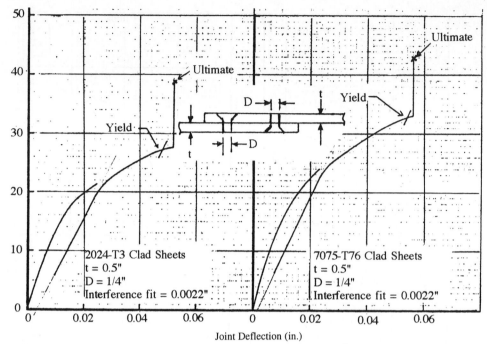

Fig. B.6.6 Load-Deflection Curve (Steel Hi-Lok HL319, $D = \frac{1}{4}"$; Aluminum, $t = 0.5"$)

APPENDIX C

Common Properties of Sections

C.1 PROPERTIES OF SECTIONS

Section Area: A (in.²); Neutral Axes: X and Y (in.); α in radians
Moments of inertia about Neutral Axis: I_x and I_y (in.⁴)

No.	Sections	A, X, Y	I_x, I_y
1	Rectangle	$A = bd$ $X = 0$ $Y = 0$	$I_x = \dfrac{bd^3}{12}$ $I_y = \dfrac{db^3}{12}$
2	Solid Circle	$A = \pi R^2$ $X = 0$ $Y = 0$	$I_x = \dfrac{\pi R^4}{4}$ $I_y = \dfrac{\pi R^4}{4}$
3	Solid Half-Circle	$A = \dfrac{\pi R^2}{2}$ $Y = 0.4244R$	$I_x = 0.1098 R^4$ $I_y = \dfrac{\pi R^4}{8}$
4	Circular Segment	$A = \dfrac{R^2(2\alpha - \sin 2\alpha)}{2}$ $X = 0$ $Y = R\left[\dfrac{4\sin^3 \alpha}{(6\alpha - 3\sin 2\alpha)} - \cos\alpha\right]$	$I_x = R^4\left\{\left[\dfrac{2\alpha - \sin 2\alpha}{8}\right]\left[1 + \dfrac{2\sin^3\alpha\cos\alpha}{\alpha - \sin\alpha\cos\alpha}\right] - \dfrac{8\sin^6\alpha}{9(2\alpha - \sin 2\alpha)}\right\}$ $I_y = R^4\left[\dfrac{(2\alpha - \sin 2\alpha)}{8} - \dfrac{(2\alpha - \sin 2\alpha)(\sin^3\alpha\cos\alpha)}{12(\alpha - \sin\alpha\cos\alpha)}\right]$

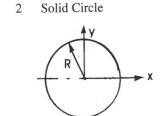

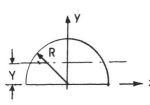

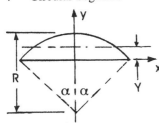

Appendix C

No.	Sections	A, X, Y	I_x, I_y
5	Circular Sector 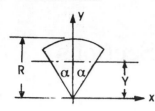	$A = \alpha R^2$ $X = 0$ $Y = \dfrac{2R \sin \alpha}{3}$	$I_x = \left[\dfrac{R^4}{4}\right]\left[\alpha + \sin \alpha \cos \alpha - \dfrac{16 \sin^2 \alpha}{9\alpha}\right]$ $I_y = \left[\dfrac{R^4}{4}\right](\alpha - \sin \alpha \cos \alpha)$
6	Hollow Square 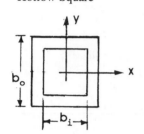	$A = b_0^2 - b_i^2$ $X = 0$ $Y = 0$	$I_x = I_y = \dfrac{(b_0^4 - b_i^4)}{12}$
7	Hollow Rectangle 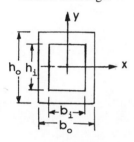	$A = b_0 h_0 - b_i h_i$ $X = 0$ $Y = 0$	$I_x = \dfrac{(b_0 h_0^3 - b_i h_i^3)}{12}$ $I_y = \dfrac{(b_0^3 h_0 - b_i^3 h_i)}{12}$
8	Hollow Square 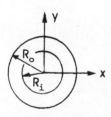	$A = \pi(R_0^2 - R_i^2)$ $X = 0$ $Y = 0$	$I_x = I_y = \dfrac{\pi(R_0^4 - R_i^4)}{4}$
9	Hollow Half-circle 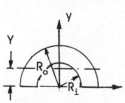	$A = \dfrac{\pi(R_0^2 - R_i^2)}{2}$ $X = 0$ $Y = \dfrac{4(R_0^3 - R_i^3)}{3\pi(R_0^2 - R_i^2)}$	$I_x = \dfrac{\pi(R_0^4 - R_i^4)}{8} - \dfrac{8(R_0^3 - R_i^3)^2}{9\pi(R_0^2 - R_i^2)}$ $I_y = \dfrac{\pi(R_0^4 - R_i^4)}{8}$

Common Properties of Sections

No.	Sections	A, X, Y	I_x, I_y
10	Hollow Ellipse 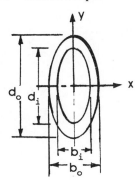	$A = \dfrac{\pi(b_0 d_0 - b_i d_i)}{4}$ $X = 0$ $Y = 0$	$I_x = \dfrac{\pi(b_0 d_0^3 - b_i d_i^3)}{64}$ $I_y = \dfrac{\pi(b_0^3 d_0 - b_i^3 d_i)}{64}$
11	Thin Annulus 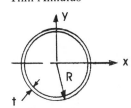	$A = 2\pi R t$ $X = 0$ $Y = 0$	$I_x = \pi R^3 t$ $I_y = \pi R^3 t$
12	Thin Half-Annulus 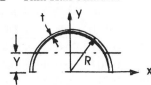	$A = \pi R t$ $X = 0$ $Y = \dfrac{2R}{\pi}$	$I_x = 0.2976 R^3 t$ $I_y = \dfrac{\pi R^3 t}{2}$
13	Thin Quarter-Annulus 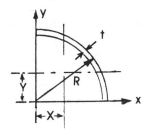	$A = \dfrac{\pi R t}{2}$ $X = Y = \dfrac{2R}{\pi}$	$I_x = 0.1488 R^3 t$ $I_x = 0.1488 R^3 t$
14	Sector of Thin Annulus 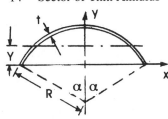	$A = 2\alpha R t$ $X = 0$ $Y = R\left[\dfrac{\sin\alpha}{\alpha} - \cos\alpha\right]$	$I_x = R^3 t \left[\alpha + \sin\alpha\cos\alpha - \dfrac{2\sin^2\alpha}{\alpha}\right]$ $I_y = R^3 t [\alpha - \sin\alpha\cos\alpha]$

Appendix C

No.	Sections	A, X, Y	I_x, I_y
15	Cruciform	$A = bH + h(B - b)$ $X = 0$ $Y = 0$	$I_x = \dfrac{bH^3 + h^3(B - b)}{12}$ $I_y = \dfrac{hB^3 + b^3(H - h)}{12}$

C.2 SECTION PROPERTIES OF ROUND TUBING

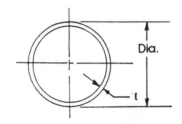

Dia. (in.)	t (in.)	Area (in.²)	I (in.⁴)	ρ (in.)	Dia. (in.)	t (in.)	Area (in.²)	I (in.⁴)	ρ (in.)
3/8	0.022	0.0244	0.000382	0.125	1-3/8	0.035	0.147	0.0331	0.474
	0.028	0.0305	0.000462	0.123		0.049	0.204	0.0449	0.469
	0.035	0.0374	0.000546	0.121		0.058	0.24	0.0521	0.466
	0.049	0.0502	0.000682	0.117		0.065	0.268	0.0575	0.464
	0.058	0.0578	0.00075	0.114		0.083	0.337	0.0706	0.458
1/2	0.028	0.0415	0.00116	0.167	1-1/2	0.035	0.161	0.0432	0.518
	0.035	0.0511	0.00139	0.165		0.049	0.223	0.0589	0.513
	0.049	0.0694	0.00179	0.16		0.058	0.263	0.0684	0.51
	0.058	0.0805	0.002	0.158		0.065	0.293	0.0756	0.508
	0.065	0.0888	0.00215	0.156		0.083	0.37	0.0931	0.502
						0.125	0.54	0.129	0.488
5/8	0.028	0.0525	0.00234	0.211		0.25	0.982	0.199	0.451
	0.035	0.0649	0.00283	0.209					
	0.049	0.0887	0.0037	0.204	1-5/8	0.035	0.175	0.0553	0.562
	0.058	0.103	0.0042	0.202		0.058	0.286	0.0878	0.554
	0.065	0.114	0.00454	0.199					
					1-3/4	0.035	0.189	0.0694	0.607
3/4	0.028	0.0635	0.00415	0.255		0.049	0.262	0.0948	0.602
	0.035	0.0786	0.00504	0.253		0.058	0.308	0.11	0.599
	0.049	0.108	0.00666	0.249		0.083	0.435	0.151	0.59
	0.058	0.126	0.0076	0.246		0.12	0.615	0.2052	0.578
	0.065	0.14	0.00828	0.243					
	0.083	0.174	0.00982	0.238					

Common Properties of Sections

Dia. (in.)	t (in.)	Area (in.²)	I (in.⁴)	ρ (in.)	Dia. (in.)	t (in.)	Area (in.²)	I (in.⁴)	ρ (in.)
7/8	0.028	0.0745	0.00669	0.3	2-0	0.035	0.216	0.104	0.695
	0.035	0.0924	0.00816	0.297		0.049	0.3	0.143	0.69
	0.049	0.127	0.0109	0.293		0.058	0.354	0.167	0.687
	0.058	0.149	0.0125	0.29		0.065	0.395	0.185	0.685
	0.065	0.165	0.0137	0.287		0.083	0.5	0.23	0.678
						0.125	0.736	0.325	0.664
1-0	0.035	0.106	0.0124	0.341		0.25	1.374	0.537	0.625
	0.049	0.146	0.0166	0.337					
	0.058	0.172	0.0191	0.334	2-1/4	0.049	0.339	0.205	0.778
	0.065	0.191	0.021	0.331		0.065	0.466	0.267	0.773
	0.083	0.239	0.0253	0.326		0.083	0.565	0.332	0.767
	0.095	0.27	0.028	0.322					
					2-1/2	0.035	0.271	0.206	0.872
1-1/8	0.035	0.12	0.0178	0.386		0.049	0.377	0.283	0.867
	0.049	0.166	0.024	0.381		0.065	0.497	0.369	0.861
	0.058	0.194	0.0278	0.378		0.083	0.63	0.461	0.855
	0.065	0.217	0.0305	0.376		0.12	0.897	0.6369	0.843
						0.125	0.933	0.659	0.841
1-1/4	0.035	0.134	0.0247	0.43		0.25	1.767	1.132	0.8
	0.049	0.185	0.0334	0.425					
	0.058	0.217	0.0387	0.422					
	0.065	0.242	0.0426	0.42					
	0.083	0.304	0.0521	0.414					

C.3 SECTION PROPERTIES OF EQUAL LEG ANGLES

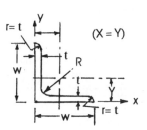

w (in.)	t (in.)	R (in.)	Area (in.²)	Y (in.)	I (in.⁴)	ρ (in.)	w (in.)	t (in.)	R (in.)	Area (in.²)	Y (in.)	I (in.⁴)	ρ (in.)
3/4	1/16	1/8	0.089	0.199	0.004	0.22	1-1/2	1/8	3/16	0.36	0.41	0.074	0.45
	3/32	1/8	0.132	0.214	0.006	0.219		3/16	3/16	0.53	0.44	0.107	0.45
	1/8	1/8	0.171	0.227	0.008	0.217		1/4	3/16	0.69	0.46	0.135	0.44
1-0	1/16	1/16	0.122	0.271	0.012	0.311	1-3/4	3/32	3/32	0.32	0.47	0.096	0.55
	3/32	1/8	0.178	0.276	0.016	0.301		1/8	3/16	0.42	0.47	0.121	0.53
	1/8	1/8	0.234	0.29	0.021	0.298		3/16	3/16	0.62	0.5	0.174	0.53
	3/16	1/8	0.339	0.314	0.029	0.293		1/4	3/16	0.81	0.52	0.223	0.52
1-1/4	3/32	3/32	0.23	0.34	0.033	0.38	2-0	1/8	1/4	0.49	0.53	0.18	0.61
	1/8	3/16	0.3	0.35	0.042	0.37		3/16	1/4	0.72	0.56	0.27	0.61
	3/16	3/16	0.43	0.37	0.059	0.37		1/4	1/4	0.94	0.58	0.34	0.6
	1/4	3/16	0.56	0.4	0.074	0.36		5/16	1/4	1.16	0.61	0.41	0.6

767

C.4 SECTION PROPERTIES OF UNEQUAL LEG ANGLES

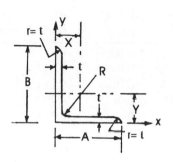

A (in.)	B (in.)	t (in.)	R (in.)	Area (in.^{2})	Y (in.)	X (in.)	I_x (in.^{4})	I_y (in.^{4})	ρ_x (in.)	ρ_y (in.)
0.75	0.5	0.063	0.063	0.0739	0.12	0.243	0.0014	0.0041	0.139	0.235
	0.625	0.063	0.063	0.0818	0.162	0.222	0.0027	0.0044	0.181	0.232
0.875	0.625	0.063	0.125	0.0922	0.15	0.269	0.0029	0.0069	0.177	0.274
	0.75	0.063	0.125	0.1	0.19	0.25	0.0049	0.0073	0.221	0.27
1.0	0.625	0.063	0.125	0.1	0.14	0.32	0.0029	0.01	0.17	0.316
	0.625	0.125	0.125	0.184	0.16	0.344	0.005	0.017	0.165	0.304
	0.75	0.063	0.125	0.108	0.178	0.299	0.0051	0.0106	0.217	0.313
	0.75	0.125	0.125	0.2	0.199	0.322	0.0086	0.0182	0.208	0.302
1.25	0.75	0.063	0.125	0.124	0.159	0.403	0.054	0.0197	0.209	0.4
	0.75	0.094	0.125	0.179	0.17	0.416	0.0074	0.0276	0.203	0.393
	0.75	0.125	0.125	0.231	0.181	0.427	0.0092	0.0346	0.2	0.387
	1.0	0.063	0.125	0.139	0.239	0.361	0.0124	0.0217	0.298	0.394
	1.0	0.094	0.125	0.202	0.25	0.373	0.0171	0.0306	0.291	0.389
1.5	0.75	0.094	0.156	0.204	0.156	0.522	0.0077	0.0463	0.194	0.476
	0.75	0.125	0.156	0.264	0.167	0.534	0.0096	0.0584	0.191	0.465
	1.0	0.094	0.156	0.228	0.228	0.472	0.0182	0.0512	0.282	0.474
	1.0	0.125	0.156	0.295	0.239	0.485	0.0226	0.0647	0.276	0.468
	1.0	0.156	0.156	0.361	0.249	0.495	0.0265	0.0772	0.271	0.463
	1.25	0.094	0.156	0.251	0.31	0.433	0.0347	0.055	0.372	0.468
1.75	1.0	0.125	0.156	0.327	0.222	0.591	0.0236	0.0998	0.269	0.553
	1.25	0.125	0.156	0.358	0.3	0.545	0.0465	0.1078	0.36	0.549
	1.5	0.125	0.156	0.389	0.383	0.506	0.0776	0.1144	0.447	0.542

C.5 SECTION PROPERTIES OF Z-SECTIONS

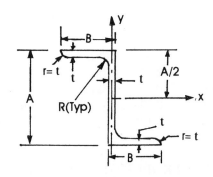

A (in.)	B (in.)	t (in.)	R (in.)	Area (in.²)	I_x (in.⁴)	I_y (in.⁴)	ρ_x (in.)	ρ_y (in.)
0.75	0.625	0.05	0.094	0.0977	0.009	0.0068	0.304	0.264
	0.625	0.063	0.125	0.123	0.011	0.0082	0.298	0.259
0.875	0.625	0.05	0.094	0.104	0.013	0.0068	0.353	0.257
	0.625	0.063	0.125	0.131	0.0158	0.0083	0.347	0.251
	0.75	0.05	0.094	0.117	0.0151	0.0122	0.36	0.323
	0.75	0.063	0.125	0.147	0.0184	0.0148	0.354	0.317
	0.75	0.078	0.125	0.177	0.0214	0.0175	0.348	0.314
1.0	0.625	0.05	0.094	0.11	0.0176	0.0068	0.4	0.249
	0.625	0.063	0.125	0.139	0.0216	0.0083	0.394	0.244
	0.75	0.05	0.094	0.123	0.0205	0.0121	0.408	0.315
	0.75	0.063	0.125	0.155	0.0251	0.0147	0..403	0.308
	0.75	0.078	0.125	0.187	0.0294	0.0175	0.396	0.307
1.25	0.75	0.063	0.125	0.17	0.0421	0.0148	0.497	0.295
	0.75	0.078	0.125	0.206	0.0497	0.0175	0.491	0.291
	1.0	0.063	0.125	0.202	0.0532	0.0366	0.514	0.428
	1.0	0.094	0.125	0.291	0.0729	0.0511	0.501	0.419
1.375	0.75	0.063	0.125	0.178	0.0562	0.0148	0.543	0.288
	1.0	0.063	0.125	0.21	0.0662	0.0367	0.562	0.418
	1.0	0.078	0.125	0.255	0.0794	0.0439	0.558	0.415
	1.0	0.094	0.125	0.303	0.0911	0.0511	0.549	0.411
	1.25	0.078	0.125	0.394	0.0951	0.0887	0.569	0.549
	1.25	0.125	0.125	0.453	0.1365	0.1308	0.549	0.537
1.5	0.75	0.078	0.125	0.226	0.0765	0.0175	0.582	0.278
	1.0	0.078	0.125	0.265	0.0961	0.0439	0.602	0.407
	1.0	0.094	0.125	0.314	0.1116	0.0511	0.596	0.403
	1.25	0.094	0.125	0.361	0.1349	0.104	0.611	0.536
1.75	0.75	0.078	0.125	0.245	0.1104	0.0175	0.671	0.267
	1.0	0.078	0.125	0.284	0.1377	0.0439	0.696	0.393
2.0	0.75	0.063	0.125	0.218	0.1272	0.0148	0.765	0.261
	1.0	0.094	0.125	0.361	0.2194	0.0511	0.779	0.376
	1.25	0.094	0.125	0.408	0.2621	0.104	0.801	0.505

Appendix C

A (in.)	B (in.)	t (in.)	R (in.)	Area (in.²)	Y (in.)	X (in.)	I_x (in.⁴)	I_y (in.⁴)	ρ_x (in.)	ρ_y (in.)
2.5	0.75	0.063	0.125	0.249	0.2172		0.0148	0.934	0.244	
	0.875	0.078	0.125	0.323	0.289		0.0288	0.945	0.298	
	0.875	0.094	0.125	0.385	0.3383		0.0333	0.938	0.294	
	1.0	0.094	0.125	0.408	0.3723		0.0512	0.955	0.354	
	1.0	0.125	0.125	0.531	0.4707		0.0636	0.941	0.346	

C.6 SECTION PROPERTIES OF FORMED HAT-SECTIONS

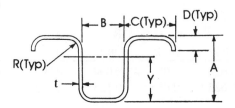

No.	A (in.)	B (in.)	C (in.)	D (in.)	t (in.)	R (in.)	Area (in.²)	Y (in.)	I_x (in.⁴)	ρ_x (in.)
1	1.25	1.0	0.88	0.25	0.045	0.16	0.2277	0.723	0.0525	0.48
2	1.25	0.76	0.74	0.25	0.063	0.13	0.2814	0.732	0.0594	0.459
3	1.25	0.82	1.04	0.3	0.063	0.16	0.3307	0.774	0.0685	0.455
4	1.25	0.82	1.04	0.3	0.071	0.16	0.3695	0.77	0.0758	0.453
5	1.25	0.8	1.08	0.35	0.08	0.19	0.421	0.777	0.0824	0.443
6	1.25	0.78	1.1	0.35	0.09	0.19	0.4704	0.779	0.0904	0.438
7	1.25	0.76	1.12	0.35	0.1	0.22	0.5158	0.78	0.0956	0.431
8	1.25	1.0	0.74	0.3	0.056	0.16	0.267	0.694	0.058	0.466
9	1.25	1.0	0.74	0.3	0.063	0.16	0.298	0.691	0.0635	0.462
10	1.19	0.76	0.74	0.2	0.045	0.13	0.1979	0.691	0.0388	0.443
11	1.18	0.76	0.74	0.25	0.063	0.13	0.2781	0.683	0.0516	0.431
12	1.25	1.0	0.765	0.26	0.032	0.19	0.156	0.704	0.036	0.479
13	1.25	1.0	0.77	0.26	0.036	0.19	0.176	0.703	0.0398	0.481
14	1.25	1.0	0.78	0.26	0.04	0.19	0.196	0.705	0.0436	0.472
15	1.25	1.0	0.79	0.3	0.045	0.19	0.223	0.709	0.0486	0.468
16	1.25	1.0	0.8	0.3	0.05	0.19	0.247	0.709	0.0536	0.466
17	1.25	1.0	0.81	0.3	0.056	0.19	0.276	0.709	0.0604	0.468
18	1.25	1.0	0.825	0.3	0.063	0.19	0.31	0.709	0.0655	0.46
19	1.25	1.0	0.84	0.34	0.071	0.19	0.355	0.71	0.0743	0.459
20	1.25	1.0	0.86	0.34	0.08	0.19	0.399	0.709	0.0821	0.454
21	1.25	1.0	0.88	0.34	0.09	0.19	0.448	0.709	0.0924	0.454
22	1.25	0.76	0.74	0.25	0.071	0.13	0.3201	0.717	0.0656	0.453

C.7 SECTION PROPERTIES OF FORMED Z-SECTIONS

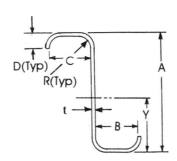

No.	A (in.)	B (in.)	C (in.)	D (in.)	t (in.)	R (in.)	Area (in.²)	Y (in.)	I_x (in.⁴)	ρ_x (in.)
1	2.46	0.9	0.9	0.33	0.081	0.19	0.3656	1.23	0.318	0.933
2	2.491	0.9	0.9	0.33	0.051	0.25	0.2281	1.246	0.203	0.944
3	2.504	0.9	0.9	0.33	0.064	0.25	0.2865	1.252	0.255	0.944
4	2.491	1.06	0.74	0.33	0.051	0.25	0.2281	1.246	0.201	0.94
5	2.504	1.09	0.71	0.33	0.064	0.25	0.2865	1.252	0.253	0.939
6	2.5	0.9	0.9	0.5	0.072	0.25	0.346	1.25	0.302	0.934
7	2.49	0.9	0.9	0.33	0.056	0.25	0.2526	1.245	0.225	0.944
8	2.5	0.9	0.9	0.5	0.081	0.25	0.3891	1.25	0.338	0.932
9	2.505	0.9	0.9	0.33	0.045	0.25	0.2034	1.253	0.184	0.951
10	2.5	0.9	0.9	0.5	0.09	0.25	0.4314	1.25	0.373	0.929

APPENDIX D

Common Formulas For Beams

Moment: $\curvearrowright +M \quad \curvearrowleft -M$ Reaction: $+R\uparrow \; -R\downarrow$ Shear: $\uparrow\boxed{+V}\downarrow \; \downarrow\boxed{-V}\uparrow$ Deflection: $+y\uparrow \; -y\downarrow$ Slope: $\curvearrowright +\theta \; \curvearrowleft -\theta$

Case	Loading and Support	Reactions and Shears	Bending Moments	Deflections and Slopes
1	Simple beam, intermediate concentrated load	$R_A = \dfrac{Pb}{L}$ $R_C = \dfrac{Pa}{L}$ $V_{A-B} = \dfrac{Pb}{L}$ $V_{B-C} = \dfrac{-Pa}{L}$	$M_{A-B} = \dfrac{Pbx}{L}$ $M_{B-C} = \dfrac{Pa(L-x)}{L}$ Max. $M_B = \dfrac{Pab}{L}$	$y_{A-B} = [\dfrac{-Pbx}{6EIL}][L^2 - b^2 - x^2]$ $y_{B-C} = \dfrac{-Pa(L-x)}{6EIL}[2bL - b^2 - (L-x)^2]$ Max. $y_x = [\dfrac{-Pab}{27EIL}][(a+2b)\sqrt{3a(a+2b)}]$ where $x = \sqrt{\dfrac{a(a+2b)}{3}}$; $a > b$ $\theta_A = \dfrac{-P(bL - \dfrac{b^3}{L})}{6EI}$; $\theta_C = \dfrac{P(2bL + \dfrac{b^3}{L} - 3b^2)}{6EI}$

Common Formulas For Beams

Case	Loading and Support	Reactions and Shears	Bending Moments	Deflections and Slopes
2	Simple beam, concentrated load at center	$R_A = \dfrac{P}{2}$ $R_C = \dfrac{P}{2}$ $V_{A-B} = \dfrac{P}{2}$ $V_{B-C} = -\dfrac{P}{2}$	$M_{A-B} = \dfrac{Px}{2}$ $M_{B-C} = \dfrac{P(L-x)}{2}$ Max. $M_B = \dfrac{PL}{4}$	$y_{A-B} = \left[\dfrac{-P}{48EI}\right][3L^2 x - 4x^3]$ Max. $y_B = \dfrac{-PL^3}{48EI}$ $\theta_A = \dfrac{-PL^2}{16EI}$; $\theta_C = \dfrac{PL^2}{16EI}$
3	Simple beam, uniform load	$R_A = R_B = \dfrac{wL}{2}$ $V_{A-B} = \dfrac{wL}{2} - wx$	$M_{A-B} = \dfrac{wLx}{2} - \dfrac{wx^2}{2}$ Max. $M_x = \dfrac{wL^2}{8}$; $x = \dfrac{L}{2}$	$y_{A-B} = \left[\dfrac{-wx}{24EI}\right](L^3 - 2Lx^2 + x^3)$ Max. $y_x = \dfrac{-5wL^4}{348EI}$; $x = \dfrac{L}{2}$ $-\theta_A = \theta_B = \dfrac{wL^3}{24EI}$
4	Simple beam, non-uniform load	$R_A = \dfrac{(2w_a + w_b)L}{6}$ $R_B = \dfrac{(2w_b + w_a)L}{6}$ $V_{A-B} = \left[\dfrac{(2w_a + w_b)L}{6}\right]$ $- w_a x - \left[\dfrac{(w_b - w_a)x^2}{2L}\right]$	$M_{A-B} = \left[\dfrac{(2w_a + w_b)Lx}{6}\right.$ $\left. - \dfrac{w_a x^2}{2}\right.$ $\left. - \dfrac{(w_b - w_a)x^3}{6L}\right]$	$\theta_A = \dfrac{-(8w_a + 7w_b)L^3}{360EI}$ $\theta_B = \dfrac{(7w_a + 8w_b)L^3}{360EI}$

Appendix D

Case	Loading and Support	Reactions and Shears	Bending Moments	Deflections and Slopes
5	Simple beam, triangular uniform load	$R_A = \dfrac{wL}{6}$ $R_B = \dfrac{wL}{3}$ $V_{A-B} = (\dfrac{wL}{2})(\dfrac{1}{3} - \dfrac{x^2}{L^2})$	$M_{A-B} = (\dfrac{wL}{6})(x - \dfrac{x^3}{L^2})$ Max. $M = 0.0642wL^2$ at $x = 0.577L$	$y_{A-B} = [\dfrac{-wx}{360EIL}](3x^4 - 10L^2x^2 + 7L^4)$ Max. $y_x = \dfrac{-0.00652wL^4}{EI}$ at $x = 0.519L$ $\theta_A = \dfrac{-7wL^3}{360EI}$ $\theta_B = \dfrac{8wL^3}{360EI}$
6	Simpel beam, double triangular uniform laod	$R_A = R_B = \dfrac{wL}{4}$ $R_{A-B} = \dfrac{wL}{4} - \dfrac{wx^2}{L}$ $V_{B-C} = \dfrac{-wL}{4} + \dfrac{w(L-x)^2}{L}$	$M_{A-B} = (\dfrac{wL}{12})(3x - \dfrac{4x^3}{L^2})$ $M_{B-C} = \dfrac{wL}{12}[3(L-x) - \dfrac{4(L-x)^3}{L^2}]$ Max. $M = \dfrac{wL^2}{12}$; $x = \dfrac{L}{2}$	$y_{A-B} = [\dfrac{-wx}{960EIL}](5L^2 - 4x^2)^2$ Max. $y_B = \dfrac{-wL^4}{120EI}$ $\theta_A = \dfrac{-5wL^3}{192EI}$ $\theta_C = \dfrac{5wL^3}{192EI}$
7	Simple beam, parabolic uniform load	$R_A = R_B = \dfrac{wL}{3}$	Max. $M = \dfrac{wL^2}{9.6}$; $x = \dfrac{L}{2}$	$y_{A-B} = [\dfrac{-wx}{90EIL^2}](3L^5 - 5L^3x^2 + 3Lx^4 - x^5)$ Max. $y_x = \dfrac{-0.01059wL^4}{EI}$ at $x = \dfrac{L}{2}$ $\theta_A = \dfrac{-wL^3}{30EI}$ $\theta_B = \dfrac{wL^3}{30EI}$

Common Formulas For Beams

Case	Loading and Support	Reactions and Shears	Bending Moments	Deflections and Slopes
8	Simple beam, end moment at point A	$R_A = -\dfrac{M_0}{L}$ $R_B = \dfrac{M_0}{L}$ $V_{A-B} = -\dfrac{M_0}{L}$	$M_{A-B} = M_0(1 - \dfrac{x}{L})$ Max. moment $M_A = M_0$	$y_{A-B} = [\dfrac{-M_0 x}{6EIL}](2L^2 - 3Lx + x^2)$ Max. $y_{x=0.423L} = \dfrac{-0.0642 M_0 L^2}{EI}$ $\theta_A = \dfrac{-M_0 L}{3EI}$; $\theta_B = \dfrac{M_0 L}{6EI}$
9	Simple beam, end moments at points A and B	$R_A = \dfrac{(M_B - M_A)}{L}$ $R_B = \dfrac{(M_A - M_B)}{L}$ $V_{A-B} = \dfrac{(M_B - M_A)}{L}$	$M_{A-B} = M_A + (\dfrac{M_B - M_A}{L})x$	$y_{A-B} = [\dfrac{x(x-L)}{EI}][\dfrac{M_A}{2} + \dfrac{(M_B - M_A)(x+L)}{6L}]$ $\theta_A = [\dfrac{-L}{6EI}](2M_A + M_B)$ $\theta_A = [\dfrac{L}{6EI}](M_A + 2M_B)$

Appendix D

Case	Loading and Support	Reactions and Shears	Bending Moments	Deflections and Slopes
10	Fixed-end beam, intermediate concentrated load	$R_A = \dfrac{Pb^2(3a+b)}{L^3}$ $R_C = \dfrac{Pa^2(3b+a)}{L^3}$	$M_A = -\dfrac{Pab^2}{L^2}$ $M_C = -\dfrac{Pa^2b}{L^2}$ $M_{A-B} = \dfrac{-Pab^2}{L^2} + R_A x$ $M_{B-C} = \dfrac{-Pab^2}{L^2} + R_A x - P(x-a)$ $M_B = \dfrac{-Pab^2}{L^2} + R_A a$	$y_{A-B} = [\dfrac{-Pb^2 x^2}{6EIL^3}](3ax+bx-3aL)$ $y_{B-C} = [\dfrac{-Pa^2(L-x)^2}{6EIL^3}][(3b+a)(L-x)-3bL]$ Max. $y_x = \dfrac{-2Pa^3 b^2}{3EI(3a+b)^2}$ at $x = \dfrac{2aL}{3a+b}$; $a > b$ Max. $y_x = \dfrac{-2Pa^2 b^3}{3EI(3b+a)^2}$ at $x = L - [\dfrac{2bL}{3b+a}]$; $a < b$
11	Fixed-end beam, concentrated load at center	$R_A = R_C = \dfrac{P}{2}$ $V_{A-B} = \dfrac{P}{2}$ $V_{B-C} = \dfrac{-P}{2}$	$M_A = M_C = \dfrac{-PL}{8}$ $M_{A-B} = \dfrac{P(4x-L)}{8}$ $M_{B-C} = \dfrac{P(3L-4x)}{8}$ $M_B = \dfrac{PL}{8}$	$y_{A-B} = [\dfrac{-P}{48EI}](3Lx^2 - 4x^3)$ Max. $y_B = \dfrac{-PL^3}{192EI}$

Common Formulas For Beams

Case	Loading and Support	Reactions and Shears	Bending Moments	Deflections and Slopes
12	Fixed-end beam, uniform load 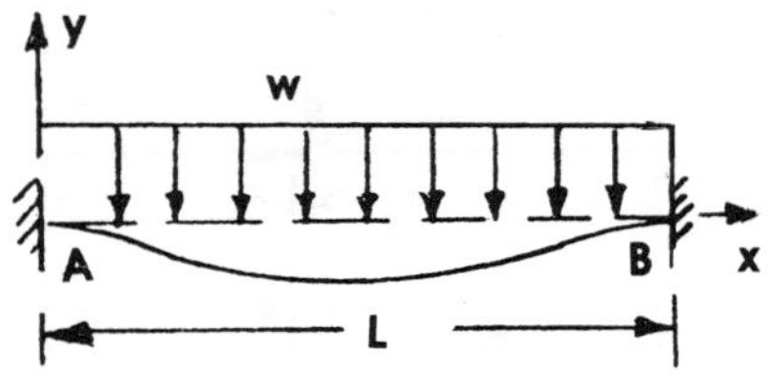	$R_A = R_B = \dfrac{wL}{2}$ $V_{A-B} = \dfrac{wL}{2} - wx$	$M_A = M_B = \dfrac{-wL^2}{12}$ $M_{A-B} = \dfrac{wL}{2}(x - \dfrac{x^2}{L} - \dfrac{L}{6})$ Max. $M = \dfrac{wL^2}{24}$; $x = \dfrac{L}{2}$	$y_{A-B} = [\dfrac{-wx^2}{24EI}](L-x)^2$ Max. $y = \dfrac{-wL^4}{384EI}$; $x = \dfrac{L}{2}$
13	Fixed-end beam, non-uniform load 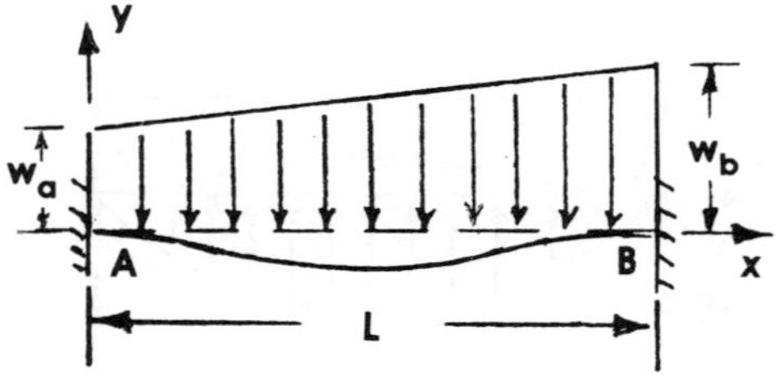	$R_A = [\dfrac{L}{20}](7w_a + 3w_b)$ $R_B = [\dfrac{L}{20}](3w_a + 7w_b)$	$M_A = [\dfrac{-L^2}{60}](3w_a + 2w_b)$ $M_B = [\dfrac{-L^2}{60}](2w_a + 3w_b)$	
14	Fixed-end beam, triangular uniform load 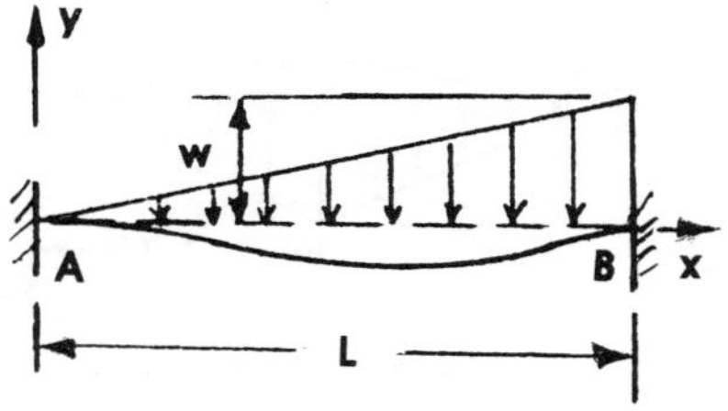	$R_A = \dfrac{3wL}{20}$ $R_B = \dfrac{7wL}{20}$ $V_{A-B} = (\dfrac{wL}{2})(\dfrac{3}{10} - \dfrac{x^2}{L^2})$	$M_{A-B} = \dfrac{wL}{2}(\dfrac{3x}{10} - \dfrac{L}{15} - \dfrac{x^3}{3L^2})$ $M_A = \dfrac{-wL^2}{30}$ $M_B = \dfrac{-wL^2}{20}$ Max. $M = 0.0215wL^2$; $x = 0.584L$	$y_{A-B} = [\dfrac{-wLx^2}{120EIL^2}](2L^3 - 3L^2x + x^3)$ Max. $y = \dfrac{-wL^4}{764EI}$; $x = 0.525L$

777

Appendix D

Case	Loading and Support	Reactions and Shears	Bending Moments	Deflections and Slopes
15	Fixed-end beam, double triangular uniform load	$R_A = R_C = \dfrac{wL}{4}$	$M_A = M_C = \dfrac{-5wL^2}{96}$ $M_B = \dfrac{wL^2}{20}$ $M_{A-B} = -M_A + \dfrac{wLx}{4} - \dfrac{2wx^3}{3L}$ Max. $M = \dfrac{wL^2}{32}$; $x = \dfrac{L}{2}$	
16	Fixed-end beam, parabolic uniform load	$R_A = R_C = \dfrac{wL}{3}$	$M_A = -M_B = \dfrac{wL^2}{15}$	
17	Fixed-end beam, intermediate moment	$-R_A = R_C = \dfrac{6aM_0}{L^3}(L-a)$	$M_A = \dfrac{M_0}{L^2}(4aL - 3a^2 - L^2)$ $M_C = \dfrac{-M_0}{L^2}(2aL - 3a^2)$ $M_{A-B} = \dfrac{M_0}{L^3}(4aL^2 - 9a^2L + 6a^3 - L^3)$ $M_{B-C} = \dfrac{M_0}{L^3}(4aL^2 - 9a^2L + 6a^3)$	$y_{A-B} = \dfrac{-(3M_A x^2 - R_A x^3)}{6EI}$ $y_{B-C} = \left[\dfrac{-1}{EI}\right][(M_0 - M_A)(3x^2 - 6Lx + 3L^2) - R_A(3L^2x - x^3 - 2L^3)]$ Max. $y_x = 0.01617\left(\dfrac{M_0 L^2}{EI}\right)$ $x = 0.565L$ (when $a = 0.767L$)

Common Formulas For Beams

Case	Loading and Support	Reactions and Shears	Bending Moments	Deflections and Slopes
18	Fixed-end beam, Δ deflection at point B	$-R_A = R_B = \dfrac{12EI\Delta}{L^3}$	$M_A = -M_B = \dfrac{6EI\Delta}{L^2}$	
19	Fixed-end beam, θ rotation at point B	$R_A = -R_B = \dfrac{6EI\theta}{L^2}$	$M_A = \dfrac{-2EI\theta}{L}$ $M_A = \dfrac{4EI\theta}{L}$	
20	Hinged-end and fixed-end beam, intermediate concentrated load	$R_A = [\dfrac{P}{2}][\dfrac{3b^2L - b^3}{L^3}]$ $R_C = P - R_A$ $V_{A-B} = R_A$ $V_{B-C} = R_A - P$	$M_{A-B} = R_A x$ $M_{B-C} = R_A x - P(x - L + b)$ $M_C = [\dfrac{-P}{2}]\left[\dfrac{b^3 + 2bL^2 - 3b^2L}{L^2}\right]$	$y_{A-B} = [\dfrac{-1}{6EI}][(R_A(x^3 - 3L^2 x) + 3Pb^2 x]$ $y_{B-C} = [\dfrac{-1}{6EI}]\{R_A(x^3 - 3L^2 x) + P[3b^2 x - (x - a)^3]\}$ $\theta_A = [\dfrac{-Pb^2}{48EI}](\dfrac{b}{L} - 1)$

Appendix D

Case	Loading and Support	Reactions and Shears	Bending Moments	Deflections and Slopes
21	Hinged-end and fixed-end beam, uniform load	$R_A = \dfrac{3wL}{8}$ $R_B = \dfrac{5wL}{8}$ $V_{A-B} = wL\left(\dfrac{3}{8} - \dfrac{x}{L}\right)$	$M_{A-B} = wL\left(\dfrac{3x}{8} - \dfrac{x^3}{2L}\right)$; Max. $M = \dfrac{9wL^2}{128}$; $x = \dfrac{3L}{8}$ $M_B = \dfrac{-wL^2}{8}$	$y_{A-B} = \left[\dfrac{-w}{48EI}\right](3Lx^3 - 2x^4 - L^3x)$ Max. $y = \dfrac{-0.0054wL^4}{EI}$; $x = 0.4215L$ $\theta_A = \dfrac{-wL^3}{48EI}$
22	Hinged-end and fixed-end beam, triangular load	$R_A = \dfrac{wL}{10}$ $R_B = \dfrac{4wL}{10}$ $V_{A-B} = \dfrac{wL}{2}\left(\dfrac{1}{5} - \dfrac{x^2}{L^2}\right)$	$M_{A-B} = \dfrac{wL}{2}\left(\dfrac{x}{5} - \dfrac{x^3}{3L^2}\right)$ Max. $M = \dfrac{0.06wL^2}{2}$; $x = 0.4474L$ $M_B = \dfrac{-wL^2}{15}$	$y_{A-B} = \left[\dfrac{-w}{120EI}\right](2Lx^3 - L^3x - \dfrac{x^5}{L})$ Max. $y = -0.00477\left(\dfrac{wL^4}{2EI}\right)$; $x = 0.477L$ $\theta_A = \dfrac{wL^3}{120EI}$
23	Hinged-end and fixed-end beam, end moment at point A	$R_A = \dfrac{-3M_0}{2L}$ $R_B = \dfrac{3M_0}{2L}$ $V_{A-B} = \dfrac{-3M_0}{2L}$	$M_{A-B} = \dfrac{M_0}{2}\left(2 - \dfrac{3x}{L}\right)$ $M_A = M_0$ $M_B = \dfrac{-M_0}{2}$	$y_{A-B} = \left[\dfrac{-M_0 x}{4EIL}\right](L - x)^2$ Max. $y = \dfrac{M_0 L^2}{27EI}$; $x = \dfrac{L}{3}$ $\theta_A = \dfrac{-M_0 L}{4EI}$

Common Formulas For Beams

Case	Loading and Support	Reactions and Shears	Bending Moments	Deflections and Slopes
24	Cantilever, intermediate load	$R_C = P$ $V_{A-B} = 0$ $V_{B-C} = -P$	$M_{A-B} = 0$ $M_{B-C} = -P(x-a)$ Max. $M_C = -Pb$	$y_{A-B} = [\frac{-Pb^2}{6EI}](3L - 3x - b)$ $y_{B-C} = [\frac{-P(L-x)^2}{6EI}](3b - L + x)$ Max. $y_A = [\frac{-P}{6EI}](3b^2L + b^3)$ $\theta_{A-B} = \frac{Pb^2}{2EI}$
25	Cantilever, end load	$R_B = P$ $V_{A-B} = -P$	$M_{A-B} = -Px$ Max. $M_B = -PL$	$y_{A-B} = [\frac{-P}{6EI}](x^3 - 3L^2x + 2L^3)$ Max. $y_A = \frac{-PL^3}{3EI}$ $\theta_A = \frac{PL^2}{2EI}$
26	Cantilever, uniform load	$R_B = wL$ $V_{A-B} = -wx$	$M_{A-B} = \frac{-wx^2}{2}$ Max. $M_B = \frac{-wL^2}{2}$	$y_{A-B} = [\frac{-w}{24EI}](x^4 - 4L^3x + 3L^4)$ Max. $y_A = \frac{-wL^4}{8EI}$ $\theta_A = \frac{wL^3}{6EI}$

Appendix D

Case	Loading and Support	Reactions and Shears	Bending Moments	Deflections and Slopes
27	Cantilever, triangular load	$R_B = \dfrac{wL}{2}$ $V_{A-B} = \dfrac{-wx^2}{2L}$	$M_{A-B} = \dfrac{-wx^3}{6L}$ Max. $M_B = \dfrac{-wL^2}{6}$	$y_{A-B} = [\dfrac{-w}{120EIL}](x^5 - 5L^4x + 4L^5)$ Max. $y_A = \dfrac{-wL^4}{30EI}$ $\theta_A = \dfrac{wL^3}{24EI}$
28	Cantilever, intermediate moment	$R_C = 0$ $V_{A-B} = 0$	$M_{A-B} = 0$ $M_{B-C} = M_0$ Max. $M_{B-C} = M_0$	$y_{A-B} = [\dfrac{M_0 a}{EI}](L - \dfrac{a}{2} - x)$ $y_{B-C} = [\dfrac{M_0}{2EI}][(x - L + a)^2 - 2a(x - L + a) + a^2]$ Max. $y_A = \dfrac{-M_0 a}{EI}(L - \dfrac{a}{2})$ $\theta_{A-B} = \dfrac{-M_0 a}{EI}$
29	Cantilever, end moment	$R_B = 0$ $V_{A-B} = 0$	$M_{A-B} = M_0$	$y_{A-B} = [\dfrac{M_0}{2EI}](L - x)^2$ Max. $y_A = \dfrac{M_0 L^2}{2EI}$ $\theta_A = \dfrac{-M_0 L}{EI}$

APPENDIX E

Computer Aided Engineering (CAE)

E.1 INTRODUCTION

In today's competitive aerospace market, it is no longer enough to simply have the best airplane or engine performance, company technology, or management programs. In order to compete for new business, an integrated product development environment is required to reduce time and product development costs, while at the same time increasing product quality and performance. This can be thought of as optimizing the design-to-certification process. Optimizing the process means bringing together various engineering disciplines, integrating them with design procedures, performing the engineering early in the design process when changes can most easily be made, and interfacing with software tools in a common IT (Information Technology) framework. This constitutes a product development process that addresses the business issues of a "faster, better, cost-effective" aerospace design-to-certification process, with the goal being to get to certification as quickly and affordably as possible.

Fig. E.1.1 shows, in simplified form, the product life cycle of a plane, from initial sale through maintenance. The life cycle of a commercial aircraft program could last for over 50 years. For airframe structures, achieving certification is required before the product can be put into use. "Product development" then becomes a process of going from design through certification, which takes place in the middle of the aircraft life cycle process.

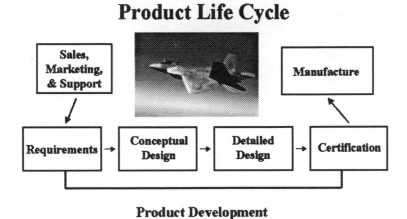

Fig. E.1.1 Product Development

E.2 DESIGN-TO-CERTIFICATION PROCESS

The structural design-to-certification process is itself comprised of three major functions, see fig. E.2.1.

Appendix E

The three main areas are:

- Define Operating Environment: In this area the mission profile and operating conditions are defined, with the outcome being the structural loads. These are called EXTERNAL LOADS because they are the operating loads that are applied to the structure. Examples include pressure on wings, temperatures and radiation on satellites, impact on the tail hook of a fighter landing on a carrier, and inertial forces resulting from vibrations in a jet engine. These loads vary with time and with angle of attack. There are typically hundreds or thousands of external load combinations that are applied in the analytical simulation. External loads are defined from flight tests, wind tunnel tests, and analytical computations (aerodynamics, for example).
- System Design and Analysis: The external loads are applied to a system-level model of the structure. This is typically a finite element model (FEM) that is capable of representing the load paths and overall system behavior. The analytical model contains the geometry and materials, and is typically created by finite element analysts. Commercial software for CAD and CAE are used in this part of the process, to build the model and to simulate system-level behavior such as structural loads, vibrations, and acoustics. The external loads are applied to this system-level model, and the INTERNAL LOADS are computed. These internal loads are the loads in each of the model's structural members.
- Detailed Design and Verification: The internal loads are then used in doing detailed stress analyses. For example, the wing used in the system-level model is usually highly idealized for efficiency. This idealization omits the small details that do not affect the internal loads calculations. The detailed model, however, retains these small details and looks now at a portion of the wing instead of the whole wing.

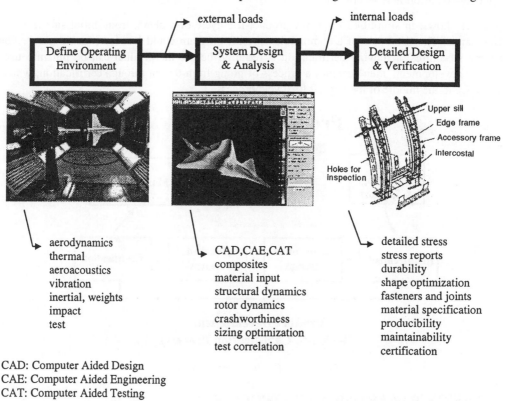

CAD: Computer Aided Design
CAE: Computer Aided Engineering
CAT: Computer Aided Testing

* Images courtesy of The MacNeal-Schwendler Corporation

Fig. E.2.1 The Structural Design-to-Certification Process

In essence, the structural design-to-certification process specifies the environment and external loads, computes the internal loads, and uses the loads to make geometry, topology, material, and tooling decisions. The overall goal is for this process to be completed in time to affect the tooling and material specification.

E.3 COMPUTER SOFTWARES

The MacNeal-Schwendler Corporation (MSC) is a leading provider of mechanical computer-aided engineering (MCAE) strategies, software, and services. MSC's core products, MSC/NASTRAN and MSC/PATRAN, apply mainly to the "System Design & Analysis" area in the design-to-certification process, and are used to analyze models to compute internal loads. The products also address areas in the "Operating Environment" area, which provides the external loads as well, such as aeroelasticity loads.

(A) MSC/NASTRAN

MSC/NASTRAN, a finite element analysis (FEA) computer program, is used by aerospace companies for aeroelasticity, heat transfer, statics, and structural dynamics analyses. A typical application, performed in the late 1970's, was coupled dynamic loads analysis in which multiple companies each worked on their part of the structure, sent mass and stiffness matrices to the "integration" company (prime contractor), and the integrator coupled them together and performed dynamic response calculations. The modeling, the matrices, and the dynamic response were all done using MSC/NASTRAN.

(B) MSC/PATRAN

MSC/PATRAN is a graphical pre- and postprocessing software product that makes MSC/NASTAN easier to use. It provides many display tools to visualize simulation results in the contour plots of maximum Von Mises stresses, a deformed plot of deflection, or an animation of dynamic response under a dynamic environment. These tools allow the engineers to quickly and accurately predict the behavior of the design. Fig. E.3.1 shows a typical usage of FEA in aerospace.

Fig. E.3.1 Typical FEA Usage in Aerospace Industry

The FEA process is a method of analyzing a part or assembly to ensure performance integrity over the product's lifetime. A geometric model is created, a finite element model is associated with the geometry, the operating environment is defined, and the structural response (deflection, stress, temperatures, etc) is computed and presented for display. If the computed response – stress, for example – is greater than the allowable or maximum design value, the structure is redesigned and re-analyzed until an acceptable design is achieved. This redesign/re-analysis cycle can be automated via structural optimization. Fig. E.3.2 summarizes the basic finite element process.

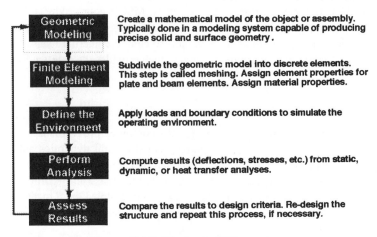

Figure E 3.2 The basic FEA process

TYPES OF FINITE ELEMENTS:

FEA programs contain numerous types of finite elements, each of which represents different physical phenomena. The choice of finite element types is important for simulating correct structural behavior. Rods and beams represent forces that act in a line; beams transmit bending moments and rods do not. These elements are used to model bridges, space trusses, aircraft stringers, and plate stiffeners.

Plate and shell elements represent forces that vary over a surface. These two- and three-dimensional elements possess both membrane (in-plane) and bending (out-of-plane) behavior. Plates and shells are used to model skin of automobiles, aircraft, and spacecraft. Fig. E.3.3 shows the applications of beams and plate elements for a three-dimensional front fuselage section.

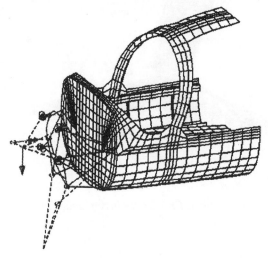

Fig. E.3.3 Mesh of A Finite Element Fuselage Section

Solid elements represent a general, three-dimensional state of stress. Solid elements are used to model engine blocks, crankshafts, bulkheads, brackets, and test fixtures. Fig. E.3.4 shows a contour plot for a three-dimensional bracket.

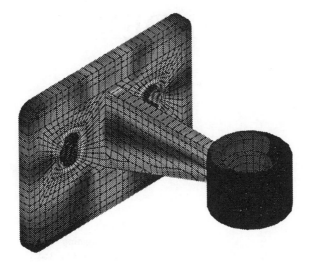

Fig. E.3.4 A Contour Plot of a Three-Dimensional Bracket

MESHING AND POSTPROCESSING

Meshing is the process of subdividing the geometry into a series of discrete elements. Meshing is done to represent complex geometry and provide more elements in regions of interest (typically, where stress gradients are greatest). Adequacy of the finite element mesh is often defined as solution convergence. "Convergence" means that as the number of elements is increased, the solution approaches a given value. With simple academic problems, this given value can be calculated from textbook solutions, and convergence is often measured in terms of these known solutions. Fig. E.3.5 shows an application of meshing and postprocessing features for a symmetric plane.

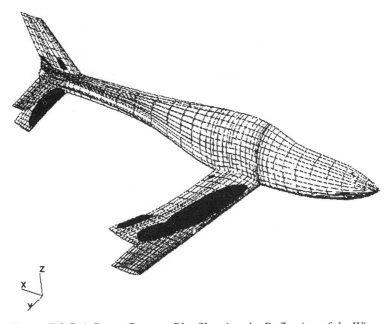

Figure E.3.5 A Stress Contour Plot Showing the Deflection of the Wing

INDEX

A

Access holes 262
Adhesive 204, 313, 320
Aerodynamic 3, 32, 784
Aileron 37
Aircraft
    – commercial 4, 29, 54, 783
    – military 4, 29
Airframe vitality 677
Airworthiness 4, 73
Allowables 67
Aluminum 60, 61, 76
Analysis 80, 128, 784
Angle of twist 169, 174
Angle fitting 343, 344, 347
Angular
    – acceleration 49, 51
    – roll acceleration 51
    – roll velocity 49, 50
    – velocity 49, 50
    – yaw acceleration 51
Approximation method 435
Assembly 13
Attachment (see Fasteners and rivets)
Average hole diameters 701-703

B

Balanced bonded joints 315
Basis 67
Bead-flanged holes 538
Beaded webs 538
Beam 87, 138, 182, 224, 670, 786
    – box 178
    – deep web beam 414
    – elastic bending 138, 139
    – I-beam 411
    – lateral buckling 411
    – plastic bending 146
    – section factor 147
    – shear 152, 157
    – tapered 164
    – torsion 169
Beam-column 418, 626, 627
    – approximation method 435

– coefficients 421, 626
– elastic supports 436
– equations 419
Beam theory 138
Bend radii (or radius) 684
Bent doubler 549
Beryllium 77
Black-box 2
Blind fasteners 279
Block loading 723
Bolt 280, 340, 344
    – bending 326
    – preload 352, 392
    – shear 321
    – tension 343, 346
Bonded joints 276, 313
    – balanced 315
    – lap 318
    – unbalanced 317
Box beam 178, 221
Brake (braking) 41, 44
Breakaway design 53, 54
Briles rivets 279
Buckling 394, 451
    – coefficients 452, 454, 459, 465, 551, 643
    – column 394
    – curved plate in-plane compression 465
    – curved plate in-plane shear 465, 508
    – edge panels of cutout 572
    – edge supports 454, 455
    – Euler equation 395
    – flat plate in-plane bending 461
    – flat plate in-plane compression 458
    – flat plate in-plane shear 460
    – initial 616, 617, 643, 647
    – inter-rivet 613
    – lateral 411
    – long rectangular plate 454
    – square plate 453
    – thin sheets 451
Buckling coefficients
    – bending (flat plates) 461
    – combined loadings 468
    – compression (curved plates) 465
    – compression (flat plates) 458, 459, 460
    – shear (curved plates) 466, 467
    – shear (flat plates) 460

Buffet 45
Bulkhead 182, 200, 345, 787
Bushing 322, 328
    – flanged 328, 699
    – plain 328
    – spherical 342
Butt welds 308
Butterfly flange 549, 553, 720
Bypass load 383

C

CAD/CAM 11, 784
Castigliano's second theorem method 99
Castigliano's theorem method 99
Casting 59, 60
Center of gravity 34
Certification 4, 783
Channel fitting 343, 344, 350
Cladding 59, 458
Clevis 342
Clips 361, 417
    – extruded angles 362, 365, 366
    – extruded T 367, 368
    – formed angles 362, 363, 364
Close-outs (panel) 209
    – square edge 209
    – tapered edge 209
Coaming stringer 573, 574
Coefficients method 425
Cold work 382, 391, 392
Column 394
    – beam-column 418
    – effective length 395
    – elastic supports 398
    – end-fixity 396
    – stepped 401
    – tapered 406
    – with elastic supports 398, 399, 400
Compatibility
    – contour 607
    – strain 85
Component tests 722
Compression panels 607
Compression test panel 737
Compression panel design curves
    – skin-extruded stringer panels 641
    – skin-formed stringer panels 642
    – sturdy integrally-stiffened panels 643
    – unflanged integrally-stiffened panels 651
Compression test panel 737
Computer-aided analysis 2, 783
Conceptual design 12
Conjugate-beam method 105
Control surfaces 342
Core 204, 206, 208
Corrosion 58, 713

Cost 11, 58, 783
    – non-recurring cost 11
    – recurring cost 11
Countersunk fasteners 277
Countersunk washers 691
Cracks 695, 741
Crack growth 75, 76, 656, 664
Crack growth stress intensity factor 666
Crippling stress 437
    – bending 445
    – extruded sections 439, 440, 444
    – formed lip criteria 439
    – formed sections 439, 440, 441
Criteria 28, 281, 439
Crushing loads 178
Curvature 178
Curved plates 464
Cut-off stress (ultimate) 657
Cutouts 530
    – corner panels 563, 593
    – cut frame 585
    – cut stringer 576, 585
    – deep shear beams 563
    – edge panels 563
    – fully framed cutout 564
    – fuselage doors 585
    – half framed cutout 570
    – quarter framed cutout 571
    – shallow beams 534, 538, 546, 548
    – transport fuselage 534
    – under axial in-plane load 573

D

Damages 695
Damage tolerant design 6, 78, 655, 687
Damage tolerant panels 655
Deep web beam 414
Deflection 418, 437, 786
Design cycles 3
Determinate structures 87
Diagonal tension factor 485, 490, 572
Diagonal tension webs 483, 484, 504
Dimensional tolerance 23
Dimpled holes 697
Displacement method 86
Donut doubler 530, 546
Doors 532, 534
Doorstop 596, 597
Double angle fitting 343, 351
Double lap joints (bonded) 315
Double-row of fasteners 704
Doublers 379, 387, 523, 549, 564, 585
    – bent doubler 548
    – corner doubler 585
    – donut doubler 530, 546
    – repair doubler 692, 720

Drawing 23
Ductility 307, 315
Dummy structures 740
Durability 78
Dynamic 45, 784
Dynamic loads 27, 45

E

Eccentricities 275, 292
Eccentric joints 292
Edge distance 277, 681, 682, 700
Edge margin 700
Edge rotational restraints 455, 459
Edge support 454
Effective column length 395
Effective width 298, 586, 610, 622
Elastic 138
Elastic limit 63
    – range 169
    – restraints 397
    – spring constants 398
    – supports 398, 399, 400, 436
Elastic center and column analogy method 107
Elastic restraints 397
Elastic supports 398, 436
Elastic-weight method 102
Elongation 66, 67
End bay 523
End-fixity coefficients 396, 397, 452, 613
Engine mount 57
Environment 58, 784
Equilibrium 85
Equivalent air speed 36, 38
Equivalent beam axial area 133
Equivalent shear web thickness 129, 266
Equivalent structures 128, 380, 381
Euler equation 395, 451
Euler-Engesser equation 396
Extrusion 59, 60, 342, 439, 440, 444

F

FAA 4
Factor of saftey 24
Fail-safe 322, 655, 668
FAR 27, 38, 52, 55, 274, 687
Farrar's efficiency factor 618, 647
Fasteners 279, 280, 625, 784
    – allowables 282-288, 337, 338
    – clusters 292
    – double-row pattern 704
    – fits 276, 390, 690
    – flexibility 690
    – hole condition 374
    – holes 382
    – knife edge 277
    – load 377, 378
    – load distribution 378, 657
    – min. spacing 377, 703
    – offset ratio 626
    – over-size fasteners 697
    – patterns 375, 704
    – spacing 623, 703
    – spring constants 267, 380
    – staggered-row pattern 704
    – symbol 278
Fatigue 58, 307, 373, 374, 382, 487, 587, 655
    – life 386, 387, 389, 656, 661
    – life utilization ratio 661
    – quality index (K) 655
Fatigue life (safe life) 655, 656, 660
Fatigue test panel 726
Feather edge (knife edge) 277
Fiberglass 207
Fillers 685, 693, 719
Fillet welds 310
Filing system 26
Finish 698
Finite element modeling (FEM) 1, 124, 379, 382, 784
First moment of area 154
Fitting 273, 321, 343, 685
    – angle fitting 343, 344, 347
    – channel fitting 343, 344, 350
    – double angle fitting 343, 351
    – factor 274, 322, 345
    – lug 321
Fitting factors 274
Flanged holes 530, 532, 535, 538, 540
Flap 36, 37, 53
Flight profiles 659
Flight-by-flight loading 723
Flutter 3, 31, 45
Forced bending 52
Force method 86
Forging 59, 60, 342
Formed section 439, 440, 443, 447
Fracture toughness 75
Framed cutouts 564, 570, 571
Frames 506, 513, 585, 586
Free-body diagrams 21
Fuel 35
    – density 49
    – head 55
    – pressure 49, 50, 55
    – slosh 55
    – tank 49, 50, 53
    – weight 35
Fuse 336
Fuselage 46, 56, 58, 200, 276, 345, 505, 507, 529, 585, 691, 717, 719, 786

G

Gap 342, 370
GAG load cycles 660
Geodesic panels 610

Grain direction 68, 323, 346
Ground-air-ground loading 660, 662, 722
Gusset joints 298
Gust 30, 37, 38, 50, 53

H

Hi-Loks 279, 345, 382
Hinge 46, 322, 342
Hole condition 382, 698
Hole filling factor 382
Holes 530
    – access holes 531
    – average hole diameters 701
    – donut 530, 546
    – flanged holes 530, 535, 539
    – lightening holes 530
    – machined flanged holes 532, 564
Honeycomb panel 203
Hooke's law 80

I

Indentation 695
Indeterminate structures 97
Inelastic limit 63
Inserts 213
Inspectability 687, 688
Inspection interval 664
Instability 514, 616, 623
Integrally-stiffened panels 610, 643, 646
Interaction curves 468
Interference fit fasteners 390
    – beam 671
    – flanged 628, 644, 645
    – panels 608, 643, 646
    – unflanged 646
Inter-rivet buckling 613, 617
Internal loads 126
Isogrid 610

J

JAR 27
Joggle 524, 682
Johnson-Euler curves 448, 449
Joints 273, 729
    – bonded 313
    – butt 308
    – eccentric 292
    – gusset 298
    – lap 315, 317, 318, 378, 379
    – supported 275
    – unsupported 275, 318
    – welded 305

K

Kink 256
Knife edge (feather edge) 277

L

Landing gear 54, 95
Landing wieght 34
Lap bonded joints 315
    – balanced double lap 315
    – single lap 318
    – unbalanced double lap 317
Lateral instability 411, 623
Leading edge 55
Least work method 99
Lift 28
Lightning strikes 3
Load 14, 27
    – analysis 32
    – block loading 723
    – brake-roll 41
    – braking 44
    – breakaway 53
    – buffet 45
    – control surfaces 31, 45, 52
    – criteria 28
    – curves 30
    – diagrams 20
    – ditching 52
    – dynamic 27, 45
    – emergency 52
    – external 27, 784
    – factor 36, 37, 49, 52, 57
    – flight 36, 54
    – flight-by-flight loading 723
    – ground 27, 38, 41, 55
    – gust 37, 38, 53
    – hand 53
    – hinge 46, 52
    – interfaces 31
    – internal 126, 784
    – landing 38, 39, 40, 41, 55
    – maneuver 27, 36
    – miscellaneous 52
    – organizations 31
    – path 1, 21, 23
    – pivoting 43
    – pressurization 46
    – reactions 20
    – rolling 37, 38
    – sequence 662
    – spectrum 658, 666, 722
    – towing 44
    – turning 42
    – yawing 37, 38, 43
Load cell 740
Load factors (jet engine mount) 57
Lockbolts 279
Lug 53, 321, 336, 342
Lumping 1, 134, 379

M

Magnesium 77
Maneuver 36, 50
Manufacturing 11, 12
Margin of safety 24, 322, 345
Material 58
    – allowables 67, 69-73
    – aluminum 69-72
    – bases 67
    – castings 59
    – extrusion 59
    – forging 59
    – processes (reworking) 681
    – properties 58
    – rating 78
    – selection 78, 656
    – sheet and plate 59
    – steel 73, 681, 689
    – sustained tensile stress 372
Meshing 787
Minimum potential energy method 99
Minor area (threaded fastener) 283
Misdrill 698
Mismatch 372
Modulus 61
    – modulus of rigidity 64
    – secant 62
    – tangent 62, 64, 396, 652
    – Young's 63
Modulus of rigidity (shear modulus) 64, 169
Mohr's circle graphic method 84, 726
Moment-area method 102
Moment-distribution method 119
Moment of inertia 139, 140
    – about an inclined axis 141
    – principal 142
Murphy's law 2, 26

N

NDI 687
Negligible damage 695
Nickel 77
Nomex core (HRH10) 208
Non-optimum effects 7
Nutplates 281
Nuts 281, 288

O

Optimum design 6, 632, 648, 786
Orthogrid 610

P

Palmgren-Miner hypothesis 660
Panels:
    – compression 607, 737
    – configurations 608
    – efficiency 608, 628
    – failure stresses 616
    – geodesic 610
    – honeycomb 203
    – integrally-stiffened 643, 646
    – isogrid 610
    – orthogrid 610
    – shear 529, 738
    – skin-stringer 616, 619, 716
    – tension 655, 726
Parting line 68
Permanent deformation 74
Pins
    – bending 326, 327
    – fuse 53, 322
    – shear 53, 321
    – tubular (hollow) 336
Plastic bending 146
Plastic limit 63
Plastic reduction factors 456
    – compression buckling 456
    – shear buckling 457
Plastic strain 620
Plate 181, 182, 451, 786
    – buckling 451
    – cantilever 186
    – curved plates 464
    – edge supports 454
    – membrane (diaphragm) 193
    – thick 183
    – thin 193
Plug 697
Poisson's ratio 65, 451
Producibility 78
Production design 12
Preloaded tension bolt 352, 392
Pressure
    – altitude 45, 48
    – cabin 47
    – differential 46, 47, 48
    – dome 199
    – flat bulkhead 200
    – fuel 49
Pressurization 46, 592
Production design 12
Principal stresses 83
Pull-through 274, 282, 286, 287, 288
Pull-up (clamp-up) 371, 372, 373
Puncture 695
Pylon 53

R

Radius of gyration 140, 396
Ramberg-Osgood 62, 652

Reduction factors
- cladding 59, 458
- cutout 577
- effective width 612
- plastic 456, 457, 620

Redundant structures 26, 97, 127
Regulations 4
Repairability 678
Repairs 677, 686, 696
- beam web 707, 720
- configurations 679
- corrosion 713
- fatigue life 687
- flush repairs 690, 691, 708, 718
- insertion 716
- min. spacing 377, 703
- non-flush repairs 690
- panel skin 707
- plugs 699
- replacement 718
- skin-stringer panels 716
- steel materials 681, 689
- temporary repairs 694

Replacement 718
Reserve fuel wieght 34
Residual strength 668, 670
Retrofit 718
Rib 55, 720
Ribbon direction 206
Rivets 280, 282, 479, 497, 709
- loads 514
- shear loads 514, 497, 522, 547
- tension loads 498, 514, 522

Roll 50, 51
Rosette strain gauge 723
Rotational restraints 400

S

Safe life (fatigue life) 656, 660
Sagging (due to diagonal tension) 523
Salvage 677, 679, 681
Salvage policy 680
Scatter factor 389, 655
Screws 280
Second moment of area 139
Section
- factor 147
- open 157
- properties 140
- symmetrical 143, 153
- unsymmetrical 145, 156

Serviceability 78
Severity factor 382, 385
Shear
- center 161, 231
- flow 20, 159, 166, 167, 226, 240
- modulus (modulus of rigidity) 64, 169
- strain 315
- stress 152, 157, 313, 315

Shear beams 534
Shear lag 225, 262, 264
Shear modulus (modulus of rigidity) 64
Shear-off 700
Shear panels 473
- diagonal tension curved web 504
- diagonal tension flat web 484
- diagonal-tension factor 485, 490
- pure diagonal tension web 483
- resistant web 474
- rivets 479

Shear test panel 739
Shear-tie 505
Shell 181, 786
- cylinderical 182, 197
- double-cylinder section 198
- hemispherical 199
- semi-spherical 200
- spherical 182

Shims 370, 373, 683, 685
Single lap joints (bonded) 318
Sign convention 19, 223
Sills 585, 586, 590, 595
Sizing 1
- detail 2
- preliminary 2, 15
- procedures 14

Skin-stringer panels 608, 616, 716
Slenderness ratio 396, 448
Slope-deflection method 115
S-N curves (or F-N curves) 373, 386, 387, 655 658
Solid rivets 279, 282
Spar 55, 720
Specifications 3
Specimen 726, 738
Speed 36, 38
Splice 274, 276, 289, 523, 527
Splice stringer 275
Spotface 346
Spring constants (fastener) 399
Stability 451
Staggered-row of fasteners 704
Steel 77, 681, 689
Stepped column 401, 402, 403
Stiffeners 478, 487, 490, 523, 527, 550, 563 565
Stiffness 3, 25, 29, 31, 86, 689
Stop-drill 695
Strain 62, 65, 80
- axial 80
- elastic 62
- plastic 62
- shear 80, 81

Strain energy method 98

Strain gauges 723, 737
Straps 719
Stress concentration 374, 385, 390, 698
Stress 62, 65, 786
    – alternating (variable) 658
    – analysis 16
    – cycles 658
    – maximum 658
    – mean 658
    – minimum 658
    – ratio 658
    – reports 16, 17
Stress-strain curves 61, 62, 63, 74
Stringers 512, 573, 585, 608, 738
    – coaming stringer 573
    – cut stringer 576, 583, 585, 591
Structural efficiency 608
Structural salvage 679
Structural index 7, 648
Structural influence coefficients (SIC) 97
Stub member 564

T

Tail 40, 56
Take-off 33, 41
Take-off gross weight 34
Tangent modulus 64
Tapered beam 164
Tapered column 402, 406
Tapered cross section 256
Tapered ends 387, 391
Tapering 378, 387, 391
Taper in width 261
Taxi 41
Tension panels 655
Tension test panel (fatigue) 726
Tests 722
    – compression 737
    – dummy structures 740
    – fatigue 726
    – load spectrum 722
    – shear 738
    – static 727, 738
    – strain gauges 723
Test setup 722
Threaded fasteners 283
Titanium 77
Torque 355
Torsion 169, 226
Torsional constant 169
    – closed sections 172, 173, 226
    – D-cell box 232, 234
    – instability 616
    – noncircular section 170
    – open section 175, 176, 226, 231
    – single-cell box 228, 236, 238
    – solid section 170, 171

    – symmetrical section 241
    – two-cell box 229, 249, 252
    – unsymmetrical section 245
Torsional shear stress 170
    – cellular sections 227
    – closed sections 172
    – solid section 171
    – open section 175
Toughness 75
Trailing edge 55
Truss 92, 786
Tube 49, 178
Tubing 446
Turbulence 54
Tubular pins (hollow) 336

U

Unbalanced bonded joint 317
Unflanged (blade) integrally-stiffened panels 646
Upright (same as stiffener)
Uranium 77

V

V-n diagram 27, 36, 38
Velocity
    – angular 49, 50
    – roll 49, 50
Vibration 3, 45, 784
Virtual work method 99
Vitality 677

W

Waffle panels 610
Warp direction 206
Washers 281, 289, 346, 353
Washer (countersunk) 691
Washer (dimpled) 697
Weight 9, 27, 32, 33, 35
    – landing 33
    – take-off gross 33
    – zero fuel 34
    – reduction 11
    – distribution 35
Welded joints 305
    – butt 308
    – fillet 310, 311
    – spot-welded 308
Wheel 40, 45
Wing 30, 35, 55, 58, 221, 257, 262, 263, 344, 529, 691, 719
Wing boxes
    – multi-cell 222
    – two-cell 221
    – single-cell 238

Wing lift 28
Wing root joints 729
Wrinkling 617, 626
Wrought forms 59

Y

Yaw 37, 43, 51
Young's modulus 63

Z

Zero fuel weight 34